TUNNELS AND UNDERGROUND CITIES: ENGINEERING AND INNOVATION MEET ARCHAEOLOGY, ARCHITECTURE AND ART

PROCEEDINGS OF THE WTC2019 ITA-AITES WORLD TUNNEL CONGRESS,
NAPLES, ITALY, 3-9 MAY, 2019

Tunnels and Underground Cities: Engineering and Innovation meet Archaeology, Architecture and Art

Volume 3: Geological and geotechnical knowledge and requirements for project implementation

Editors

Daniele Peila

Politecnico di Torino, Italy

Giulia Viggiani

University of Cambridge, UK
Università di Roma "Tor Vergata", Italy

Tarcisio Celestino

University of Sao Paulo, Brasil

CRC Press
Taylor & Francis Group
Boca Raton London New York

CRC Press is an imprint of the
Taylor & Francis Group, an **informa** business
A BALKEMA BOOK

Cover illustration:

View of Naples gulf

First published in paperback 2024

First published 2020
by CRC Press/Balkema
4 Park Square, Milton Park, Abingdon, Oxon, OX14 4RN

and by CRC Press/Balkema
2385 NW Executive Center Drive, Suite 320, Boca Raton FL 33431

CRC Press/Balkema is an imprint of the Taylor & Francis Group, an informa business

Publisher's Note
The publisher has gone to great lengths to ensure the quality of this reprint but points out that some imperfections in the original copies may be apparent.

ISBN: 978-0-367-46583-4 (hbk)
ISBN: 978-1-03-283936-3 (pbk)
ISBN: 978-1-003-02974-8 (ebk)

DOI: 10.1201/9781003029748

Typeset by Integra Software Services Pvt. Ltd., Pondicherry, India

Visit the Taylor & Francis Web site at
http://www.taylorandfrancis.com

and the CRC Press Web site at
http://www.crcpress.com

Tunnels and Underground Cities: Engineering and Innovation meet Archaeology, Architecture and Art, Volume 3: Geological and geotechnical knowledge and requirements for project implementation – Peila, Viggiani & Celestino (Eds)
© 2020 Taylor & Francis Group, London, ISBN 978-0-367-46583-4

Table of contents

Tunnels and Underground Cities: Engineering and Innovation meet Archaeology, Architecture and Art, Volume 3: Geological and geotechnical knowledge and requirements for project implementation – Peila, Viggiani & Celestino (Eds)
© 2020 Taylor & Francis Group, London, ISBN 978-0-367-46583-4

Preface

The World Tunnel Congress 2019 and the 45th General Assembly of the International Tunnelling and Underground Space Association (ITA), will be held in Naples, Italy next May.

The Italian Tunnelling Society is honored and proud to host this outstanding event of the international tunnelling community.

Hopefully hundreds of experts, engineers, architects, geologists, consultants, contractors, designers, clients, suppliers, manufacturers will come and meet together in Naples to share knowledge, experience and business, enjoying the atmosphere of culture, technology and good living of this historic city, full of marvelous natural, artistic and historical treasures together with new innovative and high standard underground infrastructures.

The city of Naples was the inspirational venue of this conference, starting from the title Tunnels and Underground cities: engineering and innovation meet Archaeology, Architecture and Art.

Naples is a cradle of underground works with an extended network of Greek and Roman tunnels and underground cavities dated to the fourth century BC, but also a vibrant and innovative city boasting a modern and efficient underground transit system, whose stations represent one of the most interesting Italian experiments on the permanent insertion of contemporary artwork in the urban context.

All this has inspired and deeply enriched the scientific contributions received from authors coming from over 50 different countries.

We have entrusted the WTC2019 proceedings to an editorial board of 3 professors skilled in the field of tunneling, engineering, geotechnics and geomechanics of soil and rocks, well known at international level. They have relied on a Scientific Committee made up of 11 Topic Coordinators and more than 100 national and international experts: they have reviewed more than 1.000 abstracts and 750 papers, to end up with the publication of about 670 papers, inserted in this WTC2019 proceedings.

According to the Scientific Board statement we believe these proceedings can be a valuable text in the development of the art and science of engineering and construction of underground works even with reference to the subject matters "Archaeology, Architecture and Art" proposed by the innovative title of the congress, which have "contaminated" and enriched many proceedings' papers.

Andrea Pigorini
SIG President

Renato Casale
Chairman of the Organizing Committee WTC2019

Tunnels and Underground Cities: Engineering and Innovation meet Archaeology, Architecture and Art, Volume 3: Geological and geotechnical knowledge and requirements for project implementation – Peila, Viggiani & Celestino (Eds)
© 2020 Taylor & Francis Group, London, ISBN 978-0-367-46583-4

Acknowledgements

REVIEWERS

The Editors wish to express their gratitude to the eleven Topic Coordinators: Lorenzo Brino, Giovanna Cassani, Alessandra De Cesaris, Pietro Jarre, Donato Ludovici, Vittorio Manassero, Matthias Neuenschwander, Moreno Pescara, Enrico Maria Pizzarotti, Tatiana Rotonda, Alessandra Sciotti and all the Scientific Committee members for their effort and valuable time.

SPONSORS

The WTC2019 Organizing Committee and the Editors wish to express their gratitude to the congress sponsors for their help and support.

Tunnels and Underground Cities: Engineering and Innovation meet Archaeology, Architecture and Art, Volume 3: Geological and geotechnical knowledge and requirements for project implementation – Peila, Viggiani & Celestino (Eds)
© 2020 Taylor & Francis Group, London, ISBN 978-0-367-46583-4

WTC 2019 Congress Organization

HONORARY ADVISORY PANEL

Pietro Lunardi, President WTC2001 Milan
Sebastiano Pelizza, ITA Past President 1996-1998
Bruno Pigorini, President WTC1986 Florence

INTERNATIONAL STEERING COMMITTEE

Giuseppe Lunardi, Italy (Coordinator)
Tarcisio Celestino, Brazil (ITA President)
Soren Eskesen, Denmark (ITA Past President)
Alexandre Gomes, Chile (ITA Vice President)
Ruth Haug, Norway (ITA Vice President)
Eric Leca, France (ITA Vice President)
Jenny Yan, China (ITA Vice President)
Felix Amberg, Switzerland
Lars Barbendererder, Germany
Arnold Dix, Australia
Randall Essex, USA
Pekka Nieminen, Finland
Dr Ooi Teik Aun, Malaysia
Chung-Sik Yoo, Korea
Davorin Kolic, Croatia
Olivier Vion, France
Miguel Fernandez-Bollo, Spain (AETOS)
Yann Leblais, France (AFTES)
Johan Mignon, Belgium (ABTUS)
Xavier Roulet, Switzerland (STS)
Joao Bilé Serra, Portugal (CPT)
Martin Bosshard, Switzerland
Luzi R. Gruber, Switzerland

EXECUTIVE COMMITTEE

Renato Casale (Organizing Committee President)
Andrea Pigorini, (SIG President)
Olivier Vion (ITA Executive Director)
Francesco Bellone
Anna Bortolussi
Massimiliano Bringiotti
Ignazio Carbone
Antonello De Risi
Anna Forciniti

Giuseppe M. Gaspari
Giuseppe Lunardi
Daniele Martinelli
Giuseppe Molisso
Daniele Peila
Enrico Maria Pizzarotti
Marco Ranieri

ORGANIZING COMMITTEE

Enrico Luigi Arini
Joseph Attias
Margherita Bellone
Claude Berenguier
Filippo Bonasso
Massimo Concilia
Matteo d'Aloja
Enrico Dal Negro
Gianluca Dati
Giovanni Giacomin
Aniello A. Giamundo
Mario Giovanni Lampiano
Pompeo Levanto
Mario Lodigiani
Maurizio Marchionni
Davide Mardegan
Paolo Mazzalai
Gian Luca Menchini
Alessandro Micheli
Cesare Salvadori
Stelvio Santarelli
Andrea Sciotti
Alberto Selleri
Patrizio Torta
Daniele Vanni

SCIENTIFIC COMMITTEE

Daniele Peila, Italy (Chair)
Giulia Viggiani, Italy (Chair)
Tarcisio Celestino, Brazil (Chair)
Lorenzo Brino, Italy
Giovanna Cassani, Italy
Alessandra De Cesaris, Italy
Pietro Jarre, Italy
Donato Ludovici, Italy
Vittorio Manassero, Italy
Matthias Neuenschwander, Switzerland
Moreno Pescara, Italy
Enrico Maria Pizzarotti, Italy
Tatiana Rotonda, Italy
Alessandra Sciotti, Italy
Han Admiraal, The Netherlands
Luisa Alfieri, Italy
Georgios Anagnostou, Switzerland
Andre Assis, Brazil
Stefano Aversa, Italy
Jonathan Baber, USA
Monica Barbero, Italy
Carlo Bardani, Italy
Mikhail Belenkiy, Russia
Paolo Berry, Italy
Adam Bezuijen, Belgium
Nhu Bilgin, Turkey
Emilio Bilotta, Italy
Nikolai Bobylev, United Kingdom
Romano Borchiellini, Italy
Martin Bosshard, Switzerland
Francesca Bozzano, Italy
Wout Broere, The Netherlands

Domenico Calcaterra, Italy
Carlo Callari, Italy
Luigi Callisto, Italy
Elena Chiriotti, France
Massimo Coli, Italy
Franco Cucchi, Italy
Paolo Cucino, Italy
Stefano De Caro, Italy
Bart De Pauw, Belgium
Michel Deffayet, France
Nicola Della Valle, Spain
Riccardo Dell'Osso, Italy
Claudio Di Prisco, Italy
Arnold Dix, Australia
Amanda Elioff, USA
Carolina Ercolani, Italy
Adriano Fava, Italy
Sebastiano Foti, Italy
Piergiuseppe Froldi, Italy
Brian Fulcher, USA
Stefano Fuoco, Italy
Robert Galler, Austria
Piergiorgio Grasso, Italy
Alessandro Graziani, Italy
Lamberto Griffini, Italy
Eivind Grov, Norway
Zhu Hehua, China
Georgios Kalamaras, Italy
Jurij Karlovsek, Australia
Donald Lamont, United Kingdom
Albino Lembo Fazio, Italy
Roland Leucker, Germany
Stefano Lo Russo, Italy
Sindre Log, USA
Robert Mair, United Kingdom
Alessandro Mandolini, Italy
Francesco Marchese, Italy
Paul Marinos, Greece
Daniele Martinelli, Italy
Antonello Martino, Italy
Alberto Meda, Italy
Davide Merlini, Switzerland
Alessandro Micheli, Italy
Salvatore Miliziano, Italy
Mike Mooney, USA
Alberto Morino, Italy
Martin Muncke, Austria
Nasri Munfah, USA
Bjørn Nilsen, Norway
Fabio Oliva, Italy
Anna Osello, Italy
Alessandro Pagliaroli, Italy
Mario Patrucco, Italy
Francesco Peduto, Italy
Giorgio Piaggio, Chile
Giovanni Plizzari, Italy
Sebastiano Rampello, Italy
Jan Rohed, Norway
Jamal Rostami, USA
Henry Russell, USA
Giampiero Russo, Italy
Gabriele Scarascia Mugnozza, Italy
Claudio Scavia, Italy
Ken Schotte, Belgium
Gerard Seingre, Switzerland
Alberto Selleri, Italy
Anna Siemińska Lewandowska, Poland
Achille Sorlini, Italy
Ray Sterling, USA
Markus Thewes, Germany
Jean-François Thimus, Belgium
Paolo Tommasi, Italy
Daniele Vanni, Italy
Francesco Venza, Italy
Luca Verrucci, Italy
Mario Virano, Italy
Harald Wagner, Thailand
Bai Yun, China
Jian Zhao, Australia
Raffaele Zurlo, Italy

Geological and geotechnical knowledge and requirements for project implementation

Tunnels and Underground Cities: Engineering and Innovation meet Archaeology, Architecture and Art, Volume 3: Geological and geotechnical knowledge and requirements for project implementation – Peila, Viggiani & Celestino (Eds)
 ISBN 978-0-367-46583-4

Analysis of the vibrations induced by a TBM to refine soil profile during tunneling: The Catania case history

G. Abate, S. Corsico, S. Grasso & M.R. Massimino
Department of Civil Engineering and Architecture, University of Catania, Italy

A. Pulejo
CMC Cooperativa Muratori e Cementisti Ravenna, Italy

ABSTRACT: The present paper deals with the new underground lines under construction in Catania (Italy). The analysed network is under construction with EPBs-TBM machine. the TBM used is a Dual Mode Machine. In order to update the tunneling method with the soil profile, a detailed geotechnical characterization is very important, in particular when a strong lithological alteration during excavation is present (such as in the analysed case history), because tunneling method can be OF (Open-Face) mode for rock formation or EPB (Earth Pressure Balance) mode for cohesive and/or incoherent soil. This paper shows a new methodology to validate the hypothesized soil profile based on a comparison between data obtained by the geotechnical survey and data achieved by HVSR tests on the vibrations induced by TBM on the surface during tunneling. A further comparison is made between data by HVSR tests and by an electric tomography.

1 INTRODUCTION

Nowadays, the need to keep levels of efficiency and environmental sustainability high has increased, looking solutions that lead to a lowering of environmental pollution generated by road traffic, improving the quality of life and reducing in costs for citizens. One way to achieve these purposes is certainly to aim at the building of underground transport lines.

Furthermore, it is important to guarantee the connections between the central urban area and the outskirts and commercial areas, often too isolated. To work around this problem, in Catania (Italy) a railway line is already present in the urban area and it is extending towards the main villages of the Catania hinterland and the "Vincenzo Bellini" airport to facilitate the movement of users through a new underground network.

The present paper deals with one of the new segments of the Catania underground, at the moment under-construction, in particular the one that connects the urban area with Misterbianco village (Abate et al. 2016, Abate & Massimino, 2017a, 2017b).

For the first time in Catania, the tunnel is being digged using a EPBs-TBM machine; the TBM used is a Dual Mode Machine (Herrenknecht S454.1, Diameter 10.6 m) able to dig in two ways: OF (Open Face) mode for rock formation; EPB (Earth Pressure Balance) mode for cohesive and/or incoherent soil. According to the OF mode the TBM machine excavates rock with disc cutters mounted in the cutter head. The disc cutters create compressive stress fractures in the rock, causing it to chip away from the tunnel face. According to the EPB mode, the TBM machine uses the excavated material to balance the pressure at the tunnel face (water pressure and earth pressure). Pressure is maintained in the cutter head by controlling the rate of extraction of spoil through the Archimedes screw and the advance rate. Additives such as bentonite, polymers and foam can be injected ahead of the face to increase the stability of the ground.

So, geological and geotechnical information of rocks and soils at the digging front is most important, in order to appropriately define the digging mode and, consequently, the front pressure to adopt in order to guarantee the digging front stability (Mohammadi 2010, Anagnostou and Kovári 1993, 1994, 1996, Carranza-Torres et al. 2013.) An error on the estimation of the front pressure can cause subsidence problems with disastrous effects in urban areas (such as Catania), for the high presence of buildings of historic and social importance (Broere 2002, Atkinson & Potts 1977, Attewell & Taylor 1984, Attewell 1978, Burland 1995).

Unfortunately, often the lithological investigation surveys carried out in the design phase could be not sufficient and have to be integrated, increasing the cost of the infrastructure. This is most important particularly when there is a strong lithological alteration during excavation, such as in the presented case-history, due to the different lava flows that characterized the history of Etna volcano in Catania. About that, the Authors in this paper propose a new simple and inexpensive method (Abate et al., 2018), based on the HVSR method (Nakamura, 1989; Nakamura, 2000), introduced by Nogoshi and Igarashi (Nogoshi & Igarashi, 1970; Nogoshi & Igarashi, 1971), using the vibration induced by the TBM on the surface, in order to update the soil profile during the tunneling. Also other authors (Barla & Pelizza, 2000; Dickmann, 2014; Dickmann & Méndez, 2017) have used different approaches to solve the problem of the uncertainty of geological formation.

The geological and geotechnical characterization of the analyzed segment performed in the design phase and updated during the tunneling is reported in this paper. Finally, the results obtained through the new method and the electric tomography test are compared, validating the method proposed by the authors.

2 THE RAILWAY NETWORK

2.1 *The case history of Catania*

The project of the new railway network of Catania consisting of electrified double-track lines with a standard gauge is part of the program to upgrade and modernize the current narrow gauge monorail.

On the basis of this development program the goal is to extend the existing railway line, confined exclusively to the urban area of Catania, to the underground and sub-metropolitan area of Misterbianco and Paternò, in the North-West direction, and towards the airport area of Fontanarossa, in the South-East direction. It will serve an estimated population of about one million inhabitants, covering an urban and suburban area of about 44 km, including 18 km in tunnels and 26 km on the surface as well as 117 km of extra-urban surface track.

Two different (currently in operation) railway lines were built in 1999: an underground double-track railway in the centre of the city (1.8 km long), and a surface monorail track (2.0 km long). Currently, two new underground railway lines are under construction (Nesima-Misterbianco and Stesicoro - “Vincenzo Bellini” Airport); the first line will connect the actual underground network with the outlying districts to the north-west of the city and with the nearby town of Misterbianco up to the town of Paternò. The second one will extend the subway line from the centre of Catania along the south-western outskirts of the city to the “Vincenzo Bellini” airport.

Figure 1 shows the whole line: the red line indicates the two railway lines, currently in operation built in 1999; the light blue line indicate two new railway lines actually under construction (Nesima-Misterbianco and Stesicoro - “Vincenzo Bellini” Airport). The dark blue lines indicate the designed railway lines, which have already been approved but which are not under construction yet. The present work regards the part of the underground railway line between the stations of Nesima and Misterbianco, which extends for a length of 1748.08 m.

Starting from the Nesima station, built in the project plan of the Borgo-Nesima segment, the route will develop in line with the ring road of Catania up to the municipal border between Catania and Misterbianco. Along the route the following stations will be present: Fontana station, Monte Pò station.

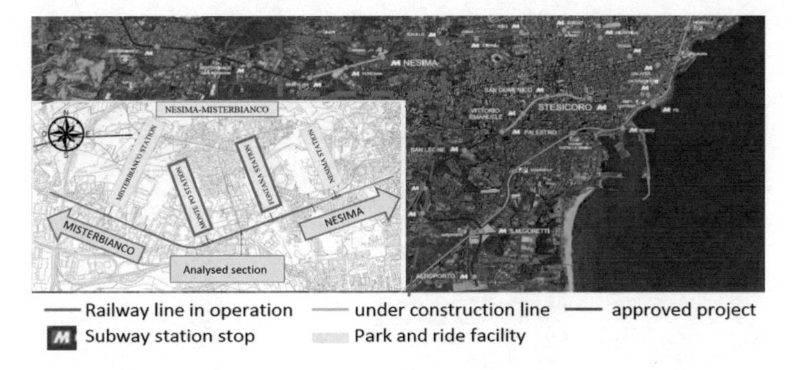

Figure 1. The underground network in Catania; in the small square the analysed segment "Nesima-Misterbianco (modified after Abate et al. 2018).

2.2 *The analysed railway line: "Nesima – Misterbianco"*

The railway line Nesima –Misterbianco is 3940 m long. The underground segment cover a length of 1748 m.

The strong heterogeneity of the geological profile of Catania's subsoil, due to Etna volcano eruptions, leads to dig in two ways, switching frequently from one to the other: Open Face (OF) mode for lava-rock and Earth Pressure Balance (EPB) mode for incoherent and soft soil, in order to balance the soil and the water pressure.

The cross section of the excavation is circular with a diameter of 10.60 m and an axis average depth of about 25.00 m. The tunnel lining consists of prefabricated reinforced concrete link with a thickness of 0.32 m and a depth of 1.50 m. Each link consists of 7 segments installed by an appropriate erector presents inside the TBM.

3 GEOLOGICAL AND GEOTECHNICAL CHARACTERIZATION OF THE ANALYSED RAILWAY LINE

In order to define the geological profile and the geotechnical parameters of the soils and rocks along the Nesima-Misterbianco segment, two different geotechnical investigations were performed: the first one in November-December in 2004 during the preliminary design and the second one in 2015 for the executive project. Furthermore, the investigations carried out for the construction of a parking area around the above mentioned Monte Pò station and the investigations performed in 1996 within the "The Catania Project" (Faccioli & Pessina, 2000) were considered.

During the geotechnical investigation survey carried out in November-December 2004 for the preliminary design of the new lines, 15 boreholes, 5 samples Q1 (UNI ENV 1997-2:2002), 3 Down-Hole tests were carried out. The most significant 3 boreholes named S2, S3, S4 were considered in the analysed Nesima-Misterbianco segment.

During the investigation survey in 2015, 17 boreholes (Si1, Si2, Si3, Si4, Si4bis, Si5, Si6, Si7, Si8, Si9, Si9bis, Si10, Si11, Si12, Si13, Si14, Si15) were performed.

In addition, boreholes named 1, 2, 3, 4, 5 performed for the construction of the parking area around Monte Pò station and boreholes named 140, 1242 and 1068 performed within the "The Catania Project" were used in order to define the soil profile as shown in Figure 2.

Generally, in the most superficial portion an anthropic layer is found; while at greater burial depths lava rock layers are observed.

The geological formations directly involved in the digging process are essentially: Volcanoclastic breasts and sand of 1669 from Si1 to Si8; Lava of Quartalaro from Si8 to Si10; Volcanoclastic breasts of 1669 from Si10 to Si11; finally, lava of Quartalaro until Si13.

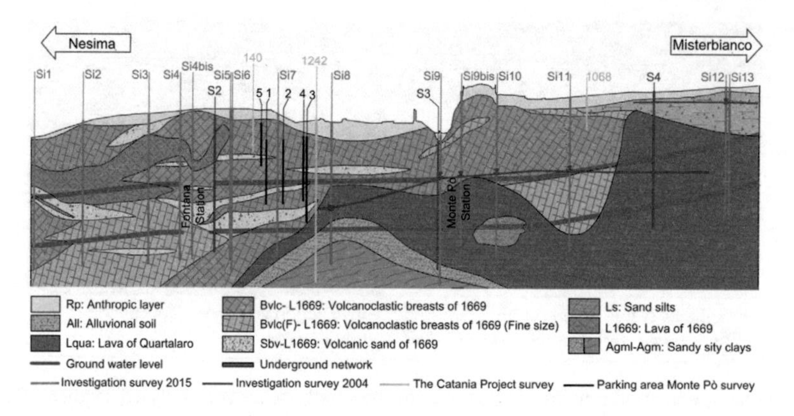

Figure 2. Soil profile and positions of the boreholes along the Nesima-Misterbianco segment established during the design phase.

Inside several boreholes shown in Figure 2 the following in-situ tests were also performed (Table 1): 2 Dilatometric Tests (DMT); 15 Standard Penetration Tests (SPT); 7 permeability Lefranc Tests (LE); 6 absorption Lugeon Tests (LU); 2 Dac-Tests; 6 Pocket Penetrometer Tests (PPT).

Furthermore, 15 Open Piezometer Tests (OP test) were performed in order to determinate the depth of the groundwater reported in Table 1.

Finally, in order to define the shear wave velocity, V_S, which is a fundamental seismic parameter, Down – Hole (DH) tests and HVSR (Horizontal-to-Vertical Spectral Ratio) tests were performed. Table 2 shows the V_S values for each geological formation.

In addition, during the investigations survey, many samples were collected. Undisturbed samples were subjected to the following laboratory tests: a) classification tests; b) tests for the determination of the resistance and deformability characteristics of the soil: edometric tests, direct shear tests, triaxial tests; c) tests for the determination of the resistance and deformability characteristics of the rock: monoaxial compression tests, point load tests, triaxial compression strength tests, indirect tensile strength tests, measurement of sonic waves. Moreover, for the geotechnical characterization of the rocks the RQD parameter was also estimated (values

Table 1. Geotechnical investigation survey.

ID	Rock Samples	Undisturbed Samples	Reshape Samples	DMT	SPT	LE	LU	Dac test	PPT	Depth Groundwater [m]
Si1	1	-	4	-	X	-	-	-	-	-
Si2	1	-	4	X	X	-	-	-	-	-
Si3	2	-	3	-	X	X	X	-	-	-
Si4	2	-	3	-	X	-	-	-	-	-
Si4bis	2	-	3	-	X	-	X	-	-	-
Si5	-	-	-	-	-	-	-	X	-	-
Si6	-	-	3	-	X	X	-	-	-	-
Si7	2	-	3	-	X	-	-	-	-	-
Si8	3	1	1	-	X	X	-	-	X	20.15
Si9	2	1	2	-	X	X	X	-	X	9.5
Si9bis	3	-	1	-	X	X	X	-	-	19.80
Si10	2	-	4	X	X	X	X	-	X	17.80
Si11	1	-	5	-	X	-	-	-	-	18.80
Si12	2	2	-	-	X	-	X	-	X	2.80
Si13	-	-	-	-	-	-	-	X	-	-
Si14	1	1	2	-	X	X	-	-	X	1.50
Si15	2	1	2	-	X	-	-	-	X	2.50

Table 2. Geotechnical parameter for each geological formation.

	γ [kN/m^3]	φ' [°]	c' [kPa]	c_u [kPa]	V_S [m/s]	E_0 [MPa]	E_u [MPa]	k [m/s]
Rp	18÷19	30÷35	0÷10	-	120÷180	70÷125		
All	18÷21	20÷23	0÷10	70÷150	200÷350		20÷45	
Sbv	18÷21	32÷40	0÷20		200÷350	200÷400		10^{-5}
Agm	20÷21	18÷27	5÷30	100÷300	400÷550	1000		
AgmL	20	18÷27	0÷15	200÷400	400÷550		55÷75	
Ls	19÷20	30÷35	0÷30	200÷400				10^{-5}
Bvlc-Bvlc(F)	18÷21	38÷45	0÷20		200÷350	250÷750		10^{-6}÷10^{-5}
Lqua-L1669	25÷27	50÷65	100÷1000		350÷800	2000÷5000		10^{-5}÷10^{-3}

between 30 and 80). Table 2 shows the geotechnical parameters for each geological formation, being: γ the unit weight; φ' the angle of shear strength and c' the cohesion obtained by the direct shear test; c_u the undrained cohesion obtained by the U-U triaxial tests and by the SPT and PPT tests; E_0 the modulus of elasticity at very small strains obtained by the V_S values; E_u the undrained modulus of elasticity obtained by the U-U triaxial tests; k the coefficient of permeability obtained by the permeability tests (LE and LU tests) and the edometric tests.

4 A NEW SIMPLE METHODOLOGY TO REFINE THE SOIL PROFILE

As mentioned earlier, in hard rock and in absence of groundwater, Open-Face mode (OF) is used, while in incoherent and/or cohesive soil, the Earth Pressure Balance mode (EPB) is used. For this reason, the knowledge of the soil is fundamental in order to choice the digging method and so essential to guarantee the safety during the tunneling.

In this paper, the authors introduce a new methodology in order to refine during the tunneling the soil profile, hypnotized in design phase. This method is based on the HVSR (Horizontal-to-Vertical Spectral Ratio) test, using the vibrations produced on the surface by the TBM.

According to the UNI9916-DIN 4150 Technical Regulation, vibrations induced by the TBM next to the buildings above or adjacent to the tunnel front must to be continuously measured in order to guarantee the safety of the buildings. The authors propose a new methodology to refine the soil profile using these necessary measurements. Thus, the proposed methodology does not involve additional costs compared to those envisaged during the design phase.

The HVSR technique, proposed by Nakamura (1989) and introduced by Nogoshi & Igarashi (1970, 1971), allow us to identify the natural frequency, f, of soil, through the spectral ratio between the Horizontal Fourier spectrum (H) and the Vertical Fourier spectrum (V) of microtremor measurements.

The peaks of H/V vs f curve indicate the natural frequencies of the soil-rock layers. HVSR technique was extensively used at various sites (e.g., Lermo and Chàvez-Garcìa 1994, Field et al. 1995, Chàvez-Garcìa and Cuenca 1995).

The goal is therefore to be able to obtain information about the stratigraphy of the soil, analyzing the frequency content, as faster as possible (about 30 m before the position of the TBM front). If the proposed methodology reveals a soil profile very similar to that established during the design phase, it is possible to confirm the soil profile given by the investigation surveys performed during the design phase. If the proposed methodology reveals a soil profile similar to that established during the design phase, it is necessary to refine the soil profile given by the investigation surveys performed during the design phase with the results achieved by the HVSR technique. If the proposed methodology reveals a soil profile different from that established during the design phase, it will be necessary to carry out further geotechnical surveys to establish the "real" profile.

In the investigated case-history, during the tunnel digging, geophones were located at the soil surface near the buildings recording a great quantity of signals in terms of velocity in three direction each day: Transversal (T), Radial (R) and Vertical (V) that exceed a fixed minimum value.

Each signal is 4 seconds long and it has been recorded with a sampling rate of 512 or 1024 Hz. Then, the average signal for each day was computed.

Then, the authors proceeded to obtain the respective accelerations for each component and the Fourier Spectra.

Finally, according to the HVSR technique, the ratio between the horizontal and vertical Fourier spectra according to the following expression (1) was carried out:

$$\frac{H}{V} = \frac{\sqrt{R^2 + T^2}}{V} \tag{1}$$

In the present paper the signals recorded between January 2017 and November 2017 were analyzed.

Figure 3a shows the natural frequency of the soil deposit below the geophone (blue line), while the red line indicates the average natural frequencies.

From Figure 3a it is possible to detect five average values of the natural frequency of the soil deposit: the first one (from section 7+000.00 to 6+870.00) is equal to 6.5 Hz; the second one (from section 6+870.00 to 6+440.00) is equal to 3.5 Hz. After the Monte Pò Station other three values of the fundamental frequency can be distinguished: the first one is equal to 8 Hz (from section 6+200.00 to 6+100.00); the second one is equal to 4 Hz (from section 6+100.00 to 5.970.00). Finally, between the section 5+970.00 and 5+870.00, a natural frequency equal to 7.5 Hz was found.

These natural frequencies were compared with the values of the natural frequencies obtained by the well-known following expression (2):

$$f = \frac{V_S}{4H} \tag{2}$$

being V_S and H, respectively, the shear wave velocity and the height of the deformable geological formations above the Lqua-L1669 (Lava of Quartalaro and Lava of 1669) formation, according to Figure 2. The values of f obtained according to the expression (2) are reported in Figure 3a. It is possible to notice a good agreement between the values of the natural frequency achieved by the geophone (red line of Figure 3a) during the tunneling and the values obtained according to the expression (2). Only in two segments, highlighted by the blue square in Figure 3, between the section 6+100.00 and 5+970.00, is possible to notice a disagreement between the numerical values and the theoretical values.

Figure 3b shows the position of the TBM front as well as on the soil profile yet reported in Figure 2.

Moreover, Figure 3c that shows the soil profile updated during the tunneling, on the base of the material actually extracted.

In the segment between the section 6+100.00 and 5+970.00, the value of the frequency obtained by the geophone was lower than the theoretical one; it is due to the presence of sandy silty clays at lower depths (Figure 3c) not present in the soil profile hypothesized in the design phase. In the segment between the section 5+970.00 and 5+870.00, the natural frequency obtained by the geophone is higher than the theoretical one; this disagreement is due to the presence of rock at a depth lower than that expected in the design phase.

The results obtained through the methodology proposed by the authors are confirmed by the Figure 3c that shows the soil profile updated during the tunneling, on the base of the material actually extracted.

This is a very important result, considering that the geophone monitoring during the tunneling is always necessary to check the performance of the buildings around the tunneling area. Then, the proposed methodology can be considered an interesting way for refining soil profile, definitely less expensive than additional geotechnical investigations.

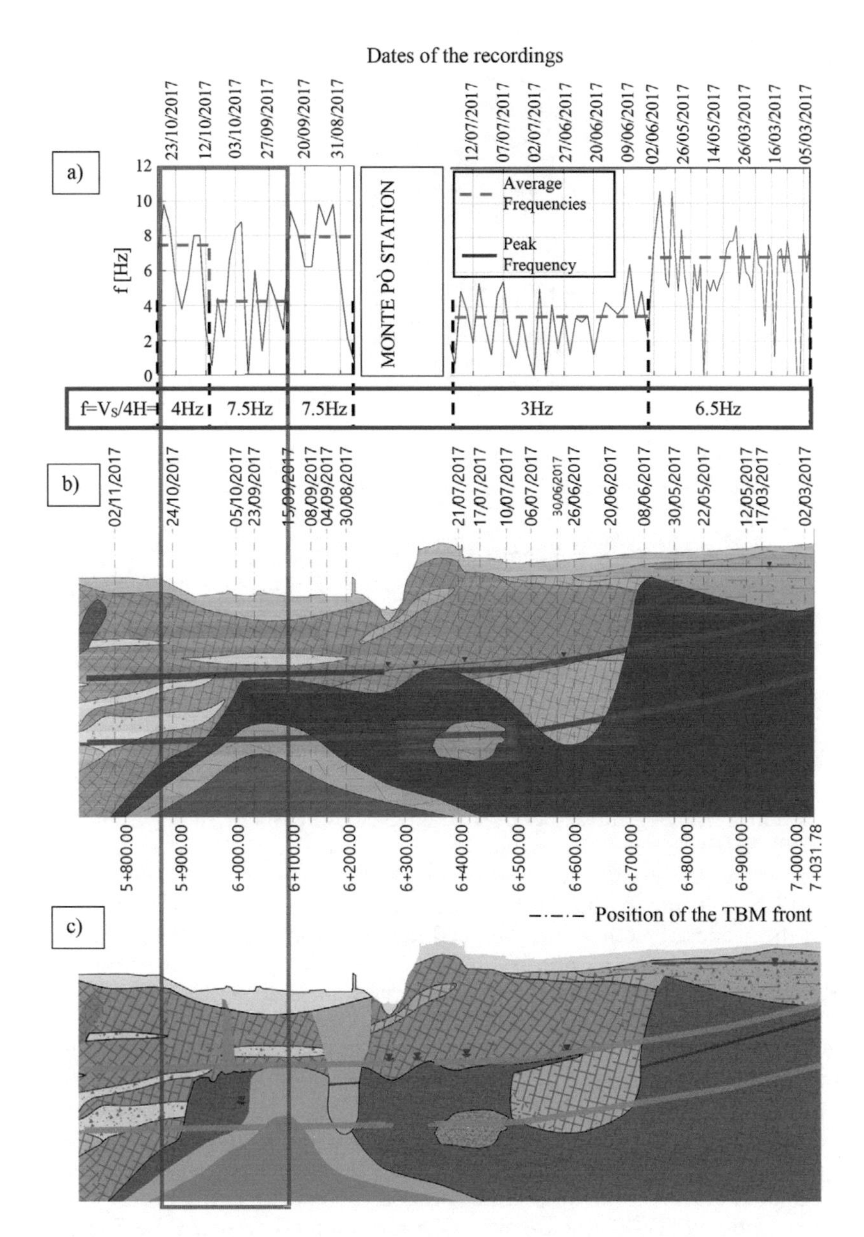

Figure 3. a) fundamental frequency of the soil deposit through HVSR technique; b) soil profile at the design phase; c) soil profile updated during the tunneling, on the base of the material actually extracted.

5 VALIDATION OF THE METHODOLOGY PROPOSED USING THE ELECTRIC TOMOGRAPHIC TEST AND TRADITIONAL HVSR TEST

The results obtained by the authors were also compared with these obtained through the Electric Tomographic Test, performed in October 2017.

In particular, three electric tomography profiles: L1, located partially above the existing tunnel, L2 and L3 located to the right and left of the L1 profile, were performed in the segment between the section 5+900.00 and 6+000.00 (Figure 3b).

Figure 4a shows the zoom of the soil profile established during the design phase (Figures 2 and 3b), Figure 4b shows the zoom of the soil profile updated during the tunneling, on the base of the material actually extracted. Finally, Figure 5 shows the results of the Electric

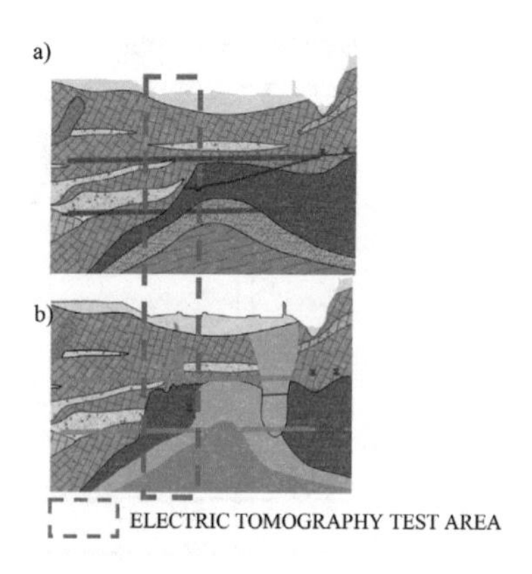

Figure 4. a) Zoom of the soil profile established during the design phase (Figures 2 and 3b); b) Zoom of the soil profile updated during the tunneling, on the base of the material actually extracted.

Tomography Test profile of the segment L1 in terms of resistivity and the new boreholes performed in October 2017, named S1T, S2T, S3T, S4T, S7T. The red square in Figure 4 shows the segment of the soil profile in which the Electric Tomographic Test was carried out. The results of the Electric Tomography Test confirm the presence of rock at a depth lower than that expected in the design phase, according to the results obtained by means of the methodology proposed by the authors (see Figure 3).

The segment investigated with the Electric Tomography Test consists of lava up to 32 m from the beginning of the profile, i.e. from section 5+900.00 to section 5+932.00 (Figure 3b). Then volcanoclastic breasts of 1669 (Bvlc - L1669) were found with a thickness equal to 30 m up to the section 5+970.00. Subsequently volcanoclastic breasts of 1669 with lower resistivity is identified (Bvlc (F) - L1669), for a thickness varying between 10 and 15 m resting on lava with high resistivity (L1669) up to the section 6+000.00.

Thus the Electric Tomographic Test confirms the value of the natural frequency (7.5 Hz) obtained by the proposed methodology applied to the measurements recorded between section 5+900.00 and 6+000.00. The natural frequency 7.5 Hz achieved with the proposed methodology is different from the value 4.0 Hz (Figure 3a) obtained according to expression (2) considering the soil profile established in the design phase (Figures 2, 3b). But, the natural frequency 7.5 Hz achieved with the proposed methodology is equal to the value of 8 Hz, obtained according to expression (2) considering the new soil profile updated thanks to the Electric Tomographic Test (Figure 4b), as well as the material actually extracted (Figure 3c).

Moreover, the frequencies obtained with the new methodology were compared with that one obtained by the traditional HVSR tests during the design phase. In particular, the frequencies achieved at the boreholes Si11 (section 6+580.00) and Si12(section 6+950.00) through the traditional HVSR test are about 4.5 Hz and 6 Hz respectively. These frequencies are in agreement with those obtained through the new methodology proposed by the Authors, achieving 4 Hz and 7 Hz at the boreholes Si11 and Si12, respectively.

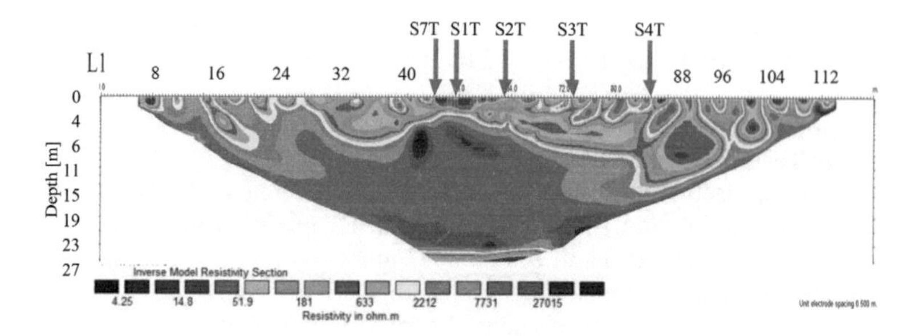

Figure 5. The Electric Tomography Test profile.

6 CONCLUSION

The present paper deals with the new underground railway that connects the urban area of Catania (Italy) with the near Misterbianco village, at the moment under-construction.

For the first time in Catania, a tunnel is being digged using a EPBs-TBM machine, able to dig in two ways: OF (Open Face) mode for rock formation and EPB (Earth Pressure Balance) mode for cohesive and/or incoherent soil. For this reason a good knowledge of the geological formations present during the digging is fundamental in order to choose the most suitable way to dig.

The authors in this paper propose a new simple and inexpensive methodology, based on the HVSR technique used for measuring the vibration induced by the TBM next to the buildings above or adjacent to the tunnel front in order to guarantee the safety of the buildings, prescribed by the UNI9916-DIN 4150 Technical Regulation.

In particular, the authors propose to use the above mentioned measurements also for a second reason: to update the soil profile during the tunneling. During the tunnel digging, a geophone was located at the soil surface approximately at a distance from the tunnel front equal to 30 m. The spectral ratios between the horizontal and the vertical component of the signals recorded day after day were computed, according to the HVSR technique.

So, the first natural frequencies of the geological formations day after day were estimated.

The new proposed methodology is validated comparing the results of the new methodology on the base of the material actually extracted during the digging and with the information obtained through an Electric Tomographic Test.

The knowledge of the natural frequency of the soil achieved with the new propose methodology showed some inaccuracies of the soil profile established in the design phase and allowed us to update the soil profile.

The authors will use this new methodology on the other segments of the underground network of Catania in order to validate ones more the methodology.

ACKNOWLEDGMENTS

The research reported in this paper was performed with the financial support of the PON - FSE-FESR - R&I 2014-2020 “Innovative PhDs with Industrial Characterization”.

REFERENCES

Abate G., Corsico S., Massimino M.R. 2016. FEM Modelling of the Seismic Behavior of a Tunnel-soil-Aboveground Building System: A Case History in Catania (Italy). *Procedia Engineering*, 158: 380–385.

Abate G. & Massimino M.R. 2017a. Numerical modelling of the seismic response of a tunnel–soil–aboveground building system in Catania (Italy). *Bull. Earthq. Eng*., 15(1): 469–491.

Abate G., Corsico S., Grasso S., Massimino M.R., Pulejo A. 2018. A new simple methodology to refine soil profile during tunnelling: the Catania case history. *16Th European Conference on Earthquake Engineering*. Thessaloniki, June.

Abate G. & Massimino M.R. 2017b. Parametric analysis of the seismic response of coupled tunnel–soil–aboveground building systems by numerical modelling. *Bull. Earthq. Eng*., 15(1): 443–467.

Anagnostou G. & Kovári K. 1993. Face stability conditions with Earth- Pressure- bassure- balanced Shields. *Tunneling and Underground Space technology*, 11(2): 165–173.

Anagnostou G. & Kovári K. 1994. The face stability of slurry- shield driven tunnels. *Tunneling and Underground Space Technology*. 9(2): 165–174.

Anagnostou G. & Kovári K. 1996. Face stability in slurry and EPB shield tunneling. *Geotechnical Aspects of Underground Construction in Soft Ground*. Rotterdam, Balkema. pp: 453–458.

Atkinson J.H. & Potts D.M. 1977. Subsidence above shallow tunnels in soft ground. *Journ. Of Geotech. Eng. Division*. ASCE GT4, pp 307–325.

Attewell P.B. & Taylor R.K. 1984. Ground movements and their effects on structures. pp 132–212.

Attewell P.B. 1978. Ground movements caused by tunnelling in soil. *Proc. Conf. on Large Ground Movements and Structures*. Cardiff, July, pp 812–948.

Barla G. & Pelizza S. 2000. TBM tunneling in difficult ground conditions. *Proceedings of the ISRM International Symposium*. Melbourne, Australia.

Broere W. 2002. Influence of excess pore pressures on the stability of the tunnel face. *Geotechnical Aspects of Underground Construction in Soft Ground*. Toulouse, France, pp 179–184.

Burland J.P. 1995. Assessment of risk of damage to buildings due to tunnelling and excavation. *First International Conference on Earthquake Geotechnical Engineering of Tokyo*. November, pp.1–12.

Carranza-Torres C., Reich T., Saftner D. 2013. Stability of shallow circular tunnels in soils using analytical and numerical models. *In Proceedings of the 61st Minnesota Annual Geotechnical Engineering Conference*. University of Minnesota, St. Paul Campus. February 22.

Chàvez-Garcìa, F.J. & Cuenca J. 1995. Site effects in Mexico City urban zone. A complementary study. *Soil Dyn, Earthquake Eng*, 15, 141–146.

Dickmann T. 2014. The Role of Tunnel Seismic prediction in Tunneling projects: best practices. *Proceedings of Indorock-2014: Fifth Indian Rock Conference*. Delhi, India.

Dickmann T. & Méndez J.H. 2017. Look-ahead seismic investigations during tunneling with shield tunnel boring machines. *The 2017 Word Congress on Advances in Structural Engineering and Mechanics*. Ilsan, Korea.

Faccioli E. & Pessina V. 2000. The Catania Project: Eartquake damage scenarios for a risk area in the Mediterranean. *E. Faccioli & V. Pessina Editors. CNR- Gruppo nazionale per la difesa dei terremoti*. ISBN: 88-900449-0-X.

Field E.H., Clement A.C., Jacob K.H., Aharonian V., Hough S.E., Friberg P.A., Babaian T.O., Karapetian S.S., Hovanessian S.M., Abramian H.A. 1995. Earthquake site response in Giumri (formerly Leninakan), Armenia, using ambient noise observations. *Bull. Seism. Soc*. Am. 85, 349–353.

Lermo J. & Chàvez-Garcìa F.J. 1994. Are microtremors useful in site response evaluation? *Bull. Seism. Soc. Am*. 84, 1350–1364.

Mohammadi J. 2010. Tunnel Face Stability Analysis in Soft Ground by EPB Method (Case Study: Tehran Metro Line 7). *M'Sc Thesis*. Tehran, Iran.

Nakamura Y. 1989. A Method for Dynamic Characteristics Estimation of Subsurface Using Microtremor on the Ground Surface. *Quarterly Report of RTRI*, vol. 30, No. 1, pp. 25–33.

Nakamura Y. 2000. Clear identification of fundamental idea of Nakamura's technique and its applications. *Proceedings of 12th World Conference on Earthquake Engineering*. New Zeland.

Nogoshi M. & Igarashi, T. 1970. On the propagation characteristics of microtremors. *J. Seism. Soc. Japan* 23, 264–280 (in Japanese with English abstract).

Nogoshi M. & Igarashi T. 1971. On the amplitude characteristics of microtremors. *J. Seism. Soc. Japan*. 24, 24–40 (in Japanese with English abstract).

UNI 9916-DIN 4150-3. 1999 "Structural vibration – Part 3: Effects of vibration on structures". www.uni.com.

Tunnels and Underground Cities: Engineering and Innovation meet Archaeology, Architecture and Art, Volume 3: Geological and geotechnical knowledge and requirements for project implementation – Peila, Viggiani & Celestino (Eds)
 ISBN 978-0-367-46583-4

The impact of saturation on the mechanical response of low porosity rocks and implications for tunnelling

M.T. Ahmed Labeid, E.L. Jaczkowski, W. Dossett & M.S. Diederichs
Department of Geological Sciences and Geological Engineering, Queen's University, Canada

J.J. Day
Department of Earth Sciences, University of New Brunswick, Canada

ABSTRACT: Unlined underground pressure tunnels and shafts have been the standard design in Norway since the 1960s, and in some cases, shotcrete is used as both temporary and permanent support. For low-porosity rocks in such an environment, the geomechanical response of the host rock with respect to water saturation under unconfined and confined loading remains ambiguous, and so does the long-term stability of these infrastructures. Researchers have investigated the effect of saturation on the unconfined strength and stiffness, but little literature exists on the confined conditions. To address this gap, this research examines the effect of saturation and confinement on the response of two low-porosity rocks: Cobourg limestone and Point du Bois granite. In this paper, stress-strain analyses are presented to investigate saturation effects in tension, compression, and confined shear. Geomechanical laboratory testing results are applied to simple numerical simulations to compare failure extent and long-term stability predictions, for unsaturated and saturated mechanical properties of these rocks, around a deep pressure tunnel.

1 INTRODUCTION

The history of hydropower in Norway originates almost 100 years ago, and currently more than 99% of the country's power is generated through hydroelectricity due to the favourable topography, stress regime and bedrock geology as well as social and environmental factors (Broch 2016). When the hydropower industry moved underground in the early 1950s, pressure shafts and tunnels were supported with steel liners; however, after successfully operating an unlined shaft at the Tafjord K3 hydropower plant in 1958 at a pressure head of 268 m, the unlined approach became the industry standard method because of its economic feasibility and design stability, as illustrated in Figure 1 (Broch 1982). According to Broch (2016), more than 200 unlined pressure tunnels and shafts are successfully operating in Norway today, and water heads exceeding 1000 m have been reached. In some tunnels and in other countries, shotcrete alone is also used as a temporary or a permanent liner with the assumption that this liner is not watertight during construction or operation. In situ water saturation and confinement conditions can significantly impact the rock mass properties and, therefore, the long-term stability of these underground structures. Thus, the objective of this study is to investigate the effect of these key environmental factors, first, on the intact rock geomechanical properties through laboratory testing, and second, on the results of stability analysis of deep unlined pressure tunnels in brittle rock with very low porosity.

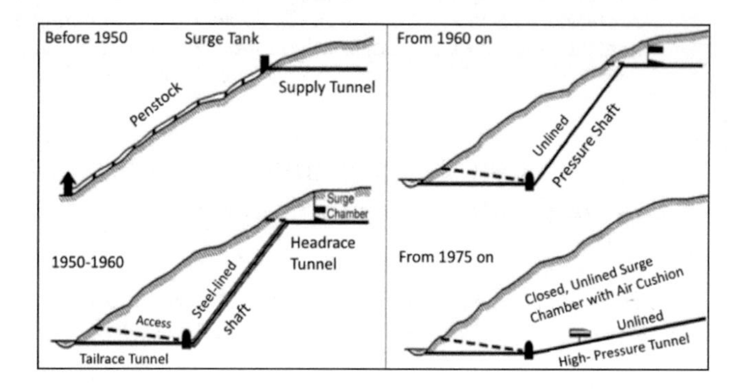

Figure 1. Development of underground hydroelectric power plants in Norway throughout the past century (Modified after Broch 2016).

2 BACKGROUND THEORY

The influence of saturation on rock strength and failure behaviour is the result of two main processes: the mechanical influence of pore water pressure and the fluid-rock chemical interaction, causing subcritical crack growth (Jaeger et al. 2007). The fluid-rock interactions highly depend on both the fluid and the rock compositions. Many researchers have shown the significant decrease in strength due to saturation in sedimentary rocks with argillaceous composition (Colback & Wild 1965; Hawkins & McConnel 1992; Vásárhelyi 2005; Erguler & Ulusay 2009; Yilmaz 2010; Li et al. 2012). However, few studies have been conducted on crystalline rocks. Lajtai et al. (1987) reported a less than 5% decrease in the short-term uniaxial compressive strength and stiffness in Lac du Bonnet granite, yet the study showed that the crack initiation threshold (stress at first damage) was significantly influenced by the presence of water (36% decrease). All these studies, for both sedimentary and crystalline rocks, focused on the rock behaviour under tension and uniaxial compression but did not investigate confined conditions, expect for Li et al. (2012) who investigated this topic in meta-sedimentary rocks. Confinement and the presence of water independently have opposing effects on the rock strength. Therefore, it is crucial to investigate their combined effects and the implication for stability analysis and predictions.

3 LABORATORY TESTING

3.1 *Testing program*

Cobourg limestone and Point du Bois granite were subjected to uniaxial compressive strength (UCS), Brazilian tensile strength (BTS), and triaxial compressive strength (TCS) tests. Multiple saturation conditions were used to vary the water content in the specimens and are classified in two general categories: saturated samples (SAT) and unsaturated or relative room humidity samples (RRH). The various saturated sample conditions will be discussed in the following sections. The number of samples tested under each condition is shown in Table 1. The limestone samples are 75 mm in diameter, while the granite samples are 47.5 mm in diameter.

3.2 *Samples description*

3.2.1 *Cobourg limestone*

The argillaceous Cobourg limestone samples were drilled from blocks retrieved from the St. Mary's Quarry in Bowmanville, Ontario, Canada. This rock has a low porosity and is composed of light grey fine-grained fossil-rich calcite nodules surrounded by dark grey clay-rich, fine-grained material, with these two phases defining the inherent heterogeneity in the samples (Figure 2 left).

Table 1. Summary of saturation testing program for Cobourg limestone and Point du Bois granite.

Rock type	Saturation condition	Number of samples by test method		
		UCS	BTS	TCS
Cobourg limestone	Relative room humidity	3	5	9
	Saturated	12	20	12
Point du Bois granite	Relative room humidity	4	7	8
	Saturated	5	5	8

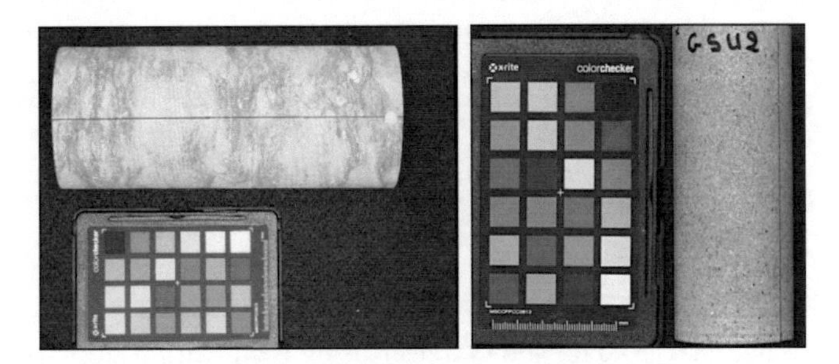

Figure 2. UCS/TCS specimens of Cobourg limestone (left) and Point du Bois granite (right).

3.2.2 *Point du Bois pink granite*

Point du Bois granite samples were retrieved from core boxes provided by Manitoba Hydro. The granite is greyish-pink in color, fine to coarse-grained and phaneritic in texture, and mainly composed of quartz, alkali-feldspars, plagioclase and biotite. Specimens used in the testing program are granular in texture and had no foliations or other visible weakness planes (Figure 2 right).

3.3 *Sample preparation*

UCS samples were prepared following the ASTM D4543 standard (ASTM 2008b) and the related ISRM suggested methods (Fairhurst & Hudson 1999). The BTS specimens were prepared following the ASTM D3967 standard (ASTM 2008a). The TCS specimens were prepared according to both the ASTM D4543 standard (ASTM 2008b) and the ISRM suggested method for triaxial testing (ISRM 1983). All BTS samples were prepared to a length to diameter (L:D) ratio of 0.5, all UCS samples had a 2.5 L:D ratio, and the TCS samples had a 2.5 L:D ratio for the granite and 2.3 L:D ratio (due to size limitations of the available 75mm Hoek cell) for the limestone.

3.3.1 *Saturation and drying procedure and water content measurement*

To examine the effect of saturation on the mechanical properties of Point du Bois granite and Cobourg limestone, different saturation procedures were used to vary the specimens' water content. For the limestone, samples were tested at room relative humidity (RRH), 1- week vacuum saturation with a synthetic pore water solution (SPW), and with 1-week, 1-month, 3-month submersions in the same solution. For the granite, only two saturation conditions were used: 1-month submersion in deionized water and relative room humidity. After mechanical testing, the samples were oven-dried in a VWR forced-air oven for a minimum of one month at 80°C and 50°C temperatures for the granite and the limestone, respectively.

3.4 *Testing procedure and equipment*

All tests were conducted using an MTS 815 servo-controlled closed-loop rock mechanics testing machine. Axial and lateral specimen deformations were measured using one circumferential chain and three 25mm axial extensometers for Cobourg limestone. Two axial and two lateral resistance strain gauges were used for Point du Bois granite specimens. Testing was performed using axial strain control. Curved bearing blocks were used to test the dually strain-gauged BTS specimens, and the standard Hoek cell was used for triaxial testing. Acoustic emissions were recorded during UCS tests to capture the microcracking and damage activity.

4 TESTING RESULTS AND ANALYSIS

4.1 *Measured water content*

Figure 3 shows the average water content values achieved in each testing program (BTS, UCS, and TCS) using different saturation methods. The vacuum saturation and submersion methods yielded the same results in the limestone as well as the 1-month and the 3-month submersion methods. Thus, only two saturation methods (relative room humidity and 1-month submersion) were used for the granite. The limestone observed water content is relatively low (less than 0.8 %), yet it is 10 times higher than that of the granite (less than 0.1 %).

4.2 *Brazilian tensile strength testing*

Brazilian tensile strength (BTS) data shows a decreasing trend with water content in the limestone; however, the data does not perfectly fit the suggested trend (R^2= 0.50), as seen in Figure 4. On the other hand, the granite Brazilian tensile strength seems unaffected by saturation.

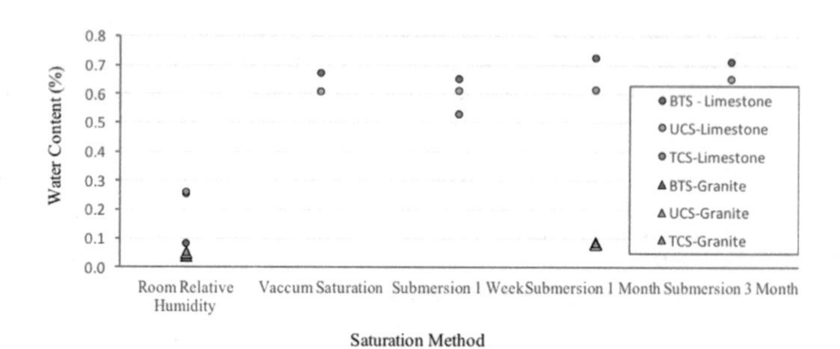

Figure 3. Average water content achieved for each testing program and the different saturation methods used for both limestone and granite.

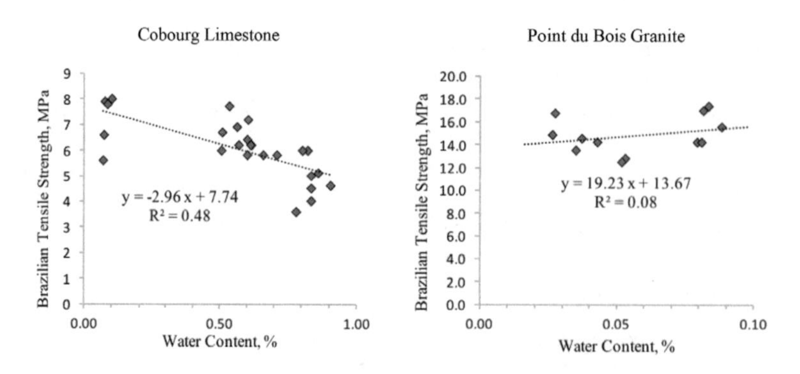

Figure 4. Effects of saturation on BTS of Cobourg limestone (left) and Point du Bois granite (right).

4.3 *Uniaxial compressive strength testing*

UCS test data was used to determine the UCS, Young's modulus, Poisson's ratio, and Crack initiation (CI) and Crack Damage (CD) thresholds. The methods used to estimate an average CI are lateral strain, crack volumetric strain, inverse tangent lateral stiffness (Ghazvinian et al. 2012), instantaneous Poisson's ratio and acoustic emissions. The methods used to determine an average CD are axial strain, volumetric strain, instantaneous Young's modulus, and acoustic emissions. In Cobourg limestone, UCS and CD show a decreasing trend with increasing the water content, while the effect on CI is minor. In Point du Bois granite, the effect of saturation is most evident on the CD threshold, while the data shows no impact of water on UCS and CI. These results are shown in Figure 5. A decreasing trend of Young's modulus with water content is present in both the limestone and the granite; however, the data does not fit the suggested trends very well (low R^2 values). On the other hand, Poisson's ratio remains insensitive to the water content increase in both rocks (Figure 6).

4.4 *Triaxial compressive strength testing*

TCS data was used to determine confined CI, CD, and peak strength in RRH and SAT conditions. Generalized Hoek-Brown envelopes (Hoek et al. 2002) were used for the three thresholds. For the Cobourg limestone, the triaxial testing results show a significant decrease of peak strength and CD envelopes with saturation, while CI remains relatively unaffected

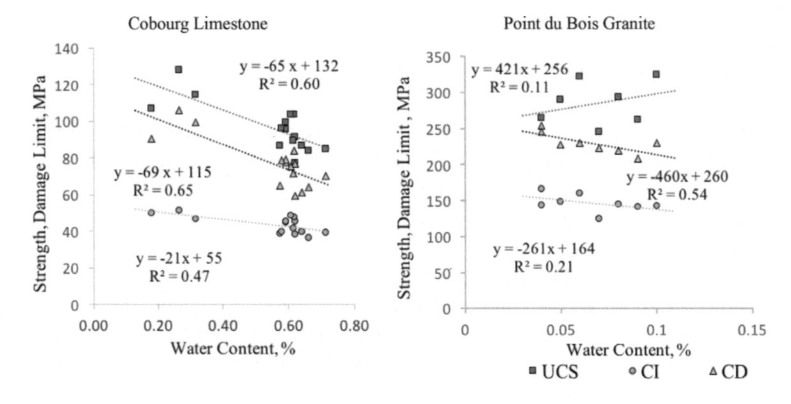

Figure 5. Strength and damage thresholds versus water content in Cobourg limestone (left) and Point du Bois granite (right).

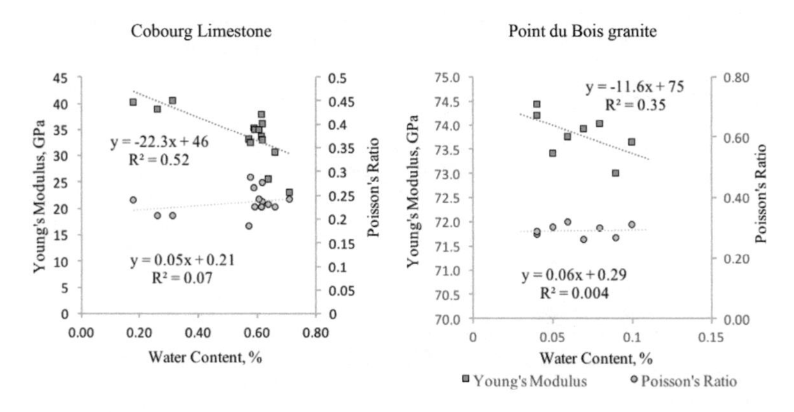

Figure 6. Stiffness versus water content for Cobourg limestone (left) and Point du Bois granite (right).

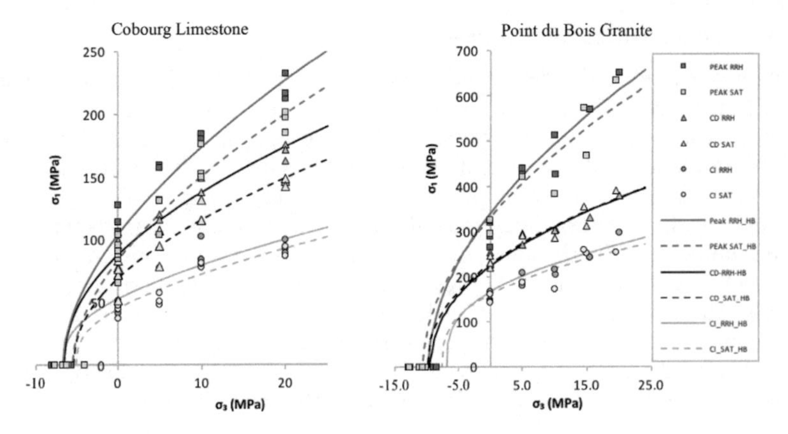

Figure 7. Effects of saturation on damage threshold and strength envelopes in Cobourg limestone (left) and Point du Bois granite (right).

(Figure 7 left). However, the data shows that saturation has no significant effect on the friction angle. Minor decreases in peak strength and CI envelopes are seen in the granite results, while the CD envelope remains nearly the same for both saturation conditions (Figure 7 right).

5 HYDROELECTRIC PRESSURE TUNNEL STABILITY ANALYSIS

This section of the study investigates the stability of a deep section of an unlined pressure tunnel in four different materials: RRH limestone, SAT limestone, RRH granite, and SAT granite.

5.1 *Model and software description*

RS2 9.0 (Rocscience 2018), a 2-dimensional continuum finite element method program, was used to examine the impact of mechanical properties on instability around a deep 10 m diameter circular pressure tunnel with a maximum depth of 1500 m (Figure 8), typical of Norwegian practice (Broch 2016). The maximum water head during tunnel operation is 400 m.

5.2 *Model settings and assumptions*

5.2.1 *Field stress*

Plane strain analysis is used in this model. A k ratio of 2 is assumed in all directions. Vertical stress is assumed to be lithostatic resulting in the stress state in Table 2.

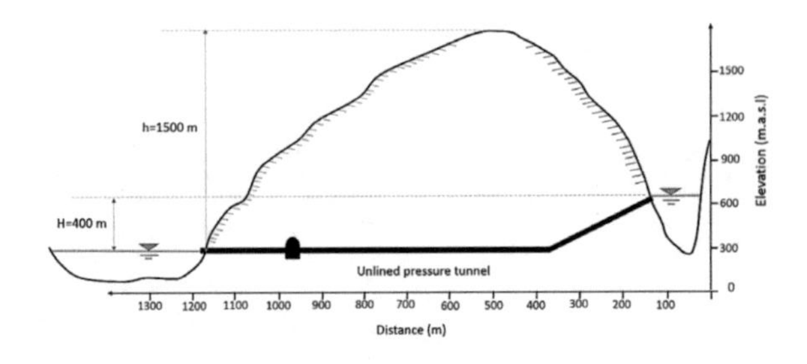

Figure 8. A schematic of the underground hydropower plant.

Table 2. Principal stresses magnitudes at tunnel depth.

Stress (MPa)	Cobourg Limestone	Point du Bois granite
Major Principal Stress, σ_1	78.2	76.8
Minor Principal Stress, σ_3	39.1	38.4
Out-of-Plane Stress, σ_z	78.2	76.8

5.2.2 *Model setup and stages*

The model consists of a 10 m diameter circular tunnel with a 100 m diameter circular external model boundary. The model is finely meshed (10,000 elements) to capture small deformations, and it is excavated in 13 stages to allow for ground relaxation and to simulate the 3D tunnel deformation. Stage 1 is unexcavated; a decreasing internal pressure is applied to the tunnel wall in stages 2 to 11, a shotcrete liner is installed in stage 12, and an internal pressure of 3.92 MPa is applied to the tunnel wall in stage 13 to simulate the hydrostatic pressure in the operating pressure tunnel.

5.2.3 *Material properties*

The Damage Initiation and Spalling Limit (DISL) model by Diederichs (2007) was used to simulate and capture the brittle rockmass behaviour around the tunnel. This model is based on the damage thresholds determined from lab testing (Table 3) and results in the model parameters shown in Table 4.

5.2.4 *Groundwater*

The groundwater condition is simulated by varying the pore water pressure in the different excavation stages. The tunnel excavation will cause a cone of depression in the water table before returning to its initial state (Figure 9). For simplicity, the local tunnel zone is assumed to be drained during construction with restoration of full pressure conditions after lining and during operation.

Table 3. Material properties from testing results used in numerical models.

Rock Type		Relative Room Humidity					Saturated				
	ρ (kg/m^3)	UCS (MPa)	CI (MPa)	T (MPa)	E (GPa)	ν	UCS (MPa)	CI (MPa)	T (MPa)	E (GPa)	ν
Limestone	2660	109.3	54.45	5.03	40	0.22	87.7	45.59	4.28	34	0.23
Granite	2610	340	167	10.4	74	0.28	332	161	11.5	73	0.30

Table 4. Generalized Hoek-Brown parameters obtained from the DISL model (as per Diederichs 2007) for both Cobourg limestone and Point du Bois granite under different saturation conditions.

Rock type	DISL Generalized Hoek-Brown Parameters	Relative Room Humidity		Saturated	
		Peak	Residual	Peak	Residual
Limestone	UCS (MPa)	109	109	88	88
	s	0.0616	0.0001	0.0730	0.0001
	m	1.338	8	1.496	8
	a	0.25	0.75	0.25	0.75
Granite	UCS (MPa)	340	340	332	332
	s	0.0582	0.0001	0.0553	0.0001
	m	1.903	10	1.595	10
	a	0.25	0.75	0.25	0.75

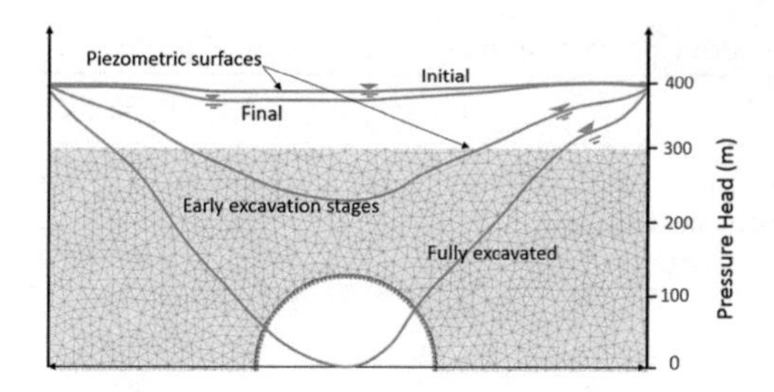

Figure 9. Effect of tunnel excavation on the piezometric surfaces.

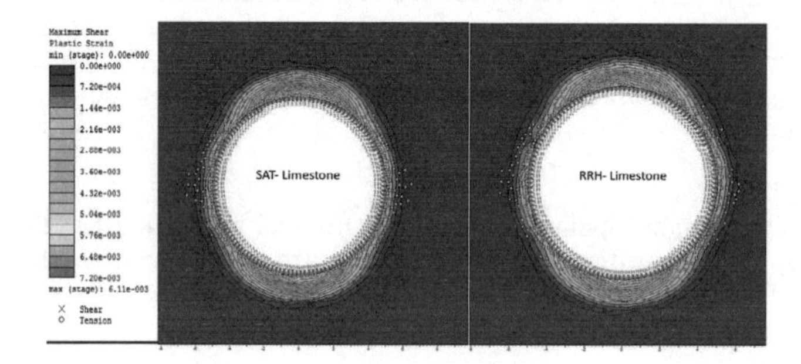

Figure 10. Maximum shear plastic strain around a pressure tunnel in saturated (SAT) limestone (left) and relative room humidity or unsaturated (RRH) limestone (right).

5.3 *Analysis and results*

To investigate the effect of rock mass saturation on the tunnel stability, yield state indicators were used to indicate internal damage within the rockmass. According to Diederichs (2007), plastic shear strain is the best indicator to assess if the rock is observably failing (as opposed to being slightly damaged). The plastic shear strain indicator was used to compare the results obtained from both saturation conditions. As shown in Figure 10, the maximum shear plastic strain in the tunnel floor and roof is higher when using SAT limestone versus RRH limestone. However, this difference is not very significant. Both tensile and shear failures are predicted around the entire excavation in both rockmasses, and the damaged area extends few meters away from the excavation boundary. This is primarily due to the high horizontal in-situ stresses, the high pore water pressure, and the seepage forces acting on the tunnel.

In the granite (Figure 11), the maximum shear strain estimated in the SAT material is three orders of magnitude higher than that of the RRH material. Tensile failure is predicted in the excavation sidewalls in both materials as well as in the roof and floor of the RRH material.

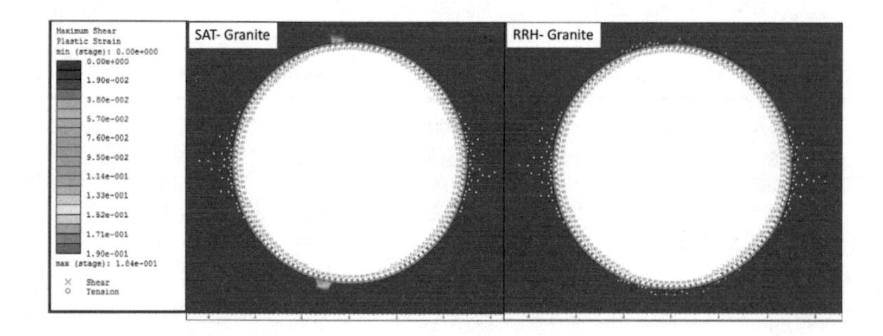

Figure 11. Maximum shear plastic strain in RRH granite (left) and SAT granite (right).

Table 5. Comparison of rockmass behaviour around the pressure tunnel for both limestone and granite (F=floor, R=roof, S=sidewalls).

Rock type	Parameter	Saturated	Relative Room Humidity
Limestone	Roof/floor Max. σ'_1 (MPa)	135	135
	Max. σ'_3 (MPa)	42	42
	Max. Total Displacement (m)	0.02	0.02
	Depth of yield (m)	5.55 (F/R) 3.0 (S)	3.5 (F/R) 2.4 (S)
Granite	Roof/floor Max. σ'_1 (MPa)	177	149
	Max. σ'_3 (MPa)	39	41
	Max. Total Displacement (m)	7.1	$7.6*10^{-3}$
	Depth of yield (m)	0.4 (F/R) 2.2 (S)	0.8 (F/R) 2.2 (S)

Other indicators were also used to compare the rock mass behaviour around the excavation in the different materials: the maximum and minimum induced effective stresses, the total displacement, and the depth of yield. For the limestone, there is no difference in these parameters for both saturation conditions, except for the depth of the yield zone. For the granite, a significantly high total displacement is predicted at the tunnel sidewalls in SAT granite. A summary of these results is shown in Table 5.

6 CONCLUSIONS

The results of this study show that a full saturation is not necessarily required to see the impact of water content on the rockmass properties, especially in limestone under both unconfined and confined short-term loading. Laboratory testing data shows a moderate to a significant negative effect of saturation on certain geomechanical properties of Cobourg limestone: UCS, CD, BTS, confined peak strength, confined CD, and Young's modulus. CI shows a minor decrease with water content. Poisson's ratio remains unaffected; the effect of saturation can be unnoticed when calculating this ratio since the hydrostatic pore water has, theoretically, the same influence on both the axial and the lateral deformations. Numerical models in the limestone show no major differences between the two saturation conditions. On the other hand, saturation has little to no effect on the granite test core; however, the RS2 models show significant differences in the plastic deformation and displacement around the tunnel between saturated and unsaturated material properties. Very high displacement is predicted around the tunnel in the saturated granite. This could potentially lead to catastrophic brittle failures in unlined pressure tunnels. This research is on-going on granite specimens, where more data is being collected to better characterize the effect of saturation on laboratory results. Different saturation technics will be included in future laboratory testing to potentially provide a wider range of water content values.

ACKNOWLEDGEMENTS

This work is supported by funding from the Natural Sciences and Engineering Research Council of Canada and by the Nuclear Waste Management Organization of Ontario.

REFERENCES

ASTM. 2008b. D4543: Standard practices for preparing rock core as cylindrical test specimens and verifying conformance to dimensional and shape tolerances. *ASTM International.*

ASTM. 2008a. D3967: Standard test method for splitting tensile strength of intact rock core specimens. *ASTM International.*

Broch, E. 1982. The Development of unlined pressure shafts and tunnels in Norway. *International Society for Rock Mechanics and Rock Engineering; Proc. ISRM International symp.*, Germany, 26–28 May 1982. Aachen: Balkema.

Broch, E. 2016. Planning and utilisation of rock caverns and tunnels in Norway. In urban underground space: a growing imperative perspectives and current research in planning and design for underground space use, *Tunnelling and Underground Space Technology* 55: 329–38.

Colback, P.S.B. & Wild, B.L. 1965. The influence of moisture content on Compressive Strength of Rock. *Proc. 3rd Canadian Rock Mechanics. Symp.*, Canada, 65–83.

Diederichs, M.S. 2007. The 2003 Canadian geotechnical colloquium: mechanistic interpretation and practical application of damage and spalling prediction criteria for deep tunnelling. *Canadian Geotechnical Journal* 44(9): 1082–1116.

Erguler, Z.A. & Ulusay, R. 2009. Water-induced variations in mechanical properties of clay-bearing rocks. *International Journal for Rock Mechanics* 46: 355–70.

Fairhurst, C.E. & Hudson, J.A. 1999. Draft ISRM suggested method for the complete stress-strain curve for intact rock in uniaxial compression. *International Journal for Rock Mechanics* 36: 279–89.

Ghazvinian, E., Diederichs, M.S. & Martin, D. 2012. Identification of crack damage thresholds in crystalline rock. *International Society for Rock Mechanics and Rock Engineering. ISRM International Symp. EUROCK*, Stockholm, Sweden, 28–30 May 2012.

Hawkins, A.B. & McConnel, B.J. 1992. Sensitivity of Sandstone Strength and Deformability to Changes in Moisture Content. *Quarterly Journal of Engineering Geology & Hydrogeology* 25: 115–30.

Hoek, E., Carranza, C., & Corkum, B. 2002. Hoek-Brown failure criterion. In *NARMS-TAC Joint Conference*. Toronto, Canada. 267–273.

ISRM. 1983. Suggested methods for determining the strength of rock materials in triaxial compression: revised version. *Int. J. of Rock Mech. and Mining Sci & Geomech. Abstracts.*

Jaeger, J.C., Cook, N.G. & Zimmerman, R.W. 2007. *Fundamentals of rock mechanics*. 4th ed. Malden, USA: Blackwell.

Lajtai, E.Z., Schmidtke, R.H. & Bielus, L.P. 1987. The effect of water on the time-dependent deformation and fracture of a granite. *Int. J. of Rock Mech. and Mining Sci & Geomech. Abstracts* 24 (4): 247–55.

Li, D., Wong, L.N.Y., Liu, G. & Zhang, X. 2012. Influence of water content and anisotropy on the strength and deformability of low porosity meta-Sedimentary rocks under triaxial compression. *Engineering Geology* 126: 46–66.

Rocscience Inc. 2018. RS2 – Phase2 Version 9.0. www.RocScience.com, Toronto, ON, Canada.

Vásárhelyi, B. 2005. Statistical analysis of the influence of water content on the strength of the Miocene limestone. *Rocks Mechanics and Rock Engineering* 38: 69–76.

Yilmaz, I. 2010. Influence of water content on the strength and deformability of gypsum. *International Journal for Rock Mechanics and Mining Sciences* 47: 342–47.

Tunnels and Underground Cities: Engineering and Innovation meet Archaeology, Architecture and Art, Volume 3: Geological and geotechnical knowledge and requirements for project implementation – Peila, Viggiani & Celestino (Eds)
 ISBN 978-0-367-46583-4

Geological and geotechnical key-factors for tunnel design of the new Naples-Bari High-Speed railway line in Southern Italy

A. Amato, G. Quarzicci, A. Sciotti, A. Pigorini & A. Nardinocchi
Italferr S.p.A., Technical Department, Rome, Italy

ABSTRACT: The Naples-Bari High-Speed railway line, in Southern Italy, has an overall length of more than 300 km, of which 120 km are currently in design or construction stage, including about 68 km of underground works (cut and cover and bored tunnels). The railway line crosses the Apennines, one of the main Italian mountain chains, mainly composed by heterogeneous soils and rock-masses, characterized by a high degree of complexity resulting from their depositional and tectonic history. The paper describes the main geological features and geotechnical factors which have represented the key-elements to develop effective design solutions for the new underground works.

1 INTRODUCTION

The "Ferrovie dello Stato Italiane" (Italian State Railways) Group is committed to the development and re-launch of the railway system in Southern Italy for national and international strategic aspects. In this context, the Naples-Bari High-Speed (HS) railway line is being designed and developed to improve the competitiveness of rail transport and to promote the integration and connection to the North of the country and Europe; the Naples-Bari route is part of Trans-European Scandinavian-Mediterranean (Helsinki-LaValletta) Corridor, which, reached the city of Naples, is divided in two direction: the first goes towards South, with the Naples-Reggio Calabria-Messina-Catania-Palermo section, the second runs South-East towards Bari (Figure 1).

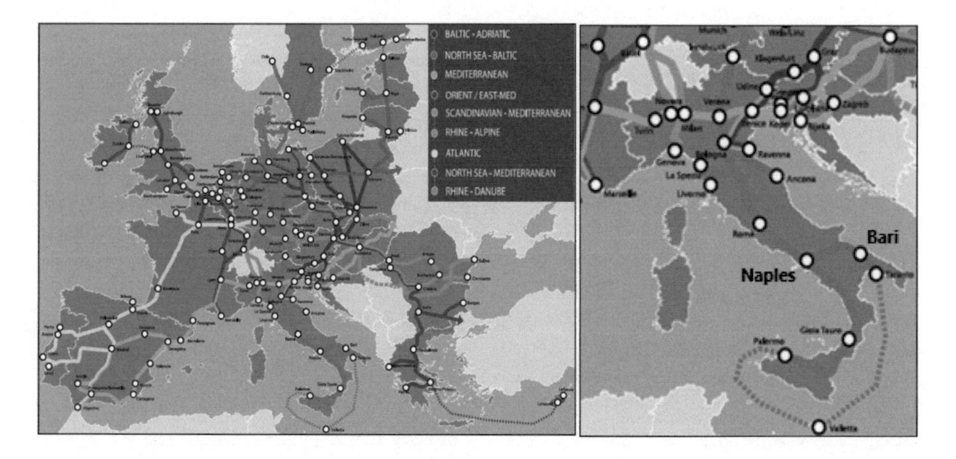

Figure 1. The Naples-Bari HS railway line along the TEN-T Corridor Scandinavian-Mediterranean.

The Naples-Bari HS railway line goes through the Southern Apennines, crossing a geological and geomorphological context extremely variable and with a high degree of complexity due to depositional and tectonic history and recent landscape evolution.

Extensive geological and geomorphological studies have been performed and a comprehensive geotechnical investigation has been carried out to define the key-factors and the critical issues to be addressed in the design of the new underground works.

2 THE NAPLES-BARI HS RAILWAY LINE

The doubling of single track sections of the existing line and new double-track routes are planned for the Naples-Bari HS line, pursuing the best solutions able to guarantee faster connections and an increase in railway transport and raising the effectiveness of the existing infrastructure through an increase in accessibility to the service in the crossed areas.

The new line has an overall length of about 180 km from Naples to Foggia and it is divided into the following sections:

1. Naples – Cancello section;
2. Cancello – Dugenta Frasso Telesino section;
3. Frasso Telesino – Vitulano section;
4. Vitulano – Apice section;
5. Apice – Hirpinia section;
6. Hirpinia – Orsara section;
7. Orsara – Bovino section;
8. Bovino – Cervaro section;
9. Foggia junction.

Three sections are in the operation stage, two sections are under construction, for two sections a public tender has been recently launched and for two sections the design stage is now underway, as represented in Figure 2.

About 120 km are going to be constructed for the completion of the line, which develops for a large part underground (68 km, bored and cut and cover tunnels (Figure 3)).

The overall cost of the investment is about Euro 6 billion and the operation of the whole line is planned in 2026.

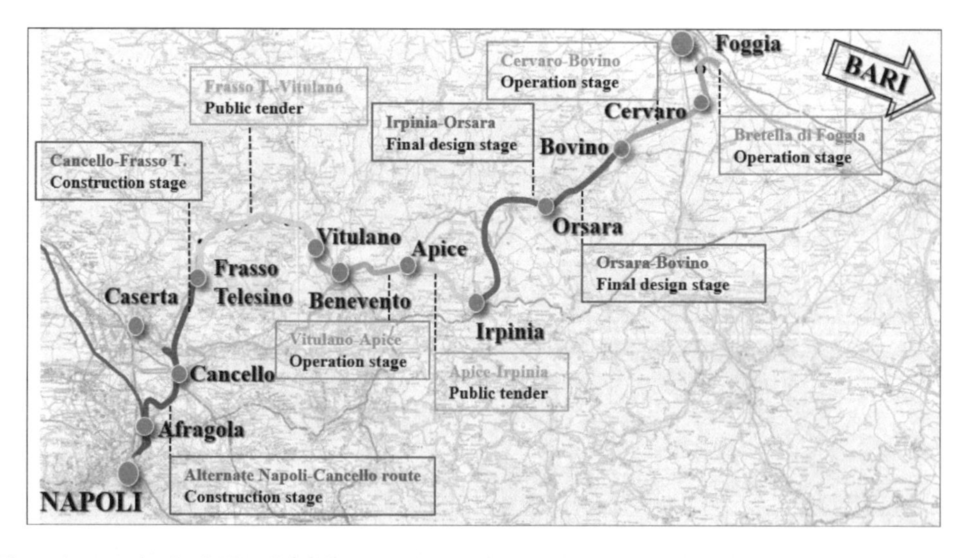

Figure 2. Naples-Bari HS railway line – project sections.

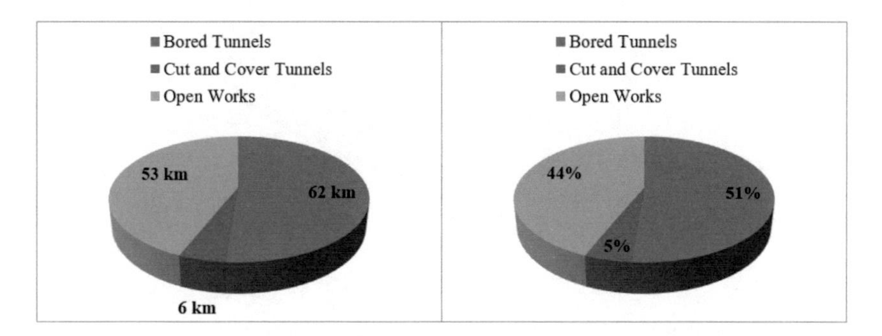

Figure 3. Infrastructures across the Naples-Bari HS Railway Line (length and percentage) (Sciotti et al., 2017).

2.1 *New Naples-Cancello section*

This section has a length of 15.5 km, of which 3.3 km will be realized by cut and cover tunnels. The Afragola HS Station, designed by the famous architect Zaha Hadid (in operation since 2017), and the cut and cover Casalnuovo Tunnel, with a length of about 2.5 km, are the most significant works built/under construction.

2.2 *Cancello-Dugenta Frasso Telesino section*

This section has an overall length of about 16.5 km, with the Monte Aglio tunnel having a length of 4.2 km and bored by conventional method. It is a double-track tunnel having intermediate safety adits every 1000 m.

2.3 *Frasso Telesino-Vitulano section*

The section is about 30.4 km long and consists of doubling the track, partly flanking the historic line and partly with a new route. The alignment is divided into 3 sub-sections. The total length of cut and cover and bored tunnels is 11 km; the bored double-track tunnels are 7 and conventional excavation is planned for them. For tunnels longer than 1 km intermediate safety adits have to be realized every 1000 m.

2.4 *Vitulano-Apice section*

The work for the doubling of the Vitulano-Apice section, about 21 km long around Benevento station, was carried out in the 1980s and the section was activated in 2008.

2.5 *Apice-Hirpinia section*

This section has a length of about 18.7 km, developing mainly underground (13 km) with 3 bored double-track tunnels (Grottaminarda tunnel, Melito tunnel, Rocchetta tunnel), which will be excavated by conventional and mechanized method. For all the tunnels intermediate safety adits have been designed.

2.6 *Hirpinia-Orsara*

This section has a length of about 28 km and develops almost completely underground across the Hirpinia tunnel having a length of about 27 km. The underground work has a twin single-track tunnels configuration with cross passages every 500 m. The tunnel will be provided of an underground safety area (Figure 4) designed as a Fire Fighting Point and to allow rescue teams access; it is realized by means of an escape tunnel between the twin tunnels, connected

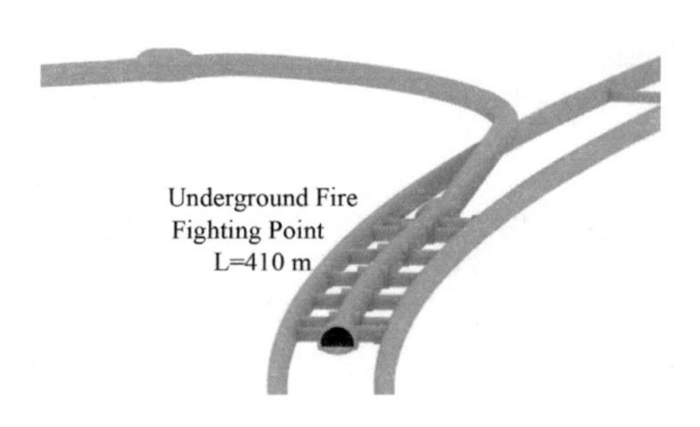

Figure 4. The underground safety area of the Hirpinia tunnel.

to them by cross-passages. The Building Information Modelling method was used for the design of all the infrastructures in this section (Figure 4). The tunnel will be excavated by conventional and mechanized method.

2.7 *Orsara-Bovino*

This section has a length of about 12.4 km, which develops almost entirely underground with the Orsara Tunnel, having a length of about 10 km. The underground work has a twin single-track tunnels configuration with cross passages every 500 m. The tunnel will be excavated by mechanized method.

2.8 *Bovino-Cervaro and Foggia Junction*

The 'Bovino-Cervaro' section, having a length of 23 km long, is in operation since June 2017; the Foggia junction allows the connection of the Naples-Foggia railway line to the existing Pescara-Bari railway line; it is in operation since July 2015.

3 GEOLOGICAL FRAMEWORK

The Naples-Bari railway line develops mainly underground from Apice to Bovino (Figure 2), passing through the Southern Apennine Chain by means of 5 tunnels.

The Southern Apennine Chain, that extends along the Campania-Calabria section of the Tyrrhenian coast, developed in two distinct stages that goes from the Late Oligocene up to the Pliocene. Three main tectonic domains can be identified: the Apennine Mountain Chain, the Apennine Foredeep Basin (or Bradanic Trough) and the Adriatic-Apulia Foreland (Figure 5).

The Apennine Chain domain is constituted by a thick sequence of thrust sheets, referred to three distinct stratigraphic and structural units, which mainly consist of marine deposits. The lower part of the sedimentary succession is composed by marine pelitic deposits and calcareous-marly sediments, with intercalation of carbonate and siliceous rocks. Towards the upper part of the sedimentary sequence, in stratigraphic unconformity, there are turbiditic formations, characterized by the alternation of sandstone or calcareous rocks with clay and clay shales. These Units have undergone tectonic movements with horizontal displacements of hundreds of kilometers from the original sedimentary basins; the displacements began in Lower Terthiary in more than one phase, first involving the sedimentary basin more distant from the Foreland.

In the Apennine Chain domain, the Units that are crossed by the Naples-Bari HS railway line are the following: 'Argille Varicolori', 'Flysch Rosso', 'Argilliti Policrome del Calaggio' and 'Argilliti con gessi di Mezzana di Forte', which are made up of clay and clay shale with a chaotic, scaly structure. The Unit called "Flysch di Faeto" is characterized by the alternation of calcareous/sandstone strata with clays layers: the investigations in the areas crossed by tunnels have shown that the lapideous component appear to be prevalent. All these units belonging to the Apennine chain domain are characterized by an intrinsic heterogeneous composition (as for turbiditic formations) and structural features (fault, joint and discontinuities) derived from the tectonic stresses.

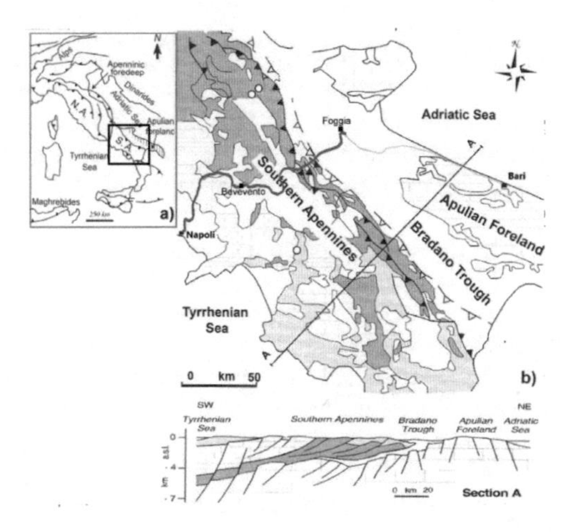

Figure 5. Layout and schematic cross-section of the three main geological domains in Southern Italy (modified from Scrocca et al., 2005).

Close to the Apennine Mountain Chain, there is the Foredeep Basin (Bradanic through), and more externally to this the Adriatic-Apulia Foreland, a region untouched by the orogenic events that generated the mountain system (Figure 5).

In the Foredeep Basin the stratigraphic series is formed by autochthonous terrigenous sediments with prevailing pelitic deposits of Plio-Pleistocene origin, only partially involved by the more recent thrust fault. At the Chain front large allochtonous masses brought about the chain are intercalated in this complex. Below the Plio-Pleistocenic deposits, there are sandy-silty deposits of the Sin-Orogenic Unit belonging to the Late Messinian. Foredeep Basin contains formations mainly composed of silty clays, marly clays and sandy clays like "Peliti di Difesa Grande" Unit and the "Vallone Meridiano" Unit.

The complex structural layout of the more external sectors of the Apennine Chain and of the transitional belt between the Foredeep Basin and the Adriatic-Apulia Foreland defines the seismic nature of Central-Southern Italy. The central zone of the Southern Apennines features the highest occurrence of seismic activity due to the many orthogonal fault systems on an approximately NW-SE axis present along the Apennine Chain that have generated its overall uplifting and the formation of important intermountain basins.

The presence of the various sedimentary successions and the numerous tectonic structures of the Southern Apennine also features considerable hydrogeological complexity, due to the lateral and vertical juxtaposition of highly permeable calcareous-marly units with less permeable sedimentary successions.

As these deposits are geologically and structurally favourable to the presence of gas, and the alternating of sediments with different permeability allows the presence of underground pockets of methane gas, studies based on the geological setting analysis and on monitoring carried out during the design phase, allowed to evaluate the risk of potential incoming flows during the tunnel excavation. The highest rank of risk has been identified in the central part of the railway alignment (Hirpinia-Orsara and Orsara-Bovino sections), which has been classified in class 2, with reference to the Interregional Note, "Underground works. Excavation in gassy terrains – Grisù 3rd edition ", which constitutes a standard reference in the Italian design practice for conventional excavation in gassy contexts.

4 GEOMORPHOLOGICAL FACTORS

The Naples-Bari HS railway line crosses a region with a high susceptibility to landslides: this feature has been indeed one of the main conditioning factor in the feasibility assessment of the whole project. Landslides and deformational phenomena are widespread and common where clay and clay shales outcrop: the slopes are gentle and characterized by geomorphological features typical of unstable area, such as an undulated surface, with depletion and accumulation zones (Figure 6). The intrinsic mechanical properties of clay and clay shales, the recent evolution history and the weathering processes, whose effects can extend to tens of metres, are the factors favouring the occurrence of instability (e.g. D'Elia et al., 1998).

Phenomena such as earthflows and slide prevail, associated with creep and slow deformations affecting the shallow and weathered part of the soil. Some landslides are at present active, many of them are characterized as quiescent, but susceptible to reactivate as a consequence of a modification of external conditions or actions. In some cases, aerial photographs have revealed ancient landslide bodies, partly reshaped or cancelled by the recent morphological evolution: these phenomena often have considerable dimensions, with thicknesses of tens of meters, affecting the slopes from the top up to the toe.

The landscape has different features where the lapideous component is predominant: steep slopes prevail and instability phenomena, such as rockfall and debris slide are less frequent.

The state of activity of the landslide phenomena along the Naples – Bari railway line has been also investigated by means of satellite radar monitoring, using the SqueeSAR™ (Ferretti 2014) differential interferometric analyses technique: superficial ground displacements are analyzed over the years estimating the displacements of radar targets present on the ground and visible from the satellite. For some active landslides, displacement rates are few millimeters per year with an average trend uniform over the years (Figure 7).

For landslide phenomena which have been identified as critical by the geological studies and SqueeSAR ™ monitoring, a proper geotechnical instrumentation has been implemented to monitor deep displacements and piezometric levels.

Figure 6. Hirpinia-Orsara section – Geomorphological map and typical landscape in clay slopes.

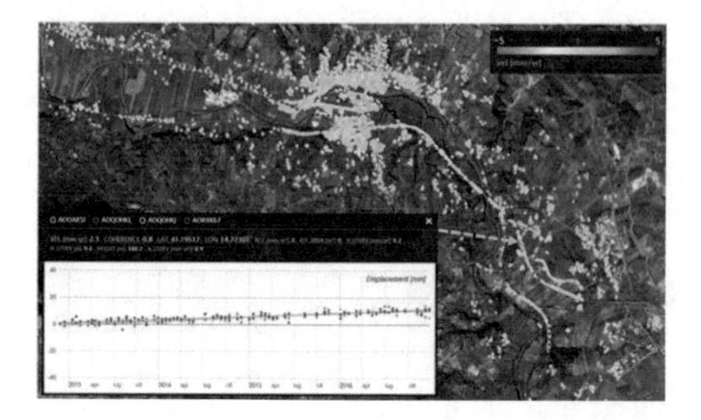

Figure 7. Satellite data with some time-depending series of monitored points in a landslide area.

5 GEOTECHNICAL FACTORS

The underground works of the Naples-Bari HS railway line cross, under variable conditions of coverage, several different geotechnical units, ranging from rock-masses (e.g.: the limestone of *Faeto Flysch*), to clay and clay shales (e.g. the '*Argille Varicolori*').

Due to their peculiar mechanical behaviour related to tunnel excavations and the significant width of outcropping along the central part of the railway alignment, a particular attention has been paid to the characterization of the so-called 'Argille Scagliose' geotechnical unit, which includes the following geological units: '*Argille Varicolori*', '*Argilliti policrome del Calaggio*', '*Argilliti con gessi di Mezzana di Forte*', '*Monte Sidone*' and '*Flysch Rosso del Frigento*'.

The 'Argille Scagliose' can be classified as silty clays and clayey silt, characterized by a high and very high fissuring intensity, with a millimetric fissure spacing and random orientation; the fissures divide the clays into lens-shaped elements ('scales' or shear lenses), having dimensions from some millimeters to few centimeters (Figure 8, left).

These scales, made up of well-oriented particles, have a high consistency, but their external surface is smooth, often slickensided, thus characterized by a low shear strength. In some cases, these structural features are partly or fully cancelled and the soil is characterized by a clayey matrix; some rare small lapideous fragments can also be identified (Figure 8, right).

These formations can be defined as "structurally complex" clayey soils and, according to Esu, 1977 and AGI, 1979, can be generally classified into 'A2′ group (sheared clay shales and shales), or as 'B3′group (disarranged layers of competent rock and clay or clay shales with a chaotic structure) for the Flysch Rosso unit, which can evolve to 'C' group (small lithorelicts or lapideous fragments surrounded by a clay matrix), as a consequence of weathering actions or gravitational deformations.

Regarding the physical properties, these clays are characterized by high value of limit liquid (LL), ranging from 40% to 90% and plasticity index (PI) from 20% to 60%; the natural water content (w_n) is generally included in a range between 15% and 30%, always close or lower than the plastic limit (PL), with values of consistency index (CI) variable between 1 and 1.3 (Figure 9). The clay fraction (CF) varies from 20% to 80% and activity is high ($A \simeq 1$). The clays can be classified as CH, according to the USCS classification.

Extensive geotechnical investigations (on-site and in laboratory) have been carried out. Namely, several geophysical tests have been performed involving mainly cross-hole and down-hole tests together with seismic refraction and MASW tests.

More than 150 undisturbed samples have been retrieved and tested; 73 triaxial compression tests, 77 drained direct shear tests were performed to assess shear strength, while 13 swelling tests under oedometric conditions and 15 Huder-Amberg oedometric tests were carried out to investigate the potential swelling behaviour.

The stress-strain curves obtained in triaxial tests show a ductile behaviour: the deviatoric stress increases along with the axial deformation, reaching an asymptotic value (Figure 10), in some cases a slight strength reduction has been observed. The triaxial specimens show different failure behaviours (Figure 10), as a consequence of the presence of fissures and shear planes, along which the overall shear surface develops, with a localization of deformations.

Figure 8. The scaly clay structure at the specimen laboratory scale.

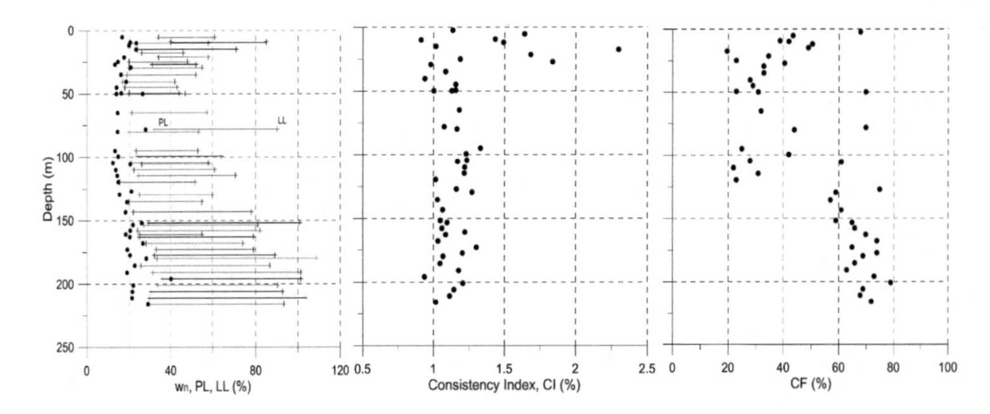

Figure 9. 'Argille Scagliose': water content, Atterberg limits, consistency index and clay fraction.

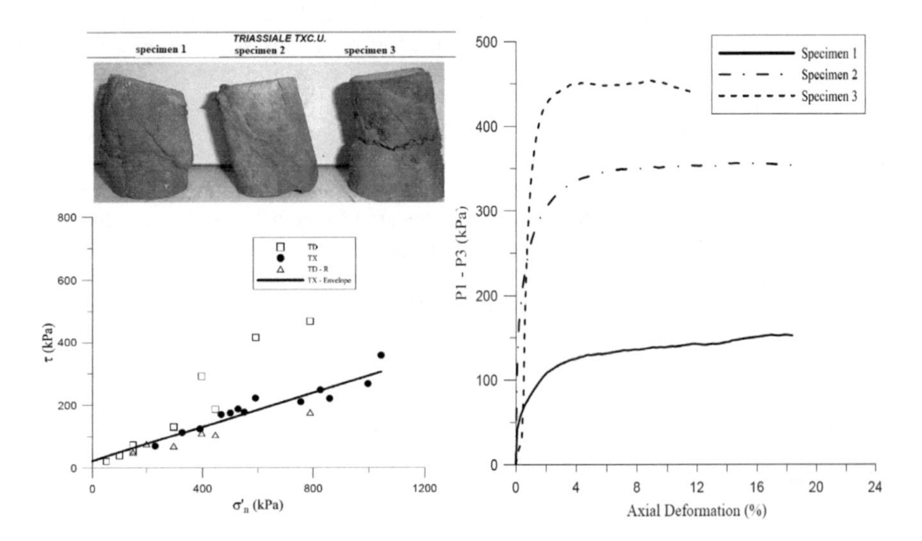

Figure 10. 'Argille Scagliose': results of triaxial compression tests and a comparison between drained direct strength (open squares), triaxial strength (solid circles), triaxial best fit (black line) and residual strength (open triangles).

The triaxial shear tests and drained direct shear tests define cohesion values (c') ranging from 20 to 40 kPa and friction angle values (φ') varying between 16 and 26°. The low values of the shear resistance, close to the residual ones, depend on the structural features of the scaly clays, as it has been widely demonstrated for such materials (e.g.: Bilotta 1987; D'Elia et al. 1998).

Shear parameters from direct shear tests are generally higher than the one derived from triaxial tests, as the shear failure developed not parallel to the shear lenses (Figure 10).

In order to identify the potential swelling behavior of the 'Argille Scagliose', oedometric tests have been carried out according to the Huder-Amberg method (Romana & Serón 2006). The values of the swelling coefficient *k* (proportional to the soil swelling capacity) derived from these tests are in the range between 1.6 – 6.5. These values are shown in Figure 11 in comparison with available literature data concerning other soils with swelling behaviour (Barla 2008; Bonini et al. 2007).

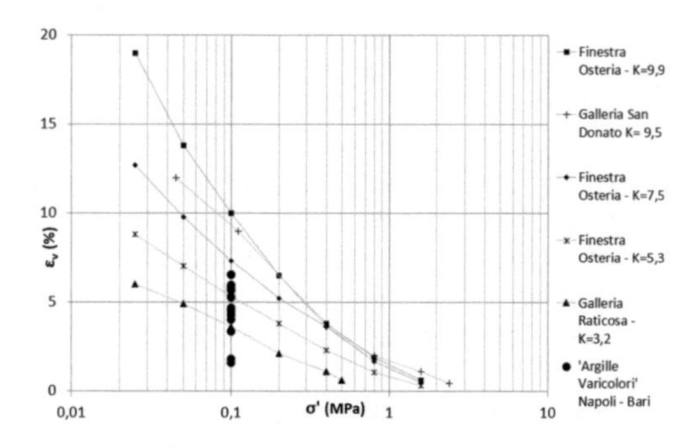

Figure 11. Comparison between k values derived from Huder-Amberg tests.

6 MAIN CRITERIA FOR TUNNEL DESIGN

Underground works along the new Naples-Bari railway line cross a geological and geomorphological context extremely variable and with a high degree of complexity due to depositional and tectonic history and recent landscape evolution.

As for geomorphological context, the widespread landslides phenomena and the potential interaction with the new infrastructures has been the main issue to address in the early stage of design. Thus, to mitigate the landslide risk the underground alignment has been considered the most effective solution, under-passing the critical areas of unstable slopes, as for Hirpinia tunnel which runs through the 'Argille Scagliose' Unit, outcropping along the railway alignment for a length of about 6 km (Figure 12).

At the adits or for shallow overburden conditions, interaction analyses have been carried out to verify if the design solutions adopted for tunnels excavation are able to avoid or minimize interaction effects between tunnel and unstable slope.

The geotechnical behaviour of most of the rocks and soil involved by tunneling is strongly affected by the heterogeneity and presence of discontinuities, both at the scale of the laboratory sample and engineering works as for the 'Argille Scagliose' Unit. The geotechnical characterization, still in progress for the last two sections currently on design stage, indeed highlighted the complexities and the difficulties in modelling and predicting the behaviour of the ground mass during excavation.

Geotechnical studies and characterization allowed to identify uniform sections along the alignment of underground works and analytical or numerical modelling helped to predict the stress-strain response of the ground in the absence of stabilization measures. The prevailing behaviours are those characterized by core-face which is stable in the short-term or unstable (category B and category C, according to the ADECO-RS definitions) (Lunardi 2008).

The feasibility of mechanized excavation in comparison to conventional method has been analyzed, taking into account different elements, such as the geological and geotechnical

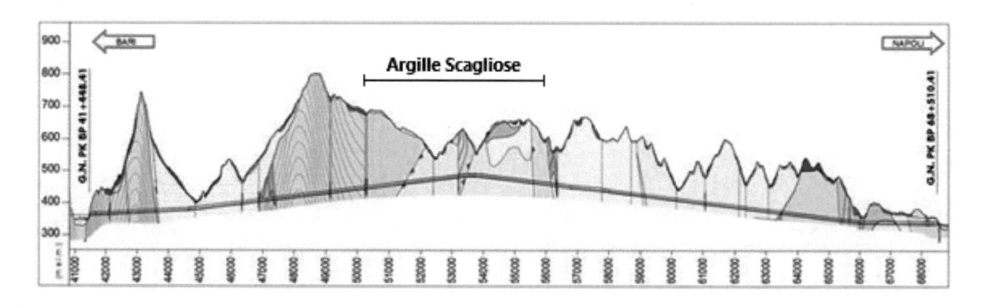

Figure 12. The Hirpinia tunnel, 27 km in length.

context, geometrical tunnels features, time and costs, safety and working environment, and a cost/benefit analysis with detailed assessment of the risks showed that the mechanized excavation is to be preferred over conventional method for all the single-track tunnels (the Hirpinia tunnel and Orsara tunnel) and for double-track tunnels having a length greater than 5 km (the Rocchetta tunnel in the Apice-Hirpinia section).

Closed EPB shield machine appears to be appropriate applying face pressure under low overburden or when the ground behaviour has been evaluated in category C.

For the excavation through 'Argille Scagliose' Unit under high depth the conventional method is adopted, due to its major flexibility and possibility of accommodating large deformations. Indeed, numerous case-histories, among which the S.Vitale tunnel on the Vitulano-Apice section, realized in 1990 (Lunardi 2015), have highlighted the peculiar behaviour of these soils, characterized by significant extrusion phenomena and radial convergences during tunnel excavation, even with long-term effects.

Full-face advancement, reinforcements in the advance core and around the cavity, and cast in place the concrete final lining close to the face are all the measures to control deformation phenomena and guarantee stable conditions. An observational approach, based on monitoring of core-face extrusion and cavity convergence, allows to face with the heterogeneities of these materials calibrating the stabilization measures within the design variabilities.

REFERENCES

AGI, 1985. Geotechnical properties and slope stability in structurally complex clay soils. *Proc. Geotechnical Engineering in Italy*, 2, 189–225.

D'Elia, B., Picarelli, L., Leroueil, S., Vaunat, J. 1998. Geotechnical characterization of slope movements in structurally complex clay soils and stiff jointed clays. *RIG*, n.3, pp. 5–32.

Esu, F. 1977. Behaviour of Slopes in Structurally Complex Formations. *General report, Session IV. Proc. Int. Symp. "The Geotechnics of Structurally Complex Formations", Capri, 2*, pp. 292–304.

Ferretti, A. 2014. Satellite inSAR Data – Reservoir Monitoring from Space. *SAGE Publications*.

Barla, M. 2008. Numerical simulation of the swelling behaviour around tunnels based on special triaxial tests. *Tunnelling and Underground Space Technology 23*, pp. 508–521. Elsevier.

Bonini, M., Debernardi, D., Barla, M., Barla, G. 2009. The Mechanical Behaviour of Clay Shales and Implications on the Design of Tunnels. *Rock Mech Rock Engng 42*: pp. 361–388.

Bilotta, E. 1987. Contributo allo studio della resistenza al taglio di argille a scaglie con prove di laboratorio. *Rivista Italiana di Geotecnica*, n.4, pp. 133–143.

Lunardi, P. 2008. *Design and construction of tunnels. Analysis of controlled deformation in rocks and soils (ADECO-RS)*. Springer.

Lunardi, P. 2015. Muir Wood Lecture 2015: Extrusion Control of the Ground Core at the Tunnel Excavation Face as a Stabilisation Instrument for the Cavity. ITA/AITES Editions.

Romana, M. & Serón, J. 2005. Characterization of swelling materials by Huder-Amberg oedometric test. *Proceedings of the international conference on soil mechanics and geotechnical engineering*. Balkema publishers, pp. 591–594.

Sciotti, A., Amato, A., Martellucci, T. 2017. Large Italian investment in railway infrastructure – The new Naples-Bari high speed/high capacity route. *Gallerie e grandi opere sotterranee 124*: 31–40.

Scrocca, D., Carminati, E., Doglioni, C. 2005. Deep structures of the southern Apennines, Italy: Thin-skinned or thick-skinned?. *Tectonics*, 24, 1–20.

Tunnels and Underground Cities: Engineering and Innovation meet Archaeology, Architecture and Art, Volume 3: Geological and geotechnical knowledge and requirements for project implementation – Peila, Viggiani & Celestino (Eds)
 ISBN 978-0-367-46583-4

In-situ stress measurements in TBM tunnels prone to rockbursts

R. Amici & G. Peach
Multiconsult, Oslo, Norway

M. Nadeem
National Development Consultants, Lahore, Pakistan

ABSTRACT: the paper details the in-situ stress measurements in twin tunnel boring machine (TBM) of the Neelum Jhelum Hydropower project, located in the Azad Kashmir region of Pakistan in foothills of the Himalayas, with overburdens up to 1870 m and high tectonic stresses. The most significant challenge during tunnel construction was the occurrence of rockbursts which, although expected, have been a constant danger to tunnel personnel and equipment. The ability to predict the likelihood, location, severity and number of rockbursts directly impacts upon the tunnel safety and daily production rates. The in-situ tests provided actual ground stresses and their orientations an indication of location along the tunnels which experienced the rockbursts. The in-situ stress results contradicted the forecasts based on the classical empirical rules combining stress analysis and rock strength. This paper details how data was interpreted and integrated to support excavation, prevention and control of rockbursts which improved tunnel safety and productivity.

1 INTRODUCTION

The Neelum Jhelum Hydroelectric project is located in Azad Jammu & Kashmir, Northeast of Pakistan. The Hydroelectric project will have the installed capacity of 969 MW with a head of 420 m. It is a long-tunnel diversion type hydropower scheme (Figure 1).

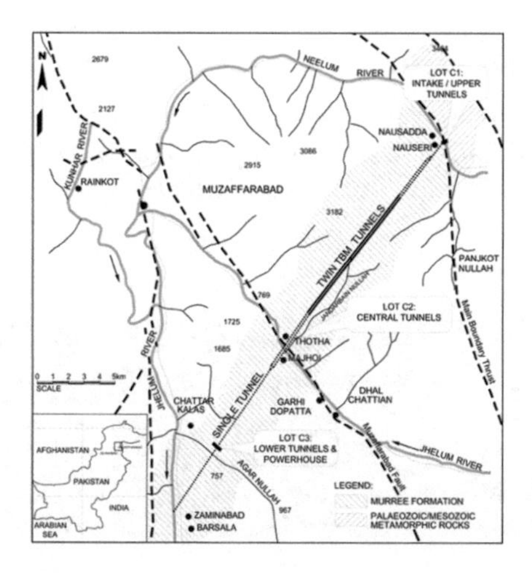

Figure 1. Project layout showing TBM Twin tunnels (in bold), major faults (dashed) and simplified alignment geology.

1.1 *Geological Rock Types, Structures and Conditions*

The geology in the TBM section is characterized by a folded, heavily tectonized, sedimentary sequence belonging to the Lower Murree Formation. The rocks encountered in the tunnels are Sandstones (SS-I), Siltstones and Silty Sandstones (SS-II), Mudstones (MST).

In addition to the expected rock types, there were also certain ground conditions which were expected to be encountered and for which the TBM was designed to manage. Also, the TBM had further facilities to detect in advance these conditions, in order to take appropriate measures to successfully negotiate the identified conditions. Structures and conditions included:

- Unstable rock zones
- Squeezing and swelling ground
- Soft ground
- High water inflows
- Extensive fault zones
- Rockbursts
- High Overburden depths (2000 m).

2 TBM SELECTION

20 km of the twin headrace tunnel was excavated by two open (gripper) 8.55 m of diameter TBMs named 696 and 697.

A significant consideration for the TBM selection was the possibility of encountering squeezing ground, with deformations of up to 500 mm on the tunnel diameter. This would exclude many types of TBM designs due to the possibility of becoming trapped within the tunnel. The open (gripper) TBM is best suited to deal with this potential condition, due to the short length of the front shield and its ability to collapse inwards various sections of the front shield, depending upon ground conditions, and still maintain the ability to excavate forward. The second ground condition which was indicated to be present in the higher overburdens and more brittle rock was rockbursting. Again, the Open Gripper TBM configuration allows for equipment to be installed to detect and mitigate potential rockbursts.

2.1 *Cutterhead & Front Shield*

The general arrangement of the open gripper TBM is show in Figure 2. The cutterhead is the only solid components and is connected to the front shield.

2.2 *Tunnel Support Equipment (L1)*

Each TBM was equipped with the follow equipment for support installation in the L1 location which was the 3.5 m immediately behind the rear of the cutter head shield. Rockbolts were installed using twin rockdrills mounted upon a circular track system which gave a radial

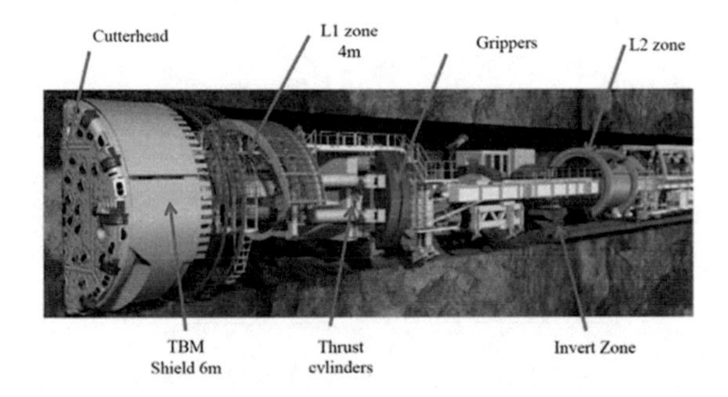

Figure 2. General arrangement of open gripper TBM.

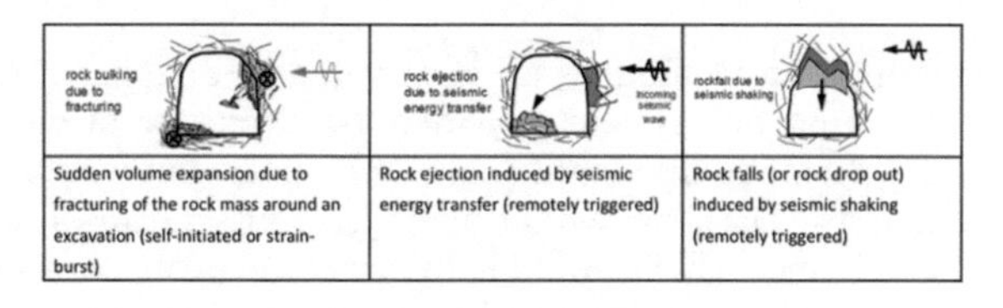

Figure 3. Rock burst damage mechanisms from Kaiser et al. 1996.

coverage for the top 260° of the tunnel peripheries. Wire mesh was installed by way of a mesh erecting device which lifted the mesh into position on the tunnel circumference and then secured by way of rockbolts. In poor ground conditions, the TBM is able to erect full circular steel rings using an erector installed within the TBM shield. Shotcrete was applied, if required, by way of a mobile shotcrete spray arm over the top 200° of the excavated tunnels.

2.2.1 *Tunnel Support Equipment (L2)*

60 m behind the L1 section was the L2 section, in this zone a duplicate set of rockbolting equipment was installed, with a slightly reduced radial coverage due to the installation of rail track. Immediately behind the rockbolting equipment was the permanent shotcrete lining equipment, this consists of two spray robots each mounted upon a longitudinal track allowing a 7 m travel distance laterally along the tunnel and a radial coverage of 270° of the tunnel crown.

3 ROCKBURST

The Canadian Rock Burst Support Handbook (Kaiser P.K., McCreath, D.R. & Tannant, D. D., 1996) defines rock bursts as damage to an excavation that occurs in a sudden or violent manner and is associated with a seismic event. This definition for rock burst was deemed too restrictive for the Neelum Jhelum project. The term “rock burst” was expanded to refer to any release of detectable energy linked with TBM excavation.

Rock bursts can be self-initiated or remotely triggered. Self-initiated rock bursts occur when tangential stresses near the excavation boundary exceed the rock mass strength, and failure proceeds in an unstable or violent manner (Kaiser et al. 1996). Once the stresses exceed the strength of the rock mass, it can violently and suddenly fracture (Figure 3).

The rockbursts which occurred during the TBM excavation are the first type, self-initiated or strain burst, due to the fracturing of the rock around the excavation. Fracturing and dilation occur when the stresses near the opening exceed the rock mass strength triggering a violent release of excess energy. The primary source of energy causing fracturing and dilation comes from the strain energy stored in the rock around the opening.

Rock bursts were first encountered in deep mines and much of the rock burst literature comes from the mining industry. Traditional mining methods do not require high concentrations of plant and equipment present near the excavation face. This contrasts with TBM excavation which continually concentrates complex equipment and manpower close to the excavation face.

4 TBM ADVANCE GROUND INVESTIGATION EQUIPMENT

The TBMs have been designed to have equipment and facilities to as far as practical manage and mitigate ground conditions.

Initially both TBMs were equipped with two separate and independent systems for carrying out advance ground investigation. These were probe drilling and Tunnel Seismic Tomography (TST) System. Later in the project when the rockburst zones were encountered, additional ground investigation systems were used in each TBM tunnel. These were rockburst prediction using Uniaxial Compression Strength (UCS) and also micro seismic monitoring.

The above four investigation techniques all gave valuable advance geological conditions in front of each TBM but no one system provided a full picture of what was ahead. The advance probe hole (30 – 40 m) could identify sandstone beds, but would not give an indication of potential rockbursts. The TST would again give an indication of changes in rock integrity up to 110 m in advance which would help target the probe locations. The rockburts prediction using UCS was time consuming and often unclear. The microseismic monitoring took a considerable time to install but would give indications of rockburts likelihood and severity.

What the advance ground investigation programme required was the measurement and information of actual ground stresses along the TBM tunnel alignment. Therefore, a fifth technique was decided upon and implemented with coring for In-situ measurement of stress.

5 ROCKBURST PREDICTION USING EMPIRICAL RULES COMBINING STRESS ANALYSIS AND ROCK STRENGTH

The prediction of severity of rock burst and depth of spalling was, first of all, attempted using empirical rules combining stress analysis and rock strength, introduced by Grimstad and Barton (1993) and Nicksiar and Martin (2013). Diederichs et. al. (2010) suggested a correlation to predict the depth of spalling.

Using the UCS from laboratory testing, the prediction of rockburst severity and depth of spalling was attempted for sandstone. This prediction shows that moderate spalling was expected for overburdens of 1000–1300 m, mainly located in the crown and invert, with a depth of spalling between 1.5–1.8 m. With overburden above 1300 m, serious rock burst incidents were predictable in the walls, with depth of spalling between 2.1–3.8 m.

The severity of rock burst is classified according to the criterion in Figure 4., using the ratio between rock UCS and tangential stress. The empirical rule is used to convey the gravitational rock stress to tangential rock stress. It can be seen that the rock burst occurrence is directly related to the overburden. The heavy rock burst was expected where overburden is greater than 1600 m. Prediction of rock burst according to Grimstad and Barton (1993) is shown in Figure 5.

b) Competent, mainly massive rock, stress problems		σ_c/σ_1	σ_θ/σ_c	SRF
F	Low Stress, near Surface, open joints	>200	<0.01	2.5
G	Medium stress, favourable stress condition	200-10	0.01-0.3	1
H	High stress, very tight structure. Usually favourable to stability. May also be unfavourable to stability dependent on the orientation of Stress compared to jointing / weakness planes*	10-5	0.3-0.4	0.5-2 2-5*
J	Moderate spalling and / or slabbing after > 1 hour in massive rock	5-3	0.5-0.65	5.50
K	Spalling or rock burst after a few minutes in massive rock	3-2	0.65-1	50-200
L	Heavy rock burst and immediate dynamic deformation in massive rock	<2	>1	200-400

Figure 4. Rockburst severity classification according to NGI Q-system.

Overburden (m)	Vertical stress σ_z (MPa)	k_0	Prediction of Rockburst in Q2 SS-1 Sandstone			
			σ_θ/σ_c	σ_θ/σ_c	Assumed Rockburst	
			Roof	Wall	Roof	Wall
1100	29.7	1.1	0.79	0.66	Slabbing/Rockburst	Slabbing/Rockburst
1200	32.4	1.1	0.87	0.72	Slabbing/Rockburst	Slabbing/Rockburst
1300	35.1	1.0	0.82	0.82	Slabbing/Rockburst	Slabbing/Rockburst
1400	37.8	1.0	0.88	0.88	Slabbing/Rockburst	Slabbing/Rockburst
1500	40.5	0.9	0.8	0.99	Slabbing/Rockburst	Slabbing/Rockburst
1600	43.2	0.9	0.85	1.05	Slabbing/Rockburst	Heavy Rockburst
1700	45.9	0.9	0.91	1.12	Slabbing/Rockburst	Heavy Rockburst
1800	48.6	0.8	0.79	1.24	Slabbing/Rockburst	Heavy Rockburst
1850	50.0	0.8	0.81	1.28	Slabbing/Rockburst	Heavy Rockburst

Figure 5. Prediction of rock burst based on Grimstad and Barton (1993) using gravitational stress and rock UCS with various assumed k_0 values and overburden.

6 PRINCIPLE OF IN-SITU STRESS TESTS BY HOLLOW INCLUSION (HI) METHOD AND BIAXIAL TESTS

The Hollow Inclusion (HI) measures strains during and after an over coring process, which is shown in Figure 6.

The elastic properties of the rock are obtained from a biaxial test on the over cored specimen. The hollow inclusion incorporates 12 strain gauges within an epoxy shell (Figure 7). This over coring process isolates the rock surrounding the cell resulting in the rock annulus changing shape in all directions according to the amount of stress that it was originally accepting and the material properties of the rock annulus. Assuming linear elastic behaviour occurs, the strain gauge readings during over coring are a function of the original in-situ rock stress.

6.1 *Instrument used for Stress Measurement*

The Hollow Inclusion (CSIRO HI) cell measures strain during, and after over coring test. During the over coring test in the bore hole, 12 strain gauges over the HI cell, read strain and the elastic properties of the rock are measured with the help of biaxial test on the over core specimen. From this strain data, the stresses can be back-calculated by the Hoek law.

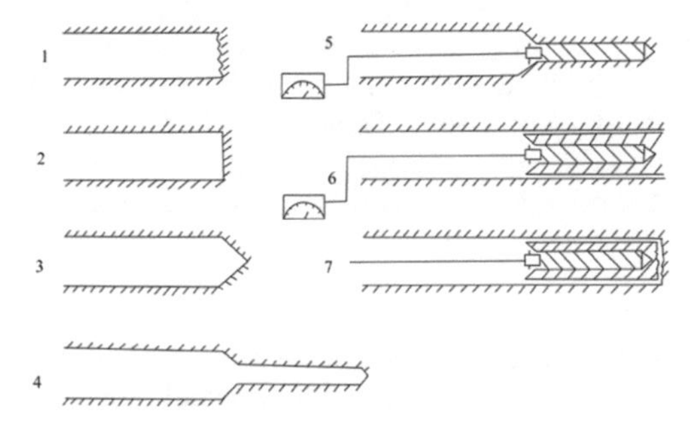

Figure 6. CSIRO HI cell test procedures for drilling, installation and measurement. 1-drill a borehole of 130 mm of diameter; 2-grind the end face of the borehole; 3-drill a trumpet on the end face of the main hole; 4-drill a borehole of 38 mm of diameter; 5-install the stress gauge; 6-overcoring the borehole and measure the strain changes in the borehole; 7-break the core and take it out of the hole.

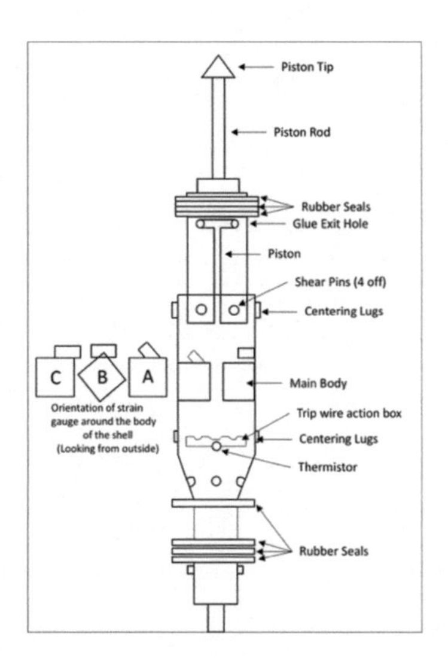

Figure 7. The different components of CSIRO HI cell.

Figure 8. Core disking in the bore hole drill at chainage 09+860.8 in TBM 696.

6.2 *Core Disking*

Core disking was regularly observed during drilling of the sandstone bore hole in both TBM's at different locations and under different overburden (Figure 8). It is a process in which drilled core are separated into disks with uniform spacing and shape. Core disking shows high stresses in the bore hole which are released during drilling of the core. It is considered that the bore holes in which core disking was observed contain more stresses as compared to the bore hole where core disking was not observed and tests were performed successfully.

Four bore holes in TBM 697 and one in TBM 696 were terminated due to core disking. In these bore holes, the core disking starts from depth ranges from 10 m to 14 m and continue up to the end of the hole. In the other bore holes, the over core sample was recovered successfully.

6.3 *Overcoring Test Results*

The results of the over coring tests are detailed in Figure 9. During the over coring tests, micro strain was recorded with the help of a cable from the bore hole. Biaxial tests were carried out on the over core sample to obtain the accurate values of elastic modulus and Poisson ratio.

Table: Over Coring Test Details in TBM 697 and TBM 696

TBM	Bore Hole Chainage	Over Burden (m)	Rock Type & Thickness (m)	Test No.	Depth Of Test in Bore Hole (m)	Length of Over Core Sample (cm)	Biaxial Test		Insitue Stresses									Normal Stresses			Shear Stresses		
							Elastic Modulus /Gpa	Poisson Ratio	σ1			σ2			σ3			X	Y	Z	X-Y	X-Z	Z-Y
									Mag. Mpa	Dip Deg.	Bearing Deg.	Mag. Mpa	Dip Deg.	Bearing Deg.	Mag. Mpa	Dip Deg.	Bearing Deg.						
TBM 697	10+881	1063	Sandstone & 40	#1	18.05 - 18.55	35	46	0.23	73.9	6	327	37.7	37	242	26	50	50	69.3	37.5	30.8	2.1	7.8	12.2
					18.75 - 19.25	42	28	0.2	108.8	11	330	35.2	47	228	23.7	41	69	94.4	40.1	33	-15.8	26.3	27.3
					19.45 - 19.95	32	25	0.2	82.1	13	339	37.7	62	224	25.8	25	75	65	42.6	37.8	-10.9	14.4	24.4
	10+935	1070	Sandstone & 57	Bore hole drill up to 15 m depth and over coring test terminate due to core discing from depth 10 to 15 m.																			
	10+936	1070		Bore hole drill up to 25 m depth and over coring test terminate due to core discing from depth 10 to 25 m.																			
	10+934	1070		Bore hole drill up to 21 m depth and over coring test terminate due to core discing from depth 10 to 21 m.																			
	09+620	1340	Sandstone & 54	#5	18.05 - 18.55	Core broken during over coring process and test fail																	
					19.50 - 20.00	45	67	0.21	88.9	2	297	54.9	60	203	11.7	30	29	86.7	24.6	43.3	2.2	3.1	11.1
					20.50 - 21.00	Core broken during over coring process and test fail																	
	08+826	1500	Sandstone & 21	#9	17.0 - 17.57	57	54	0.18	53.4	17	70	51.7	5	339	43.2	75	227	51.9	52.6	43.8	0.4	-1.1	-2.3
					18.1 - 18.5	39	49.2	0.21	50.6	7	102	48.5	13	10	40.7	75	221	50.4	48.2	41.2	0.7	0.02	-1.9
					18.6 - 18.9	30	51.6	0.19	60.7	2	95	56.4	3	5	47.6	86	211	59.3	57.8	47.6	2	-0.6	0.34
	08+422	1590	Sandstone & 19	Bore hole drill up to 19.7 m depth and over coring test terminate due to core discing from depth 14 to 19.7 m.																			
	08+287	1650	Sandstone & 31	#11	17.4 - 18.1	70	48	0.13	100.4	16	268	65.1	62	30	1.9	22	171	55.1	53.8	58.6	23.5	104	43.1
					18.2 - 18.6	40	44	0.15	141.2	22	277	85.7	68	94	37.3	1	186	106.9	63.5	93.8	17.2	1.1	42.8
					18.7 - 19.4	70	48.5	0.12	114.6	35	282	74.4	29	35	34.8	41	154	84.7	68.9	70.3	34.5	3	18.1
	08+550	1570	Mudstone & 19	Test not carried out due to poor rock condition and bore hole terminate at depth of 17.5 m.																			
	07+440	1900	Sandstone & 32	#12	17.0 - 17.6	Test failed due to water flow(20 lit/min) from the hole and core disking. Temperature of the water was 40˚C.																	
	07+339			#13	17.5 - 18.23	76	55	0.26	69.8	1	16	31.5	16	286	9.3	74	110	35.3	64.3	11	13.8	-5.9	-1.1
	07+280	1897	Siltstone & 100	#14	18.7 - 19.2	50	Biaxial test not perform due to core discing		22.6	11	48	4.8	70	169	-3	17	315	Normal and shear stresses not calculate due to core discing during over coring of HI cell					
					19.4 - 20	60			16.4	59	231	2.1	9	126	-9.2	30	31						
TBM 696	10+938.7	1060	Sandstone & 58	#6	19.05 - 19.56	51	52	0.1	118.8	31	301	44.5	16	201	17.5	54	88	92.8	41.5	46.4	2.3	42.4	6.8
					19.96 - 20.54	58	30	0.21	116.6	21	337	45	64	119	36.8	15	241	91.2	53.4	53.8	29.3	20.2	33.7
					21.0 - 21.77	77	35	0.2	102.9	32	312	48.4	47	195	24.4	37	70	82.8	37.7	55	9.8	30.7	11.3
	09+860.8	1240	Sandstone & 22	Bore hole drill up to 20 m depth and over coring test terminate due to core discing from depth 13 to 20 m.																			
	13+834	1000	Sandstone & 29	#8	23.0 - 23.57	57	42	0.23	50.7	14	102	24.3	43	206	19.5	44	358	43.3	27.6	23.6	-11.2	7	0.8
					24.3 - 25.13	83	48	0.19	54.9	30	107	27.7	13	204	21.2	57	315	43.6	30.3	29.8	-7.2	13.8	3.9
					25.9 - 26.55	65	50	0.21	50.7	39	101	21.3	22	210	15.9	43	322	32.5	24.8	30.5	-7.9	15.4	5.9

Figure 9. Over coring test results.

The in-situ stresses obtained from the over coring and biaxial test are σ1, σ2 and σ3. When the in-situ stresses located at the local coordinate of the tunnel, then the six stress component (X, Y, Z, X-Y, X-Z, Z-Y) are shown in Figure 9.

7 PRACTICAL APPLICATION OF STRESS INFORMATION

The results of the HI over coring tests show the presence of very high stress in the investigated area with maximum principal stress up to 110 MPa. In Figure 10 the field stresses are plotted versus the overburden. It can be noted that, except for the σ_2, the stresses are not strongly related to the tunnel depth. In the Figure 11 the stresses have been normalized vs. the overburden to delete the effect of depth. It's interesting to see the highest values of stress perpendicular to the tunnel X around the overburdens 800–1100 m whilst the same stress is very low in correspondence of the maximum overburden. The Average Deviation of Normal Stress is maximum at the same overburden and very low at the maximum overburdens.

It can be seen that the maximum principal stress is in the horizontal direction and normal to the tunnel axis (Figure 12). The medium and minimal principal stresses are generally in a plane that is parallel to the tunnel axis. The direction of the σ_1 can be regarded as horizontal and normal to the TBM tunnel axis; the direction of σ_2 can be regarded as parallel to the TBM tunnel axis; the direction the σ_3 can be regarded as in vertical direction.

The orientations of the principal in-situ stresses indicate that the in-situ stresses are in the in-plane direction (a plane normal to TBM tunnel axis).

It is interesting to analyze the maximum principal stress orientation along the tunnels. In Figure 13 the principal stresses are plotted in plan along the tunnels. It can be seen that the principal stress direction is almost perpendicular to the tunnels alignment.

The project area exhibits both the regional thrust and the locally developed low persistence faults. The Muzaffarabad Fault is a system with a total width of 8 km, also known as the Himalayan Frontal Thrust (HFT), has recently exhibited activity in the area resulting in catastrophic 2005 earthquake (Kaneda et al. 2008) with both strike-slip and thrust movements.

The in-situ measurements distribution along the tunnel is shown in Figure 14. It is interesting to see that the over coring tests carried out at the lower overburdens are internal to the Muzaffarabad fault system and that the lower value for the main stress is different by the others in #13 (is parallel to the tunnel) which is the location with the maximum overburden.

The Figure 15 shows the distribution of the stresses along the tunnel vs chainage and overburden. It can be seen that the maximum stresses correspond to zones were the overburden is around 800–1100 m. It can be seen that the rock burst events occurred within the Muzaffarabad Fault system in correspondence with the maximum X stress. The main stress in the XY plane (at lower part of Figure 14) show a different orientation of the stress along the tunnel. These 2D in-plane stresses are strongly related to the shape of the damage observed at the rock burst zones. (Figure 16).

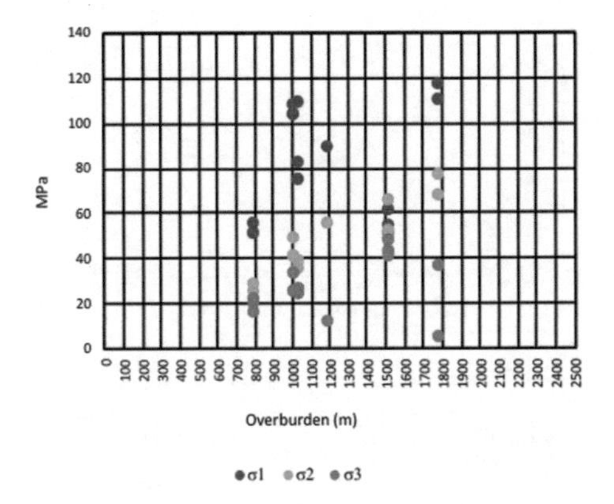

Figure 10. Magnitude of principal stresses along the tunnels Vs overburden.

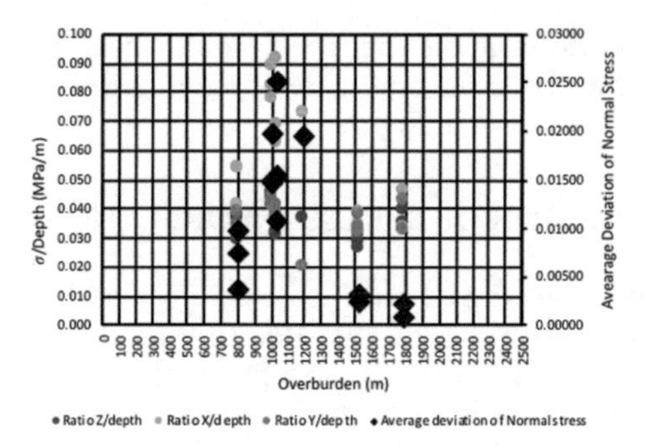

Figure 11. Normalized stresses Vs chainage with overburden and Average Deviation of Normal Stress.

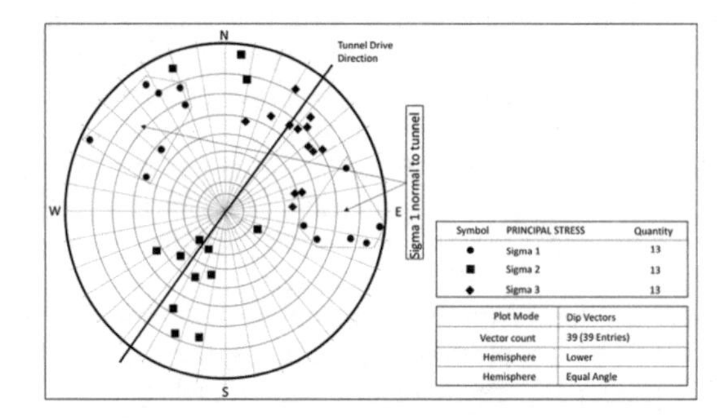

Figure 12. Pole plot showing direction of measured in-situ stresses.

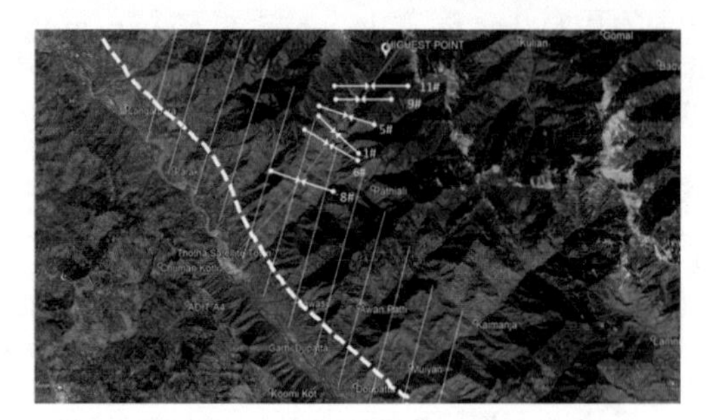

Figure 13. Principal stress direction along the TBM tunnels. White dots show the surface ruptures mapped by Kaneda et al. (2008). The white dashed area is the Muzaffarabad fault system extension.

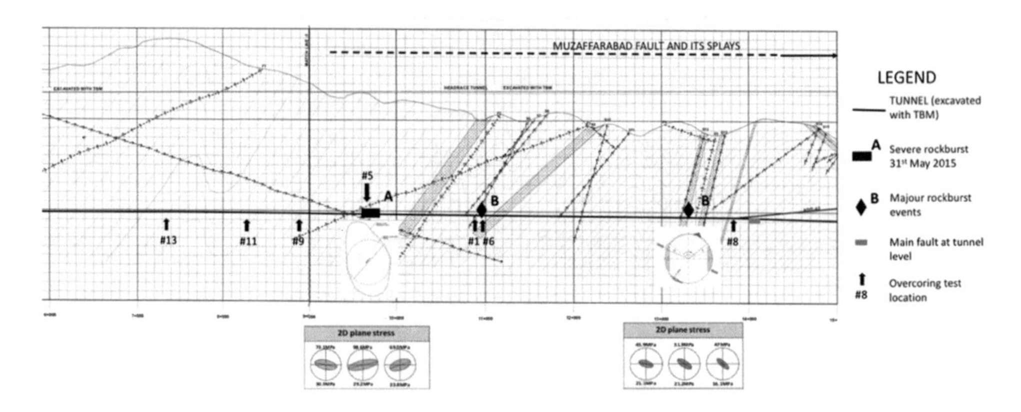

Figure 14. Main tectonic elements, rock burst events, and in-situ over coring tests along the TBM tunnel.

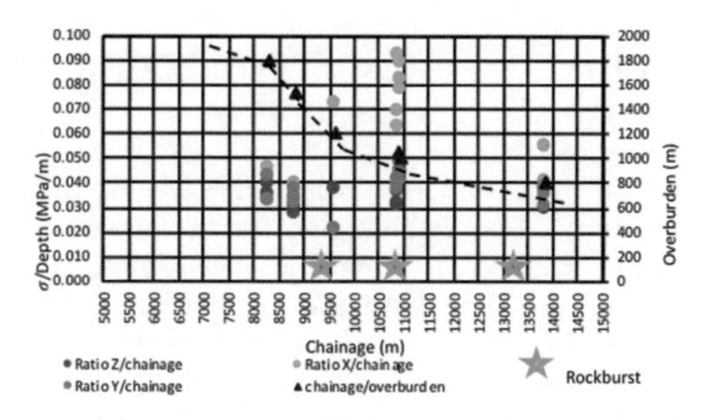

Figure 15. Normalized stresses Vs chainage at TBM tunnel. The rock burst events happened in correspondence of the maximum X stresses which is not corresponding to the maximum overburden.

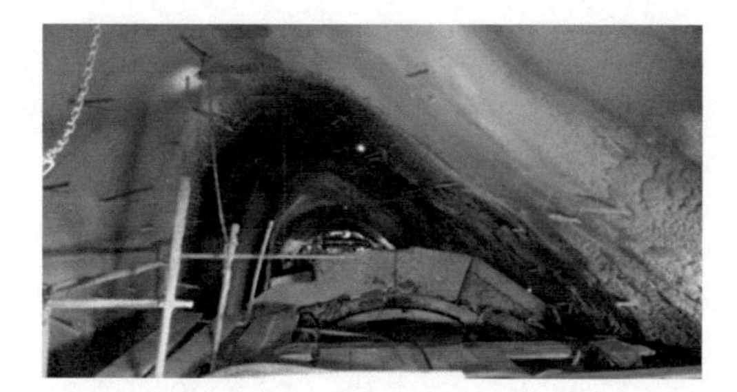

Figure 16. Rockburst zone between chainages 9+806 and 9+706 (zone A on Figure 14). It shows the orientation of maximum over break at 11-5 o'clock.

8 ROCKBURST PRECAUTIONS AND COUNTER MEASURES

The sandstone has more potential for rock burst as compared to siltstone and mudstone. Thus, it was important to identify the presence of sandstone ahead of the TBM. This was achieved by carrying out advance probe drilling up to 50 m in front of the TBM. Within the tight constraints of a TBM tunnel there are limited options for dealing with potential rockburst. What measures are available can be divided into two categories:

Preventative/Reduction

- Horizontal relief holes drilled in front of the TBM and Vertical Relief holes drilled behind the TBM shield.

The aim of both horizontal and vertical stress relief holes is to induce cracking/fracturing of the sandstone thereby eliminating or reducing the potential for a rockbursts by pushing the stress stresses deeper into the surrounding rock and away from the periphery of the excavated tunnel.

Reactive

- Full circular steel rings, heavy duty wire mesh and shotcrete applied behind the TBM shield.

All the preventative measures were aimed at containing the impact of a rockburts which may or may not occur and could influence the overall tunnel programme. With the advent of regular rockbursts and the severe rockbursts of 31st May 2015 both preventative/reduction and reactive counter measures were implemented on a daily basis and the expected highest overburdens were still some 1.1 km ahead of both TBMs.

9 IMPACT OF OVERCORING TEST RESULTS ON ROCKBURTS PRECAUTIONS AND COUNTER MEASURES

The first significant impact of the over coring test results was upon the drilling of horizontal stress relief holes. With the indication of the higher horizontal stresses within the tunnel and the cross referencing with the observed rockbursts damage around the tunnel periphery, it was decided to reduce and be more precise with the location of these stress relief holes. The number of holes were reduced from 18 in the upper 260° of the excavated tunnels to 8 in the upper 120° slightly modified to the 11 to 3 o'clock position. This reduced the drilling time from 10 hours down to 4 hours and therefore increased TBM Excavation time.

The second significant impact was upon the vertical relief holes drilled directly behind the TBM shield. The targeted application again in the 11 to 3 o'clock position resulted in further time savings and increased TBM excavation time.

The third significant impact was the indication from over coring results at 7+440, 7+339 and 7+280 that the zone of elevated horizontal stresses was coming to an end, and that the installation spacing full circular steel rings could start to increase from 0.7 m to 1.6 m.

The result of the above three significant findings resulted in a monthly production total of 180 m increasing over three months to 320 m per month.

10 CONCLUSIONS

Traditional systems to evaluate rock burst occurrence are too generic and the only show the increasing of risk with the overburden. Since the first episode of rock burst occurred in the TBM tunnel, it was clear that more detailed information was necessary to understand and manage the rock burst risk. It was decided to carry out in-situ stress measurement to understand the stress distribution and magnitude and use this information in the prediction, prevention and control of rock burst.

The results of the measurements show that the principal stress is very high and that the rock burst can occur everywhere along the tunnel.

The stress perpendicular to the tunnel shows very high values in correspondence of the lower overburden. This is related to the high tectonic stress on the rocks in the Muzaffarabad fault area, where important movements have been observed after the disruptive earthquake on 2005.

The stress orientation can be used to understand the rock burst position and the magnitude of the measured stress can be used to identify the more dangerous sections along the tunnel but cannot help whether rock burst will happen.

REFERENCES

Hoek E. and Brown E.T. 1980. *Underground excavations in rock*. Institution of Mining and Metallurgy, London, 527 p.

Diederichs, M.S., Carter, T. & Martin, C.D. 2010. *Practical rock spall prediction in tunnels*. AITES/ITA Conference, Vancouver. May.

Grimstad E. and Barton N. R. 1993. *Updating of the Q-system for NMT*. In International symposium on sprayed concrete, Fagernes, Norway. Norwegian Concrete Association, pp. 46–66

Nicksiar and Martin 2013. *Crack initiation stress in low porosity crystalline and sedimentary rocks*, Engineering Geology 2013;145;64–76.

Kaiser P.K., McCreath, D.R. & Tannant, D.D. 1996. *Canadian Rockburst Support Handbook*, Mining Research, Directorate, Sudbury, Canada.

Kaneda, H., et al 2008. *Surface Rupture of the 2005 Kashmir Pakistan Earthquake and Its Active Tectonic Implications*. Bull. of Seismological Society of America, Vol.98, No.2, pp.521–557.

Worotnicki, G., Walton, R.J. (1976) *Triaxial hollow inclusion gauges (CSIRO) for determination of rock stresses in situ*. Proc. ISRM Symp. On the investigation of stress in rock – Advances in stress measurements, Supplement, pp. 1–8, Sydney.

Tunnels and Underground Cities: Engineering and Innovation meet Archaeology, Architecture and Art, Volume 3: Geological and geotechnical knowledge and requirements for project implementation – Peila, Viggiani & Celestino (Eds)
© 2020 Taylor & Francis Group, London, ISBN 978-0-367-46583-4

Optimizing the excavation geometry using digital mapping

S. Amvrazis, K. Bergmeister & R.W. Glatzl
Brenner Base Tunnel BBT SE, Innsbruck, Austria

ABSTRACT: In conventional tunnelling, the control of blasting work by the blasting master is regarded as a key factor for successful and safe tunnel excavation. Each type of rock with its division has different excavation behaviour and makes driving difficult, even in the case of small-scale geological changes. The quality of the excavation and the cavity support is therefore highly dependent on the skills and experience of the tunnelling team. The Tunnel Control System (TCS) was developed to improve the accuracy of the excavation cross-section created by the blasting. The blasting process is optimized for this by additional information for the drill master in the arrangement of the blast holes. The geometric evaluation of present tunnel scan data with known and documented geological impacts makes it possible to predict an excavation behaviour and possible optimization of the blasting pattern.

1 INTRODUCTION

1.1 *Brenner Base Tunnel*

The Brenner Base Tunnel (BBT) consists of two single-track railway tunnels from Innsbruck (Austria) to Fortezza/Franzensfeste (Italy). The standard distance between the main tunnels is 70 m. An exploration tunnel is located 11 m below the two main tunnels. This tunnel, with an inner diameter of 6.8 m, will be excavated before the construction of the main tubes in order to obtain information about the geological and hydrogeological conditions of the mountains (see Bergmeister 2017). This preliminary investigation helps to minimise both construction costs and risks. The exploration tunnel, which will be constructed along the entire length of the main tunnel, will later be used as a drainage and service tunnel (see Insam 2018).

Approx. 50 % of the entire BBT system, 230 km long, will be excavated using conventional "drill and blast" advance and approx. 50 % using tunnel boring machines. For conventional tunnelling, normally a maximum excavation of 25 cm above the target profile, is stipulated in the construction contract for the contractor. If this limit value is exceeded, suitable measures must be taken to secure the cavity.

Within the scope of a research project, the over and under cuts and thus the inhomogeneity caused by the blasting process were documented with photogrammetric, high-resolution images and statistical evaluation of the profile data.

With these profile data and on-site face documentation, the "Tunnel Control System" was further developed and tested in recent years with the optimisation of the blasting pattern at the BBT (see Voit 2017). This optimisation method enables an optimised blasting pattern for an improved accuracy of the excavation profile.

Already in the study phase, this method improved the over-profile by 65 % through digital photogrammetric documentation and statistical prognosis.

1.2 *Documentation*

An important task in the NATM tunnelling is the geological face documentation. Conventional mapping techniques usually cover the tunnel face only. By means of tunnel scanning, a 360-degree scan can fully cover all exposed rock surfaces, face as well as walls and roof along the tunnel vault. It serves as a baseline for the determination of the driving class during the construction phase of the exploration tunnel and, in the case of the BBT, also for the implementation planning of the main tunnels. The main tunnels will be driven parallel to the exploration tunnel in the following construction lot. In addition to the conventional documentation of the heading face, both the face and the adjacent not supported tunnel section of the last blast were recorded with tunnel scanners. A tunnelling geologist from BBT SE not only prepares precise documentation, but also a geological interpretation of the recorded tunnel section.

The current technique of mapping the geology of the present rock mass during tunnelling is mainly performed by hand-drawn sketches. The difficulties in geological mapping result from the limited time available at the working face, taking into account that the assessment is always subject to the subjective impression of the respective geologist. Dust, limited lighting, little space, constant movement of personnel and vehicles as well as consistently grey motifs of the rock mass complicate the geological recording.

Precise geological documentation of the tunnel face is important for the prediction quality of the excavation behaviour.

1.3 *Tunnelscanning*

Scanning is an innovative method in tunnelling, suitable for measuring the complete tunnel geometry. The application possibilities of laser scanning are manifold and range from use as a documentation instrument for compliance with contractual specifications to the treatment of geomechanical issues (see Voit 2017).

Tunnel scanning has become increasingly important in recent years. It is an imaging, distance-based method capable of taking high-resolution, three-dimensional images (3D point clouds, see Figure 1) of the environment at different times. This allows for a multitude of visualizations and creates the prerequisite for further application possibilities.

In tunnel construction, scanner technology is mainly used to record the rock surface immediately after the excavation. This is necessary, among other things, to check the specified excavation geometry or profile (see Hofmann 2019). At the same time, the support means (steel arches, anchors and grids) of preceding excavate sections can be documented if covered by the current tunnel scan.

Based on the three-dimensional surfaces recording, the scanner technology also enables the exact spatial measurement and evaluation of geometric or geological structures (see Figure 2), which can serve as support for the geological face mapping.

The data can be used for the determination of the rock mass conditions and used as a data basis for modelling. It contains all essential information. In addition, the high-resolution

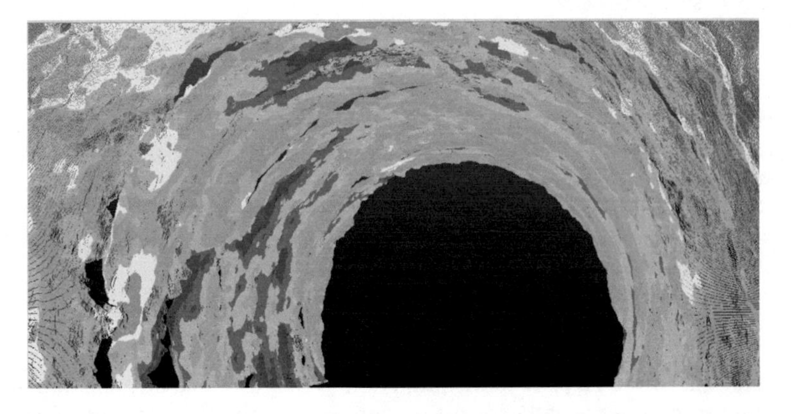

Figure 1. digital 3D point cloud reconstruction of conventional drill and blast advance.

Figure 2. combined 3D model of consecutively scanned excavations.

image allows the characterisation of the rock surface, fault planes, degree of decomposition, gap distance, roughness of the gap surfaces etc.

2 EXCAVATION OPTIMIZATION

A new evaluation procedure combines the three-dimensional measurement of the excavation geometry of previous excavations with the geological documentation of the tunnel face and enables a forecast of future excavation cross-sections or the expected over cut and under cut for future excavations in areas with comparable rock mass conditions. With the help of this information it is possible to adjust the drilling pattern (in particular the position of the drill holes) in such a way that the excavation profile is as accurate as possible.

Taking into account, the excavation surface digitally recorded by tunnel scanners, the comprehensive geological face mapping and the blasting pattern with the corresponding amount of explosive charges, the excavation profile of each round as well as its profile accuracy is stored in a database. Based on this information, the over and under cut of subsequent rounds are minimized. The database with the data of the already excavated profiles in the same lithology allows for the optimization of the drilling pattern and can also improve the construction process.

2.1 *Innovative Prognosis for Drill and Blast Excavation*

New analysis methods make it possible to estimate or forecast the expected excavation profile based on statistical evaluation of previous scans, taking into account their geological structures. In addition to the exact documentation of the over cut by means of 3D tunnel scanning, it is possible to optimize the excavation.

A method developed by author S. Amvrazis (see Amvrazis), "Tunnel Control System", was tested to minimize the over and under cut by predicting the appropriate position of the perimeter blast holes at the "Schutterstollen" (mock transport tunnel) in the Padaster valley, Austria. The smother the tunnel intrados can be excavated, the better the stability, the less shotcrete is used to fill over cut and the less trim time is required for under cut. In addition to these cost and material savings, the new method contributes to a better understanding of the behaviour of the individual types of rock mass by blasting.

2.2 *Methodology*

The excavation forecast is based on the evaluation of data already recorded (excavation geometry, geological documentation and blasting pattern with explosive charge quantities) by means of statistical regression analysis. In the regression method, the relationship between the positions of the perimeter blast holes, the excavation geometry, the geological properties and

Figure 3. rasterized profile distance on geometric area analysis lines 10 cm × 30 cm.

the profile accuracy is determined segment-wise along the reference profile (see Hofmann 2019). For modelling, the profile accuracy of the previous round is recorded as 3D point clouds of the rock surfaces directly after the excavation using a tunnel scanner (photogrammetric or laser scanner).

The profile accuracy is classified according to the geological properties, which are recorded by the geological documentation of the heading at this point. An additional comparison (see Figure 9) of the forecast profiles with the actual excavation profiles of the already predicted excavations allows an additional weighting of the data sets.

The 3D scan data is prepared by a resolution reduction: After the excavation has been recorded with a 3D point cloud, the square mean value of the over cut is determined for an area of 10 cm × 30 cm on the projected tunnel surface (10 cm in excavation direction, 30 cm along the tunnel vault) (see Figure 3). Then the averaged over cut and the radial distances of the perimeter blast hole position from the nominal profile are classified according to the position in the profile and the available geological parameters (foliation, harness, joints, fault zones) depending on type, direction of incidence and angle of incidence. Geological parameters (in the form of geological categories such as rock mass type, foliation (Sf), harness surfaces (Ha) (see Figure 6) are assigned to each rectangle in addition to the over cut spacing and entered in the database. In this way, a connection is established between the over cut distance and geological information. With the help of these correlations of previous excavations, the geological conditions at the face can be predicted for the expected over cut distances for further tunnelling.

For the prognosis, the profile accuracy is minimized as a residual. The regression function is used segment-wise to determine the perimeter blast hole positions that cause the minimum profile deviation. This is done with consideration of the geological properties in each segment that are assigned to the last tunnel face documentation. Optionally, the mean deviation of the forecast from the actual excavation of the last ten rounds can be taken into account. This will take greater account of recent changes in the outbreaks.

2.3 *Automatic Identification of joint structure in tunnel face*

In addition to the conventional face documentation, the 3D data from laser scan or photogrammetry were used in the further development to automatically assess the tunnel face for the forecast based on surface orientations and colour differences (geometric and/or chromatic discontinuities). This automatically extracts uniformly oriented surface edges and coloured contours from the 3D point cloud and the corresponding colour values (R, G, B) and categorizes them accordingly. The properties of these line characteristics serve as a further indicator for the profile forecast of the next consecutive blasting round. Similarly oriented surface edge points are used to separate surfaces and coloured contours to different types of rock.

2.4 *Geometric Modelling*

Transfer of measurement data into the TCS system takes place in the form of 3D coordinates (x, y, z) (see Markič 2019) and the corresponding colour value (R, G, B). The coordinates describe the position of surface points, and thus the geometry of the rock surface of the working face exposed by blasting. This is usually randomly distributed uneven and rough, but also, geometrically seen, interspersed with continuous uniformly oriented edges or discontinuities.

For the analysis of these geological surface edges or discontinuities, the system calculates the geometric orientations and colour contours on the face surface using the coordinate and the colour value of each surface point (see Figure 4).

Thus identified discontinuities are determined as "Points of Interest" (POIs) or "Regions of Interest" (ROIs) and marked as potential critical geological areas (see Figure 5).

From the set of POIs several line characteristics are determined. For each of them, their length, their horizontal and inclined direction and their class are calculated. The most intense geometric and chromatic discontinuities are prioritized from the set of POIs and each line is constructed with a nonlinear polynomial regression following the path of points with almost homogeneous colour information, since normally a multitude of geological structures can be recognized in this way (see Figure 6).

Figure 4. laser scan of a tunnel face with distinct fault lines.

Figure 5. laser scan of tunnel face with POIs of geological structures.

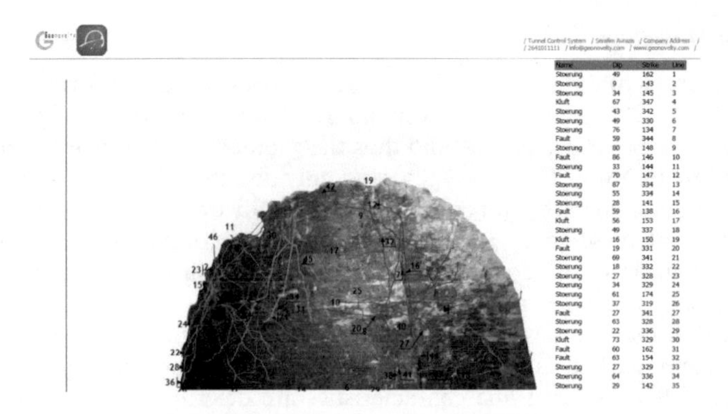

Name	Dip	Strike	Line
Stoerung	49	162	1
Stoerung	9	143	2
Stoerung	34	145	3
Kluft	67	347	4
Stoerung	43	342	5
Stoerung	49	330	6
Stoerung	76	134	7
Fault	59	344	8
Stoerung	80	148	9
Fault	86	146	10
Stoerung	33	144	11
Fault	70	147	12
Stoerung	87	334	13
Stoerung	55	334	14
Stoerung	28	141	15
Fault	59	138	16
Kluft	56	153	17
Stoerung	49	337	18
Kluft	16	150	19
Fault	19	331	20
Stoerung	69	341	21
Stoerung	18	332	22
Stoerung	27	328	23
Stoerung	34	329	24
Stoerung	61	174	25
Stoerung	37	319	26
Fault	27	341	27
Stoerung	63	328	28
Stoerung	22	336	29
Kluft	73	329	30
Fault	60	162	31
Fault	63	154	32
Stoerung	27	329	33
Stoerung	64	336	34
Stoerung	29	142	35

Figure 6. classification of geological line characteristics by means of TCS.

Figure 7. geologic and geometric modelling of discontinuities in true colour representation.

TCS uses a new mathematical method called Artificial Intelligence Surveying (AIS), in which the system learns from individual factors, e.g. different brightness conditions during a tunnel scan. The system automatically adjusts the corresponding parameters to avoid misinterpretations of the geological results.

In addition specific adjustments, based on experience or geological/geotechnical know-how, may be integrated to define the corresponding parameters. The system calculates the predicted profile of the next round (see Figure 10) based on the automatically determined line characteristics that are correlated with the finite geometry (see Figure 7 and Figure 8). This is done by statistical analysis of the finite measurements of the excavated rounds by combining the behaviour of their geometry and the geological condition defined in the system.

For each blast the average deviation per profile segment is determined. The system automatically defines segments from the areas in which different line characteristics were calculated and projects those closest to the theoretical excavation line. There is therefore a different geological line for each segment, which is projected onto the nominal profile (see Figure 9).

In addition the local effect of each perimeter blast hole is calculated from the image overlay and the average degree of excavation is displayed in a rectangular grid, also based on the geological situation.

2.5 *Database, Machine Learning*

TCS collects geometric, geological and blast specific data (blasting pattern and the quantity of explosives). A forecast for the excavation profile is calculated and correlated from these. This

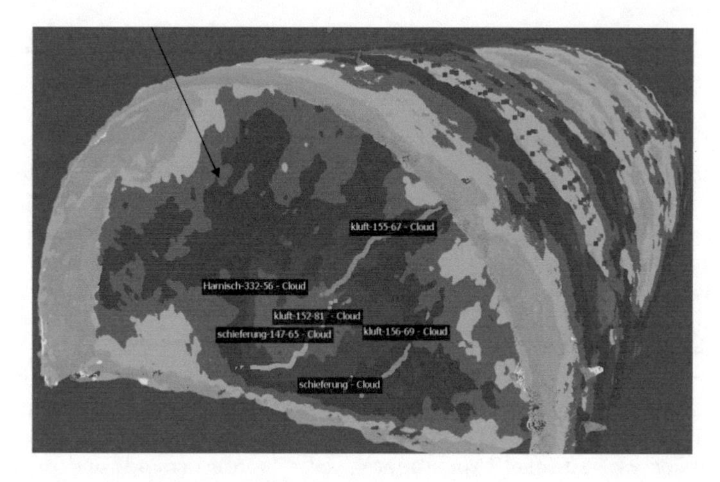

Figure 8. geologic and geometric modelling of discontinuities in false colour representation.

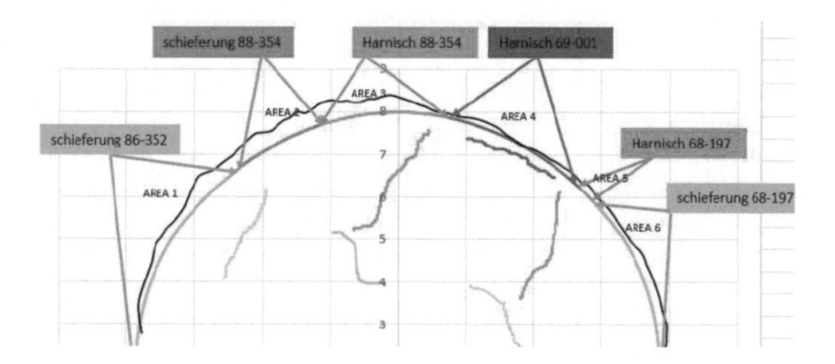

Figure 9. classification and categorisation of geological structure elements.

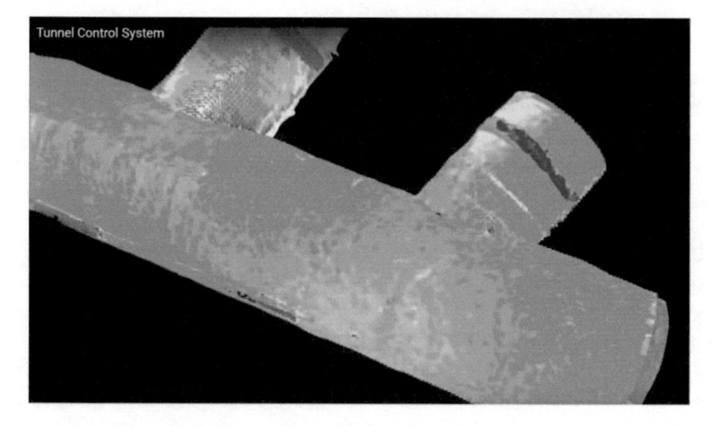

Figure 10. prediction based on previous excavation geometry and scan data.

means that the system recognizes similar geological conditions of a face in the existing data set based on the correlation and uses them for the subsequent forecast. The prediction of a new excavation profile is based on a new mathematical approach, the so-called predicted behaviour of non-linear patterns.

The use of automatically identified line characteristics in the working face allows additional measures to adapt the blast pattern for an optimized excavation profile of the following round.

The Tunnel Control System uses a database to categorize each round, containing geological-geometric elements and classify the tunnel surface geometry according to the line characteristics of the face. The database also allows for additional statistical evaluation options with regard to profile dimensions and geological conditions. The continuous recording of each blasting round with defined characteristics is a pre-condition for the first stage of the learning system. This defines the average influence of each geological condition on the geometry of the over cut, based on information already recorded.

By storing the geometric and geological properties of previous blasts, a data pool is created which can be improved incrementally with each subsequent round. It is possible to compare the predicted and actual excavation situation by scanning the area in which the predicted model was created before the excavation.

By capturing successive tunnel scans (approx. every second round) and their combination with geological documentation, it is possible to forecast following only a few rounds, provided that the rock mass conditions do not change significantly. The longer the documented tunnelling area is provided to the system, the more reliable future forecasts will become, since a larger number of matching samples result in a better hit rate. This is the pre-condition for the second stage of the learning system.

3 CONCLUSION

Excavation in tunnelling follows difficult framework conditions. It is primarily subject to the influences of the mechanical properties of the main working material — rock mass. Isotropic or anisotropic behaviour, removability, stability and geological/hydrogeological situation are further limiting factors. The arrangement of disturbances and/or separating surfaces is also decisive.

Excavation quantities, shotcrete thickness (outer shell), consumption for backfill concrete and final lining (inner shell) significantly influence the costs of the structure. Irregularities in the shotcrete lining results in inhomogeneous strain distribution and influences the load-bearing capacity and serviceability of the outer shell.

The method of excavation optimization is directed towards bringing the actual excavation geometry as close as possible to the smoothly specified theoretical excavation line — the target or nominal geometry (see Hofmann 2019).

This is done by modelling the actual excavated cavity using 3D scanners and the associated geological and structural analysis of the surrounding rock mass. Furthermore, for each documented round, the blasting pattern and the quantity of explosives is included in the model.

Different calculation modes for automatic or semi-automatic identification of line characteristics have been developed. Based on the TCS structural lines projected onto the profile, geological classification segments are created to collect reference data of previous geometries (see Amvrazis).

Thus, the input parameters "rock properties" and the influence "blasting" exerted onto the rock mass can be combined with the resulting "excavation geometry" for a certain combination. From the set of all recorded combinations, the excavation behaviour of the next subsequent round can be concluded, the same rock mass conditions provided. The prognosis created in this way serves to optimize the drilling pattern and the associated explosive charges in order to both minimize damage to the rock mass in cavity proximity and the over cut that occurs, and to possibly avoid under cut.

3.1 *Results*

The differences between different blasting rounds are recorded in the database and can be used to adjust the system to better optimize the forecast data at the second learning level. These two mixed difference patterns increase the prediction models accuracy by 85 %.

The results of the calculations were made available daily to the contractor for the planning of further measures regarding the change of the drilling pattern or the quantities of explosives used.

Overall, the over cut could be reduced by up to 65 % in parts of the examined tunnel test drives by optimizing the drill and blasting scheme (see Bergmeister 2015 and Voit 2017).

3.2 *Outlook, Future Work*

Based on the promising results of the new forecasting and prognosis method, a holistic approach will be developed (see Bergmeister 2015). By optimising the excavation geometry, shotcrete consumption can be reduced and the thickness of the shotcrete lining (outer shell) can be made more uniform. This also leads to an improvement of the tunnel geometry and thus to a more homogeneous strain distribution within the tunnel shell and the rock mass in cavity proximity. This can significantly increases the load-bearing capacity and serviceability of the tunnel structure.

Furthermore, a better knowledge of the expected rock mass behaviour can help to avoid under cut and the resulting trimming work.

Finally, a forecast of the tunnelling behaviour also makes it possible to reduce risks and increase work safety in the tunnel.

This experience-based knowledge is to be integrated into the newly developed forecasting method. All geometric data shall be imported as deterministic values. In the future, the data cloud will be developed in a stochastic model in order to achieve better results in conjunction with the in-situ evaluation of the geologist. It is to be expected that this method of digital image prediction can be used very expediently in conventional drill and blast tunnel construction.

REFERENCES

Amvrazis S., TCS Tunnel Control System, Patent number 1007395, PCT number: PCT/GR2011/00040, www.geonovelty.com.

Bergmeister, K., 2015. *Sustainable Life-Cycle Design and Innovative Skills for Building Tunnels.* Brenner Congress 2015, pp. 61–65

Bergmeister, K. & Reinhold, C., 2017. *Learning and optimization from the exploratory tunnel - Brenner Base Tunnel.* Geomechanics and Tunnelling, October, pp. 467-476.

Hofmann, M., Glatzl, R.W., Bergmeister, K., Markič, Š. & Borrmann, A., 2019. *Requirements and methods for geometric 3D modelling of tunnels.* ITA-AITES World Tunnel Congress, Naples, Italy

Insam, R., Carrera, E. & Crapp, R.: *Das Entwässerungssystem des Brenner Basistunnels.* Swiss Tunnel Congress, Luzern, 2018.

Markič, Š., Borrmann, A., Windischer, G., Glatzl, R.W., Hofmann, M. & Bergmeister, K., 2019. *Requirements for geo-locating transnational infrastructure BIM models.* ITA-AITES World Tunnel Congress, Naples, Italy

Voit, K., Amvrazis, S., Cordes, T. & Bergmeister, K., 2017. *Drill and blast excavation forecasting using 3D laser scanning.* Geomechanics and Tunnelling 10 (2017), No. 3

Tunnels and Underground Cities: Engineering and Innovation meet Archaeology, Architecture and Art, Volume 3: Geological and geotechnical knowledge and requirements for project implementation – Peila, Viggiani & Celestino (Eds)

Return of experience following the excavation of "La Maddalena Reconnaissance Gallery" – TBM performance prediction in the Ambin Massif

E. Baldovin, A. De Paola & G.L. Morelli
Geotecna Progetti s.r.l., Milan, Italy

P. Gilli & C. La Rosa
Telt sas Tunnel Euralpin Lyon Turin, Turin, Italy

ABSTRACT: The geological and geotechnical context of the rocky massif of "La Maddalena Reconnaissance Gallery" and the results of excavation with an "open type" TBM are presented, with particular reference to the stretch between ch. 3+045 and 7+020 with overburden constantly over 1000 m, up to a maximum of 2000 m. Based on the registered data of the machine and the geomechanical characteristics of the rock mass collected during the tunnel boring, the real TBM excavation progress, total and net, and the advance values predicted by some empirical models, commonly used to forecast the TBM performance on the basis of the traditional rock mass classification indexes, are compared. The consequent critical analysis of the applicability of the empirical approaches to the excavation performance parameters prediction within the Ambin Massif represents an interesting base of knowledge, which is made available for the design of future excavations in similar geomechanical contexts.

1 INTRODUCTION

"La Maddalena Reconnaissance Gallery", executed between 2013 and 2017 to collect geomechanical information for the design of the future Turin-Lyon transalpine railway base-tunnel, is 7,020 m long. Its final section crosses, for almost its entire length, the crystalline bedrock of the Ambin Massif.

The study refers to the tunnel part excavated under a high overburden, involving the geometrically and structurally deeper geologic unit of the Ambin Massif, named the Clarea Complex (CLR), mainly composed of micaschists and gneiss.

In particular, the gallery stretch of about 4,000 m excavated in the period May 2015 - February 2017, extending between ch. 3+045 and the final ch. 7+020, where the overburden is consistently over 1,000 m, up to a maximum of about 2,000 m, has been examined.

From the data collected during the construction, the net and gross advancement of the TBM have been analysed and compared with the predictions deduced from empirical methods, such as Bieniawski rock mass excavability (RME) indicator and Barton Q index.

In addition, possible correlations between the geomechanical RMR and Q indexes classifying the mass and the net progression rate recorded by the tunnel boring machine have been looked into.

2 GENERAL CONTEXT

La Maddalena Reconnaissance Gallery is a natural tunnel near La Maddalena location, within the Municipality of Chiomonte (TO), Italy. The excavation was carried out using traditional methods until ch. 0+198; further on a mechanised method was implemented, using an

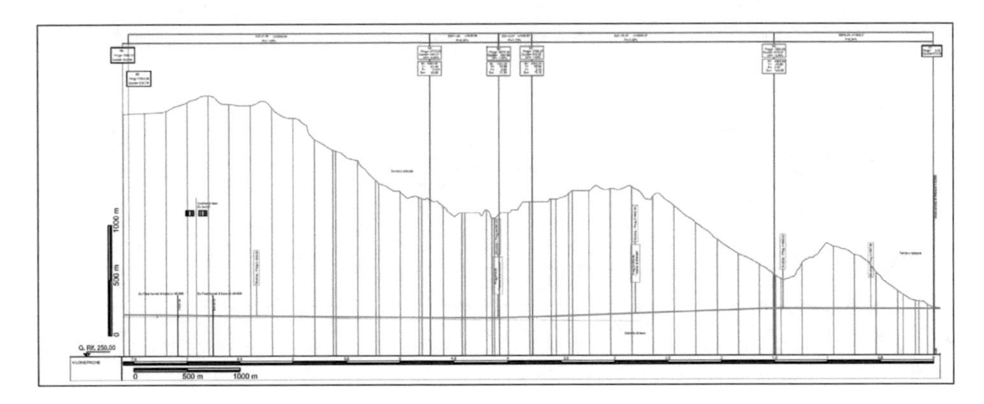

Figure 1. Elevation profile of the gallery.

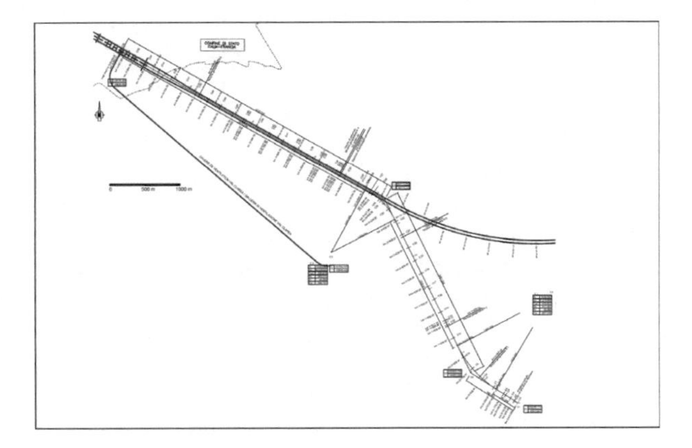

Figure 2. Planimetric layout of the gallery.

open-type TBM. The tunnel extends to ch. 7+020. The elevation of the tunnel at the entrance is about 670 m above sea level. The TBM nominal diameter of excavation is 6.30 m. The internal measurements and shapes of the tunnel were fixed during the design phase, considering the requirements during the gallery excavation, with reference to the next construction phase and the operational phase of the principal project. Within the tunnel, an accumulation basin and a redirection area for the water intercepted during excavation are located in large side chambers at ch. 4+130 and ch. 2+805, respectively. The elevation profile of the tunnel is: uphill for the first 1,500 m, with an approximate 0.34% slope, then drops downhill with a maximum slope of 3.32% for about 1,600 m. In the last 3,500 m circa, the section is uphill with a maximum slope of about 1.118% (Figure 1).

In terms of planimetric layout, the tunnel initially is straight for 300 m, followed by a bend with radius of curvature of 1,500 m, then another straight stretch of about 1,600 m and a bend with radius of curvature of 1,000 m, to finally continue straight on to the end. Starting from about ch. 3+670 the reconnaissance gallery is aligned with the two tubes of the future base tunnel (Figure 2).

3 GEOLOGICAL AND HYDROLOGICAL CONTEXT

The reconnaissance gallery for almost its entire path crosses through the crystalline bedrock belonging to the tectonostratigraphic unit of Mont d'Ambin.

Only for a limited section, near the entrance, for about the first 120 m, the excavation involves quaternary slope deposits (fluvioglacial deposits and undifferentiated glacial deposits)

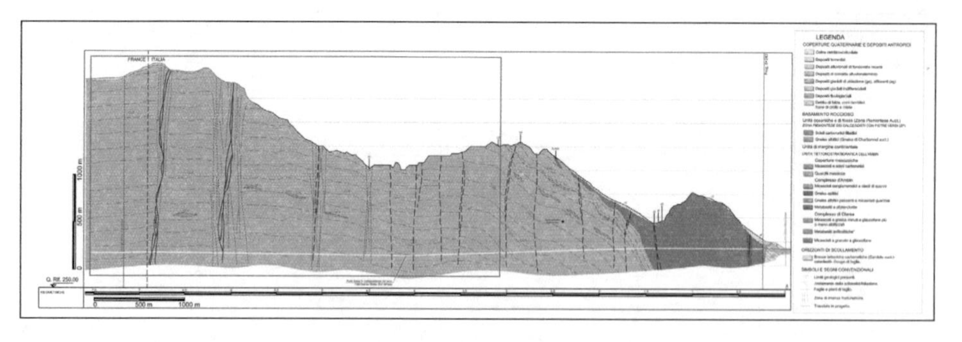

Figure 3. Geological and geomechanical profile of the tunnel.

and, subsequently, up to ch. 0+198, a series of lithotypes that can be classified as: Carniole (ca), dolomites with intercalation of talc and talcschist (do), carbonatic-phyllitic schists and micaschists with chlorite (GCC). Starting from ch. 0+198, it continues on in the Ambin Massif, first crossing aplite gneiss (AMC) with jadeite (up to ch. 1+148) and, following that, albite gneiss and quartz-conglomeratic micaschists belonging to the Ambin Massif (AMD). In the following section, from ch. 1+350 up to the end (ch. 7+020), the reconnaissance tunnel exclusively crosses the Clarea Complex (CLR) that represents geometrically and structurally the deepest unit of the Ambin Massif. This Unit consists of minute gneiss and grey-bluish micaschists with biotite, muscovite, glaucophane, garnet, quartz and albite with a layered structure, due to the presence of stretched out lenses and layers of metamorphic quartz. As anticipated, the present paper is focused on the exploratory tunnel stretch between ch. 3+045 and ch. 7+020 (the section framed in red in Figure 3), in which the overburden is consistently above 1,000 m, with a maximum of about 2,000 m in the final sector.

From the hydrogeological point of view, the entire length of the reconnaissance gallery crosses rock with an underground circulation that is substantially controlled by fracture density, interconnectivity of the joints and localised presence of persistent, permeable discontinuities. During excavation, no areas with significant concentrated water flows have been encountered, except for a few limited zones with flow along fault lines and with intense fracturing, however, with limited short-term peak flow rates.

4 TBM CHARACTERISTICS

The tunnel boring machine was designed according to the expected geological conditions and the activities requested for the investigation purposes, the stabilization of the tunnel, and to maintain high levels of productivity even in unfavourable environments. The used TBM belongs to the family of "high-performance machines in hard rock formations". It can be divided into three structural components consisting of an operational section (cutterhead, bulkhead and main beam), the gripping section (the thrust and gripping system), and the drive section (electric motors with gearboxes, gearwheel and main bearing, collection blades on the head and a conveyor belt). The potential presence of poor-quality materials, with different characteristics, such as the investigation purpose of the tunnel, have influenced the set-up of the machine in terms of the power available at the head and of its overall structure and powering. That led to the creation of two main work zones in which targeted reinforcement and geological investigations could be carried out. The machine has thus been designed with two separate working zones, named L1 (as close as possible to the head, about 6 m from the front) and L2 (about 40 m from the front). The distance between the boring front and the area where the supports were installed has determined the design choices in terms of the excavation sections. The following table lists the main characteristics of the tunnel boring machine (Table 1).

The excavation parameters were recorded by an Automatic Data Detection and Recording System able to read, display and collect the most important parameters. The Automatic

Table 1. TBM main characteristics

Technical specs	Value	Unit of measure
Boring diameter	6.30	m
N. of cutters	43	-
Max operational thrust	13,667	kN
Max thrust capacity	14,200	kN
N. of motors	7	-
Cutterhead power	2,205	kW
Cutterhead rotation speed	0–9.3	revs/min
Thrust cylinder stroke	1,830	mm
Recommended cutterhead pressure	310	bar
Nominal system pressure	345	bar

System then extracted the data collected, so it could be continuously recorded, monitored, processed and compared to the data coming from surveys and monitoring.

5 GALLERY EXCAVATION SECTIONS

On the basis of the geological and geomechanical study of the rock mass, a series of excavation and reinforcement sections that could be applied in the various scenarios according to the RMR index and tunnel overburden has been defined in the design stage, together with a criterion for their application.

Sections with occasional bolts were then adopted (sections F1 and F2), referring to the best geomechanical conditions expected along the reconnaissance gallery stretches with RMR values above 60 (class 2), and with systematic bolting (F3a-b-c and F3c_1) or single ribs (FMV) for intermediate geomechanical situations with RMR values between 51 and 60 (Class 3).

In the lower-rated geomechanical contexts with RMR values below 51 (Classes 3 and 4), where a higher fracturing degree in the mass and more evident plastic deformations are present, the sections made up of metal panels with double ribs were planned (sections F4 and F5). Lastly, to contrast rockburst phenomena, specific sections were designed (sections F4_1 and F5_1) reinforcing the shapes and increasing the frequency of the metal panels that form sections such as F4 and F5.

6 DATA COLLECTED DURING THE EXCAVATION PHASE: GEOMECHANICAL CLASSIFICATION OF THE ROCK MASS AND ADVANCE OF THE TBM

6.1 *Geomechanical classification of the massif*

In terms of geomechanical characteristics, in the stretch being examined between ch. 3+045 and ch. 7+020, the rock masses crossed through were found to be good (RMR classes 2 and 3), showing weakened characteristics only at short sections along faults or zones with intense fracturing (RMR class 4).

Figure 4 lists the subdivision into RMR classes and the trend of the relative index as the reconnaissance gallery advances.

The micaschists of the Clarea Complex (CLR) have shown average uniaxial compression strength σ_c around 100 MPa, accompanied by an accentuated mechanical anisotropy due to the typical markedly schistose texture of the lithotype.

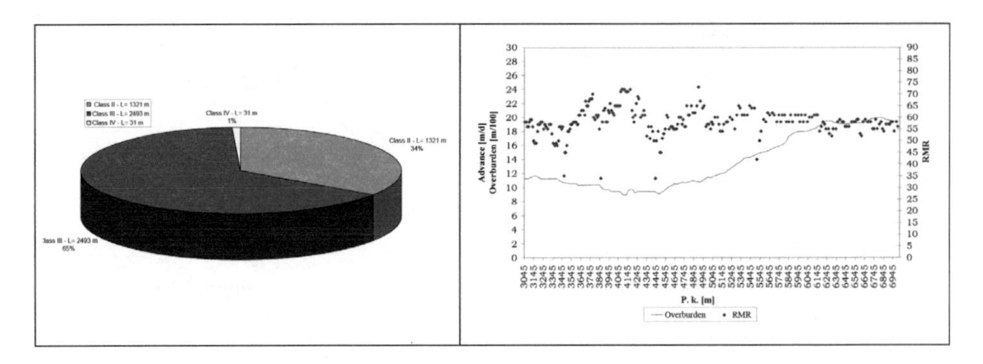

Figure 4. Rock mass quality between ch 3+045 ÷ 7+020 – Bieniawski RMR class – Percentage breakdown (a) – Sampled RMR and progression of the tunnel overburden (b).

6.2 *Progress of the tunnel boring machine - Total advance (TA) and net advance (NA).*

In the period May 2015 (ch. 3+045) - February 2017 (ch. 7+020), the tunnel boring advances were between a maximum of 280 m in April 2016 and a minimum of 45 m in August 2015, a month in which repairs were done on the TBM, with an average of about 190-200 m per month. The following figure (Figure 5) illustrates the gross monthly progress in m/month, called Total Advance (TA), which includes stoppages for maintenance, surveys, repairs of collapses, etc. which took place in the time span considered. The same figure indicates, month by month, the percentage of use (PU) of the tunnel boring machine.

An adjustment was made to the gross monthly advance TA numbers to account for the slow-downs and stoppages of the TBM which were necessary for the systematic execution of the geological studies taking place at the boring front, so as to bring the case in question to a "classic" mechanical excavation in which the effective boring time is a sum of the excavation, regripping and towing of the backup. The gross TA values were then increased by a percentage equal to the relationship between the time required for the execution of the geological studies and geotechnical investigations and the theoretical monthly time, generally about 4% on average. The average daily TA in the period of reference resulted about 6.5 m.

Starting from the aforementioned total advance (TA) values, the relative net advance (NA) numbers were calculated, multiplying the TA values for the actual machine use coefficient (subtracting the time required to carry out maintenance, repairs and surveys). The average daily net advance (NA) in the period of reference came to about 17.5 m. Below is a representation of the Total Advance TA (Figure 6) and the Net Advance NA (Figure 7) in m/day, monthly averaged, in the section studied. To be fully exhaustive, the tunnel overburden and the detected RMR values are also indicated.

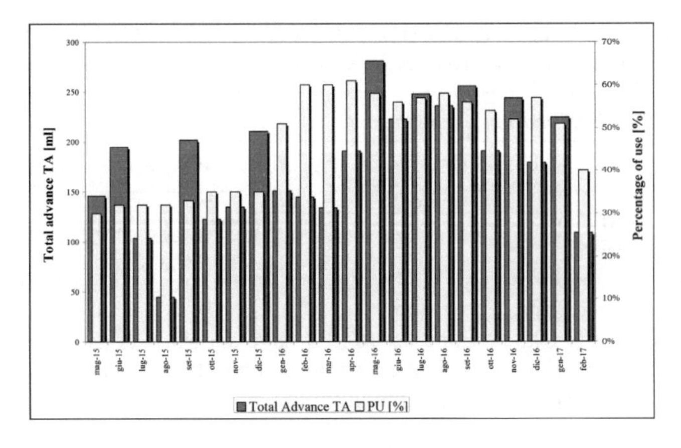

Figure 5. Monthly progress (Total advance TA) of the TBM and relative percentage of use (PU)

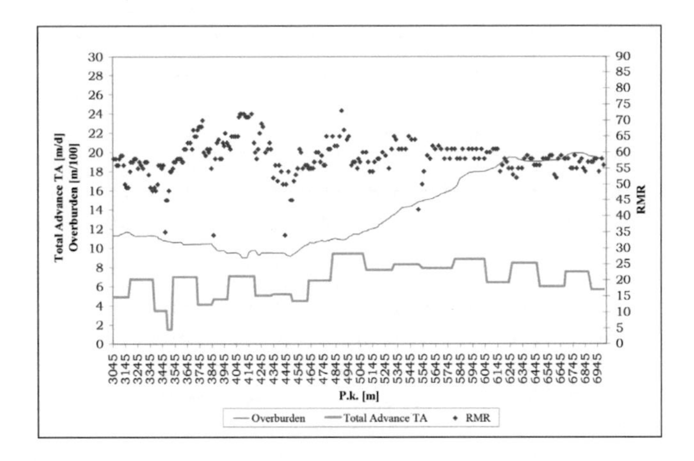

Figure 6. Total advance TA

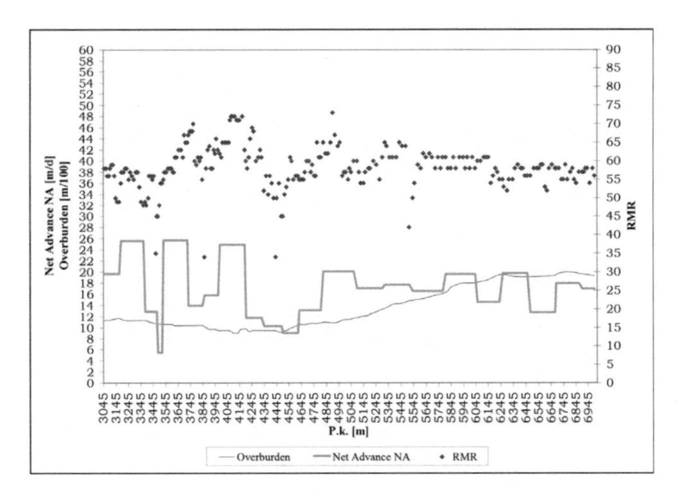

Figure 7. Net advance NA

7 THEORETICAL ADVANCE OF THE TBM BASED ON GEOMECHANICAL INDEXES AND COMPARISON WITH THE MEASURED VALUES

7.1 *Theoretical advance values calculated from Bieniawski rock mass excavability index (RME)*

During excavation, some systematic geostructural measurements were taken in the tunnel about every 10 m. The geomechanical data collected during excavations were then used to calculate Bieniawski rock mass excavability (RME) values, thereby obtaining the ARA_T (Theoretical Average Rate of Advance) and ARA_R (Real Average Rate of Advance) parameters, the latter being analogous to the Total Advance TA, which are prediction indexes of theoretical and real advancement, respectively.

Bieniawski correlations developed for open TBMs were used for this purpose:

$$ARA_T = 0.839 \times RME - 40.8 \tag{1}$$

for terrains with σ_{ci} >45 MPa and for RME<55 ARA_T=6 m/day

$$ARA_R = ARA_T \times F_E \times F_A \times F_D \tag{2}$$

where FE= Factor of Crew efficiency= 0.7÷1.2; FA= Factor of team adaption to terrain=1.0, FD= Factor of tunnel diameter = 1.59

The FE values corresponding to Bieniawski minimum and maximum values, 0.7 and 1.2 respectively, were considered. The same index was also adopted with the value of 1 (which is about the average).

The following figure illustrates the RME values and the minimum, maximum and average ARA_R values, in m/day, monthly averaged, in the section being studied. For a direct comparison, the actual Total Advance TA numbers are shown as well (Figure 8).

The average daily values of the Real Average Rate of Advance in the period of reference varied between around 7 m/day (ARA_{Rmin}) and 12 m/day (ARA_{Rmax}), compared with actual measured Total Advance TA values of about 6.5 m.

There is thus a good matching between the estimated ARA_R values and those actually measured for the Total Advance TA, especially as far as the minimum values provided by the Bieniawski correlation are concerned.

7.2 *Theoretical advances calculated based on Barton QTbm model*

The values of Barton Q parameter of the mass were calculated based on the geomechanical data collected during excavation. The Qtbm was then elaborated and used to obtain the Penetration Rate (PR) and Advance Rate (AR) parameters via the following formulas:

$$PR \approx 5 \cdot (QTbm)^{-0.2} \tag{3}$$

$$AR \approx 5 \cdot (QTbm)^{-0.2} \cdot T^m \tag{4}$$

$$T = 5 \cdot \left(\frac{L}{PR}\right)^{\left(\frac{1}{1+m}\right)} \tag{5}$$

where T = the time necessary to bore a tunnel with length L, m = the negative coefficient that represents the AR decrease over time.

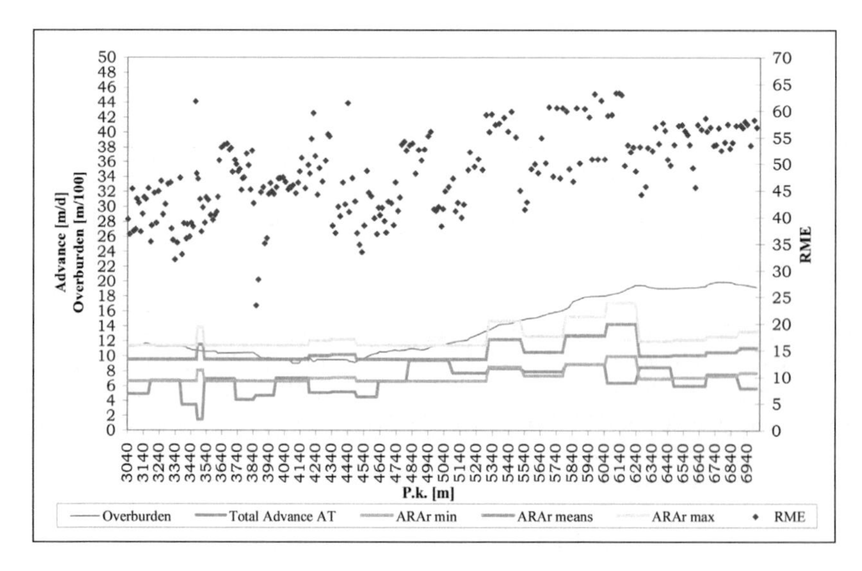

Figure 8. Monthly Real Average Rate of Advance (ARA_R), calculated based on the RME (Rock Mass Excavability) index and Total Advance TA between ch. 3+045 ÷ 7+020

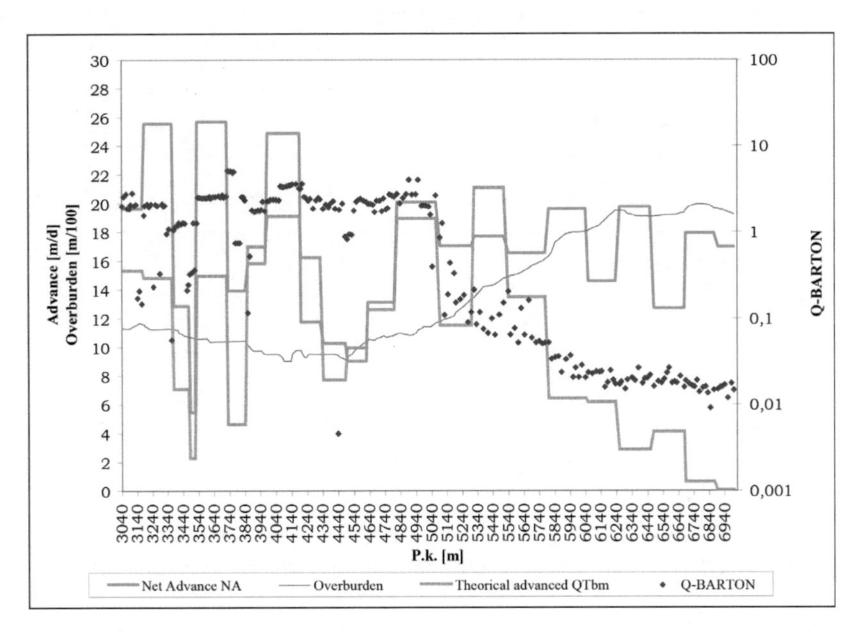

Figure 9. Theoretical monthly advances(m/d), calculated from Barton QTbm index and Net Advance (NA) between ch. 3+045 ÷ 7+020

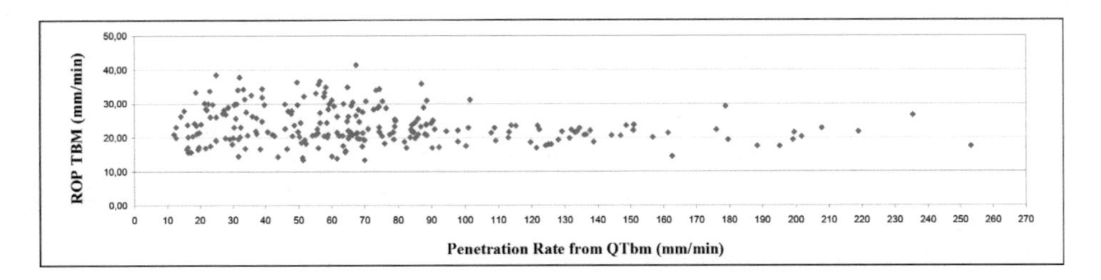

Figure 10. ROP from the TBM data and PR calculated from Barton QTbm index between ch. 3+045 ÷ 7+020

Based on the AR values, it was then possible to calculate the theoretical net average monthly advance in m/day, considering the real TBM excavation hours. In Figure 9, the values of Barton Q index and the theoretical average monthly advance numbers in m/day, calculated on the basis of the QTbm index for the section studied, a dimension analogous to the Net Advance (NA) parameter, are represented.

Figure 10 contains the graph that correlates the Penetration Rate (PR) values calculated from QTbm values and the ROP provided by the TBM data in the tunnel boring phase.

The comparison between the theoretical advance values in m/day, monthly averaged, calculated from QTbm in the studied section and the Net Advance (NA) values measured during excavation demonstrates the lack of a clear correlation. This lacking correlation appears especially as the overburden increases, where a decrease in Q linked to the tensile state (and consequently to the theoretical calculated advance) does not correspond to a clear decrease in the measured advance of the TBM. This inconsistency is also clear in the comparison of the theoretical average daily advance value in the period of reference, about 11 m, with the actual net advance (NA) number measured, which is significantly greater than 17.5 m.

Moreover, a comparison between the Penetration Rate (PR) from the QTbm index and the ROP provided by the TBM data set does not demonstrate a satisfactory correlation.

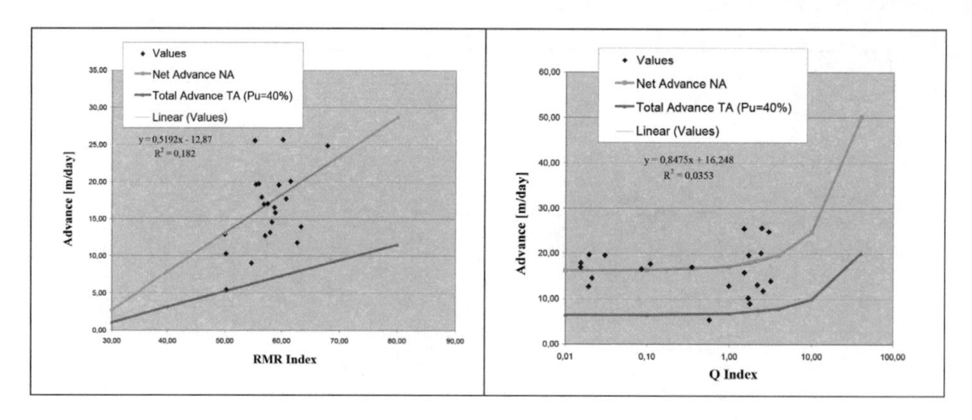

Figure 11. Correlation between Net Advance (NA) and the RMR (a) and Q (b) indexes

8 CORRELATIONS BETWEEN THE MEASURED NET ADVANCES AND THE RMR AND Q GEOMECHANICAL CLASSIFICATION INDEXES

The monthly net advance (NA) measured during the excavation were correlated with Bieniawski RMR and Barton Q geomechanical indexes as taken from the rock mass. The following empirical correlations resulted (Figure 11a and Figure 11b):

$$NA = 0.5192 \cdot RMR - 12.87\ (m/day) \tag{6}$$

$$NA = 0.8475 \cdot Q + 16.248\ (m/day) \tag{7}$$

For comparison, the same figure shows also the total advance (TA) numbers, calculated by considering an average machine use (PU) percentage of 40%.

The above correlations have generally low coefficient of determination (R^2), proving that, in the specific case being studied, factors other than the sole geomechanical quality of the mass excavated had a clearly important influence on the machine progress.

In detail, the available data show that, for RMR index values between 30 (Class 4) and 80 (Class 2), the resulting net advance (NA) is between about 2.5 m/day and 28 m/day, which corresponds to total advance TA (PU=40%) numbers between 1 m/day and about 11 m/day.

For Q index values between 0.01 (very poor rock) and 40 (good), the resulting net advance (NA) is between about 16 m/day and 50 m/day, corresponding to total Advance TA (PU=40%) values between about 6 m/day and 20 m/day. It is to be noted that for Q index values below 1, up to 0.01 (poor to very poor rock), the correlation does not highlight a decrease of the TBM advance (in line with what was actually measured), preserving the advancement almost constant, as a direct result of the elevated overburden and thus of the high stress state.

Also considering the average RMR and Q values measured in the relevant stretch, that is, 58 (Class 3) and 1.30 (poor quality rock) respectively, net advance (NA) of about 17-17.5 m/day and total advance (TA) of 6.5-7 m/day are obtained, largely in-line with the detected average advance values of the TBM in the examined section.

9 CONCLUSION

Considering the data obtained during the excavation of La Maddalena Reconnaissance Gallery, the gross and net advance values of the TBM were analysed and processed. The study has focused on the stretch of the tunnel in which the overburden was consistently in excess of 1,000 m, up to a maximum of about 2,000 m, i.e. between ch. 3+045 and the final ch. 7+020.

Then a comparison has been conducted of the measured and estimated advancement numbers, deducible from empirical methods used for the estimation of the performance of an open-type TBM in hard rock, such as Bieniawski Rock Mass Excavability (RME) index and Barton QTbm model.

The obtained numbers show a good correspondence between the predicted values of Bieniawski ARA_R and the actual TA values measured during excavation. This was particularly true in terms of the minimum values estimated using the empirical correlation proposed by Bieniawski. The result seems to confirm that the RMR index, adopted in the planning and execution phase as a base for the design and application of the support-type sections during excavation, and the Rock Mass Excavability (RME) index are sufficiently reliable to predict the mechanised excavation advance within the Clarea Complex (CLR).

The same cannot be said for the theoretical advance numbers calculated using Barton model (QTbm), which does not demonstrate a clear correlation with the advance measurements. In particular, a significant decrease in the Q index, which appears when the overburden and thus the in-situ rock stress increases, did not correspond to a significant decrease in the TBM measured advance, as the prediction model would suggest.

Lastly, possible correlations between the RMR and Q geomechanical indexes of the rock mass and the net advance rate measured during excavation were analysed. The resulting correlations, although characterised by low coefficient of determination (R^2), can still be an useful indication to consider in the planning phase for future excavations in the same geomechanical environment or in other similar contexts.

REFERENCES

Barton, N. 1999. *TBM performance estimation in rock using Q TBM*. Tunnels and Tunnelling International, Sept. 1999, pp. 30-34

Barton, N. 2009. *TBM prognoses in hard rock with faults using QTBM methods*. Keynote lecture, Inst. of Min. Metall. International Tunnelling Conf., Hong Kong.

Barton, N. 2010. *TBM prognoses for hard rock with weakness zones, using open-gripper or double-shield solutions*. Keynote paper. Int. Conf. on Underground Constructions, Transport and City Tunnels, Prague.

Barton, N. & B. Gammelsæter, 2010. *Application of the Q-system and QTBM prognosis to predict TBM tunnelling potential for the planned Oslo-Ski rail tunnels*. Nordic Rock Mechanics Conf., Kongsberg, Norway.

Barton, N. 2015. *TBM performance, prognosis and risk caused by faulting*. Plenary lecture, VIIIth S. American Rock Mech. Cong, Buenos Aires, 40

Bieniawski, Z.T., Celada, B, Galera, J.M. and Álvarez, M, 2006. *"Rock Mass Excavability (RME) Index"*. Proc. ITA World Tunnel Congress, Seoul, Korea.

Bieniawski, Z.T., Celada, B. and Galera, J.M., 2007. *"TBM Excavability and machine-Rock Interaction"* Proc. RETC, Toronto, p. 1118.

Bieniawski Z.T., Celada, B and Galera, J.M., 2007. "*Predicting TBM Excavability*". Tunnel & Tunnelling International, September, p. 25.

Bieniawski, Z.T., Celada B., Galera J.M. & Tardaguila I, 2008 – *New applications of the excavability index for selection of TBM types and predicting their performance* - World Tunnel Congress 2008 – Underground Facilities for Better Environment and Safety – India

Padovese, P., Berti A., Baldovin E., De Paola A., Morelli G.L., Ascari G., 2017 – "Innovazioni delle tecnologie di scavo meccanizzato in ammassi rocciosi caratterizzati da roccia dura con distacchi in calotta e alte coperture" – XXVI Cycle Conference of Mechanical and Engineering of rocks MIR 2017, Turin, Italy

Tunnels and Underground Cities: Engineering and Innovation meet Archaeology, Architecture and Art, Volume 3: Geological and geotechnical knowledge and requirements for project implementation – Peila, Viggiani & Celestino (Eds)
© 2020 Taylor & Francis Group, London, ISBN 978-0-367-46583-4

The use of real time monitoring systems to reduce execution risks during tunnel excavation in complex situations: The case of Stalvedro tunnel, Switzerland

G. Barbieri, M. Giani, M. Kündig & S. Motta
AF Toscano, Lugano, Switzerland

ABSTRACT: When excavating a tunnel near existing infrastructures, one of the major challenges is to minimize the risk of causing damages on them by induced deformations or vibrations. A necessary tool to be provided in these situations consists in real-time automatic monitoring systems with threshold alarms, coupled to emergency procedures and coordinated with infrastructure owners. The aim is to be continuously aware and ready for unforeseen situations and, hence, to avoid unacceptable disservices and consequent costs, such as interruption of service in adjacent infrastructures. During construction works commissioned by the Federal Road Office of Switzerland to widen existing Stalvedro tunnel along A2 highway in Switzerland, such concept was successfully applied and obtained results are herein described. Drill and blast excavation was performed dealing with adjacent highway tunnel and an upper crossing railway tunnel. Moreover, an open-cut 35 m high at the north portal was excavated in a complex geology.

1 INTRODUCTION

The Stalvedro tunnel is part of the A2 highway in the municipality of Airolo (swiss canton Ticino). It is located in close proximity to the river Ticino, in the upper Leventina valley. The existing structure is the last tunnel before the south portal of the Gotthard Road Tunnel and plays a strategic role in the north-south transportation axis of Switzerland. The Stalvedro tunnel was built in the early 1970s and comprises two separate tunnels, with the north-south tunnel featuring a length of 358.1 m and the south-north tunnel measuring 320.2 m.

Given the large volume of traffic on the axis of the A2 highway and the resulting traffic jams in front of the south portal of the Gotthard Road Tunnel, a traffic light system used to be installed at the south portal of the Stalvedro south-north tunnel to prevent the occurrence of traffic jams inside the tunnel itself. A stationary column of vehicles posed a serious safety issue for road users in the event of an incident and it was essential to prevent such a risky situation. The existing structure of the south-north tunnel did not feature a hard shoulder (emergency lane) for the emergency services or a cross passage that could act as an escape route leading to the north-south tunnel located opposite. In addition, the traffic jams limited the possibility of using the highway exit in Airolo on the northbound carriageway.

On the basis of this contour conditions, the client, i.e. the Swiss Federal Road Office (FEDRO), planned the enlargement of the northbound tunnel and the renovation of the southbound tunnel, with the aim of satisfying the latest swiss safety requirements and extending the service life of the construction. The necessary width of the new tunnel cross-section, which was conceived according to the Swiss standard VSS 640 855c for the 4+0 traffic routing specified by the client, amounts to 14 m (internal profile), having a total enlargement of about 5 m.

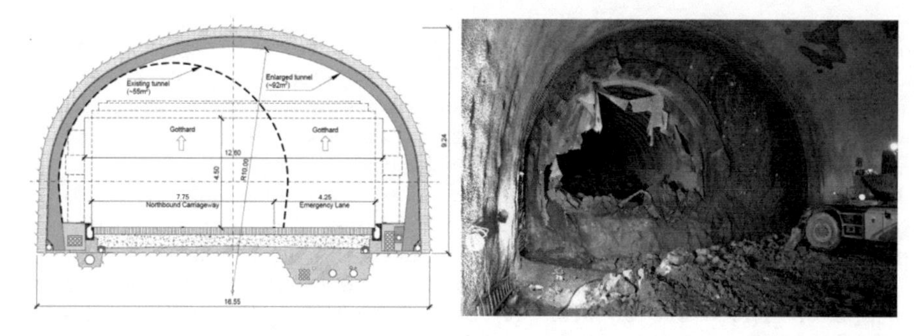

Figure 1. Enlargement of the northbound tunnel.

1.1 *Enlargement of the northbound tunnel*

The Stalvedro tunnel is predominantly located in the gneiss of the Lucomagno crystalline, which makes up the rock cliff of the Stalvedro ravine in the area of the river Ticino. The Lucomagno crystalline wedge was thrust upwards as a Lower Pennine gneiss cover via the sediments of the Gotthard Massif (Piora zone) before folding back into the massif at a steep angle. The transition between the crystalline gneiss substrate and the sediments encountered in the Triassic (predominantly cellular limestone that is partly dolomitised, subordinate quartzite) is located around 75 m from the north portal of the northbound tunnel.

For the purpose of enlarging the tunnel cross-section, the existing structure needed to be completely demolished. In addition, the native rock needed to be broken out in the area of the enlargement by means of drill and blast tunnelling, with a hydraulic breaker (jack hammer) being used in lower rock quality conditions. The lower-quality geology of the Triassic zone in the north section of the tunnel made it necessary to adapt the excavation geometry and construct an invert vault. The excavation support consisted of shotcrete or steel fibre shotcrete, rock bolts, reinforcement wire mesh and steel girders.

The new cut-and-cover tunnels that were planned for the portal zones made it necessary for the existing tunnel structure to be exposed, demolished and reconstructed.

A new pedestrian bypass between the two tunnels complete the design picture.

1.2 *Renovation of the southbound tunnel*

Based on the results of the structural inspections and the feasibility study for the enlargement of both tunnels in 2007, FEDRO concluded that the Stalvedro north-south tunnel was to be renovated without enlargement of the existing tunnel cross-section and that the remaining requirements of the standards had to be fulfilled.

Concrete restoration, new walkways, two new combined SOS/hydrant niches and a new hydrant pipeline are the main aspects of the renovation of the southbound tunnel.

2 COMPLEXITY PF THE GEOLOGIC CONTEXT

The enlargement of the Stalvedro tunnel was a complex project, having to deal with many constraints related to adjacent existing infrastructures to be kept on service and taking into account the complex geology; such aspects drove the design choices and represented the main onsite challenges.

2.1 *Nearby infrastructures*

The northbound tunnel to be widened was just at a distance of 9 m by the adjacent existing southbound highway tunnel. The latter had to be used during renewal works to assure traffic in both directions: all vehicles heading north to the Gotthard Road Tunnel were diverted through the southbound tunnel so that a bidirectional traffic management had been implemented.

During construction works, whenever a blasting phase was ready to be performed in the northbound tunnel, a short traffic interruption (less than 10 minutes) was performed into the southbound tunnel: the aim was to avoid any risk of incident inside the tunnel in service, caused by a sudden blasting in the adjacent tunnel. Moreover, it had to be avoided that accidental detachments from the existing vault could fall onto the vehicles passing by.

Moreover, the two Stalvedro tunnels pass beneath the Swiss Federal Railways (SBB) tunnel at a glancing angle and at a minimum clearance of around 8.6 m. The railway tunnel, whose commissioning dates back to 1882, had been completely excavated in the rock massif of Stalvedro and its lining consists of masonry at the portals and a shotcrete layer just at the crown. At the time of the construction works for the enlargement of the highway tunnel, which started in October 2015, the Gotthard Base Tunnel was still out of service and the railway line through the Stalvedro rocky spur guaranteed the only railway connection between the north of Switzerland and the canton Ticino through the Alps. That means that no traffic interruption was allowed and the SBB tunnel had to be meticulously checked, as well as any displacement due to the underlying tunnel excavation had to be minimized and controlled.

The Swiss Federal Railways defined specific requirements on allowable vibrations induced by blasting or mechanized operations. These requirements are summarized in Table 1.

2.2 *Geological scenario at the north portal*

The portal area in the north section is located in a challenging geological zone with a ground surface that slopes steeply in the direction of the nearby river Ticino. From a geological point of view, the planned pre-cut is partially located in the rock of the Triassic zone and is covered by the slope and settling debris. There are also fluvio-glacial deposits near to the river.

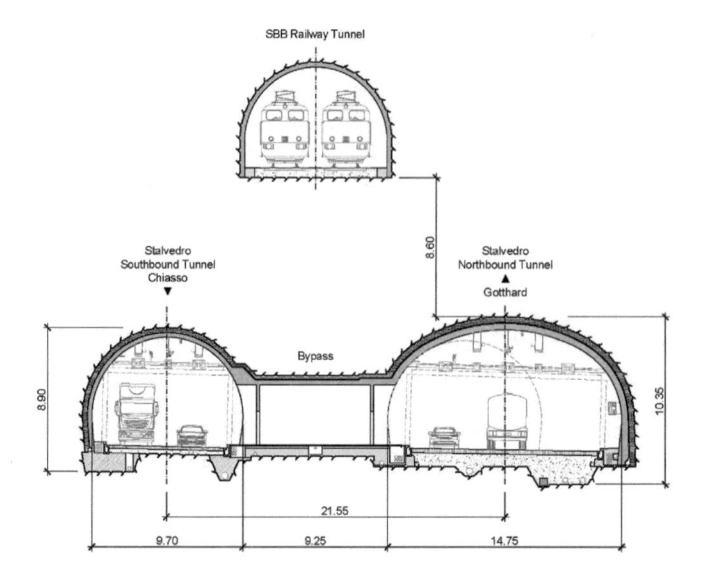

Figure 2. Cross section of Stalvedro tunnels, with the new pedestrian bypass and the SBB tunnel.

Table 1. Railway tunnel: limitation to induced vibrations (velocity).

		Frequency		
Excavation method	Railway traffic	< 30 Hz	30–60 Hz	> 60 Hz
Blasting	In service	2 mm/s	2 mm/s	2 mm/s
	Interrupted	30 mm/s	40 mm/s	60 mm/s
Jack hammer	In service	2 mm/s	2 mm/s	2 mm/s
	Interrupted	6 mm/s	8 mm/s	12 mm/s

Figure 3. Excavation trench at the north portal (2016) and historical picture of slope stabilization works.

The geologist defined the geotechnical model on the basis of individual soundings. The highly variable geological conditions and the complex morphology of the area made unavoidable a certain level of uncertainty on it. These conditions made it more difficult to design the excavation trench and to construct the new portal structure, having a height of the precut trench up to 35 m.

Moreover, according to available geological data, a quiescent landslide affects this area: historical evidence of a landslide in 1979 and the following widespread use of safety measures put the risk of a remobilization of the landside during excavation to the highest level.

Lastly, a rock boulder mass of about 140 m^3 just above the existing portal of the southbound tunnel had been identified by the geologists during onsite inspections. The design had to deal with the risk of a mobilization of this rock mass, due to vibrations possibly induced by blasting during enlargement works and by drilling during excavation of the open-cut.

As a preliminary measure, safety and reinforcement measures were provided on the slope located right above the foreseen trench. This included rockfall protection nets and surface stabilization nets, avalanches protection trestles as well as pre-tensioned anchors and rock bolts to improve stability of some boulders and outcrops. The trench was then stabilized by various levels of pre-tensioned anchors with reinforced concrete beams, shotcrete wall and rock bolts.

3 DESIGN APPROACH AND MONITORING SYSTEMS

Considering uncertainties in the geotechnical model and the complex geology of the interested area, an observational design approach was adopted, mainly consisting in two design elements:

- numerical simulations of the construction steps, based on available knowledge during design, to estimate both induced deformations and stresses and defining different support systems for different expected geological conditions;
- a proper monitoring systems for the construction phase aimed to measure real deformations, to evaluate correspondence to design assumptions/estimations and, if needed, to provide proper design adjustments during construction.

Having to deal with adjacent infrastructure in service, automatic monitoring systems were provided, aimed to a continuous check of deformation parameters, and a remote data management and alarm system was adopted to guarantee that relevant information were immediately provided to the operators. Finally, emergency procedures were defined to clarify the action flow to be followed whenever design thresholds would be overcome and, hence, to guarantee a quick action of site supervisors and designer to adapt support measures to the actual rock behaviour, in case of alert and intervention levels, and of highway and railway operators in case of extreme events (alarm levels) that would require traffic interruption in adjacent infrastructures.

3.1 *Design phase*

In the design, specific focus was put on the evaluation of displacements induced into the rock mass by the enlargement of the Stalvedro tunnel. A finite element model with a Mohr-Coulomb constitutive law were used to predict convergence of the tunnel excavation profile, displacements on the railway line above and effects on the concrete lining of the adjacent southbound tunnel on service (see Figure 4). Mohr-Coulomb parameters have been estimated from the Hoek-Brown failure criterion, by the calculation of equivalent angles of friction and cohesive strengths for each rock mass and stress range (Hoek, E. 1990).

Table 2 summarizes the results of the analysis in terms of maximum displacements estimated at the excavation profile of the enlarged tunnel and at the above railway tracks. In the same table, the displacement thresholds defined by the railway owner SBB are shown, according to the SBB I-50009 prescriptions and relevant for the present case (thresholds for a velocity v 80 km/h < v < 120 km/h).

Further analyses have been developed to investigate the response of both the slope and the adjacent southbound tunnel to the excavation of a 35 m high trench at the north portal. The excavation walls have been step by step assured by means of a shotcrete layer, 20 cm thick, about 500 nails and 100 pre-stressed ground anchors. Limit equilibrium analyses were performed to dimension support measures and numerical simulations provided estimated displacements: a 30 mm maximum displacement was predicted by the F.E. model at the bottom of the excavation trench.

3.2 *Monitoring of the southbound tunnel and SBB railway tunnel*

Monitoring of the adjacent Stalvedro southbound tunnel, as a bidirectional tunnel during the construction of the enlarged northbound one, was managed by the adoption of different systems. A preliminary inspection of the existing cracks in the concrete vault was performed and the most critical ones were injected. Further, visual checks after each blasting and the aforementioned traffic interruptions were performed and 3 multi-base extensometers were installed in the geological survey boreholes to control deformation response of the rock mass between the two tunnels. Further, 8 mini prisms were fixed to the concrete vault adjacent to the north portal open-cut in order to check displacements during excavation.

An automatic measuring system was installed inside the SBB railway tunnel, consisting of two automatic station outside the north portal of the tunnel and 13 measurement sections, each with 4 mini prisms (see Figure 5). The automatic topographic monitoring was coupled with a detailed laser-scan survey of the tunnel profile and rails, about every month.

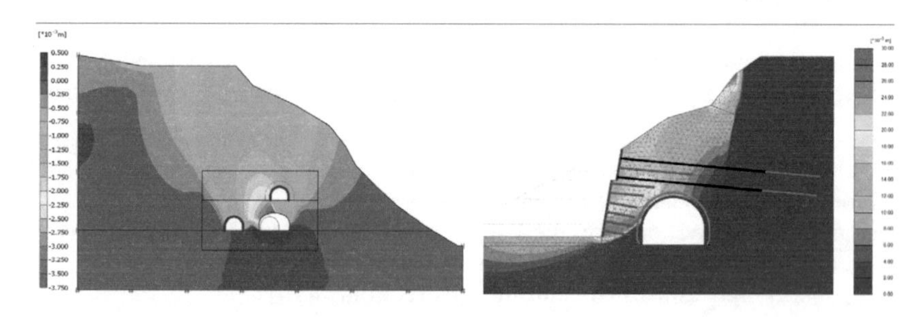

Figure 4. Total displacements (mm) of the F.E. models for the tunnel (left) and north portal trench (right).

Table 2. Displacements of the tunnel F.E. model and SBB monitoring thresholds.

Tunnel	Maximum displacement		SBB thresholds		
Excavation profile	Vertical	4.0 mm	Alert	Intervention	Alarm
	Horizontal	3.3 mm			
Railway Tunnel	Vertical	2.6 mm	6 mm	8 mm	10 mm
(lines level)	Horizontal	0.6 mm	6 mm	8 mm	10 mm

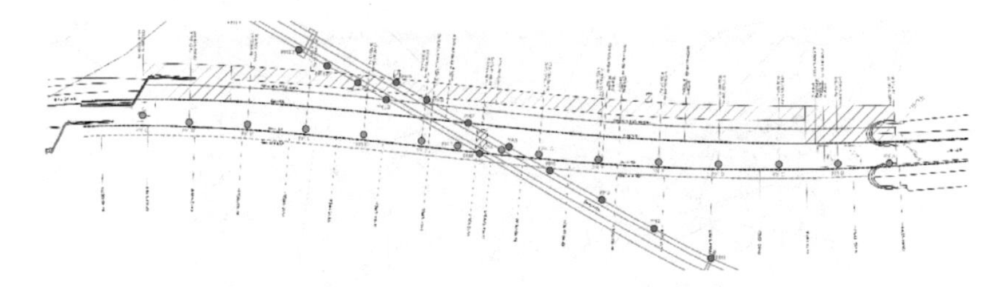

Figure 5. Plan of vibrations monitoring inside the southbound tunnel (blue) and SBB railway tunnel (red).

As already mentioned, vibrations also played a key role in the construction, as far as the strict limitations imposed by the Swiss Federal Railway are concerned (see Table 1). Hence, triaxial sensors were fixed at the tunnel abutments of both the SBB railway tunnel and the southbound tunnel. A total of 13 and 15 points of measurement respectively were considered along the tunnels, at a regular distance of 20 m one to another. Data were collected by an online connected registration unit and constantly checked by the designers and specialists. For the southbound tunnel, alert/alarm thresholds have been taken as for the railway tunnel.

At the beginning of the construction works, a careful calibration of both the quantity of explosives and the blasting scheme at the excavation front was also necessary to check the first results in terms of vibrations.

3.3 *Monitoring of the slope at the north portal*

Several topographic surface points (mini prisms) were deployed together with 3 inclinometers for the purpose of monitoring the slope area at the north portal. Moreover, load cells were installed on some anchors to control the stress load during different excavation phases and 2 sub- horizontal boreholes were drilled up to 40 m and instrumented with 2 multi-base extensometer connected to a datalogger, in order to better understand the deformation behavior of the trench wall in the loosened soil part.

A real time automatic monitoring system with threshold alarms was arranged, coupled to emergency procedures and coordinated with infrastructure owners. In case a threshold have been exceeded, an SMS would be automatically sent to the designer, the construction site supervisors and the contractor. Both an acoustic alarm and an alarm light were activated in case of alarm threshold and the risky area had to be immediately evacuated. In the following, designers, geologists and supervisors should have had to carefully check the reasons of the alarm and, in case of false alarms, restore the construction operations. Both load cells, extensometers and inclinometers were not part of the automatic alarm system: they were periodically checked once or twice a week in normal conditions, more often in case of a threshold overcoming.

Table 3 summarizes the chosen alert/intervention thresholds on the slope, whilst Figure 6 shows the distribution of the measurement points of the automatic system all over the slope above the construction site.

Table 3. North portal: monitoring thresholds.

Instrument	Measurement	Thresholds	
		Alert	Intervention
Topography (mini prisms)	Absolut displacement	± 5 mm	± 10 mm
	Relative displacement / 24 h	± 5 mm / 24 h	± 10 mm / 24 h
Extensometers	Axial deformation	5 mm	10 mm
Load cells (pre-stressed anchors)	Stress load (P_0 = initial load)	± 20% P_0	± 30% P_0

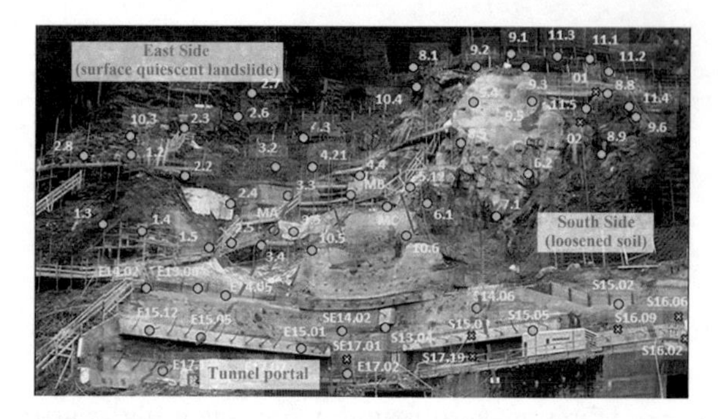

Figure 6. Distribution of measurement points all over the north portal slope.

Figure 7. Rock boulder (left), geophones (center) and vibrating wire strain gauges (right).

3.4 *Monitoring of the rock boulder at the north portal*

The presence of a big rock boulder with clear open cracks called for a special monitoring system.

In order to evaluate possible disturbance due to drilling and blasting operations, two geophones were positioned near the boulder. The sensors allowed to continuously check vibrations and to record the full accelerogram in case the trigger threshold of 0.5 mm/s was overcome.

Further, two strain gauges had been fixed at both sides of cracks, in order to check relative deformations. Topographic monitoring points had been as well integrated into the real time automatic system and the emergency procedure in case of alarm follows the same procedure described for the rest of the slope.

4 MONITORING RESULTS

The complexity of the monitoring system allowed to collect a huge amount of data, some of which are described in the following. Although the presence of the automatic alarm system, a constant check of monitoring results was performed in order to evaluate trends in addition to absolute values and anticipate possible risky events; the adopted approach and monitoring systems allowed to keep both the railway and the highway connections unaffected and, last but not least, to guarantee safety to workers, trains and motor vehicles.

4.1 *Southbound tunnel and railway tunnel: main results*

Both the tunnels did not suffer any appreciable effects of the nearby excavation. Deformations recorded by multi-base extensometers between March 2014 and December 2016 were

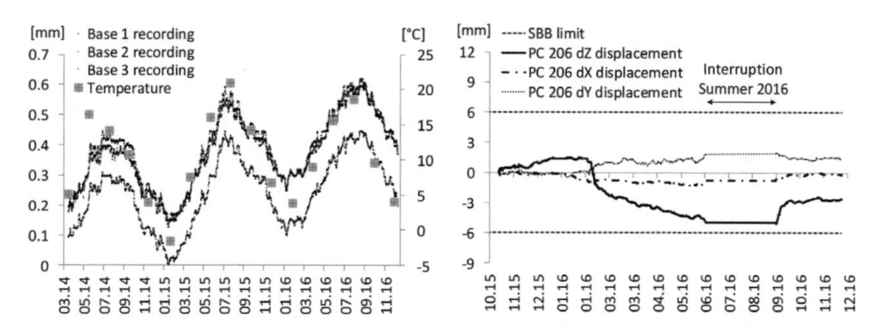

Figure 8. Multibase extensometer results (left) and example of railway tunnel topographic readings (right).

fluctuating lower than 1 mm, which was mostly due to the seasonal temperature variations, as clear in Figure 8. No relevant displacement of the southbound tunnel vault in the stretch adjacent to the north portal trench was recorded. Topographic measurements inside the SBB railway tunnel showed 3D movements during construction, but lower than the allowed limit of 6 mm (see Figure 8). Laserscan controls gave quite a similar result.

Concerning Figure 8, the lack of data in the summer 2016 corresponds to the work interruption during the same period, required by FEDRO in order to let traffic pass through the partially enlarged northbound tunnel to reduce long queues in the holiday season.

The automatic monitoring system allowed not to interrupt rail traffic also during the local collapse that occurred at the tunnel crown at the beginning of February 2016, quite near to the crossing railway tunnel: checking the continuous readings on point PC206 that was located right above the collapse position, it was possible to verify within a couple of minutes that the event had no deformation effects on the railway tunnel, hence, that no traffic interruption was necessary.

Vibrations inside both the tunnels were kept under control and generally velocity components lower than threshold limits were observed. Occasionally, alarm limits were overcome inside the railway tunnel. Thanks to the presence of many measurement points along the railway tunnel, it was possible to assess that they were only local effects, limited to just one measurement point, usually located directly above the tunnel excavation face.

4.2 *North portal trench: main results*

During excavation of the north portal trench, displacements have been observed both on the excavation wall and the slope above. Significant vertical (dZ) and transversal (dT) displacements are shown in Figure 9, together with the progressive advancement of the open-cut depth. Selected recordings of the east side of the open-cut are shown in Figure 9, whereas relevant displacements on the south side are shown in Figure 10. Location of both areas are visible in Figure 6.

In order to clarify the correlation between displacements and work activities, it should be mentioned that the advancement of the excavation (shown in the same figures) was not homogeneous over the full width of the open cut: due to the presence of the existing tunnel portal, the excavation of the material between the open-cut wall and the existing tunnel (mountain side) was delayed with respect to the rest of the trench (river side). This delayed sector was then completed together with the demolition of the existing concrete vault after the mentioned interruption of the works in summer 2016 and the operations caused additional movements, well recorded by the monitoring system.

Analyzing Figure 9, it can be seen that on the east side of the open-cut, vertical movements were more affected by the first half of trench excavation between February and April 2016, as during second excavation phase (October - November 2016) the wall was partially protected by the existing portal. Later, during excavation of the last 20 m of tunnel toward the north

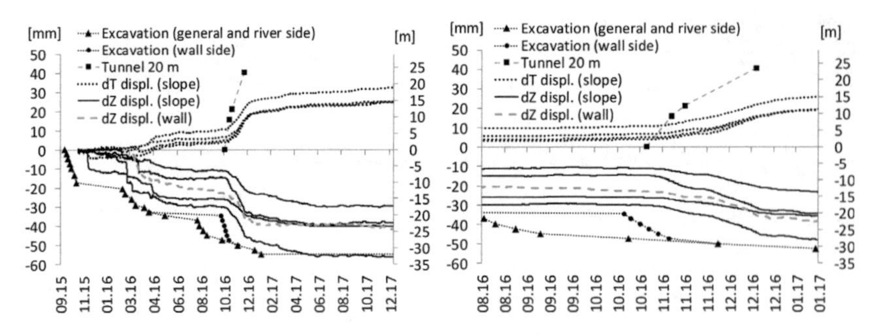

Figure 9. North portal (east side): displacements and advancement of different excavation procedures.

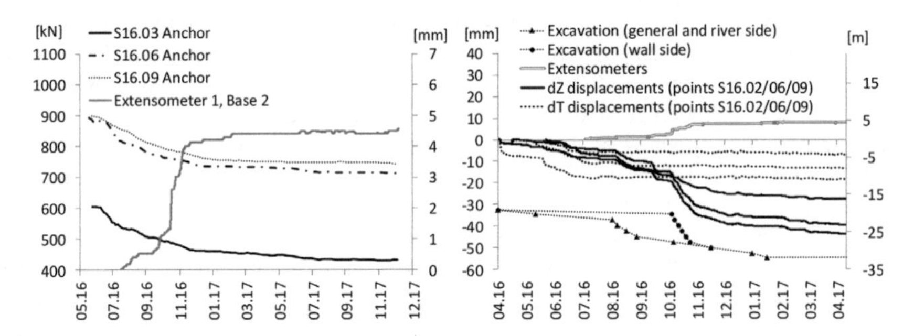

Figure 10. North portal (south side): anchors (left) and topographical readings (right).

portal, a different behavior was observed: in addition to vertical movements, also horizontal movements of the adjacent slope were recorded, showing a direct interaction between tunnel excavation and temporary landslide reactivation. Right after tunnel excavation ended, landslide movements gradually stopped as expected.

Analyzing Figure 10, a different behavior can be observed on the south side, where the trench is excavated in loosened soil: here, mainly vertical displacements were observed, related to second excavation phase as it was right in front of this side of the trench. Such high vertical movements were mainly due to the fact that the shotcrete wall was standing on loosened soil that could provide only small bearing resistance during intermediate phases, before to reach the bedrock. Moreover, load cells on anchors recorded a load reduction on them, but coupling such data to extensometer measurement (that recorded a compression deformation) it was clear that the reason of such relaxation phenomenon was due not to failure of the anchorage in rock, but to compression of the loosened soil right behind the wall and, hence, did not mean a risk of collapse.

4.3 *Rock boulder: main results*

Topographical measures showed no relevant displacement of the rock boulder both in vertical (dZ), transversal (dT) and longitudinal (dL) directions and the extensometers gave the same result, with average displacements lower than 2mm; oscillatory movements visible in the plots in Figure 11 were correlated to thermal variations thanks to the thermal sensor placed in the extensometers. As far as vibrations are concerned, during blasting inside the tunnel and drilling operations in the surrounding area, values of velocity lower than 0.5 mm/s were observed. Vibrations peaks, overcoming the design thresholds, were just correlated to local removal of small boulders and cleaning works near to the geophones.

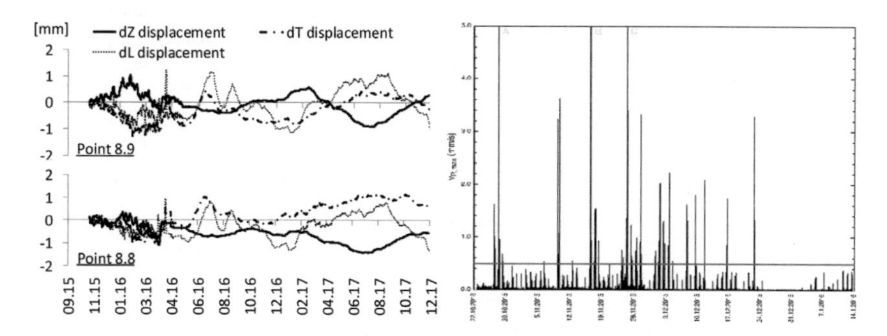

Figure 11. Rock boulder: displacements of significant points (left) and vibrations (right).

5 CONCLUSIONS

Monitoring by means of automatic reading and alarm systems allowed to carry on excavation works in safe conditions and to deal with unavoidable uncertainties in the geotechnical model and unexpected situations, such as local collapse during tunnel excavation and large displacements in different areas at the north portal, without unexpected traffic interruption. Moreover, the provided data allowed to better understand the behaviour of the rock mass and the adjacent slope and, hence, to better suit the design support measures to local geological conditions. The use of numerical analysis during design was then a useful tool to estimate displacements and provide reliable thresholds for the above monitoring systems, as well as adopting proper design solutions.

AKNOWLEDGMENTS

The authors gratefully acknowledge the Federal Road Office (FEDRO) in the name of Marco Fioroni and Gian-Mario Frei, to provide authorization to publish monitoring data and design information, as well as Studio Meier SA which implemented the automatic monitoring system and provided monitoring data and to Muttoni & Beffa SA and IFEC Ingegneria SA that were in charge of vibration monitoring systems respectively of the tunnel and of the rock boulder at the north portal during construction.

REFERENCES

Hoek, E. 1990. Estimating Mohr-Coulomb friction and cohesion values from the Hoek-Brown failure criterion), *Int. Jour. of Rock Mech., Mining Sciences & Geomechanics Abstracts*, vol.27, n. 3: 227–229.

Hoek, E. & Brown, E.T. 1988. The Hoek-Brown failure criterion - a 1988 update. Toronto: 15th Canadian Rock Mechanics Symposium.

SBB-CFF-FFS 2013. Regolamento I-50009. Monitoraggio degli impianti di tecnica ferroviaria su cantieri situati in prossimità dei binari.

Tunnels and Underground Cities: Engineering and Innovation meet Archaeology, Architecture and Art, Volume 3: Geological and geotechnical knowledge and requirements for project implementation – Peila, Viggiani & Celestino (Eds)

The role of direct investigations ahead of the tunnel face in the construction cycle of a conventional and mechanized tunnel

G. Barovero, S. Casale, H. Egger & E. Barnabei
Galleria di Base del Brennero, Brenner Basistunnel BBT SE, Fortezza-Mules, Italy

ABSTRACT: Exploration ahead of the tunnel face is extremely important in reducing geological and hydrogeological risk in tunnel excavation. In particular, direct investigations in deep-seated tunnels like the Brenner Base Tunnel are applied to verify in advance whether planning assumptions regarding the expected rock mass conditions and hydrogeology were correct. If these investigations are planned as part of a continuous Drill&Blast cycle, a sort of "macro" cycle is introduced, which includes a series of ordinary sub-cycles. In mechanized tunneling, due to the continuity of this type of excavation, exploration activities ahead of the face can be planned in parallel with other, more maintenance-oriented, activities, and therefore cause just short production interruptions. This paper examines the impacts of the direct investigation activities ahead of the tunnel face, which have so far been performed during the excavation of the BBT exploratory tunnel, on the production cycle.

1 INTRODUCTION

The results obtained by direct and indirect preliminary prospecting from the surface, aimed at assessing the geological and hydrogeological characteristics of deep tunnels such as the Brenner Base Tunnel (BBT) are not always sufficiently reliable for a detailed longitudinal geological section (profile), because it can often be necessary to complete any missing data with assumptions or interpolations that have no objective basis. Despite the extensive investigations carried out during the BBT geological mapping from the surface with state of the art borehole tests and geological-geotechnical laboratory tests on core samples, significant geological and geomechanical uncertainties still remained, especially in tectonically disturbed material such as that of the Periadriatic fault. These uncertainties include possible developments in terms of unpredictable rock mass behaviour, thus increasing the geological and hydrogeological risk during the excavation. For these reasons it is essential to plan prospecting drilling beyond the rock face in order to obtain the most reliable information on the rock mass. On the basis of this information one can assess the most appropriate excavation sections in advance, calibrating the quantity and characteristics of any support or reinforcement measures. This observational method provides for the continuous refinement of the project during construction works on the basis of guidelines that are already an integral part of the project. In order for such prospecting to be useful and convenient, its impact on the production cycle must be minimal, or in any case such that the ratio between the time required to carry it out and the impact of the result is clearly advantageous.

This paper examines the impact of advance prospecting on the production cycle, taking as an example both the stretch of the exploratory tunnel of the Brenner Base Tunnel excavated with the traditional method (D&B or percussion hammer) in the already completed Mules 1 construction lot, and the stretch excavated to date by tunnel boring machine (double shield TBM) in the Mules 2-3 construction lot which is still in progress, analysing the relationship between the time required for the execution of prospection activities and the obtained benefits in terms of information on geology and optimization of reinforcement measures.

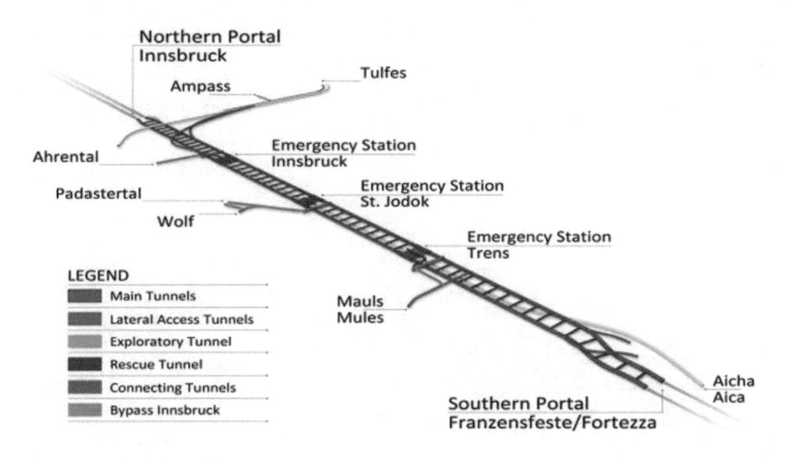

Figure 1. General layout of the BBT system.

2 THE BRENNER BASE TUNNEL

The Brenner Base Tunnel is a system of underground structures currently being excavated to build a high-speed railway connection between the town of Fortezza/Franzensfeste (Italy) and the city of Innsbruck (Austria).

The project includes the construction of two main one-track tubes characterized by a diameter of 9 m, and an approximately 6 m diameter service/exploratory tunnel, excavated in advance to explore and determine hydro-geological and geomechanical conditions, to study their response to excavation and subsequently adjust the reinforcement and support measures. The distance between the main tunnels ranges between 40 and 70 m (from axis to axis), while the exploratory tunnel runs mostly parallel to the two main tunnels, but 12 m below them. Furthermore, the main tunnels are connected transversally every 333 m for operational and safety reasons. The system also includes the already built four access tunnels (Ampass, Ahrental, Wolf and Mules/Mauls) and the three emergency stations (Innsbruck, St. Jodok and Campo di Trens/Freienfeld) located roughly 20 km apart (Figure 1).

The bypasses and the emergency stops represent the heart of the safety system for the operational phase of this line. The tunnel slopes will lie between 4.0 ‰ and 6.7 ‰ and will run under a variable overburden ranging from few meters to a maximum of 1600 m at the border between Italy and Austria.

The excavation will basically drive through four main rock types characterizing the central part of the Austroalpine area and the northern part of the southern Alps. Most of these are metamorphic rocks, consisting of Phyllite (22%), Schist with subordinate Marble (41%), Gneiss (14%), and relevant amounts of plutonic rocks, consisting of Brixner Granite and Tonalite (14%) (Figure 2).

The overall infrastructure length is about 55 km and the entire project will require the construction of about 230 km of tunnels. Once completed, the infrastructure will be directly linked to the existing 9 km long bypass tunnel connecting Innsbruck to Tulfes, making the BBT, at over 64 km, the longest underground railway connection in the world.

3 CASE STUDY 1 – CONVENTIONAL TUNNELING: EXPLORATORY TUNNEL THROUGH THE PERIADRIATIC FAULT ZONE (ITA)

3.1 *Drive description*

Case study 1 shows how the exploration measures affected the production of the roughly 1.5 km long stretch of the Exploratory Tunnel excavated through the Periadriatic Fault Zone, which showed heterogeneous and geotechnically challenging rock mass conditions.

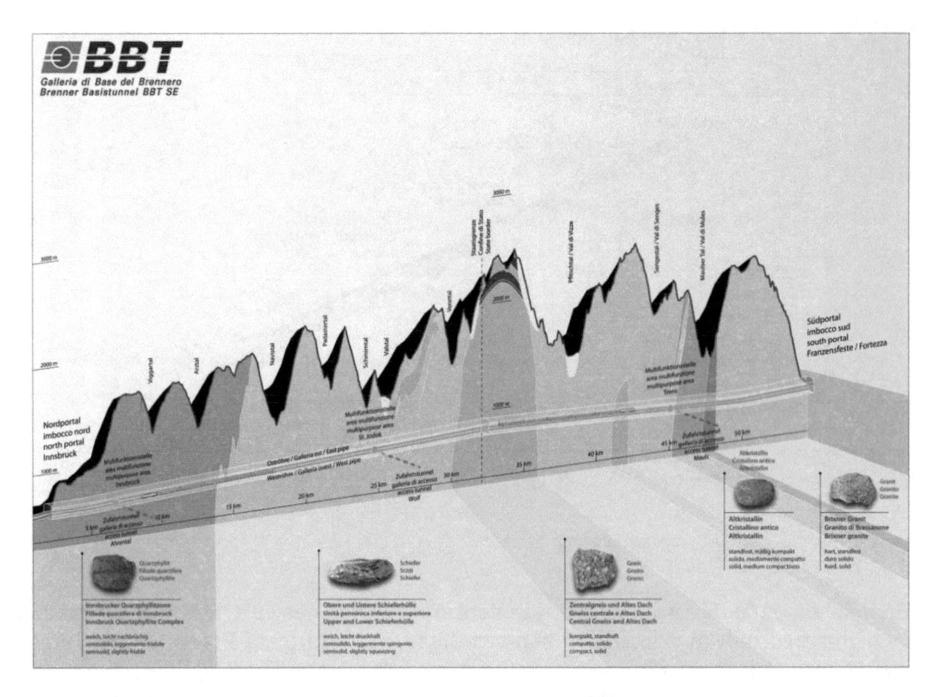

Figure 2. General overview of the main geological units crossed by the tunnel.

More than the half of this stretch, with an maximum overburden of 700 m and a diameter of about 7 m, was excavated by using hydraulic impact hammers and the remaining part by the drill & blast technique, crossing, from South to North, the following main geological units and principal faults: the Brixner Granite, a first fault zone of 30 m (Pusteria Fault), the Mules Tonalite and a second fault zone (Val di Mules Fault) with a length of 800 m.

The latter fault zone was characterized by a heterogeneous sequence of faulted metamorphic rocks, mainly phyllite, schists and cataclasite, and subordinate strata of stiffer rocks such as quartzite, paragneiss and fractured quarzitic schists. This lithological variety and the different degree of tectonic deformation clearly influenced the uniaxial compressive strength values, showing an extreme variation from about 1 MPa for the cataclasites to 125 MPa for the quartzite.

Furthermore, systematic face mapping was performed, allowing the classification of the encountered rock mass types and showing a mean RMR value (Bienawski, 1989) of 30 for the core zones and a mean RMR value of 45 for the damage zones (Skuk *et al.*, 2017).

3.2 *Aim and description of the applied exploration methods*

Systematic exploration work ahead of the tunnel face was done applying the following types of investigation:

- Rotary drilling with a data logger (RD+Drill Log)
- Horizontal core drilling and sampling (CCD)
- Percussion drilling with data logger and Increx installation for extrusion measurements (PD+Increx)

Rotary drilling with a data logger (RD+Drill Log) is a core destructive method performed by means of a drilling rig placed in a horizontal position at the tunnel face; the first 3 drillings were performed with a tricone drill bit (rotation only) and a 100 bar preventer, concurrently recording the rate of penetration, the applied thrust pressure and the torque pressure.

Continuous core drillings ahead of the face (CCD) are realized by using a core drill bit, a core lifter and proper extension tubes; as the bit cuts into the strata, it leaves a core in the central conducting tube. During the excavation of the exploratory tunnel, 6 horizontal wire-line core drillings ahead of the face were performed using a preventer device (100 bar). These drillings were of fundamental importance to understand the characteristics of the rock mass to be excavated in terms of lithologies, mechanical characteristics, jointing, alteration degree and hydraulic conditions. Furthermore, a considerable number of laboratory tests on core samples were performed to determine the physical-mechanical properties of the main lithologies.

Percussion drillings with a data logger (PD+Increx) is a core destructive method performed by means of a drilling rig placed in a horizontal position at the tunnel face as well and equipped with a down-the-hole-hammer (DTH). The drill principle consists of a combination of percussion and rotation. In the exploratory tunnel a total of 14 percussion drillings were performed by recording the rate of penetration, the applied thrust pressure and the torque pressure. An incremental extensometer (Increx) was installed in every borehole performed with the PD system for extrusion measurements, which were performed almost every day. The incremental extensometer allows measurements of the displacement along a horizontal line ahead of the face at 1 m interval. The length of the extensometer is 40÷45 m.

The aim of these prospection activities was the reduction of geological, hydrogeological and geotechnical risks and the optimization of the reinforcement measures and the lining, both for the exploratory tunnel and to obtain data for the excavation of the main tunnels.

Table 1. Exploration measures ahead of the tunnel face - Case study 1.

Advance exploration and/or measurement type		Number [-]	Total length [m]	Total time [days]	Average advance rate [m/day]	Total net time* [days]
RD+Drill Log	Rotary drilling with data logger	3	240	13	18,4	13
CCD	Continuous core drillings with preventer	6	1387	77	18	77
PD+Drill Log + Increx	Percussion drilling with data logger + Increx	14	552	7	78,8	0

**Production stop due to the execution of the referred exploration measure*

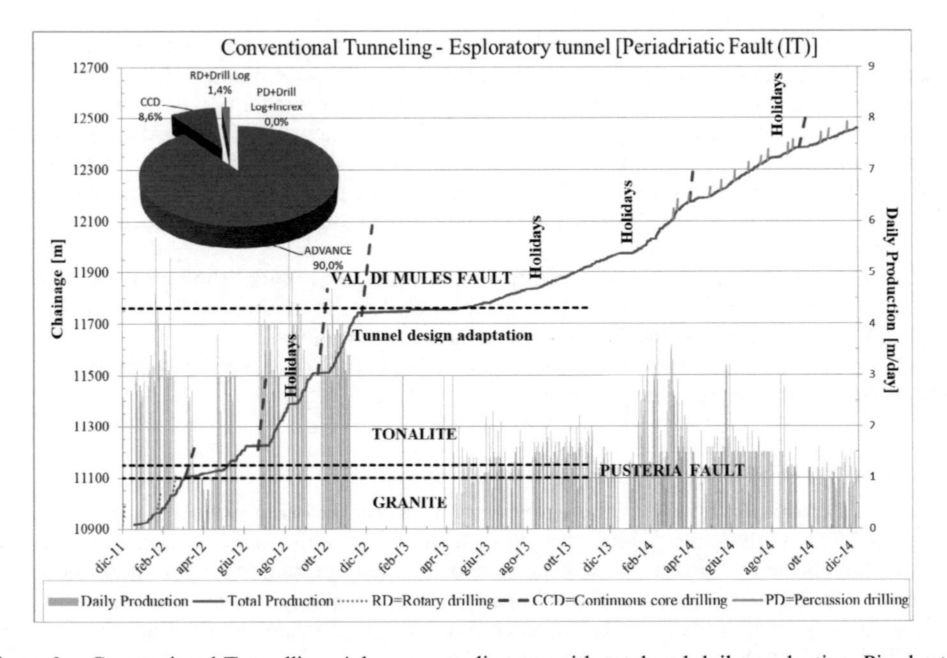

Figure 3. Conventional Tunnelling: Advance rate diagram with total and daily production; Pie chart – Exploration measures vs. Net advance.

Detailed information about the exploration measures in terms of number, total length, total exploration time and average advance rate is summarized in table 1.

3.3 *Production achieved and exploration time*

The excavation of the 1.541 m long Exploratory Tunnel began on 22.12.2011 and ended on 15.12.2014 with a total average daily production (gross advance rate) equal to about 0.89 m/day. The calculated net advance rate (gross advance time minus holidays, standbys, various stops > 1 day and minus prospection activities) was about 1.4 m/day. Figure 3 represents the main results of the advance rate (production) in terms of total and daily production, showing also the main production stops including holidays, continuous core drillings and design adaption. The pie chart compares the net advance time with the exploration time for the applied exploration concept described above (Skuk et al., 2017).

4 CASE STUDY 2 – DOUBLE SHIELD TBM: MULES – BRENNER EXPLORATORY TUNNEL (ITA)

4.1 *Drive description*

As part of the Mules 2-3 construction lot, the completion of the stretch of the exploratory tunnel in the Italian project area is also planned. Specifically, the excavations, which began at the beginning of May 2018, are being carried out with a double shield TBM and run for about 14 km up to the State border, where the tunnel will break through to the stretch of tunnel excavated in Austria.

The route crosses a particularly complex area of the Eastern Alps in terms of lithology and structural geology, as it runs along the western border of the Tauern window (mainly characterized by limestone schists and gneiss), right in the middle of the continent-to-continent collision point between Africa and Europe which caused the Alpine orogeny. Within this tectonic window we find both continental (of European origin) and oceanic units, which were pushed upwards and progressively exposed to the surface due to the tectonic denudation of the overlying strata of the Austroalpine System (African/Adriatic origin).

The following main rock types have been encountered along the stretch excavated so far (about 1,300 m): mica schists and subordinate thin layers of amphibolites of the Austroalpine domain and calcareous schists of the Tauern Window. Due to moderately fair rock mass conditions, with medium to high UCS values (40-70 MPa), the low abrasivity of the main lithologies mentioned and just a few low-volume water inflows, it was possible to achieve an average production of about 500 m/month. The drive through the predicted fault did not require any additional measurement and did not impact significantly on the daily production rate, currently between 15 and 18 m/day.

When possible, the geology was carefully documented, allowing a lithological characterization of the encountered rock mass types, since an evaluation of the RMR (Bienawski 1989) with a DS TBM type of excavation could be considered ambitious.

As the excavation work for this section began in May 2018, only the first 4 months of TBM excavation will be analysed for this paper.

4.2 *Aim and description of the applied exploration methods*

Systematic exploration ahead of the tunnel face is currently being performed in the double shield TBM excavating the exploratory tunnel, applying both direct and indirect types of investigation of the rock mass. These tests have been briefly presented and analysed as well, in order to emphasize the decidedly low impact on the excavation cycle and the substantial increase in geological knowledge of the rock mass achieved when the data are examined by crosschecking them with direct investigation outputs.

The following investigations types are performed:

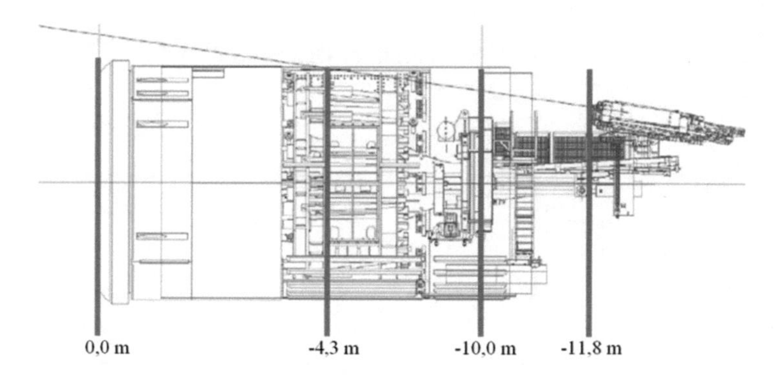

Figure 4. Longitudinal section of the Double Shield TBM with relative distances of the rotopercussion drilling (Detailed design "Mules 2-3" Lot).

- Percussion drilling with a data logger (PD+Drill Log)
- Seismic exploration using the TSP and the TSWD system
- Geoelectric investigation using the BEAM system

The rotopercussion drillings (using a Wassara-type water-powered hammer) are carried out by a device positioned behind the tailskin of the shield in the bridge 1 area, ahead of the last installed segment at an upwards angle of 12° (Figure 4). The boreholes are between 60 e 150 m long, with a maximum overlap of between 20 and 30 m. The data from the drilling logger are evaluated and the most significant drilling parameters, like advance speed, thrust, torque, penetration and specific energy are correlated with the geological-geotechnical conditions found on site and with the TBM parameters.

Furthermore, seismic tests were carried out every 100-150 m using the TSP (Tunnel Seismic Prediction) and the TSWD (Tunnel Seismic While Drilling) systems.

With regard to the TSP system, small explosive charges are shot individually in 1.5 - 2.0 m deep boreholes aligned in the tunnel wall; each shot sends out seismic wave energy which propagates through the rock mass. Variations in rock strength (acoustic impedance) occurring, for example, in fracture zones or in presence of significant lithological variations, will reflect a certain portion of the signals, whereas the remaining portion will be transmitted (Dickmann, 2005). A total of 10 measurements were performed (all through August 2018), and the result of the data processing is represented by 3D seismic models with elastic rock properties charts, deduced from the registered P and S waves velocity.

The TSWD method is based on continuous seismic monitoring of the vibration signal produced by the cutting head of the TBM by means of several geophones. In the exploratory tunnel a new geophone configuration is tested by placing them through the circular openings of the segments in direct contact to the tunnel wall.

The TBM is also equipped with the BEAM system (Bore-Tunnelling Electrical Ahead Monitoring), a geo-electric investigation method, which, based on the measurement of the induced polarization of the rock mass in terms of the percentage frequency effect (PFE) and resistivity (R), gives indications about rock mass conditions and water-inflow up to 20 m ahead of the face.

The aim of these exploration measures is to recognize in advance zones with a higher degree of fracturing or fault zones coupled with water-inflows.

Detailed information about the exploration measures in terms of number, total length, total exploration time and average advance rate is summarized in table 2.

4.3 *Production achieved and exploration time*

On August 31st, 2018, 1.329 metres of exploratory tunnel had been excavated with the TBM, corresponding to an average production of 10.8 m/day, which takes into account a learning curve and the summer break. Without considering slowdowns, the net average production is about 20 m/day (Figure 5)

Table 2. Exploration measures ahead of the tunnel face - Case study 2.

Advance exploration and/or measurement type	Number [-]	Total length [m]	Total time [h]	Average advance rate [m/h]
PD+Drill Log Percussion drillings with data logger	14	1900	183,3	10,3
TSP Tunnel seismic prediction	11	1231	-	-
TSWD Tunnel seismic while drilling	x	1295	-	-
BEAM Geoelectric investigation	x	1329	-	-

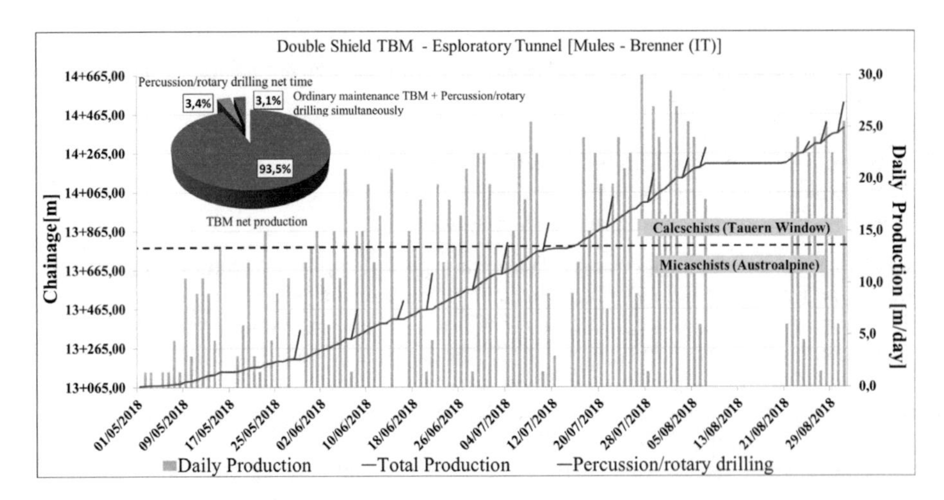

Figure 5. Double shield TBM: Advance rate diagram with total and daily production; Pie chart – Exploration measures vs. Net advance.

5 DATA ANALYSIS

5.1 *Conventional exploratory tunnel*

The table 1 presented in paragraph 3.2 shows the time required to carry out the prospecting drilling beyond the rock face, which do not take into account the actual impact that the different types of prospection activities have on the production cycle.

Tri-cone rock bit and core drillings required the total suspension of activities in the exploratory tunnel, with obvious significant repercussions on production rates. On the contrary, the short drillings (L=40÷45 m) carried out with the fast non-coring DTH method, besides being faster per se (as shown in table 1), are more flexible, as they are carried out with the same equipment, probe and workers used for the rock face and top heading reinforcement activities. Specifically, in the stretch of exploratory tunnel driven through the Periadriatic fault, preliminary reinforcement measures were carried out every 6 m of drift - known as the "excavation field" - before continuing with excavation work. Percussion drilling with a data logger and an Increx was optimized until these activities could be completed within the 20 hours required for the stabilization activities (Figure 6 and Figure 7).

5.2 *Mechanized exploratory tunnel*

With regard to the excavation of the exploratory tunnel with the double shield TBM (started on 01/05/2018), it should be noted that 6.5% of the production time is used to carry out exploratory drilling beyond the rock face. This means about 166 hours out of a total net construction period of 110 days (minus holydays - updated to 31/08/2018).

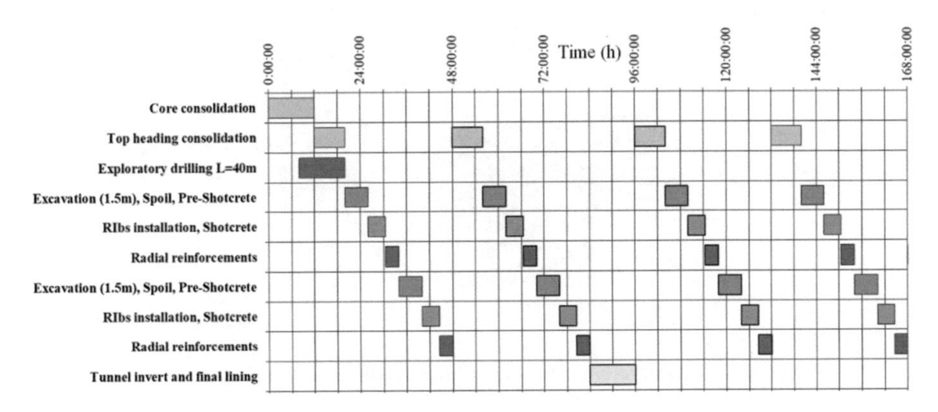

Figure 6. Standard operating cycle in the “excavation field” (L=6m), in which the exploratory drilling ahead of the excavation is carried out at the same time but without interfering with the consolidation activities (sub-cycles).

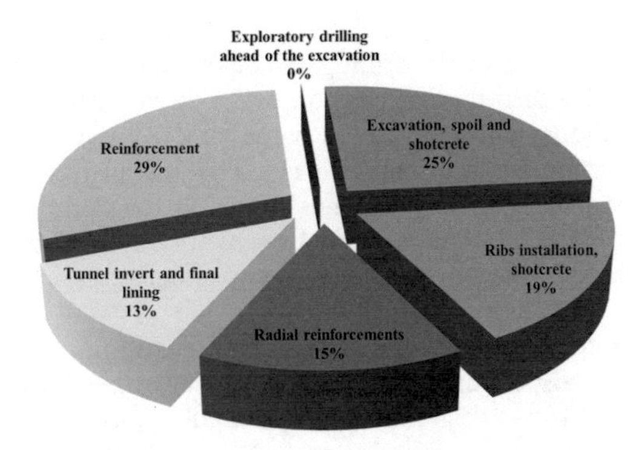

Figure 7. Pie chart showing macro-activity timing in a standard excavation cycle through the Periadriatic seam.

By moving past the learning curve and progressively improving logistics, it was possible to overlap part of the time required to carry out the exploratory drilling beyond the rock face with the routine maintenance work on the TBM. This improvement led to a reduction in the impact of such exploratory drilling on TBM productivity, from 6.5% to 3.4% (Figure 5 – Pie chart).

As regards seismic and geoelectric testing, with the technologies currently used on the TBM it is possible to carry out prospecting activities during maintenance operations, eliminating any interference with the production cycle of the TBM.

6 CONCLUSIONS

In this paper two sections of the exploratory tunnel of the Brenner Brenner Base Tunnel, excavated respectively with the conventional method (impact hammer and Drill&Blast) and mechanized excavation (TBM), were analysed and compared with regard to the impact of prospecting activities on the production cycle.

In the conventional excavated sketch of the exploratory tunnel, where systematic measures were required to consolidate the rock face with the installation of incremental extensometers (Increx) for the measurement of the face extrusion, advance drilling done by means

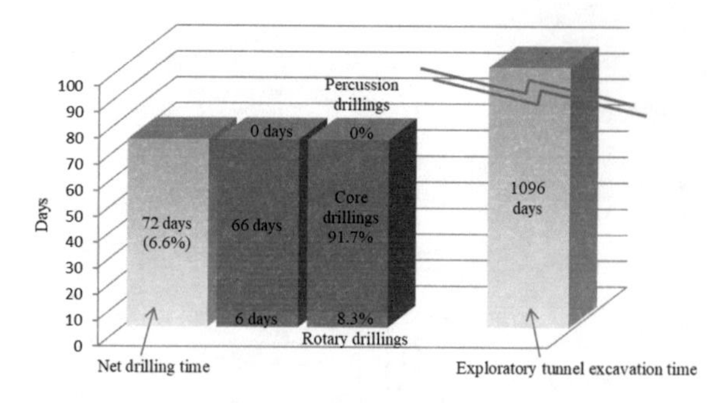

Figure 8. Actual impact of the different types of drillings on the production cycle.

of non-coring boreholes carried out with a DTH and equipped with drilling data logger has proved to be very fast and effective, as can also be seen in Figure 8, where the impact of the time taken to carry out drillings (72 days) is highlighted as compared to the time taken for excavation (1096 days).

The distribution of this timing over the three types of prospecting activities highlights the preponderant effect of continuous core drilling (66 days, equal to 91.7% of total prospection time), while it can be clearly seen that percussion drillings with automatic parameter acquisition were carried out, almost entirely, at the same time as the remaining processes, thus having a negligible impact on the production cycle. This has proved to be the most efficient prospection method, since the project guidelines prescribe the appropriate sizing of the support system based on the data obtained from prospecting drilling performed ahead of the excavation, which are then verified by means of the geological and behavioural characterization of the rock face.

With regard to the stretch excavated by TBM, which is still in progress today, in order to systematically carry out prospecting drilling prior to the excavation and according to the design guidelines, every effort has been made to continually optimise operative timing in order to reduce the impact of prospection activities on the production cycle. This has been possible thanks to the improvements in terms of technology (particularly high-performance water-powered hammers), operations (prospection activities in front of the segmental lining) and logistics, carrying out such prospecting activities in parallel with routine maintenance activities that cannot be postponed or omitted, such as the replacement of cutters and the maintenance of the drilling head, the extension and welding of the conveyor belt and the extension of medium-voltage and hydraulic services. All the planned indirect seismic and electrical tests have so far been carried out at the same time but without interfering with the execution of prospecting activities and routine maintenance; thanks to these optimizations it has therefore been possible to reduce the impact of the prospecting activities on the ordinary production cycle of the TBM to 3.4% (Figure 9).

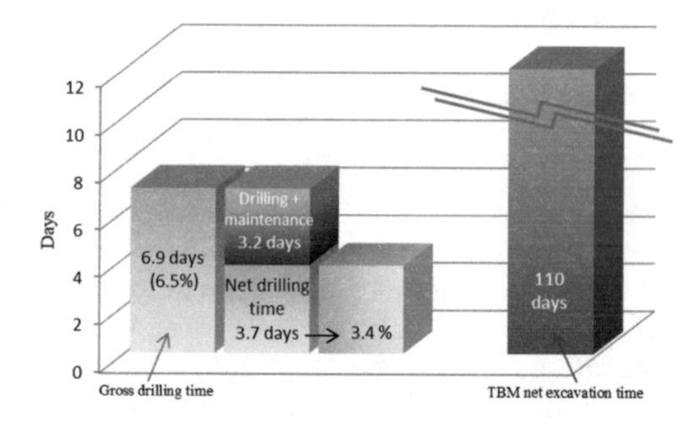

Figure 9. Effective impact of prospecting activities on the production cycle of the TBM.

The case studies of the BBT show that, both in conventional and mechanized tunnelling, exploration measures ahead of the face, if wisely implemented in the usual construction cycles/ phases, only marginally impact the production in favour of excavation safety. Therefore they should be a fixed part of the production cycle of deep-seated tunnels and become an additional measurement tool in the "observational method", since optimization of the support and consolidation methods takes place on the construction site.

REFERENCES

Chwatal, W., Radinger, A., Brückl, E., Mertl, S., Freudenthaler, A. 2011. Tunnel seismic while drilling – State of the art and new developments. In World Tunnel Congress 2011, *Proceedings*. Helsinki, 20-26 May 2011.

BBT-SE, 2016. Detailed design, *Mules 2-3 construction lot*

Dickmann, T. 2005. Seismic prediction while tunneling in hard rock. *65. Annual Meeting of the German Geophysical Society*. Graz, 23 February 2005.

Fuoco, S., Zurlo, R., Marini, D., Pigorini, A. 2016. Tunnel Excavation Solution in Highly Tectonized Zones. In World Tunnel Congress 2016, San Francisco 22-28 April.

Petronio, L., Poletto, F., Schleifer, A., Dordolo, G., Morino, A., Fabbri, B. 2010. Prima applicazione del metodo TSWD per la predizione di discontinuità davanti al fronte di scavo. *Atti del 21° Convegno Nazionale GNGTS, 09.10*.

Skuk, S., Egger, H., Schierl, H., Barovero, G. 2017. The importance of prospection beyond the tunnel face: impacts on excavation speed and benefits. In World Tunnel Congress 2017, *Proceedings: surface challenges and underground solutions*. Bergen, 9-15 June.

Tunnels and Underground Cities: Engineering and Innovation meet Archaeology, Architecture and Art, Volume 3: Geological and geotechnical knowledge and requirements for project implementation – Peila, Viggiani & Celestino (Eds)
 ISBN 978-0-367-46583-4

Geotechnical characterization of an artificially frozen soil with an advanced triaxial apparatus

M. Bartoli & F. Casini
Università degli Studi di Roma "Tor Vergata", Rome, Italy

Y. Grossi
Italferr S.p.a, Rome, Italy

ABSTRACT: The excavation of tunnels in medium to high permeability soils below the groundwater level requires the provision of temporary support and the adoption of techniques to exclude the groundwater from the excavation. Artificial Ground Freezing (AGF) is a technique that provide waterproofing and strengthening to the frozen soil body with minimal impact on the surrounding environment. This paper presents the preliminary experimental results obtained using an advanced triaxial apparatus (FROZEN), working under temperature and stress path controlled, to test a silty soil sample retrieved from a railway construction site. The apparatus allows the application of the thermal loading from a probe installed in the centre of the sample, so that the frozen front propagates from the centre of the sample to its boundary, similar to the conditions experienced in situ during the AGF implementation.

1 INTRODUCTION

The control of displacements is crucial to the viability of urban tunnelling in soft ground as ground movements transmit to adjacent structures as settlements, rotations and distortions of their foundations, which can, in turn, induce damage affecting or function, and, in the most severe cases, stability of the structure (Burland *et al.* 1977). Therefore, is necessary to adopt controllable method as support of excavation to mitigate damage on the building surrounding. Artificial Ground Freezing (AGF) was extensively adopted to stabilize the ground to support and/or exclude groundwater from the excavation during construction of Lines 1 and 6 of Napoli underground and Line C of Roma underground. (*e.g.* Viggiani & de Sanctis 2009; Viggiani & Casini 2015; Russo *et al* (2015); Russo *et al* (2017); Molè *et al* 2017).

If compared to other consolidation and water proofing techniques, AGF is suitable for any soil type of soil (from coarse to fine grained) and rocks. It is safe and environmentally friendly, as it does not release any chemical or cement mixture into the soil; on the other hands is relatively expensive.

AGF may be considered for the excavation of short tunnels in sandy soils below the water table, where traditional techniques could not be applied. AGF converts pore water to ice, the resulting frozen soils is characterized by a higher strength and it is impermeable to water flow. The soil is frozen installing the freezing pipes in the ground (Figure 1a), two concentric tubes compose them. The refrigerant fluid is injected in the inner tube and it exchanges cold energy with the surrounding soil when it returns in the outer tube. The resulting impermeable and resistant frozen body acts as temporary support and waterproofing of the excavation area until the permanent lining is installed. The pipes can be arranged with different layouts. As an example, Figure 1b shows a typical arrangement of freezing pipes completely surrounding the future excavated section of the tunnel.

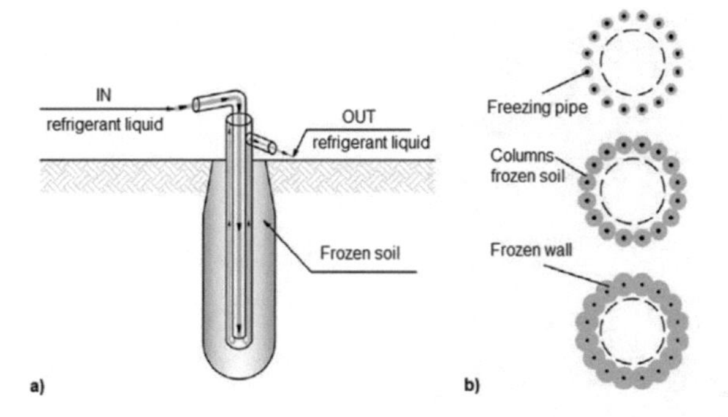

Figure 1. a) scheme of vertical frozen pipe, b) development of frozen wall for tunnelling (after Rocca 2011).

The frozen front advances mainly in the radial direction creating independent column of frozen soil around each pipe. These will eventually merge creating a so called frozen wall (*e.g.* Rocca 2011).

The thermal loading (freezing) and unloading (thawing) induce complex thermal, hydraulic and mechanical processes in the ground. These have significant mutual interactions which are still poorly understood. On freezing, the volume of the pore water increases. The interface between the liquid and ice phases of water is not flat and its curvature is sustained by surface tension balanced by suction, defined as the difference between the ice pressure and the water pressure. Suction attracts water to the frozen front, which becomes thicker, this mechanism determines the swelling observed during the freezing phase. At the same time the increasing in ice content decreases the relative permeability of the soils and increases the strength of the media.

AGF requires refrigeration of massive quantities of soil over extended periods of time, which is quite expensive. The period and the thickness of the frozen body could be optimized, if the implementation of AGF *in situ* is based on a detailed geotechnical characterization of the soil at different temperatures.

In this work are presented the preliminary experimental results obtained with an advanced triaxial apparatus (FROZEN), working with temperature below zero degree centigrade, used to characterize a silty soil retrieved *in situ* for a preliminary project with the aims of doubling the tunnel of an existing railway in Cesano (RM), Italy.

2 FROZEN: AN ADVANCED TRIAXIAL APPARATUS

Typically, the triaxial apparatuses described in the technical literature (*e.g.* Cantone *et al* 2006; Yamamoto & Springman 2014) to test frozen soils apply the thermal load with an helicoidal pipe around the specimen allowing the frozen front to advance from outside to inside the specimen.

Moreover, for water saturated sample, the volumetric strain is evaluated measuring the volume of water flowing in and out of the sample.

The main limitation of existing equipment to characterize the sample artificially frozen area: (i) the opposite direction of the frozen front advancement compared to the application of AGF *in situ*; (ii) the impossibility to obtain the volumetric strain by measuring the volume of water flowing in the specimen once the sample is frozen.

FROZEN addresses these gaps of the existing apparatus to characterize artificially frozen soils (Bartoli *et al* 2018). The prototype has been developed in collaboration between Università degli Studi di Roma Tor Vergata (DICII) and Universitat Politecnica de Catalunya (UPC, Barcelona), in the context of a project funded by Lazio region.

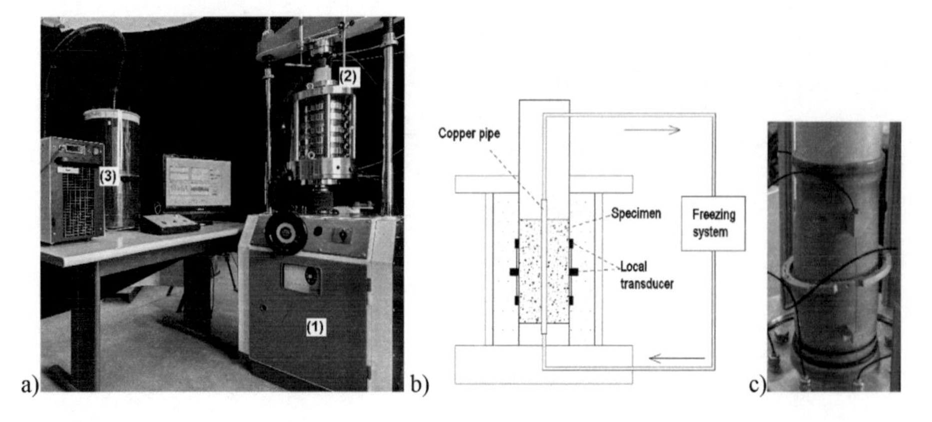

Figure 2. a) Triaxial cell, b) cooling system scheme, c) local transducers mounted on specimen.

The FROZEN apparatus is able to:

1. apply the thermal loading by means of an inner tube allowing the freezing front to advance from inside of the sample to outside, as *in situ*;
2. monitor the temperature with thermocouples installed in the sample at various distance from the inner pipe;
3. measure the radial deformation of the sample by means of local transducers.

The apparatus (Figure 2a) is composed by three principal components: (1) mechanical load frame (capacity 100 kN); (2) triaxial cell, (3) system to cool and circulate refrigerant liquid. (www.frozen.uniroma2.it)

The sample has a diameter d = 100 mm and a height h = 200 mm, which makes it possible to carry out meaningful tests also on coarse sand samples, up to a maximum dimension of the grains of the order of d_{max} =d/10=10 mm. The base pedestal and the top-cap are made of Peek, a rigid plastic material characterized by low thermal conductivity to reduce thermal dispersion.

The inner copper pipe in which the refrigerating fluid circulates with a diameter ϕ = 10 mm, is installed in the centre of the sample (Figure 2b). The circulation is bottom-up and the frozen front advances from the centre of the specimen to reproduce the same *in situ* temperature path direction. This direction of temperature path avoids the confinement of unfrozen water inside the sample.

The advancement of the frozen front is monitored with flexible thermocouples installed at various distance from the inner tube, by back analysing the temperature data is possible to calibrate the thermal conductivity of the samples at different temperature. The radial and axial deformations are measured locally using a radial belt and a pair of LVDTs (Linear Variable Displacement Transducer) mounted on the sample as shown in Figure 2c.

3 CASE STUDY

The samples have been retrieved in the area of the railway connection between Vigna di Valle and Cesano (RM) in the Lazio region (Italy), in the proximity of Bracciano lake. The design, performed by Italferr SpA, foresees the doubling and speeding up of an existing single-track railway of the Rome-Viterbo line (Figure 3).

The widening of the line will allow to increase the frequency of the train (every 15 minutes) up to Vigna di Valle. Moreover, it is planned in the next future to extend the network to Bracciano. The overall length of the designed line is approximately L=12 km and the design speed is 115 km/h, while the maximum gradient is 16‰ located in the surrounding of Anguillara.

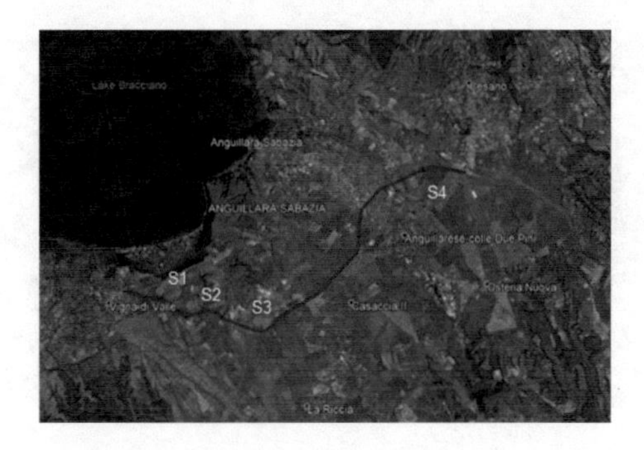

Figure 3. Position of Vigna di Valle – Cesano line and of the drilling.

The details of the new line are reported below:

Distance between the track center lines 4.00 m
Maximum operating speed of the lines 115 km/h
Minimum planimetric radius 640 m
Minimum vertical radius 3500 m
Maximum gradient 16‰
Kinematic silhouette Gabarit C
Station binary module 250 m
Electrification 3 Kv c.c

One of the supporting technique considered for the realization of the extended new line is the use of AGF. However, the technical and economical evaluations resulted from the feasibility study have highlighted that a modification of the new railway line, with the alignment fully in trench, is preferable to the original solution.

4 EXPERIMENTAL WORK

4.1 *Material and methods*

Two cored samples have been retrieved *in situ* with the following characteristics:

Table 1. Cored samples.

Acronym	Depth (m)	Date	Location
C.I. n.1 – S1	3.40-4.00	14/06/2018	Anguillara Sabazia (Rome)
C.I. n.1 – S2	16.40-17.00	18/06/2018	Anguillara Sabazia (Rome)

The specimen has a diameter d=100 mm (height h=200 mm), while the undisturbed samples have been cored with a diameter d'=90 mm. Therefore, the sample in FROZEN has been statically compacted with the material retrieved from n1-S1 reported in Figure 4.

The soil investigated is an altered material of volcanic origin, characterized by variable colour from brown to light gray, with thin layers of tuff slightly argillified and presence of leucite altered in analcime.

Particle size distribution analysis determine a well-graded soil composed by sand 45.2%, silt 42% and clay 12.4%, and uniformity coefficient U=46.8 (Figure 5). The Atterberg limits are: the liquid limit w_L=46.6%, the plastic limit w_P=30.7% and plasticity index I_P=15.9%. The specific weight γ=15.04 kN/m^3 and specific weight of solid particle γ_s=25.88 kN/m^3.

The soil has been dried in a hoven at 60°C. The soil aggregates have been broken, using a mortar, to obtain the material passing to the sieve No. 10 (2.00-mm). The sample has been statically compacted (4 layer, velocity v=1 mm/min) with a void ratio e=1.3 and a degree of saturation S_r=0.7 (Figure 6).

Figure 4. Undisturbed sample n1-S1.

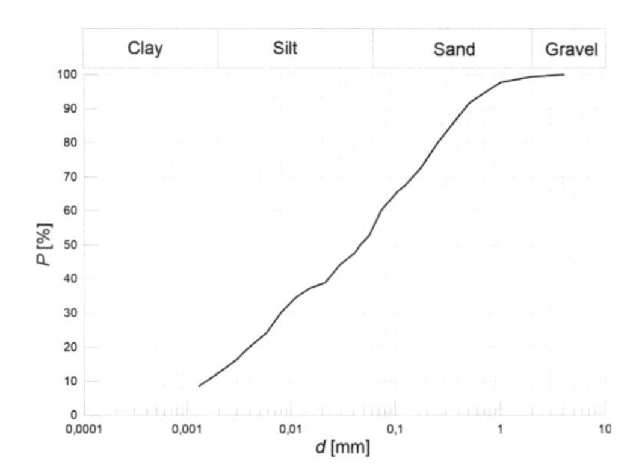

Figure 5. Grain size distribution, sample n1-S1.

The test phases are the following:

1. saturation of the sample with flushing a σ_a=σ_r=100 kPa and pore water pressure at the base u=70 kPa during the day, constant σ_a=σ_r=500 kPa and u=470 kPa during the night;
2. isotropic consolidation phase with σ_a=σ_r=700 kPa and back pressure u=500 kPa. The temperature of the cell liquid has been decreased to T~2°C;
3. freezing stage. The temperature loading has been applied circulating the refrigerating fluid with an inlet T_{in}=-19 °C, lasted for 27.5 hours;
4. axial compression phase under controlled displacement rate v=0.33mm/min up to the axial strain ε_{ax}=20%.

Figure 6. Top of specimen with filter paper and the central copper pipe.

4.2 *Experimental results*

In the following are reported only the experimental results of the freezing and of the axial loading phases. The freezing phase lasted for 27.5 hours, with an input temperature of refrigerant liquid $T \sim -19$ °C and output $T \sim -16/-17$ °C. The minimum temperature measured in the sample is T=-11.1 °C at 13.8 mm from central axis (Figure 7 point A).

Figure 7 shown temperature time evolution detected from thermocouples in different positions of the specimen: A-B at base (radius r_A=13.8 mm, r_B=19.1 mm), C-D-E on specimen border r=50 mm at 3 different height h_C=50 mm, h_D=100 mm, h_E=150 mm, F at top (r_F=19.1 mm).

After 4 hours the input temperature of refrigerant liquid is maintained constant. At 9.7 hour, when all thermocouples measure T<0 °C, the set-point of liquid cell temperature has been lowered to a temperature T_{cell}=-2.5 °C, and further lowered to T_{cell}=-4.5 °C at 22.5 hour, in order to fasten the fully freezing of the sample. This type of boundary condition also reproduces the temperature front advancing from surrounding pipes *in situ*.

The monitored temperatures have been used to reconstruct the contour in the sample at different times frame as reported in Figure 8. The contour has been obtained with a ΔT= 2°C.

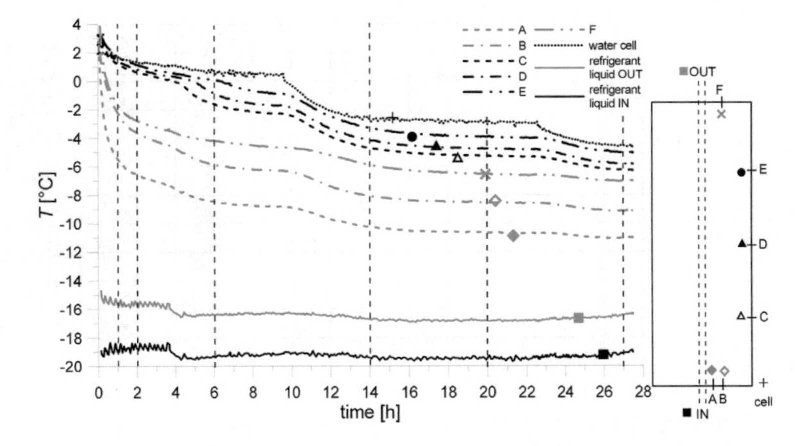

Figure 7. Temperature monitored in the sample, in the liquid cell and in the inner tube.

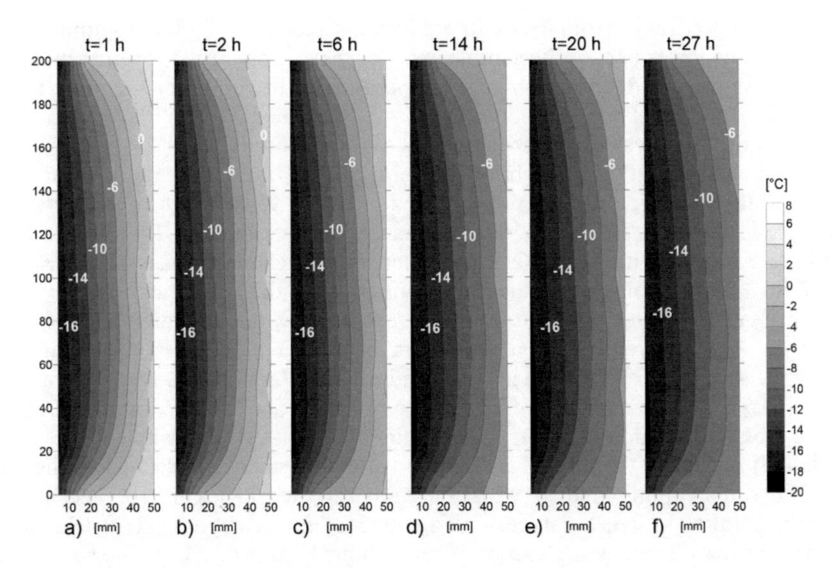

Figure 8. Temperature contours evolution inside specimen with a ΔT=2 °C.

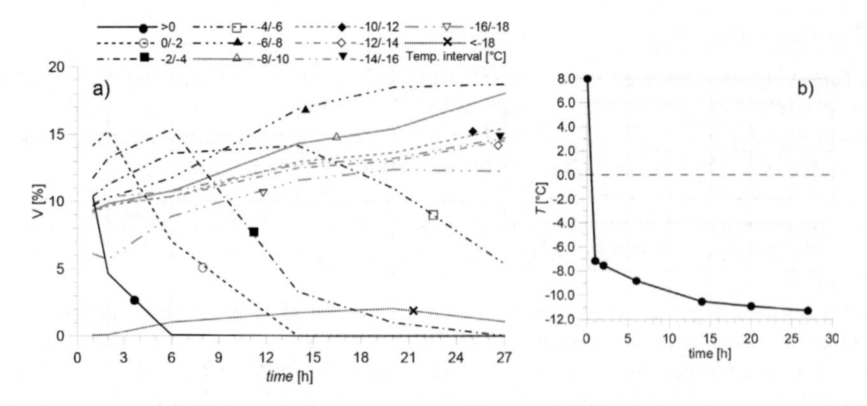

Figure 9. Evolution with time of: a) Specimen volume below temperature contour with a ΔT=2 °C, b) average temperature in the sample.

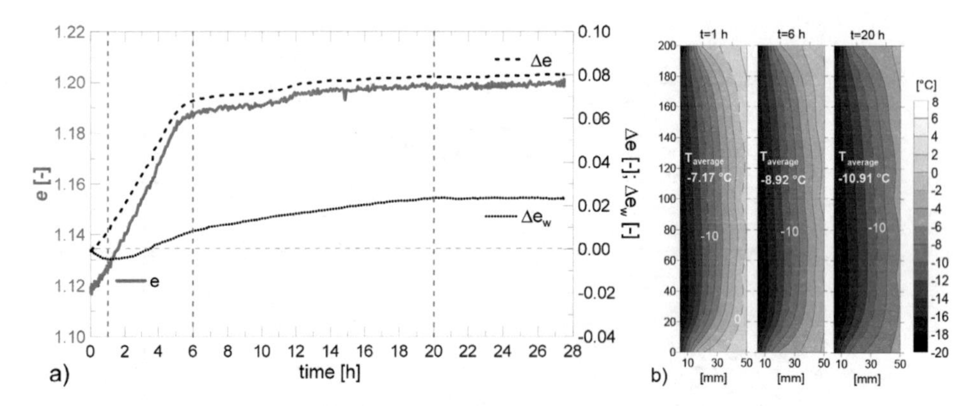

Figure 10. a) Time evolution of void ratio e (left axis); Δe and Δe_w (right axis); b) Temperature contour at three-time frame.

The frozen front advance from inside to outside, between 2h and 6h the sample has a T<0° C everywhere. The lower part of the sample experiences a temperature lower compared to the upper part. This effect is due to the higher temperature in the outlet of the inner tube higher than in the inlet of the refrigerating liquid ($\Delta T = T_{out}-T_{in}\sim$ 2.5/3.5 °C) and to the gradient in temperature in the liquid cell due to the viscosity of the cell liquid.

The volume (V) below the contours of temperature has been calculate with an interval of ΔT=2 °C, and the time evolution of V is reported in Figure 9a, while the average temperature in the sample is shown in Figure 9b. The volume in percentage with T>-6°C decreases with time, while the volume with T<-6 °C always increases. Higher increases are obtained in the range of T=-6 ÷ -8 °C compared to the other ranges. Also, the ranges with -10 °C <T<-16° C shown an increase with time characterized by a lower rate, while the interval -16 °C <T<-18° C exhibits an initial increase with a maximum at t=20.5 h followed by slight decreases. The average temperature decreases with time and is characterized by a higher rate up to 14 h ($T_{ave=14h}$= -10.5 h) and then by a lower rate up to reach a T_{avemin}=-11.3 to the end of the test.

The evolution of variation of water ratio, defined as $\Delta e_w = \Delta V_w / V_s$ where ΔV_w is the water volume (liquid) flowing in the sample and V_s the volume of solid particle; and of the void ratio $e = V_V / V_S$ (V_V voids volume) and its variation Δe is reported in Figure 10. The sample swell during all the duration of freezing stage, with an increase in void ratio Δe=1.20-1.12=0.08, while the volume of water attract by the suction in the sample is Δe_w=0.025. At the beginning of the freezing stage the water flows outside of the sample in a small amount (Δe_{wmax}= 0.005 t=1h)

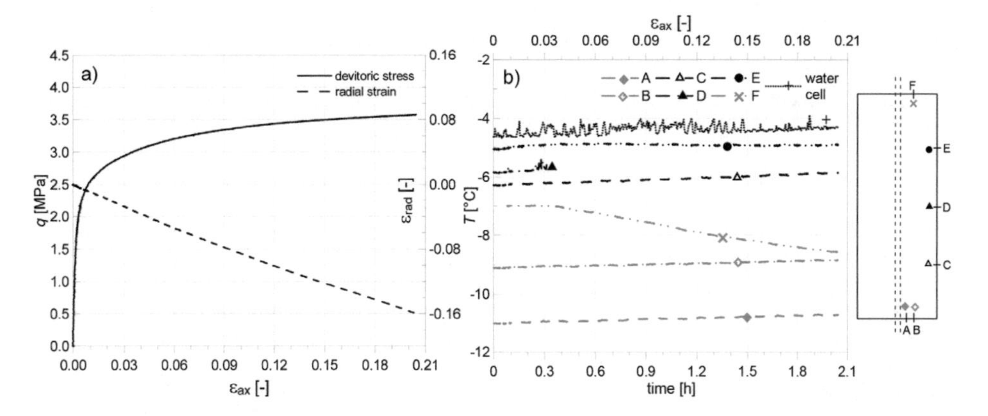

Figure 11. Axial compression phase, evolution of a) deviatoric stress and radial strain with axial strain; b) temperature with time.

followed by an inflow of liquid water as the freezing front advance (Figure 10). The void ratio *e* has a sudden increase from t=0 to t=6h, followed by a slight increase with time. Interestingly also the evolution of Δe_w experiences a change in the slope at t= 6h, followed by a linear increase up to t=20h and then remain constant due to the complete freezing of the sample.

The result of axial compression phase, at controlled displacement rate v=0.33mm/min, are reported in Figure 11a in terms of ε_a-q (left vertical axis) and ε_a-ε_r (right vertical axis). The sample exhibit a ductile behaviour over the entire range of ε_a imposed with a tangent Young Modulus $E=\delta q/\delta\varepsilon_a$=1.37 GPa and an ultimate value of q=3.57 MPa. The temperature monitored inside the sample shown a slight increase with loading ΔT~0.2°C, apart the thermocouple F positioned in the upper part of the sample which measures a T~-7°C up to t=0.3h followed by a decrease in temperature up to T~-9°C at the end of axial phase.

5 CONCLUSIONS

In this work have been presented the preliminary results of the geotechnical characterization of soil under thermal loading with temperature below zero degree centigrade. The test has been performed with an advanced triaxial apparatus, FROZEN, able to apply the thermal path from the centre of sample to outside following the same direction of AGF *in situ*.

The apparatus is able to well reproduce the advancement of frozen front, to measure the deformation during all the stages and to evaluate the strength and the stiffness during loading. The sample tested is a sand retrieved in Cesano (RM) axially loaded with a temperature of -11°C exhibit a strength of ~4 MPa and an initial tangent stiffness of ~1.25 GPa according to data reported in literature for frozen sand.

ACKNOWLEDGMENT

The financial support of LazioInnova for funding the "FROZEN" project (FILAS-RU-2014-1180) is acknowledged. The authors are grateful to Geology Department Italferr SpA for his technical support in retrieving the soil samples.

REFERENCES

Bartoli, M., Raparelli S., Casini, F. & Viggiani, G. 2018. Frozen: un'attrezzatura avanzata per prove su terreni artificialmente congelati. *IARG 2018*. Genova.

Burland, J.B., Broms, B.B. & DeMello, V.F.B. 1977. Behaviour of foundations and structures, *IX ICSMFE* 495-546. Tokio.
Cantone, A., de Sanctis, L. & Mandolini, A. 2006. Interventi di protezione degli scavi di gallerie mediante congelamento. Attività sperimentale nella stazione Municipio della Metropolitana di Napoli. *IARG 2006*. Pisa.
Casini, F., Gens, A., Olivella, S. & Viggiani, G. 2016. Artificial ground freezing of a volcanic ash: laboratory tests and modelling. *Environmental Geotechnics*: 141-154.
Molè, V., Casini, F. & Viggiani, G. 2017. Il congelamento artificiale dei terreni eseguito nei pressi di un palo di fondazione. *XXVI Convegno Nazionale di Geotecnica*. Rome.
Rocca, O. 2011. Congelamento artificiale del terreno. *Hevelius Edizioni*.
Russo, G., Corbo, A., Cavuoto, F. & Autuori, S. 2015. Artificial Ground Freezing to excavate a tunnel in sandy soil. Measurements and back analysis. *Tunneling and Underground Space Technology* 50: 226–238.
Russo, G., Corbo, A., Cavuoto, F., Manassero, V., De Risi, A. & Pigorini, A. 2017. Underground culture: Toledo station in Naples, Italy. *Proceedings of the Institution of Civil Engineers - Civil Engineering* 170(4): 161–168.
Viggiani, G.M.B. & de Sanctis, L. 2009. Geotechnical aspects of underground railway construction in the urban environment: the examples of Rome and Naples. *Geological Society, London, Engineering Geology Special Publications* 22: 215–240.
Viggiani, G. & Casini, F. 2015. Artificial Ground Freezing: from applications and case studies to fundamental research. *Geotechnical Engineering for Infrastructure and Development*. Edinburgh.
Yamamoto, Y. & Springman, S.M. 2014. Axial compression stress path tests on artificial frozen soil samples in a triaxial device at temperatures just below 0 °C. *Canadian Geotechnical Journal* 51(10): 1178–1195.

Tunnels and Underground Cities: Engineering and Innovation meet Archaeology, Architecture and Art, Volume 3: Geological and geotechnical knowledge and requirements for project implementation – Peila, Viggiani & Celestino (Eds)

Geotechnical data standardization and management to support BIM for underground infrastructures and tunnels

M. Beaufils & S. Grellet
BRGM (French Geological Survey), Orléans, France

B. Le Hello
EGIS SE, Seyssins, France

J. Lorentz
Geolithe Innov, Crolles, France

M. Beaudouin
SYSTRA, Paris, France

J. Castro Moreno
SETEC als, Lyon, France

ABSTRACT: In France, the MINnD project aims at extending Building Information Modeling (BIM) capacities for its application to infrastructures such as roads, bridges and railways. The MINnD Use Case 8 (UC8) is about Underground Infrastructures, and deals with both the description of the tunnel, its equipment parts and environment, with a main focus on pre-feasibility study to the construction phase.

This paper deals with the description of the environment part. It identifies exchange requirements in geotechnical engineering and introduce solutions for data organization and provision. This encompasses the standardization of data such as observations and measurements; geological, hydrogeological and geotechnical models; impact on anthropic and natural environment; structure sizing and proposition of construction methods and risk assessment.

In that context, propositions of reuse and extension of existing standards for geoscience data, such as GeoSciML and GroundWaterML2 from the Open Geospatial Consortium (OGC) are developed and discussed.

1 INTRODUCTION

Several definitions are often associated to the BIM acronym. While there is a consensus on saying that BI stands for Building Information, the M is often associated to multiple translations, the most popular being Model, Modeling or Management (Eastman, 2011). In brief, one can say that those three expressions respectively deal with the capacity of positioning accurately objects in space, explaining the methodology for getting it, and managing all those data during the building or infrastructure life cycle.

In the French MINnD project[1], that aims at extending BIM capacities to cover infrastructure description, an initiative was launched to deal with Underground Infrastructure, also denominated as Use Case 8 or UC8. The perimeter includes the description and organization of data associated to the design of the construction with its equipment (covered by the GC subgroup), but also addresses the topic of environmental modeling (covered by the GT subgroup).

1. Official website: www.minnd.fr/en

Geotechnics or geoengineering studies aim at collecting and combining information from the subsurface to characterize its consistence and propose solutions to enable the integration of a building or an infrastructure. This means that the data used and produced for geoengineering have a strong impact on the infrastructure design and shall be considered in BIM, as the data related to the construction or their equipment themselves.

Standardization in BIM is commonly associated to the development of the Industry Foundation Classes (IFC). Then actions on IFC Bridge for example focused on extending existing IFC or introducing new ones to be able to properly describe a bridge and its components. Following that approach, several initiatives proposed to extend IFC with geological or geotechnical concepts (China Railway BIM Alliance, 2015), (An, 2017). Such approach would offer some advantages; the most obvious is that it should facilitate compatibility with other IFCs. Yet, alternative solutions worth being considered too and a specific attention should be paid on two critical points:

First, environmental modeling follows different rules than Building Information Modeling. While BIM aims at defining the system to build and the actions to do to have it, geomodelling is about exposing what has been observed and possible interpretation that can be built from those observations. In that context, geomodeling is always associated to uncertainties. Then, to cope with them, the project coordinator relies on a risk driven management approach, focusing on the risk of delay and extra cost that each possible misinterpretation could lead to. In other words, this means that geotechnical activities do not (only) consist in the provision of a 3D map that describes the expected position, shape and properties of subsurface components, but aim at proposing infrastructure sizing and construction methods to build a sustainable infrastructure in acceptable risk conditions.

Second point is that several formats, communities and activities already exist for geoscience data standardization. Several organizations that produce environmental data even have to provide their data through specific standards according to legal rules (e.g. INSPIRE Directive in Europe). A sustainable and long-term proposition for coupling geoscience data with BIM shall integrate such dynamics and constraints, in order to avoid data duplication, limit data conversion, facilitate model updates and data maintenance in general.

Considering that context, the MINnD UC8-GT group first focused on eliciting the geotechnical activities, determining how the "geotechnical knowledge" is built and which data are used and produced to do so. Then the second activity was to study the existing standards, communities and activities around that topic. That study leads to propositions to organize geotechnical data in order to support BIM for Underground Infrastructure, with a focus on the design part of the project: this is basically the summary of that paper.

2 GEOTECHNICAL ENGINEERING

2.1 *General definition*

Geotechnical engineering is a specific part of civil engineering required to define and assess soils behavior during and after infrastructure construction. Studies are made following a detailed process from project beginning to ending in order to limit the risks associated to soils behaviors.

The geotechnical activities cover several fields of expertise as geology, geomechanics and hydrogeology and is completed by geotechnical investigations (borehole, in situ test, laboratory test). One of its main purposes is to define primary geological or hydrogeological risk of one project and propose adapted solution, requiring characterizing mechanical properties of soils and soil/structure interaction. Then, the geotechnical studies process follows classical design workflow of one infrastructure passing from prefeasibility and feasibility studies to detailed design studies, in framework of standard or specific guidelines.

2.1.1 *In France*

Several standards have been defined to ensure the quality and efficiency of French geoengineering. The French Coordination of Commissions in Geotechnical Standardization presented in (BNTRA, 2014) an overview of the applicable standards in France for the geotechnical engineering activity.

NF P 94 500 is the French standard which specifies the geotechnical studies to be done, their nature and objective, and their workflow. Geotechnical studies workflow is organized around five main missions:

- G1 is the prefeasibility study. It aims at getting an overview of the main geological, hydrological and geotechnical risk associated to the site of the project,
- G2 is the conception phase. It is separated into two main parts: the first is the identification and comparison of possible solutions for the project, the second is the development of the retained solution including costs and methodology. It ends with the preparation of tender documents to identify a constructor,
- G3 and G4 are made during the construction phase. G3 is specific to French methodology and corresponds to the geotechnical design of the company in charge of construction. G4 is the construction follow up.
- G5 is an expertise study that can be made on an existing structure.

Then NF P 94 500 is not focused on Underground Infrastructure or tunnel construction, thus can be used for any kind of infrastructure or building construction.

Yet, for tunneling construction, the French Tunneling and Underground Space Association (AFTES), proposes several guidelines in order to specialize the objective of geostechnical mission described in NF P 94 500 for tunneling project. MINnD UC8-GT group focused on the AFTES workflow and aims at describing information and data transmission from prefeasibility step to tendering step of a tunnel construction project, thus G1 and G2.

2.1.2 *In the other countries*

ISO/TC182 addresses the topic of geotechnics, yet with a focus on the description of the investigations and testing. It does not deal with the description of the general geotechnical activity processes. In that domain, international standards, especially British construction procedure differs from French standard on the fact that the company in charge of the construction do not have to perform additional geotechnical design during construction even if it is strongly advised.

Eurocode 7 is the European standard dealing with geotechnical design. It aims at describing how to design geotechnical structure and is split in two main documents (part 1 which is general rules and part 2 which is ground investigation and testing). In its two parts, but especially in part 2, the geotechnical studies procedure is described from preliminary studies to project design. It is fully consistent with G1 and G2 phases of NF P 94 500.

Then, considering the perimeter and focus of MINnD UC8 GT on the G1 and G2 phases, equivalence can be assumed for other countries as geotechnical studies and specifically design studies required for tender are similar (same objectives) inside and outside of France.

2.2 *Main topics in geotechnical works and perimeters*

2.2.1 *Decomposition of the geotechnical engineering activities*

Based on the NF P 94 500 standards, nine topics have been identified as the main subdivisions of a geotechnical engineering project all over its life.

The figure opposite (Figure 1) propose an abstract representation of the geotechnical activities and perimeter.

- RECO covers the data collection phase. It deals with the definition and execution of site investigations, measurements and laboratory tests,
- GEOL, HYDR and GTCH respectively deal with the modeling of the geological context, the hydrogeological context and the geotechnical context,
- AVOI, ENVI and RISK also deals with modeling, but respectively focus on the impact on the surrounding constructions and utility networks, the natural environment and the assessment of potential technical risks on the project.
- CALC and MECO are preconization phases that aims at modeling the interface between the infrastructure and its environment and proposing sustainable sizing and design (CALC) and construction method (MECO) to build the project.

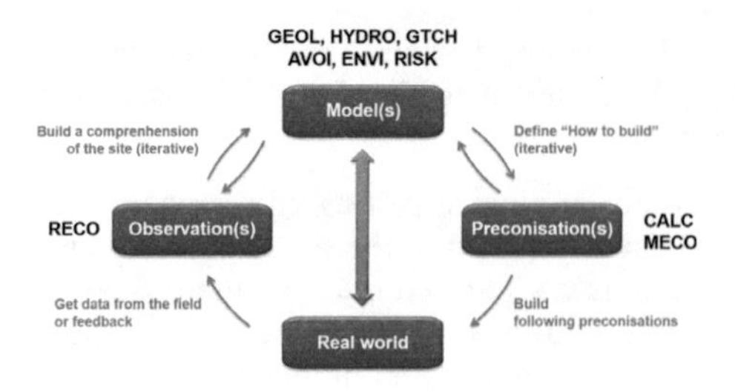

Figure 1. The main geotechnical activities.

The process starts with the collection of observations. The creation of models aims at characterizing the construction site and enable to propose solutions to integrate the infrastructure. Then feedback that would be obtained during the construction and maintenance (yet out of scope of that project) could enable to update the models and ensure the expected fidelity between the model(s) and the real world, as targeted by BIM.

2.2.2 *Information Delivery Manuals (IDM)*

In order to describe the steps, actors and formats used for geotechnical «knowledge» construction, Information Delivery Manual (IDM) were developed. Each IDM consists in a process map and a glossary.

The process map provides a graphical representation of the sequence of activities in the topic and the involved data. As recommended by ISO29481, it is based on the BPMN language. An example of process map is shown below:

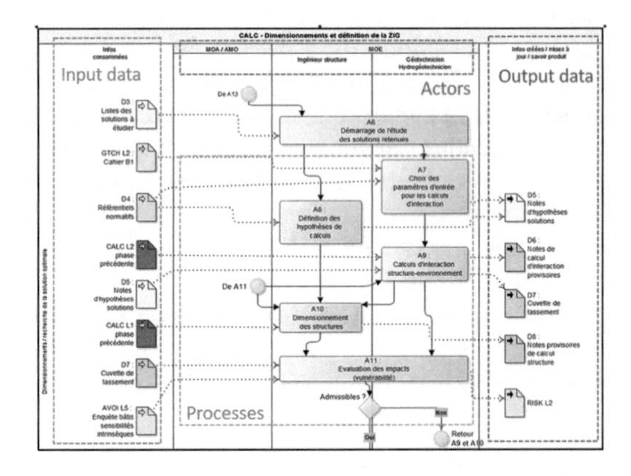

Figure 2. Extract from the CALC IDM (Infrastructure sizing from assessment of interaction with soil).

The progression of the project and thus the sequence of processes can be read from top to bottom. The realization of a process in terms of consumption/modification/production of knowledge is read from left to right.

The glossary acts as a legend for the process map and enable to provide more detailed description of both data and actions.

IDMs enable to define which data are handled, who instantiate it, who should be able to update it or just access it, thus the basic information to imagine a collaborative platform to share the data, as depicted by BIM Level 3. Such information is also crucial to define the right semantics and features to integrate in the conceptual model and the definition of model views (MVD) that would enable to fulfill the project flow.

2.3 *Geotechnical data interoperability issues and challenges*

The several IDM that were developed expose how the "geotechnical knowledge" is supposed to be built, but even if this vision is shared by the community and introduced in a standard, sometimes practices can differ and affect data exchanges.

2.3.1 *Actors and task repartition*
In some circumstances, two or more roles or jobs that have been identified in the IDM can be assumed by one unique person. Motivations for such approach are often economical and aims at reducing the data exchange (thus delays) by merging the actors. In that case, some data exchanges would not happen. Yet in order to keep the lineage, thus facilitate correction, it is strongly recommended to keep the proposed decomposition of result furniture, especially if there are specifically reused for another process.

2.3.2 *Tools interoperability issues*
Inherent to the exchanges between people having different roles is the use of different tools adapted to their job. This technical issue is increased by the presence of several software introducing their own data format. Another issue is that tenders do not always specify or enforce the format that shall be used to provide the data. For investigation contract for example, this may lead to the furniture of borehole data in very different formats: picture format (pdf), text (txt, Word, ...), spreadsheet (Excel, ...) or database (Access, AGS, ...). This can result in the best case into data conversion thus "only" time waste, yet in the worst case (e.g. image format) into the need to entry the data again, with associated mistake risk.

2.3.3 *Semantic issues*
Another issue is the use of different semantics to characterize the information. This is particularly usual as geotechnics involve several jobs (geologist, hydrogeologist, geotechnical engineer, driller, ...). Then two actors may rely on different types of classification, introducing confusion in the lecture and comprehension of the data. In the best case, this issue may be solved with value conversion. Yet if not mentioned this can lead to mistake or errors (e.g. unit format). To avoid that issue, reference to a common data dictionary or vocabulary shall be mandatory.

2.3.4 *3D modeling*
In opposition to BIM for building design, geotechnical modeling is not (for now) always associated to 3D modeling. Particularly for linear infrastructures, 2D cross sections remain a very common way (if not the most common) to map the subsurface. Until 3D geomodeling become mandatory, a solution to deal with integration of geomodels into BIM models shall consider this state and not been limited to 3D modeling.

3 GEOTECHNICAL DATA AND EXISTING STANDARDS OVERVIEW

The construction of the IDMs enabled to identify the concepts and data that compose and represent the "geotechnical knowledge". This part proposes a summary and classification of the identified concepts, as well as an overview of existing standards and initiatives that address the geotechnical topic.

3.1 *Main concepts in geotechnics*

3.1.1 *Observations and measurements*
Geological, geotechnical or hydrogeological models are built from data provided during reconnaissance surveys. These data describe different types of objects with specific properties. The main data come from observations or measurements of the soil or nearby structures. The MINnD UC8-GT group adopted the following typology:
Observations:

- Surface observations: these observations concern all the data provided during a field trip by a geologist or a geotechnical engineer. In most cases, these are diagrams, notes. These documents are accompanied by georeferenced photos of the different areas. These observations enable to provide analysis data that can take the form of an analysis of lithology and structural geology.

- Deep field observations: These observations are provided through drilling. The first observation comes from the indications of the driller during the progress of drilling. When core drilling is done, then it is possible to extract the soil to analyze it.

Measurements:

- Topography: This measure is a basic measure that serves as a shared foundation for all geotechnical data. 3D topographic data is becoming simpler to acquire thanks to easily accessible technologies such as lidar or photogrammetry.
- In situ geotechnical tests: This term includes all the tests that can be performed during a geotechnical survey. Pressuremeter or penetrometer tests are the most famous example of in situ geotechnical tests. During in situ test, the measurement of a water level is an important measure to know the state of the aquifer.
- For these geotechnical measurements, it must be ensured that the data has been validated by a geotechnical engineer based on the raw data, that is to say, the data directly obtained from the tests (pressure, deformation, . . .).
- Geophysical measurements: these measurements are different from in situ tests because geophysical measurements are non-destructive tests. The most widely used geophysical measurements are seismic, electrical or electromagnetic ones.
- Geotechnical laboratory tests. These tests concern all the geotechnical tests which are carried out not in the field, but in a laboratory. The best known are triaxial tests, oedometric tests or granulometry.
- Real time monitoring. Measurements made by sensors, for example piezometers or monitoring of nearby structures.

3.1.2 *Models*

In addition to data observed or measured on site, whatever if they are interpreted or raw data, the other type of data exchange is all analysis and design performed by specialist. This kind of data is named “model”. The principal model, the geotechnical model, is a summary of information from three main fields of expertise: geology that enables to define the general structure of the subsurface including fault presence, hydrogeology that informs the presence of water and geotechnics that determine its geomechanical properties. These model aims at synthetizing and improving data collected in useful information for construction.

Their second objective is to define the degree of uncertainty of model in order to assess the risk of construction. In addition, model could propose additional investigation to improve the quality of data measured, in order to limit the degree of uncertainty and so the risk.

According to that, each data introduced in general model (for structure design) must propose an associated degree of uncertainty and the way to limit it. This approach follows the classical workflow of geotechnical design (AFTES guidelines). Design done in prefeasibility study is a global interpretation of environment. Whereas detailed design study seeks to define more accurately the soil behavior.

Coupled to this model, impact of construction on environment (built and unbuilt) is also assessed. It is generally done (in tunnel design) by defining the expected surface settlement due to construction. This mapping is then cross-checked with existing building tolerance coming from in situ measurement assessment (AVOI). At last environmental aspect (pollutant material, excavated soil management) is studied.

3.2 *Standards for geoscience data*

An important standardization effort exists for geoscience related data. As an output several, open and non-open standards address the concepts/information needs presented in this paper. Several standardization ‘silos’ are important here.

ISO 191xx series:

- Provides the international framework under which most of the open standards are built (especially for the OGC and INSPIRE ones).

- ISO 19156 Observations & Measurements: is the key central standard framing the Observations and Sampling topics. Many domain standards are indeed specialization of it.

Open Geospatial Consortium (OGC) joint standards with the International Union of Geological Sciences Commission for Geoscience Information (CGI-IUGS)

- GeoSciML (Geoscience Markup Language) covers the domain of geology (earth materials, geological units and stratigraphy, geological time, geological structures, geomorphology, geo-chemistry) and sampling features common to the practice of geoscience, such as boreholes and geological specimens (Boisvert et al., 2017).
- GroundWater Markup Language 2 (GroundWaterML 2 or GWML2) is a data exchange standard for the groundwater domain. As part of the WaterML 2.0 standard and linked to GeoSciML, it introduces concepts such as hydrogeological units, fluid bodies, voids, flow, well, borehole construction, geological logs, management area (Brodaric et al., 2017).

INSPIRE Directive themes

- The geology theme formalizes the geology, groundwater and geophysics domains. It does have an important overlap with GeoSciML and GWML2 semantic content but also provide new structure for 'orphan' domains (ex: geophysics). Most of GeoSciML European people where involved in specifying that theme.
- Other INSPIRE themes can enrich the data panel such as 'Environmental monitoring facilities' which describes monitoring facility/networks, observation campaign and observation acquired or 'Soil'.

Industry

- The Energistics consortium propose standards RESQML and WITSML that respectively enable reservoir characterization and wits description (King, 2012). These formats are designed for the Oil & Gaz topic, yet some concepts might be reused for geotechnics.
- The AGS Data Format for geotechnics is an English initiative that provide a format to transfer ground investigation, laboratory testing and monitoring data. A project called DIGGS (Caronna, 2006) aimed providing a XML compliant format for AGS, yet this has not been officially adopted by AGS.

National standards

- Some countries also define their national data standard(s). They do not always re-use (build-on) the international initiatives previously mentioned. Even though some are willing to align with them for the same reasons as the one mentioned at the beginning of that chapter. Borehole ML, German standards on Borehole is worth mentioning here.

It is also important to understand standardization dynamics and the communities pushing them. Within the research community:

- EPOS EU Research Infrastructure made an important push in enhancing Borehole data discoverability,
- ESIP/RDA Earth, Space, and Environmental Sciences Interest Group aims to act as an 'umbrella' group trying to bring together parties from the big international research infrastructures.

Two recent activities initiated under the OGC also make a change in the standards picture in the domain. As such, the OGC Geoscience Domain Working Group gathers (if not all) the mentioned standardization 'silos' to discuss how to create bridges between the various standardization efforts. On a specific topic, the recently started OGC Borehole Interoperability Experiment (http://www.opengeospatial.org/projects/initiatives/boreholeie) builds on EPOS EU research infrastructure Borehole model and aims to test/refine it against a handful of use cases stemming from the various communities (research, Oil & Gas,, . . . and geotechnics).

4 PROPOSITION FOR GEOTECHNICAL DATA STANDARDIZATION AND MANAGEMENT

4.1 *Conceptual model*

Standards update and their implementation is a somewhat iterative approach. Thus, the MINnD UC8-GT group proposed a best approach for each of the identified concept as of today's dynamics. Needs identified in parts 2.2 and 3.1 have been confronted to the standards described in the previous part. Most (if not all) the needs identified are covered in one of the standards. The best approach proposed is the following.

Concepts and semantic for structural geology:

- Geologic units and geologic structures can be described using GeoSciML keeping in mind the flexibility offered by the concept of MappedFeature which allows to differentiate the real world object and its various representation one can encounter in various maps (different scale, levels of details,, ...).

Concepts and semantic for hydrogeology:

- GWML2 stems from the hydro Domain Working Group also with the goal to ensure the continuity in the description of surface/ground water flows. In combination with OGC HY_Feature conceptual model, it ensures a global understanding of the water cycle. It is recommended to describe underground hydrologic features and their associated flow models with those standards.

Concepts and semantic geotechnical modeling:

- Geotechnical modeling can be seen as a combination of structural geology and hydrogeology modeling with the fusion/refinement of proposed division of the subsurface in geologic units based on several properties. Thus, geotechnical modeling data exchange could be achieved using the same concepts as for the two previous domains. Yet additional geotechnics specific vocabulary might be necessary to add.

Concepts and semantic for impact on surrounding constructions:

- In the interoperability standards landscape, surrounding constructions can be described using CityGML and/or INSPIRE Building Theme. UtilityNetworkADE extension for CityGML and/or INSPIRE UtilityNetworks Theme can also be used.

Concepts and semantic for impact on the natural environment:

- INSPIRE themes 'Natural Risk Zones' and 'Area management/restriction/regulation zones and reporting units' can be a good starting point to describe the impact of a given project on its direct environment. Concepts of HazardArea, RiskZone, ExposedElements are good candidate for such a goal.

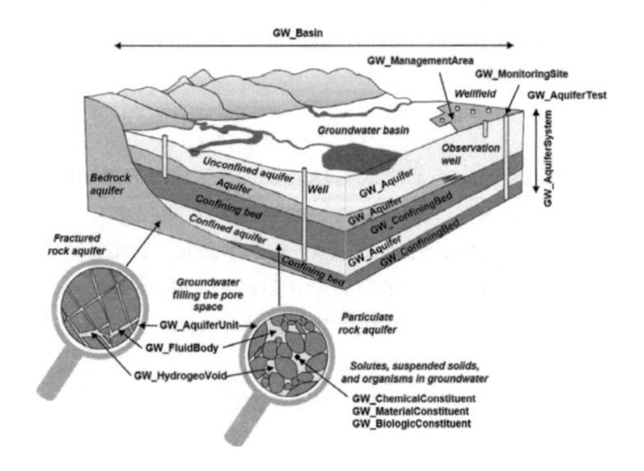

Figure 3. Some concepts from the OGC GroundWaterML2 model (Brodaric, 2017).

Observations and measurements:

- In order to describe observations collected on the field, Observations & Measurements contains all the necessary elements. The more "domain specialized" standards often propose pattern to use O&M for specific observations (ex: GeoSciML with an outcrop dedicated pattern, GWML2 with a pump test one,. . .). Logs are described in several standards. However, GWML2 takes the point of view that a log is an Observation per se, thus provide a specific O&M based pattern to do so. This is the recommended approach.
- Laboratory analysis can also be described using O&M. Depending on the types of the context, starting from GeoSciML Laboratory and Analysis package (based on O&M) might be recommended. It is a model for laboratory analytical metadata, geological sampling and specimens, and isotopic age observation results. The AGS data format is also studied as a main source of inspiration for the description of the geotechnical tests and analysis.

Survey and models:

- The concept of "Campaign" appears in the INSPIRE model, in the geology them/geophysics package. In that context, it is defined as an activity that consists in producing (geophysical) data, processing results or models. It provided the necessary hook to O&M.
- As for Geological Models, a first attempt of UML modelling also appears in INSPIRE Geology/geophysics package (Geophysical model focused). It proposes to describe them using O&M. Based on this, EPOS EU research infrastructure proposed a short summary UML description of a geological model that aims, in turn, to point back to a more thorough description using O&M. This is the approach currently pushed by the community.

4.2 *Data management*

Beside conceptual models, OGC introduces standard interfaces that enable to retrieve data and expose them on the web. Most famous standards from OGC is the Web Mapping Service (WMS) that enable to provide a map (image) representation from several GIS vector or raster data formats. Yet other OGC standards enable to expose data: Web Feature Service (WFS) to provide discrete geospatial objects, Web Coverage Service (WCS) to provide coverage, Sensor Observation Service (SOS) and Sensor Things API to provide observations and measurements.

The standardization approach concern two steps: first is the protocol to use to ask the server for specific data including filtering capacities. Second is the format and model that the result obtained conform to. In other words, such web services enable the "consumer" to build a query with a standardized syntax to get standardized data on demand without having to figure the data storage systems and format used by the provider. Use of such standards is interesting for the development of sustainable Spatial Data Infrastructures (SDI), which consists in platforms that offer users to access and share data in a collaborative way. In addition, Pub/Sub type solutions (OGC, 2013) would offer capacities to notify users of the update of some data. IDM described in the first part of that paper indicates which notification shall be transmitted to continue the geotechnical processes. Applicated to building information modeling data, such system could propose first bricks to build the collaborative data platform depicted by BIM Level 3. Use of WFS to provide IFC have even been suggested by (Schumann, 2017).

5 CONCLUSION AND PERSPECTIVES

While BIM model aims at providing the most accurate description of a projected construction, environmental modeling follows the scientific method to provide an interpretation of the subsurface organization and behavior. Then in the construction domain, the geotechnical activity consists in the realization of several models that once linked lead to the proposition of methods, infrastructure sizing that shall ensure the builders to have the project realized with acceptable risk conditions.

The MINnD UC8-GT team focused on the identification of the geotechnical data that are formally exchanged in a project. The results are represented by Information Delivery Manuals (IDM) that describe the geotechnical knowledge construction. It enabled to identify the important concepts to include in a data model proposition. Furthermore, it highlights that the interaction between the geotechnical engineering team and the civil engineering team does not (only) rely on the provision of a "map" of the subsurface. This means that current propositions of IFC extensions for geology or geotechnics that deal with the provision of representations of geological model (may it be 1D (borehole log), 2D (cross section) or 3D), only address a part of the required and formally exchanged geotechnical information.

Yet, several groups or organizations, such as INSPIRE and OGC already propose data models for geoscience data standardization. Those models are developed, supported and implemented by the geoscience community. As most concepts (if not all) needed for geotechnical data exchanges are introduced in those models, proposition from MINnD UC8-GT is to rely on them. A best approach mapping is then proposed to indicate which model or standard to use for each identified data to exchange. Implementing existing standards, especially open ones, is recommended for many reasons. With no specific order, it triggers better semantic and technical interoperability between parties involved in a project. Then tools (open source, vendors) can already be available. Those tools can range from data provision (e.g. data server), to data consumption (e.g. desktop client). In addition, in some geographical contexts, it allows to respect legal obligations (ex: INSPIRE directive in Europe - Directive 2007/2/EC). Those points have a direct impact on the costs efficiency of an approach based on those. Indeed, there is no (or less) need to adapt tools to each new customer and data context. Thus, a better Return On Investment at company/project level.

Finally, connecting building models and geomodels is a key use case for the BIM/GIS interoperability, bSI and OGC, as respective representatives of BIM and GIS, bSI and OGC data shall work together to build a common conceptual model for geotechnical concepts that would ensure the semantic interoperability. Such approach has already been successfully achieved with the alignment concept model that is identical for both bSI IFC Alignment and OGC LandInfra models (Scarponcini, 2016).

REFERENCES

An, Y. 2017 – Geological Data Extension for Subway Tunnel BIM models using a Linked Data Approach - Master Thesis. Eindhoven Technical University. March 2017

BNTRA, 2014 - Liste des normes françaises du domaine Géotechnique - Version du 31 mars 2014. Bureau de Normalisation des Transports, des Routes et de leurs Aménagements.

Boisvert, E. & et al., 2017 - Geoscience Markup Language 4.1 (GeoSciML) - Open Geospatial Consortium Standard 16-008, v4.1, 247 p.

Braeckel, A. & et al. 2013 - OGC® Publish/Subscribe Interface Standard 1.0 – Core - Open Geospatial Consortium Standard 13-131r1, v1.0, 112p.

Brodaric, B. & et al., 2017 - OGC WaterML 2: Part 4 – GroundWaterML 2 (GWML2) - Open Geospatial Consortium Standard 16-032r2, v2.2, 160p.

Caronna, S. 2006 - Implementing XML for geotechnical databases - Geo-Engineering Data: Representation and Standardisation, 10th IAEG Congress September 9th 2006 Nottingham, UK.

China Railway BIM Alliance. 2015 - CRBIM 1002-2015: Railway BIM Data Standard (Version 1.0) - China, December 29, 2015.

Eastman, C., Teicholz, P., Sacks, R., & Liston, K. 2011 - BIM handbook: a guide to building information modeling for owners, managers, designers, engineers, and contractors - Wiley.

King, M. & et al. 2012 - Reservoir Modeling: From RESCUE to RESQML - SPE Reservoir Evaluation & Engineering. Vol 15. Issue 2. 12p.

NF P 94 500 - Geotechnical engineering missions - Classification and specifications.

NF EN 1997-1/1997-2 – Eurocode 7 : Calcul géotechnique.

Scarponcini, P. & et al. 2016 - OGC® Land and Infrastructure Conceptual Model Standard (LandInfra) - OGC® Implementation Standard 15-111r1, v1.0, 315p.

Schumann, G. & et al., 2017 - Future City Pilot 1 - Recommendations on Serving IFC via WFS -OGC Public Engineering Report. 15p.

Tunnels and Underground Cities: Engineering and Innovation meet Archaeology, Architecture and Art, Volume 3: Geological and geotechnical knowledge and requirements for project implementation – Peila, Viggiani & Celestino (Eds)
 ISBN 978-0-367-46583-4

Use of TBM parameters for assessing rock mass conditions during excavation: A feedback from the Tunnel 4, Angat Water Transmission Improvement Project (Philippines)

G.W. Bianchi
EG-Team STA, Torino, Italy

I. Andreis & S. La Valle
CMC Group, Ravenna, Italy

ABSTRACT: One of the main hazards during TBM excavation in hard rock is related to unforeseen adverse conditions such as faulted rock masses. Different methods for geotechnical investigations ahead of the tunnel face are used in TBM tunneling. Many of these methods imply an interruption of excavation and their systematic application leads to significant delay. The continuous analysis of TBM excavation parameters may represent an interesting tool for monitoring of geotechnical conditions at the tunnel face and for early detection of adverse conditions. This work presents the feedback from the Tunnel 4 (Angat Water Transmission Improvement Project, Philippines) a 6.4 km long water transfer tunnel excavated by double shield TBM. During excavation, Es and FPI have been systematically analyzed and compared with RMR values. After a test section, reference values of Es and FPI were defined for each RMR-class and used for monitoring of ground conditions at the tunnel face.

1 INTRODUCTION

The use of Tunnel Boring Machines (TBMs) in the tunnel industry has undergone a continuous growth over the last years and improvements in technical features (e.g. cutter-head size, value of machine torque and machine thrust) allowed the use of TBMs in always wider range of geological and geotechnical context.

Despite of such increase in the application of TBM, there is a poor feedback from completed projects with respect to the use of machine data for assessing geological and geotechnical conditions, namely in the frame of hard rock excavation. In most of cases, the analysis of machine data is related to assessment and estimation of TBM performance, i.e. as a tool for prediction of advance rate. As a matter of fact, TBM is not a flexible method of excavation and a reliable assessment of geotechnical conditions along the tunnel alignment is a key-aspect for successful application of mechanized excavation. The choice whether using a TBM or not is therefore crucial in terms of project planning and cost estimation and this choice is often based on the expected TBM's rate of advance. In this respect, the most recognized TBM performance prediction models correspond to the CSM (Colorado School of Mines) model (e.g. Rostami and Ozdemir, 1993, Rostami, 1997) and to NTNU (Norwegian University of Science and Technology) model (e.g. Blindheim, 1979, Bruland, 1998) and adjustments factors for these two models are continuously suggested (see e.g. review in Hassanpour et al., 2011). On the other hand, there are only a few feedbacks from completed TBM projects about correlation between machine data and encountered geotechnical conditions. Bieniawski et al. (2012) suggested the use of specific energy of excavation for detecting changes in tunnelling ground conditions, based on the good correlation between this parameter and the Rock Mass Rating (RMR). Alber (1996) indicates a good relationship between specific penetration and rock mass conditions such as uniaxial compressive strength (UCS) and spacing of discontinuities. Hassanpour et al. (2011) highlight the correlation between some rock mass parameters (RQD, UCS) and the Field Penetration Index.

The present work aims to provide a first feedback from the Tunnel 4 (Angat Water Transmission Improvement Project – AWTIP) a 6.4 km long water transfer tunnel excavated by double shield TBM. During excavation, specific energy (Es) and field penetration index (FPI) have been systematically analyzed and compared with RMR values. After a test section, reference values of specific energy and field penetration index are defined for each RMR-class and currently used for monitoring of ground conditions at the tunnel face

2 PROJECT SETTING

The Tunnel 4 is a water transfer tunnel located in the Norzagaray region, about 50 km northeast of Manila. It is part of the Angat Water Transmission Improvement Project, aimed to improve the water supply systems to Metro Manila.

The tunnel is 6.4 km long with an excavation diameter of 4.94 and a precast segmental lining of 25 cm in thickness and 1.3 m in length.

2.1 *Geological setting*

The project area is characterized by occurrence of a sequence of volcanic and sedimentary rocks ranging in age from Mesozoic (Cretaceous) to Cenozoic (Plio-Pleistocene). Two main formations are identified along the tunnel alignment: the Madlum Formation and the Bayabas Formation.

The Madlum Formation of Middle Miocene age consists of two members:

- the Alagao Volcanics which corresponds to a sequence of volcanic breccia, tuff and andesite flows with local layers of claystone, mudstone and minor limestone;
- the Buenacop Limestone, a sequence of rather homogeneous, grey to light brown, massive and unweathered rocks. The basal sections of this member is a conglomerate made of pebbles and blocks of limestone and volcanic rocks (i.e. tuff, basalt and diorite) in a well cemented matrix of tuff and sand.

The Bayabas Formation is dated to be Late Eocene to Early Oligocene. This formation is composed of dark green massive basalt, tuff and volcanic breccia, with pebbles and blocks of massive basalt in a tuff-like matrix.

From the structural point of view, three main sets of faults have been identified, based on analysis of aerial images and field survey. The fault zones have been subdivided into main and minor faults according to the thickness of the core zone and to the total thickness: main faults are characterized by a thickness of the core zone greater than 5 m and a total thickness greater than 10 m; minor faults show a core zone of 1 to 3 m and a total thickness of 2 to 5 m.

2.2 *Geomechanical characterization*

Based on intact rock parameters as well as GSI and RMR values assessed during field survey and borehole analysis, the rock types along the tunnel alignment have been grouped into five geomechanical units (GU):

- Geomechanical Unit 1 (GU1) corresponds to the basalts, pyroclastites, andesite and tuff of the Bayabas Formation. This unit is characterized by hard and rather low fractured rock masses with fair to very good conditions of discontinuities. The measured parameters point to good geomechanical features in terms of rock strength and resistance parameters;
- Geomechanical Unit 2 (GU2) is represented by the volcanic rock of the Alagao Volcanics (Madlum Formation), which corresponds to a hard rock with various degree of jointing and condition of joint surfaces ranging between poor and very good. As a matter of fact,

and despite of similar values of intact rock strength with respect to the Bayabas Formation, the volcanics of the Alagao Volcanics have been considered as a different unit due to the wider range of jointing and of joint conditions (well reflected by a lower range of GSI) which may imply local different behaviour during TBM excavation with respect to GU1;

- Geomechanical Unit 3 (GU3) corresponds to interlayering of volcanic rocks and claystones, mudstones and limestones within Alagao Volcanics formation. Although the thickness of these bands may be limited to about 20–30 m, their occurrence may lead to a different behaviour of the rock mass and therefore these rocks have been classified in a different unit;
- Geomechanical Unit 4 (GU4) corresponds to the limestones and to the conglomerates of the Buenacop Limestones (Madlum Formation). Measured parameters and GSI values are consistent with a hard and poorly fractured rock. This unit includes the basal conglomerates of the Buenacop Limestone formation, which are also characterized by high rock strength and low degree of jointing;
- Geomechanical Unit 5 (GU5) is represented by disintegrated and weathered rocks occurring along fault zones and includes both core and damage zones of the faults. This unit is characterized by poor geomechanical properties with low to very values of rock strength and of GSI.

A summary of the main features of the geomechanical units and of intact rock and rock mass parameters is illustrated in Table 1.

Apart from fault zones, the data point to a rather homogeneous rock mass from the geomechanical point of view, with fair to good conditions expected along most of the tunnel alignment.

Foreseen geological and geotechnical conditions along the axis of Tunnel 4 are illustrated in Figure 1.

At the end of July 2017 the TBM has achieved 2263 m of excavation. The first 185 m were excavated within the geotechnical unit 4 corresponding to the limestones and conglomerates

Table 1. Intact rock and rock mass parameters assumed for the identified geomechanical units.

Geomechanical Unit	rock type	γ (kg/m^3)	UCS (mean, MPa)	GSI	RMR	σcm (mean, MPa)
GU1	Basalt, andesite, tuff	2680	43.0	45–80	38–66	3.5
GU2	Tuff, andesite	2640	47.1	40–80	35–71	3.79
GU3	Claystone, limestone	2520	-	35–75	32–63	-
GU4	Limestone, conglomerate	2650	42.4	55–80	47–75	10.7
GU5	Faulted, fractured rock	2400	9.8	20–40	20–33	0.2

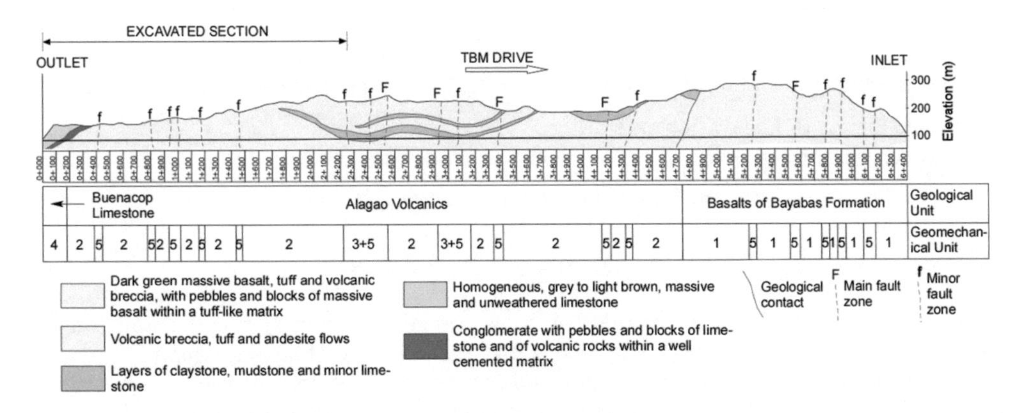

Figure 1. Longitudinal geotechnical profile along the axis of Tunnel 4.

of the Madlum Formation, in good agreement with foreseen conditions (see also Figure 1). Between chainages 185 and 2263 the geotechnical unit 2 (volcanics of the Madlum Formation) is the main encountered unit. Along the excavated section RMR values ranges between 40 and 79, with an average value of 59: most of the excavated section is characterized by RMR class III with local occurrence of RMR class II; due to the lack of real fault rocks, encountered fracture and minor fault zones led only to local decrease of RMR values but remaining within the RMR class III.

2.3 *The adopted tunnel boring machine*

The adopted TBM is a hard rock, double shield machine manufactured by SELI; total length of the shields is 10 m. The cutter head has a diameter of 4.9 m and is equipped with 31 cutters, 17" in diameter.

3 ANALYSIS OF TBM PARAMETERS

3.1 *Input data*

TBM parameters recorded during proceeding of excavation have been analysed for evaluating possible correlation with actual geotechnical conditions along the tunnel axis.

As to the excavation parameters, two main sets of data were considered:

- Machine raw-data
- Performance parameters which may provide indications of rock mass conditions as stated by the Authors mentioned above.

For assessment of ground conditions along the tunnel alignment, the RMR classification system have been adopted as indicator of geomechanical properties of the rock mass.

3.2 *TBM Parameters*

The adopted data base is formed by TBM operational parameters (machine raw data) as well as performance parameters.

The operational parameters include:

- Thrust force (kN)
- Torque (kNm)
- Rotation speed (rpm)

The performance parameters adopted in the present study correspond to:

- Penetration (mm/rev)
- Advance rate (mm/min)

These machine raw-data were used for calculating two further performance parameters:

- Specific Energy of Excavation and
- Field Penetration Index.

The Specific energy of excavation (Es) is defined as (Teale, 1965):

$$Es\ (kJ/m3) = F/A + 2\pi\ N\ T/A\ ARA \tag{1}$$

where F = total cutterhead thrust (kN); A = excavated area (m^2); N = cutterhead rotation speed (rps); T = applied torque (kNm); and ARA = average rate of advance (m/s).

Specific energy is therefore composed of two terms: the first one corresponding to thrust energy and the second corresponds to rotation energy. This parameter indicates the energy which is necessary to excavate one unit of weight of material.

The Field Penetration Index (FPI) allows to proportionate the thrust per cutter to penetrate 1 mm per revolution and it can offer therefore indications about the geomechanical quality of the rock mass. The Field Penetration Index (Tarkoy and Marconi, 1991) is defined as:

$$FPI(kN/mm/rev) = Fc/penraw \tag{2}$$

where Fc = value of thrust per cutter (kN); and Penraw = penetration raw data (mm/rev).

4 RESULTS OF THE ANALYSES

A first section of about 400 m, roughly corresponding to the learning curve, has been assumed as a test section which allowed to provide a finer tuning of the RMR measurements by the team of geologists in charge of geological mapping at the tunnel face.

With respect to the performed analysis of TBM parameters, in a first step all the operational and performance parameters listed above were correlated to geomechanical indicators of rock mass conditions, i.e. the measured RMR values.

As result, good relationships were identified between Specific Energy and Field Penetration Index on one side and RMR values on the other, as illustrated in the following. It shall be noted that the correlation is defined between two different type of data, i.e. machine data which are continuously recorded and values of a geomechanical index which cannot be provided with the same frequency.

The plot in Figure 2 illustrates the comparison between Es and RMR and highlights a rather good relationship between the pattern of these two parameters.

More in detail, and apart from the test section, it can be observed that:

- For tunnel sections characterized by RMR class II, the specific energy requested for excavation is greater than 55 MJ/m^3;
- RMR class III can be further subdivided into two main zones: the upper part of the class (i.e. with RMR values between 50 and 61) request a specific energy of about 31 to 55 MJ/m^3 for the excavation; when encountering the lower part of the class (i.e. with RMR values between 41 and 50), the values of Es decrease to about 20 to 25 MJ/m^3;
- the RMR class IV was encountered very shortly (only one RMR value) and therefore no observation can be provided with respect to Es values within this class.

As illustrated in the graphic in Figure 3, a good relationship is also observed between the pattern of RMR values and of field penetration index (FPI). Also in this case three main zones can be distinguished, apart from the test section:

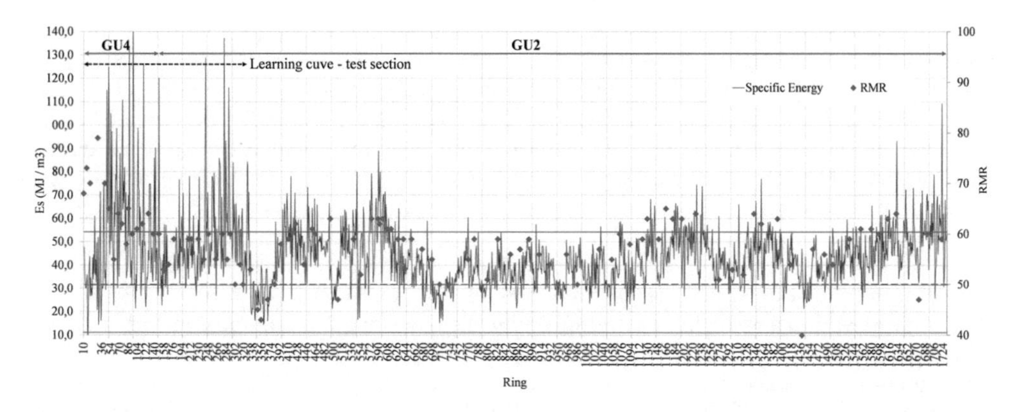

Figure 2. Diagram of measured values of specific energy of excavation (Es) and of values RMR along the Tunnel 4.

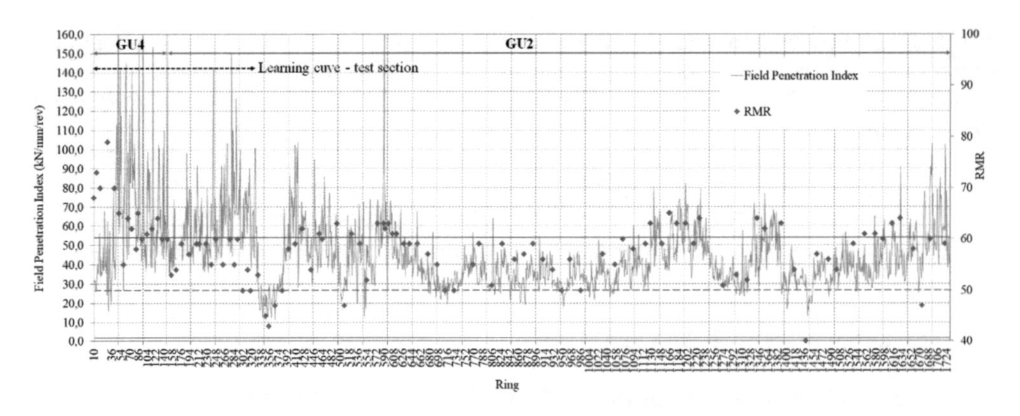

Figure 3. Diagram of measured values of field penetration index (FPI) and of values RMR along the Tunnel 4.

- For tunnel sections characterized by RMR class II, the field penetration index is greater than 53 kN/mm/rev;
- The tunnel sections excavated within the upper part of RMR class III (RMR values between 50 and 61) are characterized by values of FPI ranging between about 28 and 53 kN/mm/rev;
- A decrease of FPI below 28 kN/mm/rev is observed when encountering the lower part of the RMR class III (RMR values between 41 and 50), with lowest values of about 15 to 20 kN/mm/rev for RMR values lower than 45;
- As already stated, no reliable observation can be provided for values of FPI in correspondence of RMR class IV.

5 DISCUSSION AND CONCLUSIONS

One of the main hazards during TBM excavation in hard rock is related to occurrence of unforeseen adverse conditions such as highly fractured rock masses, often associated with water inflows. Different methods for geotechnical investigations ahead of the tunnel face are available and are commonly used in TBM tunnel projects (AFTES Recommendation GT24R2A1, 2014). Nevertheless, many investigation methods imply an interruption of excavation activity and their systematic application easily leads to significant delay in the construction schedule.

The continuous analyses of TBM excavation parameters may therefore represent an interesting tool for a continuous monitoring of geotechnical conditions at the tunnel face and for early detection of changing ground conditions and of changing degree of jointing in the rock mass. As a matter of fact, it can be observed that TBM tunnelling could be view as being a very long probe hole across the rock mass and the exploitation of excavation parameters could provide very useful informations, as it is normally done with probe holes during geotechnical investigations.

It is therefore of great interest to identify the performance parameters which display a higher sensitivity to changing ground and to geomechanical properties of the rock mass. According to previous studies (Hassanpour et al., 2011, Bieniawski et al., 2012), specific energy of excavation and Field Penetration Index displayed a good relationship with observed ground conditions.

The analysis of machine raw-data and of performance parameters from the Tunnel 4 allowed to confirm specific energy and Field Penetration Index as most sensitive parameters to changing geomechanical conditions at the tunnel face.

A continuous monitoring of these two parameters during proceeding of excavation may represent therefore a useful tool for assessing ground conditions at the tunnel face, with particular respect to joint conditions and degree of jointing. In this respect, a two-fold approach may be adopted for a continuous monitoring of Es and of FPI:

- During excavation, monitoring of Es and of FPI shall focus on the trend of these parameters: rapid decrease of Es or of FPI within a few strokes may reflect the worsening of joint conditions and/or an increase in jointing of the rock mass (Figures 1 and 2) and therefore the approaching of a fracture or of a fault zone;
- After excavation of a test section, it is possible to define reference values of Es and of FPI for each RMR-class, based on the results and on the back-analysis of data from the test section itself. The analysis of Es and of FPI may represent in this case a useful tool for identification of change in rock type at the tunnel face.

REFERENCES

AFTES Recommendation Working Group 24, 2014. Forward probing ahead of tunnel boring machines. *Tunnels et Espace Souterrain* 242, 132–169

Alber, M. 1996. Prediction of penetration and utilization for hardrock TBMs. *Proc. ISRM Int. Symp. Eurock '96*. Rotterdam, Balkema, pp721–725.

Bieniawski, R.Z.T., Celada, B., Tardaguila, I. and Rodrigues, A. 2012. Specific energy of excavation in detecting tunnelling conditions ahead of TBMs. *Tunnels & Tunneling International* pp 65–68.

Blindheim, O.T. 1979. Boreability prediction for Tunnelling. Ph.D. Thesis, Department of Geological Engineering. *The Norwegian Institute of Technology* 406.

Bruland, A. 1998. Hard Rock Tunnel Boring. Ph.D. Thesis. *Norwegian University of Science and Technology (NTNU)*. Trondheim, Norway 1–10.

Hassanpour, J., Rostami, J. and Zhao, J. 2011. A new hard rock TBM performance prediction model for project planning. *Tunnelling and Underground Space Technology* (26): 595–603.

Rostami, J. and Ozdemir, L. 1993. A new model for performance prediction of hard rock TBM. In: Bowerman, L.D. et al. (Eds), *Proceeding of RETC, Boston, MA*. 793–809.

Rostami, J. 1997. Development of a Force Estimation Model for Rock Fragmentation with Disc Cutters through Theoretical Modelling and Physical Measurement of Crushed Zone Pressure. Ph.D. Thesis. *Colorado School of Mines, Golden, Colorado*. 249.

Tarkoy, P.J. and Marconi, M. 1991. Difficult rock comminution and associated geological conditions. *In: Tunneling 91, Sixth International Symposium, London, England*. Institute of Mining and Metallurgy: 195–207.

Teale, R. 1964. The concept of specific energy in rock drilling. *Rock Mechanics Mining Science* (2): 57–73.

Tunnels and Underground Cities: Engineering and Innovation meet Archaeology, Architecture and Art, Volume 3: Geological and geotechnical knowledge and requirements for project implementation – Peila, Viggiani & Celestino (Eds)
© 2020 Taylor & Francis Group, London, ISBN 978-0-367-46583-4

Numerical method for stability analysis of rock block of tunnel with multiple geological planes

S.J. Chen, M. Xiao & J.T. Chen
State Key Laboratory of Water Resources and Hydropower Engineering Science, Wuhan City, Hubei Province, China

ABSTRACT: The multiple geological planes in tunnel tend to form rock blocks and occur local collapse. This paper proposes a numerical method for block identification and its stability analysis. Use the element reconstruction and node separation method to model multiple geological planes. The contact model considering several contact types is built to simulate the sliding and separation behavior between rock and block. The safety factor of block is calculated based on the strength reduction method. Combined with a tunnel engineering across a fault, the stability of block after excavation is analyzed. The results indicate that, the blocks are loose after excavation, and this method can better consider the influences of the initial geo-stress, excavation unloading, and rock deformation, and it can effectively identify potentially dangerous blocks. This numerical method provides an effective reference for the block stability of tunnel with multiple geological planes.

1 INTRODUCTION

During the excavation of underground tunnel, it's inevitable to meet complex geological planes (faults, joint fissures, shear zones and soft interlayers). These discontinuity planes cut the rock mass into numerous regions, and it's easy to form rock blocks and occur local collapse. Therefore, the rock block considerably affects the stability of the tunnel engineering, especially for tunnel with multiple geological planes.

For the stability analysis method of block, Goodman and Shi (1985) proposed the key block theory earlier, and evaluated the block stability by rigid limit equilibrium method traditionally. Lots of analysis program, like GeneralBlock and Unwedge, are developed to identify the unstable block in large underground excavation (Liu et al., 2004). These methods can identify the key block well, but need improvements in evaluating the block stability. The tradition limit equilibrium method treats the block as a free body and just considers its gravity stress and force on the sliding face. However, the stress from multiple contact faces of rock and block is part of the initial geo-stress field before excavation. After excavation, the rock mass unloads, and the block is subjected to stress from all contact faces, not merely the sliding face. In this process, the interactions between rock and block are important factors affecting the stability of blocks.

The simulation of the interface between rock and block has some research by numerical analysis method. Shi (1992) further proposed the discontinuous deformation analysis method (DDA), which treat the rock as discrete blocks. Though widely used in engineering, it's not accurate in simulating the deformation of surrounding rock, and it's still hard to solve the contact problems. In order to combine the advantages of continuum and discontinuum method, Beyabanaki et al. (2009) proposed a coupling analysis method between DDA and FEM. Ma et al. (2009) studied the local block collapse problem of underground cavern by integrating key block theory, DDA and software of FLAC. The existing researches show that continuum method is very mature for simulating stress redistribution and deformation of rock mass after excavation, while discontinuum method is superior in solving block problems (Cai et al., 2013). The existing methods are difficult to deal with the interface problem between rock and block. A dynamic contact force method proposed by Liu et al. (1995) is based on the explicit scheme, which has high calculation efficiency and accuracy. It is favored by researchers and well applied to engineering (Liu et al., 2018). However, this method only considers point-to-point contact, which needs to be improved to fit the large sliding problem of the block.

This paper introduces the modeling method for geological planes using element reconstruction and node separation, then builds a contact force model considering several contact types between rock and block, and finally introduces the strength reduction method to calculation the safety factor of the blocks. Combines with a tunnel engineering example, the stabilities of the blocks and rock are analyzed. It is expected to provide an effective reference for the construction and support design of tunnel with multiple geological planes.

2 MODELING METHOD FOR GEOLOGICAL PLANES

The blocks are formed by intersecting geological planes (faults, joint fissures, shear zones and soft interlayers). To build the calculation model, common modeling method may create the geometry of the block first, then mesh the inside of the block, which needs to determine the locations of all geological planes in advance. However, in practical engineering, the number and parameters of geological planes may change as the excavation progresses. At this point, the old method needs to rebuild the block geometry and re-mesh it, which has large workload and low efficiency especially when the number of new discovered geological planes is large. So, this paper introduces the element reconstruction and node separation method based on the element reconstruction method proposed by Zhang et al. (2012) which avoids the trouble of rebuilding and re-meshing. Taking a simple model for example, the basic steps are given as follows.

(1) Establish finite element model (FE model) without geological planes. The model shown in Figure 1(a) has 4 elements and 9 nodes.
(2) Obtain the information of geological planes and make the first division. The information includes the occurrence, spacing and spatial distribution of the planes. According to the extension of the model, they are divided into finite geological planes and infinite geological planes. The information can also be produced by the Monte Carlo stochastic method. As shown in Figure 1(b), the two discontinuities divide the original mesh into 10 elements.
(3) Element reconstruction is applied to change the irregular elements to regular elements. Then, use the node separation method to simulate discontinuous feature. As shown in Figure 1(c), auxiliary lines are added according to certain rules to divide the elements with poor geometry into common quadrilateral and triangular elements. Figure 1(d) shows the

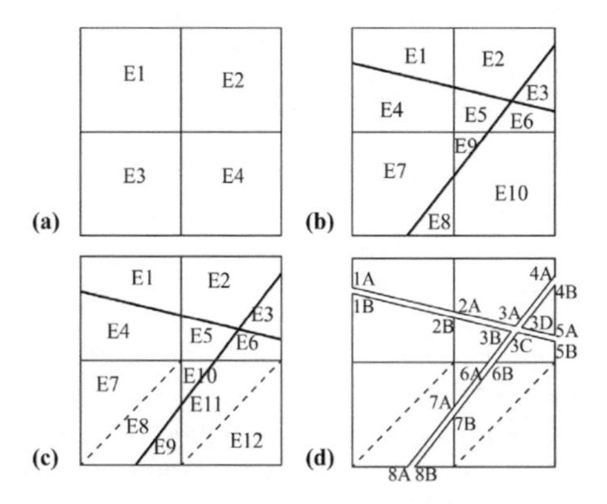

Figure 1. Simple example of element reconstruction and node separation.

FE model after node separation. A contact model on the interface will be built during calculation.

(4) Check the geometry of elements. The geometry is judged by the Jacobian matrix or element volume, and the invalid elements and nodes are deleted.

More details about the element reconstruction method can be seen in Zhang et al. (2012). The element reconstruction and node separation method in this paper can effectively simulate the multiple complex discontinuities and present the discontinuous characteristic.

3 CONTACT MODEL BETWEEN ROCK AND BLOCK

Due to the rock mass unload, the block may slid or separate from the rock. During the failure process of block, the complex contact actions are occurred on the interface between rock and block. The followings present a contact model between rock and block to simulate several contact types.

3.1 *Explicit integration format considering contact force*

On the basis of the basic equation of finite element method, the contact force of the interface and the inertia force are considered, and the overall simultaneous equation can be expressed as:

$$\boldsymbol{M}\ddot{\boldsymbol{u}} + \boldsymbol{K}\boldsymbol{u} = \boldsymbol{F} + \boldsymbol{R} \tag{1}$$

where $\boldsymbol{M}$ and $\boldsymbol{K}$ are the mass and stiffness matrixes of the interface node, respectively. $\ddot{\boldsymbol{u}}$ and $\boldsymbol{u}$ are the acceleration and displacement vectors, respectively. $\boldsymbol{F}$ is the external load vector. $\boldsymbol{R}$ is the contact force vector, and $\boldsymbol{R} = \boldsymbol{N}+\boldsymbol{T}$. $\boldsymbol{N}$ and $\boldsymbol{T}$ are the normal and tangential components of $\boldsymbol{R}$.

The velocity and acceleration of the node are calculated by central difference method, and it can be expressed as:

$$\dot{\boldsymbol{u}}^t = \frac{1}{2\Delta t}\left(\boldsymbol{u}^{t+\Delta t} - \boldsymbol{u}^{t-\Delta t}\right) \tag{2}$$

$$\ddot{\boldsymbol{u}}^t = \frac{1}{\Delta t^2}\left(\boldsymbol{u}^{t+\Delta t} - 2\boldsymbol{u}^t + \boldsymbol{u}^{t-\Delta t}\right) \tag{3}$$

Substituting (2) and (3) into (1), the node displacement at time step $t+\Delta t$ can be obtained:

$$\boldsymbol{u}^{t+\Delta t} = \bar{\boldsymbol{u}}^{t+\Delta t} + \Delta\boldsymbol{u}^{t+\Delta t} \tag{4}$$

$$\bar{\boldsymbol{u}}^{t+\Delta t} = \boldsymbol{u}^t + \Delta t\dot{\boldsymbol{u}}^t + \Delta t^2\boldsymbol{M}^{-1}(\boldsymbol{F}^t - \boldsymbol{K}\boldsymbol{u}^t)/2 \tag{5}$$

$$\Delta\boldsymbol{u}^{t+\Delta t} = \Delta t^2\boldsymbol{M}^{-1}\boldsymbol{R}^t \tag{6}$$

where t is the time. Δt is the time step size. $\dot{\boldsymbol{u}}$ is the velocity vector. $\bar{\boldsymbol{u}}^{t+\Delta t}$ is the nodal displacement vector without considering the contact force. $\Delta\boldsymbol{u}^{t+\Delta t}$ is the additional displacement vector caused by the contact force.

It can be seen from Eq. (4) to Eq. (6) that the block displacement at $t + \Delta t$ is determined by the motion state and contact force at t, which is the feature of the explicit algorithm. The motion state at t is known, while the contact force is unknown. The contact type and contact force $\boldsymbol{R}^t$ can be judged and calculated according to the contact state at t and $t+\Delta t$.

3.2 *Calculation for contact force considering several contact types*

Before the block slides, the interface between rock and block is well cemented and the contact type is point to point type. After the block slides, the interface node of the block occur relative slip with the interface node of the rock. At this time, the interface node is in contact with a certain surface of an element, which is called point to surface type. The following gives the calculation method of contact force under different contact types.

3.2.1 *Contact force calculation for point to point type*

Assume that the rock and block are in bonded state at time step $t+\Delta t$, so for the contact node pair l and l', they should satisfy the deformation coordination conditions, which means non mutual embedding in normal direction and nonrelative slip in tangent direction:

$$\boldsymbol{n}_l^T(\boldsymbol{u}_{l'}^{t+\Delta t} - \boldsymbol{u}_l^{t+\Delta t}) = \boldsymbol{0} \tag{7}$$

$$\boldsymbol{\tau}_l^T(\boldsymbol{u}_{l'}^{t+\Delta t} - \boldsymbol{u}_l^{t+\Delta t}) = \boldsymbol{\tau}_l^{\mathrm{T}}(\boldsymbol{u}_{l'}^t - \boldsymbol{u}_l^t) \tag{8}$$

where $\boldsymbol{n}_l$ is the unit normal vector of the contact node pair, pointing to l from l'. $\boldsymbol{\tau}_l$ is the corresponding unit tangent vector.

Substituting Eq. (4) to Eq. (7) and Eq. (8), and according to $\boldsymbol{R}_l{}^t = -\boldsymbol{R}_l{}^t$, the normal contact force and tangent contact force are obtained:

$$\boldsymbol{N}_l^t = \frac{2M_l M_{l'}}{(M_l + M_{l'})\Delta t^2}\Delta_{1l}\boldsymbol{n}_l \tag{9}$$

$$\boldsymbol{T}_l^t = \frac{2M_l M_{l'}}{(M_l + M_{l'})\Delta t^2}\Delta_{2l}\boldsymbol{\tau}_l \tag{10}$$

$$\Delta_{1l} = \boldsymbol{n}_l^T\left(\bar{u}_{l'}^{t+\Delta t} - \bar{u}_l^{t+\Delta t}\right) \tag{11}$$

$$\Delta_{2l} = \boldsymbol{\tau}_l^T\left[\left(\bar{\boldsymbol{u}}_{l'}^{t+\Delta t} - \bar{\boldsymbol{u}}_l^{t+\Delta t}\right) - \left(\boldsymbol{u}_{l'}^t - \boldsymbol{u}_l^t\right)\right] \tag{12}$$

where M_l and $M_{l'}$ are the lumped masses of l and l', respectively. $\boldsymbol{N}_l{}^t$ and $\boldsymbol{T}_l{}^t$ are the normal and tangential component of $\boldsymbol{R}_l{}^t(\boldsymbol{R}_l{}^t = \boldsymbol{N}_l{}^t + \boldsymbol{T}_l{}^t)$.

The above formulas are obtained under certain assumption. Therefore, it is necessary to judge the contact state of the contact point pairs and correct the contact force after each step.

If $\Delta_{1l} > 0$ and $\left\|\boldsymbol{T}_l^t\right\| > \mu_s \cdot \left\|\boldsymbol{N}_l^t\right\| + cA$, it shows the contact point is in sliding contact state, so the contact force should be:

$$\boldsymbol{T}_l^t = \mu_d \cdot \left\|\boldsymbol{N}_l^t\right\| \cdot \boldsymbol{T}_l^t / \left\|\boldsymbol{T}_l^t\right\| \tag{13}$$

If $\Delta_{1l} < 0$ and $\sqrt{(\boldsymbol{T}_l^t)^2 + (\boldsymbol{N}_l^t)^2} > cA$, it shows the contact point is in separation state, so the contact force should be:

$$\boldsymbol{N}_l^t = 0, \quad \boldsymbol{T}_l^t = \boldsymbol{0} \tag{14}$$

where μ_s and μ_d are the static friction coefficient and kinetic friction coefficient, respectively. A is control area by node l. c is the cohesion force of the interface. If the contact point haven't ever been in sliding or separation state before the step time $t+\Delta t$, then c>0. Otherwise, c=0.

3.2.2 *Contact force calculation for point to surface type*

When the contact model is point to surface type, again assume that the rock and block are in bonded state at time step $t+\Delta t$. So the contact point on the contact surface that is corresponding to the contact node l is l', and l' is in the surface belongs to element E. The displacement field $\boldsymbol{u}_{l'}^t$ and equivalent lumped mass $M_{l'}$ of node l' can be obtained by the shape function interpolation of the finite element:

$$\boldsymbol{u}_{l'}^t = \sum_j \phi_j \boldsymbol{u}_j^t \tag{15}$$

$$m_j = \phi_j M_j \Big/ \sum_i \phi_i^2, \quad M_{l'} = \sum_j m_j \tag{16}$$

where ϕ_j is the shape function value of node j in E at l'. M_j is the lumped mass of j. m_j is the mass contribution of j at l'.

According to the deformation coordination conditions and $\boldsymbol{R}_l{}^t = -\boldsymbol{R}_{l'}{}^t$, the contact force expressions of l can also be obtained in the form as Eq. (9) and Eq. (10).

It is also necessary to correct the contact state and the contact force after each step. If $\Delta_{1l} > 0$ and $\|T_l^t\| > \mu_s \cdot \|N_l^t\|$, it shows the contact point is in sliding contact state, so the contact force can be calculated by Eq. (13). If $\Delta_{1l} < 0$, it shows the contact point is in separation state, so the contact force can be calculated by Eq. (14).

3.3 *Basic steps of the contact force method*

When the contact points on the interface of block and rock break the cohesive force into sliding or separation state, the elements and element surfaces contacting with these points should be found by contact search. The contact type and contact force need to be determined at each time step. The basic steps of the contact force method at step time $t+\Delta t$ are as follows.

(1) Calculate the displacement $\bar{\boldsymbol{u}}^{t+\Delta t}$ of the contact nodes by Eq. (5) without considering contact force.
(2) Determine the contact type (point to point or point to surface) by contact search, and calculate $\boldsymbol{N}_l{}^t, \boldsymbol{T}_l{}^t, \Delta_{1l}, \Delta_{2l}$.
(3) Judge the contact state, and correct the contact force.
(4) Calculate the additional displacement $\Delta\boldsymbol{u}^{t+\Delta t}$ by Eq. (6). Then, the displacement $\boldsymbol{u}^{t+\Delta t}$ can be obtain by Eq. (4). Update the model geometry state and calculate the state of next time step.

4 SOLUTION METHOD FOR SAFETY FACTOR OF BLOCK

The strength reduction method is applied to calculate the safety factor of block, and the stability of block is quantitatively evaluated. Since the mechanical properties of the interface between rock and block play an important role in the stability of block, the critical instability state can be obtained by reducing the strength parameters of the interface:

$$c' = c/F_s, \quad \tan\varphi' = \tan\varphi/F_s \tag{17}$$

where F_s is the strength reduction factor. c and c' are the cohesion before and after the strength reduction. φ and $\varphi'$are friction angle before and after the strength reduction.

The first calculation is conducted under F_s =1. The stable block can reach a force balance state and a stable displacement value within finite steps, of which the strength reduction factor F_s is considered to be within [1,+∞]. Otherwise, the block is in an unstable state, of which F_s is within [0,1). In the corresponding interval, F_s is gradually increased from small to large until the block is in critical instability state. The critical state can be judged by displacement mutation criterion (Zhang and Mo, 2013). The reduction factor at the critical state is considered to be the safety factor of the block.

For the falling block in the top arch, the safety factor is close to zero. For the unmovable block, its safety factor approaches positive infinity. In the calculation, it is considered that if F_s <0.1, the block falls freely, and if F_s >100, the block is unmovable.

5 EXAMPLE VERIFICATION

To validate the accuracy and rationality of the contact model between rock and block and the solution method for safety factor, a simple but representative example of sliding block is used to compare the numerical results and analytical results. As shown is Figure 2, the model is consists of a sliding block and a base with variable slope (Zhao et al., 2016). The dip angel of the upper part of the slope is 40°, and the length L_1 is 10m. The dip angel of the lower part is 20°, and the length L_2 is 15m. The sliding block size is 0.5m×0.2m.

The sliding block and slope are both elastic. The Elastic modulus are both 20MPa, and the Poisson ratio is 0.3. The density of the sliding block is 2.0 Kg·m^{-3}, and the density of the slope is 2.0×10^3 Kg·m^{-3}. The bottom and lateral are fixed. The cohesion of the sliding surface is 0MPa, and the friction angle is 35°. The block slides along the slope from rest under its gravity.

The contact search is conducted at each step during calculation. The point to point contact and the point to surface contact occur alternately, and the contact force is produced on the interface between block and slope. The result of displacement time-history of the sliding block calculated by this numerical method is compared with the analytical solution, as shown in Figure 3. The block accelerates to slide down in the upper part, then slows down in the lower part, and finally stops. The result of the numerical solution agrees well with the analytical solution.

The safety factor of the sliding block is calculated by reducing the strength parameters of the sliding surface. As shown in Figure 4, when the reduction factor exceeds 0.83, the block is in the critical state. So, the safety factor of the sliding block is 0.83. The expected value of the safety factor is tan35°/tan 40° ≈ 0.834. So, this numerical method can simulate different contact types well and effectively calculate the safety factor of block.

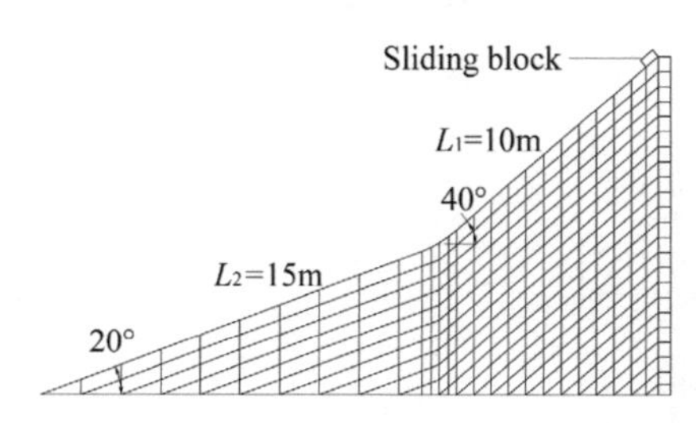

Figure 2. Sliding block model.

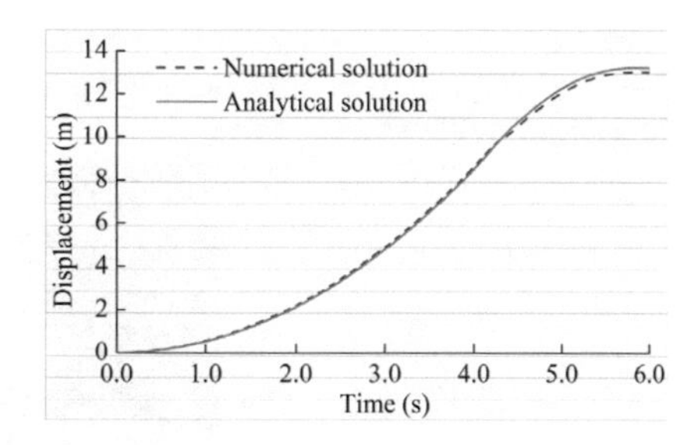

Figure 3. Displacement time-history curves of the sliding block.

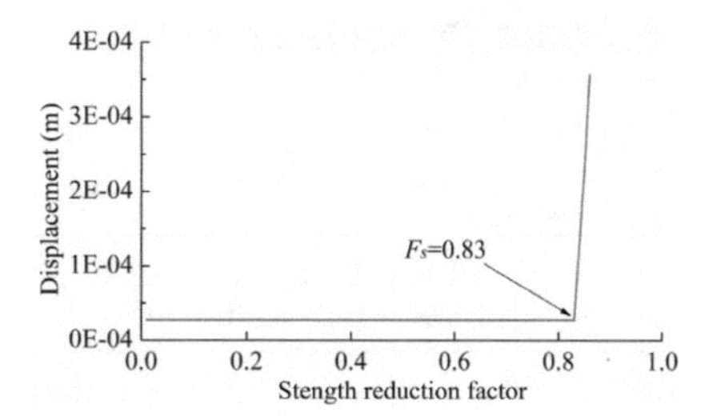

Figure 4. Displacement-reduction factor curve.

6 ENGINEERING CASE STUDY

6.1 *Calculation model and parameters*

A tunnel is crossed by a 1.5m thick fault and three large fractures. The tunnel depth is 300m. The excavation section is horseshoe shaped with a maximum width of 9.8m and a maximum height of 11.2m. The distribution of the geological planes and the calculation boundary condition are shown in Figure 5.

The tunnel Finite element model without discontinuities can be seen in Figure 6. The model size is 100m×90m (length × height), and it contains 770 elements and 1636 nodes. Then, use the element reconstruction and node separation method to build the geological planes. The final calculation model contains 992 elements and 1930 nodes. According to the block theory, four blocks are found and marked around the excavation face, which are labeled as B1, B2, B3, and B4, as shown in Figure 7.

The initial geo-stress field is obtained by stress inversion. The horizontal lateral pressure coefficient is 1.10. The maximum principle stress around the tunnel is between -8.5MPa and -9.5MPa. The mechanical parameters of the rock, fault and the interface are provided in Table 1. The materials are based on Mohr-Coulomb Criterion. The bottom of the model is fixed, and the left and right sidewall are fixed with a normal constraint. The top is a free boundary, and is applied with the gravity stress of the overlying rock mass.

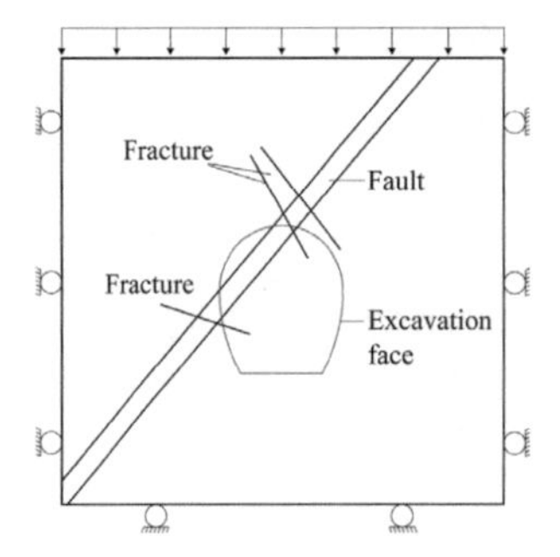

Figure 5. Distribution of geological planes and boundary conditions.

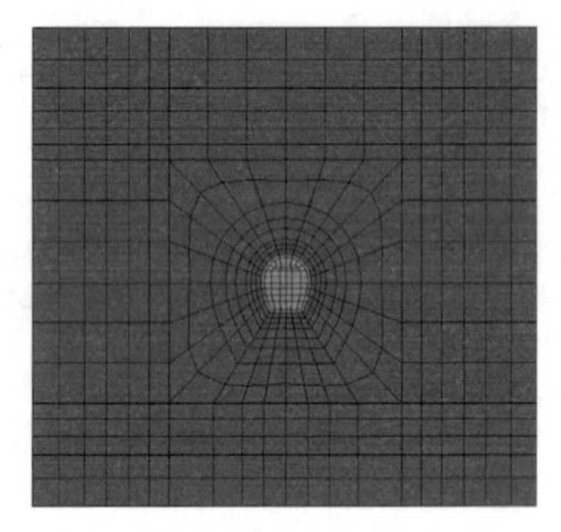

Figure 6. Initial FE model.

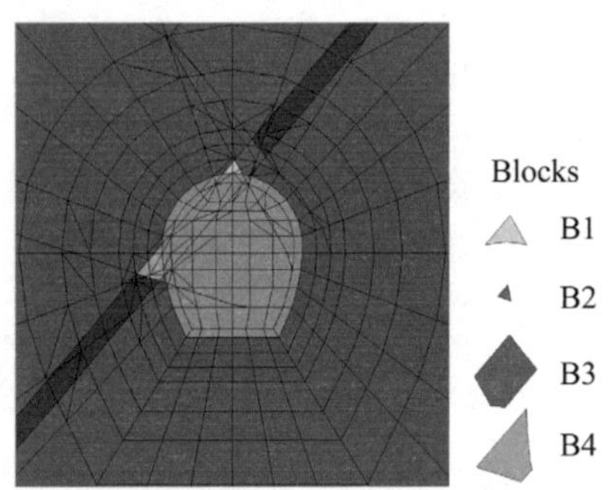

Figure 7. Reconstructed FE model and marked blocks.

Table 1. Mechanical parameters of materials

Materials	Deformation modulus (GPa)	Poisson ratio	Cohesion force (MPa)	Frictional angle (°)	Density ($g \cdot cm^{-3}$)	Tensile strength (MPa)
Rock	5.0	0.3	0.6	35.0	2.7	1.5
Fault	0.3	0.40	0.1	25.0	2.0	0.5
Interface	—	—	0.0	30.0	—	—

6.2 *Results and Analysis*

6.2.1 *Block movability and displacement of rock*

Figure 8 shows the displacement of the centroid of block. Figure 9 shows the stable and unstable block after excavation, of which the displacement is amplified by 5 times. Block B1 is falling down from the top arch of the tunnel, and its displacement and speed is continuously increased. Block B2, which has small size, slides down from the sidewall as soon as the tunnel excavation. The rock mass unloading play a major role on the failure of Block B2. The displacement of Blocks B3 and B4 reach a stable value with the support of surrounding rock. It also can be obvious seen from Figure 9 that relative movement occurs on the interface between rock and Blocks B3 and B4. As shown in Figure 10, the deformation of the rock where the fault passed is obviously large, and the deformation up and down the geological planes is discontinuous. The top arch and left sidewall have larger deformation, and the maximum value exceeds 90mm. It shows that although Blocks B3 and B4 are stable blocks, they are obviously loose and tend to slid down, which can be regarded as potentially dangerous blocks.

In the key block theory, only blocks B1 and B2 can be identified, and the displacement distribution characteristics of rock cannot be obtained. This numerical method not only can identified the key blocks, but also can show the week part and discover potentially dangerous blocks.

6.2.2 *Stress evolution of the interfaces between rock and block*

In order to analyze the stress evolution process of the interface between rock and block, the Block B3 is selected to study and three monitoring points are arranged on its three interfaces (F1, F2, and F3), as shown in Figure 9. The Figure 11 shows the normal stress and shear stress of the three interfaces. After excavation of the tunnel, the

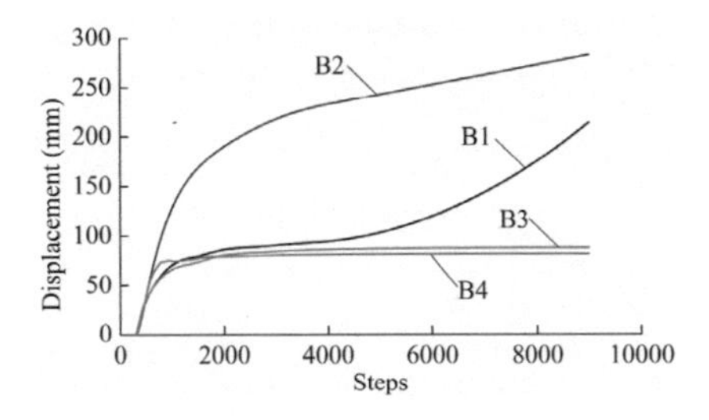

Figure 8. Monitored displacement of blocks.

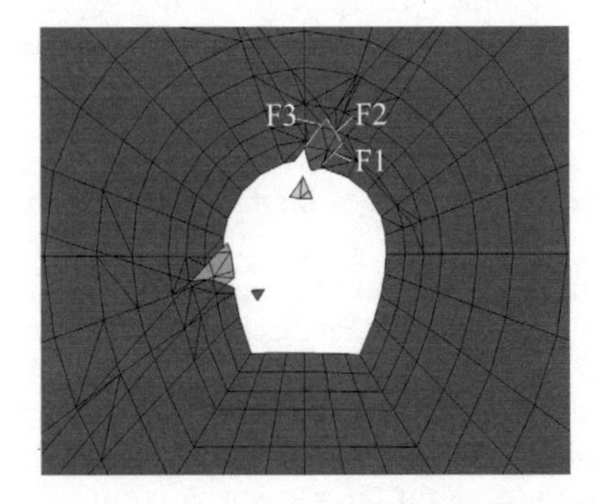

Figure 9. Unstable blocks and stable blocks (displacement amplified by 5 times).

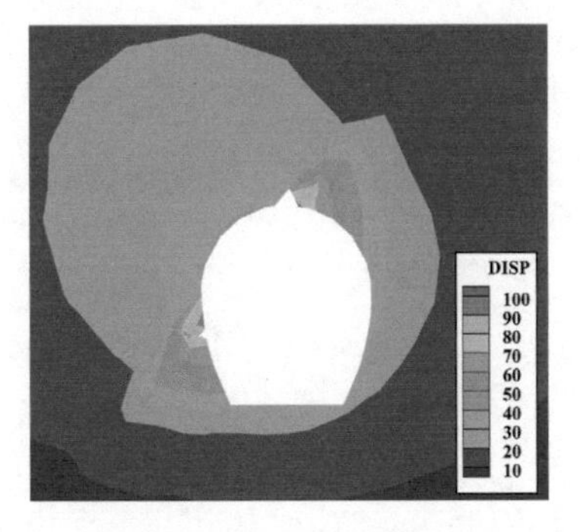

Figure 10. Deformation of rock (unit: mm).

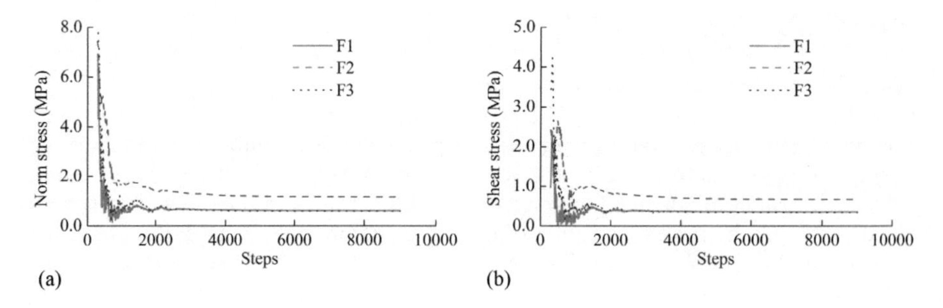

Figure 11. Stress evolution of the interfaces of Block B3 (a) Monitored normal stress (b) Monitored shear stress.

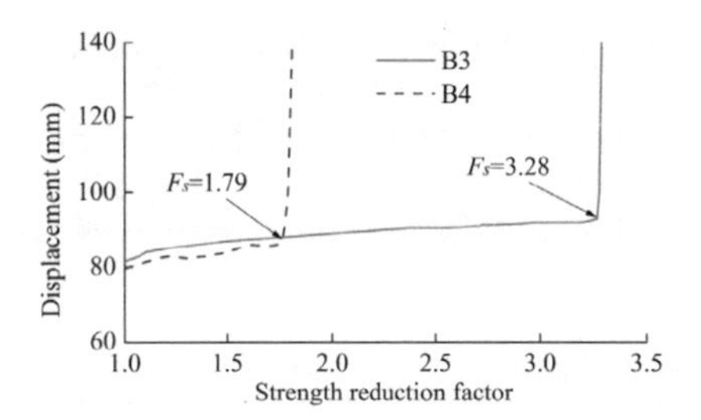

Figure 12. Safety factor of blocks B3 and B4.

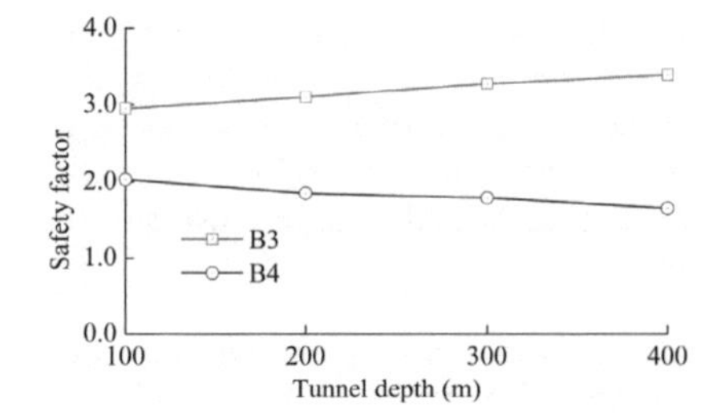

Figure 13. Safety factor under different tunnel depth.

interface stress is obviously released from its initial stress. While the deformation of the rock reach stable state, the normal stress and shear stress of the interfaces tend to be nonzero stable values, especially on the non-sliding surfaces. The stress in interface F2 is larger than that in F1 and F3.

Compared with the rigid limit equilibrium method, this numerical method better reflects the contribution of the non-sliding surfaces to the stability of blocks.

6.2.3 *Safety factor of blocks*

Through reducing the strength parameters of the interfaces, the relation curves between the displacement of blocks and strength reduction factor are obtained, as shown in Figure 12. Block B1 and B2 still have a large displacement even if the reduction face is less than 0.1. So, this two blocks fall freely after excavation. Block B1 in the top arch is mainly due to its own weight, while the Block B2 is mainly pushed down by the lateral interface stress. For block B3, the displacement mutation occurs when the strength reduction factor reaches 3.28. So, the safety factor of block B3 is 3.28. In the same way, the safety factor of block B4 is 1.79. The potential Block B3 in the top arch is relatively safer than that in the sidewall.

6.2.4 *Safety factor of blocks under different tunnel depth*

A further calculation for the safety factor of blocks B3 and B4 is made under the tunnel depth of 100m, 200m, and 400m. The safety factors of blocks under different depth are shown in Figure 13. It can be seen that as the tunnel depth increases, the safety factor of block B3 increases, while the safety factor of block B4 decreases. The deeper the tunnel, the larger the initial interface stress. The interface stress direction of block B4 points to the excavation face. So, the safety factor of B4 decreases while the tunnel depth increases. However, the interface stress of block B3 plays a positive role while the tunnel depth increases. It indicates that, the magnitude of initial geo-stress is an important factor affecting the stability of the block, and it has different effects on the blocks in different positions.

7 CONCLUSIONS

This numerical method described in this paper, first uses element reconstruction and node separation method to model multiple geological planes, then builds the contact model between rock and blocks, and the stability of blocks is studied by strength reduction method. The following conclusions are obtained:

(1) The blocks loosen, separate or slid from surrounding rock after excavation. The improved contact force method can effectively simulate the discontinuous deformation and the large sliding phenomenon of the blocks, which shows the failure process of blocks.
(2) During the excavation of the tunnel, in addition to the key blocks, there are also potentially dangerous blocks inside. This numerical method can identify the potential blocks well.
(3) The block failure is the combination effect of stress from multiple faces. The different tunnel depth has different influences on blocks in different positions. Compared to rigid limit equilibrium method, this numerical method can better consider the influences of the initial geo-stress, excavation unloading, and rock deformation, and better reflects the contribution of the non-sliding surfaces to the stability of blocks.
(4) The tunnel with multiple geological planes after excavation mainly occurs local block failure, while the surrounding rock won't have large deformation or damage. Therefore, attention should be paid more to the monitoring and analysis of complex geological conditions on site. And, a reasonable evaluation should be proposed combines with the results of the numerical method.

ACKNOWLEDGEMENTS

This study is supported by the National Key Basic Research Program (2015CB057904) of China, and the National Natural Science Foundation (51279136) of China. These supports are greatly acknowledged and appreciated.

REFERENCES

Beyabanaki, S.A.R., Jafari, A., Biabanaki, S.O.R. & Yeung, M.R. 2009. A coupling model of 3-D discontinuous deformation analysis (3-D DDA) and finite element method. *The Arabian Journal for Science and Engineering*, 34(18): 107–119.

Cai, Y., Zhu, H. & Zhuang, X. 2013. A continuous/discontinuous deformation analysis (CDDA) method based on deformable blocks for fracture modeling. *Frontiers of Structural and Civil Engineering*, 7(4): 369–378.

Goodman, R.E. & Shi, G.H. 1985. Block Theory and Its Application to Rock Engineering. *Prentice-Hall*, 26: 103–105.

Liu, G., Chen, J., Xiao, M. & Yang, Y. 2018. Dynamic Response Simulation of Lining Structure for Tunnel Portal Section under Seismic Load. *Shock and Vibration*, 2018: 1–10.

Liu, J., Li, Z. & Z, Z. 2004. Stability analysis of block in the surrounding rock mass of a large underground excavation, *Tunnelling and Underground Space Technology*, 19(1): 35–44.

Liu, J.B., Sharan, S.K., Wang, D. & Yao, L. 1995. A dynamic contact force model for contactable cracks with static and kinetic friction. *Computer Methods in Applied Mechanics and Engineering*, 1995: 287–298.

Ma, K., Xu, J., Wu, S. & Zhang, A. 2009. Research on surrounding rock stability in local collapse section of highway tunnels. *Chinese Journal of Rock and Soil Mechanics*, 10(30): 2955–2960.

Shi, G.H. 1992. Discontinuous deformation analysis: A new numerical model for the statics and dynamics of deformable block structures. *Engineering Computations*, 9(2): 157–168.

Zhang, A. & Mo, H. 2013. Improving displacement mutation criterion of slope failure in strength reduction finite element method. *Chinese Journal of Rock and Soil Mechanics*, S2(34): 332–337.

Zhang, Y., Xiao, M., Ding, X. & Wu, A. 2012. Improvement of methodology for block identification using mesh gridding technique. *Tunnelling and Underground Space Technology*, 30: 217–229.

Zhao, J., Xiao, M., Chen, J. & Li, D. 2016. Improved DDA Method Based on Explicitly Solving Contact Constraints. *Mathematical Problems in Engineering*, 2016: 1–13.

Tunnels and Underground Cities: Engineering and Innovation meet Archaeology, Architecture and Art, Volume 3: Geological and geotechnical knowledge and requirements for project implementation – Peila, Viggiani & Celestino (Eds)
 ISBN 978-0-367-46583-4

Case study of tunnel ground reaction modeling in horizontally bedded rock using continuum and fracture network models

D. Chesser, M.J. Telesnicki & J. Carvalho
Golder Associates Ltd., Mississauga, Ontario, Canada

ABSTRACT: Anisotropic, jointed rock masses can result in relatively complex conditions that are not easily approximated with numerical models using general rock mass strength parameters. In this case study, numerical models using either rock mass continua or fracture networks are examined in the context of tunnels within horizontally bedded shales with high horizontal stress. This case study demonstrates the scale and orientation effects of rock mass discontinuities on the understanding of rock mass reaction and potential failure mechanisms. A case history from the Hanlan tunnel project in Mississauga, Canada is reviewed. Continuum models for the tunnels were created using rock mass strength parameters from the Generalized Hoek-Brown Failure Envelope using laboratory testing data and Geological Strength Indices. Due to high horizontal stress and the relatively low Geological Strength Indices in the horizontally bedded rock mass, continuum models of the tunnels exhibit extensive conjugate shear failure planes through the tunnel haunches, resulting in very large zones of plasticity that are not typically observed based on local tunneling experience. Adjusting the rock mass strength parameters for a pseudo-intact rock condition and explicitly modeling fracture networks changes the model ground reaction and more accurately reflects observed behaviour.

1 INTRODUCTION

The use of continuum Finite Element Method (FEM) models to analyze tunnel excavations in bedded, jointed sedimentary rock masses in high in situ stress fields can often lead to an oversimplification and misleading results when the discontinuities are not dealt with explicitly in the models but rather are approximated by downgrading intact rock properties. To more accurately simulate rock mass failure around a tunnel excavation in bedded, jointed rock masses it is most often necessary to include the discontinuity network using joint elements rather than using a continuum model with downgraded intact rock properties to account for the discontinuities. This paper attempts to illustrate the difference between the two analytical approaches (continuum versus fracture network models) and the resulting rock mass failure mechanisms. The modelling results from the case history are then compared to observed conditions during excavation to illustrate the importance of discrete fracture modelling in bedded, jointed sedimentary rocks to accurately simulate the actual conditions.

2 BACKGROUND

A common approach, often used to model relatively simple underground excavations, such as a single tunnel or shaft, in different soil and rock units is to use a multi-material FEM model and to account for any bedding and joints by downgrading the intact rock properties using GSI (Hoek et al. 2002). One of the advantages of using a continuum model with reduced rock properties is that the model can be set-up and run very quickly which makes it relatively efficient to run multiple scenarios as sensitivity analyses. However, this approach has the disadvantage that it often does not accurately approximate the anisotropic plastic behaviour of the rock mass.

An alternative to the continuum type FEM model in bedded and jointed rock masses is to make use of a multi-material model with intact or near intact rock properties and to model the discontinuity network explicitly. Typically, in order to incorporate the behavior of discontinuities (joint networks) in FEM analysis, it is necessary to create interface (or joint) elements to represent discontinuities. The influence of joint elements on the physical behaviour is considered through constitutive laws assigned to the discontinuities. With FEM models there is no real detachment across joints when using joint elements which are limited to small displacements due to the continuum assumptions.

The first element of this kind was 'Goodman joint element' developed specifically for rock mechanics applications (Goodman et al. 1968). Early FEM models with joint elements were usually limited to a few joints due to the complex algorithms required to solve multi-joint networks. However, current commercially available FEM software packages allow for multiple joint networks with statistical distributions on the joint length, spacing, separation and joint properties. These models can be set-up relatively quickly and computation run times are now relatively fast depending on the number of joint elements in the joint network being modelled.

Therefore, in cases of relatively simple geometries and reasonable approximations of the joint networks, there is no need to make use of an isotropic plastic constitutive model to approximate a bedded and jointed rock mass when the joint network model using joint elements can more realistically simulate the anisotropic rock mass behavior due to the jointing.

3 FAILURE CRITERIA AND ROCK MASS PROPERTIES

The generalized Hoek-Brown failure criterion (Hoek et al. 2002) is widely accepted and is commonly used in numerical analysis. The Geological Strength Index or GSI (Marinos & Hoek 2000) is often used to reduce intact rock properties such as compressive strength and Young's modulus to rock mass properties in an attempt to approximate a discontinuous rock mass. The generalized Hoek-Brown failure criterion is defined using the Geological Strength Index (GSI) by:

$$\sigma_1' = \sigma_3' + \sigma_{ci}\left(m_b\frac{\sigma_3'}{\sigma_{ci}} + s\right)^a \tag{1}$$

Where σ_{ci} is the intact uniaxial compressive strength and m_b is a reduced value of the material constant m_i and is given by:

$$m_b = m_i exp\left(\frac{GSI - 100}{28 - 14D}\right) \tag{2}$$

Where *s* and *a* are constants for the rock mass given by the following:

$$s = exp\left(\frac{GSI - 100}{9 - 3D}\right) \tag{3}$$

$$a = \frac{1}{2} + \frac{1}{6}\left(e^{-GSI/15} - e^{-20/3}\right) \tag{4}$$

The parameter D is a disturbance factor which varies from 0 for undisturbed to 1 for very disturbed depending on the degree to which the rock has been subjected to blast damage and stress relaxation. The rock mass strength and elastic modulus can be approximated using the following equations. The rock mass strength (Hoek et al. 2002) is given by:

$$\sigma_{crm} = \sigma_{ci} \cdot s^{a} \tag{5}$$

And the rock mass modulus (Hoek & Diederichs 2006) is given by:

$$E_{rm} = E_i\left(0.02 + \frac{1 - D/2}{1 + e^{((60+15D-GSI)/11)}}\right) \tag{6}$$

4 MODELING METHOD AND PARAMETERS

In this case study, numerical models of the tunnels and shafts were generated using *RS2*©, a 2D elasto-plastic finite element stress analysis program. It should be noted that the shale considered in the following case history is known to swell resulting in time-dependent deformation, which is not discussed in this paper.

The pressure replacement and ground reaction curve (GRC) methods were used to estimate the amount of wall displacement that would be expected to occur behind the excavation face. An internal pressure was applied to the excavation in the model and reduced in several stages, from an initial value equal to the applied in-situ stress down to zero. The reduction in the applied pressure was used to determine the pressure corresponding to the amount of tunnel wall deformation at the time of support installation and to determine the amount of deformation that would occur after the support is installed, as well as the loading on the support elements. The final stage, modelled with an internal pressure of zero, is used to determine the extent of the failed rock (or plastic zone) that develops around the excavation.

4.1 *Continuum model parameters*

The main parameters required for continuum model include:

- excavation geometry and material boundaries for the various stratigraphic units;
- material properties for each stratigraphic unit, including Uniaxial Compressive Strength of the intact rock (UCS), intact rock constant mi, disturbance factor D, Poisson's ratio and Young's Modulus E for the intact rock. The residual properties of the rock post-failure (i.e. plastic) must also be defined – in this case all properties remained the same except D was increased from 0 to 0.1;
- GSI based on borehole data; and
- field stresses (K ratio of horizontal to vertical principal stress).

4.2 *Fracture network model parameters*

Input parameters for the fracture network models are similar to the continuum model with the exception of the following:

- joint elements are defined by orientation, spacing, and continuity to simulate the joint system within the rock mass;
- GSI values are required for the "intact" blocks of rock between the discontinuities in the model. It may be necessary to downgrade the intact rock properties somewhat to account for any minor discontinuities not included explicitly in the model; and
- joint shear strength characteristics (typically estimated from laboratory direct shear testing or from joint condition observations).

5 CASE STUDY – THE HANLAN FEEDERMAIN TUNNEL

The Hanlan Feedermain Tunnel Project in Mississauga, Ontario included a 5.8 km tunnel excavated using a main-beam, open face Tunnel Boring Machine (TBM) within Georgian Bay Formation bedrock, which is an upper Ordovician shale with minor interbeds of limestone and siltstone. The excavated diameter of the tunnel was approximately 3.7 m.

The deformation within the rock mass ahead of the excavation was approximated using pressure replacement over a number of analytical stages in the FEM model, from which a ground reaction curve was developed. Comparison of the support installation timing against the ground reaction curve provided the stress conditions in the rock mass at the time of support installation, which was used in a second model which examined the loading on the ground support.

Due to the varied geometry and rock mass quality conditions, models of several shaft and tunnel scenarios were analyzed.

Model input parameters were primarily based on data provided in the Geotechnical Baseline Report (GBR) for the project with some assumptions made, as follows:

- Uniaxial Compressive Strength (UCS) of intact rock = 13.8 MPa (measured)
- E_{ir}-Modulus of intact rock = 2,400 MPa (measured)
- E_{iw}-Modulus of intact weathered rock = 1,800 MPa (assumed)
- Intact rock constant m_i = 6 (assumed)
- Disturbance factor D = 0 for intact rock, D=0.1 for damaged rock
- Density of shale bedrock ρ_{shale} = 2.60 g/cm^3 (measured)
- Unit weight of overburden (clay) $\gamma_{overburden}$ = 21.2 kN/m^3 (measured)
- Bedding plane spacing = approx. 0.3 m average (estimated from borehole logs)
- Bedding plane conditions = planar, rough, with minor coating; friction 30° and zero cohesion (estimated from borehole logs)
- Vertical joint spacing = 2.5 to 5.0 m average (assumed from local experience)
- Field stress around the tunnel modeled with in-situ stress ratio K (σ_H/σ_V) = 6.5 (measured using United States Bureau of Mines overcore testing). Field stress for the shaft at a depth of 25 m was modeled with a horizontal major stress of 4.5 MPa and a minor stress of 3.2 MPa (measured) and a vertical stress of 0.7 MPa equal to the lithostatic pressure.
- Geological Strength Index (GSI) for the continuum models was estimated from the Rock Mass Rating (RMR) (Bieniawski 1979) using rock mass data provided in the GBR, as shown in Table 1.

The properties of the rock blocks inside the fracture network were assumed to have a GSI value of 85 for good rock. Installation of tunnel support was assumed at one tunnel radius (1.84 m) behind the excavation face, for the purposes of this study. Installation of shaft support was assumed at 1 m above the shaft bottom.

Table 1. Summary of Rock Mass Rating (RMR_{76}) Parameter Values for Continuum Models.

Rock Class	UCS [MPa]	Rating	RQD [%]	Rating	Spacing* [m]	Rating	Joint Condition	Rating	Water	Rating	GSI
Poor Rock	10	1	30	5	0.05	5	Planar, rough, coating	15	Dry	10	36
Good Rock	13.8	2	90	18	0.6	20	Planar, rough, coating	15	Dry	10	65
Shafts	13.8	2	80	16	2.5	28	Planar, rough, coating	15	Dry	10	71

* Joint spacing for the tunnel was based on the bedding joint spacing whereas the joint spacing for the vertical shafts was based on the near vertical joints.

5.1 *Tunnel models*

5.1.1 *Continuum model results*

The continuum model dimensions for the tunnel, at an obvert depth of 12 m below ground surface, are illustrated in Figure 1.

The results of the ground reaction model are shown at the stage of support installation in Figure 2. The area of failed rock, referred to as the plastic zone, is defined by the yielded model elements shown in the figure. The plastic zone shows a classic butterfly shape that occurs in high differential stress scenarios.

Using the results of the GRC model, the remaining ground pressure at the time of support installation is used in a new model with support installed. The typical temporary support for this project comprised steel ribs with an I-beam section (section depth 106 mm and weight 19.3 kg/m) on a 1.5 m spacing with timber lagging. Results of this model are shown in Figure 3 with the axial force distribution on the rib.

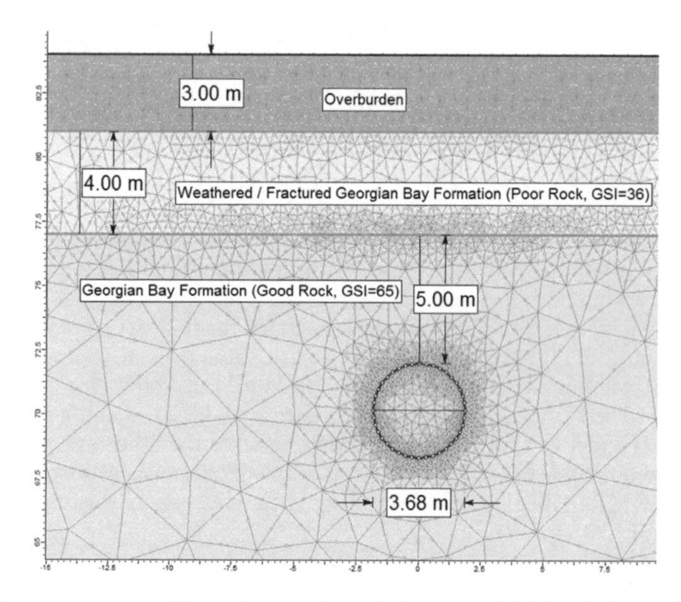

Figure 1. Continuum tunnel model dimensions. Tunnel obvert depth is 12 m below ground surface.

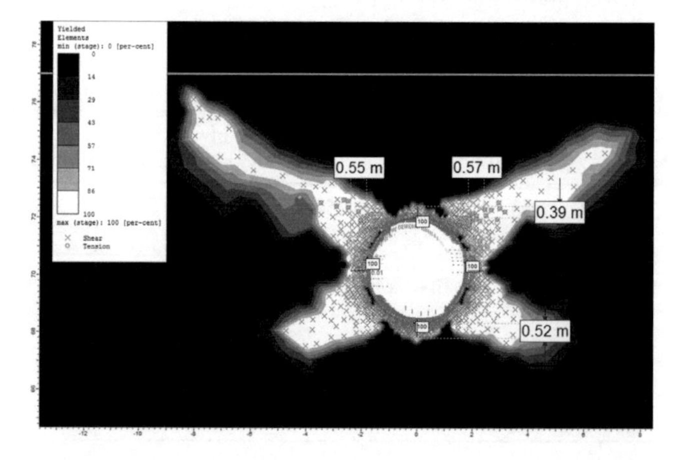

Figure 2. Model results at the support installation stage showing yielded elements. Dimensions of plastic zone are shown (shown above the tunnel: springline left and springline right; to the right: obvert and invert).

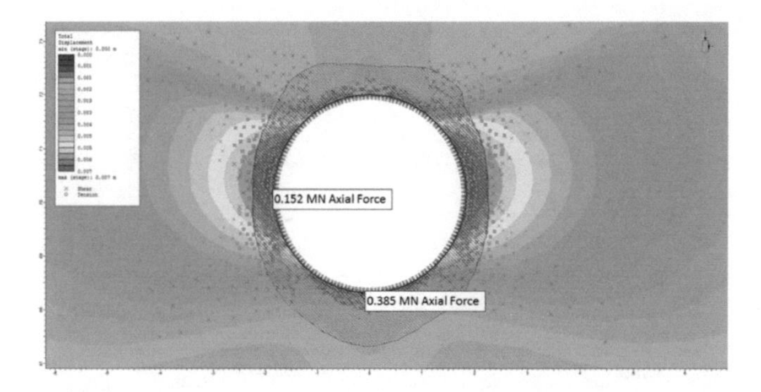

Figure 3. Results of the continuum model with support installed. Axial force acting on the tunnel lining is shown in the dark blue hatch, with minimum and maximum values displayed at their locations.

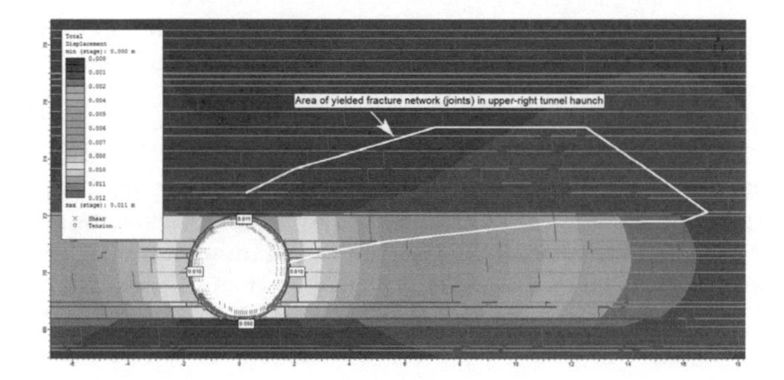

Figure 4. Results of the fracture network model showing displacement in the coloured contours, and yielded joints in the fracture network as red lines.

5.1.2 *Fracture network model results*

Modeling using a fracture network (i.e. joint elements) to account for the discontinuities in the rock mass was also carried out. For the "Good Rock" mass, a GSI of 85 was used for the blocks between the fracture network discontinuities in order to account for some bias in the UCS results since weaker shale samples are typically difficult to prepare for testing without breakage and also to account for less persistent, minor joints not represented in the fracture network. The model properties were otherwise the same as the continuum model; both used the same stress gradient.

The model geometry is shown in Figure 4. Instead of yielded elements within the rock mass, the joints elements of the fracture network have yielded and are accompanied by much fewer yielded rock mass elements localized around the tunnel excavation. As expected, the greatest extent of the area of yielded joints is along bedding planes located at the top and bottom of the tunnel opening. The displacement along the joint elements reduces the stresses transferred across the joints resulting less stress concentration in the crown and invert of the tunnel. Results of the model with the steel rib support installed are shown in Figure 5 with the axial force distribution on the rib.

5.2 *Shaft models*

Models for the shaft were 7.3 m diameter, excavated to a depth of 25.2 m below ground surface.

5.2.1 *Continuum results*

Similar to the tunnel, a GRC model was used to estimate loads on temporary support due to excavation convergence. The plastic zone for this model was up to 2.3 m thick. Welded wire mesh pinned with 3 m long fully grouted 22 mm diameter rock bolts on a 1.5 m spacing was

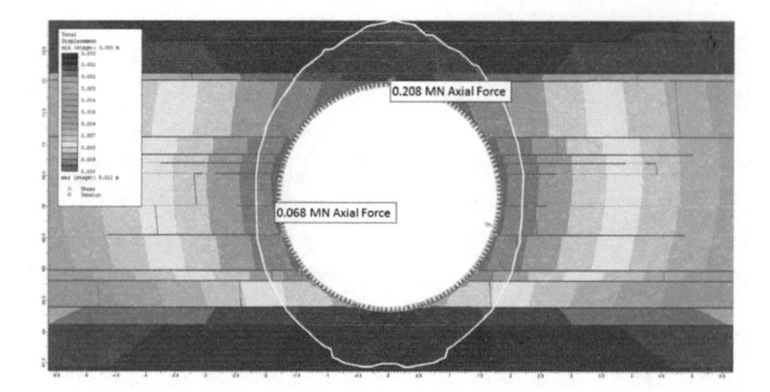

Figure 5. Results of the fracture network model with support installed. Axial force acting on the tunnel lining is shown in the white hatch, with minimum and maximum values displayed at their locations.

used for temporary shaft support (installed within 1 m of the shaft bottom) and the results of this support model are shown in Figure 6, with axial force plotted along one of the bolts. Note the axial force in the bolt reaches 0.19 MN, in comparison to the bolt capacity of 0.2 MN.

5.2.2 *Fracture network results*

The GSI of the rock mass was increased to 85 and a fracture network added to the model geometry for the shaft with the results showing much smaller plastic zones concentrated around the intersections of joints at oblique angles to the shaft walls, as shown in Figure 7. Axial force within the bolts is generally lower than the continuum model, typically up to 0.05 MN, however the worst case bolt is subject to 0.16 MN when traversing the corner of one of the smaller blocks that experienced plastic failure.

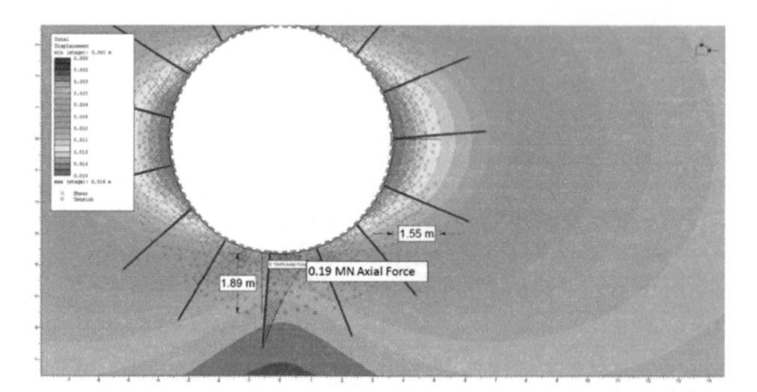

Figure 6. Results of the continuum model for the shaft with support installed. Axial force acting on a rock bolt is shown in the dark blue hatch.

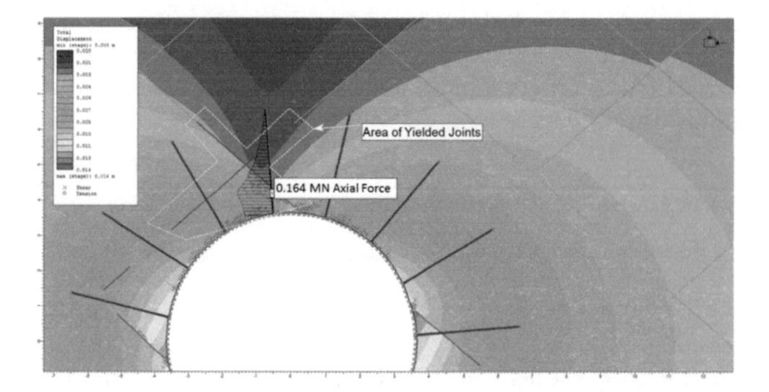

Figure 7. Results of the fracture network shaft model showing displacement in the coloured contours, and yielded joints as red lines. Axial force on the worst-case bolt is shown as black hatch.

5.3 *Discussion of continuum and fracture network results*

The resulting loads on the temporary tunnel support in the models are summarized in Figure 8 and for the bolts in the shaft in Figure 9. These figures are illustrative of the potential impact of the model method on the loads used in the structural design of the temporary excavation support. These differences will likely become more pronounced with larger excavations where support may be installed closer to the face, making it important to model the failure mechanism properly.

5.4 *Observed rock mass conditions*

Site visits to the main tunnel leading from the shaft were carried out in July 2014 and at that time 12 m of the tunnel had been excavated. Tunnel support comprised steel rib rings fabricated from steel W4×13 I-beams on a 1.5 m spacing with staggered hardwood lagging. While the rock crown in the tunnel was mostly obscured by the lagging, visible areas appeared to show some delamination along bedding planes but little or no failure of intact rock. Photographs from inspections are shown in Figures 10 and 11. The figures show the delamination of thin shale beds from the tunnel obvert but otherwise relatively intact bedrock surrounding the

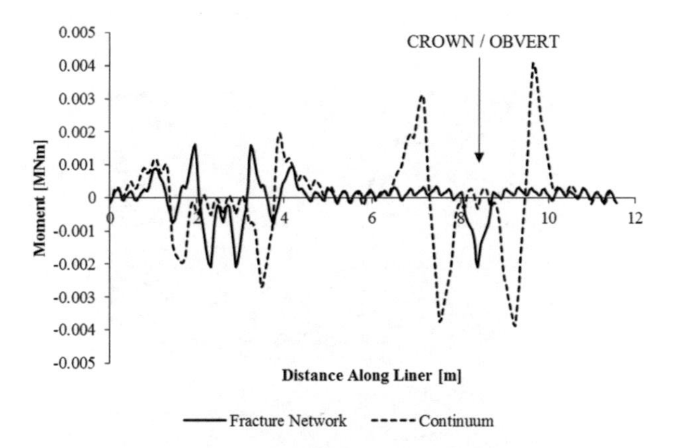

Figure 8. Comparison of bending moment generated on the steel I-beam rib between the fracture network and continuum models for the Hanlan tunnel.

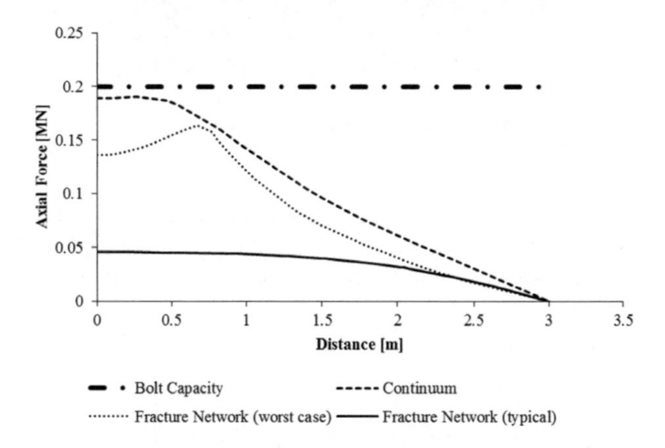

Figure 9. Comparison of axial force along rock bolts installed in the shaft between the continuum and fracture network models.

Figure 10. View from within the shaft of the brow of the starter tunnel (photograph taken before the TBM started excavating – the starter tunnel was excavated using a roadheader).

Figure 11. View of the sidewall of the starter tunnel near the shaft.

tunnel eye, consistent with the bedrock failure patterns shown in the fracture network model. It is reasonable to assume the rock mass at the shaft-tunnel intersection would incur more failure in comparison to mid-tunnel conditions and therefore any failure of the rock mass would be evident in this area.

6 CONCLUSIONS

Observations of actual tunnel conditions within the shale bedrock show that the primary mode of rock mass failure is shearing and subsequent delamination of the bedding layers in the crown, consistent with the fracture network model. No evidence was observed to indicate failed rock in excess of 2 m thick around shafts as shown in the continuum models. While failure of the tunnel haunches in a "butterfly" pattern as shown in the continuum model is possible under certain conditions with high horizontal stress, observed behaviour does not suggest that this is the case for the scenarios examined in this study. The fracture network model is considered a better analytical method for estimating the anticipated tunnel failure mode and lining loads.

ACKNOWLEDGEMENTS

The authors would like to thank Bill Turner and the Regional Municipality of Peel for allowing us to present data from the Hanlan Water Project, and to Gabriele Mellies for her work on the project.

REFERENCES

Bieniawski, Z.T. 1979. The Geomechanics Classification in Rock Engineering Applications, Proc. 4th International Congress for Rock Mechanics, ISRM, Montreux, vol. 2, pp. 41–48.

Goodman, R.E., Taylor, R.L. & Brekke, T.L. 1968. A Model for the Mechanics of Jointed Rock. Journal of the Soil Mechanics and Foundation Division, Vol. 94, Issue 3, pp. 637–660

Hoek, E., Carranza-Torres, C. & Corkum, B. 2002. Hoek-Brown Failure Criterion – 2002 Edition. 5th North American Rock Mechanics Symposium and 17th Tunneling Association of Canada Conference: NARMS-TAC, pp. 267–271

Hoek, E. & Diederichs, M.S. 2006. Empirical Estimation of Rock Mass Modulus. International Journal of Rock Mechanics and Mining Sciences, 43, 203–215

Marinos, P. & Hoek, E. 2000. GSI – A Geologically Friendly Tool for Rock Mass Strength Estimation, Proceedings GeoEng2000 Conference, Melbourne, pp. 1422–1442

Tunnels and Underground Cities: Engineering and Innovation meet Archaeology, Architecture and Art, Volume 3: Geological and geotechnical knowledge and requirements for project implementation – Peila, Viggiani & Celestino (Eds)

Probabilistic Fault Displacement Hazard Analysis (PFDHA) of a Bhairabi tunnel in North East India

R.L. Chhangte
Research Scholar, Department of Civil Engineering, NIT Silchar, Assam, India

T. Rahman
Associate Professor, Department of Civil Engineering, NIT Silchar, Assam, India

A.I. Laskar
Professor, Department of Civil Engineering, NIT Silchar, Assam, India

ABSTRACT: In this study, we have estimated Seismic Hazard Analysis of the Bhairabi Tunnel based on Probabilistic Fault Displacement Approach. We have proposed a new methodology for estimating site specific seismic hazard using Probabilistic Analysis of Fault Displacement method. The methodology for this approach is based on the Probabilistic Seismic Hazard Analysis (PSHA) with the ground motion prediction equation replaced by the Fault Displacement Prediction equation. The methodology for Probabilistic Fault Displacement Hazard Analysis (PFDHA) was developed for a strike-slip faulting environment and the probability distributions used in this study is based on the regional site specific tectonic set up and all others earthquakes related parameters. We have prepared a Fault Map around the Bhairabi Tunnel and based on this tectonic set up, we have estimated ground shaking parameters at the Bhairabi Tunnel located in the North East India (NEI) using PFDHA. The estimated ground shaking parameters will be very much useful in design the engineering structures like the Bairabi Tunnel.

1 INTRODUCTION

Bhairabi Tunnel is located is situated in the state of Mizoram in the North-Eastern part of India (NEI) which is shown in Figure 1. It is the eastern most region of the country and connected to the mainland India only through a small landmark. NEI has been assigned under seismic zone five, which is known to be the most severe seismic zone in India. NEI has covers the areas with the highest risks zone where several devastating earthquakes occurred in the past (Tiwari et al, 2015).

However, contrasting to this high seismic risk in NEI, there have been insufficient earth-shaking measuring device network and no proper measure have been taken to mitigate or reduce the effects of ground shaking hazard due to earthquake in NEI, particularly in the state of Mizoram in India. Hence it is of utmost importance to analyze the possible seismic hazard which can be cause by the future earthquake. For this reason, in this paper, the fault displacement hazard analysis of Bhairabi Tunnel in Mizoram have been undertaken by us.

Probabilistic fault displacement hazard analysis (PFDHA) is a relatively new methodology which is derived from Probabilistic Seismic Hazard Analysis (PSHA). PFDHA is used to determine the occurrence probability of surface fault displacement for a specific region based on the nearby fault parameters (Coppersmith & Youngs, 2000). This methodology is useful for designs of new structures and for analysis of existing structures where an active faults lie in close proximity. PFDHA result may be interpreted as a probability of exceedance of fault displacement at a specified return period instead of a probability of exceedance of ground shaking as in PSHA (Wells & Kulkarni, 2014).

There has been many application of PFDHA for various types of facilities, especially underground structures e.g. pipelines, train tunnels, water and sewer tunnel as well as critical facilities e.g. dams, nuclear power plants. One of the most interesting application of PFDHA was performed in the seismic hazard analysis of the Yucca Mountain Nevada, USA where nuclear fuel and high toxic waste are deposited. This study is still considered as one of the most complex and comprehensive analyses of seismic ground shaking hazard and the first PFDHA carried out (Stepp et al, 2001).

In general, all the procedure followed in PFDHA is more or less similar to the PSHA methodology. Technique followed in PFDHA for different types of fault slip (strike-slip, normal or reverse slip) have been studied and published by various researchers in the field of seismology and engineering structures design. The basic framework of PFDHA are (1) the determination of earthquake magnitude and rate of occurrence of earthquakes on a faults for a specific region or site, (2) the location of rupture along the fault, and (3) the expected amount and distribution of fault displacement at the fault and its vicinity, (4) to calculated the probability of exceedance of specified displacement (Wells & Kulkarni, 2014).

As mentioned earlier, PFDHA was first developed by Stepp et al (2001) for normal faults encountered at the proposed nuclear waste disposal at Yucca Mountain, Nevada, USA. The same fault displacement approach was later developed by Braun (2000) for Wasatch fault in Central Utah, USA. Petersen et al (2011) and Moss and Ross (2011) extended the probabilistic fault displacement approach for different types of fault such as strike-slip and reverse faults respectively in California, USA. Recently, Shantz (2013) compiled and formulated a simplified procedure of PFDHA developed by the California Department of Transportation (Caltrans) to determine fault displacement on strike-slip faults in California.

There are two fundamental approaches for analysis of fault displacement. The first one is called the earthquake approach. It follows the basic PSHA framework for ground shaking hazard and the probability of occurrence of displacement at a region or site at or near the ground surface especially to the occurrence of earthquakes (fault slip at depth) in the specific region. The second approach is called the displacement approach. This approach utilizes the characteristics of fault displacement observed at the site of interest to quantify the hazard without determining a specific mechanism for the cause of earthquake in the region (Youngs et al, 2003).

Tunnels built in earthquake-prone areas must be designed to withstand both seismic and static loading. The tunnel may have to be constructed over or in the vicinity of a fault or fault zone as it is not always possible to avoid engineering structures construction over the active faults. In such situations, earthquake-triggered fault movement may subject the tunnel to structural displacements and generate stress concentration, which can result in the partial or complete failure of the structures (Wang et al, 2012). Therefore, fault displacement analysis for an underground structure in an earthquake prone areas is of utmost importance. Hence, in this study, the fault displacement hazard analysis based on earthquake approach is carried out to analyse the seismic hazard of Bhairabi tunnel and its surrounding region.

2 SEISMICITY AND GEOLOGICAL STATUS OF NEI

NEI is very highly active seismic region and is considered to be one of the six most active region in the world. The other five are Mexico, Taiwan, California, Japan and Turkey (Tiwari, 2002). NEI is located in the seismic zone V in India. As per Indian Standard code of practice IS 1893: 2016, seismic zone V is considered to be the most severe seismic zone in India. The seismotectonic map of NEI is shown in Figure 1. The region include the Eastern Himalaya, Arakan-Yoma Belt including the folded belt of Tripura, Irrawaddy basin, Shillong Plateau where Dauki fault is located and the northern Bengal basin (Kayal, 1996; Nandy, 2001). The anomalous gravity and seismicity are assumed to be caused by large scale tectonic movements that have taken place in the area mostly during the Cretaceous and Cenozoic

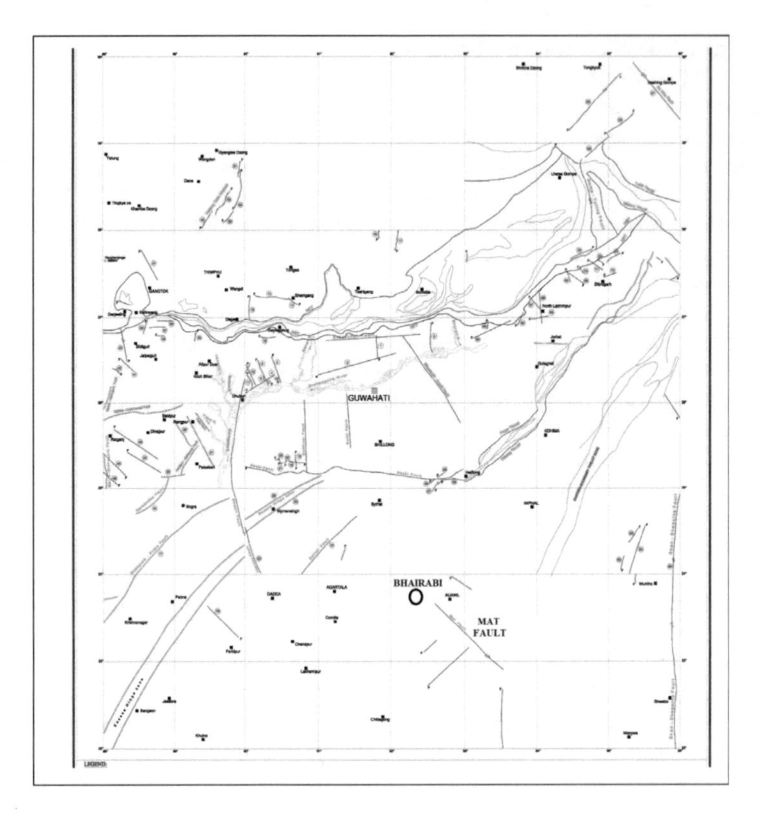

Figure 1. Location of Bhairabi and Mat fault from NEI fault map.

times, due to interplate movements of the Indian, Tibetan, and Burmese plates. The high seismic activity in NEI indicates that the plate's movement are continuing (Ni et al, 1989, Verma et al, 1976).

Bhairabi is located in the state of Mizoram, one of states of India. Mat fault is the most prominent fault in Mizoram (Jaishi et al, 2013 & 2014) and the data for other fault (such as Kaladana fault, Churachandpur-Mao fault, etc.) in the proximity of Bhairabi tunnel are scarce. For this reason, Mat fault is selected for fault displacement hazard analysis of Bhairabi tunnel and its surrounding region. The Mat fault is located on the central part of the outer wedge of the Indo-Burmese Region (IBR) (Tiwari et al, 2015). IBR is a part of Surma Basin and this region is an area of folded sediments and is characterized by westerly convex, structural ridges and valley. There are many trending NE-SW and NW-SE faults located within this basin. Bengal basin demarcates the north-western boundary of the Surma basins. There are many faults in NEI at the southern part of Surma basin (Mizoram) which have the strike-slip displacement of the folded axes along them (Tiwari, 2002).

The seismicity in the Surma Basin is quite complex and a simple seismic analysis may lead to inaccurate results. Earthquakes in this region are confined at shallow to intermediate depth level and majority of them occur at depth greater than 70 km (Tiwari et al, 2015). This means that the region's earthquakes comprise of shallow interface subduction earthquake as well as the intermediate depth intraslab subduction earthquake. The location of the nodal planes of these earthquakes is consistent with the geological fabric of the Indian plate (Kundu and Gahalaut, 2012).

The region of Mat fault is not highly active in terms of seismicity as compare to the other prominent fault in NEI such as Dauki fault. For the past 50 years no earthquake of more than moment magnitude 5.5 has occurred within 100 km proximity of Mat fault (Sailo et al, 2011). However, an earthquake of moment magnitude 6.1 has occurred in southern part of

Mizoram on November 21, 1997. The epicentre for this earthquake is recorded to be at a distance more than 100 km south-west from Mat fault. As discussed earlier, all these earthquakes are of intra-slab type, as they occurred within the Indian Plate (Kundu and Gahalaut, 2012). Even other smaller fault in Mizoram, such as Kaladan fault located in the western part of Mizoram does not exhibit high seismic activity. Till date, it is uncertain whether Mat fault is responsible for smaller magnitude earthquake occurring in the state of Mizoram. Unfortunately, there is no local earthquake measuring network in Mizoram to confirm this fact (Tiwari et al, 2015).

3 PROCEDURE FOR FAULT DISPLACEMENT HAZARD ANALYSIS

In this study, we have used a simplified procedure of PFDHA which has been formulated by the California Department of Transportation (Caltrans). In this method, a single fault, called as the principal fault has been selected for the analysis. The selected fault is considered to have maximum effects on the nearby site and structure. In our case, we have selected Bhairabi tunnel as a case study to estimate the fault displacement seismic hazard. The reason for selecting Mat fault as principal fault for the analysis is because it is nearest and largest fault in the proximity of Bhairabi tunnel and may cause severe seismic hazard at the tunnel site particularly during fault-triggered earthquake.

Different fault parameters of Mat fault and its seismic activity are presented in Table 2. The simplified procedure of PFDHA by Caltrans has been initially developed for strike-slip faults in the state of California, USA for the faults having a high slip rate (Li et al, 2015). This confirm that Caltrans procedure is suitable for our analysis as Mat fault has high slip rate (29-36 mm/year). In this study, we have used the Caltrans procedure, maximum moment magnitude scaling relations and the average fault displacement prediction equation for obtaining the seismic hazard curve of PFDHA.

Various steps involved in our PFDHA study are presented below:

In this method, we have estimated the characteristic earthquake magnitude (preferably moment magnitude, M_w) for the fault of interest using known or estimated fault parameters such as fault length and area. Earthquake magnitudes are determined using empirical scaling relations of earthquake magnitude.

Recurrence interval of the earthquake is estimated using the equation shown below:

$$T_r = (M_o/0.8m_o) \tag{1}$$

Where m_o is the moment rate and M_o is the seismic moment. The moment rate is determined using the equation $m_o=\mu AS$, where μ is the crustal rigidity, A is the effective fault area, and S is the fault slip rate. The seismic moment is estimated using the relation developed by Hanks and Kanamori (1979).

We have also estimated the lognormal distribution of the average surface rupture displacement value for the characteristic earthquake using an empirical average fault displacement prediction equation.

We can calculated the fault rupture hazard curve using the probability equation proposed by Youngs et al, (2000) for PFDHA earthquake approaches:

$$\nu_k(d) = \sum \alpha_n(m^o) \int_{m^o}^{m} f_n(m)[\int_0^{\infty} f_{kn}(r|m).P_{kn}{}^*(D>d|m,r).dr].dm \tag{2}$$

where $\alpha_n(m^o)$ is the rate of all earthquakes on source n above a minimum magnitude of structural engineering significance, $f_n(m)$ is the probability density of earthquake size between minimum magnitude (m^o) and the maximum magnitude (m^u) that source n can produced. f_{kn} $(r|m)$ is the conditional probability density function of distance from site k to an earthquake of magnitude m occurring on the source n. $P_{kn}{}^*(D>d|m,r)$ is fault displacement's 'attenuation function' at or near the ground surface. The conditional probability is split up into two components as shown below:

$$P_{kn}^{*}(D>d|m,r) = P_{kn}(slip|m,r).P_{kn}(D>d|m,r,slip) \quad (3)$$

where P_{kn}(slip| m,r) is the conditional probability that some amount of displacements occurs at site k as a result of an earthquake on source n of magnitude m with rupture at a distance r from the site. P_{kn} (D>d| m,r,slip) defines the conditional probability of the amount of fault displacement given that fault slip occurs.

Scaling relation of maximum moment magnitude is the basic concept which is needed in fault displacement hazard analysis. Sometime, there is a requirement to estimate the earthquake magnitude for a specific fault. In such case, a scaling relation is used to estimate the earthquake magnitude from fault parameters e.g. fault length, width and area. Different empirical earthquake scaling relations have been developed in different part of the world based on the data collected from each specific site and region, in the form of varying regression equations. Caltrans procedure used equations developed by Hanks and Bakun (2008) scaling relations. Hanks and Bakun (2008) relations is most suitable for strike-slip type fault located on a fast moving plate boundaries. Another scaling relations for determining the earthquake magnitude of strike-slip fault had been developed by Mark Leonard (2010). Leonard propose scaling relations to determine the earthquake magnitude for all type of faults from seismic moment, rupture area, length, width and average displacement which predict linear log-log relationships among the various parameters. Leonard (2014) updated his previous scaling relations to determine the earthquake magnitude for more complex Stable Continental Region (SCR) strike-slip faults.

The fundamental concept of fault displacement hazard analysis is to establish prediction equation for average fault displacement, which represents the displacement of ground surface lying above a fault, which ruptured under earthquakes with different range of magnitudes. For this purpose, historical or previous rupture data of the specific fault are to be collect along with the fault parameters. From this data, empirical relations will be developed for the specific fault relating fault displacement and earthquake magnitude. For the prediction equation of average fault displacement, two equations are considered: a) Wells and Coppersmith (1994) and b) Hecker et al (2013). Both of the mentioned equations are suitable for all types of fault.

Wells and Coppersmith (1994) compiled a sources parameters for historical earthquakes worldwide for developing an empirical relationships among various faults parameters. They compiled the earthquake data to develop a series of empirical relationships between earthquake magnitude, surface and subsurface rupture length and width, rupture area and average and maximum displacement for each earthquake events. They have brought up the following conclusion from their study: (1) the length of surface rupture is generally about 75% of the subsurface rupture length, (2) the average surface displacement for each event is about one-half of the maximum surface displacement, (3) the average subsurface displacement on the fault plane is less than the maximum surface displacement but more than the average surface displacement.

Hecker et al (2013) investigated the nature of earthquake-magnitude distributions on faults. They compared the interevent variability of surface displacement at a point on a fault from composite global data set of paleoseismic observations using truncated exponential and characteristic earthquake model. The characteristic earthquake model produces coefficient variation (CV) values consistent with the data (CV~0.5) only if the variability for a given earthquake is small, which indicate that rupture patterns on a fault are stable. The truncated exponential distribution model give higher values of CV than their respective value obtained from empirical constraint.

4 GROUND MOTION DATA

The ground motion (earthquake) data for this study is taken for a 100 km proximity of Bhairabi tunnel in all direction as the distance beyond 100 km will not contribute

Table 1. Ground motion data for Bhairabi tunnel and its surrounding region.

Date	Location (N, E)	Magnitude (M_w)
2016-01-03	24.804, 93.650	6.7
1997-05-08	24.894, 92.250	6.0
1984-12-30	24.641, 92.891	6.0
1984-05-06	24.257, 93.545	6.0
1958-03-22	23.515, 93.842	6.0
1957-07-01	24.310, 93.894	6.2
1950-08-15	25.343, 92.944	6.0

much to the ground shaking hazard (Wong et al, 1996) for a specific fault. The earthquake data from range of moment magnitude 6.0 – 8.0 is taken, following the pattern of Petersen et al, 2011 for the strike slip fault hazard analysis. An earthquake of moment magnitude 6.3 – 7.9 is taken for the fault displacement hazard by Petersen et al (2011). The recorded earthquake data available for the site-specific Bhairabi tunnel are very few for thorough analysis and development of PFDHA. In total, there are 7 strong motion records available which are presented in Table 1. Most of these strong motion records are available in Cosmos Virtual Data Center, International Seismological Center, Raghukanth and Somala (2009) and Bora et al (2014). These strong motion database are also crossed checked from the records in National Earthquake Information Center (NEIC) maintains by US Geological Survey (https://www.usgs.gov).

5 FAULT PARAMETERS OF MAT FAULT

It is to be highlighted here that the Mat fault which is very nearer to the proposed project site Bairabi Tunnel. The various parameters of Mat Fault are obtained from different sources such as the work done by B.P Tiwari et al (2015), H.P Jaishi et al (2013) and H.P Jaishi et al (2014). The empirical equations from Wells and Coppersmith (1994) and Hecker et al (2013) are also used for the average fault displacement prediction. Maximum Moment Magnitude Scaling Relations by Hanks and Bakun (2008) and Leonard (2010) are used to determine the characteristic moment magnitude of Mat fault. The data obtained for fault parameters are quite contrasting to each other in terms of their numerical values. Hence, for this reason, we will take the normalized value as well as the upper and lower limit value of the fault parameters for our hazard analysis. The fault parameters are as given below:

Table 2. Mat fault parameters.

Parameters	Normal Value	Lower Value	Upper Value
Fault Length (km)	37	32	43
Fault Width (km)	5.5	4	7
Dip Angle (^{o})	84	79	89
Fault Locking Depth (km)	20	16	25
Slip rate (mm/year)	33	29	36
α	0.0012		
β	0.0017		
Characteristic Moment Magnitude (Mw)	6.6	6.4	6.7

6 RESULTS

The main concept of PFDHA is acquiring the hazard curve which shows the probability of annual exceedance of a fault displacement exceeding specified displacement of a site or region. The fault displacement hazard curve for Bhairabi Tunnel due to Mat fault is shown in Figures 2–7. The characteristic moment magnitude for the region has been established as 6.6 from the scaling relations. The established earthquake magnitude has a return period of 2,475 years. However, unlike other study, it may not be safe to neglect lesser return period like 100, 475 years and so on. The reason is that there are only few seismological study in and around Mizoram state by the researchers and so, the specified return period of 2,100 years cannot be truly relied upon based on the minimal site specific studies.

Here, the sensitivity parametric study of PFDHA based on the fault parameters, scaling relations and displacement prediction are performed. The parametric study based on fault parameters are analysed for fault length, dip angle, fault locking depth and slip rate. Since the fault parameters data obtained from different studies are somewhat different, the normalized value as well as the upper limit and lower limit value are fixed for each fault parameters. The results obtained are more or less equivalent for the three limit values. However for the slip rate, the obtained lower limit displacement curve has distinctive lesser value as compared to the normalized and upper limit values.

The parametric study for earthquake magnitude scaling relations is based on Hanks and Bakun (2008) and Leonard (2010). For the parametric study of average fault displacement prediction equation, Wells and Coppersmith (1994) and Hecker et al (2013) are used. The fault displacement hazard curve obtained for both parametric study are almost similar and no distinctive lesser values.

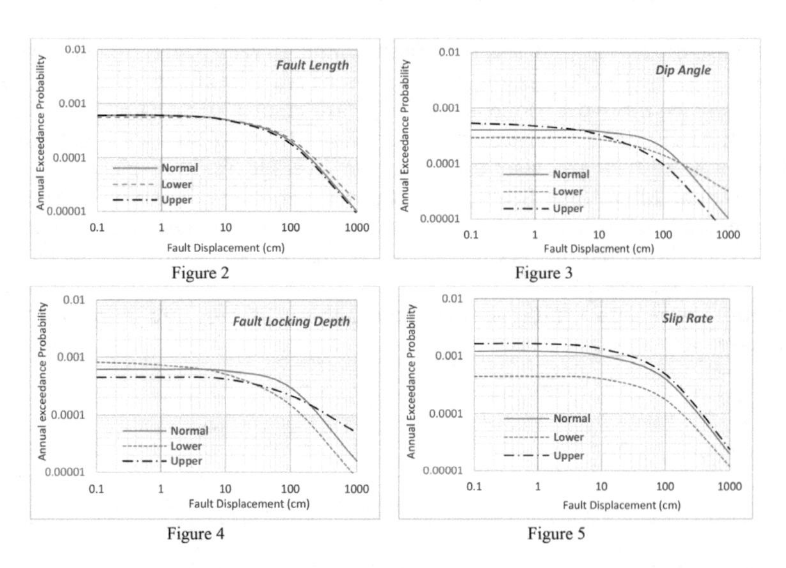

Figures 2–5. Sensitivity of fault displacement hazard curve for Mat fault with respect to (2) Fault length, (3) Dip angle, (4) Fault Locking Depth and (5) Slip rate.

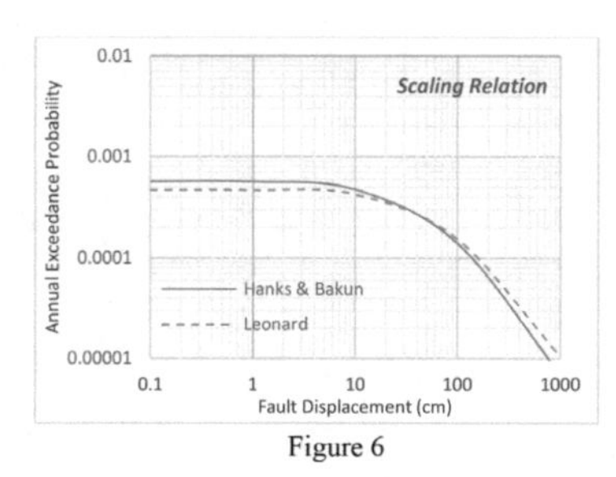

Figure 6

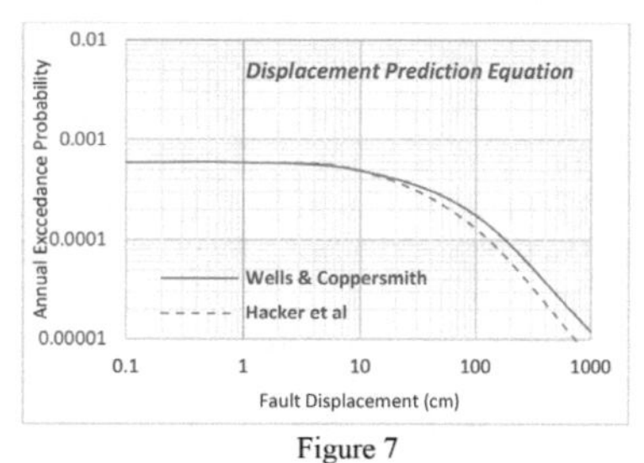

Figure 7

Figures 6–7. Sensitivity of fault displacement hazard curve with respect to (6) magnitude scaling relations and (7) average displacement prediction equation.

7 CONCLUSIONS

Caltrans procedure developed by Shantz (2013) has been used to analyse Mat fault displacement for the probabilistic displacement hazard analysis of Bhairabi tunnel in Mizoram. The fault hazard curve are prepared using Caltrans methods. The parametric study of the sensitivity of fault displacement hazard curve are performed for different fault parameters as well as maximum earthquake magnitude scaling relations and average fault displacement prediction equation. The maximum characteristic moment magnitude (M_w) relevant for Bhairabi and its region is also determined.

The main limitation of this study is that only one fault which is nearest to the site is assumed to be the contributor of the displacement hazard at the region. Observing from the north-eastern India tectonic map, there are many fault which may contribute ground shaking at the project site which can be estimated using the displacement hazard in the region. However, this study will be useful for the designing of engineering structures, particularly underground structures such as tunnels, pipelines, sewer lines, etc. located over or near the faults. The results obtained also signify that the hazard curve are sensitive to the fault parameter as well as scaling relations and displacement prediction equation. Therefore, care should be taken while selecting the fault parameters data and scaling relations and displacement prediction equation while performing PFDHA for the designs of engineering structures.

Another important factor to consider in PFDHA is the return period of certain earthquake for which the characteristic magnitude are based upon. While it may seem reasonable to consider a shorter return period such as 475 years for designs of structures in ground motion analysis, longer return period should always be preferred for the case of fault displacement hazard analysis. The reason behind this is the hazard and damages caused by the earthquake in fault displacement maybe much higher than the ground shaking hazard.

REFERENCES

Bora, D.K., Skolov, V.Y. & Wenzel, F. 2016. Validation of strong-motion stochastic model using observed ground motion records in North-east India. *Geomatics, natural hazards and risks*. 7(2): 565–585.

Braun, J.B. 2000. Probabilistic fault displacement hazards of the Wasatch fault. *Master's Thesis, Department of Geology and Geophysics*, University of Utah, Salt Lake City, Utah.

Coppersmith, K.J. & Youngs, R.R., 2000. Data needs for probabilistic displacement hazard analysis. *Journal of geodynamics*. 29(2000): 329–343.

Hanks, T.C. & Kanamori, H. 1979. A moment magnitude scale. *Journal of geophysical research*. 84(B5): 2348–2350.

Hanks, T.C. & Bakun, W.H. 2008. M-log A observations for recent large earthquakes. *Bulletin of the seismological society of America*. 98(1): 490–494.

Hecker, S., Abrahamson, N.A. & Wooddell, K.E. 2013. Variability of Displacement at a Point: Implications for Earthquake-Size Distribution and Rupture Hazard on Faults. *Bulletin of the seismological society of America*. 103(2A): 651–674.

Jaishi, H.P., Singh, S., Tiwari, R.P. & Tiwari, R.C. 2013. Radon and Thoron anomalies along Mat fault in Mizoram, India. *Journal of earth system science*. 122(6): 1507–1513.

Jaishi, H.P., Singh, S., Tiwari, R.P. & Tiwari, R.C. 2014. Correlation of radon anomalies with seismic events along Mat fault in Serchhip district, Mizoram, India. *Applied radiation and isotopes*. 86: 79–84.

Kayal, J.R. 1996. Earthquake source process in North-east India: A review. Himalaya Geology. 17, 53–69.

Kun, M. & Onargan, T., 2013. Influence of the fault zone in shallow tunneling: a case study of Izmir metro tunnel. *Tunneling and underground space technology*. 33(2013): 34–45.

Kundu, B. & Gahalaut, V.K. 2012. Earthquake occurrence processes in the Indo-Burmese wedge and Sagaing Fault region. *Tectonophysics*. 524: 135–146. doi: 10.1016/j.tecto.2011.12.031.

Leonard, M. 2010. Earthquake fault scaling: self-consistent relating of rupture length, width, average displacement and moment release. *Bulletin of seismological society of America*. 100(5A): 1971–1988.

Leonard, M. 2014. Self-consistent earthquake fault-scaling relations: update and extension to stable continental strike-slip faults. *Bulletin of seismological society of America*. 104(6): 2953–2965.

Li, F., Hsiao, E., Lifton, Z. & Hull, A. 2015. A simplified probabilistic fault displacement hazard analysis procedure: applications to the Seattle fault zone, Washington, USA. *6th International conference on earthquake geotechnical engineering*, Christchurch, New Zealand, 1–4 November, 2015.

Nandy, D.R. 2001. Geodynamics of North-eastern India and the adjoining region. ABC Publications, Calcutta, p.209.

Moss, R.S. & Ross, Z.E. 2011. Probabilistic Fault Displacement Hazard Analysis for Reverse Faults. *Bulletin of the seismological society of America*. 101(4): 1542–1553.

Ni, J.F., M. Guzman-Speziale, M. Bevis, W.E. Holt, T.C. Wallace, & Seager, W.R. 1989. Accretionary tectonics of Burma and the three dimensional geometry of the Burma subduction zone. *Geology*. 17, 68–71.

Petersen, M.D., Dawson, T.E., Chen, R., Cao, T., Wills, C. J., Schwartz, D.P., & Frankel, A.D. 2011. Fault displacement hazard for strike-slip faults. *Bulletin of seismological society of America*. 101(2): 805–825.

Raghukanth, S.T.G., & Somala, S.N., 2009. Modeling of strong-motion data in northeastern India: Q, Stress Drop, and Site Amplification. *Bulletin of seismological society of America*. 99: 705–725.

Sailo, S., Tiwari, R.P., & Baruah, S. 2011. Seismotectonics of Surma Basin with special reference to Sylhet and Mat faults. *Geological society of India memoir*. 77: 185–193.

Shantz, T. 2013. Caltrans procedures for calculation of fault rupture hazard. *http://www.dot.ca.gov/newtech/structures/peer_lifeline_program/docs/Caltrans_Procedures_for_Fault_Rupture_Hazard_Calculation.pdf*, accessed 29/7/2018.

Stepp, J.C., Wong, I.V., Quittmeyer, R. & Toro, G. 2001. Probabilistic seismic hazard analyses for ground motions and fault displacement at Yucca mountain, Nevada. *Earthquake spectra*. 17(1): 113–151.

Stirling, M., Goded, T., Kelvin, B. & Nicola, L. 2013. Selection of earthquake scaling relationships for seismic-hazard analysis. *Bulletin of seismological society of America*. 103(6): 2993–3011.

Tiwari, R.P., Gahalaut, V.K., Rao, Ch.U.B., Lalsawta, C., Kundu, B. & Malsawmtluanga. 2015. No evidence for shallow shear motion on the Mat fault, a prominent strike slip fault in the Indo-Burmese wedge. *Journal of earth system science*. 124(5): 1039–1046.

Tiwari, R.P. 2002. Status of seismicity in the Northeastern India and earthquakes disaster mitigation. *ENVIS Bulletin*. 10–11.

Verma, R. K., Mukhopadhyay, M. & Ahluwalia, M. S., 1976. Seismicity, gravity and tectonics of northeast India and northern Burma. *Bulletin of the seismological society of America*. 66(5): 1683–1694.

Wang, Z.Z., Zhang, Z. & Gao, B. 2012. The seismic behavior of the tunnel across active fault. *Proceedings of the 15th world conference on earthquake engineering* Lisbon, Portugal, 30th August, 2012.

Wells, D.L. & Coppersmith, K.J. 1994. New empirical relationships among magnitude, rupture length, rupture width, rupture area and surface displacement. *Bulletin of the seismological society of America*. 88(4): 974–1002.

Wells, D.L. & Kulkarni, V.S. 2014. Probabilistic fault displacement hazard analysis – sensitivity analyses and recommended practices for developing design fault displacements. *Proceedings of the 10th national conference in earthquake engineering*, Earthquake engineering research institute, Anchorage, AK, 2014.

Wong, I.G., Pezzopane, S.K., Menges, C.M., Green, R.K. & Quittmeyer, R.C. 1996. Probabilistic seismic hazard anal;ysis of the Exploratory studies facility at Yucca Mountain, Nevada, Proceedings, Methods of seismic hazard evaluations, Focus '95. *American nuclear society*, 51–63.

Youngs, R.R., Arabasz W.J., Anderson, R.E., Ramelli, A.R., Ake, J.P., Slemmons, D.B., et al. 2003. A methodology for probabilistic fault displacement hazard analysis (PFDHA). *Earthquake spectra*. 19 (1): 191–219.

Tunnels and Underground Cities: Engineering and Innovation meet Archaeology, Architecture and Art, Volume 3: Geological and geotechnical knowledge and requirements for project implementation – Peila, Viggiani & Celestino (Eds)
© 2020 Taylor & Francis Group, London, ISBN 978-0-367-46583-4

Drone based deformation monitoring at the Zentrum am Berg tunnel project, Austria. Results and Findings 2017–2019

K. Chmelina
Geodata group, Leoben, Austria

A. Gaich
3GSM GmbH, Graz, Austria

M. Keuschnig & R. Delleske
Georesearch Forschungsgesellschaft mbH, Wals, Austria

R. Wenighofer & R. Galler
Chair of Subsurface Engineering, Montanuniversität Leoben, Austria

ABSTRACT: In the paper we present a drone based deformation monitoring system that is developed by the Austrian partners Geodata Group, 3GSM GmbH and Georesearch Forschungsgesellschaft mbH in the frame of the EUREKA research project DefDrone_3D. Since May 2017 the system is used at the Austrian tunnel project Zentrum am Berg for monitoring the 3d deformation of rock slopes and the ground surface at the portal area of two parallel road tunnels under construction. The contribution describes the developed system, the monitoring missions performed at the tunnel project, the developed data processing technique and the achieved results and findings.

1 INTRODUCTION

For several years now unmanned aerial vehicles (UAVs), commonly known as drones, are successfully used for surveying tasks in different areas including (infrastructure) inspection, construction and mining. For these areas new products (drones, cameras and data processing software) have been pushed to market in the last few years offering interesting new features. Add-ons like multispectral, thermal, methane gas detection, and also LiDAR add flexibility and functionality to drones. Besides, extra-long flight times, direct georeferencing, high precision navigation, automatic starting and landing, swarming, short post-processing times and resistance to harsh environmental conditions make them a good choice for surveying applications in manifold environments. UAVs are more and more becoming an industry standard.

Consequently, drones already start to be applied also in tunnelling projects on a routine basis, e.g. for site documentation, generation of 3D terrain models and orthophoto maps, inspection and volumetric evaluations (e.g. for earth works, landfills).

In the EUREKA research project DefDrone_3D the partners Geodata Group, 3GSM GmbH and Georesearch Forschungsgesellschaft mbH started 2017 with the use of UAVs for deformation monitoring. The goal is to develop a drone based deformation monitoring system able to monitor structures such as embankments, dams and rock slopes/rockwalls that are affected by tunnel construction. The system is supposed to enable the periodic/repeated measurement of critical, dangerous or inaccessible areas and to deliver deformation results of very high accuracy. It is to become a valuable tool for geotechnical data interpretation and early warning.

To achieve these goals, the drone based deformation monitoring system has been designed, developed and tested at the Austrian tunnel project Zentrum am Berg (ZaB). The following chapters present the system, the tunnel project with its research area, the performed monitoring missions and the achieved results and findings between 2017 and 2018. It is to note that the system, particularly its data processing, is still under development and final results can be expected in 2019.

2 THE MONITORING SYSTEM DEFDRONE_3D

2.1 *General design*

Our monitoring system consists of:

- a UAV (multicopter drone) including remote control device, vision system, communication system, GNSS navigation system, collision avoidance system, accessories (charger, batteries, transport box/bag) and remote control software,
- a data acquisition system consisting of a camera system (camera sensor, camera lenses), a camera mounting (gimbal), a data storage device and the raw data produced (= the aerial images),
- a set of Ground Control Points (GCPs) including marking material,
- a mission planning and control software,
- a data evaluation software (= image processing software) and
- a data management software (= monitoring information system).

2.2 *UAV(s) and data acquisition system(s)*

In our research project we tested below combinations of drones and cameras/sensors (Figure 1).

1. DJI Phantom 4 Pro with integrated 1"CMOS camera, 20 MP, 24mm equivalent,
2. DJI Mavic Pro with integrated camera, 12 MP, 24mm equivalent (used in swarm mode with 5 drones),
3. DJI S1000 with Sony A7R, 36 MP, 35mm lens,
4. DJI M600 with Phase One iXM-100, 100 MP, 80mm lens,
5. DJI M600 with Phase One iXU-1000, 100 MP, 55mm lens,
6. DJI M600 with Hasselblad A6d, 100 MP, 55mm lens and
7. DJI M210rtk with Zenmuse X5S, 20.8 MP, 15mm lens (30mm equivalent).

The different combinations allowed the study of effort and time for preparation and handling of equipment, flight performance, data quality and other parameters relevant for the given purpose of deformation monitoring. As can be imagined, the selection of the UAV - camera combination is a compromise. For achieving highest data quality and precision of results, of course, it

Figure 1. UAVs, cameras and pilots at the study site Zentrum am Berg, numbers refer to the above list.

must be used a powerful high resolution/high end camera. However, due to the then higher camera weight it is also required a bigger drone offering a higher payload but causing extra effort for transport, mounting, calibration etc. Besides, it is more costly. Taking a smaller/cheaper drone with an integrated camera (e.g. DJI Phantom 4 Pro) makes your mission an easy job but, however, the quality (esp. precision) of the results might not be the highest.

2.3 *Ground Control Points (GCPs)*

For georeferencing different kinds of Ground Control Points have been designed, manufactured, installed and tested in the course of the project (Figure 2). They range from simple bireflex targets (Figure 2 upper left) to prism targets of bigger diameter (Figure 2 lower right). They can be selected with regard to flight height/ground distance and image resolution. Their specific advantage is that they can be measured with mm-accuracy by conventional surveying using a total station. Therefor, they can be rotated to view into any direction required like a typical surveying target but, at the same time, remaining measurable from above by a drone. In cases where GCPs cannot be installed in stable area they are to be remeasured prior to each drone mission. The GCPs can be installed on an adapter (Leica type) for easy removal without loss of point identity. Various quick installation techniques can be chosen depending on the type of ground.

2.4 *Mission planning and control software*

The generation of accurate elevation data with high ground resolution requires extensive mission planning. In our case it is of particular importance to keep camera-object-distances constant throughout the entire area of interest. For this purpose, an initial digital elevation model (DEM) is required as input data for flight planning (Figure 3). The required resolution of the DEM depends on the complexity of the terrain and the required ground sampling distance (GSD) of the 3d model to be processed. For flat areas and a higher GSD, a 30m DEM can be sufficient. However, for a constant and very low GSD of 1 cm in complex topographies such as steep alpine terrains, an up to date DEM with 1 m resolution is recommended.

To achieve the targeted high quality results, our monitoring system requires a mission planning software able to import and use a high resolution DEM for flight planning. If not available in advance, the DEM is produced from data of an initial flight. Furthermore, it is convenient to also have the possibility of importing orthoimages of the area of interest.

Each of these requirements is met by the software product UgCS that allows to plan and fly drone survey missions for many UAV platforms including DJI drones. After importing the

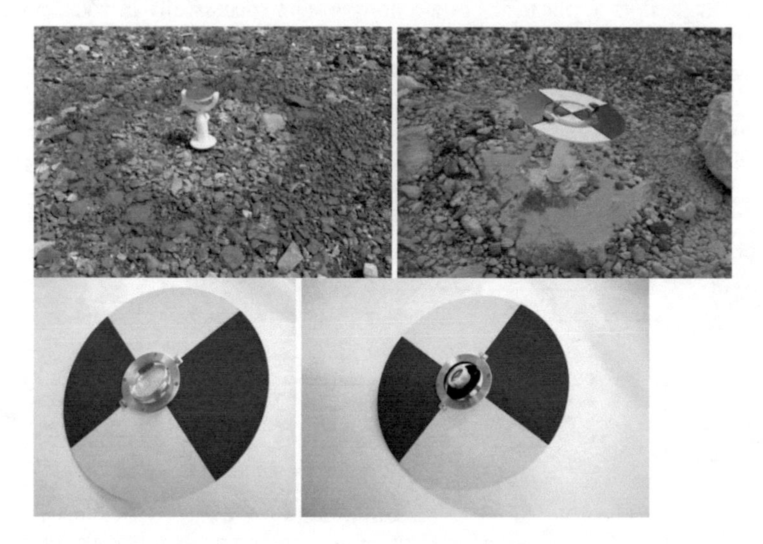

Figure 2. Different Ground Control Points based on bireflex or prism targets.

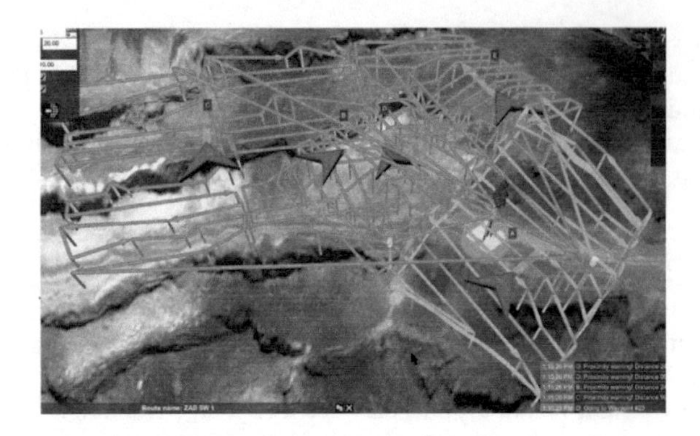

Figure 3. Flightplan based on a 0.5m DEM involving 5 UAVs for a GSD of 0.5 cm.

DEM, the orthoimages and defining the optimal camera direction, the flight directions and camera angles can be defined. Finally, an optimal mission with different flight plans can be designed and executed.

2.5 *Data evaluation software (image processing software)*

Image processing comprises the following main steps: (i) generation of an accurate 3D model of the area of interest and (ii) producing a map that displays the change between 3D models taken at different points in time (Figure 4).

The meaning of *accurate* is twofold in this context: (i) the intrinsic accuracy of the 3D model, i.e. how comprehensive the surface in question is described (model accuracy) and (ii) the positional accuracy of the 3D model in terms of the correct localization in a superior coordinate system. It is straightforward that both accuracies need to be better than the extent of displacements to be detected.

2.5.1 *3D model generation software*

The image processing software uses latest algorithms from photogrammetry to generate 3D models automatically. The used principle mainly follows the so-called Structure-from-Motion approach which is a photogrammetric range imaging technique for estimating three-dimensional structures from two-dimensional image sequences (Snavely et al. 2008). The technique has been used in various applications and among others successfully in rock mass characterization and blast design (Gaich et al. 2017).

The resulting 3D models are available in a proprietary format as used by the ShapeMetriX system (3GSM 2018) but are exportable to standard formats such as Wavefront OBJ (OBJ 2018).

Figure 5 displays a 3D model as generated with the used software. The model shows 2.4 million points (adjustable) leading to a point spacing of 18 cm. It bases on 422 photographs and is

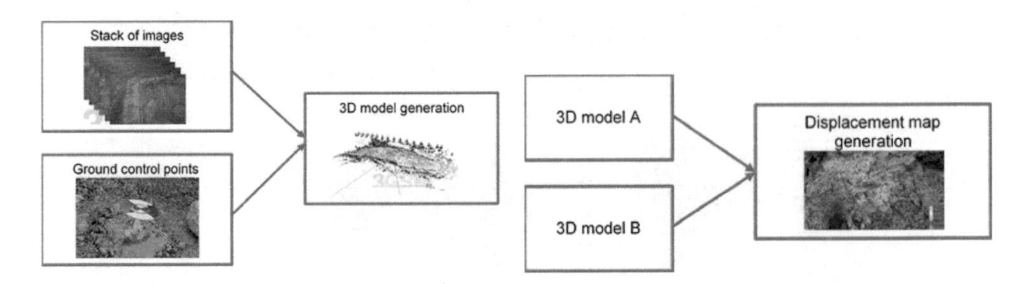

Figure 4. Principle workflow in the image processing software: from a set of images and surveyed ground control points a 3D model is generated using modern photogrammetry algorithms; 2 different models are then analyzed in order to quantify any changes between the 3D models.

Figure 5. 3D model of the research area as computed with the used photogrammetry software.

computed in 3 hours on a laptop computer. The positional accuracy is 5.9 cm shown by the residuals at 11 ground control points. The computation of the model requires a minimum on user intervention, namely specifying the images and file names, as well as assigning co-ordinates to ground control points.

In order to investigate the model accuracy, i.e. how well a surface is described by the model, several tests using different image matching techniques have been applied. Among others algorithms as described in (Hirschmüller 2008, Bleyer et al. 2011, Schöneberger 2016) have been implemented and their behavior analyzed.

2.5.2 *Displacement map generation*

The comparison of 3D models requires them in the same co-ordinate system. A software component named ModelComparator has been implemented to handle 3D models and visualize their differences. This experimental workbench allows the application of different algorithms for the computation of displacements hence enabling to test the quality and runtime behavior of the generated displacement maps. Investigated methods for the displacement computation include:

- "Shooting" rays from the surface in direction of the cameras (this model viewing rays)
- Surface normal
- Vectors in direction of definable axes (X, Y, Z, custom)
- Using ICP (iterative closest points) in order to identify corresponding points between the surfaces
- 2D image matching techniques to identify corresponding points between the surfaces
- 3D descriptor matching techniques to identify corresponding points between the surfaces

Figure 6. Screenshot of the software component *ModelComparator*.

Each of the mentioned algorithms features several parameters, so there is a vast number of possibilities to determine the displacement maps. Work on that topic is still in progress. Figure 6 shows a screenshot of the *ModelComparator* software. The analyzed models have an area of about 150 x 150 m and consist of 35 million points reducing point spacing down to 2.7 cm; referencing accuracy: 2.4 cm.

3 THE TUNNEL SITE "ZENTRUM AM BERG"

The monitoring system is currently tested at the Austrian tunnel project Zentrum am Berg (=ZaB) which is an underground research, development and training center of the University of Leoben. The center reflects the fact that underground construction is an indispensable component of modern infrastructure for energy supply and traffic which is inextricably linked with future economic progress. As the reliance of traffic on tunnels increases the development of safety technology, education of emergency services organizations and training of maintenance personnel needs to include testing at representative tunnel sites. The ZaB tries to tackle the requirement of realistic conditions for this purpose and is located in an altitude of 1100 m on a terrain level called Dreikönig in a disused part of the Erzberg mountain in Styria which is the largest open pit mining area of Austria. The ZaB will encompass two two-lane road tunnels (Figure 7) and two single-track railway tunnels when in full operation. The tunnels represent the current state of the art cross sections regarding design and size and are interlinked at a Y-node. On this terrain level several galleries exist from former underground mining. They are integrated in the ZaB and will expand the functional range of the testing area (Figure 8). Especially the road tunnels traverse a section where several galleries had to be crossed as indicated in Figure 8 left showing a crossing gallery underneath the research area. These former transport galleries and shafts mainly influenced the alignment of the tunnel axis to avoid coming across unexpected chambers. Moreover, the gradients were adapted to connect properly into the galleries.

In September 2017 the contractor Swietelsky Tunnelbau GmbH commenced the main construction phase of the ZaB excavating the rail tunnels in the siliceous Blasseneck porphyroid. The main characterization of the porphyroid is its rhyolitic mineralogic composition and its degree of fragmentation by numerous joints in the face. In June 2018 the excavation of the road tunnels has started. The surrounding rock consists of the calcareous Sauberger limestone partially containing significant contents of ore. The limestone is characterized by its intense jointing and its layers dipping out of the face in the portal area. Consequently, starting the excavation of the road tunnels was preceded by extensive preparation of the terraced former open pit area which was a vital step for the work safety and is continued as can be seen in Figure 7 right. Therein the portals are protected by depositing a layer of excavation material onto the shotcrete arcs to safeguard the area underneath from rock fall.

Figure 7. Left: Aerial view of research area (= portal of two two-lane road tunnels); Right: Excavator filling up the portal area by excavation material to keep the area safe from falling rocks.

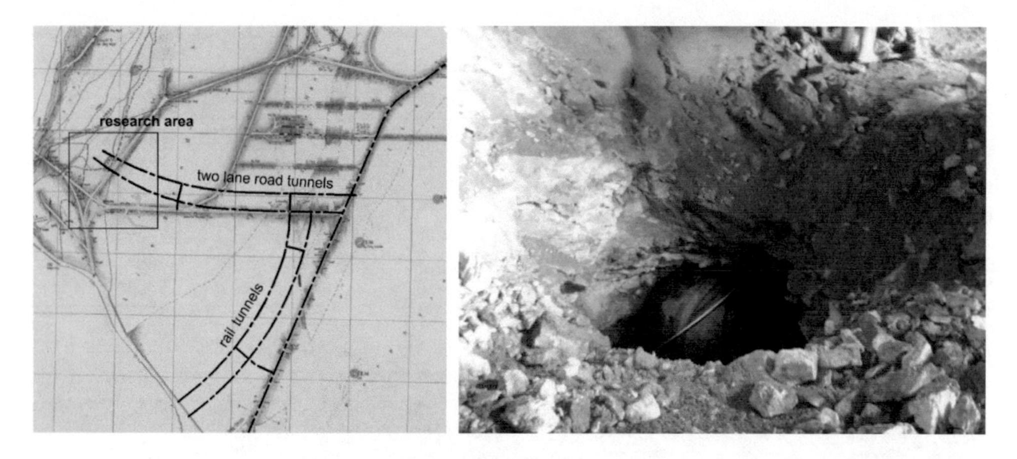

Figure 8. Left: layout of the Zentrum am Berg (ZaB) indicating research area at the portal of the two two-lane road tunnels. Black: tunnels in construction. Colored: galleries on the terrain level called "Dreikönig"; Right: crossing gallery seen from excavation of the top heading.

As of August 2018, 180 m (East) and 213 m (West) of the rail tunnels as well as 345 m (North) and 133 m (South) of the two-lane road tunnels are excavated. The construction phase regarding tunneling is expected to end in autumn 2018 when the subsequent installation of interior tunnel equipment will begin.

4 THE MONITORING MISSIONS

4.1 *Regular missions*

To detect surface changes at the ZaB research area data acquisition started on 10th of May 2017 with the first flight. Since then we conducted seven missions with the DJI Phantom 4 Pro (Table 1). Within this time we changed the settings of the flights to optimize the amount of data, data quality and flight time. From the 7th of Nov. 2017 on the flightplans (Figure 9) have not changed but the DEM was constantly updated. The flights have been carried out with a camera tilt of 60° to the rockwalls, an overlapping of 80 percent and flight heights of approx. 20-30m.

4.2 *Special missions and investigations*

In addition to the regular missions for detecting coarse volumetric changes of the area by processing model differences (Figure 10 right) special missions have been carried out such as flights into the tunnel and flights where small deformations have been provoked/simulated.

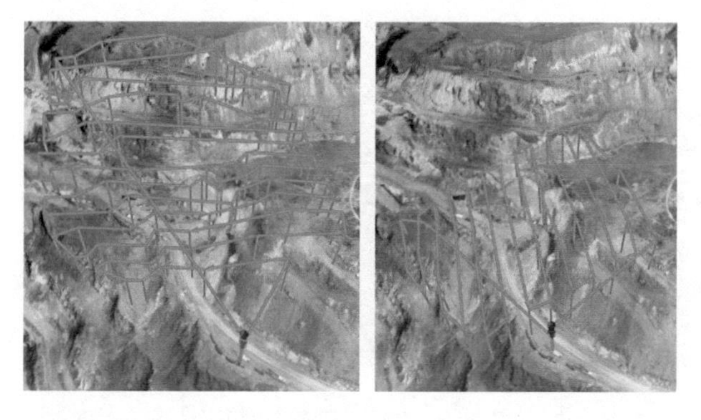

Figure 9. The two flightplans executed at a regular mission.

Table 1. Mission statistics.

Date	nr. of Images	Flight Time (min)	Nr. of Flights	GSD (cm)	Angle
10.05.2017	443	40	2	1	45°
16.05.2017	850	90	4	1	Nadir, 30°, 60°
05.09.2017	1200	120	5	1	Nadir, 30°, 60°
07.11.2017	820	70	3	1	60°
23.05.2018	448	40	2	1	60°
21.06.2018	410	40	2	1	60°
20.07.2018	650	50	2	0,5	60°

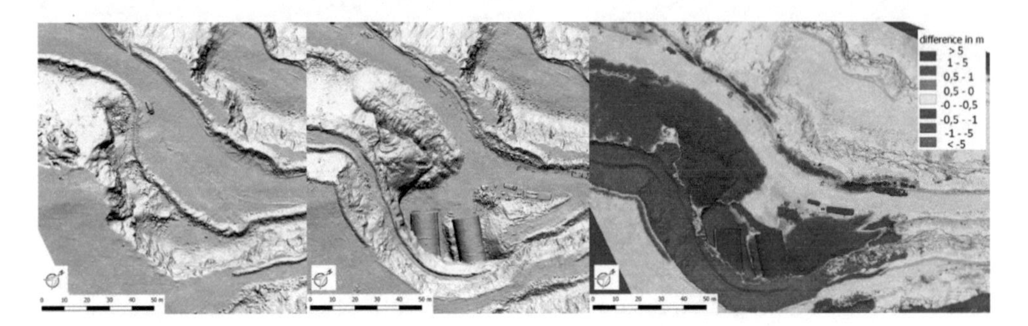

Figure 10. Left: Hillshade of the first flight in May 2017; Center: Hillshade of the flight in May 2018 showing road tunnels; Right: model difference in discrete coloring showing surface changes.

4.2.1 *Tunnel flight test*

A manually flown mission has been carried out in a 40m long tunnel section using a DJI Mavic AIR drone (Figure 11-12). The flight has been done in several lines parallel to the tunnel axis, one for each tunnel side wall, one for the tunnel crown (with camera upwards) and one for the invert/ground (with camera downwards). The tunnel face has been extra-acquired in three horizontal lines perpendicular to the axis just a few meters in front of the tunnel face.

4.2.2 *Deformation simulation test*

To produce reference data for the image processing software a simulation test has been done. Wooden plates with rock texture and each equipped with three 3D targets (Figure 13) have been laid out in the research area, a flight done, the plates moved in different ways (tilted, shifted) and the next flight done. Each time the displacements of the plates, resp. 3D targets, have been measured with a total station. The plates have been measured two times, (i) during the regular missions and (ii) from closer distance in a higher resolution during a special mission.

Figure 11. Left: DJI Mavic AIR drone at the tunnel face; Right: 3D tunnel model generated from underground drone imagery in conjunction with 3D model generated from aerial imagery outside the tunnel.

Figure 12. Side wall of tunnel as reconstructed from underground drone imagery; GSD is 5 mm/pixel.

Figure 13. Left: Wooden plates laid out in the research area; Right: Each plate has three geodetic targets.

5 THE DATA PROCESSING & RESULTS

The software components as described in chapter 2.5 are applied to imagery from the test site. The 3D model of the entire area shows a point spacing of 18 cm. For the detection/quantification of displacements the model description is made even more detailed. Additional images from the areas of interest are taken with large GSD. The overview model shows a GSD of 0.8 cm, the detailed view features 0.2 cm. This allows to reduce point spacing down to 1.0 cm.

Figure 14 left displays a result from the image processing software obtained in the deformation simulation test described in chapter 4.2.2. The areas where changes of the wooden plates occurred are highlighted in the displacement map. The left plate has been shifted laterally along the surface resulting in larger displacements (in red) in the upper part where the plate has been removed. The right plate has been tilted which manifests in increasing displacements in the lower part of the plate (colour changes from green over yellow to red). Figure 14 right shows a close-up of the right plate in conjunction with some colored dots on the surface. Again the colour of the dots refer to the amount of detected displacements.

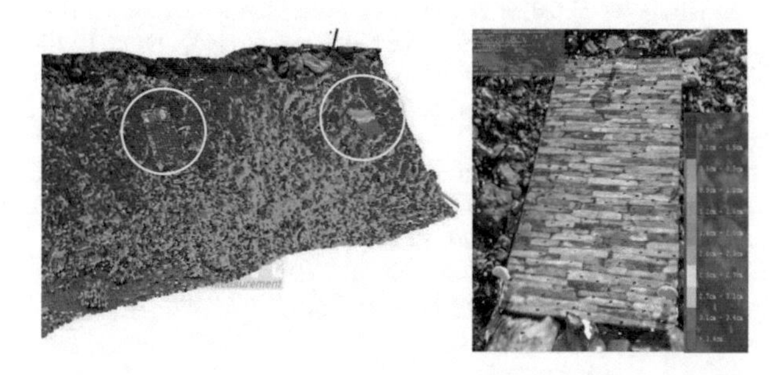

Figure 14. Left: 3D models with displacement map and two highlighted areas where displacements occurred; Right: Detailed view of 3D model inside the ModelComparator software; the colour of the pins indicates the amount of changes on the surface.

6 CONCLUSIONS

In the research project DefDrone_3D a drone based monitoring system is currently developed and investigated. Objective of the system is to detect and quantify the 3D deformation of objects such as embankments, slopes and rockwalls from repeatedly performed flight missions with cm- accuracy. It is considered useful especially in cases where other monitoring methods and systems cannot be applied, e.g. when inaccessible areas are to be monitored.

The system consists of a commercial multicopter drone (or several drones operated in swarm mode if desired), a suitable high resolution camera and software packages for flight mission planning and control, data evaluation and data management. Data evaluation is based on advanced image processing techniques applying Structure-from-Motion technology.

Since May 2017 the system is tested at the Austrian tunnel project Zentrum am Berg. Until August 2018 seven regular and several special flight missions have been carried out to test and validate the developed system.

So far it is concluded that the chosen hardware, the DEM based mission planning method and the data evaluation strategy processing the displacements turned out useful. 3D displacements in the range of about 1 cm can be determined under good conditions. The strategy is to compute from coarse to fine, i.e. to identify those areas where changes occurred and then increase the level of detail at those areas. This prevents from generating massive amounts of data that do not gain information. Several methods for quantifying the displacements have been applied so far and there is still space for more investigations.

Our research work is still ongoing currently focusing on further improving the data evaluation software to finally allow for the generation 3D displacement vector maps. In the near future also automatic starting, landing and battery charging/changing seem feasible. Our vision is having a drone or even drone swarm performing autonomous survey missions in a tunnel project including all data evaluation without human intervention.

In the frame of the project tests have also been carried out at a Chilean dam project that will be reported at a later stage.

ACKNOWLEDGEMENTS

The described research is funded by the Austrian Research Promotion Agency (FFG) and carried out in the frame of the EUREKA research project Defdrone_3D.

REFERENCES

3GSM, ShapeMetriX UAV. https://3gsm.at/produkte/shape-metrix-uav/, accessed on August 23, 2018.

Bleyer, M., Rhemann, C. Rother, C. 2011. PatchMatch Stereo - Stereo Matching with Slanted Support Windows. In Jesse Hoey, Stephen McKenna and Emanuele Trucco, *Proceedings of the British Machine Vision Conference*, pages 14.1-14.11. BMVA Press, September 2011. http://dx.doi.org/10.5244/C.25.14

Gaich, A., Pötsch, M. & Schubert, W. 2017. Digital rock mass characterization 2017. Where are we now? - What comes next? *Geomechanics and Tunnelling 10*, No.5, pp. 561-567. DOI: 10.1002/geot.201700036.

Hirschmüller, H. 2008. Stereo Processing by Semi-Global Matching and Mutual Information. IEEE Transactions on Pattern Analysis and Machine Intelligence. Volume: 30, Issue: 2, pp. 328-341. DOI: 10.1109/TPAMI.2007.1166.

OBJ, OBJ file fomat. https://www.cs.cmu.edu/~mbz/personal/graphics/obj.html, accessed on August 23, 2018.

Schöneberger, J.L. 2016. Structure-from-Motion Revisited. *IEEE Conference on Computer Vision and Pattern Recognition (CVPR)*. DOI: 10.1109/CVPR.2016.445.

Snavely, N., Seitz, S.M., Szeliski, R. 2008. Modeling the World from Internet Photo Collections. *International Journal of Computer Vision*: Volume 80, Issue 2, pp. 189-210.

Tunnels and Underground Cities: Engineering and Innovation meet Archaeology, Architecture and Art, Volume 3: Geological and geotechnical knowledge and requirements for project implementation – Peila, Viggiani & Celestino (Eds)

Necessary geological and geotechnical information for a metro project in an historical and urbanised city area. The case of "Metro Bucharest, Line 4 Extension"

V. Ciugudean Toma & C. Vajaeac
Metroul SA, Bucureşti, Romania

A. Arigoni & S.S. Saviani
Amberg Engineering AG, Regensdorf-Watt, Switzerland

ABSTRACT: All the underground infrastructures require a specific level of geological and geotechnical knowledge usually based on the common practice, on the Country laws and on the Design Level. The actual level of data recommended for a metro project in an urbanised area is an important technical and financial issue to be analysed in comparison with the already available information and the needs of the design level. Besides the geotechnical investigations carried out can considerably decrease all the geological and geotechnical potential risks. The comparison of the amount of investigations carried out before the tender, during the design and in the construction phase reflects well the efforts to ensuring a safe and economical tunnel construction process. The analysis of the Rumanian standard requirements for the geotechnical investigations and the case history of the engineering geology and geotechnical assessment particularly for "Metro Bucharest, Line 4 Extension" Project is discussed.

1 INTRODUCTION

All the underground infrastructures require a specific level of geological and geotechnical knowledge. Design experiences clearly show that the actual level of data and geo-studies recommended for a metro project in an urbanised area is an important technical and financial issue that has to be analysed in comparison with the already available information and with the needs of the design level. Besides the geotechnical investigations carried out can considerably decrease all the geological and geotechnical potential risks.

The case history of the "Metro Bucharest, Line 4 Extension" Project is presented. In this specific project, the existing Metro line 4 Gara de Nord – Străuleşti will be extended with a new metro line that could be fully underground or partially at ground level. If the construction is completely underground and crossing the downtown area, twin-tube tunnels having an inner diameter of 5.70 m will be built for a length of about 11.2 km with a number of stations actually foreseen of 15, constructed with cut and cover method. For this project, it is possible to rely both on the already available investigations and results of the geotechnical assessments carried out for the existing metro lines and for private projects and on the data of the additional investigation campaigns done specifically for this.

2 THE COMPLEXITY OF GEO-ASPECTS IN BUCHAREST'S URBAN ENVIRONMENT

Geologically, Bucharest is located in the Romanian Plain. The problems concern on the one hand its bedding and on the other hand the lithological and stratigraphic composition of the newer sedimentary formations. The bedding of the area is a rigid basement consisting of crystalline shale of various types situated at a depth of 5000 m - 6000 m. This fact points out that the city is located in an area with an unstable ground thus warning of the reaction to earthquakes due to the lack of a rigid bedding close to the surface. The sedimentary cover gradually extends through different erosion phases and accumulation of marine, lake or continental sediments ending with the quaternary age. In its entirety, the quaternary is a complex stage, in which a very large amount of deposits were genetically and lithofacially formed. The stratification of the land is not constant in relation to the sedimentation conditions in which it was established.

The appearance of the land surface was defined during the modernization of the city, when some existing drops were filled with local materials, resulting from the systematization of street networks and constructions, or excavated following the construction of the Parliament Palace, so they are now characterized by absolute levels ranging from 72.00 m to 87.60 m.

As other capital cities, Bucharest has firstly developed along rivers (with affluents, meanders, small islands, hills). The city developed at first along Dâmboviţa river and later also along Colentina river where the necessary conditions for constructing buildings were more suitable.

The number of hills in Bucharest, the network of brooks, puddles, ponds etc. – all together indicate the difficult foundation conditions that characterize from a geological, hydrogeological and geotechnical point of view the sedimentary deposits that constitute the foundation ground of the city. At present, the morphology of Bucharest is a plain with microrelief resulted from erosion and sedimentary processes which extended along valley of Dâmboviţa River to the south and valley of Colentina River to the north.

The two streams are almost parallel with a NW-SE trend, with old large meanders and small depressions, in which surface waters and the water from the springs situated at the bottom of the slopes have been accumulated. Marshes were formed, that are now hydro-technically drained.

Bucharest has three typical geomorphological zones: 1) the lower meadows on the riversides; 2) the field between the rivers, with altitudes up to 90 m, in the N-W, and up to 75 m above the Black Sea level in S-E; 3) the higher plains, extending towards S and N. The downtown is extending in the interfluvium area on an almost plane surface with elevation around 80 m.

From a geological point of view, due to neo-tectonic movements, the specific structural frame of the Romanian Plain, where Bucharest is situated, is the one of a syncline with a general character of subsidence development, oriented towards SW- NE, on the background of which the neogen and quaternary deposits accumulated as in Figure 1.

This highlights the character of the region's lability, warning about the reaction to earthquakes, due the lack of a rigid foundation close to the surface or hard rocks. Indeed, with

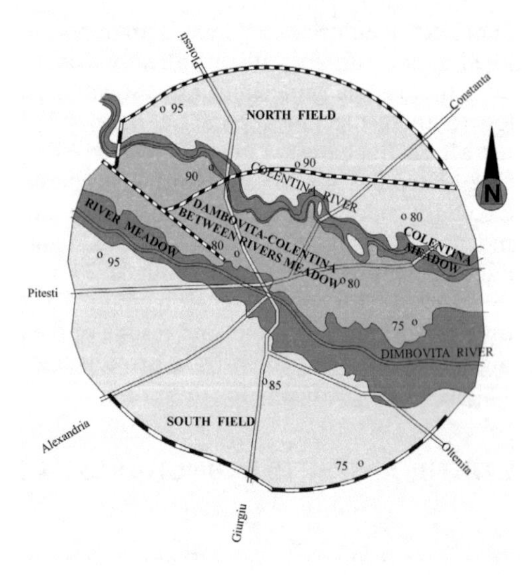

Figure 1. Geological Map of Bucharest (The Technical Economic Memo on Metro construction in Bucharest, 1953).

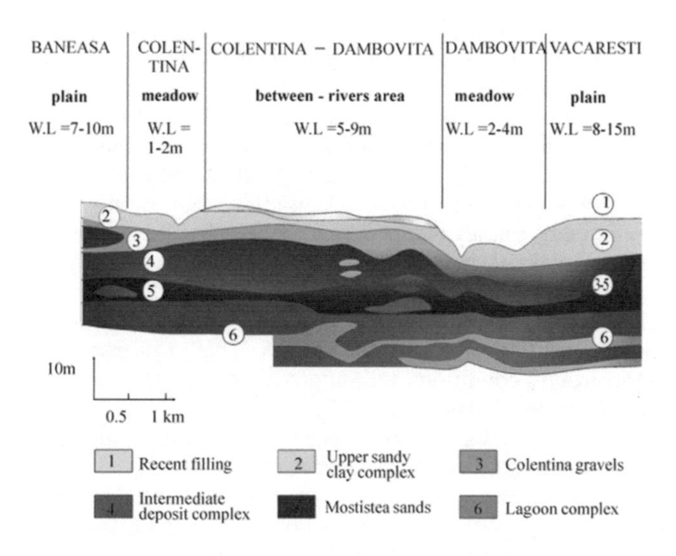

Figure 2. Geological section (The Technical Economic Memo on Metro construction in Bucharest, 1953).

about 2 million inhabitants and approx. 110,000 buildings Bucharest is the biggest city with the highest seismic risk in Europe, with soft soil conditions and with an high fragility of buildings. In the last two centuries the city experienced five significant earthquakes characterized by a moment magnitude ranging from 6.9 to 7.9; the last one being in 1990. The new hazard map, according to SR11100/1-1993 gives a design peak ground acceleration of 0.30g (MSK intensity scale with a 50 years average return period).

The lithological succession of Bucharest, completed during the quaternary period, is formed by an alternation of lithological formations, consisting of cohesive soil, sometimes with interlaces of sandy sub-horizons, and also non-cohesive soil such as gravel sands or sands only, with intersections of clay sub-horizons. For design purposes, the following lithological formations, called "Complexes" or "Typical Layers", are identified:

- Embankments (Vegetable Soil and Anthropic Fillings),
- Upper Sandy-Loam Complex,
- Colentina Gravels Complex,
- Intermediate Clay Complex,
- Mostistea Sand Complex,
- Loamy Complex (Lacustrine; Coconi),
- Lower Gravels Complex (of Fratesti).

The hydrogeological conditions in the area highlight the existence, within the quaternary deposits, of three aquifer horizons characteristic of the Roman plain: macrogranular horizon in type 3 layer, Mostistea sands horizon in type 5 layer and Fratesti stratum horizon in layer 7.

3 ROMANIAN LAW REQUIREMENTS AND URBAN CONSTRAINTS

All the infrastructures require a specific level of geological and geotechnical knowledge that is usually based for each design level on the country laws and on the common practice.

3.1 *European and Romanian legislative requirements*

Eurocode 7 (EC7) stipulates the need to develop two types of geotechnical documentation for a new project, namely the Ground Investigation Report and the Geotechnical Design Report.

The results of the previous desk studies and geotechnical investigations shall be compiled in the Ground Investigation Report which shall form a part of the Geotechnical Design Report.

Particularly in Romania, the correct and complete application of EC7 is actually important in the current period, when various infrastructure works with a very complex geotechnical component are designed and executed, most of them being public investments with European funds.

According to the Romanian and European legislation, the soil in metro projects is investigated since the preliminary phases of the design, basing on the geotechnical category of the works and the geotechnical risk induced by them. The volume of the geotechnical investigations is also recommended in the norms.

Besides according to EC7, the assignment of a project (or part of it) to one of 3 geotechnical categories and their association with geotechnical risk is based primarily on the complexity of the structure, on the ground conditions and on groundwater, taken into account in a qualitative manner and involving local experiences.

Differently from EC7, the Romanian norm NP074 introduces a quantitative method for the selection of the geotechnical category. In particular, a level of risk is associated to each category, that is the result of a geotechnical risk assessment that takes into account quantitatively the ground conditions, the groundwater, the construction classification, the possible impacts on the existing structures and the seismicity. This preliminary framing of a work into one of the three geotechnical categories is done prior to the investigation of the construction site and can be subsequently changed as a result of the new supplementary data that field investigation may supply.

The following step is the definition of the new investigation campaigns in accordance with the requirements of NP074 and in compliance with EC7, basing on the assessed category and focusing on the design needs and on the field investigation (in situ and laboratory) stages.

Essentially, the Romanian legislation in the field of geotechnics transposes the requirements of the European legislation and at the same time is even more exhaustive, giving more detailed guidelines and procedures to be followed in the decisional process (e.g. campaign investigation and design procedures selection based on quantitative geotechnical category definition).

Furthermore, the importance of the local experience of the specialists with their specific level of attestation in the field is explicitly considered, not only for the calculation of the geotechnical category. As per Law 10/1995, the verification of the projects and the execution of the works must be done by specialists approved by the Romanian Authority. In Romania the examination of the specialists in order to obtain the attestation as "Verifier" and "Expert" in the geotechnical field "Af", requires special preparation and examination, with the revision of the attestation every 5 years. This is a further example that general good practice is to involve every time local experts since the early design stages.

3.2 *Urban constraints: the complexity of Bucharest urban environment*

Because of the occurred numerous floods, fires, earthquakes and the tragic communist era of demolitions, blocks building and chaotic public networks development, the city can be considered a real trap, both considering its natural conditions and its anthropic location.

Over time, building underground and on the ground constructions have required extensive works with direct implications in the morphology and hydrogeology of the site, due to excavations and fillings, surface and underground concrete, pumping phreatic aquifers or under pressure aquifers etc.

The complex architectural and urban city context and the constraints due to undersized streets in relation to the population density in central and peripheral areas increase the difficulty of designing underground structures in the urban environment and implicitly increase also the difficulty of positioning and executing geotechnical investigations in situ.

The difficulties encountered in designing open air/underground structures consist of:

- natural factors, as geological, hydrogeological, geotechnical and seismic conditions, characteristic of a site whose ground is represented by an alternation of soft sedimentary deposits of very large thicknesses (clay & sand under water) and the lack of a rigid bedding;
- urban anthropogenic factors:

- the existence of large underground works in operation (public networks as collectors, aqueducts) or new works under construction,
- the location of metro structures close to old or damaged existing buildings & infrastructures, especially in the central area (for about 60-80% of the Line 4 Extension corridor), which requires according to Law 10/1995, structural and geotechnical expertise,
- "the unknown" of the old town of Bucharest underground.

Based on what above, all these constraints of Bucharest's urban environment have been assessed since the planning stage of the geotechnical investigation locations.

A special issue concerns the available spaces for the drilling and positioning of the site installations, particularly for their legal regime, the restrictions, the possible neighbourhood with protected areas, the historical monuments, the hydro-technical development of the Dâmbovița river, etc. Other constraints are related to the existing interferences, such as public infrastructures, electric networks, etc. Besides, the environmental constraints are becoming increasingly restrictive, in requiring more compact workspaces, noise reduction, continuity of traffic, avoiding damage and discomfort to the owners of the buildings near which the works are executed.

Another aspect that can negatively affect the execution of the complex geotechnical studies and might lead to delays is the complex bureaucratic process, in particular for obtaining the necessary permits for executing the drilling works in Bucharest.

4 THE ROLE OF THE EXISTING GEOTECHNICAL DATABASE

In Bucharest metro design, the assessment of the optimal transportation solution and then of the optimal structural solutions in terms of performances, time and costs requires a complete approach since the early stage, starting from defining the geotechnical parameters resulting from in situ investigations and laboratory tests, to the design and monitoring of geotechnical works.

For this reason, the geotechnical database of Bucharest is the first important available resource to obtain the geotechnical information necessary for the design. During all the previous metro design activities (started more than 50 years ago) and in a city area of 20 km of diameter, important volumes of geotechnical information were accumulated and archived, taking into account that there are approx. 80 km of metro in operation, as per Figure 3. For example, an important part of the graphic database was taken from the Bucharest Institute of Design (ex. Hydrogeological Study Bucharest) and in that case the surface of the maps exceeds even 4 m^2.

In order to better identify the difficult areas for the construction of the metro structures, in a first stage all the information provided by the city geological maps, by the maps with the anthropic filling areas (which were built in order to rehabilitate the building pits used in the

Figure 3. Bucharest city map with metro network (Metroul).

interwar period in brick factories), by the maps with hydro-isohyets that give the real image of the direction of the underground water flow, by paleohydrographic maps (which highlight the meandering course of the rivers that crossed Bucharest), etc. were used. Furthermore also the hydrogeological profiles made in the 20th century by specialized institutions were very useful.

In particular, the design of the Metro Line 4 Extension at the prefeasibility (PFS) and feasibility study (FS) phases benefits of the geotechnical information resulting from the execution in the city area close to the corridor of at least 120 previous geotechnical drillings, with over 3000 m of drilled columns and 1000 m of complex drilling sheets, from archive data.

Following what above, at prefeasibility and feasibility study stages, all the geological, hydrogeological and geotechnical data were assessed using governmental and non-governmental institutions and the optimal planning of the new geotechnical investigation locations campaigns was defined in order to complete the information where there was not a sufficient database.

5 OPTIMAL PLANNING CRITERIA FOR THE GEOTECHNICAL INVESTIGATIONS

Linear transport infrastructures, partially or fully underground, involve deep excavations and structures with strong interactions with soil and a high risk due to the neighbouring surroundings. The planning of the geotechnical campaigns was done based on the above described frame by design engineers specialized in underground constructions, geotechnical engineers, geologists and hydro-geologists to allow the geotechnical and hydro-geological studies to meet the requirements of geotechnical design (in terms of structural framing, dewatering, monitoring, drainage works, compliance to existing norms, risks reduction etc.).

Based on the archive data, in accordance with the Romanian legislation, following the analysis of the design team with local experience and based on tunnelling and geotechnical specialists' expertise, the criteria for geotechnical investigations planning in PFS and FS have been defined basing on the following:

1. Site conditions assessment:
   - natural site conditions (presence of rivers, interception of meadow areas, interception of aquifers under pressure, uniformity/strata heterogeneity, etc.),
   - anthropic conditions of the site (presence of large-scale fillings, etc.),
   - conditions imposed by neighbouring surroundings (existence of buildings with different structural framings, bridges, tunnels, etc.).
2. Short term and long term design conditions:
   Short term and long term design conditions must be considered so as to cover all the conditions that may occur during construction and operation. Particularly, the first main criterion for taking into account the time factor in defining the design situation is the permeability of the soils; the second one is to verify the undrained properties of the soils wherever applicable.
3. TBM Type
   Selection of possible different TBM types to better face the variable geological conditions (EPB shield, Hydro shield, Multimode TBM, Variable Density TBM).
4. Risk of occurrence of dangerous phenomena during construction and operation phases:
   - "barrier phenomenon", manifested when the metro structures are positioned perpendicularly to the direction of the main groundwater flow,
   - subsidence phenomena, in case of hydro-works accident (during dewatering, drainage, ..) or for TBM operational problems.
5. Design of geotechnical works (auxiliary to basic structures):
   - ground improvement measures (waterproofing of the bottom of the excavations, conventional tunnels excavation, ..),
   - specific systems for water supply by deep wells,
   - dewatering systems for excavation works of stations or shafts.

6. Design criteria of main structures and related calculations methods:
   - most probable collapsing mechanisms,
   - geotechnical parameters for numerical analyses (soil behaviour, constitutive laws, ..),
   - use of numerical methods for tunnels and geotechnical works.
7. Creating a geotechnical and hydrogeological database as comparison and baseline for long-term monitoring of the behaviour of the metro and the neighbouring structures and of the public wells for water supply.

Finally, with specific reference to the Metro Line 4 Extension corridor, specific locations were identified to be studied in details:

- the area close to the Parliament Building and to the Romanian Orthodox Cathedral,
- the interferences and crossings of existing metro lines in operation (M1 line, Pasaj Hasdeu-Uranus and M2 line undercrossing),
- the probable planned interferences with future metro lines (undercrossing of M5 new line, M4 Chirigiu station over M7, M4 Toporasi station over M8, M4 connection to the existing M4 line with M4 in operation),
- the crossings of underground roads,
- the undercrossing of Dâmboviţa channel.

6 BRIEF GEOTECHNICAL DESCRIPTION OF THE METRO LINE 4 CORRIDOR SITE

The examination of the strategic options done in the PFS identified different transportation scenarios (bus, tram, light rail, commuter rail, metro) and by the comparative analysis of these modes of transport it was shown that each of these modes would operate along the same corridor between Gara de Nord and Gara Progresul. As in Figure 4 the corridor is about 10 km long and the alignment of Line 4 Extension crosses geographically, morphologically and hydrographically the Dâmboviţa River meadow area on both banks and also the plain area both to the north and south of the Dâmboviţa River.

The hydrographic network is poorly represented, the main watercourse being Dâmboviţa. The lithological succession intercepted along the corridor is considered in relation to the

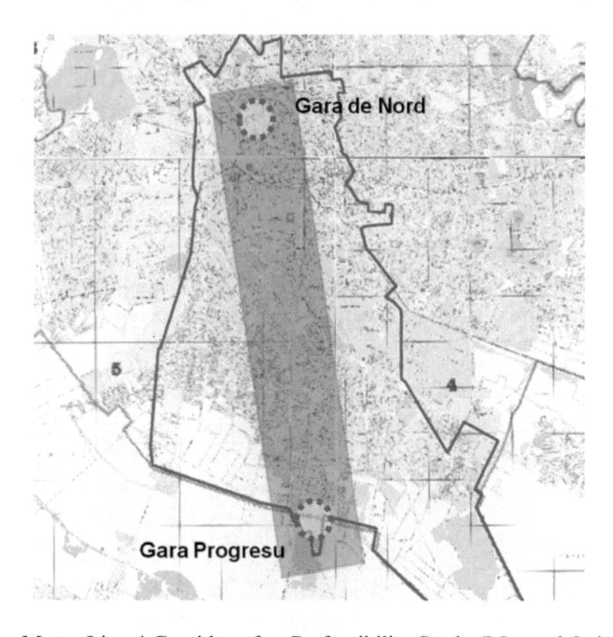

Figure 4. Bucharest Metro Line 4 Corridor after Prefeasibility Study (Metroul & Amberg 2018)

sedimentary conditions in which the meadow area and the field area were formed, by the uniformity/heterogeneity of strata, in terms of vertical and horizontal development, granulometric structure, presence of aquifers etc.

From the hydrogeological point of view, the considered corridor is characterized by the presence of three distinct hydrogeological structures: depth, medium and deep aquifers. They have been analysed and assessed through investigations already carried out for the hydrogeological design of metro works (dewatering, drainage, deep wells) with which the Line 4 will interact, such as Gara de Nord Station, Eroilor Station 1 and 2, Izvor Station, Eroii Revolutiei Station.

In the north of Dâmbovița river field area, the specific lithological sequence is characterized by an alternation of impermeable and permeable deposits, with different thicknesses and extension but uniform in sedimentation. Underground water was intercepted at depths ranging from 4 m to 5 m from ground surface.

In the meadow area, the land stratification is lenticular, sometimes incomplete as type strikes (erosion of the surface clay, clay complex, type 2 layer) with great variations in the granularity of the material both vertically and horizontally, highlighting the sinuous route of the old course of the river Dâmbovița. The underground water was intercepted at depths ranging from 1.7 m to 3 m from ground surface. For this area, where the underground route will under-cross the course of the Dâmbovița river, the already collected data from the pumping tests of the works for Metro Line 5 (used for the design of the dewatering systems) were also considered (Metroul 2011).

In the southern field area of Dâmbovița river, the characteristic feature of the field area is the uniformity of the sedimentation process, over which was deposited the formation type layer 2, consisting of deposits with small granulometric variations in the area of the clayey and silty fractions, with a significant vertical development. The stabilized hydrostatic level is at depths ranging from 4.5 m to 14 m, on a approx. 7 km long route, between the Natiunile Unite Boulevard, Drumul Găzarului, Progresul Station. In order to identify the permeability characteristics of the strata, in the Southern area, at Progresul Station, drillings for hydrogeological tests were carried out (pumping tests and recordings by piezometers in observation drills).

Also, the geotechnical main parameters were defined based mainly on the available database with checks carried out by the new investigations. These parameters were defined for the 3 different areas of Bucharest explained above (summary below in Table 1).

7 THE IMPORTANCE OF AN ACCURATE GEOTECHNICAL STUDY

Geotechnical constraints underlie fundamental decisions on the metro project (alignment selection, construction methods, TBM types, ground improvement methods, mitigation measures). The success of underground works design in general, but especially for metro projects in intensively urbanised area, is ensured largely by the quality of the geotechnical study.

At the beginning of the design, the purpose of field investigations should be determining the main relevant information to confirm the most important design assumptions, to estimate the technical difficulties and to identify the risks in order to confirm the feasibility, the location and the cost-effectiveness of the project. Then the development of a sound detailed design including the optimization of the costs becomes important in defining the geotechnical investigations.

The soft soils of Bucharest with their lithological succession along the metro corridor may sometimes differ even on distances of 20 m. Given the variability over small distances, the need

Table 1. Bucharest Metro Line 4 Corridor – Main geotechnical properties.

N.	Layer type	γ [kN/m^3]	φ [°]	c [kPa]	Ic [-]	E [MPa]	ν [-]	K_0 [-]
1	Filling	17.5	20	2	-	8	-	0.53-0.80
2	Clayey-sandy	19-20.5	14-20	28-40	0.55-0.9	0.7-14	0.4	0.49-0.60
2′	Clayey-sandy	19	20-24	10-15	-	9-12	0.3	0.40-0.55
3	Macrogranular	19-20	25-40	0	-	15-30	0.3	0.49-0.55
4	Intermediate clays	20	10-18	40-65	0.65-0.9	10-18	0.4	0.43-0.55
4′-5	Mostiștea sands	20	22-30	0	-	15-20	0.3	0.49-0.60

of adequate definition of the geological and geotechnical conditions along the alignment, by a detailed desk study and then by an adequate additional investigations plan, to avoid all the risks due to an insufficient research and knowledge of the site, is evident since the early stages.

Again at the very early stage more importance should be given to the collection and interpretation of quantitative results from geotechnical drilling and associated laboratory tests compared to qualitative information to define the physico-mechanical characteristics of the soils.

Besides in a urbanised city, with the project corridor close to important old and often damaged buildings and historical monuments, an important issue is the analysis of the presence of the phreatic and pressure aquifers. They require a detailed knowledge of the variation in the hydrostatic level and of the hydrogeological parameters underlying all the hydraulic and geotechnical calculation to guarantee all the information for the design of the dewatering systems, for TBM choice and operation and for the specific mitigation measures for the structures.

For Metro Line 4 Extension, a very detailed desk study with two additional investigation campaigns in Pre-Feasibility study (PFS) and in the Feasibility Study (FS) for Preliminary Design were used to confirm the main assumptions for the project (geology along the alignment with additional boreholes, soil conditions in the station locations, hydrogeological properties of the soils, ground conditions to define TBM types and soil conditioning system, ...). For what explained, it is clear that the investigations campaigns must be based on the overall analysis of the available information of the city area and must be defined in order to provide a general check along the corridor and detailed verifications in specific locations.

Particularly the additional investigations plan of the PFS phase was defined in order to achieve at least the frequency of the in situ investigation of 1 borehole every 500 m route length. The soil investigations were performed to a depth that exceeds the final elevation of the excavation by at least 10 m (approx. 1.5 tunnel diameter). Even taking into account the knowledge of Bucharest city area, only for the PFS, at the end of the investigations the geotechnical study of the area used approx. 40 boreholes (total length approx. 1000 m) including the ones done in '80ies and the new ones done in the 2017, with a final average distance between them of about 300 m along the metro alignment. For the ongoing FS phase for the Preliminary Design, the average drilling density is increased and for each metro station a minimum of 2 or 3 drillings will be available. By the end of this phase more than other 100 new boreholes will be available (total length more than 3000 m). This shows that the preliminary design and the detailed design of the works will be based at least on a hydro-geological and geotechnical model developed using investigations carried out with an average distance between two boreholes of about 80m.

Even though there is not an international accepted standard to define the boreholes campaigns, as per Table 2 it is clear that some Laws and Authors provided instructions for the investigations to define the geotechnical model. A reasonable application of all the instructions in Table 2 and in the related references allows to define a sound investigations campaign; obviously the local experience on site is always essential.

Table 2. Boreholes recommended spacing (in borehole/distance or in borehole length/tunnel length)

References	Instructions & Recommendations
EC7 (2007)	1/20 - 200 m (Linear structures)
BS 5930 (2015)	1/20 - 200 m (Linear structures)
AASHTO 88 (1988)	Soft Ground: 1/15 - 30 m (Adverse cond.) & 1/90 - 150 m (Favorable cond.)
	Mixed Face: 1/8 - 15 m (Adverse cond.) & 1/15 - 23 m (Favorable cond.)
USNC (1984)	L = 1.5 borehole linear feet per route foot of alignment
	1/125 m (wide tunnels database)
	1/80 m (database without deep tunnels)
IGS (2016)	1/60 - 100 m (Natural alluvial deposits)
Look B. G. (2007)	*1/25 - 50 m*
Elfatih M. A. A. (2014)	*Feasibility phase: 1/200 - 400 m*
	Preliminary Design phase: 1/50 - 100 m
	Detailed Design phase: 1/30 - 100 m

**in Italic recommendations from case histories.*

The comparison of Table 2 and the planned campaigns shows that the geotechnical model of this project will be based on more information than what usually foreseen in the major part of the available recommendations. Particularly, with reference to the tunnels and already at the end of the Preliminary Design stage, the investigation campaign will be fully in line with EC7 (spacing between 20-200 m) and AASHTO 88 for soft ground excavation in medium conditions (90-150 m), even though these laws refer to all the investigations including detailed design phase.

Finally it is worth to highlight that the level of exploratory borings, expressed in term of drilled borehole total length divided per alignment length (based on the planned investigation) is 0.36, similar to the "common practice value" 0.42, as in US NCTT 1984. Actually the current value is lower than the recommended target value (1.5) but this discrepancy is considered reasonable due to the fact that the general recommendation does not take into account the already available wide knowledge of the project site and that besides the present information level will be increased during the last design stage.

Taking into account that the "common practice value", as per US NCTT 1984, was defined more than 30 years ago to have the best picture of the geotechnical conditions in order to avoid any costs increasing during construction, this value could be considered an high target value (Parker 2004). In our opinion for a city metro project (not high depth), if there is a detailed basic knowledge, after a desk study by other case histories, and in collaboration with other investigation methods (geophysics) nowadays widely applied, this value could be reduced to 0.5-0.75.

For information, the average cost of geotechnical studies in Romania for the design of each metro line varies between 0.12 - 0.16% of the total investment value. Cost differences are due both to natural site conditions or to differences in route lengths. Due to specific national conditions (equipment costs, workers rates, …), the costs cannot be compared with the known values from world-wide common practice (0.5-3% of the total investment value).

8 CONCLUSIONS

The Metro Bucharest Line 4 Extension Project is a good example in which a sound and reliable practice is applied to define the geological and geotechnical model of a linear infrastructural project in an urbanised area. Particularly the importance of the use of all the available information and experiences and of the planning of new investigations campaigns fitted on the specific needs and locations since the very early design stages allow to have a complete picture of the main technical issues, thus supporting the most critical choices and optimizing at the same time the total investigation efforts in accordance with the most reliable international standards.

REFERENCES

AASHTO 1988. *Manual on subsurface investigations.*

Bala, A., Hannich, D., Ritter, J.R.R. & Ciugudean Toma, V. 2011. *Geological and geophysical model of the quaternary layers based on in situ measurements in Bucharest.* Romanian Report in Physics, Vol. 63-1.

Bancila, I., Florea, M. & Moldoveanu, T. 1980. *Engineering geology*, vol. I, 487-526.

Ciugudean Toma, V. 1985–2018. *Geotechnical studies and projects* elaborated at S.C. Metroul S.A

Cotet, P. 2001. *Data regarding geomorphology of Bucharest city*, Geographical problems, VX

Law 10 1995. *Quality in Construction.*

Metroul S.A. & Amberg Engineering 2018. *Line 4 Pre-Feasibility Study & Geotechnical Report.*

Metroul S.A. 2011. *Line 5 Geotechnical Study.*

NP 074 2014. *Geotechnical Documentation Norms.*

Parker, H.W. 2004. *Planning and Site Investigation in Tunnelling*, 1′ Congresso Brasilero de Túneis e Estruturas Subterráneas, Seminárió Internacional South American Tunnelling

SR EN 1997-2 2007. *Eurocode 7: Geotechnical design.*

The Technical Economic Memo on Metro construction in Bucharest, 1953. *Geotechnical and Hydrogeological Studies*, vol. 3, Chapter II.

U.S. National Committee on Tunneling Technology 1984. *Geotechnical Site Investigations for Underground Projects.* National Research Council.

Tunnels and Underground Cities: Engineering and Innovation meet Archaeology, Architecture and Art, Volume 3: Geological and geotechnical knowledge and requirements for project implementation – Peila, Viggiani & Celestino (Eds)
© 2020 Taylor & Francis Group, London, ISBN 978-0-367-46583-4

Ante-operam v/s as-built data ARA_T for a 15.965 m TBM-EPB, A1 Highway, Italy

M. Coli & E. Livi
Department of Earth Science, Florence University, Italy

A. Selleri
ASPI Autostrade per l'Italia, Roma, Italy

S. Comi
SPEA Ingegneria Europea S.P.A., Milano, Italy

ABSTRACT: The 7,548 m long and 15.965 m wide Santa Lucia tunnel, close to and north of Florence (Italy), is in execution by means of a TBM-EPB with high overburden and water head, and methane is presence. The tunnel crosses a highly tectonized shale complex and a limestone formation set in a series of tight anticlines and synclines. In the design phase, the GSI index was defined for the entire tunnel length; lab tests allowed determining USC and drillability. On this basis, during the design stage the ARA_T (theoretical Average Rate of Advance, m/d) was computed according to additional methods, such as for instance RME (Rock Mass Excavability) and FPI (Field Penetration Index, kN/cutter/mm/rev). The calculated ARA_T ranges, according to the methods and the tunnel chainage, is around 15 m/d. This figure is compared to the real excavation ARA_T and matched to the TBM data related to the employed energy in excavation.

1 INTRODUCTION

The case history we present regards a new highway tunnel 7,548 m long, with an excavation cross section of 15.965 m (three lanes). This tunnel, the Santa Lucia tunnel, is in the Northern Apennine range, close to and north of Florence (Italy), and is excavated by means of an EPB-TBM, under a high overburden and water head.

The TBM-EPB used, built by Herrenknecht AG, represents one of the biggest TBM used in the world for an entire tunnel and presents an innovative engine system based on 16 electric engines placed on the circumference of the cutting head and working synchronously; the TBM also provides for water and methane proofing.

Many authors have addressed the topic of defining TBM advancing rate since the advent of the TBM and recently too (from Barton 1999, 2015 to Rispoli et al. 2018). Recently, Yüksel & Bilgin (2014) categorized the factors affecting TBM performances into three groups

a) Mechanical factors, related to the TBM “hardware”.
b) Geo-factors, related to the geological and geomechanical setting.
c) Worksite factors, related to the worksite organization and staff experiences.

TBM advancing rate can be divided into ARA_T and the ARA_R:

- ARA_T (theoretical average rate of advance, m/d) only considers the net daily advancing span without any kind of stop or delay,
- ARA_R (real average rate of advance, m/d) takes into account all the stop and the delay related to the worksite organization, like staff experience, material supplying, mucking, and so on, which relate to the c) factors.

In this paper, we focus our attention to the ARA_T, which is linked to the a) and b) factors; moreover, because the TBM hardware (a) factor) is fixed,\the main variable to be consider into evaluating the ARA_T is the b) factor which is the geological and geomechanical setting of the rock-masses along the tunnel.

2 GEOLOGICAL SETTING

The preliminary geological field surveys and in situ and laboratory investigation allowed thorough defining of the geological setting of the area. The Santa Lucia tunnel passes through the shales of the Sillano Frm. and the bedded limestones of the Monte Morello Frm. (Figure 1).

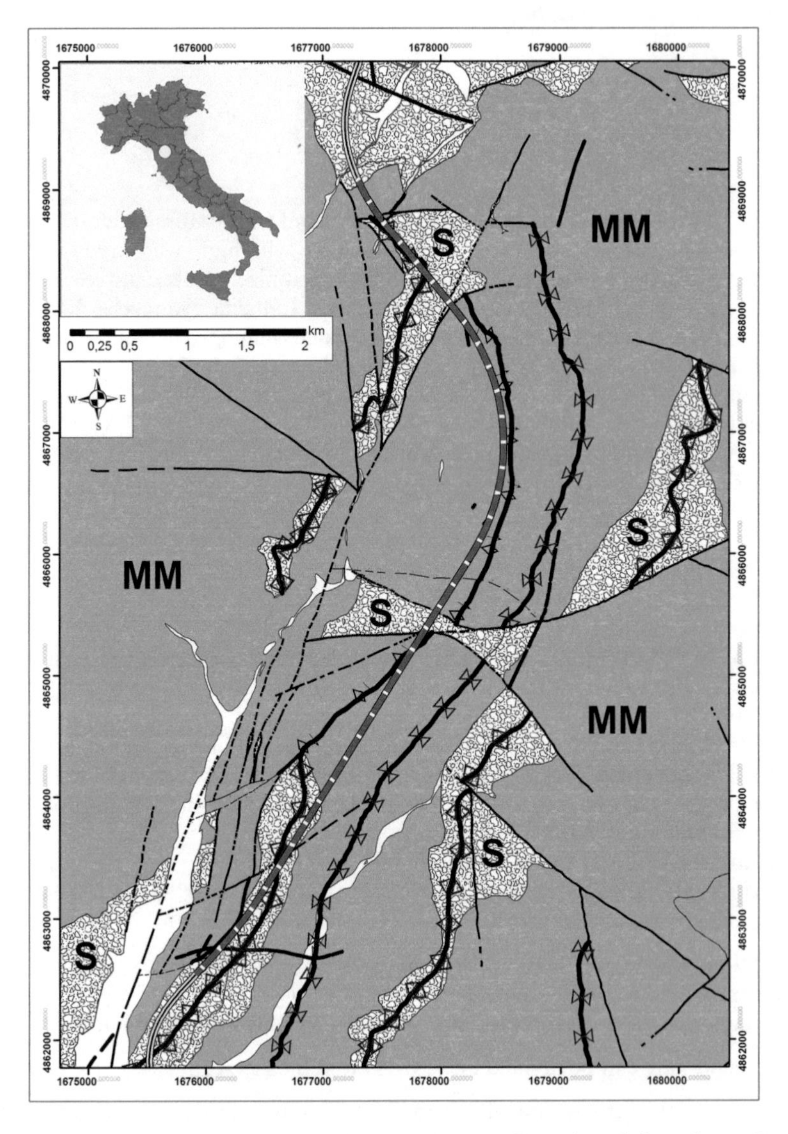

Figure 1. Geological map of the Santa Lucia tunnel area. Grey dotted line: Santa Lucia tunnel; MM: Monte Morello Frm., S: Sillano Frm; anticline and syncline axial-plane traces and normal faults are reported.

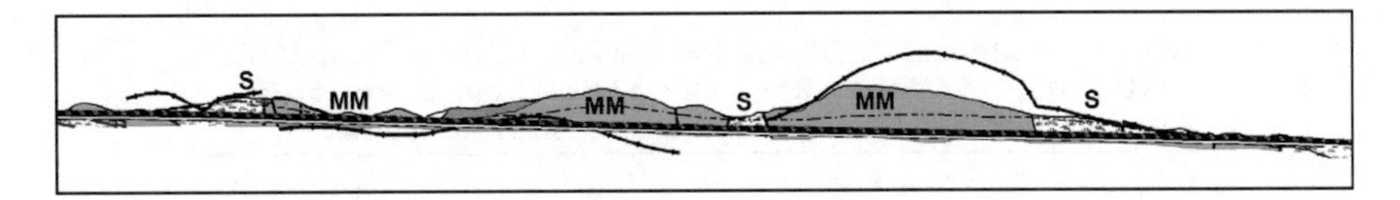

Figure 2. Geological cross-section along the Santa Lucia tunnel. MM: Monte Morello Frm.; S: Sillano Frm.; faults and the main anticline axial-plane trace are reported; dotted line: water table.

The Sillano and Monte Morello formations are part of the alloctonous Ligurian Units, a stack of the tectonic pile of the Northern Apennines. The Monte Morello Frm. is in sedimentary sequence over the Sillano Frm. These formations constitute the Calvana Supergroup, a sedimentary unit that resembles the cover of the European margin towards the Piedmont-Ligurian Ocean, now tectonically unrooted and thrusted onto the more external continental units, the Tuscan Nappe (Boccaletti et al. 1980, Coli and Fazzuoli 1983, Cicali and Pranzini 1986, Abbate 1992).

In the area of interest, these two formations are set in a series of tight anticlines and synclines NE-SW trending and SE verging, in places with local thrusts, with the Sillano Frm. plastically deformed and the Monte Morello Frm. brittle deformed.

The Santa Lucia tunnel runs almost parallel to the main tectonic structures and intercepts one of the main anticline axial plane (Figure 2).

The geological setting is within the fold axial-planes dipping towards west of about 40° and the bedding planes also dipping towards west of about 50° on the normal limb, and of about 30° on the reverse limb.

The Sillano Frm., Late Cretaceous (Bortolotti 1982), consists of black shales with rare inter-beds or levels of thin-bedded quartz-calcareous turbidites; very often, the tectonic processes highly disrupted the Sillano Frm. reducing it to broken formation and scaly-complex.

The Monte Morello Frm., Paleocene-middle Eocene (Coli and Fazzuoli 1983, Bortolotti 1982), is a calcareous flysch with alternating beds of limestone, marly limestone and marls. The beds are regularly set and with a thickness up to one meter; there are two sets of ubiquitous joints regularly spaced every few tens of centimetres. Close to faults or in the hinge zone of folds this assemblage is highly fractured.

The Monte Morello Frm. and the competent beds of the Sillano Frm., if not disrupted, present two sets of orthogonal joints, normal to bedding, and trending about N-S (*bc* joints) and W-E (*ac* joints), respectively.

3 GEOMECHANICAL SETTING

The field surveys and boreholes analysis led to dividing the rock-mass to be crossed into more RMZs (ISRM 1978) and thanks to 72 geostructural-geomechanical survey stations, uniformly distributed along the tunnel trace and the RMZs, it was possible to define the rock-mass settings.

The GSI index (Hoek et al. 1998, Hoek et al. 2000, Marinos and Hoek 2000) was defined for both the Sillano and the Monte Morello Frms, whereas, was not possible to define the RMR (Bieniawski 1989) for the highly tectonized Sillano Frm. (Table 1).

Laboratory tests regarded the UCS, abrasivity CAI (CERCHAR 1986) and the drillability (Bruland 1998), of course, for the Monte Morello Frm. only (Table 2), because of the poor consistence of the Sillano Frm. shales. These data deeply influence the TBM performance and, therefore, their availability drives the type of analysis executable.

Table 1. Synthesis of the GSI and RMR indexes for the different RMZs. MML-1, MML-2, MML3 = RMZs for the Monte Morello Frm.; SIL = RMZs for the Sillano Frm., ar: arenaceous, mc: marly-calcareous.

RMZ	RMR	GSI	UCS (MPa)
MML-1	56 - 76	40 - 65	82
MML-2	49 - 60	28 - 45	82
MML-3	35 - 50	20 - 55	82
SIL	–	18 - 25	78
SIL-ar	–	20 - 45	46
SIL-mc	–	20 - 40	70

Table 2. Results of the laboratory tests for UCS, abrasivity (CAI) and drillability for the intact rocks.

RMZ	UCS (MPa)	CAI	DRI
MML	82	1.04	68
SIL	78	0.63	66
SIL-ar	46	0.60	75
SIL-mc	70	1.00	68

4 ADVANCING RATE PREDICTION

The most performable tools for the prediction of TBM advance are the RME (Bieniawski et al. 2006, 2008), the Q-TBM (Barton 2000) and the FPI (Hassanpour et al. 2010).

The RME allows calculating the ARA_T on the base of the geo-factors and taking into account the TBM-factors in an indirect way through abacus referenced to different types of TBM, and the ARA_R is then calculated including the worksite-factors.

The FPI gives the ARA_T, it takes into account the TBM-factors on the base of the RMR, GSI or Q index, which resemble the geo-factors, the worksite-factors is included in a very simple way.

The Q-TBM gives the ARA_T and considers into the process the TBM- and geo-factors.

The Santa Lucia tunnell was divided into several RMZs according to the local geological-geomechanical setting. The field survey allowed having the GSI index for all the tunnel stretches corresponding to each RMZ, and the RMR index only for the tunnel stretches into the Monte Morello Frm. RMZs. For the tunnel stretches into the Sillano Frm. RMZs, the RMR index was indirectly inferred from the GSI though the correlation by Hoek et al. (1995).

The Q index was not surveyed in the field, due to the type of rock-mass; anyway, the Q index had been derived by the RMR index through the equations by Bieniawski (1984).

The use of the corresponding equations allowed having the distribution of these three indexes all along the Santa Lucia; their distribution is strictly related to the lithological setting, bedding and joint attitude and spacing, faults and axial planes occurrence (Figure 3).

Other data, from laboratory tests, were available during the design stage for inferring the expected advancing rate of the TBM was the UCS, abrasivity (CAI) and drillability.

For the TBM technical data (thrust, power, cutter head, RPM, …) we refer to the TBM-EPB, 15,40 m in diameter equally water and methane proofed, used in the last years for the close Sparvo tunnel in a similar geological context. The main difference between the two TBMs being the engine: hydraulic with a central torque beam for the Sparvo tunnel, 16 electric engines placed on the circumference of the cutting head and working synchronously for the Santa Lucia tunnel, 15.965 m in diameter. The data we

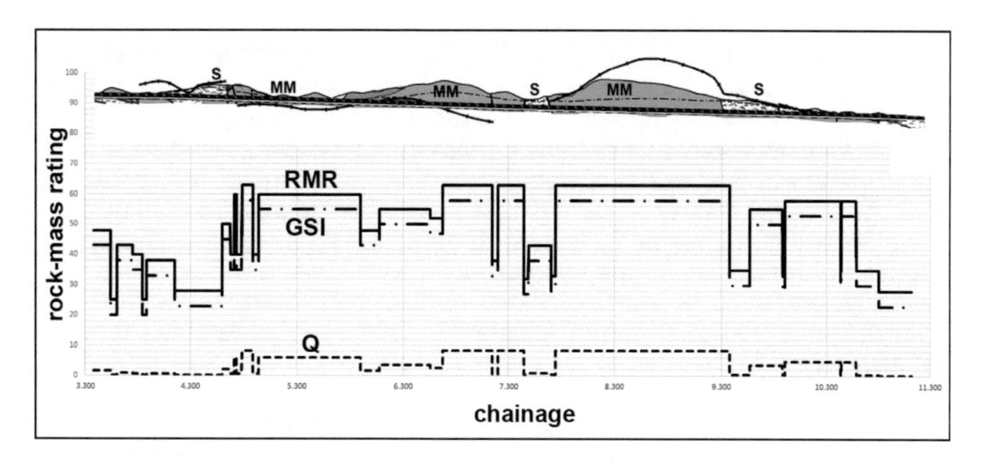

Figure 3. GSI, RMR and Q indexes plotted in respect to the geological setting as by Figure 2.

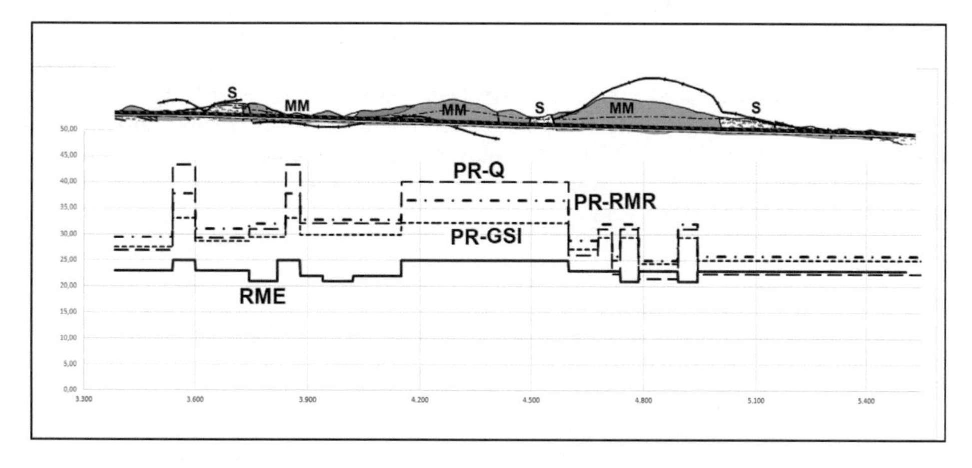

Figure 4. Average advancing rate computed for PR by FPI (according from RMR, GSI and Q index) and for RME approaches plotted in respect to the geological setting as by Figure 2.

used for transforming the FPI into PR (Penetration Rate, m/d).were 1 rpm and a thrust of 250 kN/cutter.

The elaboration of these input data allowed predicting the expected ARA_T for the Santa Lucia TBM, according to the several RMZs, faults and main tectonic features crossed by the tunnel. The prediction of the expected TBM advancing rate has been performed (Figure 4) by the approaches best fitting with the available data: FPI (Hassanpour et al 2010) and RME (Bieniawski et al. 2008).

In our analysis we did not used the Q-TBM because it requires some specific data not available at that project stage.

- PR by FPI: the analysis brings to an ARA_T with the values derived through the RMR being in average 26.82 m/d and standard deviation 2.78, those from the GSI with average 26.44 m/d and standard deviation 6.76, and from the Q index equal to 28.47 m/d and standard deviation 4.06. In our opinion, these values, because of previous experience, are overestimated, and in order to have more realistic values the computed ARA_T has been reduced to the 60%, this leads to a value around 15 m/d.

– RME: the analysis bring to an ARA_T with an average value of 23 m/d and standard deviation of less than 1 m. Because the referring abacus were made basing on data from TBM up to 12 m in diameter, equal to about 113 m^3/m and because our excavated volume is about 177 m^3/m (113/177= 0.64), we decided to conservatively reduce the expected daily rate of 1/3, to 15 m/d.

The resulting referring values of about 15/d derived independently by the two approaches are fully acceptable for a TBM of such a large diameter working in terrains so different, with high water head and gas, too; for the 7.500 m of the tunnel, the expected working days were computed in about 18 months of effective work.

5 AS-BUILT DATA

After some delays due to permits releasing, the Santa Lucia TBM started the excavation work on 2017 May. At the date of closing this paper (2018.09.01), due to several unpredictable hitches, for about 450 days the TBM only worked for 250 days (55%). As-built data for this tunnel stretch of about 2 km are reported in Figure 5.

The TBM had excavated about 2 km of the 7.5 of the tunnel from 2017 May 11th to 2018 August 2nd; which very clumsily corresponds to an ARA_R value of about 4.5 m/d, with ARA_T equal to 8 m/d. After each stop, the restart of the TBM implies low advancing rate; advancing the work, TBM stops decreased and the average TBM velocity increased pointing to 15 m/d.

Regarding the technical data, the most suitable available one is the energy consumption for segment (2.2 m length) (Figure 6), the minimum energy for segment being 3.545 kW, max 111.781 kW, mode 8.730 kW and average 13.387 kW with standard deviation of 7.103. The trend line resembles the medium values.

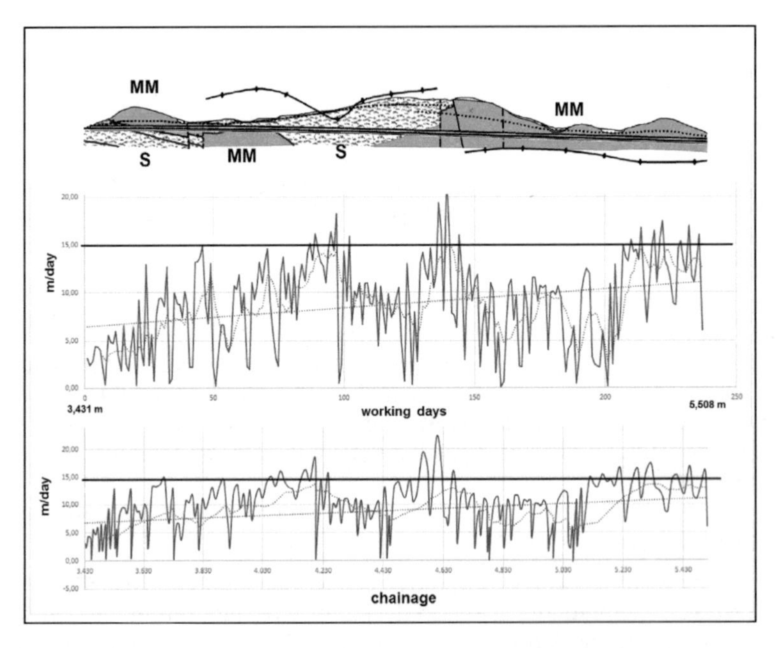

Figure 5. a) TBM advancing rate for the first 2 km of tunnel; dashed black line is the tendency, dotted black line is the moving average advancing rate computed every 15 m, to be compared with the geological setting of this tunnel stretch. b) TBM advancing rate for the first 2 km of tunnel; dashed black line is the tendency, dotted black line is the moving average advancing rate computed every 7 days, to be compared with the geological setting of this tunnel stretch. MM = Monte Morello Frm., S = Sillano Frm., anticline axial plane trace and water table (dotted) are reported.

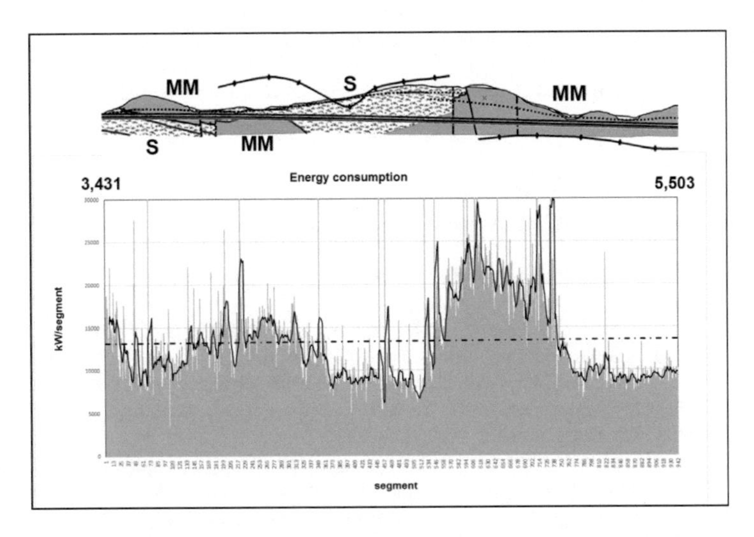

Figure 6. TBM energy consumption for the first 2 km of tunnel; dashed-dotted black line is the tendency, that is constant to about 13.000 kW/segment, to be compared with the geological setting of this tunnel stretch. MM = Monte Morello Frm., S = Sillano Frm., anticline axial plane trace and water table (dotted) are reported.

6 DISCUSSION

Due to the several stops, and the short stretch of the as-built data available, it is not simple to analyse these as built data, but some considerations can be made:

- if we consider the time laps during which the TBM worked with more continuity we can note that the expected ARA_T of 15 m/d is not far from the as built data,
- there is not any clear correlation between the TBM advancing rate and the geological setting, even if the advancing rate seems higher in the shale of the Sillano Frm. and lower in the limestone of the Monte Morello Frm.
- energy consumption for segment is lower in the shale of the Sillano Frm, and higher in the limestone of the Monte Morello Frm., in particular where there is a high overburden.
- there seems to be an inverse correlation between energy consumption and advancing rate: more energy-low advance/less energy-more advance; this point needs more data that we are expecting from the rest of the tunnel now expected to be excavated with more continuity.

7 CONCLUSIONS

Our aim to analyse the as-built performances of the Santa Lucia tunnel large TBM-EPB (15.965 m) in comparison to the geological setting and the design predicted advancing rate has been partially impeded by many yard and external factors that delayed the TBM start and its progression.

Presently, after the excavation of a stretch of 2 km of tunnel on 7.5 km of the entire tunnel (26%), a few preliminary considerations can anyway be made:

- The expected ARA_T of about 15 m/d is not far from the as-built average value of advancing.
- Until now the yard factors, plus external ones, heavily played against the TBM advancing rate, at now the ARA_R of 4 m/d.
- The excavation in the bedded hard limestone of the Monte Morello Frm. requires more energy for segment advancing; lower energy for advancing in the shale of the Sillano Frm.

This preliminary data from a so large TBM-EPB confirm our prediction in the design stage and encourage the application of these approaches to others tunnels under design.

REFERENCES

Abbate, E. 1992. Introduzione all'escursione pre-congresso. In *Guida alla traversata dell'Appennino Settentrionale, Soc. Geol. It.*: 1–14.

Barton, N. 1999. TBM performance estimation in rock using Q-TBM. *Tunnels & Tunneling International*, September 1999: 30–34.

Barton, N. 2000. *TBM Tunneling in Jointed and Faulted Rock*. AA Balkema. CRC Press.

Barton, N. 2015. TBM performance, prognosis and risk caused by faulting. In A.O. Sfriso et al. (eds), *Invited lectures of the 15th Pan-American Conference on Soil Mechanics and Geotechnical Engineering and 8th South American Congress on Rock Mechanics, Buenos Aires, Argentina, 15–18 November 2015*: 80–117. Amsterdam: IOS Press.

Bieniawski, Z. T., Celada, B., Galera, J.M. 2008. New applications of the excavability index for selection of TBM types and predicting their performance. *Proceedings of the* ITA-AITES *World Tunnel Congress 2008, Underground Facilities for Better Environment and Safety, Agra, India, 19–25 September 2008*: 1618–1629. Rotterdam: Balkema.

Bieniawski, Z.T. 1984. *Rock mechanics design in mining and tunnelling*. Rotterdam: Balkema.

Bieniawski, Z.T. 1989. *Engineering Rock Mass Classifications*. New York: John Wiley & Sons.

Bieniawski, Z.T., Celada, B, Galera, J.M., Álvares, M. 2006. Rock Mass Excavability (RME) index. *Proceedings of the ITA World Tunnelling Congress, Seoul, Korea, 22–27 April 2006*, Paper no. PITA06-254.

Boccaletti, M., Coli, M., Decandia, F.A., Giannini, E., Lazzaretto, A. 1980. Evoluzione dell'Appennino Settentrionale secondo un nuovo modello strutturale. *Mem. Soc. Geol. It.* 21: 359–373.

Bortolotti, V. 1982. Domini paleogeografici precoci. *In Guide Geologiche Regionali Appennino Tosco-Emiliano* 4: 16–18. BE-MA (Ed.).

Bruland, A. 1998. Hard rock tunnel boring. *Drillability test methods, Project Report 13A-98, NTN Anleggsdrift.*

CERCHAR. 1986. The Cerchar Abrasiveness Index. *Cerchar, Centre d'Etudes et Recherches de Charbonnages de France.12 S., Verneuil.*

Cicali, F., Pranzini, G. 1986. Il rapporto portata delle sorgenti-deflusso di base dei torrenti nel bilancio drogeologico di un rilievo carbonatico (Monti della Calvana - Firenze). *Geol. Appl. e Idrogeologia* 21: 155–172.

Coli, M., Fazzuoli, M. 1983. Assetto strutturale della Formazione di Monte Morello nei dintorni di Firenze. *Mem. Soc. Geol. It.* 26: 543–551.

Hassanpour, J., Rostami, J., Khamehchiyan, M., Bruland, A., Tavakoli, H.R. (2010), TBM performance analysis in pyroclastic rocks: a case history of Karaj water conveyance tunnel. *Rock Mech. Rock Eng.* 43(4): 427–445.

Hoek, E, Marinos, P, Benissi, M. 1998. Applicability of the geological strength index (GSI) classification for weak and sheared rock masses-the case of the Athens schist formation. *Bull. Eng. Geol. Env.* 57(2): 151–160.

Hoek, E., Kaiser, P.K., Bawden, W.F. 1995. *Support of Underground Excavations in Hard Rock*. Rotterdam: Balkema.

ISRM (International Society for Rock Mechanics). 1978. Suggested Methods for the Quantitative Description of Discontinuities in Rock Masses. *International Journal of Rock Mechanics and Mining Sciences & Geomechanics Abstracts* 15: 319–368.

Marinos, P., Hoek, E. 2000. GSI – A geologically friendly tool for rock mass strength estimation. *Proceedings of the International Conference on Geotechnical and Geological Engineering (GeoEng2000), Melbourne, 19–24 November 2000*: 1422–1446.

Rispoli, A., Ferrero, A. M., Cardu, M., Farinetti, A. 2018. TBM performance assessment of an exploratory tunnel in hard rock. *Proceedings of the 2018 ISRM International Symposium, Geomechanics and Geodynamics of Rock Masses, St. Pietroburgo, Russia May 2018*: 1287–1296.

Yüksel, A., Bilgin, N. 2014. The effect of geomechanical properties of rock on the performance of tunnel boring machines employed in metro tunnels. *Yerbilimleri/Earth Sciences* 35(1): 17–36.

Tunnels and Underground Cities: Engineering and Innovation meet Archaeology, Architecture and Art, Volume 3: Geological and geotechnical knowledge and requirements for project implementation – Peila, Viggiani & Celestino (Eds)
© 2020 Taylor & Francis Group, London, ISBN 978-0-367-46583-4

Mixed transitional ground impact on a TBM. The OCIT Experience in Akron, Ohio, USA

E. Comis & W. Gyorgak
McMillen Jacobs Associates, Mayfield Heights, Ohio, USA

D. Chastka
Kenny/Obayashi Joint Venture, Akron, Ohio, USA

ABSTRACT: For the first time in North America a dual mode Rock/EPB TBM, 9.26m diameter bore was used to excavate the Ohio Canal Interceptor Tunnel (OCIT) in the downtown Akron area, USA. Tunneling was carried-out at a uniform slope of 0.15 percent through ground conditions that consisted of soft ground, mixed face soft ground over bedrock, and bedrock. The Mixed Transitional Ground (MTG) resulted in the biggest challenge for the TBM. This paper specifically addresses the influence of the MTG on the TBM Utilization Factor and present the counter-measures implanted in these special geological conditions.

1 INTRODUCTION

The nominal or installed operational values for a Tunnel Boring Machine (TBM) are: thrust, torque, power, and rotational speed. These are highly related to the expected geology and the tunnel diameter and will directly affect the TBM penetration rate (PR), which is the TBM progress in terms of feet/hour (m/hr) during the boring process.

However, the TBM utilization (U), which reflects the amount of time that the TBM actually excavates, has to be determined for a comprehensive TBM performance analysis. Another parameter, that is often cited as part of performance prediction, is the cutter life. This parameter is typically expressed in terms of average cutter life in hours, meter travelled on the face, cutters per meters of tunnel, or cutters per cubic meter of excavated rock.

The primary focus of performance prediction studies is the prediction of the TBM's advance rate (AR) often expressed as the amount of daily advance expressed in feet per day (m/day), feet per week (m/week). These studies have mainly tried to find out the relationship between the rock mass characteristics and the machine performance. Few publications can be found on utilization records for TBMs working in soft ground while typical utilization rates in rock are reported from 5%, for very difficult and complex geologies with poor site management, to around 55%, in perfect working conditions for an open type TBM in moderately strong rock with no ground support requirements. However, the most common range of TBM utilization is reported to be in the 20 to 30% range (Rostami, 2016).

Machine utilization is very sensitive to ground condition. In rock tunnels bad ground could cause face collapse and cutterhead jamming. Ground convergence and squeezing could also become an issue for shielded TBMs and water bearing formations where the inflow of water can interrupt the operations and cause long delays for dewatering and drainage. Often measures offered to mitigate the ground related issues, such as probe drilling or pre-excavation grouting, cause their own delays and can become source of inaccuracies in predicting machine utilization.

The impacts on machine utilization of factors such as contractor and crew experience, management approach, labor issues, site arrangement, logistical issues with supplies, repair and spare parts, electricity and power supply, transportation and site access limit muck haulage, availability of local workforce, and so on, are also very important but extremely difficult to quantify.

TBM utilization has a direct impact on TBM advance rate (AR) and ultimately project schedule and costs. The majority of case histories report the penetration rate while the advance rate or utilization, are not reported as often, or it is not clear if they are based on available time, boring days, working days, or calendar days.

The Ohio Canal Interceptor Tunnel (OCIT) Project involved the construction of a 6,200-foot-long (1,890m) conveyance and storage tunnel with a finished inside diameter of 27 feet (8.23m) to control combined sewer overflows for several regulators in the downtown Akron area. The OCIT Project was awarded to Kenny/Obayashi Joint Venture with a Notice-to-Proceed on November 4, 2015.

The section of the GBR related to the OCIT identified three major reaches that were defined as distinctly different ground conditions:

- Reach 1 which primarily consisted of soft ground, Silty Sand interbedded with Silt underlain by Glacial Till
- Reach 2 which was a transitionary zone with soft ground overlying bedrock (consisted of Shale and Siltstone), defined as mixed ground conditions
- Reach 3 which was comprised of bedrock with two sections of low rock cover.

The tunnel was excavated using a 30.38-foot (9.26m) diameter dual mode type "Crossover" (XRE) Rock/EPB Tunnel Boring Machine (TBM), manufactured and supplied by The Robbins Company.

The TBM was launched in October 2017 and completed the drive in August 2018. Reach 1 and 2, for a total of 810 ft (247 m), presented a challenge for the project as characterized by low penetration rate, high cutter consumption, difficulties in maintaining face pressure, and low TBM utilization. Reach 3, for a total of 5,400 ft (1,647 m), was distinguished by an exceptional high TBM utilization and steady advance rates.

2 IMPACT OF MIXED TRANSITIONAL GROUND ON TBM PERFORMANCE

Because of alignment restrictions and increase in demand for tunnel with larger diameters, more and more tunnel boring machines are working in mixed face conditions. As mentioned by several authors and later summarized by Oliveira and Diederichs (2016), mixed face conditions still present a challenging scenario for TBMs.

Several issues that have been reported from tunneling projects excavated along mixed ground conditions (Della Valle, 2001; Thewes 2004, Zhao et al., 2007; Shirlaw, 2016; Comulada et al., 2016; Silva et al. 2017) such as:

- Ground loss, settlements and sinkholes
- Slow rates of tunneling
- Rapid tool wear, especially of the disc cutters and screw conveyor
- Damage to tools, mixing arms and other parts of the TBM
- Very frequent and lengthy interventions
- Clogging
- High temperatures inside the excavation chamber

Reach 2 of the OCIT was described in the GBR as a transitionary zone with soft ground overlying bedrock. Soft ground consisted of silty sand containing interbedded layers of silt with a variable layer of glacial till deposits following the top of the bedrock. Bedrock consisted primarily of shale with varying layers of siltstone. The top of the bedrock was identified as highly weathered at depths up to 10 feet (3 m). Groundwater level was 4 feet (1.2 m) from the crown of the tunnel at the start of the reach moving to the crown of the tunnel at the end.

This geological condition can be denominated mixed transitional ground (MTG) as defined by Oliveira and Diederichs (2016).

Langmaack and Lee (2016) emphasized that a successful TBM drive in general and especially in difficult ground conditions can only be reached by combining adequate mechanical solutions together with the use of suitable chemicals as well as working with experienced TBM operators.

Especially in EPB tunneling, the correct choice and use of well adapted soil conditioners can make a considerable difference for the success of a tunneling project; both in highly permeable grounds as well as in sticky clays. Soil conditioning is determined by the following parameters defined in EFNARC (2005); Foam Expansion Ratio (FER), the ratio between the volume of foam at working pressure and the volume of the solution (with a specified the range between 5 and 30), Foam Injection Ratio (FIR), the ratio between the injected volume of foam at working pressure and the banked volume of ground (with specified range between 10% and 80%), and Concentration of surfactant (C_f), the concentration of used surfactant in the preparation of the foaming liquid (with a specified range between 0.5% and 5%).

The TBM design features adopted for the expected operating conditions were described by Comis and Chastka in 2017. Cutterhead design for proficient boring in both rock and soil conditions, adjustable main drive speed with an over-speed mode for operation in hard rock, and special screw conveyor wear protection features were addressed for these specific ground conditions. TBM specifications are listed in Table 1.

A soil conditioning assessment was also conducted prior to the construction of the OCIT. Testing varied according to soil type. For the sandy soils, testing for mobility was made using slump. For the silty soils, testing included adhesion (using a variation of the Langmaack adhesion test (NAT 2000, Langmaack)), slump, static flow, dynamic flow, and rheometry. Based on these tests the recommendation was a polymer foaming agent for Reach 1, with FER and FIR to be adjusted with actual ground conditions ant the use of an anti-clay polymer if more than expected portion of clay was found in the excavated soil. For the mixed face conditions (Reach 2) it was suggested the same polymer foaming agent with particular attention to the change of rock/soil ratio. If the ratio of fine particles in the ground increased, water could be additionally injected into the chamber in order to increase fluidity and to adjust viscosity. If the ratio of fine particles in the ground decreased, foaming agent, polymer, or bentonite could be additionally injected into the chamber in order to keep necessary fluidity and viscosity. Liquid type polymers were also considered suitable for this purpose. Final set-up for Reach 2 ground conditions were considered sufficient for the bedrock zone, Reach 3.

The OCIT TBM weekly production for the whole drive is shown in Figure 1. An average weekly advance rate of 34.6 ft/week (10.5 m/week) has been recorded for Reaches 1 and 2, and 242 ft/week (73.8 m/week) for Reach 3. Reach 1 and 2 have been characterized by low penetration rate, high cutter consumption, and difficulties in maintaining face pressure.

Lack of production (high thrust with low penetration rate) prompted several cutterhead inspections which revealed an almost a full face of dry sand and showed material packed in

Table 1. TBM Specifications

TBM type	Crossover Rock/EPB TBM
Bore Diameter (m)	9.26
Cutting tools (17" Cutters/Knives)	56
Drive System	12 x 190kW VFD electric motors
Maximum Rotational Speed (rpm)	3.5
Rated Hydrostatic Pressure (bar)	3.5
Shield length (m)	12.19
Maximum Cutterhead Torque (kNm)	23,270
Maximum Thrust Force (kN)	65,900
Screw Conveyor Capacity (m^3/h)	773 @ 16 rpm
Primary Voltage (V)	13,200

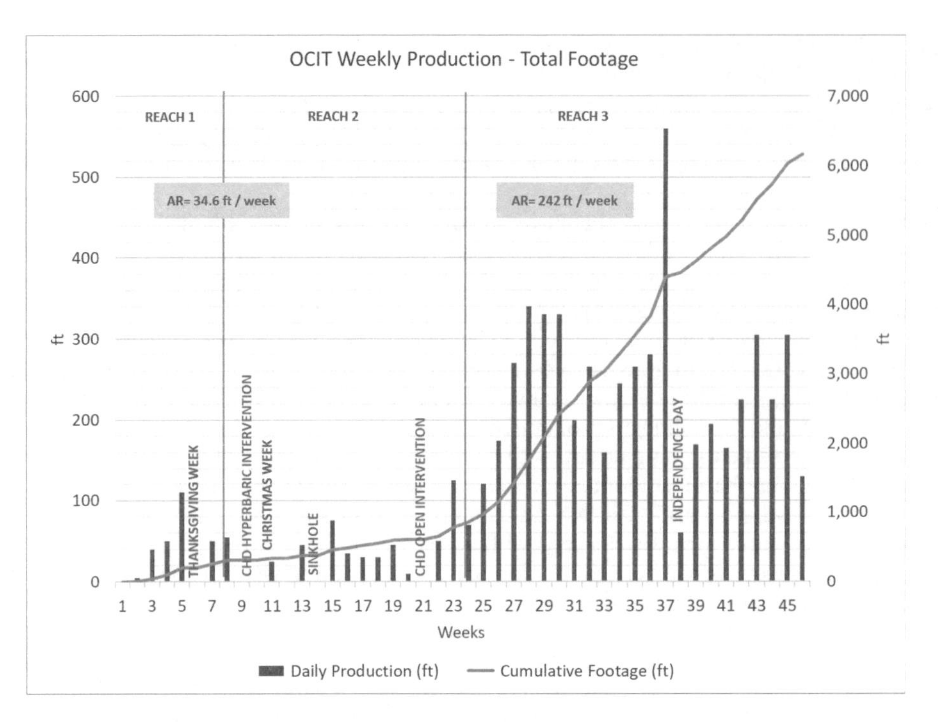

Figure 1. OCIT Weekly Production.

Figure 2. Different types of cutters being tested.

the rear of the housings as well as the buckets. Buckets were filled of clay/sand, easily removable with a pressure washer. Most of the outer radial buckets were hard packed and needed the use of pneumatic tools to remove material. The dryer conditions at the face were thought to be related to the dewatering carried out at the mining site for the portal construction.

Material was packing into the cutter housings preventing the cutters to rotate. Three different cutter types have been tested (Figure 2): standard disk, traction cutters, and carbide tip. Standard and traction cutters both had issues with the housings clogging and then stopping the rotation of the cutter which then led to excessive wear. Carbide tip cutters provided more traction and allowed the cutter to rotate even after the housing were clogged. The original dressing of the cutterhead had carbide-tipped cutters at the gauge position while the remaining cutters were all standard disks. During the first intervention, a decision was made to switch the transition cutters over to carbide-tipped as well. The cutterhead continued to clog developing high temperature in the plenum, baking the clay and sand material and ultimately steaming muck which was conveyed outside the tunnel (Figure 3).

During the last intervention in closed mode some grill bars were removed as well, hoping to mitigate the cutterhead clogging by increasing its opening ratio that was designed in consideration that 75% of the drive was going to be in rock. Unfortunately, the penetration rate did not improve remarkably.

Figure 3. Clogging inside the TBM cutterhead and steam in the tunnel.

Attention was brought to the soil conditioning as well. Initial foam produced a low-quality result, a liquid consistency with no stand time, when checked at the foam generators on the TBM. Foam supplier ran multiple tests to see if it was an issue with the product or the ratios being used getting good results in their lab. The medium in the generators were then evaluated and changed from the initial set-up of loosely packed larger metal cylinders to a tightly packed plastic top hat and ultimately to a tightly packed steel wool. The final result was a foam of good consistency and quality. Despite getting the desired output, the cutter-head still experienced clogging.

Focus was finally shifted to foam parameters adjustment. The initial soil conditioning evaluation suggested an FER of 15 with C_f between 3% and 5%. However, even with a FIR that has been as high as 200–300% the clogging was not mitigated. The FER was reduced to 2 and the FIR set between 60% and 80%. This finally allowed the completion of Reach 2. The penetration rate increased from 5mm/min to 15mm/min suggesting that the cutterhead was progressively starting to clear-up, concurrently with the increase of rock ratio at the face of excavation (Figure 4).

Average Instantaneous Penetration Rate (PR) and Cutter Consumption are shown in Figure 5 for each single reach. The mixed transitional ground resulted in an average instantaneous penetration rate of 8.52mm/min with a cutter consumption of 1.8m/cutter. This data highlights that mixed transitional ground has to be considered a difficult ground condition by the industry.

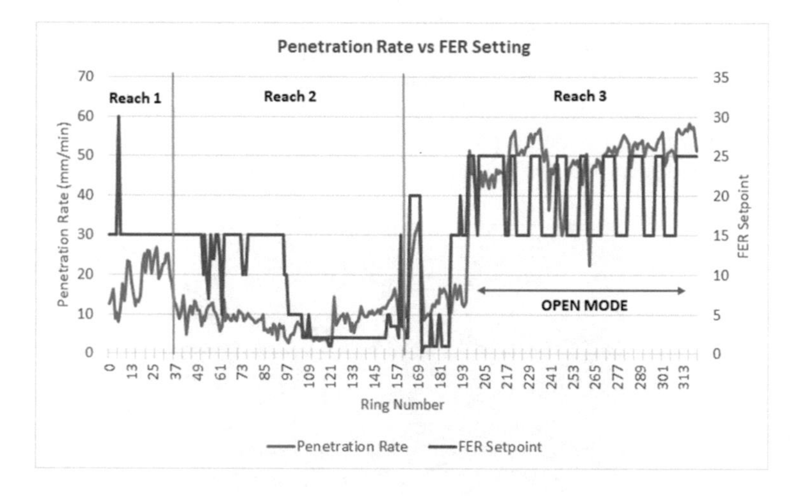

Figure 4. TBM penetration rate vs. FER setting.

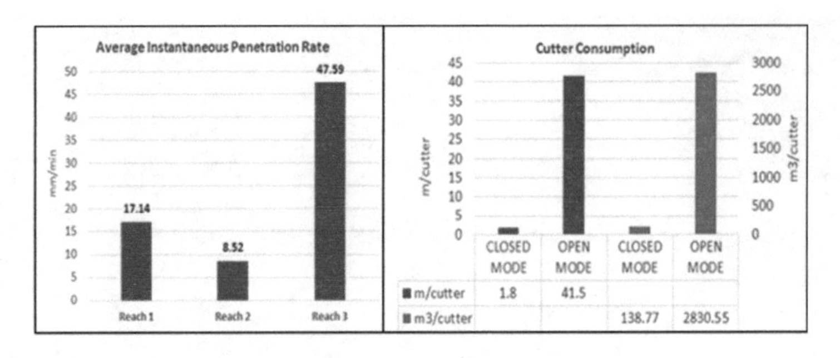

Figure 5. Average instantaneous penetration rate and cutter consumption.

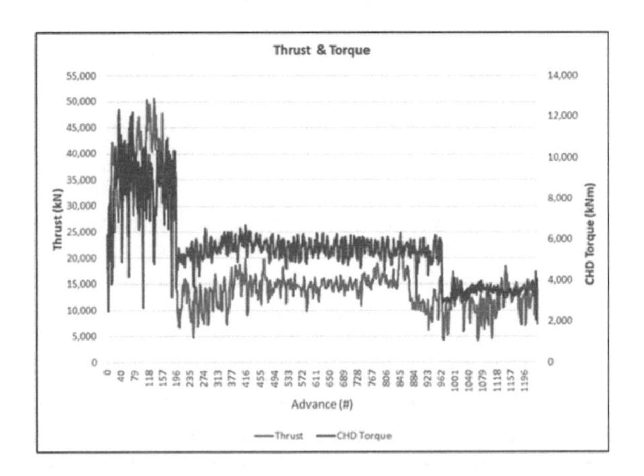

Figure 6. Torque and Thrust for the whole drive.

Reach 3 had instantaneous penetration rates as high as 63.1mm/min with an average of 47.6mm/min. The average PR was affected by limitation in torque and thrust for the last 1,200 feet (366 m) due to failure of two main drive motors and one of the screw motors (Figure 6).

Reach 1 and 2 were characterized by high torque and thrust levels as described above.

According to Comis and Chastka (2017), the main challenge in Reach 3 was expected to be the screw conveyor wear. Screw conveyor speed was in fact increased up to 14 RPM in order to increase its efficiency in rock. Special wear prevention features were adopted in the screw conveyor design. A wear monitoring plan was defined in the *Tunnel Excavation Plan* and carried-out during the mining operation with periodic visual inspection and was casing thickness measurement. The 5,400 feet (1,650 m) boring in rock did not result in measurable wear. As shown in Figure 7, the chromium carbide plates installed at the auger flights and at the casing and the hard facing in a crosshatch pattern resulted just polished by the mix of muck, soil conditioning and bentonite injected in the plenum.

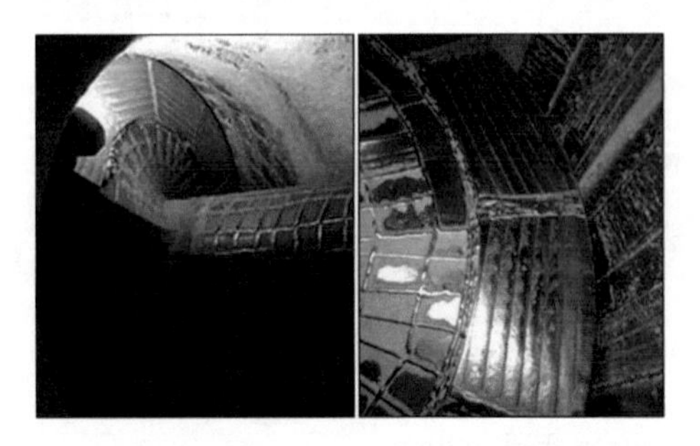

Figure 7. Screw wear visual inspection.

3 TBM UTILIZATION

A mechanized tunneling project schedule and cost estimate relied on performance prediction of TBMs. This process involved a thorough understanding of the complexities in the site geology, machine specification, and site management.

Performance analysis for the OCIT TBM was carried-out from data collected from both the TBM data logger and contractor Shift Reports and categorized as shown in Table 2. The working week consisted usually of six days, five of mining and one dedicated to maintenance of the TBM and ancillary systems. The overall Utilization for the length of the drive has been 45% as shown in Figure 8. Major delays were observed for the cutterhead maintenance (15%) and non-machine related (18%). The Utilization was analyzed for each tunnel Reach as identified in the GBR.

3.1 *Reach 1*

Reach 1 utilization was 37% (Figure 9) with major downtimes observed in the TBM and non-machine related delays. This can be considered a common distribution, as it reflects the start-up of the job, with site final set-up for mining still to be implemented and TBM commissioning activities still on-going.

3.2 *Reach 2*

Reach 2 utilization was 31% (Figure 10) with major downtimes due to cutterhead maintenance (29%) and TBM break-downs (17%). The non-machine delays were reduced to 15% as the site was finally configured for full mining operation. The combined effect of learning a new TBM, addressing constantly changing ground conditions in the first two reaches, and a low level of experienced workers led to poor production, increased downtime, and a greater level of difficulty troubleshooting any issues that occurred.

Table 2. Performance categories.

Operation	TBM Delays	Non-TBM Delays
Safety	Cutterhead Maintenance	Probe Drilling
Boring	Ordinary Maintenance	Tunnel Conveyor System Delays
Ring Building	TBM Downtimes	Supply Chain Delays
	Back-Up Downtimes	Surveying
		Utility Extension
		Transport
		Power Outage

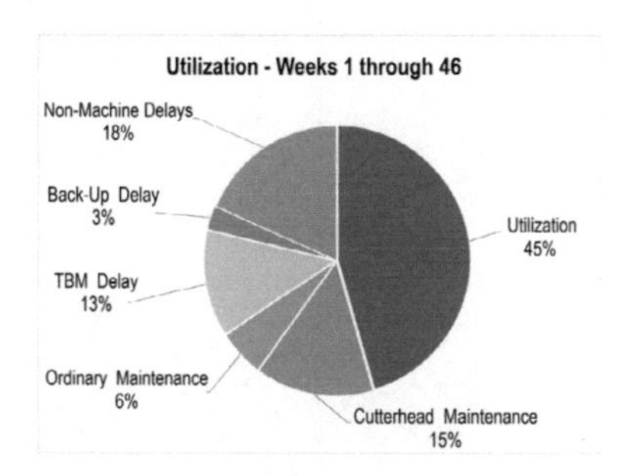

Figure 8. Total Utilization for the whole drive.

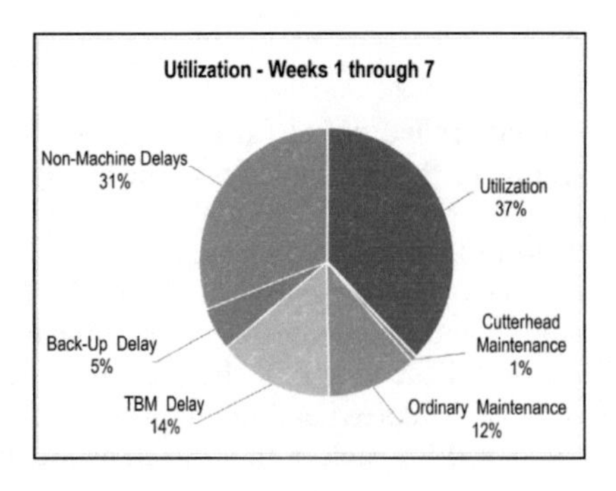

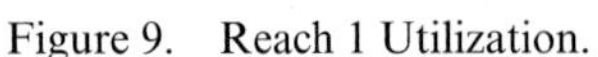

Figure 9. Reach 1 Utilization.

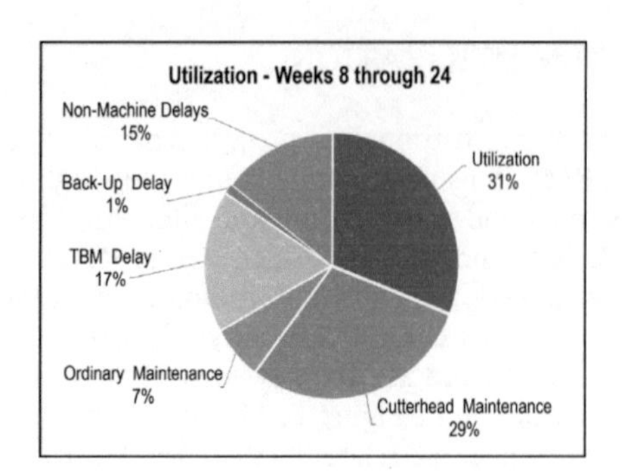

Figure 10. Reach 2 Utilization.

The OCIT project requirement to hire 35% of local workforce made the project more difficult from a labor perspective. Akron had not had any TBM tunnels prior to this project thus it limited the workforce experience on the machine. The end result was a much longer learning curve due to extensive training and turnover.

The ring building time learning curve can be an example of these difficulties (Figure 11), especially in combination with the difficult ground conditions. Ring building is different on every machine as the operator and supporting staff need to understand the capabilities of the erector and the process of setting the segments. Furthermore, a relationship exists between production and ring build time. The natural assumption that during low production periods, setting of only two rings in a shift would lead to an efficient ring building with high construction quality is often misleading. The reality was in fact that in closed mode, when production was low, the ring build were around 60 to 80 minutes with gap and step issues. Almost instantaneously when the TBM switched to open mode the production increased dramatically and the ring build went to less than 40 minutes with quality issues almost negligible. This can be considered a direct result of more practice with the equipment but also the crew finding a good rhythm with a consistent cycle.

3.3 *Reach 3*

Reach 3 Utilization was 60% over 20 mining weeks. This result is remarkable, especially considering that muck is removed from the mixing chamber with a screw conveyor (less efficient than a belt conveyor that is usually adopted for rock TBMs). The overall performance in this Reach was affected by the non-machine delays. The utilization analysis identified the weakness of the system as being the tunnel conveyor with 45.5% of the downtimes (Figure 12).

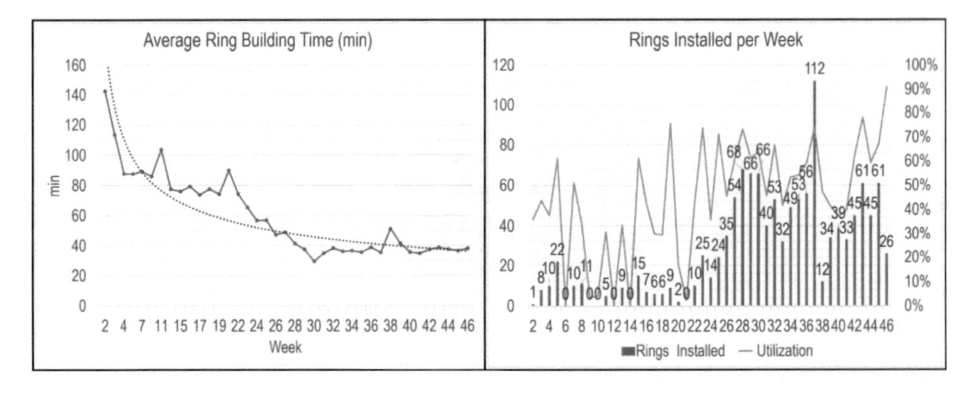

Figure 11. Ring Building Time Learning Curve and Weekly Production.

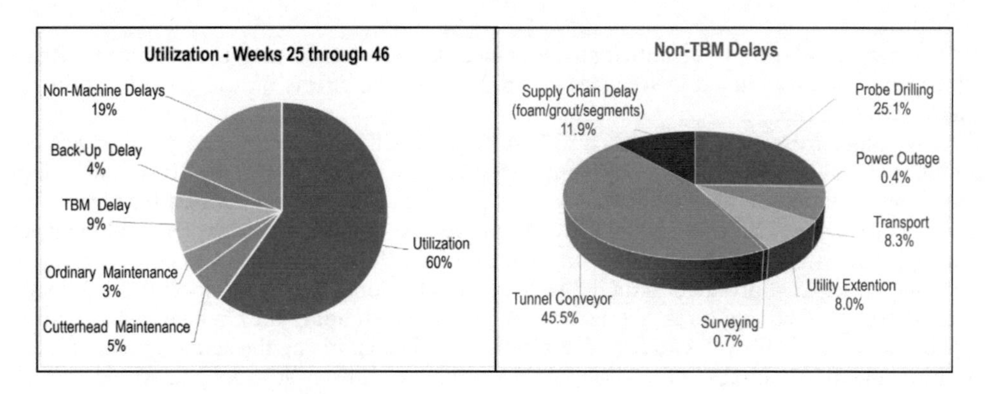

Figure 12. Reach 3 TBM Utilization.

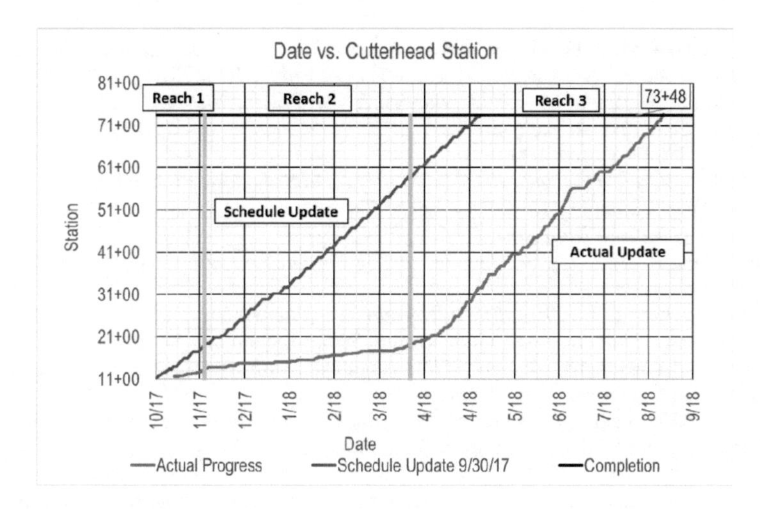

Figure 13. Estimated mining schedule vs real mining schedule.

Explanation can be found in the attention required at the three transfer points in surface and respective hoppers (tunnel conveyor to overland, overland to stacker, and stacker to muck pile). Increases in amount of muck or changes in consistency of the material caused clogging of transfer points and large downtime to clean the system. The TBM penetration rate was also adjusted accordingly, in order to limit the muck loaded on the conveyor. Differently, probe drilling was an unavoidable contractual requirement has to be considered during future advance rate estimates.

4 CONCLUSIONS

The OCIT Project experienced the most challenging portion of the excavation in the first two Reaches of the tunnel. Although this was where the challenges were anticipated, the type and severity were under-estimated. The natural assumption of initial slower advance rates was made in consideration of inexperienced workers, typical learning curve on a TBM, and adjusting ground conditions. However, these factors combined with others such as cutterhead design, foam generation, soil conditioning, and constantly changing geological conditions resulted in an initial inefficiency of the system and a longer learning curve for the crew.

As shown in Figure 13, the anticipated advance rates were demonstrated possible by Reach 3 but were under-estimated in Reaches 1 and 2 due to the effects of a lower utilization rate and penetration rate.

As we move forward as an industry, TBM projects will only continue to grow in size and complexity. As cities grow and infrastructure expands, the demand for larger TBMs and excavation in more difficult and varying geology will increase. The need for tunnels throughout the world is increasing at an exponential rate. This leaves the limited pool of experienced managers and engineers with the challenge of finding and training a TBM tunnels experienced workforce.

Basing anticipated production on TBM advance rate alone ignores the specific complexities of each project. Defining a real Utilization Factor, understanding the challenges and realistic capabilities of available workforce and equipment, and categorizing the actual causes of inefficiency can help to compile a database useful to make the industry advance as a whole.

REFERENCES

Comis, E. and Chastka, D. 2017. *Design and Implementation of a Large-Diameter, Dual-Mode "Crossover" TBM for the Akron Ohio Canal Interceptor Tunnel*; Proc. RETC 2017: pp. 488–497.

Comulada, M., Maidl, U., Silva, M.A.P., Aguiar, G. & Ferreira, A. 2016. *Experiences Gained in Heterogeneous Ground Conditions at the Twin-Tube EPB Shield Tunnels in São Paulo Metro Line 5*; Proc. ITA 2016 WTC, San Francisco: 1-11.

Della Valle, N. 2001. *Boring Through a Rock-Soil Interface in Singapore*; Proc. RETC 2001: pp. 633–645.

EFNARC, 2005. *Specification and Guidelines for the Use of Specialist Products for Mechanized Tunneling (TBM) in Soft Ground and Hard Rock.*

Girard, D. and Chen, R. 2018. *Lessons Learned in Dry Ground Excavation Using an EPBM*; Proc. NAT 2018: pp. 908–914.

Langmaack, L. and Lee, K.F. 2016. *Difficult ground conditions? Use the Right Chemicals! Chances–Limits–Requirements.* Tunnelling and Underground Space Technology 57: pp. 112–121.

Oliveira, D.G.G. and Diederichs, M. 2016. *TBM Interaction with Soil-Rock Transitional Ground*; Proc. TAC 2016, Annual Conference, Ottawa: pp. 1-8.

Rostami, J. 2016. *Performance Prediction of Hard Rock Tunnel Boring Machines (TBMs) in Difficult Ground.* Tunnelling and Underground Space Technology 57: pp. 173–182.

Shirlaw, N. 2016. *Pressurized TBM Tunnelling in Mixed-Face Conditions Resulting from Tropical Weathering of Igneous Rock.* Tunnelling and Underground Space Technology 57: pp. 225–240.

Silva, M. A. A. P., Katayama, L. T., Leyser, F. G., Aguiar, G. and Ferreira, A. A. 2017. *Twin Tunnels Excavated in Mixed-Face Conditions.* Proc. WTC 2017, Bergen, Norway.

Thewes, M. and Burger, W. 2004. *Clogging Risks for TBM Drives in Clay.* Tunnels & Tunnelling International, June 2004: pp. 28–31.

Zhao, J., Gong, Q.M., and Eisensten, Z. 2007. *Tunnelling Through a Frequently Changing and Mixed Ground: a Case History in Singapore.* Tunnelling and Underground Space Technology 22: pp. 388–400.

Tunnels and Underground Cities: Engineering and Innovation meet Archaeology, Architecture and Art, Volume 3: Geological and geotechnical knowledge and requirements for project implementation – Peila, Viggiani & Celestino (Eds)

Geological and geotechnical main design aspects of two tunnels on Fortezza-Verona High Speed Line in Northern Italy

A. Corbo, A. Sciotti & S. Vagnozzi
Italferr S.p.A. – Tunnelling Unit, Rome, Italy

S. Rodani
Italferr S.p.A. – Geology Unit, Rome, Italy

ABSTRACT: The two single track railway tunnels of Ponte Gardena along Fortezza-Verona High Speed line (North of Italy) are designed to be excavated by EPB-TBM. Its alignment runs on the left bank of the Isarco River. Combined interpretation of geological and geotechnical data led to crucial information during the design process, allowing a deeper knowledge on the geomorphological complexity of the area. These studies together with the contribution of the satellite radar monitoring technique permitted to define the borders of a large-scale landslide through thick debris deposits overlying a fractured rock basement. The railway alignment was optimized to prevent any influence of the tunnel construction on the landslide. Based on the new track layout, design solutions were carried out to solve the interaction issues related to the underpassing of an existing main road and to mitigate effects produced by the tunnel excavation. Specific numerical analyses were conducted to evaluate the effectiveness of the selected design choices.

1 INTRODUCTION

The planned Verona-Fortezza High Speed (Italy) Southern access railway line to the Brenner Base Tunnel between Austria and Italy (currently under construction) is part of the TEN-T Scandinavian-Mediterranean European Core Corridor (Figure 1), stretching from Helsinki (Finland) to La Valletta (Malta). From Verona to Fortezza the project consists in the quadrupling of the current railway (about 180 km long), of which the first section in design phase is from Fortezza to Ponte Gardena: this section is characterized by two tunnels, named Scaleres and Gardena for a total length of 22 km, excavated both with conventional and mechanized methods.

They are separated by a viaduct crossing over the Isarco River that spans for 250 m across the Isarco Valley. Furthermore, the project includes the connection of the new line to the existing one through two single track railway tunnels (excavated by EPB-TBM) located on the river left bank (Figure 2), approximately between the towns of Ponte Gardena, Laion and Albions (Bolzano). In this area, a large-scale landslide has been recognized through the thick debris deposits overlying the fractured rock basement. Besides, these two railway tracks are designed to run under an existing main road infrastructure connecting Italy and Northern Europe. Indeed, this peculiar scenario required a detailed geological and geotechnical characterization of the area involved by tunnelling.

In the paper, design criteria are reported for defining the Ponte Gardena tunnels alignment avoiding interaction with the landslide, together with the description of the adopted solution for safely underpassing the highway with a tunnel cover limited to few meters from its foundations.

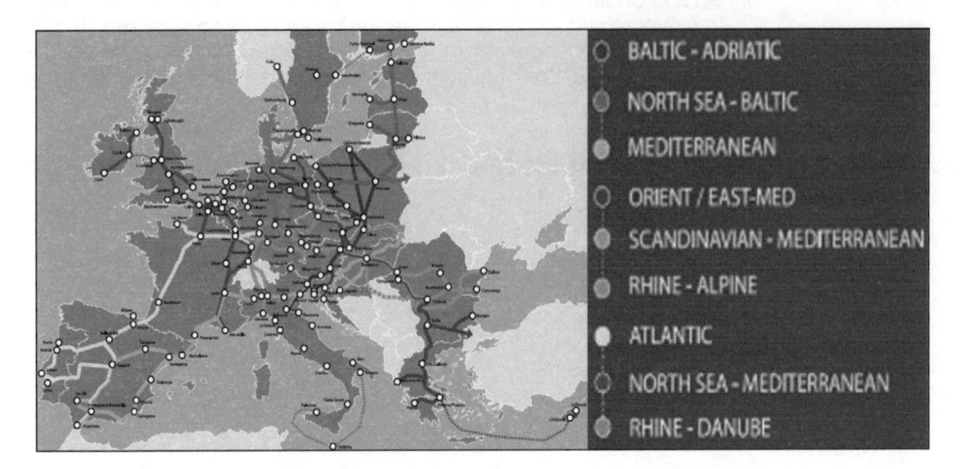

Figure 1. TEN-T Core Network Corridors.

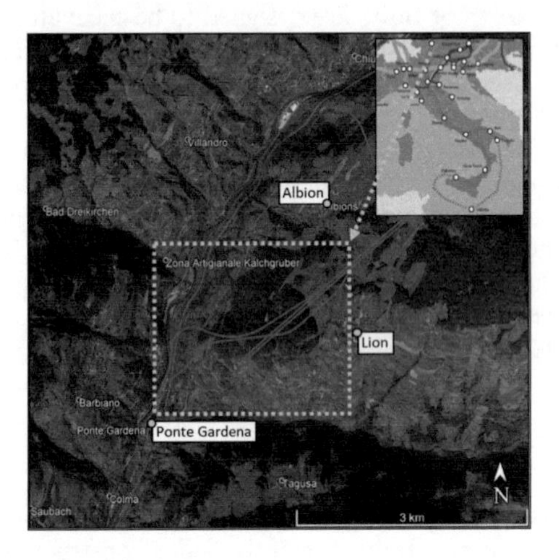

Figure 2. Ponte Gardena railway tunnels area.

2 GEOLOGICAL AND GEOTECHNICAL STUDIES

Geological studies defined two different metamorphic units supposed to be in tectonic contact through a roughly east-west oriented thrust plane. The upper unit lies in the southern part of the area of interest, near Laion and Ponte Gardena, and consists of a sequence of paragneiss and quartzites with a massive structure, often intensely fractured, interbedded with quartziferous phyllites and quartz levels. The lower unit is mainly composed by a sequence of phyllites with variable petrophysical properties from quartz-phyllites to biotite. Furthermore, the Laion area is characterized by an intense fracturing due to joints with high or sub-vertical inclinations, probably associated to a fault system oblique to foliation. Wide variety of quaternary debris covers overlapping the bedrock is the product of recent glacial, post-glacial and alluvial dynamics.

Indeed, landslides are the main geomorphological elements identified in the area. These phenomena remarkably affect the slope on the hydrographic left bank of the Isarco River between Albions and Ponte Gardena, where an extensive landslide has been recognized. It involves a slope portion of more than 350 meters in altitude from the bottom of the Isarco Valley up to 850 meters a.s.l. where a fluvial-glacial terrace is located. Main causes of displacements are referred to the post-glacial relaxing phase after the glacier retreat and subsequent gravitational collapse. The post-glacial relaxation in combination with the fragile structural framework of the area affected by a pervasive fractures system, seems to have led to the described local geomorphological instability. However, the possible erosion at the base of the slope induced in the past by the Isarco River must be considered, which might have forced to a clear deviation towards south-east from a tributary on the hydrographic right bank.

To study this landslide, multi-scale and multi-temporal analyses of aerial photos and orthophotos were carried out. Analysis of digital terrain model from Lidar (at 2.5 meters resolution) was also of remarkable support. Data boreholes confirmed the complex nature of the slope deformation and the boundaries assumed according to geological data-field; at the same time, they helped to define the maximum landslide thickness of about 100 m. It spreads laterally over 1 km and consists of a complex system of bodies having different kinematical features.

Data comparison and integration of both surface geological surveys and subsurface geognostic (direct and indirect) activities were of strategic importance in order to understand the structure of the potential complex slope movement, together with the support of the Interferometric Synthetic Aperture Radar technique (InSAR) for surface deformation monitoring; all the information was collected in a GIS database.

Then, the most reliable geological model was reconstructed representing the structural features of the crystalline basement and the quaternary deposits, which form the gravitational morphologies insisting on the area. In Figure 3 the main geological and structural features and geomorphological elements are summarized.

Geotechnical investigations were carried out consisting in both in-situ and laboratory tests on retrieved samples and an intense monitoring programme was actuated in the area involved by the landslide. In particular, no. 19 core drilling boreholes for recovery samples and no. 2 destructive drilling were executed, reaching a maximum depth of 415 m. Several in situ tests were performed in the boreholes such as Standard Penetration Tests, pressuremeter and dilatometer tests respectively in soils and rocks, Lefranc or Lugeon permeability tests. The geophysical exploration involved mainly 2D electrical resistivity and seismic refraction surveys. Piezometers and inclinometers have been also installed in the landslide area. Laboratory tests were performed both on retrieved rock and undisturbed soil samples.

Site investigations and laboratory tests permitted to characterize the geotechnical model involved by the two tunnels to be excavated along the heterogeneous Isarco Valley left bank, constituted by: a) debris deposits consisting of stone blocks loosens from the weaker paragneiss layer mixed with glacial sandy-clayey layers; b) paragneiss bedrock characterized by the presence of an intense system of fractures; c) sand-gravel alluvial deposits of the Isarco River at the base of the slope. Piezometric measurements indicated a total head generally at the river level. The following physical and mechanical properties of these soils and rock layers are reported in Table 1.

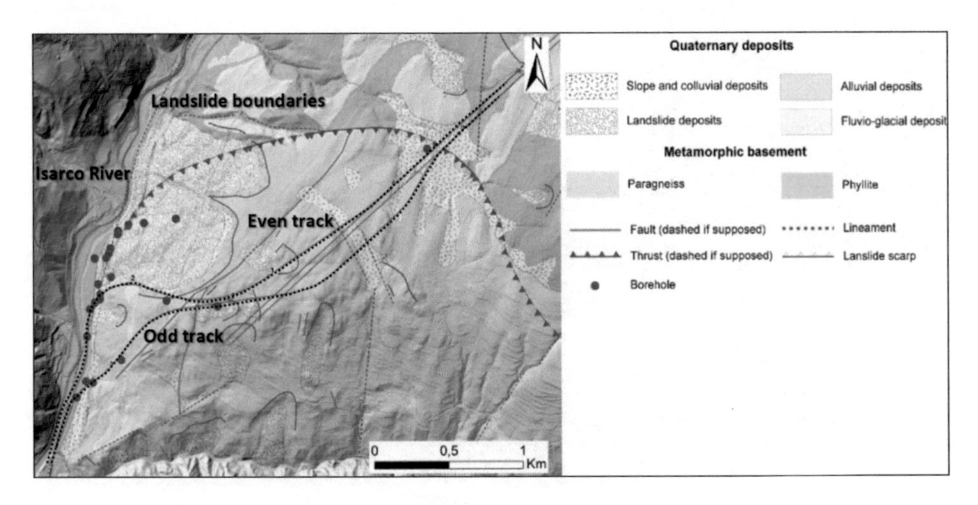

Figure 3. Geological-geomorphological simplified map of the Ponte Gardena area.

Table 1. Parameters assigned to rocks and soils.

		Debris deposits	Paragneiss	Alluvial deposits
γ	[kN/m^3]	24	27	20
E	[MPa]	250	1800	60
c	[kPa]	50	170	0
φ	[°]	39	46	37
ν	[-]	0.3	0.3	0.3

3 SATELLITE REMOTE-SENSING TECHNIQUE

Related to the monitoring plan adopted during the design phase, a fundamental contribution was given by the satellite radar technique, permitting to obtain further information on the deformational phenomenon above described. These analyses allow not only to estimate the current deformation behavior, but also to reconstruct its evolution over time through processing the long time-series of satellite radar interferometry data related to different satellite constellations and different observational periods: ERS (1992–2000), RADARSAT (2003–2011) and COSMO-SkyMed (2012–2017). In this paper, dataset images in the period between July 2012 and May 2017 are discussed from the COSMO-SkyMed satellite (CSK) in descending orbit.

The interferometric technique for remote and high-precision monitoring of earth surface deformation phenomena allows to observe land displacement by satellite data analysis without the installation of measurement points on the ground since surface deformations are monitored by estimating displacements at radar targets already present on the ground (Ferretti 2014). Each displacement measurement is one-dimensional, along the direction of the satellite. The identified measurement points in the studied area are displayed in Figure 4 with an average annual velocity (mm/year) calculated over the entire period of analysis and represented through a color-coded scale: yellow, orange and red colors identify movements away from the satellite radar sensor, green color identifies substantially stable points (values between -1 and 1 mm/year) while movements towards to the radar sensor are characterized by different gradations of blue.

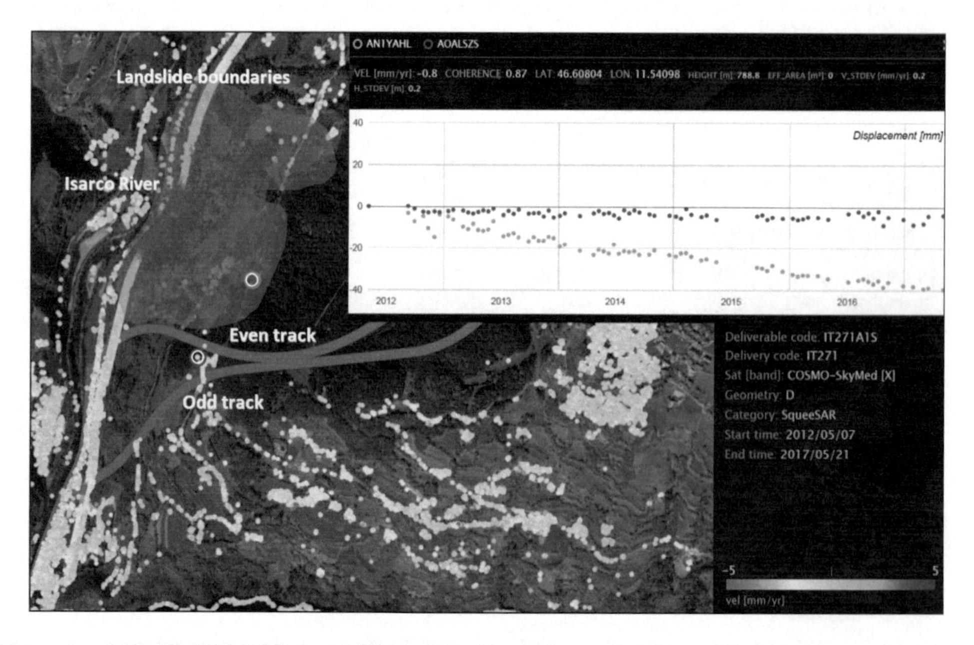

Figure 4. COSMO-SkyMed data satellite and time depending series of two monitored points.

Satellite data show that the landslide (highlighted in blue in Figure 4) affecting the western edge of the slope has movements of the order of 7 mm/year with zone exceeding even 10 mm/year. Comparing velocity time-series of points outside the landslide area (in purple in the graph in Figure 4), substantially stable, with time-series of points in motion inside the detected area (in blue in the same graph), it is possible to clearly see their different trend during time.

4 ADOPTED DESIGN SOLUTIONS

Geological-geomorphological studies together with the satellite radar monitoring allowed to acquire a deeper and accurate knowledge on the geomorphological structure of the valley, defining the boundaries of the unstable area and highlighting at the same time its complex nature.

Tunnel excavation usually leads to stress changes around the cavity due to soil/rock removal and due to the excess of volume extraction to leave enough room to fit the lining, representing in this case, a possible risk for further mobilization of the surrounding unstable slope.

As a consequence of the mentioned studies the railway alignment was modified and optimized (as reported in Figure 4) to prevent any influence of the tunnel construction on the landslide and, on the other hand, to avoid possible structural impacts in the tunnel lining. Therefore, the first portion of the even track has been planned to run outside the boundaries of the area affected by the slope deformation, then moving significantly away in the second portion with the rapid increase of the cover proceeding within the slope. The odd track, located much further south of the even track, is very far from the area affected by the landslide.

Adapted the railway alignment to the results of studies and investigations, *ad hoc* design solutions were developed to solve the interaction between tunnels excavation and the overlying existing highway (Figure 5).

The interaction between the two single track railway tunnels and the highway consists of the even track running by a side of the highway near one of its viaduct piers (pier "A") underpassing then a second viaduct pier foundation (pier "B") while the odd track is located to underpass the highway embankment connected to the mentioned adjacent viaduct. To study the interaction, geometrical-typological characteristics of embankment and viaduct's foundations and their structural scheme were analysed together with the topographic surveys carried out.

Therefore, due to materials to be bored, limited covers and interfering structures, it was decided to use the mechanized tunnelling (EPB Shield excavation), together with preventive protection measures, to guarantee conditions of stability and mitigate as much as possible subsidence: structural elements (barrier of contiguous micropiles) with the aim of modifying and limiting the field of induced displacements; rock/soil grouting to reduce deformations around the tunnel section to be excavated.

FEM numerical analyses using Plaxis 2D were performed to study the tunnel excavation/highway interaction. Tunnel construction was simulated in the hypothesis of plane strain condition by imposing a volume loss V_L in an expected excavation scenario (0.2% in rock and 0.5% in soils) and in the worst excavation scenario (0.4% in rock and 1.0% in soil) through the application of a

Figure 5. Slope instability boundaries with, on the background, the highway section to be underpassed.

uniform radial contraction to the tunnel boundary. The analyses were carried out to study the design solutions effects on the stress-strain variation induced by the tunnel excavation.

4.1 *Even tunnel track interaction with the highway viaduct pier "A"*

The even tunnel track moves away from the highway next to the viaduct pier "A": the minimum distance between its foundation (481.50 m a.s.l.) and the tunnel (crown at 474.00 m a.s.l.) is about 3.2 m (Figure 6).

In order to contain ground displacements caused by the tunnel excavation, a barrier of contiguous bored micropiles between the pier "A" and the tunnel is designed to be pre-installed, before the EPB-TBM passage. Mitigation barriers of bored piles, micropiles, or jet-grouting columns have been demonstrated to be effective in reducing ground movements produced by tunnel construction processes and consequently decreasing the level of damage eventually induced (Bilotta & Russo 2011, Rampello et al. 2016).

The choice of a micropiles barrier is connected to the versatility of execution in consideration of the heterogeneity that characterizes the slope debris involved by the works in the section interfering with the pier "A". This material is, in fact, composed by an alternation of metamorphic rock blocks of the order of 1–2 m and transported loose sediments. Small diameter piles can be performed both in soil and in rock, as well as allowing the minimization of installation effects in terms of vibrations and displacements, compared to medium or large diameter piles. Aim of the intervention is to limit the field of variation of stress-strain state produced by the tunnel construction through the presence of the micropiles. The barrier, linearly extending for about 30 m, consists of Ø300/900 micropiles on two rows 78 cm spaced, having a distance between structural elements of 1.5D. The vertical length is about 35 m in order to maximize the efficiency of this type of intervention. In fact, according to technical literature, the most convenient length, extended below the tunnel depth, is $L = z_0 + D$, with z_0 the tunnel axis depth and D its diameter. Higher values usually determinate no significant advantages.

In this case, numerical simulations have been conducted referring to the orthogonal section to the tunnel axis characterized by a cover of 18 m. As mentioned, here the tunnel crosses the slope debris deposits; towards the valley there is the transition with the sandy-gravelly deposits at a distance of 30 m from the tunnel axis. To the right of the tunnel (about 10 m from its axis) there is the passage to the paragneiss (Figure 6).

The caisson foundation pier "A" of 15x11m, elliptical in plant, was simulated in the plane model as a 15 m wide parallelepiped (to maximize effects in terms of differential subsidence) 10.30 m high, with linear elastic behavior characterized by stiffness according to the concrete strength class. As for the unit weight, this was modified to take into account the three-dimensional effect of the foundation shape in the two-dimensional calculation domain, thus determining a reduction of g, therefore equal to 14 kN/m^3. Span and pier weights plus the overloads were simulated by the application of a concentrated load of 37500 kN spread over the foundation length in the plane orthogonal to the calculation. Micropiles wall has been modeled as linear element with elastic behavior and equivalent stiffness function of geometric

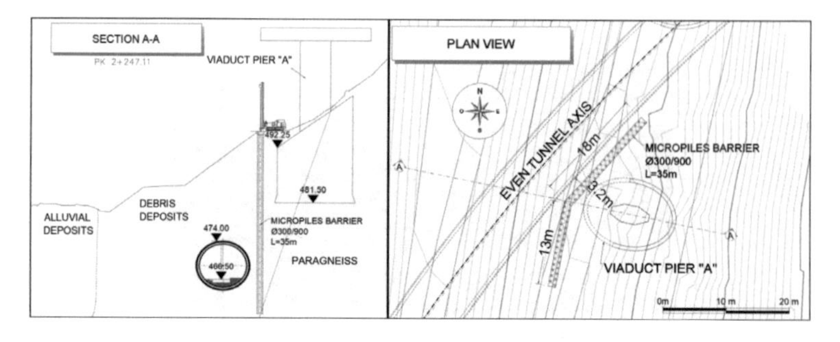

Figure 6. Geotechnical cross-section and plan view of the intervention next to the viaduct pier "A".

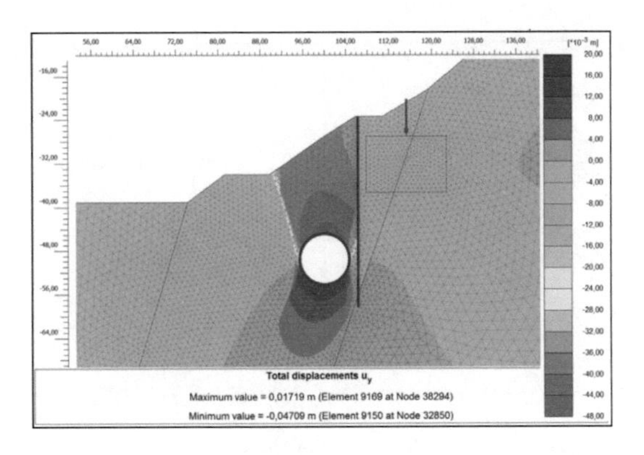

Figure 7. Settlements produced by the excavation (V_L=1.0%) in presence of micropiles barrier.

Table 2. Max settlement at the base of the viaduct pier "A" foundation.

	V_L=1.0% without mitigation intervention	V_L=1.0% with mitigation intervention
w_{max} [mm]	14	2

characteristics, concrete properties and micropiles spacing. Previous studies highlighted the importance of the barrier/soil interface on numerical results regarding vertical barriers simulations to prevent ground movements produced by tunneling (Bilotta 2008). Sensitivity analyses were performed to value its influence on the mobilised strength along the micropiles and on the displacement field induced by the excavation. Hence, a fully rough interface was assumed.

Calculated settlements are reported in Figure 7 referring to the worst scenario (V_L=1.0%), while in Table 2 is reported the maximum settlement at the foundation base, referring to the TBM passage calculation phase, simulated both without micropiles and in presence of the mitigation barrier.

Results confirmed that the design solution of a micropile barrier executed before the TBM passage between the foundation of the existing pier and the tunnel track allows to reduce effects of interaction between the tunnel and the viaduct caisson foundation "A". A quantitative evaluation of the mitigation effect is provided by comparing the maximum vertical displacement calculated at the base of the pier foundation, in presence and absence of the micropiles, in terms of local efficiency:

$$\eta_{loc} = 1 - \frac{w_{pier\ with\ micropiles}}{w_{pier\ without\ micropiles}} \tag{1}$$

thus, highlighting how the presence of the micropile barrier interposed between tunnel and foundation to be protected allows significant settlement reduction of about 90%.

4.2 *Odd tunnel track interaction with the highway embankment*

Interaction of the odd tunnel track with the highway is represented by the underpassing of the embankment located next to the viaduct previously described, close to the Southern shoulder, with a cover of around 16 m from the road rolling plan. Here, cement injections are designed around the tunnel perimeter composed by sand-gravel alluvial deposits and part of the loose materials which constitute the embankment body. Scope of the treatment is the increasing of stiffness and strength of the soils around the tunnel section for a thickness of about 4 m, performed by means of sub-horizontal drilling and subsequent injection of ternary cement mixture at controlled grouting volumes and pressures through tubes *à manchettes* before the arrival of the TBM-EPB (geometries shown in Figure 8).

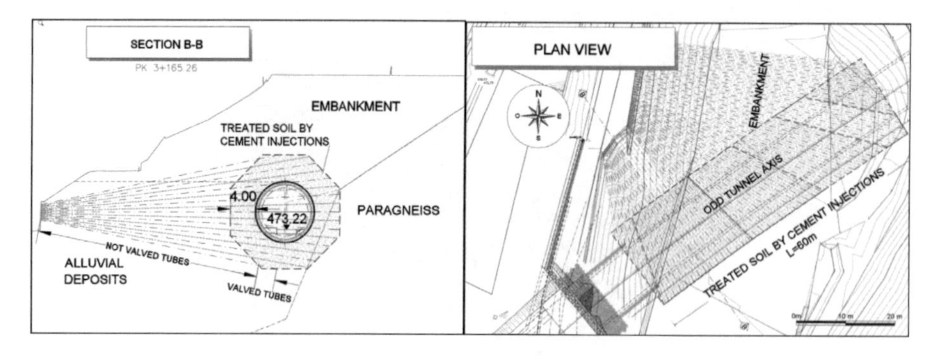

Figure 8. Geotechnical cross-section and plan view of the intervention below the embankment.

Numerical simulations referring to the orthogonal section to the tunnel axis have been carried out in which the tunnel crosses the sandy-gravelly deposits below the loose materials of the embankment body. To the right of the tunnel, about 7 m from the tunnel axis the transition to paragneiss is defined (Figure 8). The soil treated by cement injections was modeled by simple activation (wished in place), assigning new mechanical characteristics to the materials affected by the treatment, considering an increase in stiffness and cohesion. A distributed load of 35 kN/m^2 was applied on the top of the embankment, for a length of 22 m equal to the carriageways width in order to schematically simulate the highway load on the embankment.

Figure 9 shows the calculated displacements contour map in the analysis in case of V_L=1.0% while in the Table 3 is reported the maximum settlement at the top of the embankment (calculation phase of the TBM passage, simulated both without the treated soils and in presence of the mitigation works). Also in this case, results confirmed that the design solution regarding the grouting injections executed before the TBM passage allows to reduce effects of interaction between the tunnel and the embankment. The quantitative evaluation of the mitigation effect at the top of the embankment, in presence and absence of the treated soils, expressed as:

$$\eta_{loc} = 1 - \frac{w_{embankment\ with\ injections}}{w_{embankment\ without\ injections}} \tag{2}$$

points out how the presence of the injected soils around the tunnel perimeter determines a significant settlement reduction of about 70% on the top of the embankment.

4.3 *Even tunnel track interaction with the highway viaduct pier "B"*

The even tunnel track underpasses the highway below the "B" pier with a distance of 11 m, measured between the tunnel extrados and the base foundation (484.80 m a.s.l.) as shown in Figure 10.

Numerical simulations have been conducted referring to the orthogonal section to the tunnel axis characterized by a cover of 26 m. In this section (Figure 10), the tunnel will be excavated in the basement represented by the fractured paragneiss below a layer of debris deposits with a variable thickness between 5 and 15 m. This simulation has been focused on effects of TBM excavation in the fractured rock respect to the foundation pier.

The caisson foundation pier "B" was modelled with the same approach described before, but having a height of 7.65 m. In Table 4 is reported the maximum settlement at the base of the caisson foundation, referring to the calculation phase of the TBM passage regarding the worst case (V_L=0.4%) while Figure 11 shows vertical displacements calculated in the analysis. In this scenario, the calculated differential settlement is of the order of millimeters below the foundation. Therefore, the adoption of mechanized excavation will allow to reduce effects of interaction with the foundation of the viaduct pier.

Anyway, referring to the randomly distributed system of fractures present in the weak and fractured rock, a treatment of high penetrability cementitious injections around the tunnel

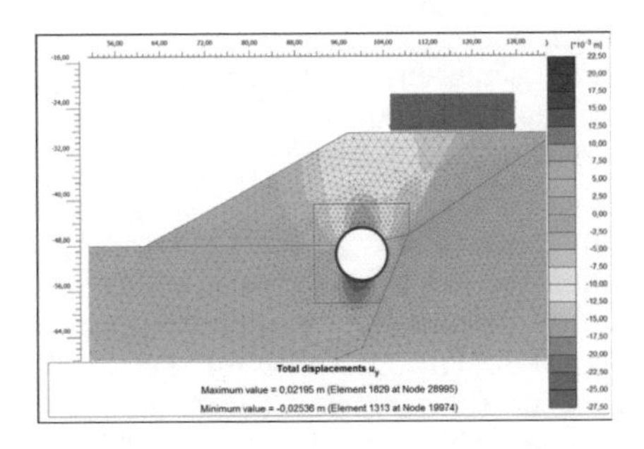

Figure 9. Settlements produced by the excavation (V_L=1.0%) in presence of injected soils.

Table 3. Max settlement at the top of the embankment.

	V_L=1.0% without mitigation intervention	V_L=1.0% with mitigation intervention
w_{max} [mm]	28	9

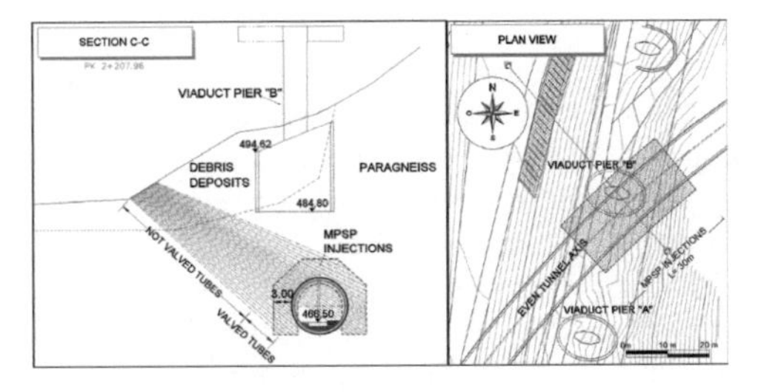

Figure 10. Geotechnical cross-section and plan view of the intervention below the viaduct pier "B".

Table 4. Max settlement at the base of the viaduct pier "B" foundation.

	V_L=0.4%
w_{max} [mm]	7

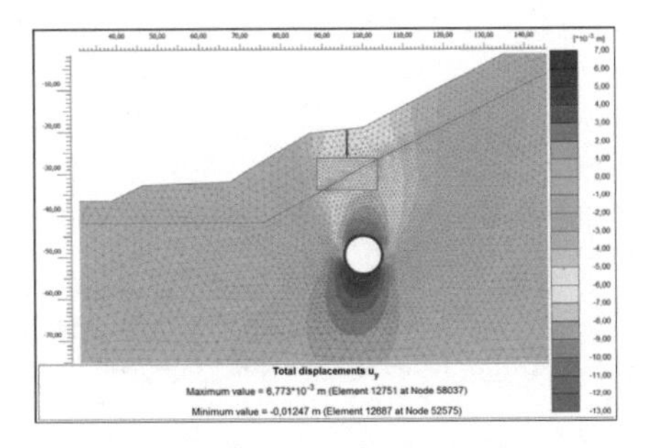

Figure 11. Settlements produced by the TBM excavation (V_L=0.4%).

perimeter has been designed. The grouting method adopted for a thickness of 3 m around the tunnel perimeter is the Multi Packer Sleeved Pipe (MPSP) carried out before the arrival of the EPB-TBM. MPSP system consists essentially in the installation inside boreholes of a plastic pipe equipped at regular intervals with rubber sleeves and bag packers, expanded through grout injection against the hole walls, to isolate the sections to be grouted (Manassero & Di Salvo 2015). The main role of the grouting is to improve the mechanical properties and secondly to reduce the hydraulic permeability of the fractured rock, filling the system of fractures with high penetrability cement-based grouts giving homogeneity to the rock mass (locally fractured or very fractured) and thus mitigating deformation effects induced by the excavation. Figure 10 shows the geometry of the intervention, extended along the tunnel axis for a total length of about 30 m below the foundation pier.

Furthermore, regarding the interaction between adjacent piers, the angular distortion between two extreme slab points (35 m long isostatic structure supported by the viaduct piers) has been calculated respect to the obtained settlements from numerical simulations. In the scientific literature (Viggiani et al. 2011) limit values of the angular distortion beyond which manifestations of damages are possible in case of multiple span bridges are reported. This value, equal to 1/250 can also be referred to the viaduct object of intervention. Distortions achieved from the calculated settlements are well below this limit with a maximum value of around 1/3500 as result of the designed deformation mitigation interventions. They are also below the limit reported in "Eurocode 7 - Geotechnical Design" equal to 1/2000 beyond which effects on structures can occur.

5 CONCLUSION

In the described project, the combined interpretation of geological and geomorphological data led to crucial information during design process, allowing a deeper and accurate knowledge on the structural and geomorphological complexity of the area on the left bank of the Isarco River, and helping to define the correct choice regarding the Ponte Gardena railway tunnels alignment on Fortezza-Verona High Speed line.

Then, different design solutions have been carried out to solve the interaction issues related to the highway underpassing, in function of different geotechnical contexts in order to mitigate effects produced by the tunnel excavation. Specific numerical analyses have been conducted to predict the deformation behavior due to tunnelling and to evaluate the effectiveness of each design choice. Therefore, micropiles located adjacent to a first viaduct pier, effectively modifies the displacement field caused by tunnelling, acting as a barrier against the propagation of the deformations around the tunnel section, while in case of the embankment underpass, the stiffness and strength soil increase by means of cement injections around the tunnel perimeter positively reduces deformations and plastic zones induced by its construction. The mechanized excavation in weak fractured rock at the base of a second viaduct pier shows a favorable scenario in terms of reduction of deformations induced by tunnelling together with the aid of the high penetrability cement-base injections for filling the fracture system potentially present in the rock.

REFERENCES

Bilotta, E. 2008. Use of diaphragm walls to mitigate ground movements induced by tunnelling. *Géotechnique*, 58(2): 143-155. London: ICE Publishing.

Bilotta, E. & Russo, G. 2011. Use of a Line of Piles to Prevent Damages Induced by Tunnel Excavation. *Journal of Geotechnical and Geoenvironmental Engineering*, 137(3): 254-262. Reston (VA): ASCE:

Ferretti, A. 2014. *Satellite InSAR Data, Reservoir Monitoring from Space*, 9: 159. The Netherlands: EAGE Publications.

Manassero, V. & Di Salvo, G. 2015. The application of grouting technique to volcanic rocks and soils to solve two difficult tunnelling problems. In *Proc. of The International Workshop on Volcanic Rocks and Soils*: 399-408. Ischia Island: AGI.

Rampello, S., Fantera, L. & Masini, L. 2016. Diaphragm wall as a mitigation technique to reduce ground settlements induced by tunnelling. In *Geotechnics for Sustainable Infrastructure Development*: 325-333. Phung: Geotec Hanoi (edt).

Viggiani, C., Mandolini, A. & Russo, G. 2011. *Piles and Pile Foundations*. 278 pp. London: CRC Press, Taylor & Francis.

Tunnels and Underground Cities: Engineering and Innovation meet Archaeology, Architecture and Art, Volume 3: Geological and geotechnical knowledge and requirements for project implementation – Peila, Viggiani & Celestino (Eds)

Torino Metro Line 1 extension: Working solutions

A. Damiani & L. Mancinelli
Lombardi Ingegneria srl, Milan, Italy

R. Crova & E. Avitabile
Infratrasporti.To Srl, Turin, Italy

ABSTRACT: The paper describes relevant design aspects for the extension of Torino Metro Line 1, work that includes a double track circular single tunnel (total length equal to 1733 m), two underground stations (Bengasi and Italia '61-Palazzo Regione), three shafts (PB22, PB23 and P24). The tunnel, 6.88 m in (internal) diameter, has been excavated by using an EPB TBM, with the contemporary disposition of a precast lining composed by 5+1 r. c. segments, 1.4 m long and 0.3 m thick. Intervention area is strongly urbanized, reasons why working choices and procedures have been defined to obtain an adequate level of robustness, also taking into account the difficult underground scenario, extremely heterogeneous and in presence of water. Technological aspects and particular solutions are discussed and described. In detail, the effect of the design choices about the definition of the correct TBM confinement pressure on the tunnel face is analyzed, by comparing settlement predictions and monitoring results.

1 INTRODUCTION

Turin Metro Line 1 is in an Southward extension phase running along Via Nizza, with the construction of Bengasi and Italia'61-Palazzo Regione stations, of end-line shaft PB24 and of mechanized double track tunnel (single tube), in connection with the existing station Lingotto (end of the currently working line). The new system is completed by two ventilation shafts (PB22 e PB23) located in the middle of the two new metro stretches (Lingotto-Italia'61 and Italia '61-Bengasi).

The area of intervention is heavily urbanized, with buildings and roads, aspect that, in association with the geotechnical context, obliged to a particular design approach, with robust solutions like diaphragm walls sustained by temporary and permanent struts and/or slabs for shaft and stations. For the tunnel a mechanized system was chosen.

A monitoring program, uses as a tool to check the evolution of settlement regimes on a daily/weekly/monthly basis and to make comparisons with thresholds and predefined limits, was also available. Figure 2 represents the tunnel after its realization, before the disposition of internal elements related to traffic (exercise configuration of the infrastructure).

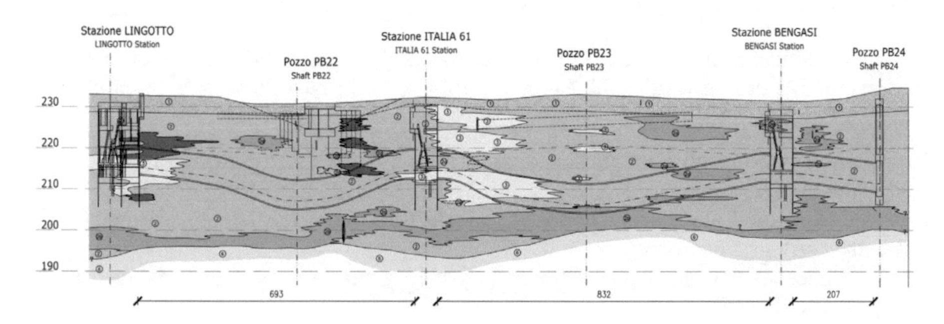

Figure 1. Turin M1 extension scheme (units in m).

Figure 2. Natural tunnel: Bengasi Station–Shaft P24 stretch.

2 CHOICE OF THE EXCAVATION SYSTEM

The natural tunnel has been realized by an EPB (earth pressure balance) Herrenknecht machine, having a weight of 400t and a total length of about 100m; the excavation diameter was 7.76m. The tunnel is 1733m long with overburden between 12.3m and 21.4m. The internal lining, a precast ring composed by 6 r. c. segments, is 30 cm thick, its external diameter is 7.48m for a length of 1.40m. The production system was equipped with train transportation for muck and segments.

The choice of the method and of the best tool to face off the different situations was made considering advantages and disadvantages of available possibilities. EPB was then the identified way, after the following main important aspects emerged in a value engineering analysis:

- high control and limited quantity of conditioning material to add (and extract) for the excavation;
- best face control, obtained with low active pressure levels (never higher than approx. 2.5 bar);
- low environmental impact;
- high performance in the expected context;
- good ratio between costs and expected production.

In addition, the experience already matured in Turin for other underground lines were used to analyze this new work and to confirm EPB technology, as one of the best for application in urban field. Other aspects that influenced the choice were the possibility to proceed with technical modifications and optimization of the excavation tools (head configuration) during the construction process, as well as a good versatility and compatibility with the break-in and break-out activities, always present and important in metro works.

3 GEOTECHNICAL CONTEXT

From a geotechnical point of view, the context shows the presence of three main lithographic formations, with a design water table laying at approx. 225 m asl. Proceeding from the top we find:

- fluvial and fluvio-glacial Rissian deposits, formed by gravels, cobbles and bars of sand in sandy-silty matrix;
- lacustrine and fluvio-lacustrine deposits (inferior/superior Pleistocene), formed by clayey silts with sandy-gravelly bars;
- Neritic (Pliocene) marine deposits formed by clayey silts, sandy silts and blue-grey sands with (including) fossils.

Table 1 summarizes the geotechnical parameters assumed for the design.

Table 1. Geotechnical parameters.

Description	Unit	γ [kN/m^3]	E [MPa]	ϕ' [°]	c' [kPa]	ν [-]
Superficial layer	0	19	10	28	0	0.25
Gravel and sand, loose to slightly cemented	1	19	15-150	28-35	0-5	0.35
Gravel and sand, slightly to medium cemented	2	21	170	35	20	0.35
Silt (sandy to slightly clayey)	3	19	60	26	25	0.30

4 TBM FACE PRESSURE DEFINITION

The definition of the face pressure for the whole work has been done on the basis of a series of analyses, carried out for this purpose. In particular some scenarios have been identified to proceed with engineering studies constituted by

- empirical evaluation following Anagnostu-Kovari approach;
- finite elements calculation on cross section to identify convergence regime (plane strain conditions);
- finite elements calculation to identify face extrusion regime (axisymmetric conditions).

The main situations are described below.

- Scenario S1: zone at half stretch Bengasi station-P24 Shaft, overburden approx. 14 m, homogeneous face in sand/gravel (unit 1).
- Scenario S2: maximum overburden, approx. 21.0 m, between Bengasi and Italia 61 stations, near PB2 shaft, homogeneous face in sand/gravel (unit 1), partially occupied by a silty portion.
- Scenario S3: low overburden, approx. 13.5 m, near Italia 61 station and new Regione Piemonte building, mix face conditions with sand/gravel (unit 2) and silty sand (unit 3).
- Scenario S4: underpassing the Lingotto urban road tunnel, nominal overburden approx. 18 m, reduced to approx. 8 m because of the structure's presence, homogeneous face in sand/gravel (unit 1).
- Scenario S5: minimum overburden approx. 12.5 m, near Lingotto station, homogeneous face in slightly cemented sand/gravel (unit 2).

For every scenario, Table 2 shows the watertable conditions, where H_w represents the height of water from the axis of the tunnel and p_w the corresponding pressure in bar (1bar=100 kPa).

The results of the empirical evaluations are summarized in Table 3 (where s represent the total stress at tunnel axis level). Those values, as per hypotheses of the formulation, represent the pressure that guarantees the stability of the face without any filtration process.

Table 2. Hydraulic heights and pressures.

Scenario	S1	S2	S3	S4	S5
H_w [m]	7.3	15.0	6.8	11.2	5.8
p_w [bar]	0.73	1.50	0.68	1.12	0.58

Table 3. Empirical analysis results.

Scenario	s [bar]
S1	0.85
S2	1.60
S3	0.85
S4	1.25
S5	0.70

Table 4. Loss of volume vs. effective face pressures.

Scenario s' [bar]	S1	S2	S3 V_p [%]	S4	S5
0.25	0.42	0.32	1.21	0.31	0.20
0.50	0.24	0.27	0.89	0.23	0.14
0.75	0.16	0.22	0.64	0.17	0.10
1.00	0.10	0.19	0.47	0.12	0.07
1.25	0.06	0.15	0.35	0.09	0.03
1.50	0.03	0.11	0.23	0.06	0.0
1.75	0.00	0.08	0.08	0.01	0.0

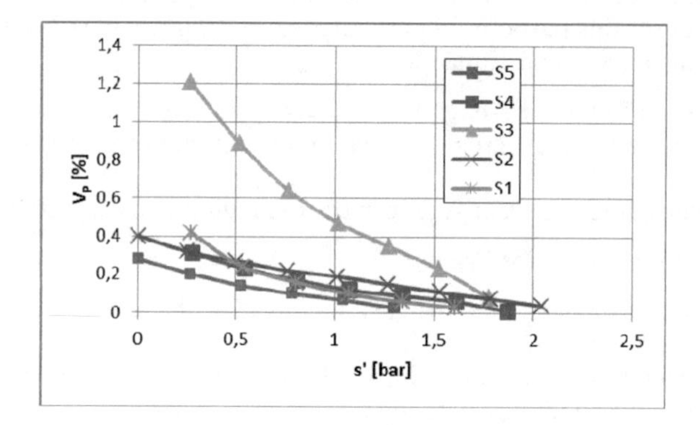

Figure 3. Loss of volume vs. effective face pressures.

It is clear that they must be equal or higher than the water pressure p_w and, for design purposes, are assumed to be the minimum face pressure to be adopted.

While the method doesn't give any information about the deformation regime connected to this stability, it is necessary to use something else to catch this very important aspect and FEM (finite elements method) represents a very good tool in this sense. Its application in plane strain and axisymmetric conditions give, as a result, the loss of volume. For the design purposes, those studies have been performed in a parametric manner, applying to the model pressures from 0.0 to 2.0 bar (in effective stress regime).

Table 4 and Figure 3 summarize the results, in terms of loss of volume V_p (%) vs. pressures s' (bar). V_p incorporates the effects of face extrusion (deduced form axisymmetric analyses) and cavity convergence (deduced from plane strain analyses).

The results offer an important vision of the situation, where Scenario 3 appears clearly most critical. This particular situation, characterized by low overburden and water, need an important face pressure control to limit the volume loss and, consequently, the manifestation of a deterimental settlement regimes. The other scenarios seem to be similar between them and less demanding.

Operative TBM face pressures s (total stress) are defined by the following formula

$$s = s' + p_w \tag{1}$$

where s' represents the effective stress, necessary to maintain the loss of volume in a range 0.2%÷0.4% (as a function of the risks), and p_w the water pressure. Table 5 shows the results obtained in terms of s. All values must be intended at tunnel axis level.

Table 5. Face pressure.

Scenario	S1	S2	S3	S4	S5
$s=s'+p_w$ [bar]	1.20	1.90	1.60	1.50	1.10

5 MONITORING AND BACK ANALYSIS

The development of the whole work has been associated with a real time control of productions and monitoring data. With this aim, a dynamic database, web-GIS system has been adopted and configured with apposite thresholds for alert and alarm situations. Great attention in this field was dedicated to muck weight, injected volumes, face and backfilling pressures, impacts at ground level and on existing structures, for the eventual definition of corrective actions on every process. Available data were then used to verify design hypotheses on the basis of back analyses. An example of this activity, on a representative portion of tunnel (approx. 400 m between Bengasi and Italia '61), is here summarized. In particular the following images represent the reconstruction of some subsidence cross sections (transversal to tunnel axis at ground level) and the statistical synthesis for k parameter and volume loss of.

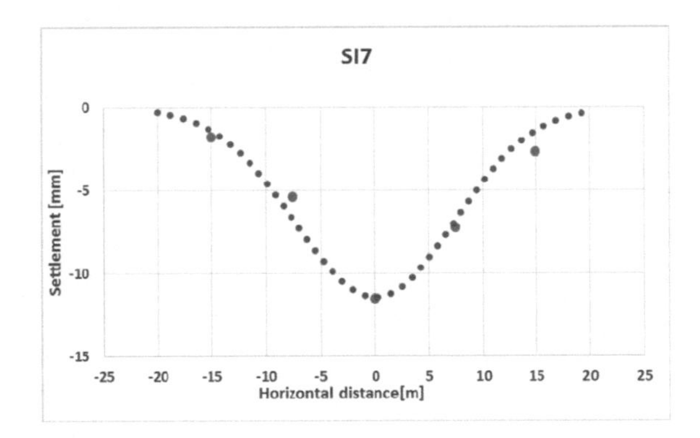

Figure 4. Back analysis for subsidence cross section "SI7".

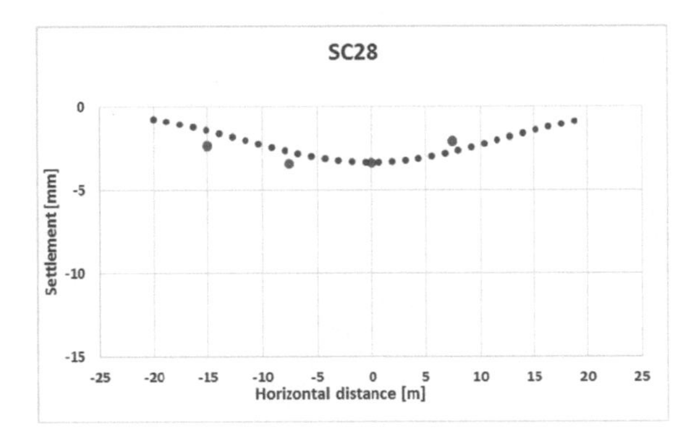

Figure 5. Back analysis for subsidence cross section "SC28".

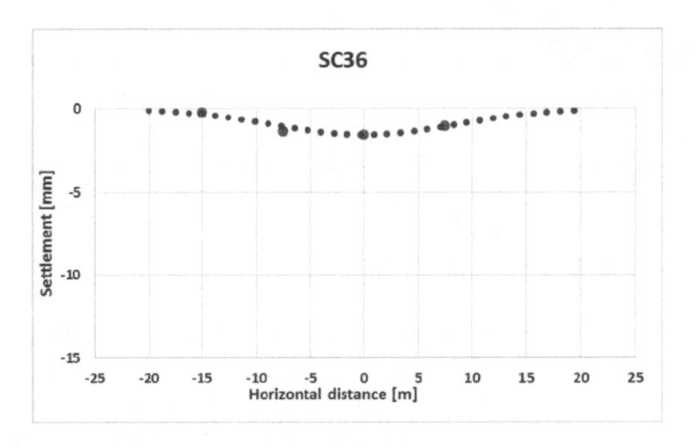

Figure 6. Back analysis for subsidence cross section “SC36”.

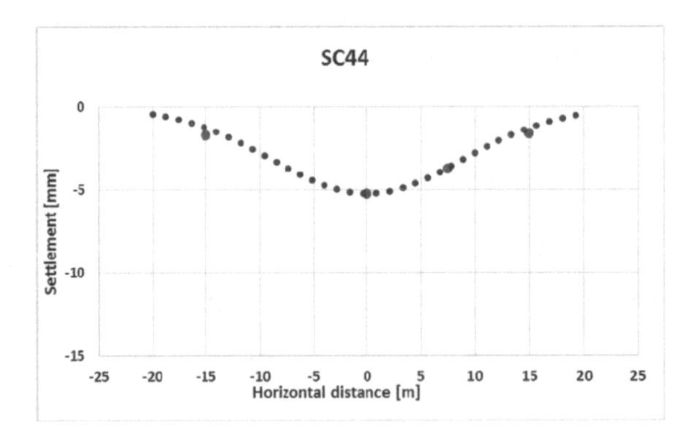

Figure 7. Back analysis for subsidence cross section “SC44”.

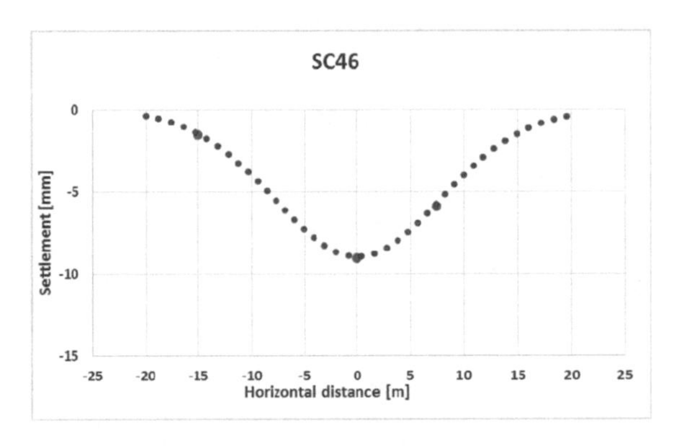

Figure 8. Back analysis for subsidence cross section “SC46”.

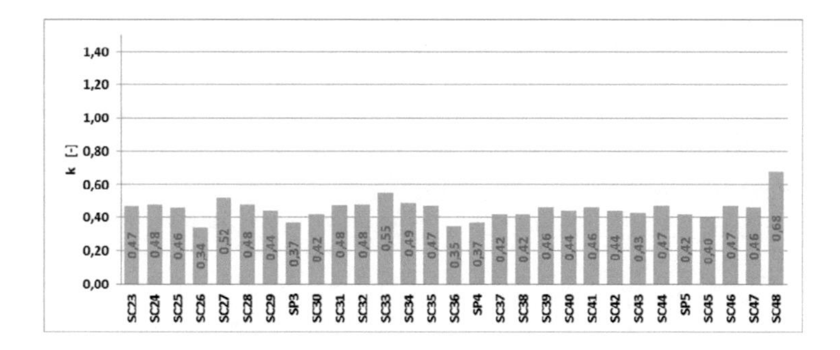

Figure 9. k parameter: back analysis results.

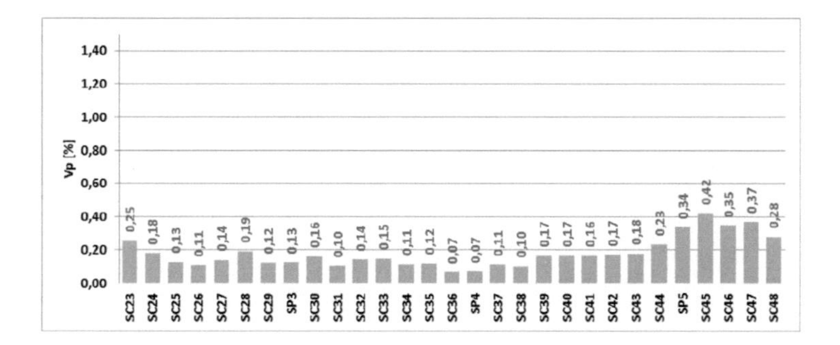

Figure 10. Volume loss of: back analysis results.

The results in terms of k parameter are particularly interesting. Geotechnical formations in the area of works point, at design stage on the basis of literature and previous experiences, to the assumption of k=0.4. Field results allowed to a fine tuning, obtaining an average value of k=0.44. This should be considered as a confirmation of an adequate design approach. We remind here that the k parameter is the coefficient that relates, in Mair's formulation, the width of the subsidence profile to the depth of the tunnel axis.

The volume loss V_p shows values in the range between 0.1 % and 0.4 %, with an average value of 0.2%.

6 CONCLUSIONS

The extension of Torino Metro Line 1 has represented an important step in the direction of mechanized underground works in Italy. The large amount of information recorded during the activities and the back analysis performed on these data have given new support to studies and knowledge of applied geotechnics in the area. This contribution will help designer in their work, also in different context, for approach method and contents.

REFERENCES

Anagnostou G., Kovári K. 1996. Tunneling and Underground Space Technology. *Face stability condition with Earth-Pressure-Balance shields.*

Cassani G., Mancinelli L. 2005. IACMAG Torino. *Monitoring surface subsidence for low overburden TBM tunnel excavation: computational aids for driving tunnels.*

Damiani A., Crova R., Mancinelli L., Avitabile E. 2018. Strade e Autostrade - n° 128. *I lavori di prolungamento della Linea 1 della Metropolitana di Torino verso Bengasi.*
Mair R. J., Taylor R. N., Burland J. B. 1996. Proc. Int. Symposium on Geotechnical Aspects of Underground Construction in Soft Ground, London (eds R. J. Mair and R. N. Taylor), pp. 713–718. *Prediction of ground movements and assessment of risk of building damage due to bored tunneling*. Balkema.
Mancinelli L. 2005. Geotechnical and geological engineering - Vol. 3, n.3 – pp. 263÷271. *Evaluation of superficial settlements in low overburden tunnel TBM excavation: numerical approaches*

Tunnels and Underground Cities: Engineering and Innovation meet Archaeology, Architecture and Art, Volume 3: Geological and geotechnical knowledge and requirements for project implementation – Peila, Viggiani & Celestino (Eds)
 ISBN 978-0-367-46583-4

2D and 3D geological-geomechanical GIS model for underground projects: Collection, storage and analysis of geological and geomechanical data during the design and construction phases

R. De Paoli & F. Gobbi
AF TOSCANO SA, Mesocco, Switzerland

L. Thum
AF TOSCANO SA, Lausanne, Switzerland

F. de Martino
AF CONSULT ITALY SRL, Milan, Italy

ABSTRACT: The accuracy of a geological and geomechanical model is a key factor in the implementation of infrastructure projects and in particular, for underground infrastructures. The reduction of risks and optimization of costs are highly influenced by the quality of the geological model. The role of geologist in the construction phase is to collect and interpret field data, and to validate the hypothesis assumed during the design phase. A digital GIS-based method has been developed to facilitate the interpretation of data. GIS models consist of a geodatabase in which the individual items can be inspected, either visually in a 3D model or in tables. A tablet PC is used to digitally collect soil information. Data are then automatically integrated into the model. A PDV (Parallel Data Viewer) graph, parallel to the excavation axis, allows one to automatically plot and control data associated with the excavation, such as geomechanical parameters, etc.

1 INTRODUCTION

The construction of tunnels and underground works in general is constrained by the geological and geotechnical conditions of the site. Many authors, such as Bieniawski (1989), Hoek (1982), Kvartsberg (2013), Loew et al. (2010), Parker (2004) have highlighted the importance and necessity of having a good knowledge of the geological structure of the construction site.

In the design phase, the rock masses are described geologically and hydrogeologically, affiliated to tectonic units and finally, geomechanically characterized. The types of rock mass and the site structure allow segments of a uniform nature and behaviour to be defined, to which fitted support measures will be applied.

Current design software programs, drilling technology for geognostic and geophysical surveys nowadays allow the representation of information sampled in the field simultaneously in a three-dimensional model. Such procedure results in a greater precision in the geolocalization of the individual geological structures (faults, lithological limits, ...) and the possibility of interpreting, connecting and projecting these structures three-dimensionally. The fit of the resulting geological-geotechnical model is more easily controlled.

During execution, an appropriate geological-geotechnical survey makes validating the suitability of the rock support possible. This constant surveying is of critical importance, particularly in the case of tunnels with a high overburden, as unexpected rock behaviour (squeezing conditions or instability) and faults could be encountered despite careful investigations (Loew et al. 2010, Pozzorini 2017). An extensive geological dataset is required to manage and analyse the collected geological field data during the geological survey. These data are the basis for the verification and forecasting of rock support measures. This database includes, in addition to geological and geomechanical data sampled during the execution, also the geolocalised representation of the interpretation of geognostic drilling at the tunnel heading, thus allowing the prediction of the geological and geomechanical rock conditions in a ca. 100 m perimeter zone around the tunnel face.

In recent decades, authors such as Kim et al. (2005), Fuerlinger et al. (2008), Choi et al. (2009), Zheng et al. (2010), Aldiss et al. (2012) and Thum & De Paoli (2015) have reiterated the importance of geological-geotechnical monitoring during excavations as a support for the provision of rock-support measures to be taken and for geological prediction, potentiality in the use of digital supports for the acquisition and processing of sampled geological data and of the three-dimensional analysis of such information.

The developed method responds in particular to the technical requirements in the case of large underground works in heterogeneous rock masses, allowing a rapid collection and synthesis of large quantities of fragmented data. The technique was developed in parallel to the geological survey of the Ceneri Base Tunnel (CBT) (Ticino, Switzerland, AlpTransit Gotthard AG Media Office 2012) with the aim of optimizing and facilitating the collection, management and the analysis of quantities of georeferenced geological/geomechanical data. It provides a synthetic overview of geological information that is very difficult to gasp with isolated non-georeferenced data.

2 SOFTWARE, GEO-DATABASE ORGANISATION

By analogy to the technologies used for data management during the planning and construction of complex buildings (BIM), a georeferenced database has been developed as part of underground works that allows the acquisition, storage, visualization and the analysis of geological and geomechanical data.

The geological and geomechanical characterization of rock masses in both the design and construction phases, i.e. the parameters taken into account, refer to the recommendations contained in the SIA standards (Swiss Society of Engineers and Architects).

The georeferenced database in a geographic information system (GIS) allows geological data to be collected on lithological unit, rock discontinuities (type, dip-direction, dip, spacing and persistence of discontinuities), hydrogeological conditions (water inflow wand initial flow rate) and geotechnical information such as GSI, RQD, UCS, … locating them in a model with geographic coordinates.

In the construction phase, to complement the above-mentioned information, excavation-related data can be collected, including geomechanical test values, wedges volumes, location or heights of overprofiles and convergence values.

In addition to consultation and analysis of the data in tabular form, the GIS system allows them to be visualised in a geo-referenced 3D model or in a horizontal cross-section.

The technique presented was developed using the GIS software proposed by Esri. Esri's ArcMap and ArcCatalog 10.2 with a "Basic" license level are used to create and manage the databases and maps. The "3D Analyst" extension is required to build the 3D model in ArcScene. Data are separated in thematic file geodatabases (.gdb) and stored in feature classes by attribute type to make the data easier to handle and to simplify the display of specific layers. A number of useful site-related data such as site plans, geological/hydrogeological maps, cross-sections, boreholes, digital elevation models (DEM), topographical maps and land registries are gathered in the geodatabase (Figure 1).

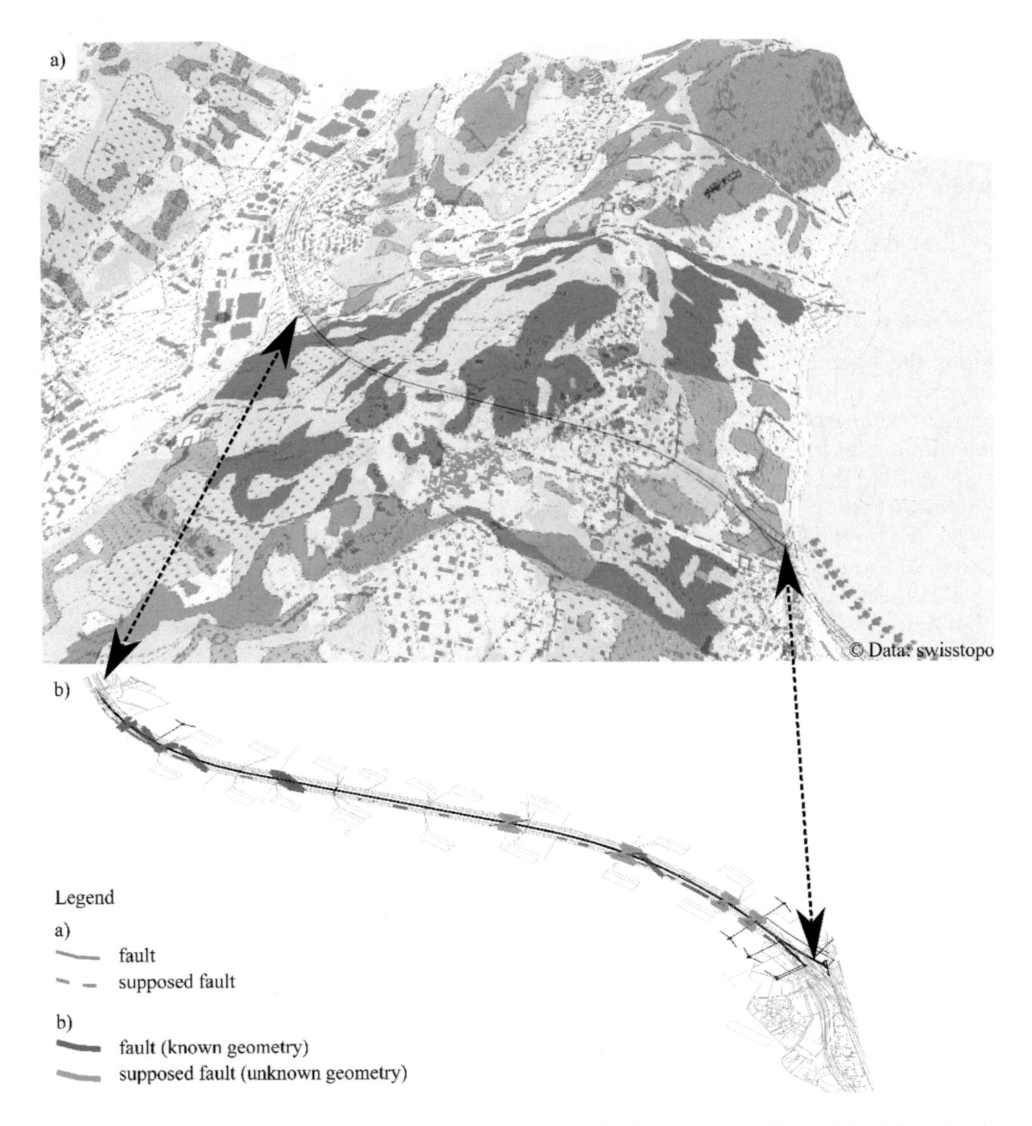

Figure 1. Project map with geological information. a) Geological map in a 3D model. b) Localization of faults mapped by geological survey of technical tunnel excavation.

3 DIGITAL DATA ACQUISITION AND GEOLOGICAL INTERPRETATION

For the acquisition of geological data, this is done separately for the project phase and the construction phase. In both cases the graphic representation of the collected rock discontinuities as faults or lithological boundaries is carried out using the tables of attributes associated with each object.

3.1 *Creation of the geological model of the project in the design phase*

In the design phase, the design of rock reinforcements is based on the geological/geomechanical model. The collected geological data are stored in a feature class/tabledatabase and will be used to generate the underground 3D model of the geological discontinuities (lithological boundaries, faults, etc,...). In particular, the persistence, dip-direction and dip attributes of the discontinuities will be used as parameters for the creation of single entities. The interpretation of geognostic and geophysical surveys are geo-referenced and integrated into the 3D model. 3D objects are created using specific tools created in ArcGIS. The combination of the surface geological map in the 3D model and the interpretation of the geological investigations (facilitates the connection/interpretation of the individual geological structures in the underground and validation of the geological model of the project.

3.2 *Geological survey during excavations*

During the execution stage, as an alternative to the traditional hand-drawn manual acquisition of data of the geological surveys, the mapping is carried out digitally. A tablet PC equipped with ArcGIS allows the digital mapping of the heading face and/or walls on a GIS map with a local coordinate system. The spatial reference system for this "face map" is vertical and parallel to the face of the surface.

A set of Python scripts implementing trigonometric functions convert the coordinates of any feature on the "face map" into geographic coordinates. Once their coordinates have been converted, features can be visualized at their real location on the horizontal site map as well as in a 3D model. For each structure/features the attribute table is compiled.

For a detailed explanation of the technique of geological site survey of the excavations with digital support, please refer to Thum & De Paoli (2015).

4 APPLICATION AND DISCUSSION

4.1 *Design phase*

To illustrate the interest of the method below, a three-dimensional analysis is carried out for the project "Third-lane extension of the N2 motorway section Lugano sud-Mendrisio - general project for the construction of the third tube of the tunnel San Salvatore between Melide and Grancia" (Federal Roads Office, FEDRO), nowadays at the stage of studying different tunnel alignment solutions.

The basis for the set-up of the structural geology model is the national geological map (sheet CH 1353), the detailed geological maps of the tunnel portal sector and the geological surveys from the technical tunnel excavated between the two current motorway tunnels (Figure 1).

The source of the geological data is therefore fragmentary and discontinuous. The information given on the surface geological map is disturbed by the morphology of the territory and by the surrounding vegetation. The mapped discontinuities are sometimes real, sometimes assumed like a lithological boundary of a presumed tectonic nature according to geological senses or identified on the basis of geomorphological analysis.

The purpose of the study is the verification of the persistence of surface-mapped geologic structures and of those detected during excavation of the tunnel. The three-dimensional analysis allows the study of faults, verifying their real spatial extension, their planarity and the order of fault families (Figure 2).

With reference to the study of different tunnel alignments, the graphic representation of the main tectonic features in a 3D model allows the immediate localization of these underground structures. Consequently, it is possible to identify their interference with all tunnel route variants under examination without any reworking of geological data such as the creation of geological cross-section prognosis for each route variant (Figure 2a).

4.2 *Geological survey during excavations*

The proposed method for georeferenced acquisition and archiving of geological-geomechanical data during geological surveys of the tunnel heading face is particularly suitable for

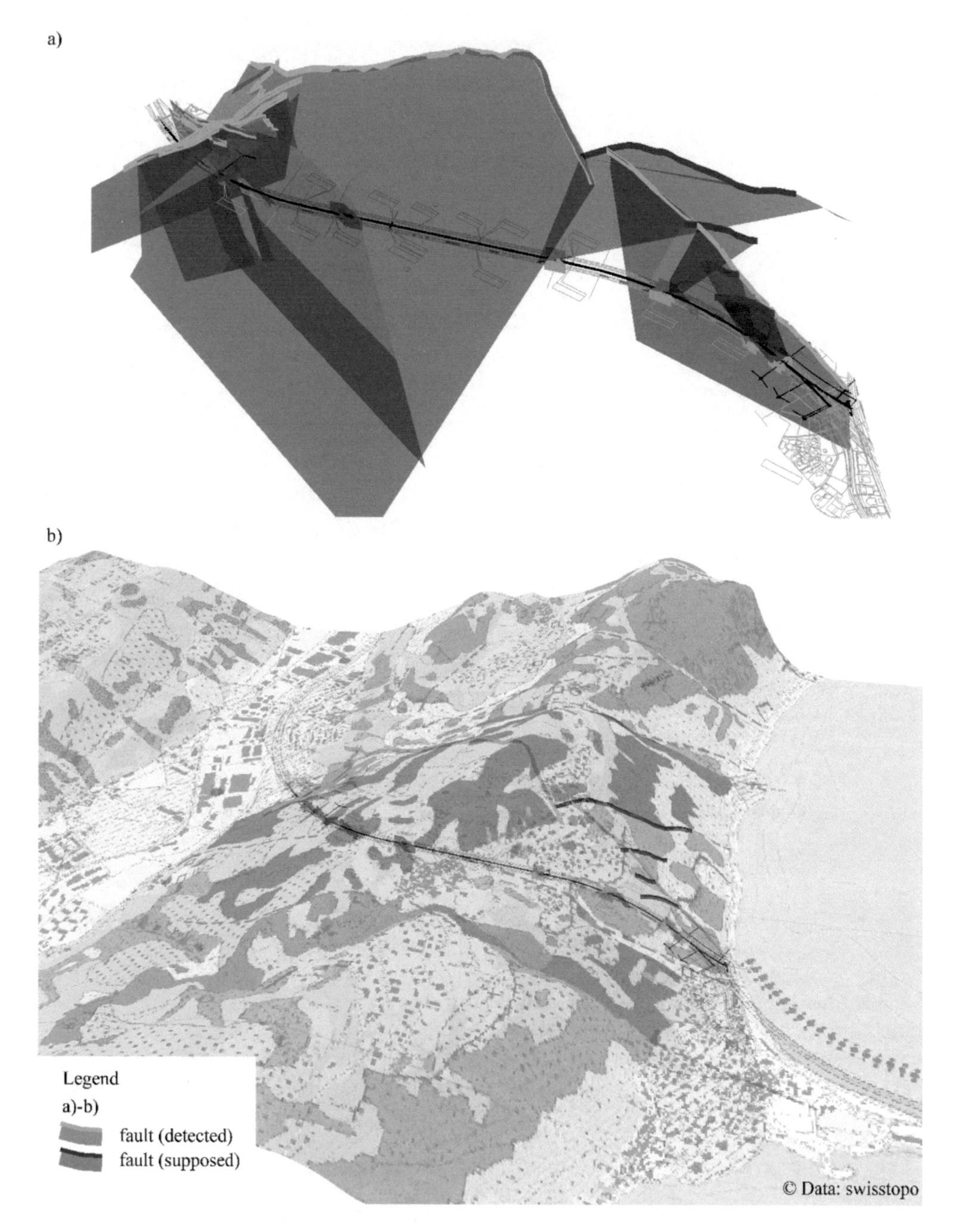

Figure 2. Project geological model, 3D analysis of the main faults: a) Underground projection of mapped (red) and supposed (blue) fault. b) Comparison of fault underground model and geological map.

large underground projects with high overburden and complex geological contexts where the understanding of the geomechanical context has important economic repercussions on the costs of rock-support measures. Knowledge of the geological conditions, of the rock mass behaviour and of their spatial variations reduces the risk of unforeseen problems during excavation.

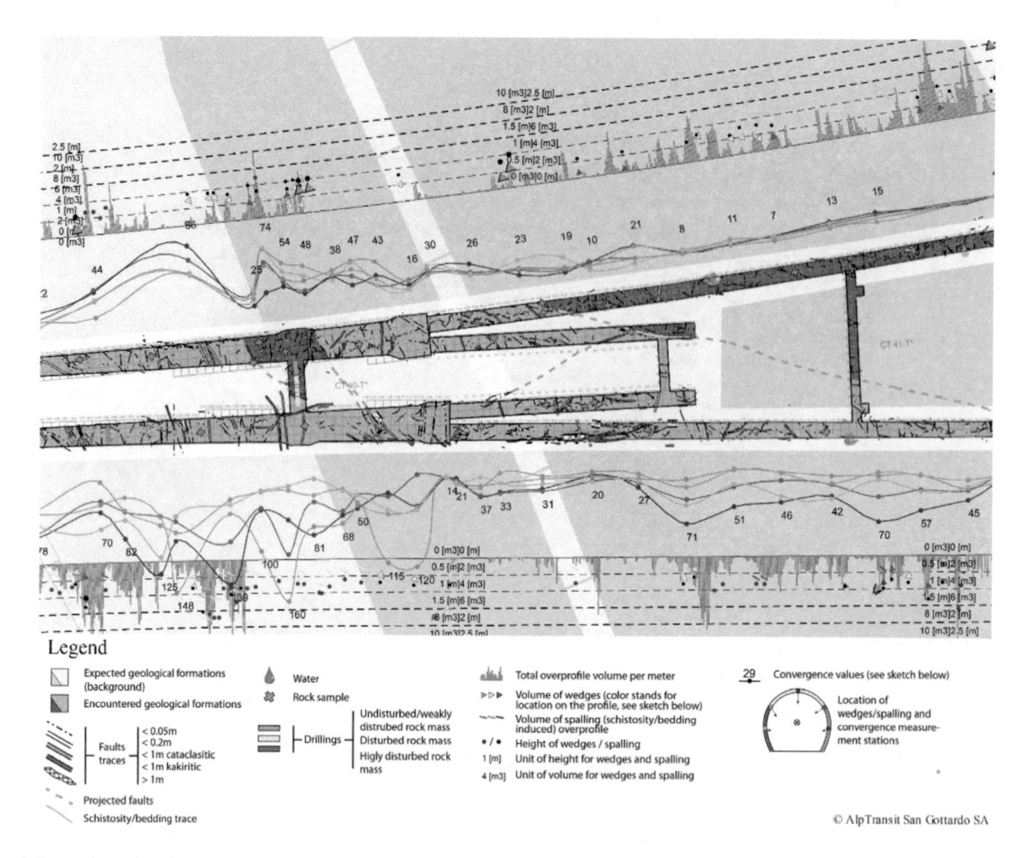

Figure 3. Geological and geomechanical synthesis plan of the sector Saré Cavern, CBT Sud.

Figure 3 shows an extract of the geological-geomechanical synthesis plan of the CBT Tunnel, Saré cavern sector. In addition to the geological feedback model, the results of the geotechnical monitoring (convergence measurements) and of the wedges (overprofiles) observed during the excavations are reported. Thanks to the synthesis model shown in Figure 4, the lithological contacts could be foreseen at the excavation phase, optimizing and verifying suitability of the rock reinforcements adopted. The back analysis takes place visually in the first instance, then, if a statistical analysis becomes necessary, it is possible to isolate data to be analysed from the tables of attributes.

The precise geolocation of geological information in the 3D model allows the approximations of the geological prognosis around the tunnel face to be reduced. Figure 4 shows a borehole of approx. 600 m in length which has a relatively large curvature of the drilling axis. In this sector coverage above the CBT track reaches 800 m. The correct location of boreholes implies a drastic decrease in the error of the geological prognosis in terms of location and inclination of geological structures. In the case of fault 15 B, the integration in the 3D model of the feedback geology and the interpretation of the borehole allowed the prognosis and the verification of the orientation, of the nature and of the extension along the route of the aforementioned fault/disturbed zone. This procedure allowed the engineers to plan the fault crossing in detail.

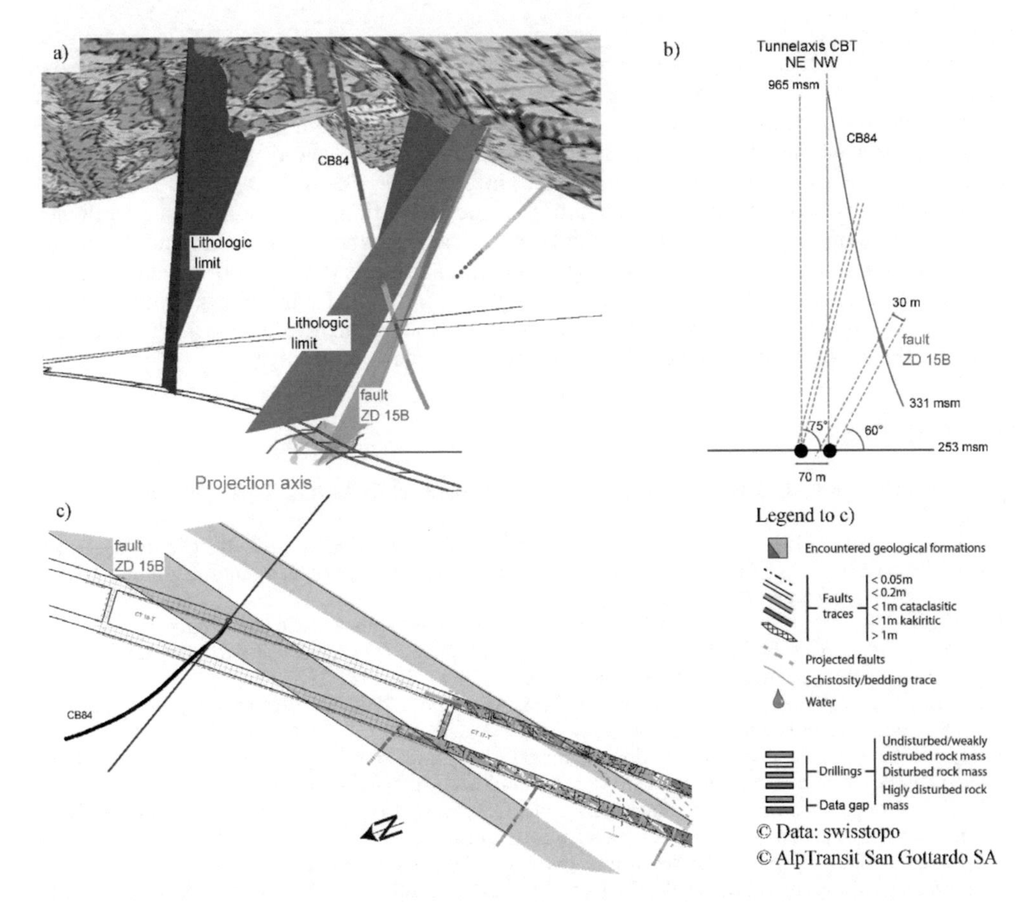

Figure 4. 3D model of 15B fault by CBT railway tunnel: a) 3D geological model of fault and lithological limits. b) Cross-section perpendicular to CBT road with location of drilling axis, CGT roads and interpretation of the fault 15B. c) Geological survey map with prognosis of extension of the 15B fault in the tunnel sector.

5 CONCLUSIONS

Geological mapping of both surface and underground during the geological survey of underground excavations provides an important amount of fragmented geological data. This information is the basis on which the geomechanical engineer measures the infrastructure project.

The proposed GIS solution is implemented with the "Basic" license of Esri ArcGIS Desktop with the addition of the "3D Analyst" extension. More details can be obtained at http://www.geol-gis.ch. The aforementioned program is compatible with CAD files and DEM terrain models.

The proposed method requires, in the geological modelling of the underground, the use of the possibilities offered by technological development in the three-dimensional drawing and in the management of the databases.

Thanks to special tools, developed in ArcGIS, data coming from the single geomechanical surveys are managed, analysed and geo-located. Geo-localized 2 D and 3 D visualization of the collected information allows for a better understanding of the systematicity, persistence, frequency of geological discontinuities such as faults or lithological contacts, thus facilitating the understanding of the site's geological structure even to non-specialists.

In the design phase, the creation of a 3D geological model of the underground according to the exposed technique takes place by first representing the geological structures in the underground, and then the extraction of the geological cross-sections. In this way the approximation errors that result from interpreting the data by interpolation of maps and two-dimensional geological sections are reduced.

In the construction phase of the project, the simultaneous view of geological conditions with technical-constructive (excavation-related) information, like the extent and frequency of danger scenarios (wedges, cracks, deformation of the excavation...) allows for an overall view of the stability of the excavation and the interaction between the underground and infrastructure. The tables of attributes associated with individual objects (features) of a geological and geomechanical nature allow the statistical analysis of individual parameters. Data sampling can be done directly by selecting objects in the 2D plane or in the 3D model. Proceeding in this way, the discernment of the objects of interest is simple and intuitive.

REFERENCES

AlpTransit Gotthard AG Media Office 2012. Galleria di Base del Ceneri/Geologia. *AlpTransit Gotthard AG*, 14–15/18–19. Luzern.

Aldiss, D.T., Black, M.G., Entwisle, D.C., Page, D.P. & Terrington, R.L. 2012. Benefits of a 3D geological model for major tunnelling works: an example from Farringdon, eastcentral London, UK. *Q. J. Eng. Geol. Hydrogeol.* 45: 405–414.

Bieniawski, Z.T. 1989. Engineering rock mass classification. Wiley-Interscience, New York.

Choi, Y., Yoon, S.-Y. & Park, H.-D. 2009. Tunneling Analyst: a 3D GIS extension for rock mass classification and fault zone analysis in tunneling. Comput. Geosci. 35: 1322–1333.

Fuerlinger, W., Weichenberger, F., Leblhuber P., & Amann, G. 2008. GIS-Anwendung und baugeologische Erfahrungen. *Felsbau Mag.* 324–330.

Hoek, E. 1982. Geotechnical considerations in tunnel design and contract preparation. In 17th Sir Julius Wernher Memorial lecture, Trans. Inst. Min. Metall. (Sect. A: Min. industry) 91: 101–109.

Kim, C.Y., Hong, S.W., Kim, K.Y., Baek, S.H., Bae, G.J., Han, B.H. & Jue, K.S. 2005. GIS-based application and intelligent management of geotechnical information and construction data in tunnelling. In Erdem, Y., Solak, T. (Eds.), Underground space use. Analysis of the past and lessons for the future, Proceedings of the International World Tunnel Congress and the 31st ITA General Assembly. Taylor & Francis Group. Istanbul 197–204.

Kvartsberg, S. 2013. Review of the use of engineering geological information and design methods in underground rock construction. Department of Civil and Environmental Engineering, Division of GeoEngineering, Chalmers University of Technology. Gothenburg 2013:3.

Loew, S., Barla, G. & Diederichs, M. 2010. Engineering geology of Alpine tunnels: past, present and future. In Proceedings of the 11th IAEG Congress, Taylor & Francis Group. Auckland, New Zealand 201–253.

Parker, H.W. 2004. Planning and site investigation in tunneling. In 1° Congresso Brasileiro de Túneis e Estruturas Subterrâneas, Seminário Internacional South American Tunnelling, Brazilian Association of Soil Mechanics and Geotechnical Engineering. Brazil.

Pozzorini, D. 2017. Ceneri-Basistunnel – Umgang mit prognostizierten und unerwarteten Störungen: Fallbeispiel von schleifend zur Tunnelachse streichenden Störzonen bei komplexen Baugrundverhältnissen, Swiss Bulletin for Applied Geology 22(2): 21–31.

Swiss Society of Engineers and Architects (SIA) 1998. 199 Erfassen des Gebirges im Untertagbau, SIA Empfehlung. Zurich.

Thum, L. & De Paoli, R. 2015. 2D and 3D GIS-based geological and geomechanical survey during tunnel excavation, Engineering Geology 192: 19–25.

Zheng, K., Zhou, F., Liu, P. & Kan, P. 2010. Study on 3D geological model of highway tunnels modeling method., J. Geogr. Inf. Syst. 2: 6–10.

Tunnels and Underground Cities: Engineering and Innovation meet Archaeology, Architecture and Art, Volume 3: Geological and geotechnical knowledge and requirements for project implementation – Peila, Viggiani & Celestino (Eds)

Contribution of continuous geophysical measurements to the success of tunnelling

T. Dickmann, D. Krueger & J. Hecht-Méndez
Amberg Technologies AG, Regensdorf, Switzerland

ABSTRACT: Today, many geophysical methods are available, which support the exploration of a tunnel route before tunnelling commences. Some have also been used during tunnelling for many years. Among them, seismic methods account for the largest share in both the number of tunnel applications worldwide and the success rate of contributing to geological exploration.

Since deep and long tunnels harbour considerable geological uncertainties, continuous measurements are increasingly becoming urgent to obtain a complete forecast. It is precisely these tunnels that are being driven by more powerful TBMs, which require a smooth and fast course of all measurements on the job site. Moreover, flexible operation and analysis of results is required. Modern systems are mastering these requirements more and more. Case studies are presented in which operational as well as data-analytic aspects are compared and discussed. Likewise, new trends in the interaction between tunnelling and geophysics will be presented.

1 INTRODUCTION

Site investigations ahead of the tunnel face by means of geophysical methods are increasingly becoming an essential part of the risk management process for the last 20 years. The tunnelling industry has already identified the potential of these usually non-destructive methods that valuably contributes to the assessment of the ground conditions and to the provision of an interpretative reporting.

Among available methods being used nowadays, two groups can be identified: alternative geophysical techniques emerging in the tunnelling sector, mostly electromagnetic techniques, and techniques already mature for the sector and successfully used in many working sites. Seismic methods have established as the frontrunner offering a wide range of data acquisition techniques, data processing types and result visualization tools. Tunnel seismic has undergo a recurrent development accompanying the tunnel industry for several years and adapting to its demanding requirement. With the advent of the industry 4.0, new horizons are being opened, hence more effective data collection, higher amount and faster data transfer, and fancy data analysis technique are expected. Therefore, existing modern seismic technologies should cope with newer requirements as well as with new trends in the tunnelling industry. For instance, in operational terms, the use of TBMs for long and deep tunnels is more and more preferred. In terms of data analysis, a reasonable question would be how we can get more useful information from all data that are being collected. These situations pose new challenges to tunnel seismic.

Continuous seismic measurements look for optimising both data acquisition process as part of the operative work flow while tunnelling and the rapidness in providing reliable geological predictions for the next tens of meters ahead of the face. In the case of TBM drivage an additional issue is related to the type of seismic source being used. Due to technical and operative constrains, the use of explosive might be restricted. In that case, impact sources surge as an alternative to be considered. Beside this operational aspect, continuous measurements allow for gradual detection of relevant geological target coming ahead. Such targets however, may not always lay directly in front of the drivage but they may run sub-parallel to the tunnel excavation. If seismic data from continuous measurement is available, it should be possible to process and evaluate this data in a special fashion to prospect not only targets ahead but laterally located around the tunnel excavation.

2 OVERVIEW OF SEISMIC METHODS

This paper focuses on seismic methods, because seismic reflection imaging is the most effective prediction method due to its large prediction range, high resolution and ease of application on a tunnel construction site. When using the information of the full seismic wave field propagating through the ground, seismic properties such as seismic velocities and their derived elastic parameters such as Poisson's ratio or stiffness present valuable information to characterise the ground.

There are nowadays several seismic methods being applied during mechanised tunnelling. The operation of the Integrated Seismic Prediction (ISP) methods is possible during normal TBM operation; meaning measurement preparation occurs while the TBM advances and measurement itself is done during ring-building and stand-by times avoiding long TBM downtimes. Here, an impact source generates both body and surface waves, whereas the surface waves is being converted at the tunnel face to an S-wave propagating towards discontinuities in the ground and being reflected (Borm et. al. 2003).

The TSWD-method has been developed for seismic exploration ahead of the tunnel face during TBM tunnelling where the cutting process of the TBM itself is used as the source of seismic waves ensuring a continuous seismic monitoring without hindering the drilling and driving operations (Petronio et al. 2003).

A meaningful alternative is Tunnel Seismic Prediction (TSP) - a rapid, non-destructive and highly sophisticated measuring method and system especially designed for underground construction works. The TSP method was firstly introduced to the underground construction market in 1994. With the use of the latest technology of the TSP 303 system, true 3-D data processing tools and presenting parameters of rock characterisation ahead of the face in three dimensions is available.

2.1 *Continuous seismic methods*

Continuous seismic methods can be an effective component of many site characterisation investigations. One of the primary benefits of continuous seismic measurements is to increase spatial sampling density so that background and anomalous conditions can be identified early in the investigation. As with all geophysical methods, seismic methods are limited in depth penetration, i.e. the range of the target depth. As the target becomes deeper meaning further away, the resolution of geophysical measurements decreases. Due to larger volume sampling or wave signal attenuation, the contrast between the target and surrounding materials needs to be even greater. At some point, a discrete localised target, such as a cavity, may be more difficult or impossible to detect at distances greater than 100 meters ahead of the face. (Fig. 1). However, as the excavation advances, the target gets closer and a better resolution or imaging will be possible by continuous measurements.

Recently, a new concept for continuous seismic exploration while excavation has been introduced, Tunnel Seismic Prediction while Excavation, TSPwE® (Dickmann et al. 2018). By deploying three pairs of receivers along each tunnel wall and blasting a minimum number of shots for a given face position (Fig. 2), 3-D images are generated and updated every 10 to 15 meters, depending on the advance rate. This methodology brings some advantages when prospecting geological features as the cavity example mentioned above. In this case, detection shortcomings due to the distance of the target at early stages, such as wave signal attenuation, penetration depth and lateral resolution are counteracted by continuously calibrating the 3-D results while getting closer to the target. Moreover, since the largest reflection signals are based on the cavity (largest contrast), higher accuracy should be expected during automatic data processing.

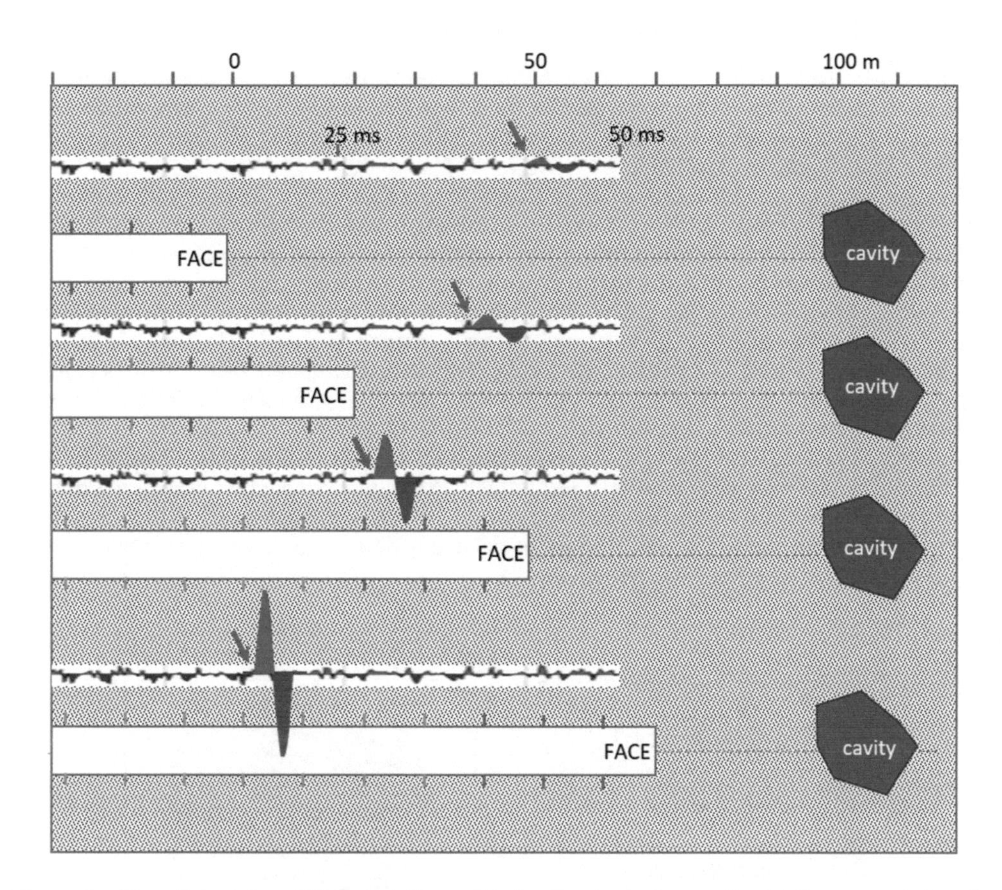

Figure 1. Tunneling at four different stations while heading towards a cavity. At the upper station, the cavity is about 100 m ahead of the face and the seismic signal reading of the reflection at the cavity is hardly recognizable (refer to red arrow at approx. 40 ms). When approaching the cavity while measuring seismic reflection signals from source points behind the tunnel wall, the signal becomes stronger and significant.

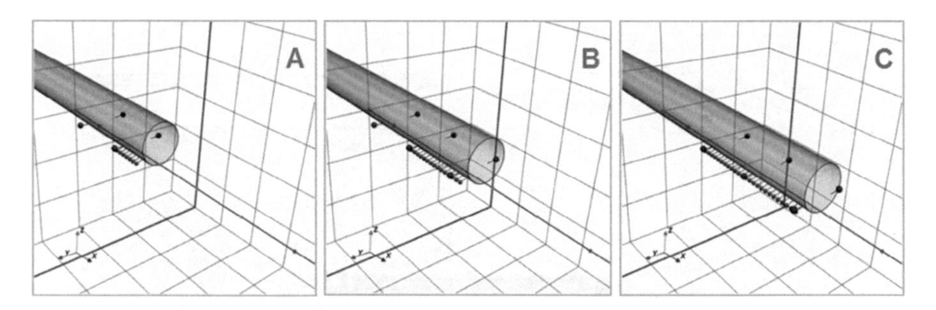

Figure 2. Concept of TSPwE®: tunnel seismic prediction commences after deployment of four receivers (blue dots) (A). Shooting in small boreholes along the side wall happens along with heading. (red dots in A, B and C). After about 10 to 15 meters a third receiver pair is deployed (B). After 20 shots along the side wall, the rear receivers are being deployed as front receivers (C).

2.1.1 *Addressing operational aspects using continuous seismic*

Continuous seismic measurements can be done in both conventional and mechanised tunnelling. In conventional headings, explosives are being commonly used as the seismic source. In mechanised tunnelling however, the use of explosive as a seismic source poses an important restriction since it is not necessarily available at site or not available at all. Additionally, due to licensing restrictions and safety regulations, the use of explosive or selection of conventional tunnelling as the excavation method is slowly decreasing in some countries, as for instance in China. Under this scenario, the use of TBM is expected to increase in the upcoming years.

To cope with this scenario, the use of an alternative impact source together with continuous measurements is foreseen. A sledge hammer, an electronically controlled mechanical hammer or actuator can then be employed as the impact source. Similar as explosive sources, impact sources also generate body waves that travel through the rock mass and are reflected at discontinuities. The signal transferred to the medium by this type of source presents different characteristics with regards to signal amplitude and frequency range and compared to those generated by explosives. In addition, the total energy transferred into the rock mass is much lower. Consequently, the penetration depth and the achievable seismic resolution for detecting a given target must be evaluated.

Figure 3 shows the TSP layout and comparison of 3-D distribution of the P-wave velocity for single measurements performed in a test tunnel. TSP data was acquired using a mechanical hammer at 18 "shot" positions (Fig. 3a). In total, 103 blows were performed (5 to 6 blows) per shot position. Hammer data was stacked at each position to increase the Signal-to-Noise

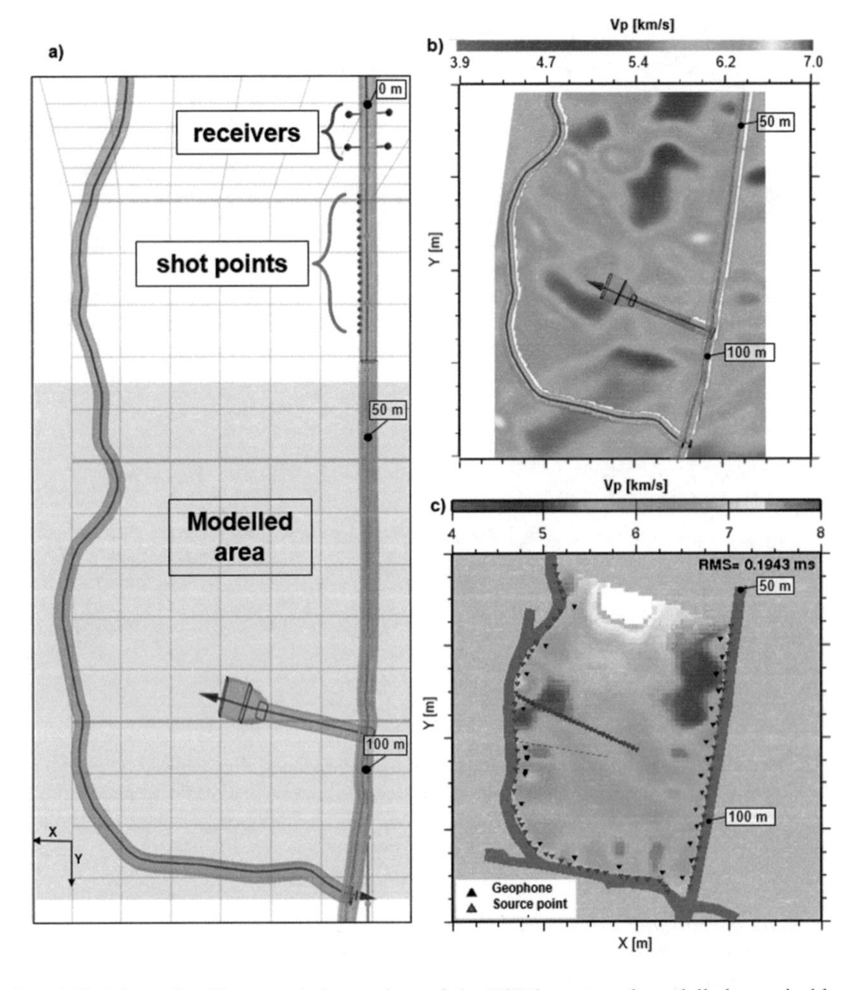

Figure 3. a) Position of receivers and shot points of the TSP layout and modelled area in blue. b) 3-D P-wave velocity distribution obtained from the TSP data on modelled area in plane-view. c) 2-D P-wave distribution obtained from seismic tomography (modified from Richter, H. 2010).

Ratio (SNR). Stacked data was then processed by the standard processing workflow of the TSP software. The control data correspond to seismic tomography acquired along three surrounding galleries as shown in Figure 3c. In this case, seismic tomography is set as the benchmark since resolution achievable by this method is expected to be higher. It must be point out that the 2-D tomography results used for the comparison already existed from former surveys. TSP data was processed independently, i.e. with own input parameters for seismic data processing and modelling.

Figures 3b and 3c depict the P-wave velocity distribution for the TSP results and seismic tomography, respectively, throughout the modelled area as indicated by the blue square in Figure 3a. The area starts approximately 10 meters ahead of the last shot point of the TSP layout and extends about 80 meters in the longitudinal direction (Y). In the lateral direction (X), the model spans about 25 meters from the tunnel axis where the seismic measurement was done. In general, the results are in satisfactory agreement, the prevailing modelled P-wave velocity estimated by both techniques varies around 6 km/s (green). Although some differences are present, important features or velocity anomalies are evident in both results. For instance, the high velocity zones at the top/centre and surrounding the left tube in the tomography are also noticeable in the TSP results. Similarly, the largest low velocity zone (blue) at the top, surrounding the right tube, are also reproduced by the TSP results although slightly shifted to the left. Towards the bottom gallery, results differ slightly more. A relevant aspect is the magnitude of the contrast in each result. In the seismic tomography results, the velocity contrast between the rock prevailing velocity and the anomalies is apparently higher. In turn, although the spatial distribution of the anomalies indicated by the TSP results match the tomography, the contrast is somehow lower particularly for the high velocity zones. This might be due to different input parameters in data processing of each methodology as for instance the initial value assigned to the average P-wave velocity for each model, 5.4 km/s and 5.7 km/s for the tomography and the TSP, respectively. Influence of other input parameters can also not be disregarded.

This benchmark attempt indicates that the TSP results coming from data acquired using hammer blows is reliable for the first tens of meters ahead of the seismic layout, in this example up to around 80 m. Considering this, the use of mechanical hammer for continuous measurements is feasible. As mentioned above, the TSPwE concept implies updating 3-D results in short spatial intervals, every 10 to 15 meters ahead of the face which will lay within a reasonable range for this type of source. Moreover, since the data density will increase by using up to 6 receivers and optionally by using two shot lines, the accuracy and reliability of models coming from data acquired using impact sources should considerably improve.

2.1.2 *Extending data analytics in continuous seismic*

One advantage of using continuous seismic relies on the gradual detection of targets as the tunnelling progress approaches to it. Certainly, it is preferable to obtain early indicators of the presence of unfavourable structures to further explore them and the risks associated to them. In many cases, unfavourable structures are not ahead and vertically oriented but rather they run parallel or subparallel to the tunnel, i.e. horizontally layered. In the worst scenario, this feature will intersect the tunnel at some future stationing resulting in possible hazardous situations, e.g. water ingress, collapses from the crown, etc.

Seismic data acquired within a tunnel contain useful information coming from a 3-D space around the tunnel. Depending on the type of data analysis, structures ahead of the tunnel face or surrounding the tunnel can be modelled. Such data analysis can be referred to as processing to tunnel side or side processing. Side processing of TSP data coming from standard acquisition procedure, i.e. sporadic or periodic measurements, present limitations related to the lateral resolution due to the length of the seismic layout. Hence, only a limited portion of the potential layer (reflector) will be detected compared to the actual length of the geological feature extending horizontally. By using continues measurements such limitation can be partially overcome because each portion of the target feature are continuously imaged (Fig. 4).

The surface above a given tunnel with low overburden, e.g. with subsea tunnels, gives an ideal situation for validating the applicability of side processing to TSP data acquired using

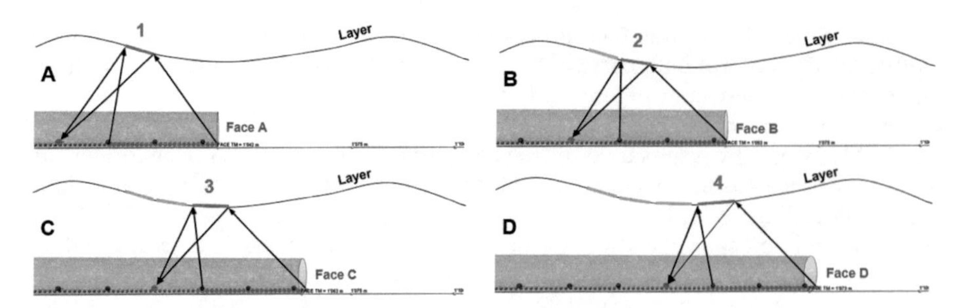

Figure 4. Schematic representation of side processing of data acquired using the TSPwE concept. The capital letters A, B, C and D indicate various tunneling stages with their respective face positions (advance from left to right). The red and green segments lying on the layer represents the modelled segment by current and previous side investigations, respectively. As the continuous data acquisition advances, the layer is continuously modelled.

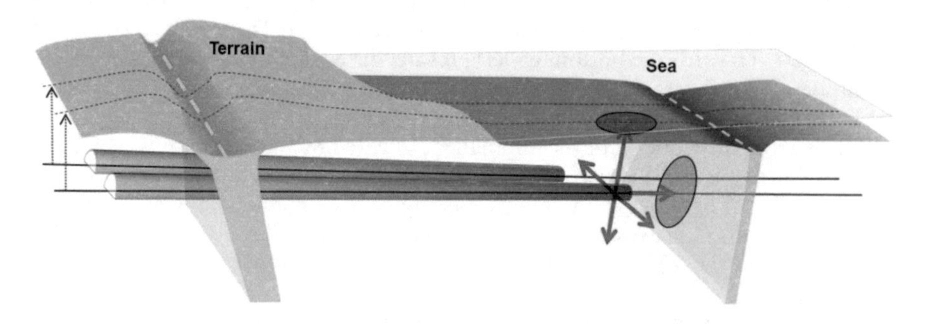

Figure 5. Schematic representation of subsea tunnels showing the terrain and sea surface as well as the seabed. Surface delineation (blue dotted lines) as well as detecting other structures (red arrows with imaging planes) are of interest for the project.

the continuous approach. Figure 5 schematically shows such a situation in a subsea tunnel. The surface may also be deemed as a formation change, a discontinuity, or any other significant geological structure presenting sufficient rock physical property contrast. As depicted in this picture, the surface above the two tubes will be imaged by seismic measurements done within the tunnels.

Figure 6 shows the longitudinal view of the 3-D migrated sections from two single TSP measurements done in parallel tubes for such a scenario upon the geological forecast. Although the measurements were carried out in different tubes and not strictly following the TSPwE concept, it resembles very well the significance of continuous measurements. For the sake of simplicity both results are projected into the same plane located at the middle of the tunnel axes.

A migrated section represents the real spatial position or distribution of reflective rock zones along the section, hence the intensity of the colouring (red or blue) indicate zones of high reflectivity or in other words areas with significant rock condition changes. White colouring denotes no reflectivity. In section Face A (left), two major features are identified, the most significant corresponds to a strong reflectivity zone smearing vertically up to the surface. This zone matches very well the weakness zone inferred in the geological forecast (dotted red line behind the section). Towards the surface, the reflection zones tend to bend following the contact between the hard rock and the deposit material. In this example, the proximity and magnitude of the weakness zone strongly dominates the migrated section. In turn, section Face B shows the highest reflectivity directly around a section of the tunnel, possibly due to the presence of a significant Excavation Damaged Zone (EDZ), and at a small portion nearby the

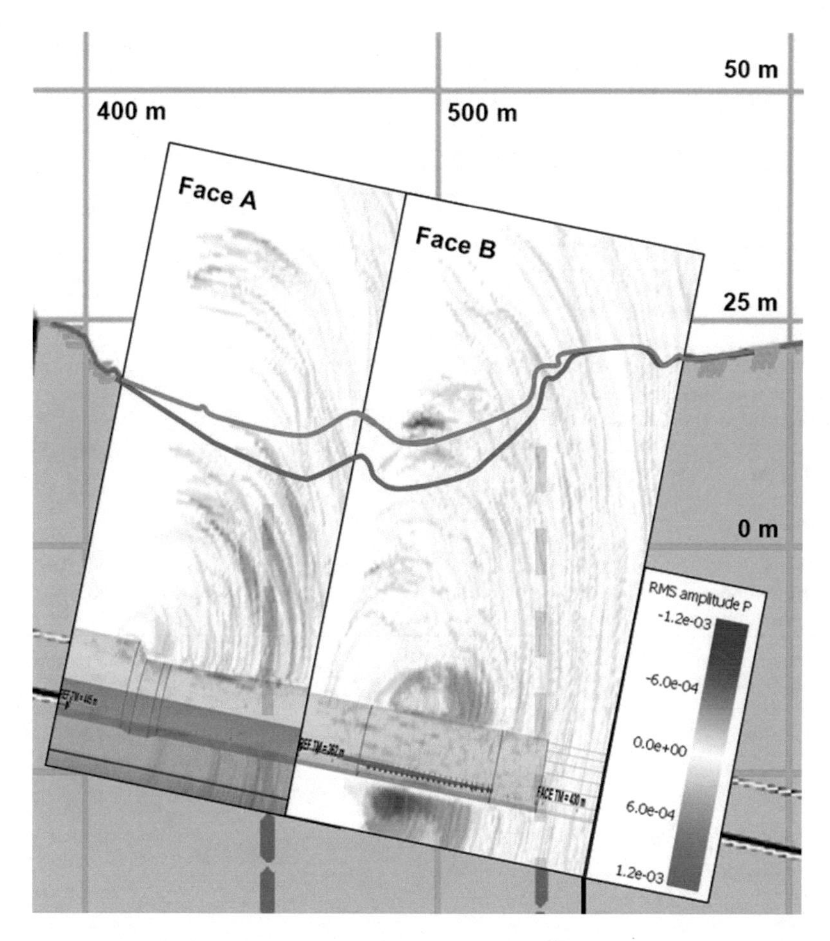

Figure 6. Two migrated sections of the P-wave component analysis upon geological forecast of seismic investigations in a subsea tunnel. Each section was obtained from periodical seismic measurements. Red and green lines represent the contact between hard rock and deposit material and the surface, respectively.

surface. Because of the upward dip of the surface towards the right, unfortunately, it is not possible any longer to obtain reflections that make possible to image the contacts along the entire section. In any case, evaluation of these periodic seismic measurements already gives insights into the potential of using a combined approach of continuous investigation and special data analysis as side processing.

2.2 *Benefits and contribution of tunnel seismic*

Considering the common high uncertainty of the geological forecast for a tunnel project, any variant of tunnel seismic together with probe drilling and geological documentation will help in validating and refining the geological forecast. Selection of the more suitable seismic technique for a given project should be done following an evaluation of the capabilities and limitation of each methodology available and site requirements. If TSP is considered as the right option, various measurement schemes can be followed: sporadic, periodic and continuous (Fig. 7). Then, it is the tunnel engineer's decision on the right selection of the best approach for the project.

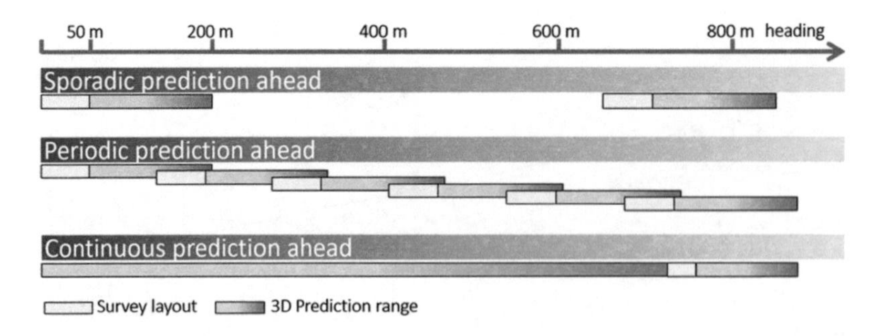

Figure 7. Top: Sporadic measurements may be carried out, when only specific sections of the tunnel course shall be investigated. Middle: A better than sporadic approach is the periodic prediction, where a new measurement is to be done before the previous investigation range exceeds. Bottom: a continuous prediction delivers data while excavating and updates a predictive model quasi continuously in short intervals of about 10 meters.

Table 1. Comparison of the three types of data measurement available for TSP. Excavation methods: mechanised or conventional. Production cycle means the workflow plan for the tunnel excavation and how TSP can be implemented.

Type of measurement	Excavation method	Seismic source	Production cycle	Downtime (min)	Prediction range (m)
Sporadic	all	explosive mechanical	independent	~60*	120 – 150 80 – 120
Periodic	all	explosive mechanical	integrated	~60*	120 – 150 80 – 120
Continuous	all	explosive mechanical	part of it	< 10	120 – 150 80 – 120

* Standard acquisition, if Multiple Shot Recording is used downtime reduces by 1/3.

Table 1 summarizes major differences and benefits for the three types of measurements available for TSP. The table may be considered as a support decision tool for the selection and implementation of TSP as the seismic prospection method.

3 NEW TRENDS OF GEOPHYSICS IN TUNNELLING

There are not seldom many methods used in underground construction projects, including geo-technical sounding and sampling, geophysics, environmental sampling and testing, hydrogeology and geotechnical laboratory testing. One of the biggest challenges is to make use of all the data during the interpretation and the modelling. The traditional way includes a lot of drawings, plots, diagrams, tables spread on desktops, walls, floors, screens etc. (Cracknell & Reading 2014). The digital world has evolved during the last 30 years and it does find a peak in industry 4.0 where data exchange has never been more essential than before and hence opens many new opportunities for joint interpretation. The GeoBIM concept suggests how to make use of these opportunities available today, state of the art geotechnical and geophysical data handling and workflow (Svenson 2017).

Once data exchange and joint interpretation becomes possible, a further step to artificial intelligence such as machine learning algorithms is made. Machine learning algorithms use an automatic inductive approach to recognize patterns in data. Once learned, pattern relationships are applied to other similar data to generate predictive models. Hence, there is much scope for the application of machine learning algorithms to the rapidly increasing volumes of geophysical data obtained by continuous measurements for geological mapping problems.

Machine learning approaches can be applied at different stages throughout the geophysical evaluation, ranging from raw data analysis by classifying and filling gaps, processing parameter adjustment (data mining) and ultimately on automatic interpretation which comprises predicting rock properties and extracting discontinuities from seismic data (faults, layers, etc). However, there are challenges for application of this technique in geophysical data mainly related to the amount, quality and non-uniqueness of geophysical data, the homogeneity of collected data and the dependency of the involved variables. Nevertheless, with increasing computational power and the experience already gained in application of machine learning in different field of geoscience, adaptation and rapid development for the tunnel industry is possible.

4 CONCLUSIONS

Beside the operative benefits provide by the execution of continuous seismic measurements, further data analysis can be done from a single dataset. While prediction ahead allows to tackle continuously geological structures ahead of the face that may pose a hazard to the excavation, specialized processing to tunnel side helps investigating subparallel structures that may intersect the tunnel along as the excavation advances. Since the same dataset is being used for both type of predictions: ahead and side, no extra job must be carried out in the tunnel once a measurement has been completed. The lateral resolution limitation of side processing due to the length of the TSP layout is until some extent overcome by employing the continuous measurement approach. This was demonstrated by analysing two datasets of single TSP measurements with the task of detecting the surface in a subsea tunnel.

The use of impact hammer for continuous measurements is also feasible. This is particularly important for TBM drivage where the use of explosive is restricted. By comparing TSP results of a sporadic measurement against results of high resolution seismic tomography in a test gallery, important insights about the penetration depth and resolution of the TSP results was possible. In this example, reliable results were obtained until about 50 meters away from the seismic layout. Considering that 3D results of continuous measurements are update every 10 to 15 meters as the excavation progress, the penetration depth achievable when using impact source should be fair enough for the integration of this type of source into the continuous seismic approach. Certainly, the use of a seismic source as an impact hammer in a TBM drivage should be well prepared. The impact hammer can be mounted in a suitable place of the TBM structure or could be mounted in an additional vehicle if there is enough space for accessing.

With these extended capabilities, tunnel seismic technologies like TSP, shows its versatility and recurrent development adapting to the new requirements and trends of the tunnelling industry. As these trends grow, implementation of alternative geophysical prospection technologies in tunnelling is expected to increase in the following years. Certainly, a combination of geophysical techniques may also be an attractive approach as a tool for further mitigate the geological risk still coming ahead or from the sides while tunnelling.

REFERENCES

Borm, G., Giese, R., Otto, P., Amberg, F. & Dickmann, T. 2003. Integrated Seismic Imaging System for Geological Prediction During Tunnel Construction. *ISRM-2003 Technology roadmap for rock mechanics, South African Institute of Mining and Metallurgy, Symposium Series S33*: 137-142.

Cracknell M.J. & Reading A.M. 2014. Geological mapping using remote sensing data: A comparison of five machine learning algorithms, their response to variations in the spatial distribution of training data and the use of explicit spatial information. *Computers & Geosciences* 63: 22-33.

Dickmann, T., Krueger, D. & Hecht-Méndez, J. 2018. Optimization of Tunnel Seismic operations for fast and continuous investigations ahead of the face. *Proc. of the World Tunneling Congress*, Dubai, June 2018.

Petronio, L., Poletto, F., Schleifer, A. & Morino, A. 2003. Geology prediction ahead of the excavation front by Tunnel-Seismic-While-Drilling (TSWD) method. *73rd Annual International Meeting*. Society of Exploration Geophysicists, Expanded Abstracts.
Richter, H. 2010. Hochauflösende seismische Tomographie zur Charakterisierung eines Gebirgsblocks im Lehr- und Forschungsbergwerk „Reiche Zeche"- Freiberg unter Verwendung von Strecken- und Bohrlochmessdaten, *Diploma thesis*, Universität Potsdam.
Svensson, M. 2017. GeoBIM for Infrastructure Planning. *Proceedings of the 23rd European Meeting of Environmental and Engineering Geophysics*. 3-7 September 2017, Malmö, Sweden.

Tunnels and Underground Cities: Engineering and Innovation meet Archaeology, Architecture and Art, Volume 3: Geological and geotechnical knowledge and requirements for project implementation – Peila, Viggiani & Celestino (Eds)
© 2020 Taylor & Francis Group, London, ISBN 978-0-367-46583-4

Study on judgment of tunnel face failure in soft surrounding rock and failure control measures

J. Du, Z. Mei, H. Gao, B. Zhang & W. Yuan
China Railway Southwest Research Institute, Co., Ltd., Chengdu, Sichuan, P. R. China

ABSTRACT: In recent years, with the increase of tunnel construction, the problems existed in the construction of tunnel with soft surrounding rock are getting more and more tricky. The soft surrounding rocks which have poor engineering properties can be easily influenced by excavation disturbance which could lead to obvious deformation. Though many studies are focusing on the failure mechanism of soft surrounding rock and its control measures, there is yet no consensus. In this thesis, according to strength reduction theory, combined with displacement mutation and plastic strain energy mutation, a judging method for stability of tunnel face has been developed by using safety factor as a criterion. Then, the pre-reinforcement measures against instability have been categorized. The stability of tunnel face has been studied under different pre-reinforcement conditions by using safety factor as a criterion. At last, suggestions for pre-reinforcement have been given.

1 GENERAL INSTRUCTIONS

In recent years, the numbers and mileages of tunnels in railway, highway, urban rail transit and municipal engineering have shown a high-speed development trend in China. Hong (2015) believe that the 21st century is the golden season for tunnels and underground projects in China. The construction of tunnel in China has made great progress, but it also faces more severe challenges. There will be more and more bad geological tunnels, such as rock burst, soft rock deformation, high ground temperature, active fault, high-cold, karst, gas and so on (Zhao, 2016). For the weak surrounding rock tunnel, there are more and more problems, which not only bring safety, quality, construction period, investment and other impacts on the tunnel construction, but also adversely affect the surrounding environment of the tunnel, such as: large deformation, tunnel face collapse, distortion of the arch, initial cracking, surface collapse and so on (Wang, 2004). The soft surrounding rocks which have poor engineering properties can be easily influenced by excavation disturbance which could lead to obvious deformation. Though many studies are focusing on the failure mechanism of soft surrounding rock and its control measures, there is yet no consensus (Sun et al. 1991). In this thesis, according to strength reduction theory, a judging method for stability of tunnel face has been developed by using safety factor as a criterion. Then, the pre-reinforcement measures against instability have been categorized. The stability of tunnel face has been studied under different pre-reinforcement conditions.

2 JUDGING METHOD

2.1 *Strength reduction theory and safety factor*

The strength reduction theory is initially applied to the slope stability analysis. The finite element safety factor method developed by it is one of the important methods to calculate the safety factor of slope stability. This method can not only determine the safety factor of the slope. It can also automatically find the potential sliding surface of the slope, so it has been widely used in slope stability analysis. In recent years, Academician Zheng Yingren et al. (2012) applied the finite element strength reduction method to the stability analysis of surrounding rock, which provided a new analysis method for the stability analysis of the tunnel face.

In this method, the surrounding rock strength parameters c and φ are simultaneously divided by the reduction coefficient K_i, and the new parameters c' and φ' are substituted into the model calculation, and the above steps are repeated in sequence until the model is unstable. The reduction coefficient K_i at this time is Safety factor.

$$c' = c/K_i \tag{1}$$

$$\varphi' = \arctan(\tan\varphi/K_i) \tag{2}$$

2.2 *Catastrophe theory and cusp catastrophe model*

In 1972, the theory of catastrophe was first proposed in the "Structural Stability and Morphogenesis" published by the French mathematician Thom. The main starting point is the bifurcation theory and the singularity theory, as well as the concept of structural stability. The catastrophe theory mainly explains how the nonlinear system changes from the continuous gradual state to the mutation of the system property, that is, how the continuous change of parameters leads to the discontinuity (Ling. 1987). The cusp catastrophe model is a kind of catastrophe theory, which is proposed by Zeeman (1974). Its potential function V is a two-parameter function (two control variables u and v) whose state variable is x.

$$V = x^4 + ux^2 + vx \tag{3}$$

The corresponding equilibrium position meets:

$$\frac{\partial V}{\partial x} = 4x^3 + 2ux + v = 0 \tag{4}$$

Through a series of conversions after applying the catastrophe theory, the bifurcation set equation is finally obtained:

$$D = 8u^3 + 27v^2 \tag{5}$$

The above formula is a sufficient criterion for the mutation. The eigenvalue D can be used as the distance between the steady state of the surrounding rock and the critical state: when $D>0$, the surrounding rock is in a stable state; when $D\leq 0$, it is only possible for the system to mutate across the bifurcation set (Ma et al. 2010).

2.3 *Judging method*

Firstly, based on strength reduction theory, the numerical simulation method is used to reduce the cohesion c and the internal friction angle φ of the tunnel surrounding rock in the numerical model, and draw the relationship between the reduction coefficient K_i and different criterion factors. Secondly, based on catastrophe theory, different cusp mutation models are established for different criterion factors. Finally, the mutation point on the curve is found and the safety factor FV is determined.

Specific steps are as follows:

1. Establish a numerical model;
2. Determine the reduction coefficient K_i;
3. The cohesive force c and the internal friction angle φ of the surrounding rock are reduced;

4. Different criterion factor $F(K_i)$ are selected, such as key point displacement, plastic strain energy, plastic zone volume, unbalanced force, and so on;
5. Draw the relationship curves between different criterion factor $F(K_i)$ and the reduction coefficient K_i;
6. The polynomial fitting is performed on the relationship between the criterion factor $F(K_i)$ and the reduction coefficient K_i to obtain a fitting equation;

$$F(K_i) = a_4 K_i^4 + a_3 K_i^3 + a_2 K_i^2 + a_1 K_i + a_0 \tag{6}$$

7. Establish a cusp mutation model.

$$V(K_i) = K_i^4 + u K_i^2 + v K_i \tag{7}$$

$$u = \frac{a_2}{a_4} - \frac{3 a_3^2}{8 a_4^2} \tag{8}$$

$$\nu = \frac{a_1}{a_4} - \frac{a_2 a_3}{2 a_4^2} + \frac{a_3^3}{8 a_4^3} \tag{9}$$

   From the catastrophe theory, $\Delta = 8u^3 + 27\nu^2$ is the mutation characteristic equation, which can be used as the judging equation of the system mutation. Only when D≤0, the system will undergo mutation.
8. Find the point of mutation and determine the safety factor FV.

3 NUMERICAL SIMULATION

A numerical simulation model is shown in Figure 1. The values of the surrounding rock parameters refer to the design codes, as shown in Table 1. The surrounding rock constitutive model uses the Mohr-Coulomb strength criterion. The fixed constraint is applied to the bottom surface, the horizontal constraint is applied to the four sides, and the top is a free surface. The initial ground stress is applied by gravity balance. The buried depth of the tunnel is selected as 1D (D-the width of tunnel), and the numerical model is shown in Figure 2.

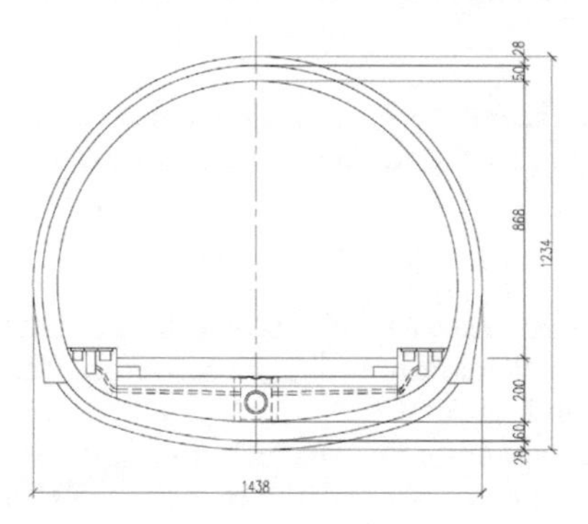

Figure 1. Cross-section of tunnel (unit: cm).

Table 1. Field geotechnical parameters.

Severe γ (kN/m^3)	Elastic modulus E (GPa)	Poisson's ratio μ	Internal friction angle φ (°)	Cohesion c (MPa)
15	0.2	0.415	11	0.005

Table 2. Relationship between the reduction coefficient and the cohesion c and the internal friction angle φ.

reduction coefficient	cohesion c' /Pa	internal friction angle φ' /°
0.05	100000.00	75.57
1.00	5000.00	11.00
2.00	2500.00	5.55
3.00	1666.67	3.71
4.00	1250.00	2.78
5.00	1000.00	2.23
6.00	833.33	1.86
7.00	714.29	1.59
8.00	625.00	1.39
9.00	555.56	1.24
10.00	500.00	1.11

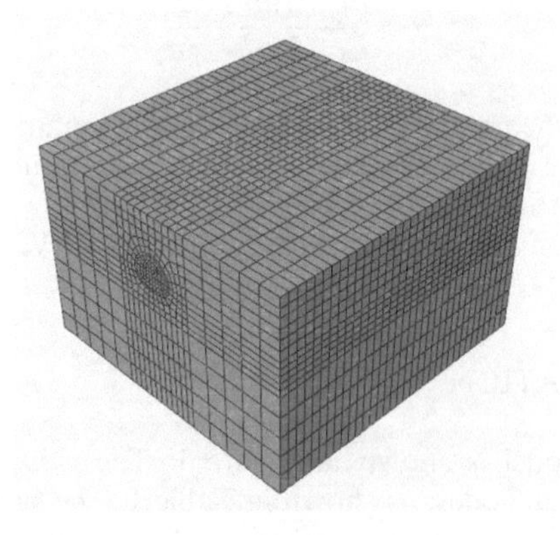

Figure 2. Numerical model.

4 FAILURE CRITERION ANALYSIS OF TUNNEL FACE IN SOFT SURROUNDING ROCK

According to Equation 1 and Equation 2, the cohesive force c and the internal friction angle φ of the surrounding rock are reduced, and the reduction process is set as follows:

4.1 *Displacement mutation*

As the reduction coefficient K_i increases, the strength parameters of the surrounding rock gradually decrease. Until the calculation no longer converges, the resulting displacement contour of tunnel face is shown in Figure 3.

Taking the point A of the maximum displacement in Figure 3 as the key point, the relationship between the key point displacement and the reduction coefficient K_i is plotted, as shown in Figure 4.

u and v can be obtained by using equations (8) and (9) by fitting the coefficients in the four expansions of the Taylor series under different reduction coefficient K_i. Then, u and v are substituted into the mutation characteristic equation, and the displacement mutation characteristic value Δ of the key point A corresponding to the different reduction coefficient K_i can be obtained, as shown in Table 3.

It can be seen from Table 3 that when the reduction coefficient $K_i = 0.367$, $\Delta = 5.73\times10^{-5}>0$, there is no mutation at the key point A; when the reduction coefficient $K_i = 0.374$, $\Delta = -5.53\times10^{-5}<0$, the key point A is abrupt, and the tunnel face is unstable, that is to say the safety

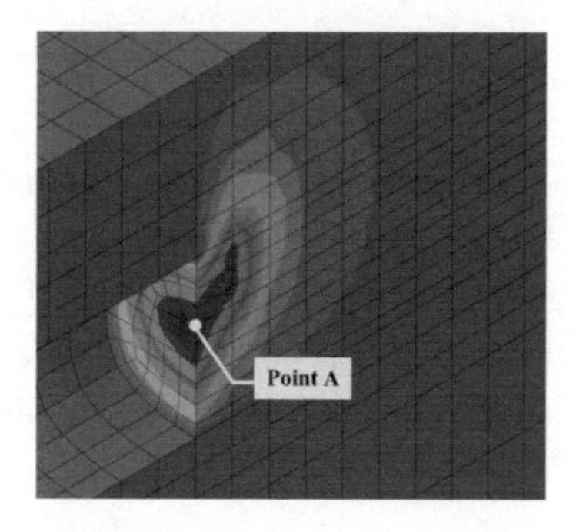

Figure 3. Displacement contour of tunnel face when the calculation does not converge.

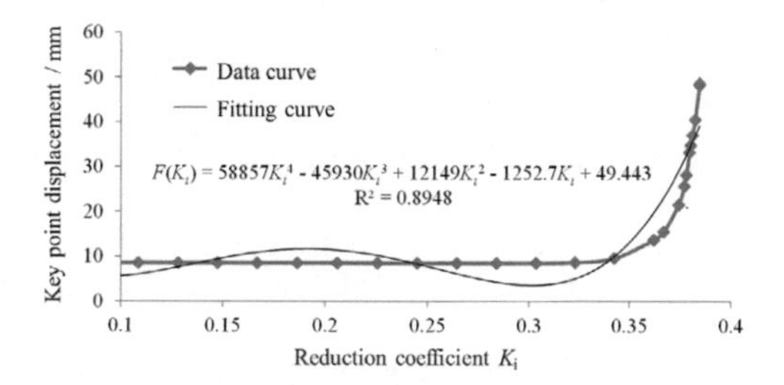

Figure 4. Fitting curve of displacement and reduction coefficient at key point.

Table 3. Displacement mutation judgement at key point.

K_i	u	v	Δ	status
0.362	-1.77×10^{-2}	-1.19×10^{-3}	3.75×10^{-3}	No mutation
0.367	-1.81×10^{-2}	-1.97×10^{-3}	5.73×10^{-5}	No mutation
0.374	-1.91×10^{-2}	-1.04×10^{-4}	-5.53×10^{-5}	Mutation

factor is between 0.367~0.374. In general, the subdivision solution is no longer continued. It is determined that the safety factor FV is equal to K_{i-1}, that is, the safety factor FV1=0.367.

4.2 *Plastic strain energy mutation*

The numerical simulation software has built-in calculation formula for plastic strain energy, which can be directly evaluated. The relationship between the plastic strain energy and the reduction coefficient K_i is shown in Figure 5.

The process of fitting the curve to solve the safety factor is the same as the operation procedure in the displacement mutation. After calculation, the safety factor FV2 is equal to 0.367.

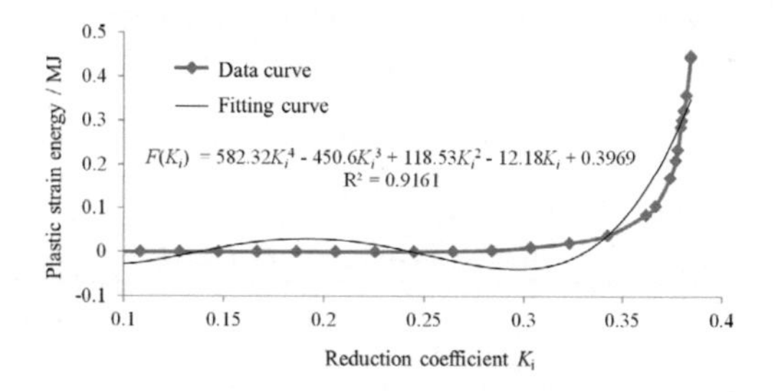

Figure 5. Fitting curve of plastic strain energy and reduction coefficient.

4.3 *Unbalanced force mutation*

The unbalanced force mutation is based on the numerical simulation calculation non-convergence as the judgment condition. In this example, the corresponding reduction coefficient K_i is 0.384, that is, the safety factor FV3 is equal to 0.384.

The final calculation result of the above three criteria is FV1=FV2 <FV3. Obviously, the unbalanced force has the largest safety factor. Because it considers fewer factors and is affected by the convergence conditions, the error is large and will not be used. The other two criteria, although different in discriminating methods, have obtained consistent results and can be used as criterion for instability of the tunnel face.

5 STUDY ON COUNTERMEASURES OF TUNNEL FACE STABILITY IN SOFT SURROUNDING ROCK

5.1 *The types of tunnel face reinforcement*

The pre-reinforcement of the tunnel face is mainly carried out by means of protecting or reinforcing the face.

One is to directly strengthen the tunnel face by measures such as front anchor, horizontal swirling, advanced grouting, freezing to improve the strength and suppress the deformation of the surrounding rock. This article refers to collectively as reinforced reinforcement.

The other is to protect the tunnel face by measures such as advanced long pipe shed, horizontal jet grouting pile and pre-cutting lining in the arch of the face, and try to cut off the contact between the tunnel face and the surrounding area, and reduce disturbance to surrounding rock. It is called protected reinforcement.

If both methods are combined, it is called integrated reinforcement.

5.2 *Take reinforced reinforcement to control the stability of tunnel face*

Only the strength parameters of the reinforced reinforcement are increased, and the other surrounding rock parameters are unchanged, as shown in Table 4.

According to judging method for stability of tunnel face by using safety factor as a criterion, set the different reinforced reinforcement lengths, such as 0D, 0.2D, 0.4D, 0.6D, 0.8D, 1.0D, 1.2D, 1.4D, 1.6D, 1.8D and 2D. The fitting curve of reduction coefficient and plastic strain energy can be obtained under the different reinforced reinforcement lengths. Only the calculation results at the working condition of 0.6D are listed, as shown in Figure 6. The fitting curve of reduction coefficient and key point displacement can be obtained under the different reinforced reinforcement lengths. Only the calculation results at the working condition of 0.6D are listed, as shown in Figure 7.

The safety factor of the surrounding rock of the tunnel face under different reinforced reinforcement lengths can be obtained, as shown in Table 5.

It can be seen from Figure 8 that with the increase of the length of the reinforced reinforcement, the safety factor of the surrounding rock of the tunnel is gradually increased. When the length is increased to 1D, the safety factor no longer increases. The effective reinforcement length of the reinforced reinforcement is about 1D. The length greater than 1D does not increase the safety factor, but it can be used as a safety reserve.

Table 4. The parameters of reinforced reinforcement.

cohesion c' /Pa	internal friction angle φ' /°
1.0×10^4	22

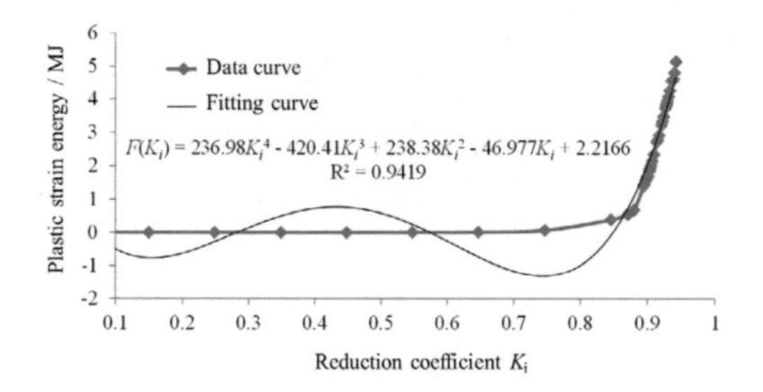

Figure 6. Fitting curve of reduction coefficient and plastic strain energy under the reinforced reinforcement lengths of 0.6D.

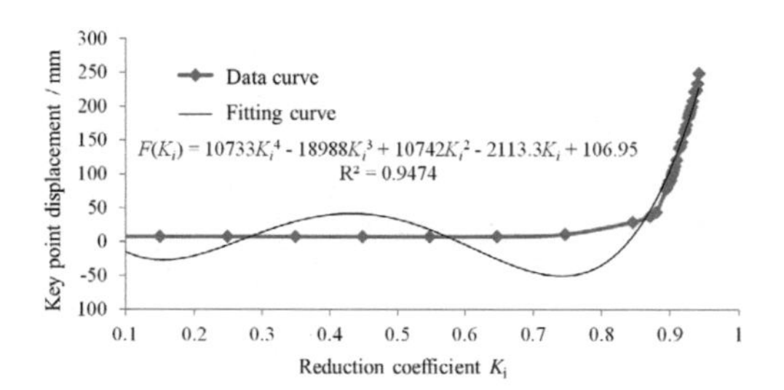

Figure 7. Fitting curve of reduction coefficient and key point displacement under the reinforced reinforcement lengths of 0.6D.

Table 5. Safety factor of surrounding rock in tunnel face under different reinforced reinforcement lengths.

Length of reinforced reinforcement	0D	0.2D	0.4D	0.6D	0.8D	1D	1.2D	1.4D	1.6D	1.8D	2D
Safety factor	0.367	0.448	0.647	0.880	0.922	0.943	0.943	0.943	0.943	0.943	0.943

5.3 *Take protected reinforcement to control the stability of tunnel face*

The range of protected reinforcement of the tunnel face is set to the extent of the dome above the arch line. Its thickness is 2m, as shown in Figure 9. The parameters of protected reinforcement are shown in Table 6.

According to judging method for stability of tunnel face by using safety factor as a criterion, set the different protected reinforcement lengths, such as 0D, 0.2D, 0.4D, 0.6D, 0.8D, 1.0D, 1.2D, 1.4D, 1.6D, 1.8D and 2D. The safety factor of the surrounding rock of the tunnel face under different protected reinforcement lengths can be obtained, as shown in Table 7.

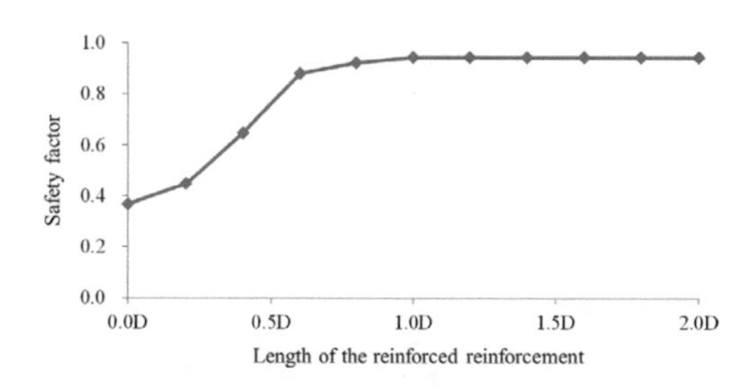

Figure 8. Ralationship of safety factor and reinforced reinforcement length.

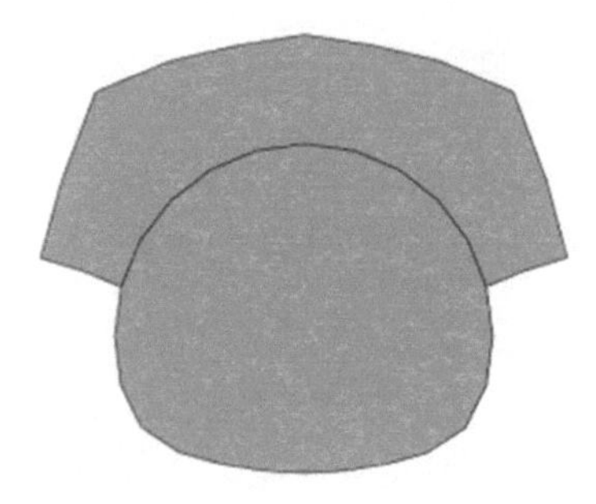

Figure 9. Protected reinforcement sketch map.

Table 6. The parameters of protected reinforcement.

Severe γ /kN	Elastic modulus E /Pa	Poisson's ratio ν /	Cohesion c' /Pa	Internal friction angle φ' /°
18	3×10^9	0.2	1.0×10^4	22

Table 7. Safety factor of surrounding rock in tunnel face with different length of protected reinforcement.

Length of protected reinforcement	0D	0.2D	0.4D	0.6D	0.8D	1D	1.2D	1.4D	1.6D	1.8D	2D
Safety factor	0.367	0.637	0.655	0.698	0.718	0.737	0.766	0.766	0.766	0.766	0.766

It can be seen from Figure 10 that with the increase of the length of the protected reinforcement, the safety factor of the surrounding rock of the tunnel is gradually increased. When the length is increased to 1.2D, the safety factor no longer increases. The effective reinforcement length of the reinforced reinforcement is about 1.2D. The length greater than 1.2D does not increase the safety factor, but it can be used as a safety reserve.

5.4 *Take integrated reinforcement to control the stability of tunnel face*

The integrated reinforcement is a combination of two types of pre-reinforcement methods: reinforced reinforcement and protected reinforcement. The combination of the two methods makes the method of controlling the stability of the tunnel face more flexible. Combining the reinforcement parameters of two pre-reinforcement methods in Sections 5.2 and 5.3, the stability of the integrated reinforcement control tunnel face is studied. Among them, the

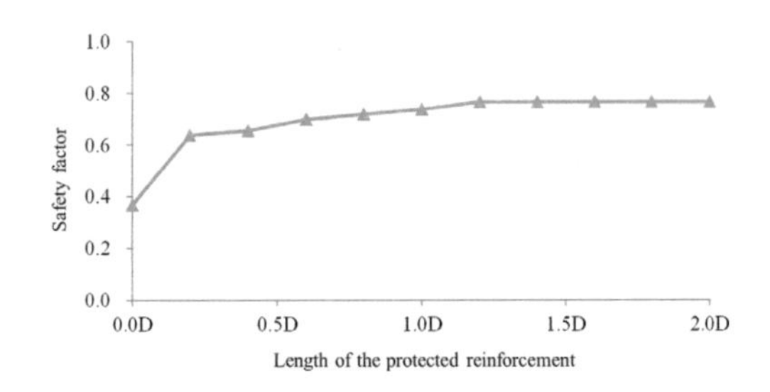

Figure 10. Relationship of safety factor and length of protected reinforcement.

Table 8. Safety factor of surrounding rock in tunnel face with integrated reinforcement when length of protected reinforcement is 0D.

Length of reinforced reinforcement	0D	0.2D	0.4D	0.6D	0.8D	1D	1.2D	1.4D	1.6D	1.8D	2D
Safety factor	0.367	0.508	0.647	0.880	0.922	0.943	0.943	0.943	0.943	0.943	0.943

Table 9. Safety factor of surrounding rock in tunnel face with integrated reinforcement when length of protected reinforcement is 0.6D.

Length of reinforced reinforcement	0D	0.2D	0.4D	0.6D	0.8D	1D	1.2D	1.4D	1.6D	1.8D	2D
Safety factor	0.698	0.768	0.863	0.945	1.016	1.104	1.104	1.104	1.104	1.104	1.104

Table 10. Safety factor of surrounding rock in tunnel face with integrated reinforcement when length of protected reinforcement is 1.2D.

Length of reinforced reinforcement	0D	0.2D	0.4D	0.6D	0.8D	1D	1.2D	1.4D	1.6D	1.8D	2D
Safety factor	0.766	0.848	0.974	1.117	1.134	1.174	1.174	1.174	1.174	1.174	1.174

reinforcement parameters of the reinforced reinforcement are shown in Table 4, and the reinforcement parameters of the protection reinforcement are shown in Table 6.

Safety factor of surrounding rock in tunnel face with integrated reinforcement can be obtained when length of protected reinforcement is 0D, 0.6D or 1.2D, and length of reinforced reinforcement is 0D, 0.2D, 0.4D, 0.6D, 0.8D, 1.0D, 1.2D, 1.4D, 1.6D, 1.8D or 2D, as shown in Table 8 to Table 10.

It can be seen from Figure 11 that with the increase of the length of the integrated reinforcement, the safety factor of the surrounding rock of the tunnel is gradually increased. When the length of the protected reinforcement is constant, the safety factor increases with the length of the reinforced reinforcement. Compared with a single reinforced reinforcement, the safety factor also no longer increases after the length is increased to 1D.

5.5 *The selection of reinforcing type*

No matter what reinforcement method is adopted, the ultimate goal is to ensure the stability of the tunnel face. Taking the reinforcement parameters of Section 5.2 (Tables 4) and Section 5.3 (Tables 6) as the benchmark, take the safety factor equal to 1 as the criterion. If the safety factor is greater than 1, the tunnel face is considered to be stable when the safety margin is not considered.

1. Reinforced reinforcement

If only reinforced reinforcement is used, the safety factor is 0.943, which does not meet the requirements. A method of increasing the cohesion can be adopted, and if it increases to 15 kPa, the safety factor is 1.045. Or increase the internal friction angle, and if it increases to 33°,

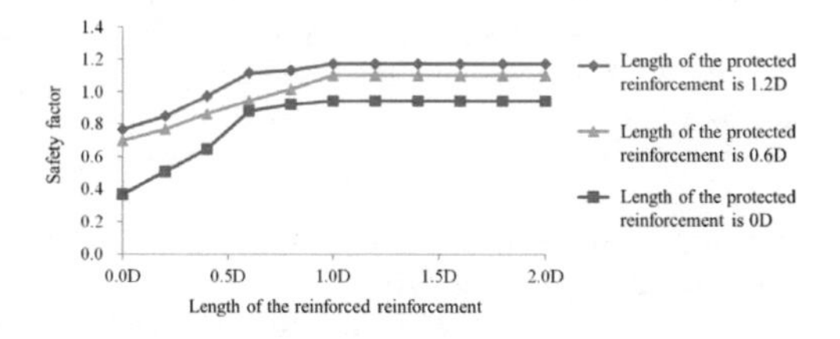

Figure 11. Relationship of safety factor and length of integrated reinforcement.

the safety factor is 1.045. Or increase the cohesion and internal friction angle at the same time, so that the safety factor is greater than 1, to ensure the stability of the tunnel face.

Note: The calculation process is no longer explained.

2. Protected reinforcement

If only protected reinforcement is used, the safety factor is 0.766, which does not meet the requirements. Taking the increase of cohesion and internal friction angle, if increased to 25kPa and 33°respectively, the safety factor is 0.827, but still can not meet the requirements. Because the surrounding rocks of the tunnel face are too poor and the increase of the strength parameter cannot stabilize it.

3. Integrated reinforcement

When the length of the protected reinforcement is 0.6D and the length of the reinforced reinforcement is 0.8D with integrated reinforcement, the safety factor is 1.016 (Table 9), which satisfies the requirements. When the length of the protected reinforcement is 1.2D and the length of the reinforced reinforcement is 0.6D with integrated reinforcement, the safety factor is 1.117 (Table 10), which satisfies the requirements.

Therefore, based on the above three analyses, it can be considered that a single reinforcement mode or an integrated pre-reinforcement mode can be selected to control the stability of the tunnel face. The final implementation plan can be determined by combining the surrounding rock conditions, construction difficulty, and project cost during the specific design and construction.

6 CONCLUSIONS

In this thesis, according to strength reduction theory, a judging method for stability of tunnel face has been developed by using safety factor as a criterion. Then, the pre-reinforcement measures against instability have been categorized. The stability of tunnel face has been studied under different pre-reinforcement conditions. The conclusion is as follows:

1. According to strength reduction theory, cusp catastrophe theory, combined with displacement mutation and plastic strain energy mutation, a judging method for stability of tunnel face has been developed by using safety factor as a criterion.
2. The pre-reinforcement measures against instability have been categorized. The stability of tunnel face has been studied under different pre-reinforcement conditions by using safety factor as a criterion. Suggestions for pre-reinforcement have been given.
3. It can be considered that a single reinforcement mode or an integrated pre-reinforcement mode can be selected to control the stability of the tunnel face. The final implementation plan can be determined by combining the surrounding rock conditions, construction difficulty, and project cost during the specific design and construction.

REFERENCES

Hong, K.R. 2015. State-of-art and prospect of tunnels and underground works in china. *Tunnel Construction*, 35(2): 95–107.

Ling, F.H. 1987. Catastrophe theory and its application. Shanghai: Shanghai Jiao Tong University Press.

Ma, S. & Xiao, M. 2010. Judgment method for stability of underground cavern based on catastrophe theory and monitoring displacement. *Chinese Journal of Rock Mechanics and Engineering* 29(supp. 2): 3812–3819.

Sun, J. & Hou X.Y. 1991. Underground structure. Beijing: Science Press.

Wang, M.S. 2004. General theory of shallow buried and excavation technology for underground engineering. Hefei: Anhui Education Press.

Zeeman, E. 1974. On the unstable behavior of stock exchanges. *Journal of Mathematical Economics* 1(1): 39–49.

Zhao, Y. 2016. Development and innovation of china railway tunnel. *China tunnel and underground engineering conference, Chengdu, 22–27 October 2016.*

Zheng, Y.R., Zhu, H.H. & Fang Z.C. 2012. Surrounding rock stability analysis and design theory of underground engineering. Beijing: China Communications Press.

Tunnels and Underground Cities: Engineering and Innovation meet Archaeology, Architecture and Art, Volume 3: Geological and geotechnical knowledge and requirements for project implementation – Peila, Viggiani & Celestino (Eds)
 ISBN 978-0-367-46583-4

BIM-FEM interoperability for the modelling of a traditional excavated tunnel

S. Fabozzi
National Research Council, Institute of Environmental Geology and Geoengineering, Rome, Italy

G. Cipolletta
A-squared Studio, London, UK, formerly University of Naples Federico II, Naples, Italy

E. Capano, D. Asprone, G. Dell'Acqua & E. Bilotta
University of Naples Federico II, Naples, Italy

ABSTRACT: The Building Information Modelling approach started to be implemented recently for the infrastructure design such as roads, railways, bridges, tunnels, although the BIM model not always integrates the digital terrain model that requires the availability of site surveys and specific tools to interpolate the available data into a 3D model.

The present work recognizes the possible development of this innovative design approach compared with the possible integration of the geological-geotechnical information for the definition of the digital terrain model and its potential applicability to infrastructures and, more in general, to geotechnical structures and works (e.g. excavations, foundations, soil reinforcement, dams, underground structures). Focusing on these aspects, a possible procedure to implement geotechnical information into a BIM model has tested for a benchmark case history of tunnelling with conventional heading in the urban area of Naples.

1 INTRODUCTION

The rapid development of Information Technologies in Architecture, Engineering and Construction Industry (AEC) puts in continuous evolution the definition and application of the Building Information Model/Modeling (BIM). This is a novel and innovative design and management approach that has undergone to a growing interest in the civil engineering field in the last ten years, with particular attention to the structures. This approach combines and integrates each information of the structure under design (e.g. geometry, dimensions, type of materials and their properties, foundation, installations) into a three-dimensional digital and computerized model that becomes the core of this 'intelligent' project generated into a full integrated BIM software. The main benefits of this approach is to guarantee higher quality to the project and to reduce the time and costs of the manufacture since it is able to control the project during all its life cycle.

This innovative approach needed a very performing dedicated software, managing and exchanging data systems able to manage the entire design and construction industry. These technological instruments have seen in the last years an impressing growth thanks to which the structures in particular and the service utilities have reached a very high level of details (*LoD*) that represent the different levels of abstraction required throughout the planning progress, from the *Concept* to the *As built*.

Very recently, the BIM approach started to be implemented for the design, construction and management of infrastructures (I-BIM, *Jack et al., 2016*) dedicated to the energy and public transportation (*Dell'Acqua et al. 2018*), utilities and hydraulic systems. This field of application is rather complex since some different disciplines such as geology and geotechnics should be involved and integrated into a unique digital model. This observation holds particularly for the case of tunnels, since they are completely immersed in the ground.

Overall, for tunnel design two issues can be identified to be implemented into a I-BIM model: 1) the geometrical definition of the tunnel alignment (plano-altimetric profile) and structure (tunnel lining); 2) the definition of the geotechnical problem.

According to a review of the technical literature (*Borrmann et al. 2014, Borrmann et al. 2015, Ninic & Koch 2017*) and the projects of the international design companies adopting Digital Construction (DC) tools, the first issue is addressed with a rather high level of definition while the second one suffers a lower level of BIM methodology knowledge. In fact, the available software tools are still insufficient for an efficient BIM implementation. Moreover, unlike the aboveground structures, the underground 'geometry' and properties may be known only after the field investigations.

The definition of a soil model integrated into a BIM tool, requires that a 3D digital terrain model should be generated starting from the ground investigations and constantly refined during the full site investigation process. Furthermore, the results of site and laboratory tests should be available into a common sharing environment to properly calibrate the geotechnical model for the specific problem that in turn re-enter in the BIM flow.

For tunnels that run underground for many kilometres, geological field data and geotechnical design integration into a BIM strategy could potentially improve the entire design and construction process in a cost- and time-saving manner for the reduction of unforeseen problems which are highlighted by the definition and update of the BIM model.

An important advantage to design an underground tunnel with a BIM approach is to import the 3D model (ground model and tunnel structure) into numerical codes commonly adopted for study different geotechnical problems such as, for example, the tunnel-soil interaction during the excavation process in urbanized areas. Actually, the interoperability between a BIM authoring software and geotechnical 3D numerical codes, that is the capability to exchange data to automate workflows, still depends on file formats for geometry (Drawing eXchange Format, DXF, for example), while they not yet support data model for the interchange of object model such as the Industry Foundation Classes, IFCs, (*buildingSMART 2013*) or *aecXML*. This is a gap that could be bridged only by improving the technological computerization of such aspects, integrating the geotechnical engineering within digital model.

The present work recognizes the potential of such innovative method for the possible integration of the geological-geotechnical information to define the digital terrain model and its potential applicability to infrastructures and, more in general, to geotechnical structures and works (e.g. excavations, foundations, soil reinforcement, dams, underground structures and others).

Focusing on these aspects, a possible procedure to implement geotechnical information into a BIM model has been adopted, as an exercise, referring to a new stretch of the underground in Naples that is currently under construction, for which a certain amount of data and information were available to the authors. Different commercial BIM tools have been used to define the I-BIM model of the case of study and its interoperability has been experienced with the finite element geotechnical software Plaxis 3D.

2 CASE OF STUDY

The stretch of tunnel used as benchmark is part of the underground network in the area of Naples and is currently under construction. It spans between two stations in the north-eastern area of the city along the route shown in Figure 1. This stretch is about 1 kilometre long, with an initial sub-horizontal progress, a subsequent slope of 2.6‰ until the end station where the alignment becomes sub-horizontal again. The tunnel is quite shallow, with a minimum and maximum tunnel cover equal to 10 and 30 m respectively, crossing a loose dry sandy layer.

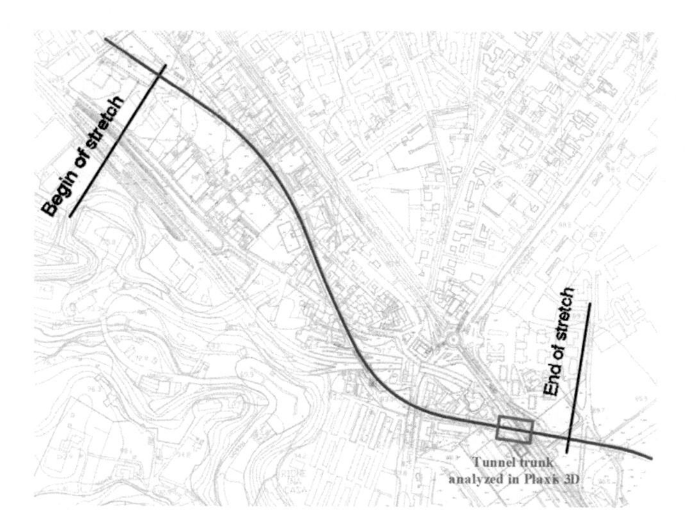

Figure 1. Tunnel stretch under consideration.

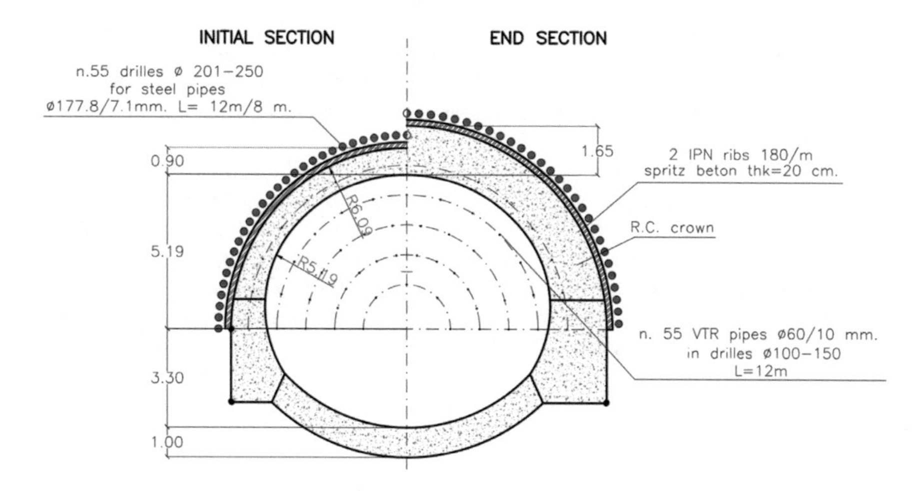

Figure 2. Typical section of the running tunnel.

Along the considered stretch, the main tunnel section is that shown in Figure 2. It is a conical trunk section 8 m long, with an internal diameter from the center-line equal to 5.19 and an external diameter of the initial and end section equal to 6.09 m and 6.84 m respectively, from the center-line. The section has a crown thickness equal to 0.9 m and 1.65 m in correspondence of the initial and final section while the invert is 1 m thick. Ground reinforcement was foreseen in order to guarantee the excavation front stability. To this end, 55 VTR pipes with a diameter equal to 60 mm and 12 m long are installed in correspondence of the front. Furthermore, forepolings are installed over the crown creating a steel pipe umbrella arch. They are 12 m long with a diameter equal to 177.8 mm.

3 BIM-BASED PROJECT

As introduced before (§1), a BIM-based road/rail tunnel project is a very complex product resulting from the integration of different aspects such as 1) tunnel structure, 2) tunnel alignment, 3) topography, 4) soil stratigraphy, 5) interferences with other infrastructure, 6) class detection, 7) quantification of excavated volumes, 8) quantification of structural materials volumes, and many others. In this paper, only the first four aspects have been included in the BIM-based project of the tunnel stretch since it focuses mainly on the geotechnical problem concerning the tunnel excavation.

To this aim, a scheme of the adopted BIM-work-flow is shown in Figure 3. It consists of four main phases defined as follows:

- *Phase 1*: Definition of the surface topography;
- *Phase 2*: Geological – Geotechnical data management;
- *Phase 3*: Definition of the tunnel BIM model;
- *Phase 4*: Definition of the FE tunnel model.

These phases have been implemented into two different BIM software authoring, Autodesk Civil 3D (blue box in Figure 3) and Bentley OpenRail Designer (red box in Figure 3) respectively, exploiting BIM tools dedicated to geotechnical aspects, made available by both software. Autodesk Civil 3D and Bentley OpenRail Designer represent actually the most advanced BIM software dedicated to infrastructures. Civil 3D is used both for rail and road tunnels while Bentley has developed two different dedicated software, OpenRail and OpenRoads Designer, for rail and road tunnels respectively.

As shown in Figure 3, each phase of the BIM-work-flow goes in parallel during the BIM tunnel project implemented in Autodesk and Bentley software, with very similar procedures.

Phase 1 consists of the definition of the topographic surface; in a preliminary design step, it has been defined through the support of the Digital Elevation Model, DTM, available online (e.g. Microsoft Bing Map) and, in a more advanced design phase, through available cartography of the area of interest.

Phase 2 deals with the management of the geological and geotechnical data coming from site investigations, including logs, data and reports, integrated into a Geographic Information System, GIS. *HoleBASE software*, developed by Keynetix, and *gINT* software made available by Bentley, have been adopted for the geological and geotechnical data managing in Autodesk and Bentley BIM software respectively. Civil 3D and OpenRail Designer for instance, import directly the analyzed soil data from the geological-geotechnical interoperable software, defining, in the Phase 3 of the BIM-work-flow, the soil stratigraphy trough the dedicated geotechnical tool.

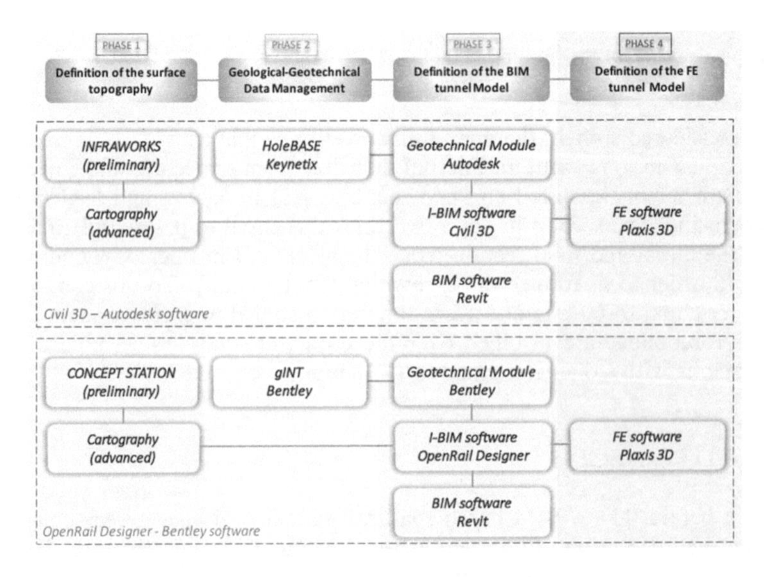

Figure 3. BIM-work-flow adopted in this exercise.

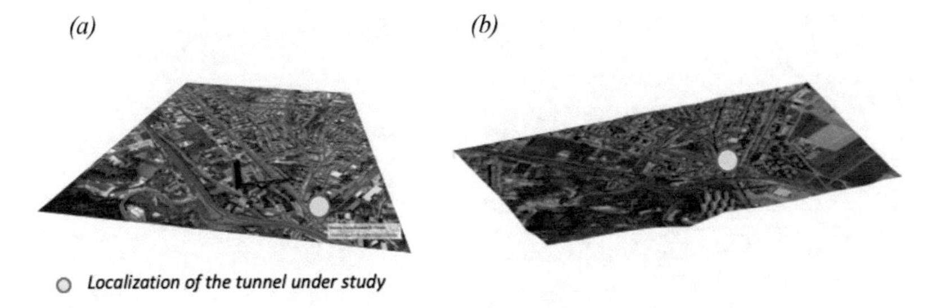

Figure 4. Topographical surface imported from (a) InfraWorks360 and (b) Concept Station.

Phase 3. Once defined the 3D digital terrain model, the BIM tunnel model is completed in this phase by means of the definition of the 3D tunnel alignment and the parametric lining structure. The latter in particular, has been defined in Revit, a BIM software dedicated to the structures, and imported both in Civil 3D and OperRail Designer to define the entire stretch under study.

Phase 4. The geometry of the soil layering and of the tunnel has been imported in Plaxis 3D to define the numerical FE model of a part of the studied stretch to verify the effect of the excavation in the urban area.

A detailed description of each phase of the followed BIM-work-flow is proposed below.

3.1 *Topography and stratigraphy definition*

The three-dimensional cartographical base has been defined, in a preliminary design phase, through *InfraWorks360* and *Concept Station,* for both BIM models defined in Autodesk Civil 3D and Bentley OpenRail respectively. They use the background data of Microsoft Bing Map to create the surface terrain starting from level curves. Figure 4a-b for instance, shows the shared data with Autodesk and Bentley software respectively, to define the topography.

In a more advanced design phase, the cartographical base of Naples Metropolitan Authority has been adopted to redefined the base map. Starting from COGO (COordinate GeOmetry) points, TIN (Triangulated Irregular Network) surfaces were created as cartographical base.

Once defined the topography, the stratigraphy was defined starting from the available field investigation data. These latter were firstly managed and catalogued into a Geographic Information System, GIS. *HoleBASE software*, developed by Keynetix, and *gINT* software made available by Bentley, were adopted for the geological and geotechnical data managing in Autodesk and Bentley BIM software respectively. These softwares allow to stay in control of the geotechnical project data and streamline reporting during every stage of a site investigation.

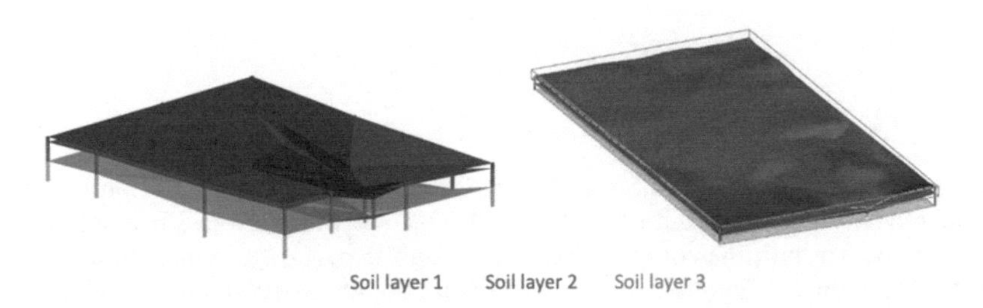

Figure 5. Soil stratigraphy defined trough the geotechnical module of (a) Autodesk Civil 3D and (b) Bentley OpenRail Designer.

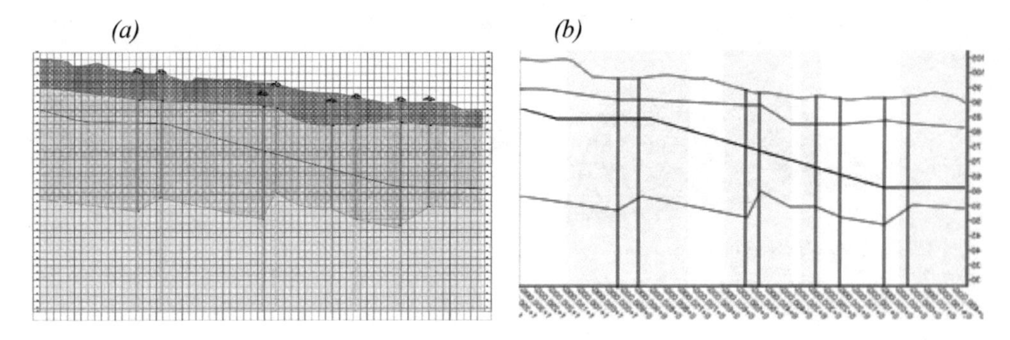

Figure 6. Plano-altimetric tunnel profile generated in (a) Civil 3D and (b) Bentley OpenRail Designer.

Data log managed in HoleBASE and gINT were directly imported in Autodesk Civil 3D and Bentley OpenRail Designer respectively, and analyzed trough dedicated geotechnical tools (*Geotechnical Module*) defining, in the Phase 3 of the BIM-work-flow, the soil stratigraphy of the model. Figure 5 for instance, shows the stratigraphy obtained with Civil 3D (Figure 5a) and OpenRail Designer (Figure 5b).

3.2 *Tunnel model definition*

This part of the BIM-work-flow concerns the definition of the tunnel BIM model, including the tunnel axis 3D profile and its structure.

The plano-altimetric profile has been firstly defined into Civil 3D and OpenRail Designer based on the available documentation of the project of the tunnel stretch under study. Figure 6 shows the obtained results in both software. Starting from the defined profiles, the solid tunnel is generated through the *Modeller* available both in Civil 3D and OpenRail Designer.

Referring to the tunnel geometry descripted in §2 (see Figure 2), two different tunnel sections have been modelled in Civil 3D and OpenRail Designer, a simplified model (Figure 7) and a real model (Figure 8). The simplified model has directly defined into the adopted BIM software while the detailed one has been defined in Autodesk REVIT BIM software, creating a parametric family object and then imported into Civil3D and OpenRail Designer, before to be defined in Plaxis 3D.

The main differences between these two models is that in the first case, the tunnel section is constant along the longitudinal axis and corresponding to the central section of the tunnel segment, in the second case instead, the entire conical trunk section has been reproduced including the forepolings over the crown creating a steel pipe umbrella arch.

Furthermore, while the simplified tunnel section type is assigned automatically in BIM software along all the tunnel alignment (Figure 7), the detailed one needs some manual operation to assign the imported tunnel section to the tunnel route. As example, Figure 7 shows the 3D solid tunnel importing the simplified tunnel section in Civil 3D and OpenRail Designer.

4 BIM-FEM INTEROPERABILITY

The last phase of the BIM-work-flow of Figure 3 consisted in the importation of the ground and tunnel model, defined trough two different I-BIM software, into a FEM software, Plaxis 3D in this study. To limit the computational load, only a part of the entire tunnel stretch has been studied, the portion highlighted in red in Figure 2b, about 100m long. This choice comes from the need to study a particular interaction condition of the excavated tunnel with a pier of a crossing bridge in this area.

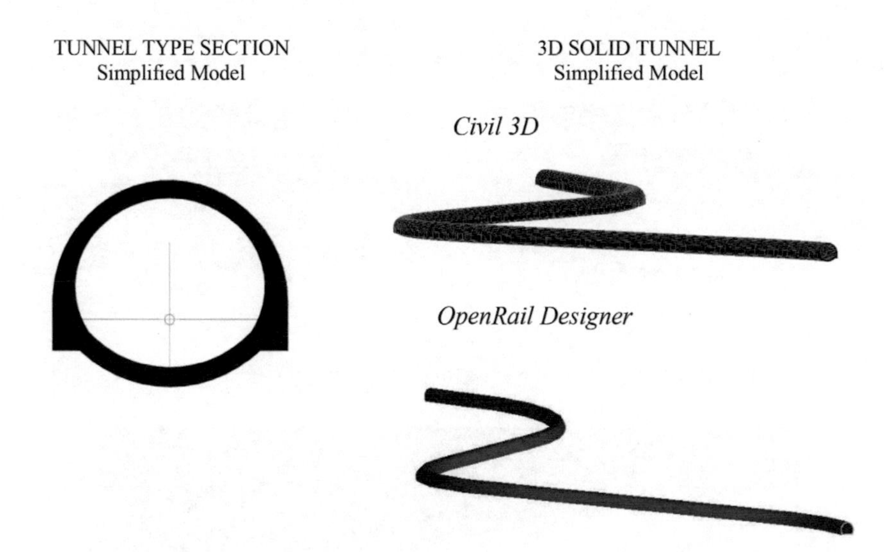

Figure 7. Simplified Model: Tunnel type section and 3D solid tunnel created in Civil 3D and OpenRail Designer.

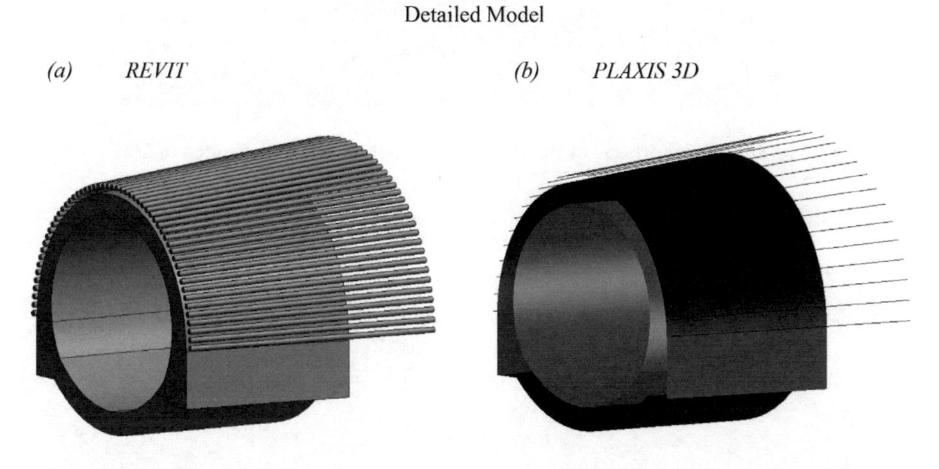

Figure 8. Detailed Model: (a) Tunnel type section created in REVIT and (b) imported in Plaxis 3D.

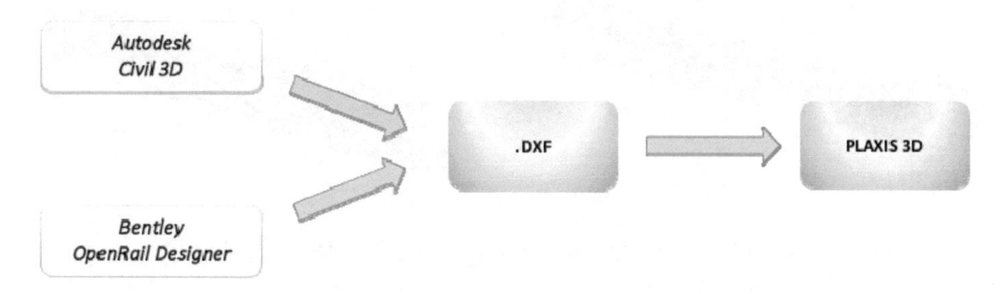

Figure 9. BIM-FEM interoperability: work-flow.

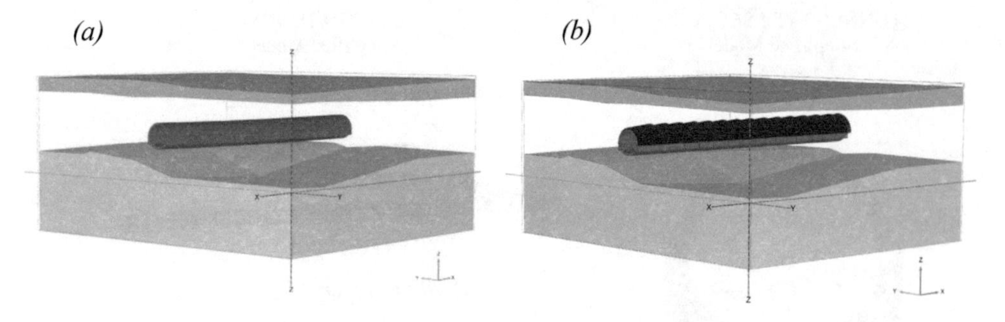

Figure 10. Geometric model for (a) simplified and (b) detailed tunnel model implemented in Plaxis 3D.

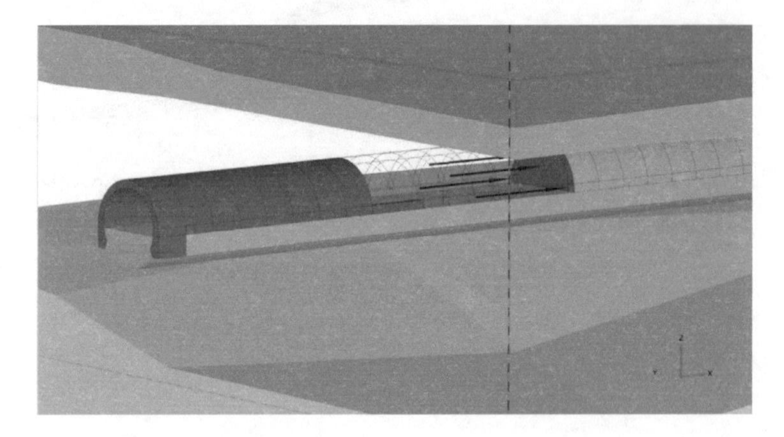

Figure 11. NATM excavation process of the simplified tunnel model implemented in Plaxis 3D.

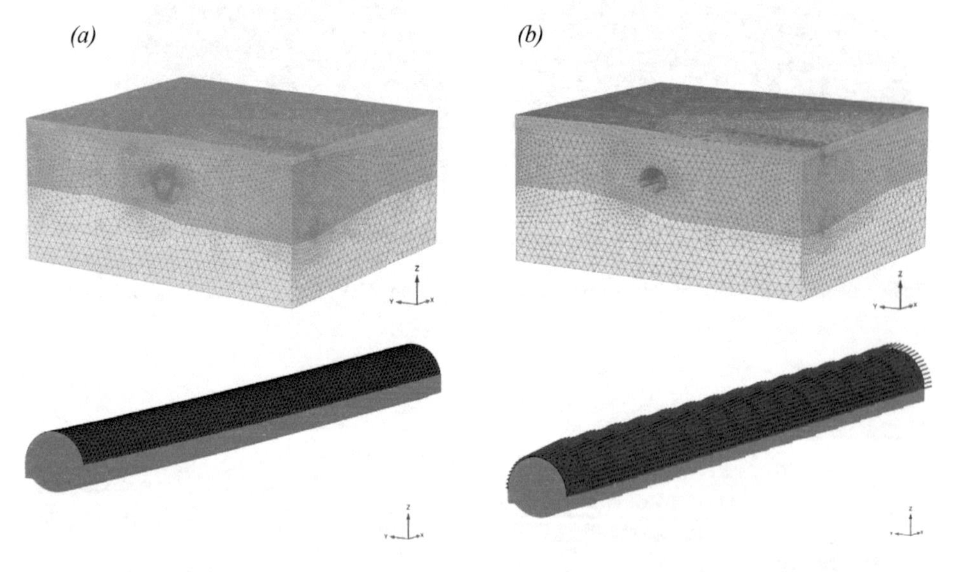

Figure 12. Mesh model for (a) simplified and (b) detailed tunnel model implemented in Plaxis 3D.

Since FEM codes like Plaxis does not yet support not-proprietary formats, currently BIM-FEM interoperability consists of an exchange of geometric information. Following the workflow of Figure 9 in fact, it has been possible to import (1) the geometry of the stratigraphy and (2) the tunnel geometry; for both cases (1) and (2) the imported geometric entities were surfaces only. In Figure 10 are shown the geometric models carried out for the simplified (Figure 10a) and real (Figure 10b) tunnel model.

Note that (Figure 11) the tunnel section has been divided into crown, abutment and invert portions in order to simulate the NATM excavation process. The tunnel lining is simulated by means of concrete volume, the temporary support lining as a plate structural element available in Plaxis 3D while, in correspondence of the front, a normal load surface has been added to stabilize the front during the advancement. The steel pipes umbrella arch in the detailed model were modelled as embedded beam structural elements to simulate the contact friction between the steel and the surrounding soil.

During automatic meshing, the geometric model is converted into a numerical mesh. Figure 12 for instance, shows the mesh obtained for the simplified and detailed tunnel model, zooming in the mesh of the tunnel structure.

The definition of the mesh model is this final step of the BIM-work-flow planned for the project of the Metro Line 1 of Naples. A series of numerical analyses and parametric comparisons have been then carried out for both simplified and detailed tunnel models which results are out of the scope of this paper and will be discussed elsewhere.

5 CONCLUSIONS

The present work explores the possible development of a BIM design approach that includes the geological and geotechnical information for the definition of the digital terrain model and its potential applicability to infrastructures like tunnels.

To define a ground model integrated into a BIM tool, a 3D digital terrain model should be generated starting from the ground investigations and constantly refined. Furthermore, the results of site and laboratory tests should be available into a common sharing environment to define properly a suitable geotechnical model for the specific problem, that in turn re-enter in the BIM flow.

A possible procedure to implement geotechnical information into an I-BIM model has been tested against a benchmark case history of tunneling with conventional heading in the urban area of Naples. Two different BIM authoring software were used for the scope, Autodesk Civil 3D and Bentley OpenRail Designer, together with some dedicated geotechnical tools made available by both software house, that represent actually the most advanced BIM software dedicated to infrastructure.

The so defined BIM models have imported into FEM commercial code, PLAXIS 3D, to exploit their interoperability. Actually, the interoperability between a BIM authoring software and geotechnical 3D numerical codes, that is the capability to exchange data to automate workflows, still depends on file formats for geometry (Drawing eXchange Format, DXF, for example), while they not yet support *data model* for the interchange of *object model* such as the *Industry Foundation Classes* (*IFCs*) or *aecXML*. This is a limitation that could be filled only improving the technological computerization of such aspects, integrating the geotechnical engineering within digital model.

REFERENCES

AutoCAD Civil 3D (2018), AutoCAD Civil 3D Manual and Tutorial Manual.

Borrmann A., Flurl M., Jubierre J.R., Mundani R.P., Ernst Rank E. Synchronous collaborative tunnel design based on consistency-preserving multi-scale models. Advanced Engineering Informatics 28 (2014) 499–517.

Borrmann A., T.H. Kolbe, A. Donaubauer, H. Steuer, J.R. Jubierre & M. Flurl. 2015. Multi-Scale Geometric-Semantic Modeling of Shield Tunnels for GIS and BIM Applications. Computer-Aided Civil and Infrastructure Engineering 30 (2015) 263–281.

buildingSMART. 2013. Industry Fondation Classes IFC4 Official Release, buildingSMART http://www.buildingsmart-tech.org.

Dell'Acqua G., Guerra de Oliveira S., Biancardo S.A. 2018. Railway-BIM: analytical review, data standard and overall perspective, Ingegneria Ferroviaria n. 11, (2018).
Infraworks 360 (2018), Infraworks 360. Tutorials
Jack C.P. Cheng, Qiqi Lu, Yichuan. 2016. Deng Analytical review and evaluation of civil information modeling, Automation in Construction 67 (2016) 31–47.
Ninic J. & Koch C. 2017. Parametric multi-level tunnel modelling for design support and numerical analysis. EUROTUN 2017: Innsbruck University, Austria.
Plaxis 3D Software Delft (2017), Reference and Manual.

Tunnels and Underground Cities: Engineering and Innovation meet Archaeology, Architecture and Art, Volume 3: Geological and geotechnical knowledge and requirements for project implementation – Peila, Viggiani & Celestino (Eds)
 ISBN 978-0-367-46583-4

Double-shield TBM performance analysis in clay formation: A case study in Iran

M. Fallahpour
Department of Mining and Metallurgical Engineering, Amirkabir University of Technology, Tehran, Iran

M. Karami
Department of Mining and Petroleum Engineering, Shahrood University of Technology, Shahrood, Iran

T. Sherizadeh
Department of Mining and Nuclear Engineering, Missouri University of Science and Technology, Rolla, MO, USA

ABSTRACT: The Konjancham conveyance tunnel is a part of the tropical water conveyance plan that is designed to transfer water to both Kermanshah and Ilam provinces in Iran. The second part of this tunnel (KT-2) is a 12 kilometers long with a diameter of 5.56 meters that is being bored using a double-shield TBM (DS-TBM) machine. The tunnel is excavated in both Aghajari and Lahbari formations, including claystone, siltstone, and sandstone. In this study, an overview of the basics of the project, including geology, engineering geology, hydrogeology, as well as the characteristics of the tunnel and the TBM, have been reviewed, firstly. Then, the field performance of the TBM is analyzed by focusing on the geological conditions and the functional parameters of the machine such as thrust force and RPM. Finally, the wear of the disc cutters was analyzed as one of the important issues in the boring process.

1 INTRODUCTION

In mechanized tunneling, due to high advance rate, a long tunnel can be constructed at a faster rate and safer compared to other methods. Shielded TBMs are among the most technically sophisticated excavation machines in use by tunneling projects within weak or hard rocks. The use of shields around the TBM gives the machine the capability to pass through weak grounds and adverse geological conditions. In tunneling projects, Performance of a TBM depends on machine specifications, ground characteristics, and operating parameters. Accurate prediction and evaluation of rate of penetration (ROP), advance rate (AR), utilization factor (UF), and cutter cost/life is necessary for the successful use of a TBM on any tunneling project.

Several prediction methods have been offered throughout the years to determine boreability of rock masses and its relation to machine performance (e.g. Blindheim 1979, Bruland 1998, Rostami & Ozdemir 1993, Rostami 1997, Barton 2000). In recent years, due to growing use of TBMs and the necessity to accurately evaluate and predict performance of machines in different ground conditions, many researchers have worked to develop new evaluation and prediction models for the common existing models (e.g. Yagiz 2007, Gong & Zhao 2009, Hassanpour 2009, Hassanpour et al. 2009).

This paper presents an evaluation of TBM performance by focusing on the geological and geotechnical parameters of the Konjancham water conveyance tunnel (KT-2) in Iran as a case study. Accordingly, machine operational parameters obtained from the bored section of the KT-2 project were used to compare the results of the prediction models such as NTH, Q_{TBM} and RME with actual machine performance in an attempt to evaluate accuracy of each

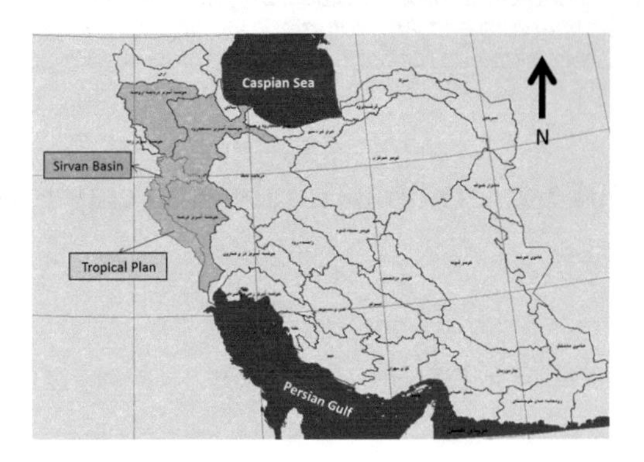

Figure 1. Location of Garmsiri (tropical) plan in Iran (Karami et al. 2018).

model. Furthermore, disc cutters wear has been evaluated as one of the crucial indices for assessing the rock-machine interaction in mechanized tunneling because of the major role it plays on total time and cost of TBM tunneling projects.

2 KONJANCHAM WATER CONVEYANCE TUNNEL (KT-2 PROJECT)

2.1 *Project description*

The Garmsiri (tropical) plan has been studied with the aim of transferring water from Sirvan basin to tropical regions of both Kermanshah and Ilam provinces in the west of Iran (Figure 1). In this plan, the water of Nosoud tunnel is transferred through channels, pipelines, dams, and tunnels and then, it is distributed based on requirements (Karami et al. 2018).

Konjancham water conveyance tunnel is located in Ilam province. The second part of the tunnel (KT-2) is 12 kilometers long with a diameter of 5.55 meters that is being bored using a double-shield TBM machine. As of March 2018, the tunnel has been bored to about 5 km from the outlet (i.e. chainage 12+000 to 7+000).

2.2 *Geology and hydrogeology of KT-2 project*

The case study area is located in the southern margin of Zagros Mountains. The geology of this area, in addition to quaternary deposits (QL), includes Aghajari and Lahbari formations. Aghajary formation is made of mudstone (siltstone, claystone, silty-claystone, clay-siltstone). In general, only 10 to 15 % of tunnel route is composed of sandstone layers. Lahbari formation includes sandstone (silty-sandstone, clay-sandstone). KT2 passes through Agajari-Lahbari formation, and there are not any adverse geological conditions along the tunnel's route (Karami et al. 2018). Table 1 lists the stratigraphic units encountered along the tunnel route. The summary of the rock mass properties within the tunnel are presented in Table 2. Also, in Figure 2, the geological plan and profile of the tunnel show the distribution of rock units along the bored section of the tunnel (IMN Company, 2015).

Average geomechanical characteristics of the engineering geological units in the area were studied using some empirical rock mass classification systems, such as RQD (Deere et al. 1967), RMR (Bieniawski, 1989), and the Q-system (Barton et al. 1974). Summary results are listed in Table 3.

According to the collected data from geotechnical boreholes and seep 2D numerical simulations, the maximum water inflow rate into tunnel face and total length of the tunnel were estimated as 0.20 L/s and 109 L/s (with a safety factor of 1.5), respectively. Almost 85 percent of permeability tests performed in the boreholes drilled in different geological units, yielded Lugeon between 0-3, indicating low permeability in the bedrock (IMN Company, 2015).

Table 1. Stratigraphic and geological unit symbols identified along the tunnel (IMN Company, 2015).

No.	Stratigraphic units	Engineering geological units	Lithology	Description
1	Aghajary and lahbari formations	Aj	Mudstone (Siltstone, Claystone, Silty-Claystone, Clay-Siltstone)	Slightly weak,Very blocky, Weathered, Broken and unstable
2	Aghajary and lahbari formations	Ajs	Sandstone (Silty-Sandstone, Clay-Sandstone)	Moderately strong to Slightly weak,Very blocky, Weathered, Broken and unstable

Table 2. Physical and mechanical properties of rock masses within KT2 (IMN Company, 2015).

Parameter	Value	Unit
UCS	4	MPa
Unit weight(γ)	2440	kg/m^3
Elastic modulus (E)	1.5	GPa
Poisson's ratio(ν)	0.32	-
Cohesion	0.11	MPa
Friction angle	31	°

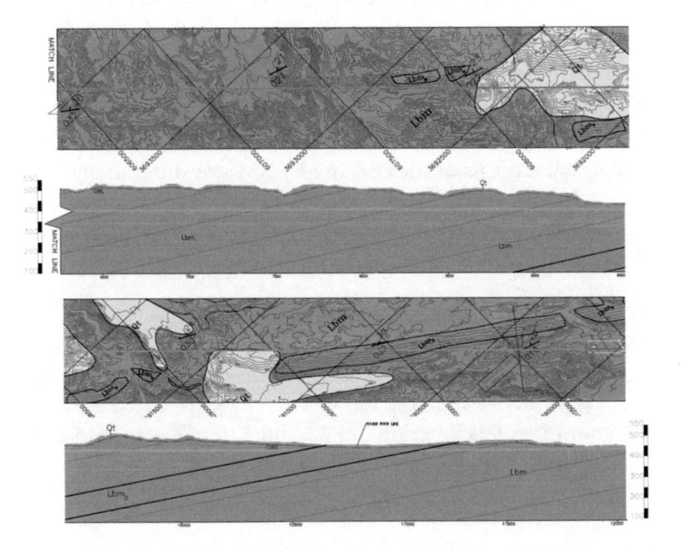

Figure 2. Geological plan and profile along the KT-2 (from chainage 12+000 to 7+000) (IMN Company, 2015).

Table 3. Summary results of the rock mass classification in the project area (IMN Company, 2015).

Formation	RQD (%)	RMR (Basic)	RMR (Modify)	GSI	Q
Aghajari	65–75	46–49	42–45	45–55	1.5
Lahbari	85–95	49–52	44–47	50–60	1.3

Figure 3. DS-TBM employed in KT-2 under assembly.

Table 4. Specifications of DS-TBM used in KT2.

Parameter	Value
Cutter-head diameter	5.56 m
Disc-Cutter diameter	432 mm
Outer lining diameter	5.3 m
Segment width	1.5 m
Segment thickness	25 cm
Maximum thrust force	32500 KN
Cutterhead Torque (nominal)	4210 KN.m
Cutterhead Power	6×350=2100 KW
Cutterhead speed	0-9.5 rpm
TBM weight (approx.)	1178 ton
TBM Length	315 m

2.3 *Technical specifications of DS-TBM*

Based on the geological and geotechnical characterizations, a double shield hard rock TBM (DS-TBM) has been recommended and designed for use in the excavation of the KT2 (Figure 3). The main specifications of the designed TBM are summarized in Table 4.

3 TBM FIELD PERFORMANE EVALUATION

In this section, TBM field performance are investigated based on the actual operational data obtained from boring about 5 km of the tunnel. Figure 4 shows the weekly and cumulative rate of TBM boring. As shown in Figure 4, about 5 km of the tunnel route was completed in 28 full working weeks (about 187 working days) at an average rate of about 171 m/week. The minimum and maximum rate of excavation are 42 and 225 m/week, respectively.

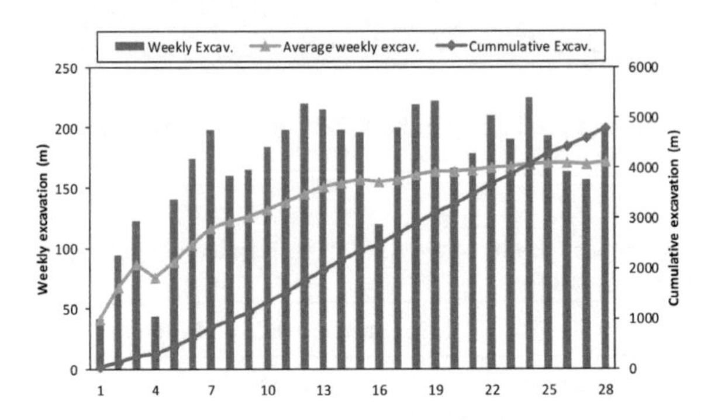

Figure 4. Weekly and cumulative rate of excavation.

The weekly average of the penetration rate (ROP) ranges between 1.5 and 6 m/h with an average of 4 m/h (Figure 5). Also, the weekly average of the advanced rate (AR) ranges between 6 and 32 m/day with an average of 24.5 m/day (Figure 6).

The weekly average of utilization factor (UF) ranges between 9.2 and 37.7% with an average of 24.7 % (Figure 7).

Furthermore, the results of analysis of daily performance of the machine show that maximum TBM advance of 42 m/day, rate of penetration of 8.8 m/h and utilization factor of 49% were experienced during the 5 Km excavation of KT-2 project.

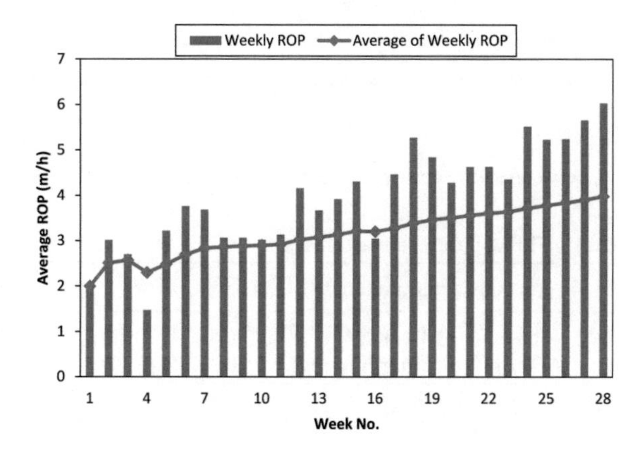

Figure 5. Weekly average of TBM rate of penetration.

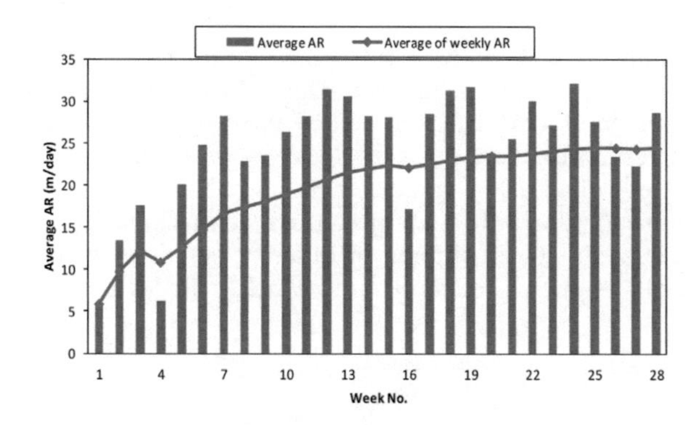

Figure 6. Weekly average of TBM advanced rate.

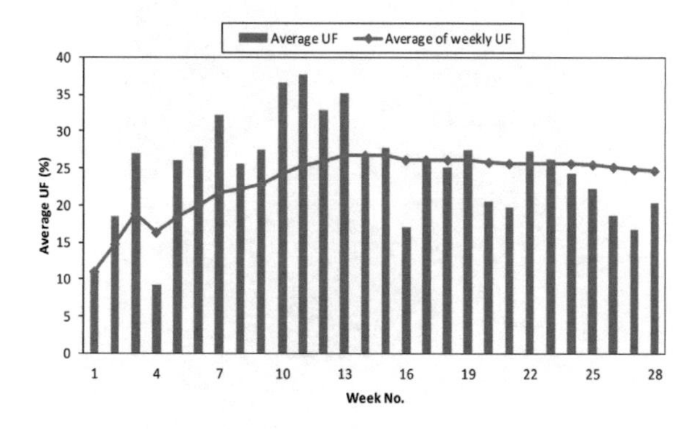

Figure 7. Weekly average of TBM utilization factor.

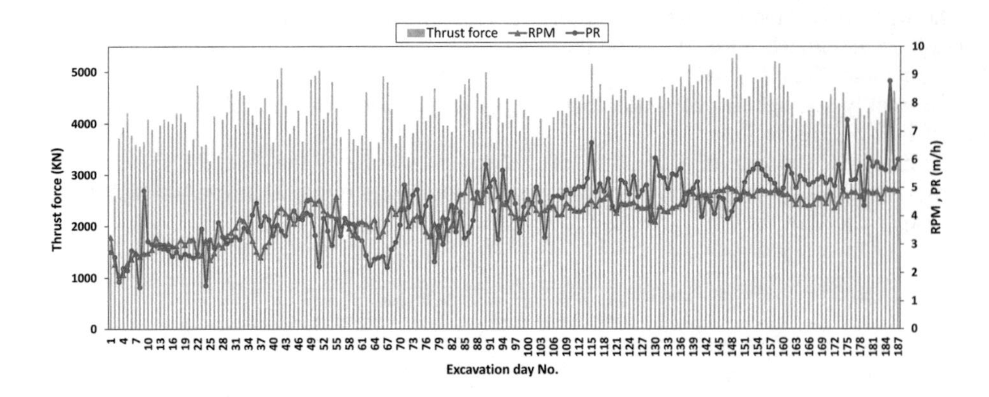

Figure 8. Relationships between TBM performance parameters in selected tunnel section (daily).

Table 5. Summary results of TBM parameters evaluation.

	Performance parameters					Operational parameters		
	AR (m/day)	ROP (m/h)	PR (mm/rev)	FPI (KN/ Cutter/mm/rev)	U (%)	Thrust (KN)	Torque (KN.m)	RPM
Weekly average	6–32	1.5–6	8.7–20.7	4.2–8.5	9.2–37.7	2230–4907	96–1130	1.6–4.9
Total average	24.5	4.0	16.3	6.2	24.7	4010	911	3.8

The daily relationships between TBM performance parameters such as thrust force, RPM, and rate of penetration are shown in Figure 8. Daily thrust force ranges between 1650 and 5353 KN with an average of 4251 KN and daily RPM ranges between 1.9 and 5.3 with an average of 4.0. The results of the daily analysis show maximum daily ROP (8.78 m/h) is occurred at a thrust force of 4577 KN and cutter rotational speed of 4.9. The summary results of TBM performance evaluation in KT-2 project are listed in Table 5.

Figure 9 shows average time distribution of different activities in the tunnel for the first 5 km of KT-2 tunnel. As shown, advance time which includes required times for boring,

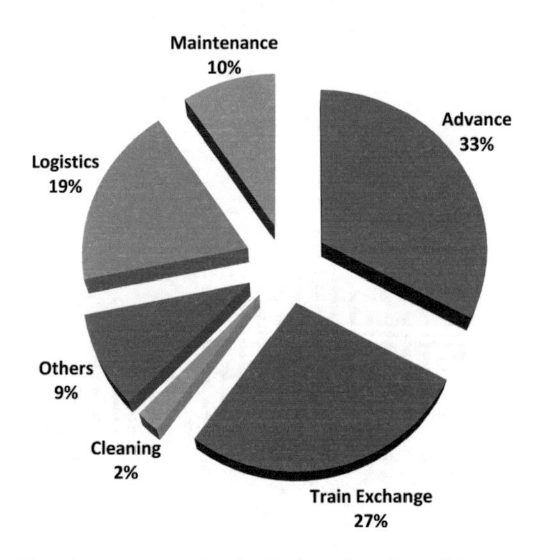

Figure 9. Average tunneling time consumption in the bored section of the tunnel (about 5 Km).

Table 6. Comparison between prediction models and actual field parameters (average RPM=3.8).

Prediction Model	PR(mm/rev)	PR(m/h)	AR(m/day)
NTH	5.9	1.4	7.4
Q_{TBM}	24.8	5.6	32.4
RME	17.7	3.2	18.8
Actual	16.3	3.9	24.5

segment installation and re-gripping comprises 33% of all activities. In the bored section of the tunnel, the main downtime for the advance of TBMs was as follows: train exchange 27 %, delays due to maintenance 10 %, delays related to site management and logistics 19 %, delays due to cleaning 2 %. Breakdown and shift change were also a part of the downtime that influenced the TBMs usage and accounted for 10 % of total time.

Table 6 summarizes the results between the prediction models of the TBM performance and the actual field parameters. NTH (Bruland, 1998), Q_{TBM} (Barton 1999, Barton 2000b), and RME (Bieniawski 2006, Bieniawski 2007, Alber 2000) are the prediction models that were used for the TBM performance analysis. The results depicts the RME prediction model results to have good correlation with the actual performance parameters from KT-2.

4 WEAR OF DISC-CUTTERS

Disc cutter lifetime can be expressed in various definitions. A frequently used unit is the rolling distance (in km) before a disc has to be replaced. In this concept, the penetration rate and the position of each cutter should be known (Hassanpour et al. 2015). Disc lifetime can also be expressed as the length of tunnel mined or volume of rock excavated per cutter, shown as $H_m$and H_f, respectively (Bruland, 1992):

$$H_m\ (\mathrm{m/cutter}) = \frac{L(\mathrm{m})}{N_{TBM}} \tag{1}$$

$$H_f\ \left(\mathrm{m^3/cutter}\right) = \left(\frac{H_m\, \pi\, d_{TBM}^2}{4}\right) \tag{2}$$

Where H_m is the average length of tunnel bored and H_f is the volume of rock excavated for each cutter change, respectively. N_{TBM} is the total number of disc cutters changed and L is tunnel length excavated for each full dressing of the head.

The TBM cutterhead of KT-2 is equipped with 39 disc cutters, including 4 double-ring center, 20 single-ring face, and 11 single-ring gauge cutters of diameter 17 in. or 432 mm. The cutter number increases with the distance to the center of the cutterhead. Figure 10 shows the cutter arrangement on the cutterhead with the cutter numbers.

Figure 11 shows the rolled distance and number of replacement of any disc-cutter. As shown, during boring about 5 km of the KT-2, 62 disc-cutters including 4 center, 29 face, and 29 gauge discs were replaced. Normal wear, oil leakage, and cutter jamming are the main reasons for disc-cutters replacement (Figure 12). A sample of normal wear of cutter ring in KT-2 project is shown in Figure 13.

A comparison analysis was conducted between required disc-cutters for the 5 Km excavation in KT-2 and predicted values from NTH model (Bruland, 1998) and VHNR model (Hassanpour et al. 2015).

Bruland (1998), on the basis of the CLI parameter, obtained the following expression, which can be used to estimate the net mean time (in hours) between one substitution of a disc and the next one (H_h):

$$H_h = \frac{H_o . K_D . K_Q . K_{RPM} . K_N}{N_{TBM}} \tag{3}$$

Where, $H_o$is the basic disc lifetime (in hours) related to CLI parameter. K_D, K_Q, K_{RPM}, and K_N are the corrective coefficients that take into account the diameter of the TBM, the quartz contents of the rock, the rotation velocity of the excavation head and the number of discs, respectively.

It should be noted that in these calculations, values of quartz contents, uniaxial compressive strength (UCS), Q number, spacing between disc-cutters, and max force applied by the tool in the direction perpendicular to the excavation face are considered to be 15%, 16 MPa, 1/1, 69 mm, and 220 KN, respectively. Also, all calculations were done based on the average RPM

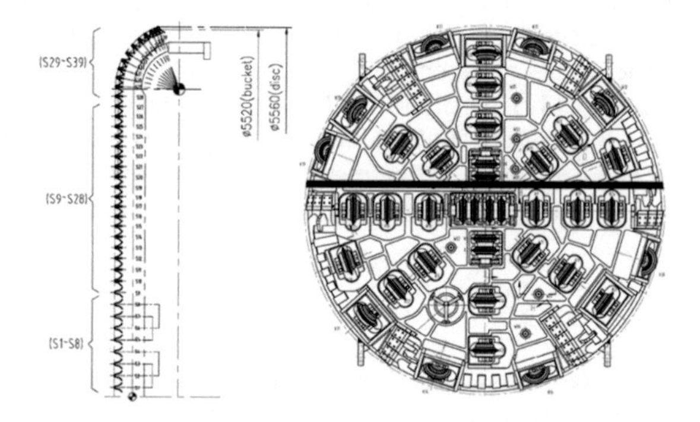

Figure 10. View of DS-TBM cutterhead and cutter arrangement on it in the KT-2.

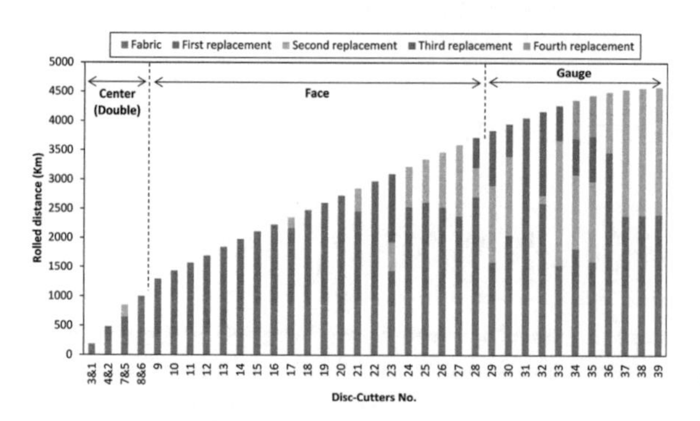

Figure 11. Rolled distance and number of disc-cutter's replacement.

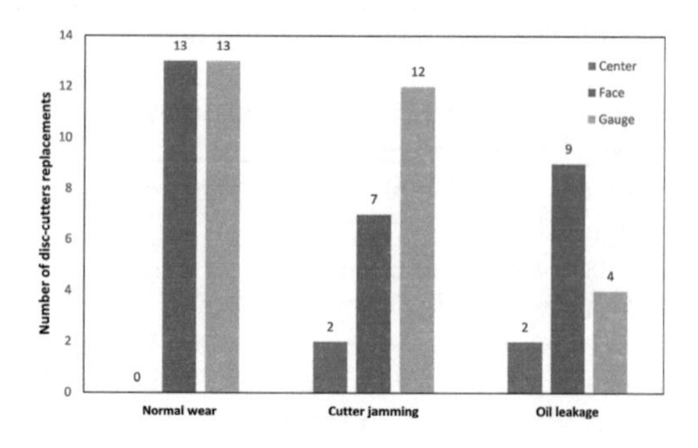

Figure 12. Reasons and number of replacement for every three type of disc-cutter.

Figure 13. A view of disc-cutters in KT-2 a) a new cutter, b) and c) normal wear.

Table 7. Coefficients of NTH model for estimation of required cutters in KT-2.

Coefficient	H_O	K_D	K_Q	K_N	K_{RPM}
Value	130–150	1.4	1.5	0.94	2.4

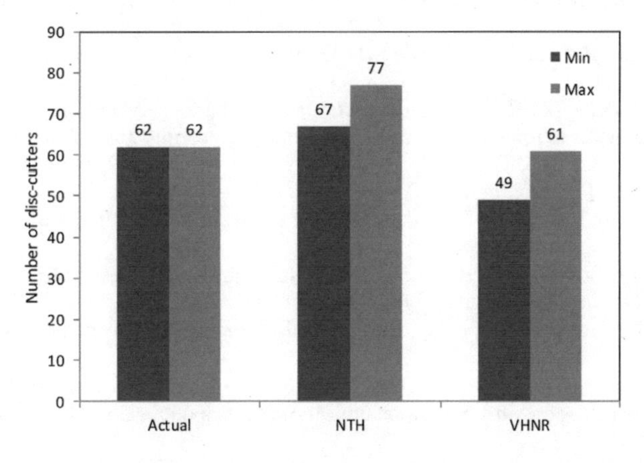

Figure 14. Comparison between the number of disc-cutters used in KT-2 and values of some prediction models.

3.8 and average ROP 4 m/h. The coefficients of NTH model for estimation of required cutters in KT-2 are shown in Table 7.

Also, in the VHNR method, simple empirical equations for cutter life prediction and the relationship between actual cutter life or H_f (m^3/cutter) are developed based on different ground parameters including intact rock properties (UCS, quartz content, and Vickers Hardness Number or VHNR) and fracturing degree or rock structural properties (joint count number or J_v, RQD, Basic RMR, and GSI). To estimate actual cutter life along the tunnels, a procedure explained by Bruland (1998) was used. In Figure 14, a comparison was made between the number of disc-cutters used in KT-2 with values of some prediction models.

5 CONCLUSIONS

Accurate Estimation and evaluation of TBM operational parameters, such as rate of penetration, advance rate, utilization factor and wear of disc-cutters is necessary to achieve high efficiency in mechanized tunneling. This study aimed to analyze the TBM performance parameters in the Konjancham water conveyance tunnel in Iran as a case study. Based on the data obtained from an approximate 5 Km of bored tunnel in clay and silt rocks types within the Aghajari and Lahbari formations, an evaluation of actual field parameters was conducted based on daily and weekly operations results. Also, a comparison was made between results of the three commonly used TBM performance prediction models and actual performance of the TBM for this project. The results indicate that the actual performance parameters in KT-2 correlates very well with the RME prediction model. The wear of disc-cutters was also analyzed as it was noted to be an important index in determining the success of the project. Normal wear, oil leakage, and cutter jamming were the main reasons of disc-cutters substitution. The results indicated, 62 disc-cutters including 4 center, 29 face, and 29 gauge were consumed in the excavated section of the KT-2 project.

REFERENCES

Alber, M. 2000. Advance rates of hard rock TBMs and their effects on project economics. Tunnelling and underground space technology 15(1): 55-64.

Barton, N.R. 2000. TBM tunnelling in jointed and faulted rock. CRC Press.

Barton, N., Lien, R. & Lunde, J. 1974. Engineering classification of rock masses for the design of tunnel support. Rock mechanics 6(4): 189-236.

Barton, N. 1999. TBM perfomance estimation in rock using QTBM. Tunnel and Tunnelling International: 30-34.

Barton, N. & Abrahao, R.A. 2003. Employing the QTBM prognosis model. Tunnels & Tunnelling International 35(12): 20-23.

Bieniawski, Z.T. & Bieniawski, Z.T. 1989. Engineering rock mass classifications: a complete manual for engineers and geologists in mining, civil, and petroleum engineering. John Wiley & Sons.

Bieniawski, Z. T., Caleda, B. & Galera, J. M. & Alvares, M. H. 2006. Rock mass excavability (RME) index. In ITA World Tunnel Congress (Paper no. PITA06-254).

Bieniawski, Z.T. & Grandori, R. 2007. Predicting TBM excavability-part II. Tunnels & Tunnelling International: 15-18

Blindheim, O.T. 1979. Boreability Predictions for Tunneling. Ph.D. Thesis, Department of Geological Engineering, the Norwegian Institute of Technology.

Bruland, A. 1998. Hard rock tunnel boring. Ph.D. Thesis, Norwegian University of Science and Technology (NTNU).

Deere, D.U., Hendron, A.J. & Patton, F.D. & Cording, E.J. 1966. Design of surface and near-surface construction in rock. In The 8th US symposium on rock mechanics (USRMS). American Rock Mechanics Association.

Gong, Q. & Zhao, J. 2009. Development of a rock mass characteristics model for TBM penetration rate prediction. International journal of Rock mechanics and mining sciences 46(1): 8-18.

Hassanpour, J. 2009. Investigation of the Effect of Engineering Geological Parameters on TBM Performance and Modifications to Existing Prediction Models. Ph.D. Thesis, Tarbiat Modares University, Tehran, Iran.

Hassanpour, J., Rostami, J., Khamehchiyan, M. & Bruland, A. 2009. Development new equations for performance prediction. Geo Mechanics and Geoengineering: An International Journal 4(4): 287–297.

Hassanpour, J., Rostami, J., Zhao, J. & Azali, S.T. 2015. TBM performance and disc cutter wear prediction based on ten years experience of TBM tunnelling in Iran. Geomechanics and Tunnelling 8(3): 239-247.

IMN Company, 2014-2015. Geological and engineering geological reports for Konjancham Water Conveyance Tunnel project.

Karami, M., Fallahpour, M. & Sherizadeh, T. 2018. Technical and Practical Analysis of Contact Grouting in Mechanized Tunneling: A Case Study in Iran. In 52nd US Rock Mechanics/Geomechanics Symposium. American Rock Mechanics Association.

Rostami, J. 1997. Development of a force estimation model for rock fragmentation with disc cutters through theoretical modeling and physical measurement of crushed zone pressure. Doctoral dissertation, Colorado School of Mines.

Rostami, J. & Ozdemir, L. 1993. A new model for performance prediction of hard rock TBM. In: Bowerman LD et al (eds) Proceedings of RETC: 793–809.

Yagiz, S. 2007. Utilizing rock mass properties for predicting TBM performance in hard rock condition. Tunnelling and Underground Space Technology 23 (3): 326–339.

Tunnels and Underground Cities: Engineering and Innovation meet Archaeology, Architecture and Art, Volume 3: Geological and geotechnical knowledge and requirements for project implementation – Peila, Viggiani & Celestino (Eds)

The coseismic faulting of the San Benedetto tunnel (2016, Mw 6.6 central Italy earthquake)

P. Galli
Dipartimento Protezione Civile, Rome, Italy
CNR-IGAG, Rome, Italy

A. Galderisi
CNR-IGAG, Roma
UNICH, University of Chieti-Pescara, Italy

M. Martino
ANAS, Rome, Italy

G. Scarascia Mugnozza & F. Bozzano
Sapienza University of Rome, Rome, Italy

ABSTRACT: in this paper it is hypothesized that coseismic faulting caused the huge damage recorded by the San Benedetto tunnel during the October 30, 2016 earthquake in Central Italy. The results of the survey of the damage suffered within a short sector of the tunnel, coupled with both geological observations carried out at surface, and the available seismological data, strengthen this hypothesis.

1 INTRODUCTION

On August 24, 2016 a Mw 6.2 earthquake (RCMT 2016) caused the death of 299 persons under the rubbles of their old, stone-masonry houses at the border among Lazio, Umbria and Marche Regions (central Italian Apennines). This was the first event of a long seismic sequence which is still ongoing now (late 2018), after two years. Another Mw 6.1 event occurred on October 26, whereas the largest one (Mw 6.6) was on 30 October (see Mw and beach-balls in Figure 1). Earthquakes have been sourced by one of the normal fault dissecting the fold-and-thrust Apennine chain, the Mt. Vettore fault system (MVFS in Figure 1). The resulting mesoseismic area was almost entirely stretched in the hanging-wall of the MVFS, reaching the highest macroseismic intensities (11 MCS degree; Galli et al. 2017) ever assigned in Italy since the catastrophic Fucino earthquake in 1915 (Mw 7.0).

During the Mw 6.6 event, the abrupt rupture of the 30-km-long MVFS caused impressive surface faulting phenomena all along the fault trace, with coseismic scarps as-high-as 2 m. In addition to the master fault, several synthetic and antithetic fault segments ruptured at surface, with offsets ranging from 1 m to few centimeters. Here we maintain that one of these affected at depth the San Benedetto tunnel (National Road 685, connecting Lazio and Marche to Umbria), causing a ~10–20 cm displacement of this single-tube tunnel, and its immediate and long closure to the traffic (Figure 1).

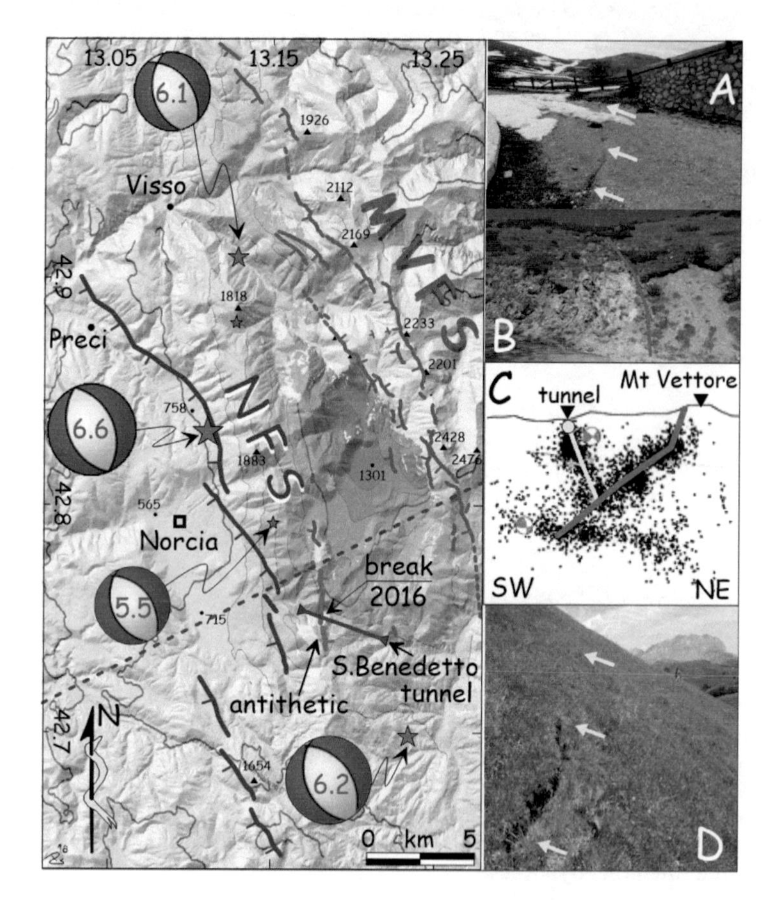

Figure 1. Left, digital elevation model of the Mt Vettore earthquake area (Mw 6.6; bold brick-red line indicates the fault rupture; blue areas suggest the downthrown of the hanging wall, as resulting from ALOS-2/PALSAR-2 Interferogram; isolines every 20 cm. GSI source. NFS, Norcia Fault system). The bold blue line indicates the tunnel track. The bold red line indicates the path of superficial cracks identified at surface along the antithetic fault, as those represented in the right panels of this figure (A, D). Panel B is the geological fault outcropping along the cracks (Corniola Formation vs Marne del Serrone). Panel C is the hypocentral distribution of aftershocks (from Chiaraluce et al., 2017), depicting the antithetic fault (yellow line) that likely affected the tunnel.

2 SURFACE FAULTING OF TUNNELS

Surface faulting is a rather rare event, always associated to high magnitude crustal earthquakes, usually larger than Mw 6. The probability that coseismic faulting might cut an anthropic artifact is even more rare; however, due to the dense network of lifelines that since Roman times have been designed across the entire Mediterranean region, this is not the first time that we observe the faulting of an artifact, as accounted also by several paleoseismological and archaeoseismological studies published in the past 30 years (e.g. Galli et al. 2008).

Also during the 2009 L'Aquila earthquake (Mw 6.3, central Italy), the Paganica-San Demetrio fault rupture damaged several buildings and cut the Gran Sasso aqueduct, causing its displacement and huge damage in the surroundings (Galli et al. 2010; 2011).

However, while we are aware of existing faulted pipelines, roads, houses and canals, only few cases have been reported worldwide concerning the faulting of road tunnels, as the one investigated here. Among these, those of Wrights (California earthquake of 1906, Mw 7.7,

Figure 2. Detail of TBM Wirth tb11 360h with full cross section cutterhead used for the "pilot" tunnel (extract from Ricci 1986).

Prentice & Ponti 1997) and Kern County, also in California (earthquake of 1952, Mw 7.5, SCEDC 2017).

The San Benedetto tunnel, in central Italy, is located along the Italian National Road 685. It is a two-way tunnel, about 4400 m long, with a WNW-ESE direction, connecting the provinces of Perugia in Umbria region (Norcia tunnel entrance) and Ascoli Piceno in the Marche region (Arquata del Tronto tunnel entrance).

The Norcia side entrance is located at 1001 m a.s.l., while the Arquata del Tronto side entrance is at 1014 m a.s.l., with an average inclination of 0.29 % and a maximum overburden of 600 m. The nominal internal radius of the tunnel is 5.1 m.

The construction of the tunnel started in 1986, with a pilot tunnel performed by TBM (diameter 3.6 m; Figure 2), successively enlarged by using blasting techniques. The first stage lining consists of radial bolts, electrowelded mesh and shotcrete, while the final lining was made of non-reinforced concrete; between the two linings a PVC coating was put in place. In the crown, there is also a suspended ceiling slab for the ventilation system (Figure 2).

A relevant feature of the tunnel is the absence of the inverted arch. As a matter of facts, under the concrete roadway surface we have found a thin stratum of rockfill laying over the bedrock. During construction, abundant water inflow was encountered at around 3700 m from the Norcia side entrance, which yielded the installation of a drain pipe under the road platform.

3 TECHNICAL-STRUCTURAL ANALYSIS OF THE DAMAGED TUNNEL

Following the 30 October 2016 mainshock (Mw 6.6), the tunnel suffered damage and deformation of the roadway that led to the prompt traffic closure. Damage focused between progressive 10 + 956 km and 10 + 976 km, consisting basically in the downthrown of the eastern section (Arquata del Tronto side). This mainly occurred along an oblique joint, with a rough NNE-SSW trending, associated with several breaks and offsets of the wall and of the crown (Figure 3), even with lateral slip (Figure 4).

Inside the tunnel, we carried out a technical-structural analysis in order to depict the joint pattern along the wall and the crown. The fractures pattern was reported on a log with the aid of a laser total station, and it was successively integrated and corrected through photographic documentation (Figure 5, bottom).

In the immediate aftermath of the mainshock, i.e., before any work in the rupture zone (e.g., the opening of the explorative pit in Figure 3, #8), we made a detailed topographic leveling of the road surface using an aluminum triplometer. Starting from the 940 m progressive, we performed 1-m-spaced leveling, except in the rupture zone where we took measures every

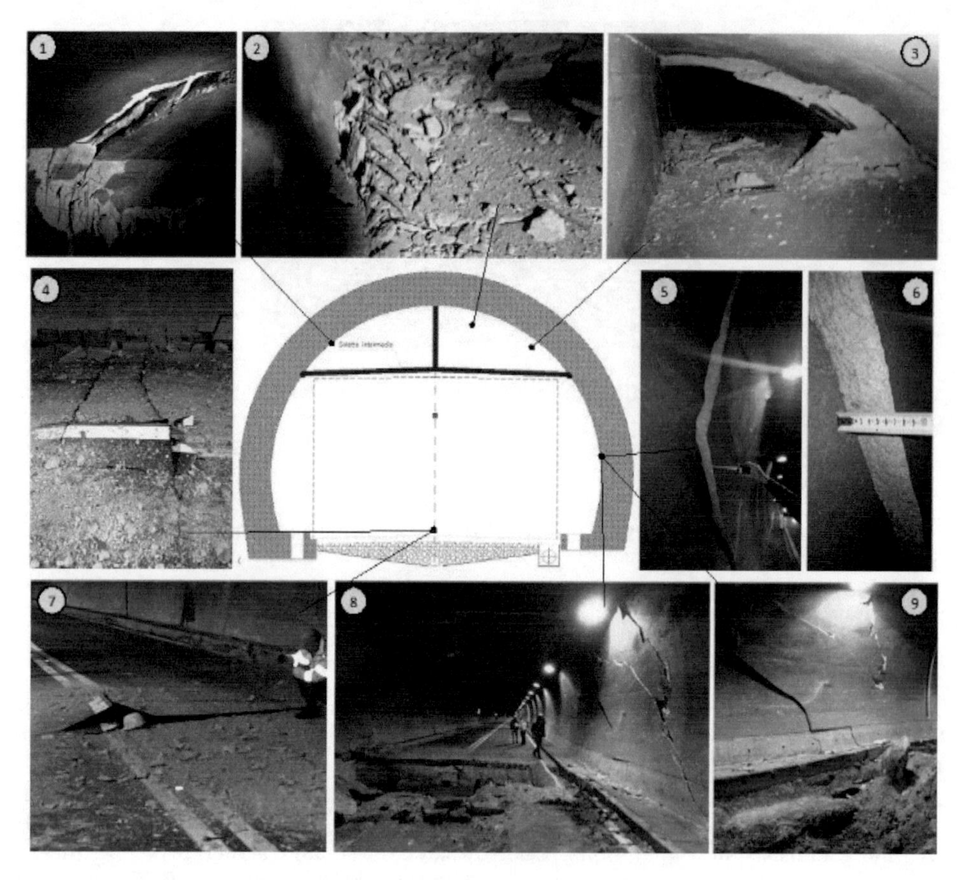

Figure 3. Cracks opened in the tunnel. Photos 1–3 and 8 were taken looking SE. Photos 5–7 and 9 looking NW, toward Norcia. Photo 4 is perpendicular to the road, with Norcia to the left. 1–3 were taken in the vaults, and show the vertical displacement of the concrete lining. Photo 8 shows also the pit opened for investigating the presence of the inverted arch.

0.1 m. This made it possible to calculate the relative offset between the Norcia side block and the Arquata side block.

4 DATA ANALYSIS

The deformation zone inside the tunnel extends for about 15 m slantwise with respect to the road axis. In particular, on the southern side of the tunnel, the deformation was focused between the progressive 10 + 962 and 10 + 976 km while on the northern side was between 10 + 956 and 10 + 966 km.

On the one hand, the tunnel shows a left offset of ~13 cm, which was well detectable looking the edge of the sidewalk (Figure 4). In addition, and more consistently, the road surface was affected by an inflection/scarp about 2-m-wide, associated to an overall 20 cm downthrown of the eastern side, and a 5-cm high bulge in the western side (see profile Figure 5a). The edges of the inflection were accompanied by a complex network of cracks, likely controlled by the ductile behavior of the asphalt carpet. Remembering that the tunnel has not an inverted arch, the offset of the road should

Figure 4. Early view, looking west, of the southern wall of the tunnel which clearly shows the lateral component of the offset. Here we measured ~13 cm of left offset, associated to ~20-cm of vertical offset accounted by the scarp in the middle of the image.

be considered the direct expression of the rock-rupture below the asphalt (i.e., around the tunnel), in terms of 1) trend, 2) amount of vertical and horizontal offset, and thus 3) fault kinematics.

5 DATA DISCUSSIONS

Leaving aside the strong coseismic shaking of the concrete structure, the effects of which should have been visible along the whole structure, and not only within a short section, from all the above, it is possible to make some hypotheses concerning the cause of the tunnel break.

First of all, it is worth to remember that we surveyed at surface several sets of antithetic faults and fractures (Figure 1) that roughly match the area of the tunnel break, all with associated coseismic slip on 30 October. In turn, these surficial breaks match an antithetic geological fault (Figure 1B) that was seen also through the hypocentral distribution of the aftershocks of the 30 October event (Figure 1C). Then, inside the tunnel, the offset on the road surface is oblique (about 40°) with respect to the axis of the tunnel itself, and it matches the set of fractures surveyed along the northern and southern wall, besides in the vault itself (Figure 3). Here fractures can be grouped into 2 sets, one with 45° west-dipping planes and another with roughly vertical break (Figure 5c). The first set is clearly evident on the southern side of the tunnel, and branches off at the fracture intersection with the road surface. It is here where we measured both the vertical 20-cm offset and the 13-cm left offset (Figure 4).

Therefore, we suggest that the net vertical and horizontal offset observed across the road (Figure 5a) and across the vault (Figure 3, panels 1–3) is the direct expression of coseismic faulting of the tunnel (Figure 6), while the fractures along the concrete vault and wall, showing an oblique dislocation (e.g., 45°), can be considered the brittle response of the concrete lining to the downthrown of the eastern block.

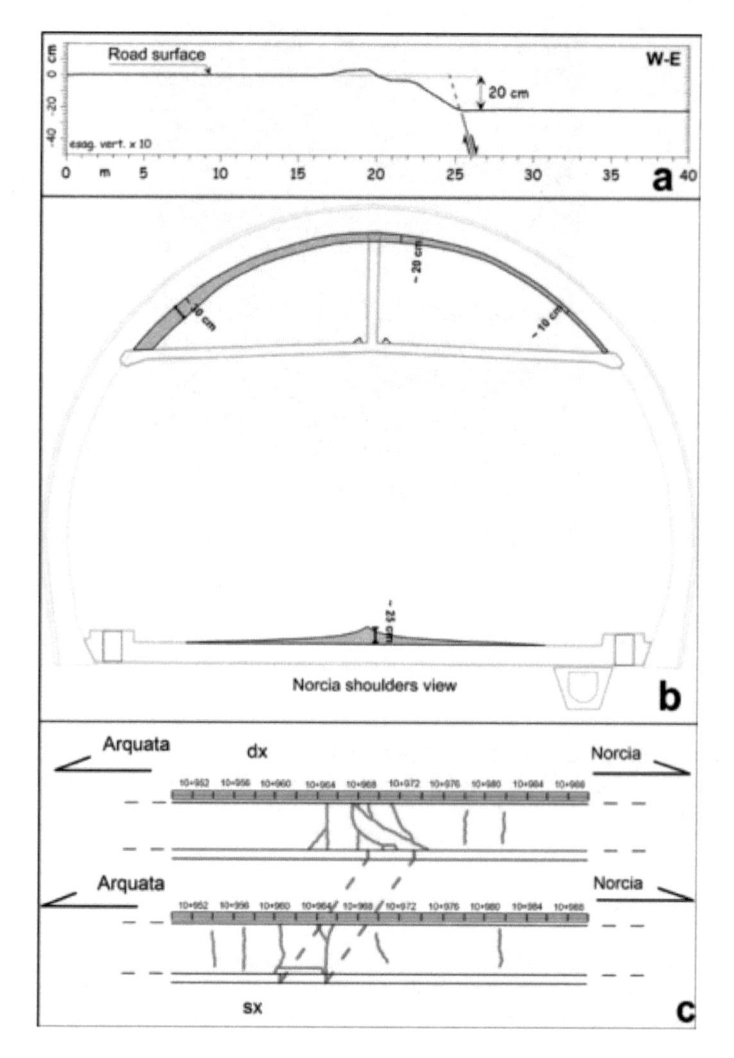

Figure 5. a) Topographic profile of the road surface, accounting for a vertical offset of 20 cm. b) Schematic section of the tunnel. The areas in pink represent the deformation that occurred after the 30 October earthquake. c) Log of the left and right flanks with associated fractures. The red dashed area indicates the 20-cm-high scarp formed in the road.

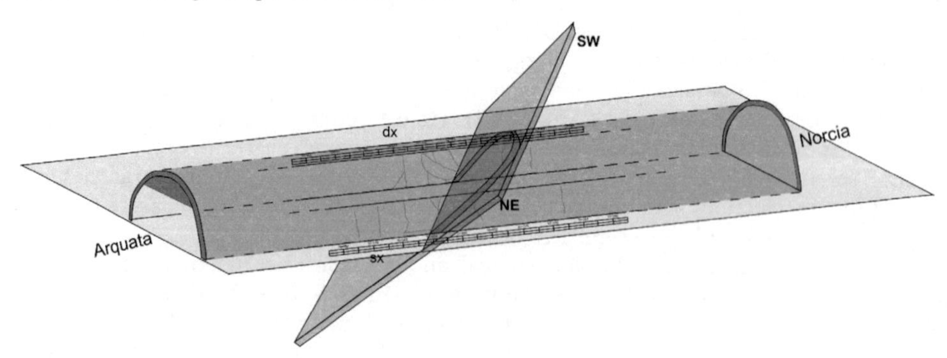

Figure 6. Simplified sketch of the tunnel at the intersection with the fault (pink plane), showing all the visible fractures along the sides of the tunnel (green lines). The NE-SW-trending plane (in red) is the supposed left, transtensional fault plane that cut the tunnel on October 30th. Note that the fractures along the sides of the tunnel are mostly arranged at 45° with respect to the fault plane.

6 CONCLUSIONS

If our hypothesis is correct, the coseismic offset of the San Benedetto tunnel is the first case of faulting of this typology of structure in Europe, and one of the few in the world. The fault plane responsible for this displacement is part of a segmented, roughly N-S, antithetic structure of the Mt Vettore fault system (MVFS), characterized by a main normal kinematics, although with a left component. This kinematics, observed within the tunnel, fits the extensional stress field of the region, which has a σ_3 NW-SE trending.

As a concluding remark, we strongly maintain that also during the design of road/railways tunnels or other underground linear structure (gas/oil/water pipeline) it should be mandatory to perform microzonation investigations, and in particular regional studies on active faults capable for rupturing at surface, and thus to cut and destroy temporary or definitely any strategic structure.

REFERENCES

Chiaraluce, L., Di Stefano, R., Tinti, E., Scognamiglio, L., Michele, M., Casarotti, E., Cattaneo, M., De Gori, P., Chiarabba, C., Monachesi, G., Lombardi, A., Valoroso, L., Latorre, D., Marzorati, S. 2017. The 2016 Central Italy seismic sequence: a first look at the mainshocks, aftershocks, and source models, *Seismological Research Letters* 88: 757–771.

Galli, P., Galadini, F. & Pantosti, D. 2008. Twenty years of paleoseismology in Italy. *Earth Sciences Reviews* 88: 89–117.

Galli, P., Giaccio, B. & Messina, P. 2010. The 2009 central Italy earthquake seen through 0.5 myr-long tectonic history of the L'aquila faults system. *Quaternary Science Reviews* 29: 3768–3789.

Galli, P., Giaccio, B., Messina, P. & Peronace, E. 2011. Paleoseismology of the L'Aquila faults (central Italy, 2009 Mw 6.3 earthquake). Clues on active fault linkage. *Geophysical Journal International* 187: 1119–1134.

Galli, P., Castenetto, S. & Peronace, E. 2017. The macroseismic intensity distribution of the 30 October 2016 earthquake in Central Italy (Mw 6.6): Seismotectonic implications. *Tectonics* 36: 1–13.

Kontogianni, V. & Stiros, S. 2003. Earthquakes and Seismic Faulting: Effects on Tunnels. *Turkish Journal of Earth Sciences* 12: 1–27.

Prentice, C. & Ponti, D. 1997. Coseismic deformation of the Wrights tunnel during the 1906 San Francisco earthquake: A key to understanding 1906 fault slip and 1989 surface ruptures in the southern Santa Cruz Mountains, California. *Journal of Geophysical Research* 102: 635–648.

RCMT 2016. European-Mediterranean RCMT Catalog. Website: http://www.bo.ingv.it/ (accessed on 12 September 2018).

Ricci, P. 1986. Rassegna dei Lavori Pubblici del Maggio 1986, Ing. Paolo Ricci, Ferrocemento SpA.

SCEDC 2017. Kern County Earthquake. Website: http://scedc.caltech.edu/significant/kern1952.html (accessed on 12 September 2018).

Tunnels and Underground Cities: Engineering and Innovation meet Archaeology, Architecture and Art, Volume 3: Geological and geotechnical knowledge and requirements for project implementation – Peila, Viggiani & Celestino (Eds)
 ISBN 978-0-367-46583-4

Geotechnical monitoring: Application in a case history of Azienda Nazionale Autonoma delle Strade (the National Roads Authority)

R. Gandolfo, S. Majetta & M. Martino
Anas S.p.A., *Rome, Italy*

ABSTRACT: According to new Italian Building Code, ANAS (Italian national agency for state roads) developed a new monitoring system, in order to control the geological aspects for road infrastructures (tunnels, bridges, embankments…). This paper describe a case history of geotechnical monitoring design located in a geomorphologically complex area. Geotechnical monitoring is the driving engine of the design for double aspects: safety and control, from the design stage until the operating phase

1 INTRODUCTION

In the last years, Anas SpA (hereafter Anas) is developing the territory control and its interaction with the infrastructures (Tunnels, Bridges, Embankment, etc.).

Following is described the SS652 project plan, for which the results of the geotechnical monitoring carried-out allowed the civil works designers to identify main problems and to find the optimal route in accord to the territory nature. The section under construction is 5.3 km long and consider the realization of:

- a tunnel long about 2.5 km (excavation diameter 14.80 m) mined with traditional methodology;
- an emergency tunnel long about 2.5 km (excavation diameter 4.0 m) mined with using TBM;
- five viaducts long 1.2 km.

2 SS 652 FONDO VALLE SANGRO

2.1 *Short Project history*

The SS652 Fondo Valle Sangro road, also known as "Strada a Scorrimento Veloce (SSV) Sangritana", is an Italian national road that connects the Molise region to the Adriatic coast. It crosses the Sangro valley (Abruzzo region) and ends in the SS16 "Adriatica" national road (Figure 1). The design project involved the variation of the last road section, about 5 km long, between Gamberale station and Quadri town. In 2013 Anas drawn up a Feasibility Study where alternatives routes were studied.

During the elaboration of the Definitive Project based on the approved Preliminary Project, Anas carried out a study on the interaction between the geomorphological aspects and the infrastructure area.

The geomorphological problems that emerged in the study led to a project modification. It was evaluated the realization of a natural tunnel section (about 730 m long) and different supporting structures such as bulkhead, retaining wall and counterscarp for consolidation of the landslide bodies crossed by the road.

Figure 1. Localization of the project, in red the new "SS652" plan.

The results obtained from the realization of further geotechnical investigations have highlighted a more complex geomorphological and stratigraphic scenario compared with the data available up to that time. Furthermore, in January 2015, the installed geotechnical monitoring probes highlighted the reactivation of quiescent landslide bodies.

These new evidences led to a further project review aimed to simplify the structures and improve their safety. The review foresee the realization of natural tunnel (total development around 2,500 m long) that cross the most critical areas.

2.2 *Site general information*

2.2.1 *Project description*

The project road plan develops for about 5.7 km on the left bank of the Sangro River inside the regional territory of Abruzzo (Chieti province) starting from 34 + 400 km of the SS652, in correspondence of the existing viaduct on the Sangro River, close to Gamberale – Sant'Angelo railway station.

Passing the existing viaduct, the road deviates into the Abruzzo side continuing on the left side of the Sangro River and cross the "Sangritana railway" with a new viaduct (Viaduct 1, 350 m long). The exit ramp for "Pizzoferrato" and "Gamberale" starts from the new viaduct and ends on a new roundabout that connect the existing roads ("1", in Figure 2).

Following starts a tunnel approximately 2.5 km long. The tunnel, flanked by a safety tunnel along the mountain side, present a succession of circular curves with radius respectively 1.050 m and 1.100 m and finish with a straight line ("2", in Figure 2). After the tunnel, starts a viaduct (120 m long) and an alternation of bulkheads and walls.

Close to Quadri town, the route continues over the recently modernized road of the 2nd lot, 2nd shred, 1st section (known as "Variante di Quadri") and pass over the railway and the watercourse with another viaduct (600 m long). In this last section is planned the junction of "Quadri" ("3", in Figure 2).

Summarizing, the project consider the construction of a natural tunnel, five viaducts and a series of minor works and the stabilization of a landslide activated during the design period and located close to Mincolavilla area, next to the entrance of the natural tunnel.

2.2.2 *Geological setting*

Marine lithologies with superficial deposits of the continental succession are present in the area. In particular, there are clayish-marl-arenaceous lithologies (Agnone flysch) and grayish and reddish marl-clays lithofacies (Gamberale-Pizzoferrato formation). In this geological context are frequent landslides due to the reduced mechanical characteristics of these terrains that are particularly altered by exogenous agents (rains, river erosion, etc.)

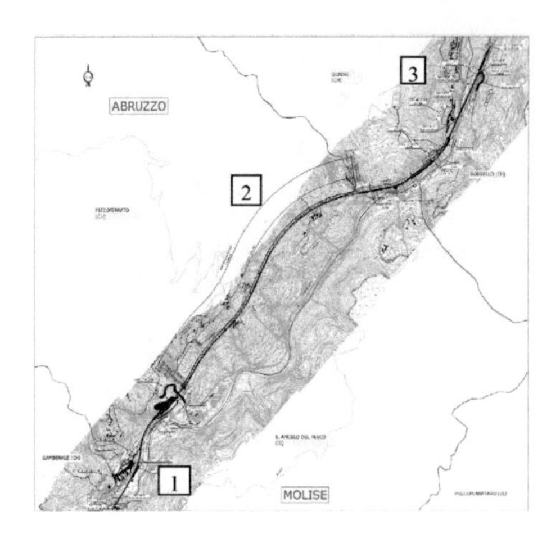

Figure 2. Design plan.

2.3 *Monitoring Plan*

The road infrastructure insist in a particular geological and geomorphological territory that requires constant monitoring of landslide phenomena. Due to the emerged geological complexity, Anas developed a specific monitoring plan in order to have an additional and important tool for the control of the territory and its infrastructures.

In the project economic framework was included a specific task "Geotechnical monitoring" (task cost item: € 2.414.140). A new tender, separated from the main contract and directly managed by the Works Director, has been done to assign the execution.

The monitoring plan has been splitted in two parts, one external and one internal the working activities, and includes the following controls:

- Structures: walls, bulkheads, buildings, temporary and final tunnels cladding, foundations, etc.;
- Land: countryside plan, landslide bodies, embankments, surveys, excavation fronts.

The following instrumentation is going to be installed:

- Traditional inclinometers;
- Automated inclinometers;
- Topographic benchmark;
- Levelling rod;
- Accelerometers;
- Open tube type piezometers and/or Casagrande cells;
- Toroidal load cells for tie-rods;
- Strain gauges;
- Load cells at the foot center;
- Optical objectives.

In addition to the traditional instruments, the following equipment is planned to be used to monitor the landslide areas:

- Rain gauges;
- Robotic topographic station;
- Terrestrial interferometer;
- Laser scanner;
- Automatic inclinometer to monitor the different depths.

In these areas is also planned the utilization of an automated control program to record continuously:

- The underground horizontal movements for the identification of potential slipping planes;
- The surface movements of the ground;
- Subsidence phenomena;
- Groundwater level;
- Interstitial soil pressures and their variations in time.

2.4 *Innovative techniques for landslide control*

The problems encountered during the different project design phases have been overcome with the definition of a new route that cross the most critical areas through the realization of a natural tunnel (development 2,500 m long – Figure 3).

Depending on the geological-geomorphological conditions, the environmental control was extended throughout the tunnel section.

The traditional monitoring systems was integrated with innovative instruments able to acquire and transmit data continuously, such as: inclinometers and terrestrial radar interferometers.

This methodology was applied during the executive design to control the gallery entrance area (Gamberale side) affected by an active landslide (Figure 4). The particular geomorphological conditions didn't allow to perform measurements on the landslide body in classical mode due to the quick deformation of instruments installed (the instruments were deformed between the zero measurement and the next one).

A vertical inclinometer chain (S7_16_in), composed of 50 nodes installed at a distance of 70 cm and a total length of 35 m, was used for the measurements. The instrument is a chain with nodes, inside which different types of sensors can be inserted, in order to collect different types of data on the same vertical. The results of the study showed not only the nature and the shape of the landslide body, but also the movement typology. In this case were used two types of displacement sensors:

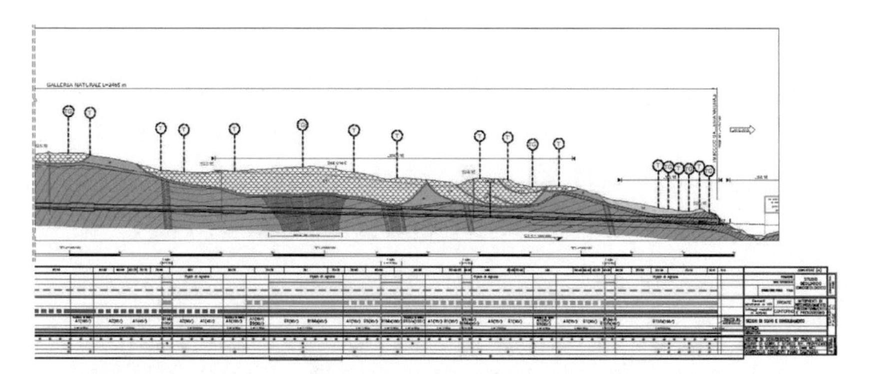

Figure 3. Tunnel profile with the location of the monitoring sections.

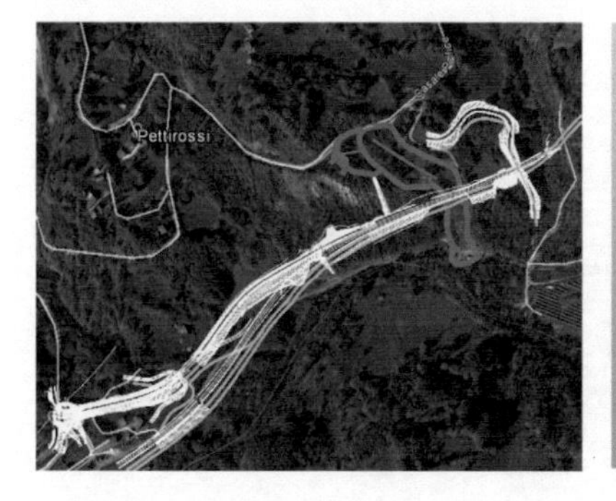

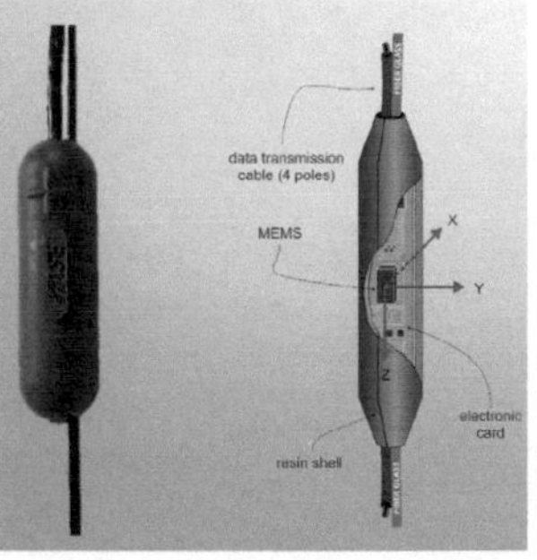

Figure 4. On the left: the landslide near the tunnel entrance. On the right: a MEMS sensor.

- MEMS sensor (Micro Electro Mechanical Systems – Figure 4) equipped with a three-dimensional accelerometer, a magnetometer (measuring the gravitational field in the three orthogonal directions) and a temperature probe;
- MEMS sensor equipped with a temperature probe and an electrolytic cell that measures the inclination of the node with respect to a horizontal plane;

Each node of the chain, equipped with both sensors, allows a high quality data collection thanks to the MEMS versatility, to the accuracy of the electrolytic cell and to the high number of measurement collected at each depth.

In this case, the recording time was set with an hourly rate and the beginning date (zero), considered as starting point for elaborations, was the day after the instrumental installation (19 November 2016, 00:43).

The inclinometer sensors registered a clear deformation phenomenon starting from the night between 23rd and 24th January at a depth included between 5.5 and 6.5 m and with a surface movement included between 0.7 and 1.5 m depth (Figure 5). In particular, the local displacement along the sliding surface shows a movement of 6.64 cm in a month (azimuth equal to 158.5 °) highlighting a clear acceleration (Figure 6) compared to the previous trend (ref.: ASE ltd, Monitoring Report for the period 19 November 2016 – 07 March 2017 – Parma 10/03/2017).

The last measure valid for the whole inclinometer chain was recorded on 27 February 2017 at 11:45 cause in the following reading (12:45) all the nodes below the sliding surface were out of order due to the electric cable break and the consequent interruption of the signal. The results of the measurement show that the sliding surface is strongly localized in few centimeters.

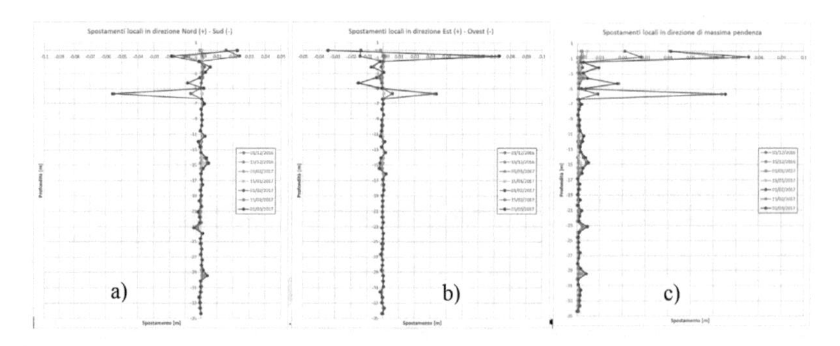

Figure 5. data registered by the inclinometer sensors along the vertical.a) displacements on the N-S axis; b) displacements on the E-W axis; c) total displacement on the maximum sope.

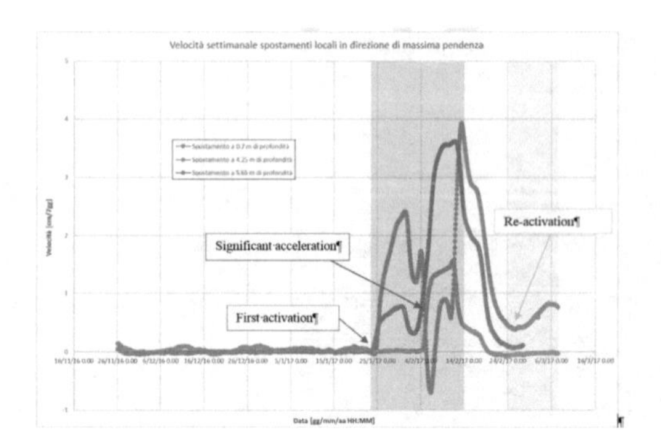

Figure 6. Speed of movements over time.

3 CONCLUSIONS

The monitoring system integrates the control planned for the main works and is designed to provide the necessary information for a correct assessment of the activities to be carried out during construction.

The monitoring system allow to be able to adopt any corrective actions in case of variation from the original project, monitoring the interference effects with the pre-existences conditions (landslides, infrastructures, buildings, etc.) and the new structures in construction. The monitoring plan also allows to check the maintenance of safety conditions during the work construction and during the exercise performing a direct control of structures (tunnels, bridges, minor works), slopes, infrastructures.

REFERENCES

AGI – Associazione Geotecnica Italiana 1997. Raccomandazioni sulla programmazione ed esecuzione delle indagini geotecniche. http://people.unica.it/fabiomariasoccodato/files/2013/10/AGI77.pdf

AGI, Associazione Geotecnica Italiana 1994. Raccomandazioni sulle prove geotecniche di laboratorio. http://www.associazionegeotecnica.it/pubblicazioni/raccomandazioni-sulle-prove-geotecniche-di-laboratorio

ANAS, 2017. Linee Guida in materia di Monitoraggio Geotecnico. Pubblicazione portale Anas http://portaleanas/ComunicazionidiServizioDocumenti/LineeGuidaMonitoraggioGeotecnico.pdf

NTC 2018, 2018. Aggiornamento delle Norme Tecniche per le costruzioni. http://www.gazzettaufficiale.it/eli/gu/2018/02/20/42/so/8/sg/pdf

A. S. Maxwell, A.M. Pearson, J. McCallum – *Guidelines for the Implementation of Data Management Systems on Tunnelling Projects* – Institute of Materials, Minerals and Mining – Proceedings of the underground design and construction conference – 2015

Angus Maxwell, S. Vasagavijayan, S. Satkunaseelan – The Contribution of Computer Systems to the Management of Geotechnical and Construction Risk on the Klang Valley MRT SBK Line Underground Works – International Conference and Exhibition of Tunnelling and Underground Space IEM Kuala Lumpur 3–5 March 2015

Angus Maxwell, Chong Poh Ting, Michael Chin Yong Kok – Thrusting in the same direction. Client-contractor collaboration for integrated TBM process and geotechnical control – Powergrid Cable Tunnels Singapore – International Conference and Exhibition of Tunnelling and Underground Space IEM Kuala Lumpur 3–5 March 2015

Angus Maxwell and Anmol Bedi – *Mission Control; Monitoring Temporary Works in Tunnels* – British Tunnelling Society, Evening Meeting report – March 2016

Brunetti A., Mazzanti P., Moretto S., Scancella S. – *L'Interferometria SAR Terrestre per il monitoraggio geotecnico e strutturale* – Ingenio-web, n.56. – 2017

Mazzanti P., Bozzano F., Brunetti A., Esposito C., Martino S., Prestininzi A., Rocca A. & Mugnozza G. S. – *Terrestrial SAR Interferometry Monitoring of Natural Slopes and Man-Made Structures.* – In Engineering Geology for Society and Territory-Volume 5 (pp. 189–194) – Springer International Publishing – 2017.

Mazzanti P., Bozzano F., Cipriani I., Prestininzi A. – *New insights into the temporal prediction of landslides by a terrestrial SAR interferometry monitoring case study* – Landslides, 12(1), 55–68. DOI 10.1007/s10346-014-0469-x – 2014.

Mazzanti P. – *Displacement Monitoring by Terrestrial SAR Interferometry for Geotechnical Purposes.* – Geotechnical News, 29(2), 25–28 ISSN: 0823-650X – 2014.

Moretto S., Bozzano F., Brunetti A., Della Seta M., Majetta S., Mazzanti P., Rocca A., Valiante M. – *The 2015 Scillato Landslide (Sicily, Italy): deformational behavior inferred from Satellite & Terrestrial SAR Interferometry* – 10th International Symposium on Field Measurements in Geomechanics July, 17 – 20, 2018, Rio de Janeiro, Brazil – 2018.

Segalini A, 2013, ASE S.r.l. Advanced Slope Engineering s.r.l. – Spin Off company of the University of Parma, Italy

Tunnels and Underground Cities: Engineering and Innovation meet Archaeology, Architecture and Art, Volume 3: Geological and geotechnical knowledge and requirements for project implementation – Peila, Viggiani & Celestino (Eds)
© 2020 Taylor & Francis Group, London, ISBN 978-0-367-46583-4

Geotechnical monitoring: The control of infrastructure projects

R. Gandolfo, M. Martino & A. Umiliaco
ANAS S.p.A., Design Department, Rome, Italy

ABSTRACT: The Italian territory is mostly constituted by outcropping soils that are defined in geotechnical terms as "complex structure soils". The territory is overall characterized by new formations, related to the young geological age of the Alpine chain and the Apennine ridge, and by many landslides and instability phenomena (which cover 6.6% of the whole national territory). The need to realize the infrastructural network, in such a complex geological and geotechnical background, has led all the Italian production chain to develop a deep know-how in the whole construction process. According to the new Italian Building Code and considering the strong impact of the infrastructures on a geologically complex territory, ANAS started a new geological and geotechnical monitoring service investing 25 million euro in the period 2018–2020. This paper aims to highlight the evolution of geotechnical monitoring and the new strategies prepared by ANAS.

1 GEOLOGY OF ITALY OVERLAPPING THE INFRASCTRUCTURAL NETWORK

The Italian territory, despite of its small surface, shows a significant geological variability. Two main tectonic events formed the peninsula: the orogeny of the Alpine chain and the orogeny of the Apennine ridge.

More in detail, the territory can be divided into five different outcropping formations (Figure 1):

1) the Alpine chain, outcropping in the Northern side and in the Calabria region;
2) the Apennine chain, that is the skeleton of Italian Peninsula from north to south up to the Sicily;
3) the continental deposits of Sardinia;
4) the volcanic deposits;
5) the Plio-Pleistocene sedimentary deposits.

Such a geological variability is also associated to a great geotechnical heterogeneity which implied over time the rise of the Italian geology and infrastructural engineering schools, whose some major exponents are well known all over the world.

Italian geotechnical engineering, as an example, has often faced cases of large diameter tunnel excavations in "complex structure" rocky formations, which are usually considered the most difficult geomaterial to deal with. In the geotechnical classifications, such peculiar formations are neither rocks nor soils, thus, their mechanical characterization is very hard to conduct by the classical approaches of both the rock mechanics, as well as, the soil mechanics.

In this geological framework, it should be also considered that the combination of a seismic territory and of severe hydrogeological conditions triggers the development of very different kinds of landslides. According to the Italian statistics, firstly

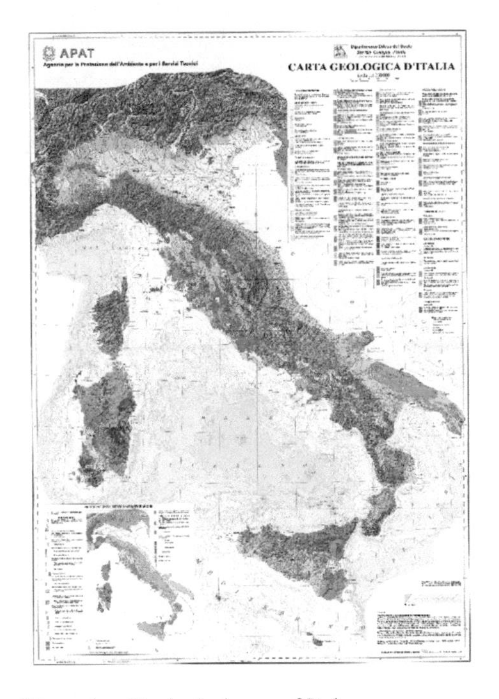

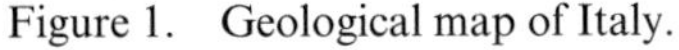

Figure 1. Geological map of Italy.

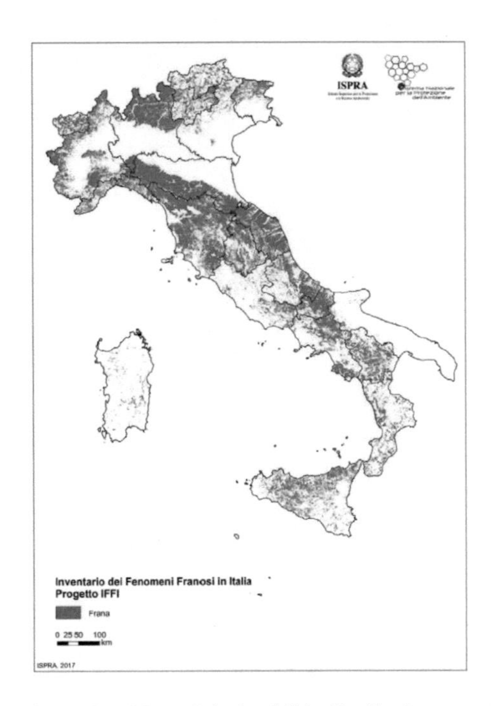

Figure 2. Map of the landslide distribution.

earthquakes, and just secondly the landslides represent the natural disasters which causes major damages to the population, infrastructures, environmental, historical and cultural heritage.

The *IFFI Project* (database of Italian landslides), realized by ISPRA (Istituto Superiore Per la Ricerca Ambientale) and by autonomous regions and districts, provides a detailed framework about the distribution of landslides on the Italian territory (Figure 2). The database, up to now, has counted more than 499,511 landslides, on a surface of 21,182 Km^2.

ISPRA data reveal that 6180 critical points are affecting the main Italian infrastructural network (720 of which along the highways network and 1862 along the railways network) related to the reactivation of different kinds of landslides (fall, topple, slide, flows, etc.), already mapped and surveyed by IFFI Project.

Nowadays, the national existing road network, managed and controlled by ANAS (National Road Agency), interferes with thousands of slope instability situations (Figure 3). The same issues also affect the new road networks during both the design stage and during the construction phases.

To worsen the risk situation, there is the susceptibility of the road network to extreme seismic events which regard the major part of the Italian territory. Figure 4 shows the Italian seismic hazard map, edited by INGV (National Institute for Geophysics and Volcanology), based on the maximum expected acceleration for a rigid and horizontal ground.

Italy is divided into 4 hazard zones: low, medium, high and very high hazard. The map shows that a great part of Italy has a seismic hazard between high and very high, in particular on the Apennine ridge (center and South) an on the North-Eastern part of the Alpine chain.

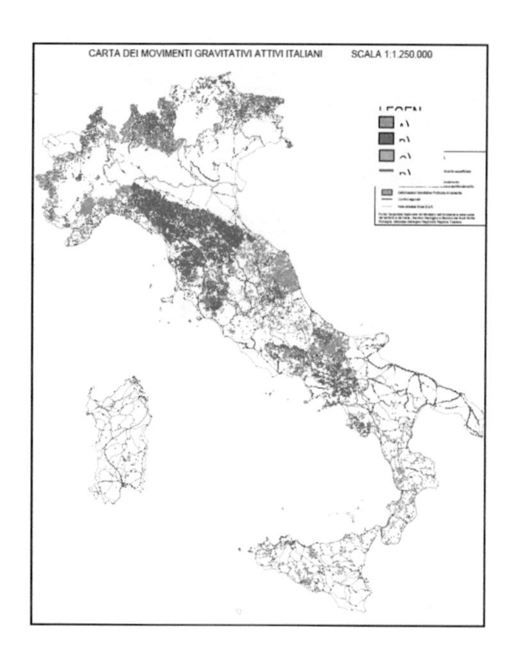

Figure 3. Map of landslides of ANAS road network. A) Superficial landslides: creep, flows, superficial translational slides; B) Deep slides: fall/topple, rotational slides, deep translational slides, lateral spreads, sinkholes, etc; C) Deep seated gravitational slope deformations; D) ANAS road network.

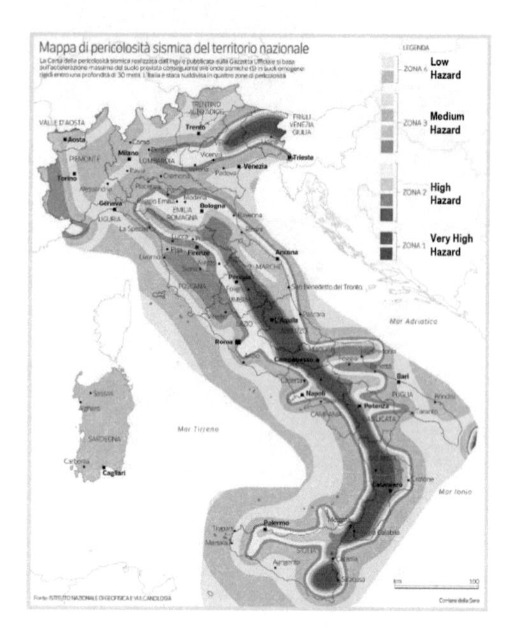

Figure 4. Seismic hazard map of Italy.

2 HISTORICAL BACKGROUND

The Roman Empire was the first civilization to realize an infrastructural network for eight centuries, since 300 B.C. up to the fall of the Western Roman Empire (476 A.D.). In particular, Romans were the greatest road constructors of history and developed a road network, in the period of Emperor Domitian, of 372 main roads for an extension of almost 80.000 Kilometers (Appia – 312 B.C., Aurelia – 241 B.C., Flaminia – 220 B.C., Emilia – 187 B.C.) (Figure 5).

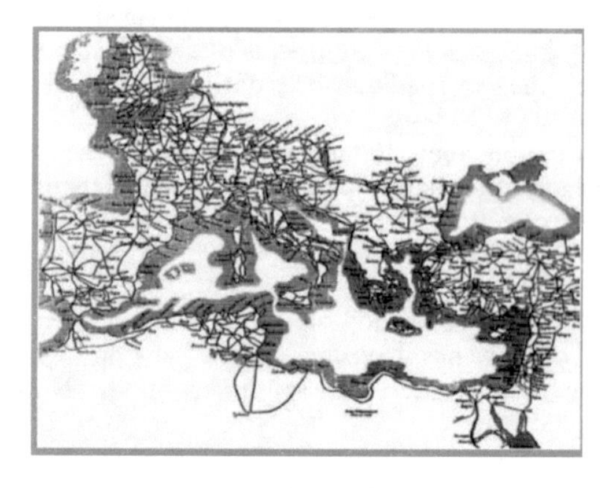

Figure 5. The Roman Empire road network in the period of his maximum expansion.

Figure 6. View of Crypta Neapolitana. During the roman period, the tunnel was dark and narrow but later it was enlarged and lowered at the floor. Today, it is between 4–5 m wide, and 5–20 m tall. At the time, it was lighted by torches and by two light wells. The tunnel orientation is such as the sun is aligned with the two entrances, during the equinox, at the sunrise and at the sunset.

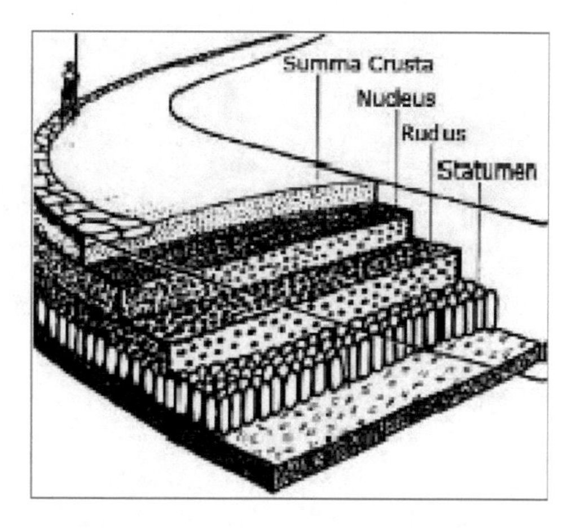

Figure 7. Scheme of a typical roman road pavement.

These roads had already some technical characteristics of modern infrastructures. The geometric design of the ancient roman roads, in fact, was planned in order to insert the roads in the surrounding territory through the introduction of tunnels, bridges and viaducts. An example is the "Crypta Neapolitana", a road tunnel of 700m length, built under the hill of Posillipo, Naples, in 37 B.C. (Figure 6).

Romans built road pavements depending on the bearing capacity of the soils and with different functional stone layers (Figure 7). On a big stones layer (*statumen*), a second gravelly layer (*ruderatio*) was placed, on which the third layer of rubble (*nucleus*) was placed; on the top, the final pavement of stone slab was placed for the major roads, or a layer of gravel (*summa crusta*) for the minor roads. The most common material used for pavements was flint, for his hardness and resistance, otherwise, limestone, granite or sandstone.

More recently, for the development of the steam engine and the industrialization, a new linear infrastructure was born: the railroad. The Napoli-Portici railway was the first railroad built in Italy, in the 'Due Sicilie' kingdom, that was characterized by a double track line of 7.25 Km length.

On June 19, 1836, it was signed the agreement for the construction of the Napoli-Portici railway and the line was opened on October 3, 1839, with solemnity by the King Ferdinando II di Borbone.

2.1 *Roads changing in times*

Since 1928, year of foundation of A.A.S.S (Azienda Autonoma Statale della Strada; in 1946 A.A.S.S. was replaced by ANAS), Italy is working for the continuous improvement of the existing road heritage and for the development of infrastructures that allow the connection between the Northern and the Southern parts of the whole peninsula. In the first years, the greatest effort was to improve the service conditions of the existing network, as shown in the images in Figure 8.

Figure 8. The "Cassia" road connecting Rome and Viterbo towns. Comparison between before and after the rectification works, finished in 1932

Nowadays, ANAS road network includes more than 30,000 Km of roads, with a heritage of 1700 operating tunnels, for a total length of 856 Km distributed in 500 Km of double tubes tunnels and 356 Km of single tube.

3 MONITORING SYSTEM

Due to geological and geomorphological framework affecting the Italian road network, the new Italian Building Code highlights the role of geotechnical monitoring with the aim to control the safety of the civil works, as specified at paragraph 6.2.6 of the new update of "Norme tecniche per le costruzioni 2018" (NTC 2018).

Following the guidelines of the NTC 2018, ANAS produced specific "Guidelines" ('Linee Guida in materia di monitoraggio geotecnico') to lead the designers to draw up detailed monitoring documents. The NTC 2018 imposes to adopt a monitoring plan for the projects and also indicates to put in light the aims of the monitoring. Following the requirements, it must be provided a monitoring system for the soil-structure system, before, during and after the construction, with specific control parameters relevant to represent the overall behavior of the soil-structure interaction. Thus, the monitoring system has to provide the measurement of significant physical parameters, such as displacement, strain, strength and water pressure, during the different phases of the project; moreover, the installation of specific tools for a correct data acquisition and interpretation is needed to reduce the risk of damage or collapse of the structure.

An effective monitoring allows:

- To check the correspondence between the design hypothesis and the physical take-over;
- To confirm the validity of the design solution or, on the contrary, to identify a better one;
- To control the functionality of the structure during its lifetime.

Summarizing, monitoring ensures:

- A support in the design phase by the characterization of the area (e.g. the ground water level variability, rock masses stability, subsidence etc...);
- The analysis of the risks and, eventually, a specific action in order to ensure the safety of the area against landslides, floods, earthquakes, etc.;
- A support to the construction phase by checking the initial assumptions on the behavior of the structure;
- The improvement of the quality of the structure. Geotechnical monitoring tools can improve the correct construction of the work;
- An additional tool for a legal protection. Monitoring data can be used in case of dispute with third parties involved, who declare themselves damaged by the construction;
- A Support in the management of the project to reduce the risk of damage or collapse during the construction or the operating phases.

Geotechnical monitoring is a primary issue for the application of the "OBSERVATIONAL METHOD". The NTC2018, at paragraph 6.2.5, requires using the observational method "*when*

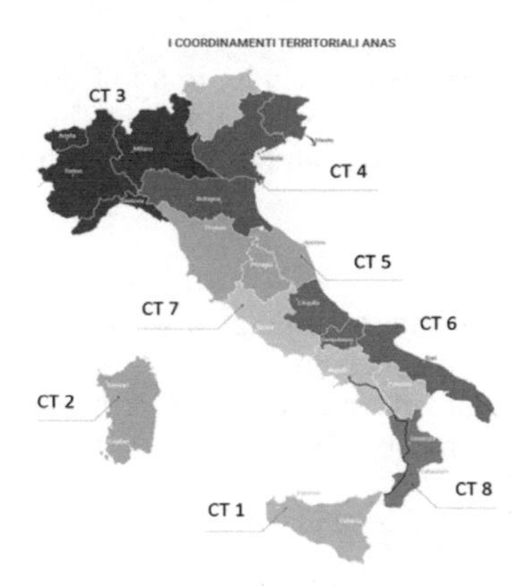

Figure 9. Anas Coordination Offices of the territory.

due to the specific complexity of geotechnical framework and the relevance of the project, after having applied extensive and deep investigations, remain documented reasons of uncertainty".

Designers can apply the observational methods whenever the geotechnical context do not allow to define uniquely the design solution.

In the light of the new reference Italian Building Code, ANAS stipulated some Service Contracts in order to provide the geotechnical monitoring service, for a total investment of 25 million euro in the period 2018–2020, distributed on the eight Coordination Offices (Figure 9) which locally manage the road national network in Italy.

The monitoring services are activated during several stages: *ante operam*, during the construction and *post operam*, for each type of ANAS works (tunnels, viaducts, embanKments, building, slope safety). For each project a monitoring plan is drawn up with specific documents (Figure 10), such as the position and type of the instrumental devices, frequency of the data acquisition, etc; such documents are integral part of the monitoring service contract.

Two types of monitoring service contracts are provided:

1) Supplying, installation, measurement, restitution and interpretation of data acquired by geotechnical-geomorphological-topographic monitoring tools, as planned in the design documents attached to the contract.
2) Verification of the monitoring data, provided by third parties (company, owners, etc), of measurements, data restitution and interpretation, reports, etc.

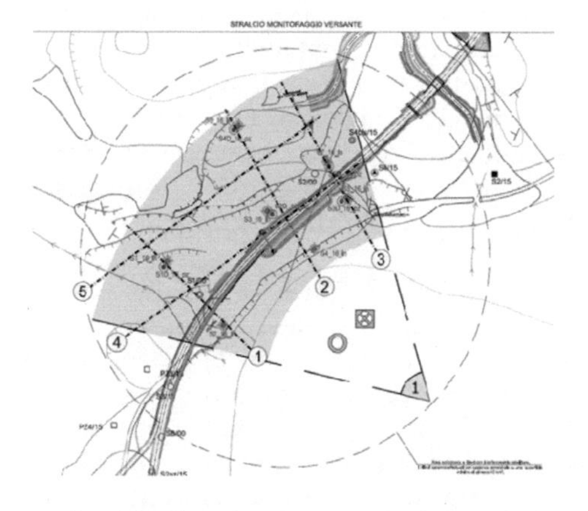

Figure 10. A typical scheme of monitoring plan for a landslide interfering with a new road design. In this case the adopted monitoring systems are: terrestrial interferometer, inclinometers, piezometers, rain gauge stations.

Summarizing, each design process will include, in addition to the monitoring executed by the construction company, a monitoring project aimed at checking and controlling the infrastructures, directly managed by Anas through the Service Contract.

3.1 *Tunnel monitoring*

In the design and construction of underground civil works there are a lot of factors that can produce a potential risk. To ensure an adequate level of safety, during the construction phase, the monitoring plays an important role in order to promptly verify conditions of possible risk for the work, for the staff, or for buildings and other structures surrounding the areas involved in the excavations.

As mentioned before, the monitoring allows to:

- Verify that the monitored parameters reflect the values established in the design;
- Verify that the work remains in safety conditions during its use;
- Check that the parameters relating to the environmental working conditions respect the standard thresholds.

It is possible to monitor different cases, such as:

- Cracks in the rock around the excavation;
- Movements along joints or faults;
- Convergence of two points around the excavation;
- Extrusion of the excavation front;
- Subsidence, or other effects related to structures or territory;
- Water level oscillations in ground;
- Strains changing in the rock or in the support structures;
- Energy emission as seismic or microseismic waves (vibrations).

In addition to the control of safety conditions, by means of the adoption of new technologies (e.g. Laser scanner, Georadar, etc), it is also possible to monitor the compliance of the work in progress with the project specifications. Through these tools it is now possible to evaluate: extra thicknesses or reduced thicknesses of the linings, presence of voids in the back of the final linings, spacing of the ribs. These tools let the Works Management constantly check the quality characteristics of the works in progress.

4 CONCLUSIONS

Because of the development of the instrument technology, the increased importance of safety and the consequent adaptation of national building codes, geotechnical monitoring has acquired an increasingly central role as an important tool to control numerous aspects of the construction and the management of an engineering system. The continuous evolution of the instrumentation, characterized by an increasing speed in monitoring and interpreting the data, the sophisticated technology of transmission and visualization of the data in real time and in any part of the world, allows increasing the safety factors of any work.

In road infrastructure sector, ANAS, which is managing almost 30,000 Km of existing roads and is planning to realize approximately 92 Km of underground excavations in the three years 2018–2020, has developed a new approach to control the infrastructures through the systematic use of geotechnical monitoring. Through the activation of specific Service Contracts, ANAS has defined criteria for the design and selection of instrumentation, storage, processing and transmission of data, not least the management of the information flow and the professional figures involved in the operational monitoring structure.

REFERENCES

A.G.I. 1979.Some Italian experiences on the mechanical characterization of structurally complex Formations. Proc. IV I.C.R.M. 1, 827–846. Montreaux.

ANAS 2017. Linee Guida in materia di Monitoraggio Geotecnico. Anas. Roma. Pubblicazione portale Anas http://portaleanas/ComunicazionidiServizioDocumenti/LineeGuidaMonitoraggioGeotecnico.pdf.

INGV 2004. Istituto Nazionale di Geofisica e Vulcanologia - Mappa della pericolosità sismica del territorio nazionale - http://zonesismiche.mi.ingv.it/.

ISPRA. Dipartimento difesa del suolo, Servizio geologico - Carta geologica d'Italia, scala 1:250.000. http://www.isprambiente.gov.it/it/cartografia/carte-geologiche-e-geotematiche/carta-geologica-dei-mari-italiani-alla-scala-1-a-250000

ISPRA 2018. Dissesto idrogeologico in Italia: pericolosità e indicatori di rischio. http://www.isprambiente.gov.it/it/pubblicazioni/rapporti/dissesto-idrogeologico-in-italia-pericolosita-e-indicatori-di-rischio-edizione-2018

ISPRA 2007. Progetto IFFI "Inventario dei Fenomeni Franosi in Italia" Metodologia e risultati. APAT. Roma. http://www.isprambiente.gov.it/it/progetti/suolo-e-territorio-1/iffi-inventario-dei-fenomeni-franosi-in-italia

NTC 2018 – Aggiornamento delle Norme Tecniche per le costruzioni – 17/01/2018 - http://www.gazzettaufficiale.it/eli/gu/2018/02/20/42/so/8/sg/pdf

Picarelli L. 1986. Caratterizzazione geotecnica dei terreni strutturalmente complessi nei problemi di stabilità dei pendii. Proc. XVI Convegno Nazionale Geotecnica, Napoli.

Trigila A., Iadanza C., Bussettini M., Lastoria B. 2018. Dissesto idrogeologico in Italia: pericolosità e indicatori di rischio. ISPRA, Rapporti 287/2018 - http://www.isprambiente.gov.it/it/pubblicazioni/rapporti/dissesto-idrogeologico-in-italia-pericolosita-e-indicatori-di-rischio-edizione-2018.

Tunnels and Underground Cities: Engineering and Innovation meet Archaeology, Architecture and Art, Volume 3: Geological and geotechnical knowledge and requirements for project implementation – Peila, Viggiani & Celestino (Eds)
© 2020 Taylor & Francis Group, London, ISBN 978-0-367-46583-4

Characterization of underground rock masses employing structure from motion: Application to a real case

R. García-Luna, S. Senent, R. Jurado-Piña & R. Jimenez
Technical University of Madrid, Madrid, Spain

ABSTRACT: We propose an inexpensive and easy-to-use methodology for remote characterization of underground rock masses at the tunnel face. The method is based on the joint use of the Structure from Motion (SfM) photogrammetric technique and a Discontinuity Set Extractor (DSE) software, in which a high-density 3D point cloud is created using a commercial SfM software (Agisoft PhotoScan). To maintain the real features of the tunnel face, such digital model must be orientated and scaled; therefore, a "portable orientation template" is employed. The point cloud is then introduced in DSE, where discontinuity orientation data are "extracted" and analyzed to identify joint sets. The measurements obtained with the SfM + DSE approach have been compared with those obtained with a traditional analysis based on manual compass measurements. Both techniques identify the same number of discontinuity sets with orientation differences within the uncertainty range associated to manual measurements.

1 INTRODUCTION

A good knowledge of the discontinuity network in a rock mass is fundamental for the design and control of tunnels under construction, as discontinuities are surfaces of weakness that control the occurrence of unstable blocks in the tunnel and the loads imposed by the blocks on the excavation support (Goodman 1976, Hudson & Harrison 1997). However, characterizing these discontinuities using manual methods (i.e. direct measurements made with compass) can impose a risk on the safety of the operators who take the data, as the (unsupported) tunnel face is the area of the tunnel with a higher risk of collapse (Slama et al. 1980, Health and Safety Executive 1996). In addition, the results obtained manually are often very subjective, depending on factors such as the access conditions to the outcrop, the number of measurements made, the worker´s ability, the time available and the scale of the problem.

Due to these difficulties, remote "non-contact" techniques —such as digital photogrammetry and laser scanner methods— are becoming increasingly common to obtain digital 3D models of tunnel supports (Roncella et al. 2012, Chaiyasarn et al. 2015) and to identify and characterize the underground rock mass discontinuities (Buyer & Schubert 2016, 2017). However, these works do not discuss in detail the performance of the remote technique.

In this paper we introduce an inexpensive and easy-to-use methodology for remote identification of discontinuity sets at underground rock masses under real tunneling conditions. The proposed methodology has been applied to one tunnel under construction in Northern Spain, comparing the results obtained using the remote SfM technique with those obtained using traditional data collection methods with a compass, and describing the photographic conditions for acquisition of the 3D digital model, emphasizing the importance of using auxiliary processing tools that allow us to visualize jointly the main discontinuity sets identified by DSE and the real tunnel scenario as a 3D point cloud.

2 METHODOLOGY

This process employed to introduce and validate the methodology is divided into three parts. First, we characterize the main discontinuities sets present in the tunnel face using traditional manual measurements (with compass) and a DIPS analysis (Rocscience 2017). The next step is to generate a photogrammetric 3D model from several overlapping photographs of the tunnel face with the commercial software Agisoft PhotoScan (LLC 2016), and using the structure from motion technique combined with a portable orientation template to orientate and scale the model. Then once the 3D model is ready, it is introduced in the open-source software DSE (Riquelme et al. 2014), where the different discontinuity sets data are "extracted" and analyzed. Finally, the results provided by both methods are validated through two different ways: (i) comparing the results obtained with DIPS (in terms of Dipdir/Dip) directly with the measurements obtained with the SfM + DSE approach, and (ii) making a combined 3D visual inspection of the point cloud classified with DSE in superposition with the original unclassified point cloud.

2.1 *Portable orientation template*

One of the main disadvantages of the SfM photogrammetric technique, in contrast to traditional photogrammetry, is that it generates digital 3D models within an arbitrary reference system without scale or orientation. Therefore, to geo-reference the 3D model, it is necessary to include several ground control points (GCPs) of known 3D coordinates (x-y-z), within the scene. Singularities or traces easily recognizable in the images, may also be useful for this purpose (Dowling et al. 2009, Dandois & Ellis 2010).

To solve this problem, we have developed a "portable orientation template" that avoids the need to use topographic control devices (Figure 1). Its operation is inspired by a conventional compass, but on a larger scale. The template contains five GCPs labeled inside a rigid square built with EVA foam; a central one that serves as the origin of the system of coordinates (local) and other four located at the corners. The template incorporates an arrow that can be aligned with the north using a compass and a spirit level can be used to place it horizontally. As the distance between the points is known, the template serves as reference plane and we can use it to orientate and scale the scene in a single step, using the relative coordinates provided by its points, within a local reference system.

Figure 1. Portable orientation template.

2.2 *Description of the tunnel*

The methodology has been tested in a real tunnel under construction in the Santa Barbara Foundation research area (FSB 2017), located in Northern Spain. The rock mass is constituted by limolites, lutites and sandstones (RMR 40–60), and the tunnel had an excavated cross-section of 60 m2 and a length of approximately 295 m when this study was conducted. The tunnel is constructed using drill and blast methods, with full-face excavation and it is supported using rock bolts and shotcrete. Shotcrete was present in the lower parts of the tunnel face, hiding the rock mass surface and obstructing the real discontinuities.

2.3 *Manual characterization of discontinuity sets*

With the purpose of validating and comparing the results provided by the SfM + DSE methodology, we decided to identify first the main discontinuity sets manually using a traditional compass measurements and a DIPS analysis. We measured the orientations of 33 discontinuities found at the tunnel face with a Freiberger-type geological compass, taking about 30 minutes to complete and register all the measurements. Figure 2 shows the results provided by DIPS (pole density) and the three main discontinuity sets identified using such measurements.

The statistical analysis of the pole vectors of all discontinuities measured with the compass (Figure 2) identifies three main discontinuity sets: F1 (240/79), F2 (131/71) and F3 (4/10). The two first (F1-F2) outcrop clearly at the tunnel face, being cut at 45° by a N-S vertical sampling surface (note that the direction of advance of the tunnel at the time of data collection was W-E). Both families (F1-F2) present conjugate orientations (ΔDipdir = 109°) with sub-vertical dip angles (β = 79–71°). The last family (F3) presents a sub-horizontal dip angle (β = 10°) and a more variable dip direction (α = 12–273°). The orientation of F3 family, combined with the orientation of the tunnel face, make this discontinuity set to accumulate shotcrete, especially in the lower parts of the tunnel face. This complicates the measurement of F3 orientations, which are better measured at reverse faces (wall), which only appear clearly in the upper area of the face, inaccessible for manual measurements.

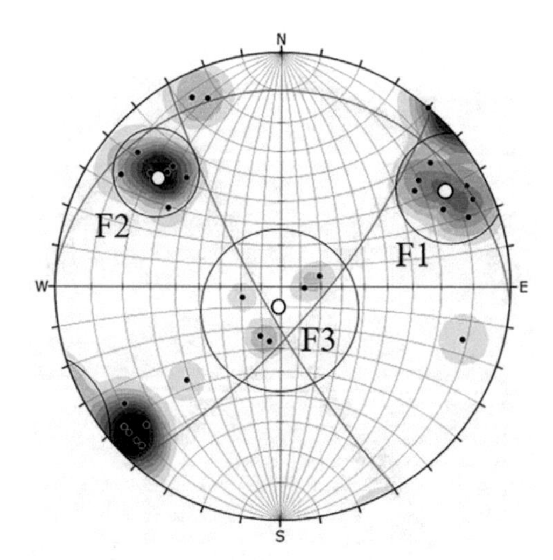

Figure 2. Stereographic projection (lower hemisphere, equal angle) of the pole vectors of discontinuity orientations measured with the compass (black dots) and the three main discontinuity sets (F1-F2-F3) identified by DIPS (white dots).

2.4 *Photographic equipment and data acquisition*

To generate the photogrammetric 3D model, we employed an un-calibrated Sony Digital Single-Lens Reflex camera (model α230) with a common lens Sony DT 18–55 mm (F 3.5–5.6). There are some challenges for 3D SfM reconstruction in underground spaces, such as the need to work in low-luminosity conditions while maintaining low exposition times (of < 1/4") to avoid the effect of micro-vibrations in the field. To overcome them, we decided to install the camera on a professional heavy weigh tripod (model Yashica YT-300) and to activate the camera by remote control.

The same lighting conditions as in a real tunnel under construction have been used in this work. The tunnel face was illuminated by a drilling jumbo positioned 4 m away from the face, which provided a luminosity of about 100–130 LUX (measured at the face). In addition, we use an extra portable illumination source situated directly above the camera (LED torch, model Yongnuo YN-216 LEDs) using an L-Bracket plate on the tripod. This torch provided an extra illumination of the rock surface of about 10–30 LUX (depending on the distance from the tunnel face) that pointed directly at the same direction of the camera, helping us to attenuate the presence of "shadow areas" that can produce occlusions zones in the final 3D model.

We make a full-scan from the tunnel face, taking 206 photographs from 18 different positions (Figure 3), achieving a high level of superposition between each picture (> 60%). All the photographs were taken with the highest quality settings available in the camera (10 MP, RAW-JPEG format) and with fixed camera adjustments (manual mode, AF, focal length 18 mm, aperture F/3.5, shutter speed 1/4" and ISO 100). These parameters are defined before photo shooting according to the available lighting conditions of the tunnel face; in this way, the number of maneuvers between photographs is minimized (point-stabilize-shoot), helping us to streamline the process of data collection.

In order to be able to geo-reference and scale the 3D model correctly, the portable orientation template was located directly on the ground (oriented and levelled), at the lower part of the tunnel face before starting the shooting. In addition, seven precision targets were distributed along the surface of the tunnel face to evaluate the accuracy of the scale of the 3D model provided by the portable orientation template (comparing the distances between them, measured "in situ" with measuring tape with the corresponding distances computed from the scaled 3D model).

Figure 3. Front view of the photogrammetric 3D model (textured mesh) and camera positions (blue squares).

The total time spent in the preparation and acquisition of photogrammetric data was 30 minutes, or about the same time employed in the field campaign for the manual characterization with compass. Although these times may seem very high, it was decided to perform an over-characterization of the tunnel face to ensure the generation of a solid 3D model.

2.5 *Generation of the 3D model*

The 3D model was constructed with the professional version of Agisoft PhotoScan photogrammetric software, using the structure from motion approach. (Note that the standard version does not allow to work with reference points.) The surface of the whole 3D model has a total area of 192 m2 and was generated from all the available photographs (206). The total number of points was 6,058,037, with a density of 2.5 points/cm2. The resulting 3D point cloud was finally scaled and oriented, according to a local reference system, based on the five GCPs included within the portable orientation template. The errors of the relative coordinates (x-y-z) of the reference points provided by the template are listed in Table 1. Table 2 summarizes the distance errors for precision targets (7) installed on the tunnel face.

Generation the photogrammetric 3D model in Agisoft PhotoScan took a computation time of about 25 h, including the time to align and orient the initial "disperse" point cloud (1h 46 min) and the time to process the second high-density point cloud (23 h 45 min) using MVS techniques. (All computations were performed using a single core of an Intel Core i7–6700 computer [3.4GHz processor, 16 GB RAM] running the Windows operating system.)

Finally, so that the results are not affected by the presence of shotcrete in the tunnel, the zones of the model where its presence is greater have been eliminated, keeping only the central area of the front, where the rock mass is not covered with shotcrete. In addition, an automatic filtering (subsample) of the point cloud was conducted, establishing a minimum distance of 1 cm between the points, in order to avoid that the areas of the model with higher density (points/cm2) can affect the DSE analysis (Riquelme 2015). Both the edition of the cloud and the subsampling have been performed with the 3D CloudCompare software (Girardeau-Montaut 2016.). In this way the final "Region of interest" of the model (ROI), with a total of 295,705 points, is defined.

Table 1. Errors of (x-y-z) coordinates of template's reference points within the 3D model.

GCPs	x	y	z
	mm	mm	mm
1	-2.8	1.4	1.7
2	-3.8	3.1	2.5
3	-0.3	0.3	1.9
4	-3.8	-2.5	2.1
5	-6.7	0.4	2.4

Table 2. Distance errors for precision targets.

Scale Bars	Real distance*	3D distance	Error
	mm	mm	mm
target 1 – target 2	441	451.0	10.0
target 3 – target 4	755	776.0	21.0
target 5 – target 6	547	561.1	14.1
target 5 – target 7	598	611.7	13.7
target 6 – target 7	350	359.1	9.1

*Measured with tape.

3 RESULTS AND DISCUSSION

The study was developed in two simultaneous phases: (i) an initial phase in which the 3D model was analyzed in its entirety, validating the manual results (identifying F1, F2 and F3); and (ii) a secondary phase that focused on a smaller upper sector of the tunnel, resulting in the characterization of a new accessory family, associated to F2 and named F2B, which had not been detected in the initial analysis of manual measurements with DIPS. This case has also been analyzed in more detail in García-Luna et al. (2019).

3.1 *Identification of discontinuity sets with DSE*

Figure 4 and Table 3 show the results provided by the Discontinuity Set Extractor analysis of the high-density point cloud of the ROI (295,705 points) generated by Agisoft PhotoScan and subsampled by CloudCompare. In particular, Figure 4 shows a lower-hemisphere equal-angle stereological projection of the density deviations of the plotted points normal vectors recognized by DSE using all planes identified within the ROI. Table 3 summarizes the representative orientations values (in terms of Dipdir/Dip) of the principal poles of each identified discontinuity set. In total 117,026 points were classified and 178,679 points have not been assigned.

Altogether, three main discontinuity sets were identified with DSE: F1 (249/82), F2 (132/73) and F3 (352/19). Note that they are approximately equal to those identified with DIPS using the compass measurements: two sub-vertical and conjugated discontinuities (F1-F2, with ΔDipdir = 117°) and one sub-horizontal (F3), with maximum differences between their

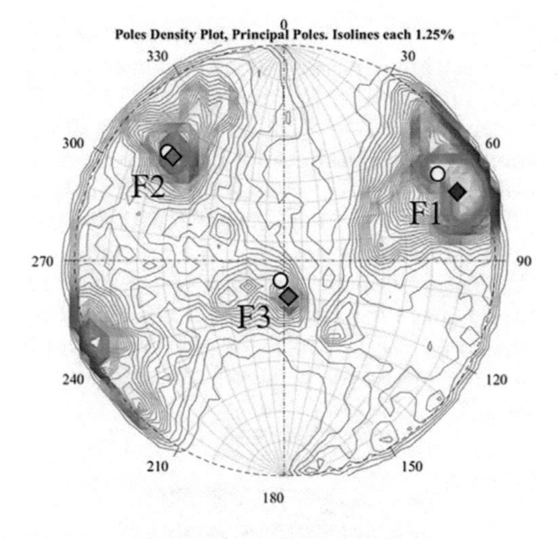

Figure 4. Stereographic projection (lower hemisphere, equal angle) of the plotted points normal vectors (density deviations) and the three main discontinuity sets (F1-F2-F3) represented with diamond symbols, both calculated by DSE. Results of the manual characterization with compass are also provided (white dots).

Table 3. Representative orientations provided by DSE.

Discontinuity sets	Dip Dir.	Dip	Number of assigned points*
	[°]	[°]	[%]
F1	249	82	27.60
F2	132	73	11.01
F3	352	19	1.95

*Over the total number of points

Table 4. Differences of orientation between the discontinuity sets identified with DIPS and DSE.

Discontinuity sets	DIPS		DSE		Differences
	Dip Dir. [°]	Dip [°]	Dip Dir. [°]	Dip [°]	[°]
F1	240	79	249	82	9
F2	131	71	132	73	2
F3	4	10	352	19	9

corresponding orientations of only 9°, 2° and 9°, respectively. Table 4 shows the differences between the orientations computed with both approaches, using the difference between the acute angle formed by their unit normal vectors (Jimenez-Rodriguez & Sitar 2006).

The angular differences shown in Table 4 are within the error range associated to manual measurements for these types of planes, of about 6–11° depending on the dip angle of the joint (Robertson 1970, Windsor & Robertson 1994).

3.2 *Visual inspection of the results and detail analysis*

One main advantage offered by the use of digital 3D models for the analysis of discontinuities is the wide range of modelling tools associated with the SfM technique. For instance, the 3D point cloud processing software CloudCompare allow us to represent the planes associated to each discontinuity set identified with DSE (F1-F2-F3) back into the original 3D model developed with Agisoft PhotoScan (see Figure 5).

Figure 5 shows that joint-sets F1 and F2 were identified clearly along a great percentage of the face, whereas the F3 set appears less often. However, in the upper area of the 3D model (where several blocks produced by the intersections of the discontinuities planes F1, F2 and F3 can be clearly seen), the sub-vertical F2 set is not well represented. This observation introduces doubts about the discontinuity F2 itself and made us to conduct an alternative analysis of that area of the tunnel face. Therefore, we created a new detailed ROI of this region of approximately 20 m2 (1/3 of the front surface) with 40,509 points.

The results provided by DSE for this detailed area are shown in Figure 6b, where two main families with a NE-SW component are observed: F2A (130/80) and F2B (157/77). The first one corresponds to the same family previously identified as F1 (132/73) with a slight increase in its dip angle (7°), while the 3D visualization of the F2B planes (Figure 6a) confirms that they represent a "new" discontinuity family. This new family

Figure 5. Spatial distribution of surfaces associated to the main discontinuity sets (FI-F2-F3) identified with DSE, projected back into the original 3D model.

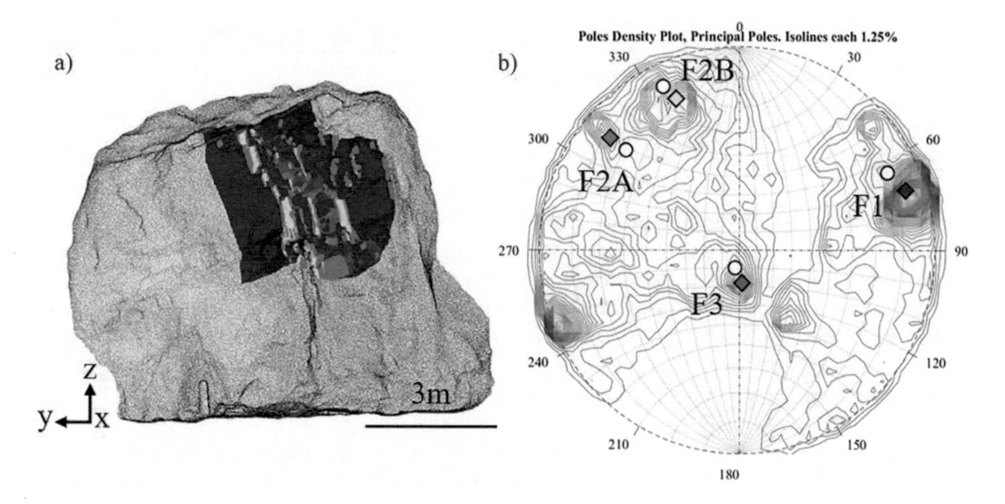

Figure 6. (a) Spatial distribution of surfaces associated to the main discontinuity sets (FI-F2A-F2B-F3) identified with DSE, projected back into the new ROI of the upper area. (b) Stereographic projection (lower hemisphere, equal angle) of the plotted points normal vectors (density deviations) and the four main discontinuity sets (F1-F2A-F2B-F3) represented with diamond symbols, both calculated by DSE. Results of the manual characterization with compass are also provided (white dots).

Figure 7. Spatial distribution of surfaces associated to the main discontinuity sets (FI-F2A-F2B-F3) identified with DSE, projected back into the original 3D model.

(F2B) is able to delimit correctly the contour of the blocks for which the first classification had left doubts around the F2 family.

A more detailed analysis of the DIPS results (Figure 2) confirms the presence of this family within the stereonet F2B (156/83), as it was registered in two compass measurements (158/82 and 154/84) that where initially discarded when F2 was characterized.

Once this measure has been fixed it is possible to return to the original model (ROI 295,705 points), to introduce the new discontinuity F2B and to proceed to the joint visualization of the four discontinuity sets: F1, F2A, F2B and F3 (Figure 7). It can be observed that the family F2A (in green) is distributed along the entire face, although it appears more often in the lower areas, where compass measurements are possible (0–2 m). However, the family F2B (in yellow) mainly appears in the upper area of the tunnel face (unreachable for direct measurements) and its presence in the lower parts is rare, which justifies that only two manual measurements of this set were registered.

4 CONCLUSIONS

Compared with traditional ground photography, underground characterization of rock masses using digital photogrammetry is challenging due to a dark environment that demands a greater care for shooting preparation, ensuring the stability of the sensor.

In this work, however, has been proved that the generation of 3D models of rock masses in underground environments with real working lighting conditions (provided by drilling jumbos) is possible using a regular camera installed on a tripod and without having advanced photographic knowledge; hence becoming a suitable methodology for real construction projects.

In addition, the use of a portable orientation template has been shown to be a good solution to orientate and scale the model without the need to employ plenty of time and resources.

The analysis performed with the methodology proposed in this paper (SfM + DSE) identifies the same number of families as the analysis performed with traditional methods, showing only slight angular differences that are within the stablished uncertainty range associated to manual compass measurements. Moreover, the use of this technique allows to analyze inaccessible areas of the tunnel face usually not registered by operators (above 0–2 m), also becoming a safer worker environment since the hazards associated to collapses are reduced.

Finally, we have proved that the use of alternative tools to filter and fragment the tunnel face into smaller regions is a good solution to focus the analysis on the most important outcrop areas, where the discontinuities surfaces are not hidden by shotcrete accumulation. This process also reduces processing times and decrease the weight of digital files.

REFERENCES

Buyer, A. & Schubert, W. 2016. Extraction of discontinuity orientations in point clouds. *Rock Mechanics and Rock Engineering: From the Past to the Future* 2: 1133–1138.

Buyer, A. & Schubert, W. 2017. Calculation the Spacing of Discontinuities from 3D Point Clouds. *Procedia engineering* 191: 270–278.

Chaiyasarn, K., Kim, T., Viola, F., Cipolla, R. & Soga, K. 2015. Distortion-free image mosaicing for tunnel inspection based on robust cylindrical surface estimation through structure from motion. *Journal of Computing in Civil Engineering* 30 (3): 04015045.

Dandois, J.P. & Ellis, E.C. 2010. Remote sensing of vegetation structure using computer vision. *Remote Sensing* 2 (4): 1157–1176.

Dowling, T., Read, A. & Gallant, J. 2009. Very high resolution DEM acquisition at low cost using a digital camera and free software 2479–2485.

FSB, Fundación Santa Bárbara. http://www.fsbarbara.com.

García-Luna, R., Senent, S., Jurado-Piña, R. & Jimenez, R. 2019. Structure from Motion photogrammetry to characterize underground rock masses: experiences from two real tunnels. *Tunnelling and Underground Space Technology* 83: 262–273.

Girardeau-Montaut, D. 2016. Cloud Compare: 3D point cloud and mesh processing software, open-source project. Version 2.6.

Goodman, R.E. 1976. *Methods of Geological Engineering in Discontinuous Rocks*. St. Paul: West Publishing Company.

Health and Safety Executive. 1996. *Safety of New Australian Tunnelling Method (NATM) Tunnels. A review of sprayed concrete tunnels with particular reference to London clay*. Sudbury: (HSE) Books.

Hudson, J.A. & Harrison, J.P. (1 ed.) 1997. *Engineering Rock Mechanics: An Introduction to the Principles*. Tarrytown: Pergamon.

Jimenez-Rodriguez, R. & Sitar, N. 2006. A spectral method for clustering of rock discontinuity sets. *International Journal of Rock Mechanics and Mining Sciences* 43 (7): 1052–1061.

Llc, A. 2016. Agisoft PhotoScan Professional Edition. Version 1.2.

Riquelme, A.J., Abellán, A., Tomás, R. & Jaboyedoff, M. 2014. A new approach for semi-automatic rock mass joints recognition from 3D point clouds. *Computers & Geosciences* 68: 38–52.
Riquelme, A.J. 2015. Uso de nubes de puntos 3D para identificación y caracterización de familias de discontinuidades planas en afloramientos rocosos y evaluación de la calidad geomecánica (PhD Thesis). *Universidad de Alicante.*
Rocscience Inc. 2017. Dips User Manual. Version 7.0.
Roncella, R., Umili, G. & Forlani, G. 2012. A novel image acquisition and processing procedure for fast tunnel DSM production. *International Archives of the Photogrammetry, Remote Sensing and Spatial Information Sciences* 297–302.
Slama, C.C., Theurer, C. & Henriksen, S.W. 1980. Manual of Photogrammetry. *American Society of photog rammetry*. Falls Church, Va., United States.

Tunnels and Underground Cities: Engineering and Innovation meet Archaeology, Architecture and Art, Volume 3: Geological and geotechnical knowledge and requirements for project implementation – Peila, Viggiani & Celestino (Eds)
 ISBN 978-0-367-46583-4

Design and construction of TBM tunnel lining crossing active fault in Thessaloniki Metro Project, Greece

E. Gavrielatou, D. Alifragkis & E. Pergantis
Structural Engineering Department, ATTIKO METRO S.A., Athens, Greece

ABSTRACT: During construction of the Thessaloniki Metro Project, the TBM-bored tunnels had to cross an active geological fault. This case study presents the design and construction of a special segmental lining that was implemented. The possible movements of the fault were evaluated according to geological investigation data. After examining alternative solutions, the Projects Authority ATTIKO METRO S.A. concluded to proceed with using an innovative system of seismic joints between the tunnels' concrete rings. 3D FEM analysis was used to obtain tunnel stresses and displacements due to the estimated fault movements. It was shown that boring the tunnels through the fault with EPB-TBMs was feasible but required certain modifications of the segments' steel reinforcement and the use of special seismic connectors. Extensive laboratory tests verified the proper behavior of the proposed system. Fabrication and installation of the special segments in the tunnels have been successfully completed.

1 INTRODUCTION

The city of Thessaloniki is the second largest city in Greece behind the capital Athens and has a population of about one million people in its greater area. The necessity of an underground transport system (metro) had been evident since decades. After a long period of back-and-forth tendering efforts, the construction of the Main Metro Project commenced in 2006, under the administration and supervision of the Project Owner Authority ATTIKO METRO S.A. The Project comprises one Metro Line 9.5 km long that covers mainly the city center, thirteen (13) modern center-platform stations, and a 50,000 square meters Train Depot. The Line consists of two independent single-track tunnels constructed mostly (7.7 km) by means of two Tunnel Boring Machines (TBM) and secondarily (1.8 km) by the Cut and Cover method. In 2013 this Metro Line was decided to be extended, under a separate Project Contract, towards the suburb of Kalamaria by an additional length of 4.8 km twin single-track tunnels, as well as with another five (5) new Stations.

As the geological study by Pitilakis, (2008), had shown, the TBM bored twin tunnels were to come across an active fault "Pylaia Fault" at two locations, one at the Main Project and the other at the Kalamaria Extension. During the design stage of both Projects, a thorough geological investigation and assessment study was executed. The investigation plan consisted of a set of boreholes along the Line, in situ and laboratory tests, in situ measurement of water pressure and water table.

Strong earthquakes, something not rare at the region, are associated with movements of faults which may cause failure or severe problems on the Project structures made of reinforced concrete such as widespread cracking, leakage of ground water and tracks misalignment.

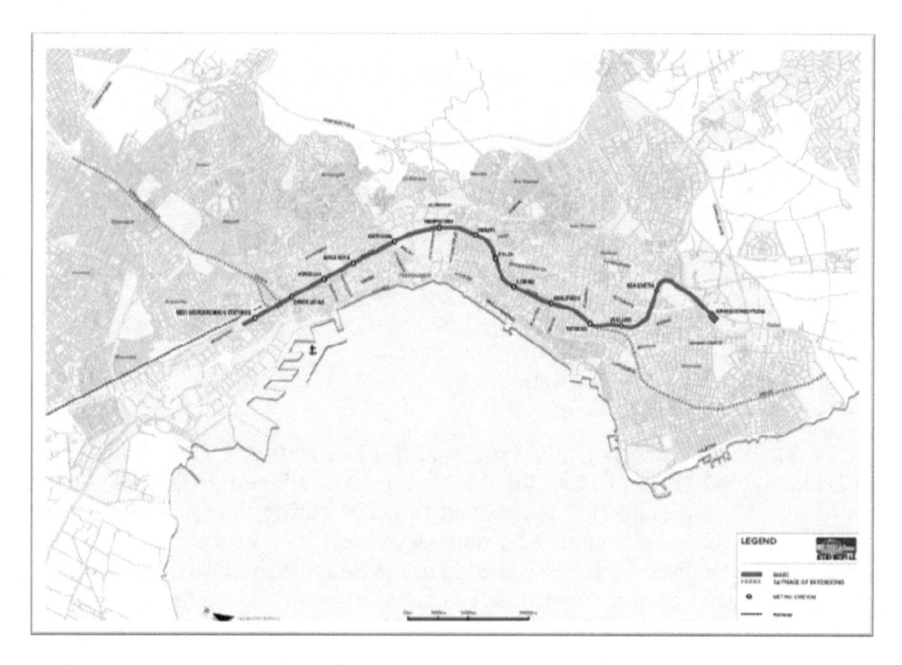

Figure 1. Plan view of Thessaloniki Metro Main Project and Kalamaria Extension.

2 GEOTECHNICAL CONDITIONS

“Pylaia Fault” is located at the Eastern part of the Thessaloniki city and is normal with E-W direction, has a total length of 9 km (Figure 2). This fault is not associated with any known earthquake in the past but it is characterized as active by Zervopoulou and Pavlides (2008) on the basis of seismotectonic data.

Based on empirical relationships Wells and Coppersmith (1994) the magnitude of the mean expected earthquake in case of a surface rupture of 6 km length is Mw=6.0, while in case of activation of each individual segment of length 3 km, which is considered the most likely, the mean expected magnitude of earthquake is Mw=5.6.

ATTIKO METRO S.A. commissioned an extended study to Pitilakis (2008), according to which the potential fault plane (discontinuity) was defined in terms of slope, dip and geometry; the fault plane is assumed to cross the longitudinal tunnel axis perpendicularly, under a

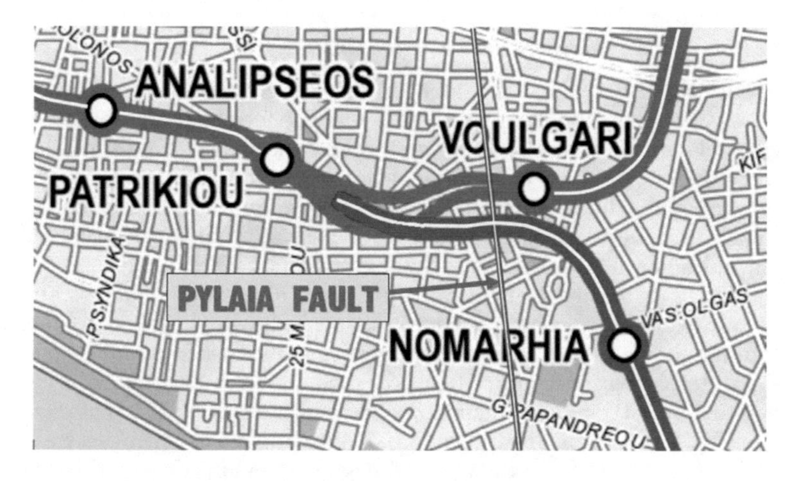

Figure 2. Location of Pylaia Fault on the plan view of the Projects.

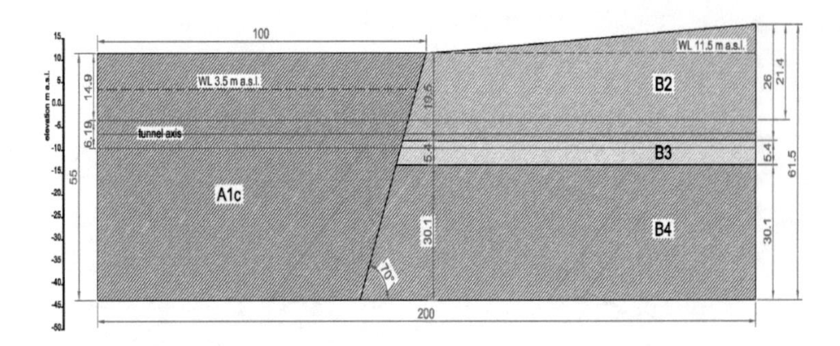

Figure 3. Stratigraphy around Pylaia Fault.

dip angle of 60° or 70°. The potential coseismic displacement is 15cm for the Operational Design Earthquake (ODE) and 25cm for the Maximum Credible Earthquake (MCE). ODE corresponds to a seismic event that is expected to occur during the lifetime of the structure (100 years), while MCE corresponds to a quite severe earthquake scenario of higher return period. It should be emphasized that, according to the established knowledge of faults tectonics, these displacements correspond to the estimated average slip of the fault and that the offset at the ground surface may be much lower and in some cases it may be not visible at all. There are several cases where the slip at the fault has not manifested at the surface as shown for example in the 1995 Kozani, Greece earthquake.

The stratigraphy of the ground around the fault can be simplified as shown in Figure 3. The first layer of 2m depth is Man Made Deposits (MMD). This unit covers both sides and has been neglected due to its small thickness and the fact that it is not involved in tunnel construction. Unit A1c consists of sandy clays (CL) of low plasticity, with clayey/silty sands (SC-SM) with gravels. Unit B2 consists of lacustrine phase Neogene formations of medium to high plasticity and sandy silt (ML), as well as sandy flexible silt (MH), which exhibit strength reaching the levels of very weak siltstone, at certain positions. Unit B3, found under Unit B2, is mainly calcareous silty/clayey sand (SC-SM) and dense to very dense calcareous silt (ML) with sand, of low plasticity. Finally, Unit B4 is formed by semi - rocky formations, such as claystone and siltstone. This Unit is assumed to be the bedrock of the tunnel section investigated. The ground water level is encountered at about 10 m above tunnel axis before the "Pylaia" active fault, and at about 17 m above tunnel axis after the fault towards Voulgari Station.

3 PROBLEM DEFINITION - ALTERNATIVE SOLUTIONS

3.1 *Modification of Project layout - Tunnel lining data*

According to the initial design of the Main Project, the geological discontinuity-potentially seismic fault–was crossing "Voulgari" station. To avoid intractable problems, for example severe structural damages, in case of a fault movement, the station had to be relocated to the "stable" part of the discontinuity (Unit B in Figure 3). To this end, the discontinuity (Pylaia fault) was now crossing, almost perpendicularly, the twin tunnels connected with the station. The twin tunnels are of circular cross-section with excavation diameter equal to 6.19m; the precast segmental lining has an internal diameter of 5.3m and the segments are 0.3m thick (thus external lining diameter equal to 5.9m); the backfill (grouting void) is about 0.15cm; second phase trackwork invert is not structurally connected with the precast segments.

Each ring of the tunnel lining consists of five segments plus the key and is 1.50m long. The lining segments are of precast reinforced concrete, class C40/50 per EN 206-1, EN 1992-1-1 and the Greek Concrete Technology Standard while reinforcement consists of reinforcing bars class B500C per the Greek Standard ELOT 1421-3. Watertight elastomer gaskets are provided all around each segment to ensure the required degree of waterproofing while specific limits are set - and lab tested - with regard to offset, gap and max water pressure. Three

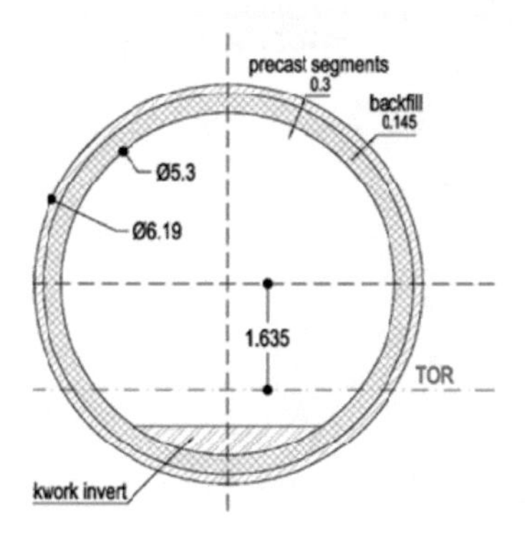

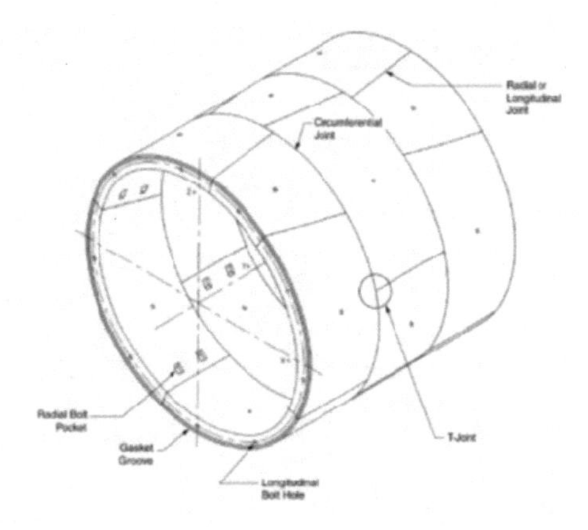

Figure 4. Schematic sections of tunnel.

circumferential joint bolts for each segment are used to fasten and keep the joint between two adjacent rings closed, while two bolts in radial joints ensure the gasket required compression during installation (Figure 4).

The initial design provided for the tunnels to be mined by the use of Earth Pressure Balanced Tunnelling Boring Machine (EPB-TBM). The two tunnels were to be driven with the same direction of advance, from "Voulgari" Station to "Patrikiou" Station. (Figure 1). After the relocation of Voulgari station, the two TBM tunnels could experience considerable displacements due to a potential activation of the fault. If an offset occurs along the fault influence zone, the tunnel lining will potentially suffer damages which will require repairs, and the degree of damages depend on the total offset amount and on the width of the zone where it develops. Besides, one should bear in mind that after such an incident, not only will the tunnel lining be damaged, but the railway tracks will be misaligned too.

3.2 *Alternative tunnel construction*

Following common practice, ATTIKO METRO S.A first investigated the possibility, instead of boring the tunnels by the TBMs as was the original plan, to construct two enlarged section NATM tunnels at the North side of "Voulgari" Station which would cover the length of the fault influence zone (the term "NATM" is herein used in a simplified manner and stands for tunnel mining method with conventional mechanical means).

The internal diameter of these NATM tunnels would be equal to 6.50m in order to include, with an adequate clearance, the inner TBM tunnel section (internal diameter=5.30m). They would be designed to promote a relatively discrete structural break during fault displacement development and to minimize the volume of damaged lining section to be repaired, and the tracks could be easily rearranged (Figure 5). Considering the great increase in the construction time and cost for the enlarged NATM tunnels and its results on the Projects Schedule, as well as the small expected fault displacements, ATTIKO METRO S.A decided to examine other alternatives.

3.3 *Finally chosen solution*

According to FHWA-HRT-05-067 (2004) if fault movements are small and/or distributed over a relatively wide zone, it is possible that a tunnel segmental lining may be designed to accommodate the fault displacements by providing articulation of the precast segments using ductile joints. The idea is to allow the tunnel to distort into an S-shape through the fault zone without rupture and with repairable damage. The closer the joint spacing, the better the performance of the liner. This approach is feasible for relatively small fault displacements. Design of a lining to accommodate fault displacement becomes more

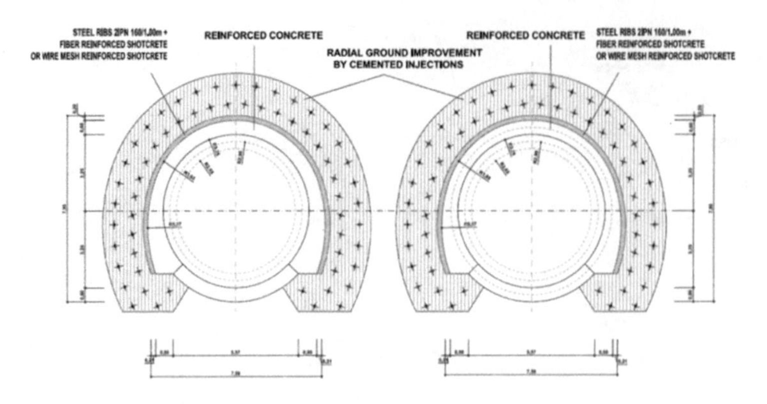

Figure 5. Enlarged NATM tunnel section with final liner to accommodate fault displacement.

feasible in soft soils where the tunnel lining can more effectively redistribute the displacements. The requirements for tunnel water-tightness must also be evaluated when considering the use of movement joints.

With these considerations ATTIKO METRO S.A. commissioned the Contractors to undertake a detailed investigation on the use of the already applied segmental lining, but with special seismic joints and reinforcement.

4 SEISMIC ANALYSIS & RESULTS

With the aim to obtain realistic stresses and displacements of the tunnel, three-dimensional numerical FEM analyses were carried out for each project. The analyses were performed by the Contractors' design companies, i.e. ROCKSOIL S.p.A., Italy, for the Main Project and OMIKRONKAPPA Consulting, Greece, for Kalamaria Extension.

Soil was modelled with 10-nodes tetrahedrons solid elements with a Mohr-Coulomb constitutive model; tunnel was modelled with 6-nodes triangular plate elements with an elastic constitutive model; fault plane was modelled with 12-nodes triangular interface elements. Some important aspects were investigated with several preliminary FEM analyses, which were the influence of the resistance to slip of the active fault plane, the influence of the interaction between tunnel and soil and the efficiency of a deformable backfill between tunnel and soil.

Tunnel was modelled as a continuous pipe and, to take into account the presence of radial joint between two adjacent segments, a reduction of the thickness of the segments, as suggested by Wood (1975) was applied; to take into account the presence of circumferential joint between two adjacent rings a reduction of the stiffness of tunnel in longitudinal direction, adopting an anisotropic plate model, was made. In order to obtain a representative range of results, four different values for stiffness were considered. Results of these analyses showed that the fault movement had effect in both predominant directions:

- a general increment of stresses occurs in transversal direction for a total length of about 60m
- an inflection of the longitudinal axis of the tunnel occurs that produce mainly tensile and compressive axial force in longitudinal direction.

These results allowed us to conclude that the realization of the tunnel with precast segmental lining through "Pylaia" active fault was feasible, but would request certain modifications of the segments.

Regarding the first aspect, the increment of stresses may be dealt with increasing the segment reinforcement that will ensure safety in regard to collapse and ensure the shape of the transversal section of the tunnel. The main transversal reinforcement resulted to be about 350% increased, while the cross reinforcement did not demand any modification. (Table 1)

Table 1. Increase of segments reinforcement in section between Voulgari and Patrikiou Stations.

Region	Main Reinforcement B500C (inner and outer reinforcement)	Cross Reinforcement B500C (inner and outer reinforcement)
Regular segments from Voulgari to Patrikiou Stations (expect fault zone)	8 Ø 18	15 Ø 10
Special segments in fault zone (tunnel length approx. 60.00m)	12 Ø 28	15 Ø 10

As regards the second aspect - and particularly the tensile forces - the best approach was to permit a displacement between adjacent rings through the use of special seismic joints.

Structural Analysis concluded that structural failure is avoided under both seismic scenarios (ODE and MCE). Regarding water inflow into the tunnel, this can be secured - in ODE scenario only - following a simple reparability methodology.

For the segments to be installed at the area of the Pylaia Fault, a new system of longitudinal connections was designed while the radial bolts do not need to be modified. During the possible seismic events, both in ODE and in MCE, the two ground masses crossed by the tunnels slip along the fault at predetermined values of displacement and, therefore, the lining is required to be able to bear those induced displacements.

5 SPECIAL JOINTS IN SEGMENTAL LINING

Due to the fact that the standard circumferential joint bolt connection is not able to bear the required displacements, a composed scheme of seismic devices and connectors has been developed. Limit values of the gap and the offset between two adjacent rings have been proposed as follows: a) in ODE case, a gap equal to 10 mm and an offset equal to 15 mm, and b) in MCE case, a gap equal to 18.5 mm and an offset equal to 30 mm. The bearing displacement scheme between the rings of the tunnel lining is realized by the following steel elements:

- a device of a ductile tensioner linked to four springs system that bear the tension longitudinal stress allowing the design displacements of the joints foreseen in ODE and MCE (Figure 6),
- a pinned plate that spreads the compressive force induced by the ductile tensioner on the segment (Figure 7, 8),
- a seismic shear connector that allows the shear stress transition between two adjacent rings. This device is able to support shear in ODE and MCE as foreseen in the design without interferences with the calculated displacements of the seismic devices with springs (Figure 9),
- a centering pin that reduces installation errors and does not bear any kind of stress in seismic and static conditions (Figure 10).

The above mentioned scheme was manufactured by FAMA S.p.A., Italy, a specialized designer and manufacturer of accessories for TBM tunnels. This scheme was exclusively designed for the specific Project of Thessaloniki Metro. Anti-corrosion protection of the metallic elements is to be applied by appropriate galvanization coating.

6 LABORATORY TESTING OF THE SCHEME ELEMENTS

In order to verify the proposed scheme, ATTIKO METRO SA. in 2016 demanded a program of laboratory tests for the certification of the device (ductile tensioner and springs system) as well as of the seismic shear connector The testing of the device included two phases. The first phase was performed in the I.S.I.S. Malignani Laboratory and aimed to test the steel device under tensile force only and furthermore to tensile force combined with the displacements foreseen in the ODE and MCE design conditions. The goal of these test was to define the geometry and dimensions of the elements and to identify their technical specifications. In the

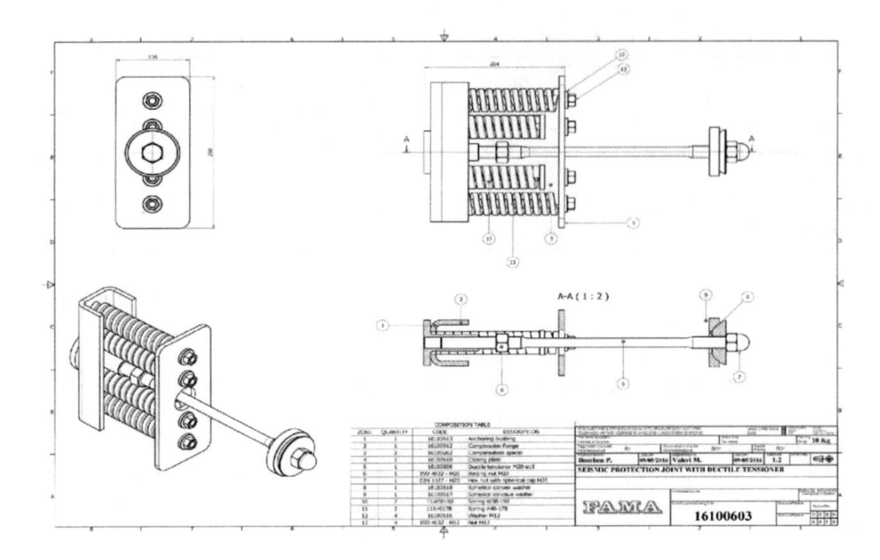

Figure 6. Ductile tensioner and springs system.

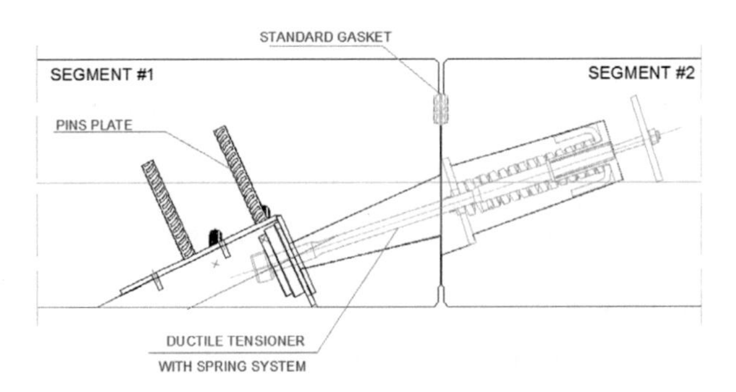

Figure 7. Transversal section between two rings.

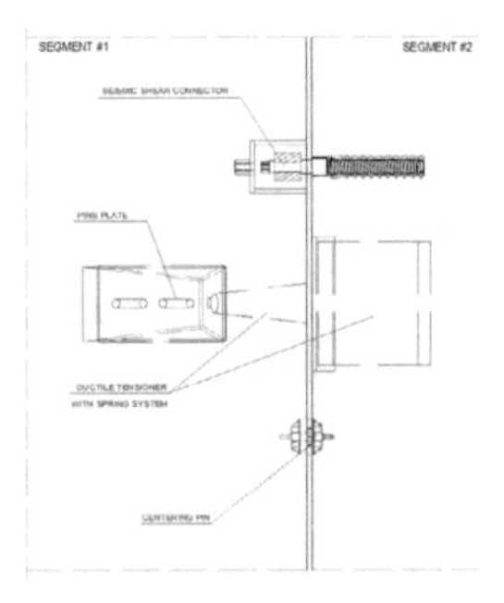

Figure 8. Longitudinal section.

second phase tests, in order to simulate the real conditions, the device was embedded in concrete blocks having reinforcement detailing and concrete properties similar to the precast segments used for the tunnel lining (Figure 11). Two target level of gap and offset were defined: a) in ODE: offset equal to 15 mm and gap equal to 10 mm, and b) in MCE: offset equal to 30 mm and gap equal to 18.5 mm. The tests were performed at the Laboratory of the Tunneling Engineering Research Centre of the University of Rome “Tor Vergata” and the results are presented in Figures 12 and 13.

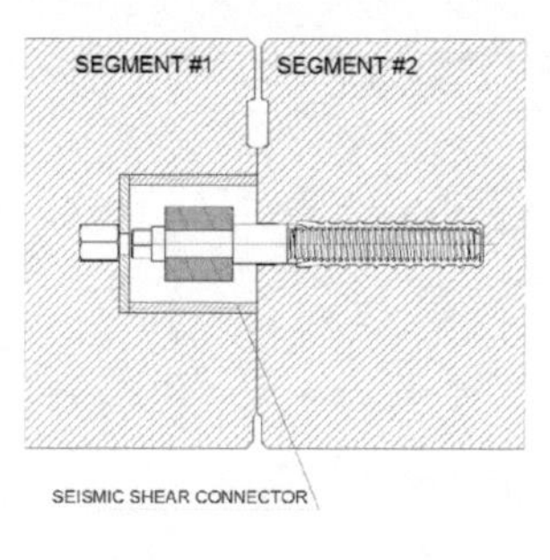

Figure 9. Seismic shear connector.

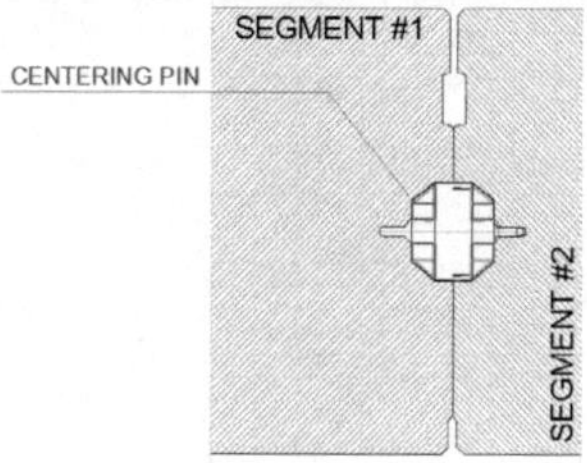

Figure 10. Centering pin.

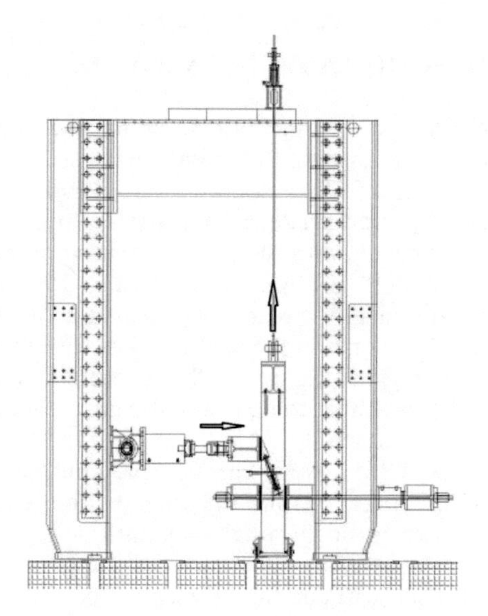

Figure 11. Second phase Test set-up.

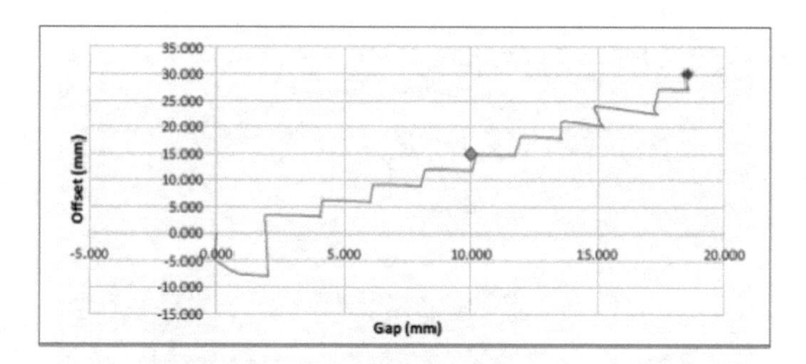

Figure 12. Experimental curve of applied offsets-gaps (targets: green-ODE, red-MCE).

The shear connector was tested also in CATAS S.p.A laboratory under ODE and CDE conditions.

All the above tests showed that the elements of the proposed scheme could safety undertake the anticipated fault displacements.

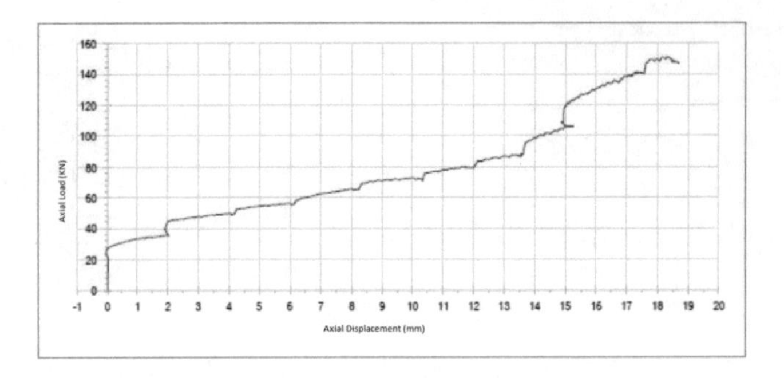

Figure 13. Axial load vs. axial displacement (gap) curve.

7 SEGMNENT CONSTRUCTION AND INSTALLATION

The above mentioned scheme was manufactured by FAMA S.p.A., Italy, a specialized designer and manufacturer of accessories for TBM tunnels. This scheme was exclusively designed for the specific Project of Thessaloniki Metro. Anti-corrosion protection of the metallic elements was applied by appropriate galvanization coating.

Production of the special segments with the seismic joints took place at the prefabrication factory ARMOS PROKATASKEVES AE, Thessaloniki, for both Projects in 2016-17 (Figure 14). This factory produced all the regular segments for the Project as well. One set of moulds - out of the eight available ones in the factory - was transformed to accommodate the seismic joints for the special segments (Figure 15). The process was reversible in the sense that this transformed set of moulds was able to be restored back to normal segments production. Indeed, one mould set was transformed (in FAMA S.p.A., Italy) for the first Project Contract in 2016 and restored after the TBM passing through the fault, and the same set was again transformed (in the Thessaloniki factory) for the second Project Contract in 2017 and put again in normal production after passing through the fault.

These special segments underwent all the prescribed factory tests successfully and were afterwards installed in the tunnels without any problems. The whole length of the tunnels for both Project Contracts at the area of the fault is already completed.

Figure 14. Production of special segments at ARMOS PROKATASKEVES AE factory.

Figure 15. Seismic joints installed in the mould at ARMOS PROKATASKEVES AE factory.

8 CONCLUSION

This paper provides an overview of the development of a set of special segments for passing "Pylaia" active fault in Thessaloniki Metro Projects (Main Project and Kalamaria Extension).

Based on geological evaluation, the predicted values of the tunnel displacements due to possible fault movements were established under two possible earthquake scenarios. After examining alternative solutions, the responsible authority ATTIKO METRO SA concluded that tunneling through the fault and constructing a segmental lining was feasible and preferable in terms of construction time and cost.

However, instead of the regular segments used along the Projects tunnels, the analysis showed that the segments to be used across the fault would require increased transversal reinforcement in order to ensure structural safety and secure the tunnel's transversal shape. Additionally, special seismic joints had to be introduced to permit displacements between adjacent lining rings. These joints were designed, manufactured and tested exclusively for these specific Projects taking into account the geometry of the tunnels, the site geology and the fault characteristics.

Good cooperation between the authority, the contractors, the designers and the manufacturer led to timely decisions and the satisfactory implementation of a successful solution for a challenging problem. The experience gained from these Projects will certainly be an asset for the future when planning tunnels under similar geological conditions.

ACKNOWLEDGMENTS

The authors would like to thank ATTIKO METRO S.A., both the administration and colleague engineers, for the encouragement and support during their drafting the present paper.

REFERENCES

Caputo, Chatzipetros, Pavlides, Sboras, 2012. The Greek database of seismogenic sources (GreDaSS): state of the art for northern Greece, Annals of Geophysics

Muir Wood, 1975. The circular tunnel in elastic ground, Geotechnique, 1

Pitilakis, 2008. Evaluation of the spatial distribution of the permanent ground displacements at the level of the twin tunnels due to potential fault crossing.

USA Dep't of Transportation, 2004. Seismic Retrofitting Manual for Highway structures: Part 2- Retaining structures, slopes, tunnels, culverts and Road ways, FHWA-HRT-05-067.

Wells and Coppersmith, 1994. New empirical relationships among magnitude, rupture length, rupture width, rupture area and surface displacements, Seism. Soc. Am. Bull.

Zervopoulou and Pavlides, 2008. Neotectonic faults in Thessaloniki urban complex, 3rd Greek Congress of Seismic Mechanics.

Tunnels and Underground Cities: Engineering and Innovation meet Archaeology, Architecture and Art, Volume 3: Geological and geotechnical knowledge and requirements for project implementation – Peila, Viggiani & Celestino (Eds)

3D geological modelling for the design of complex underground works

F. Giovacchini, M. Vendramini, L. Soldo, M. Merlo, D. Marchisio, G. Ricci & A. Eusebio
Geodata Engineering, Turin, Italy

ABSTRACT: Modern planning and design for complex infrastructures provides the conception of a robust geological conceptual model that accounts for all the involved parameters, their intrinsic variability and associated risks. Geodata Engineering has undertaken an innovative approach by systematically implementing 3D geological models to suit this demand. An implicit modelling engine developed for the Civil Engineering industry is used to find a unique solution that honors both the provided dataset and the geoscientist interpretation while mimicking natural geological geometries. This paper analyses methods and workflows used to integrate 3D implicit modelling and 2.5D explicit modelling. The key advantages of this approach are examined through relevant examples. Better visualization of available data brings improved understanding, robust analysis and dynamic interpretation. The modelling tool of choice not only expedites the first iteration of a 3D geological model, but it allows a faster updating process and the construction of multiple versions to test several interpretations. This process leads to improved precision and a reduced geological risk, becoming a milestone of the Geodata Risk Analysis Based Design (GRBD) approach. In addition, it is explained how 3D geological modelling provides an innovative communication tool for internal and external collaboration among professionals with a focus on the integration of 3D geological models with engineering designs from BIM.

1 INTRODUCTION

Geologic conditions are the greatest source of unknowns prior to actual construction of underground excavations, especially for deep and large tunnels in rock. It can, therefore, be argued that the geological model has a commanding impact on the entire design and risk evaluation process. It is therefore necessary to conceive a robust Geological Conceptual Reference Model to minimize risks and costs.

Several authors (Venturini et al., 2001; Knill, 2002; IAEG Commission C25, 2014; Soldo et al., 2014, Riella A. et al., 2015) have progressively proposed the concept of Geological Reference Model (GRM) as a framework capable to fulfil these necessities. The GRM is built progressively, with the contribution of different specialties (e.g. geomorphology, structural geology, etc.), it is based on natural laws with experts' judgement. In many projects 3D models can be very useful in assessing baseline geological and geotechnical reference conditions, hazards, risks and in defining the proper mitigation measures.

The Geological 3D modelling is an industry standard in mining and oil & gas, but it is just starting to grow in the civil engineering industry.

Subsurface information is intrinsically incomplete since sampling is not as dense as necessary to solve uncertainties and underground conditions are heterogeneous. It is therefore necessary to use specific software to construct 3D models. This must be able to give 'geological sense' to geometries using custom algorithms. A 3D model is a tool to store and process the input data, to understand, define, quantify, visualize, or simulate a certain aspect of geological conditions. It requires interpretation and refinement of the input data as well as a defined workflow and must be built by a geologist. 3D models allow to see how the layers interact in 3D which the human brain cannot process from a 2D geological map.

One of the main purposes of a 3D geological model is the conception of a robust geological model that leads to improved precision and reliability of the reference base model for underground works design.

2 SOFTWARE

3D geological modelling software needs to be capable of handling complex geological information, from irregularly spaced data sets (e.g. from boreholes and/or surface exposures).

The use of Geological Modelling tools and methodologies in building the Design Geological & Geotechnical Model stands among the foundation milestones of the design procedures in Geodata Engineering.

Recently, the deployment of Leapfrog Works by Seequent has brought interesting outcomes. The software's Implicit Modelling Engine is based on a specifically designed algorithm (FastRBF™, Fast Radial Basis Functions).

2.1 *Basic Principles*

There are two methods that are used in geological 3D modelling software; Implicit and Explicit modelling. *Explicit Modelling* provides that the modeler defines geological structures such as lithologies, faults, folds and veins by drawing them on regularly spaced sections and joining them. The data offers constrains for the structures, for example in the form of borehole stratigraphic data projected onto the cross section, but the process is fundamentally based on drawing and the reasoning is based on 2D elements that are then joined together in a 3D space.

In *Implicit Modelling* computer algorithms directly create the model from a combination of measured data and user interpretation. Input data is evaluated against a mathematical function, by following geologically meaningful rules and parameters that are set up by the user for a specific geological setting. The interpolant is suitable to construct surfaces using discrete variables such as lithologies as well as continuous variables such as geotechnical parameters. The result is a unique solution that honors both the provided dataset and the geoscientist interpretation while mimicking natural geological geometries.

The advantages of implicit modelling compared to explicit modelling are multiple (Cowan et al. 2002, Cowan et al., 2004):

- the result of the interpolation is the smoothest surface available that is the most suitable solution for geological modelling;
- just a limited amount of digitization is necessary;
- evaluating thousands of points takes seconds therefore incorporating a huge amount of contact points is not a time-consuming process;
- editing is easy and fast, it entails digitization of polylines, points and measurement discs over existing surfaces or chosen sections with any orientation;
- geometrical constraints in the form of rules as a fault terminating against another are easily applicable.
- the improved speed of the process compared to explicit modelling leaves more time for the actual data conditioning and interpretation process.
- the model building and updating process becomes an iterative process;
- updates or production of multiple versions are sped up significantly.

The used interpolant is based on FastRBF™. Radial Basis Functions describe the predicted value at a point X as a function that depends on the distance from the point where the value is known. RBFs are a group of global interpolation methods where the interpolant is dependent on all data points. Fast RBF interpolant uses advanced algorithms to solve the combination of Radial Basis Functions and evaluate the estimates quickly. The interpolation of surfaces depends on the distance from the points where values are known.

2.2 *Input Data and Interpretation*

Leapfrog Works can import several forms of input data that are listed below.

- Boreholes and every form of associated data as interval data and point data; this can include rock type distribution, natural defects, values from in situ or lab tests or calculated indexes such as Core Recovery, RQD, GSI, RMR, etc;
- Elevation grids;
- Surface data from geological surveys (geological contacts, discontinuity and foliations trend or orientation);
- Remotely sensed data and ortophotos
- Geophysical data;
- Scanned Cross Sections;
- GIS data as shapefiles;
- CAD data as .dxf o .dwg.

3 KEY POINTS OF THE MODELING APPROACH

When building a geological model, there are a few key rules to keep in mind. It is fundamental to have a clear final purpose. A good workflow does not involve all the available data, instead it focuses on modelling the key elements to clarify doubts and concerns that have arisen in the preliminary evaluations.

Another key element of a successful modelling process is to keep the complexity to the minimum to have a fast and agile elaboration process. It is also important to model only over the area of interest. Radial Basis Functions tend to "flatten out" in the areas where constraints from data of interpretation are not present making the model less geologically reliable.

A basic version of a geological model in Leapfrog Works can be built based on a small amount of input data as a geological map and surface elevation detail; this can guide the first evaluation on the area and help in further investigations planning.

4 3D MODELLING CASE HISTORY IN GEODATA ENGINEERING

In the following paragraphs some GDE case histories are described.

4.1 *Example A – 3D Geological Modelling - Hymalaian base Tunnel (India)*

The first case history is a long base tunnel (approximately 12 km long), part of the improvement of the service and safety condition of the Manali – Sarchu Road, connecting Lahaul district, in Himachal Pradesh, to Ladakh in Jammu and Kashmir, India. The tunnel is located underneath one of the highest Himalayan passes (el. 4890m a.s.l.), yearly closed for long periods because of the extreme climatic conditions.

Figure 1. The planning for the next decade provides the construction of several long base tunnels in the extreme Himalayan environment, requiring an appropriate approach to the identification of the Design Geological and Geotechnical Reference Model (photo Geodata).

The bedrock complexity is the result of the juxtaposition of geologic features - formed during a multi-phase depositional and deformation history - including variable sedimentological structures, unconformities, folds, faults and joints, both at regional and local scale. The state-of-the-art approach entails the geological study and it has been supported by *Leapfrog Works* 3D modelling.

A key item for the project is the understanding of the complex relationships and geometry of some major faults. The fractured rock masses reveal a complex deformation history related to the interaction of a pervasive ductile and brittle deformation phases that took place in different temperature (T) and pressure (P) conditions.

The rock masses exhibit an intense degree of fracturing, somewhere concentrated, without large outcrops of fault related rocks (gouges or cataclasites). The complex and scattered geometry of this faulting-related fracturing, together with the extent of the areas covered by Quaternary deposits made the field recognition and characterization of the fault zones difficult. This is particularly relevant for the slope debris fan and moraines that cover most of the valley downstream the region where literature references map the presence of thrust planes.

The use of 3D analysis has been of paramount importance providing a powerful stand-alone environment for data integration, cross-section construction and 3D model building, making possible the understanding and validation of tectonic geometries otherwise hidden. The following figures show some examples of the analyses on some relevant structures.

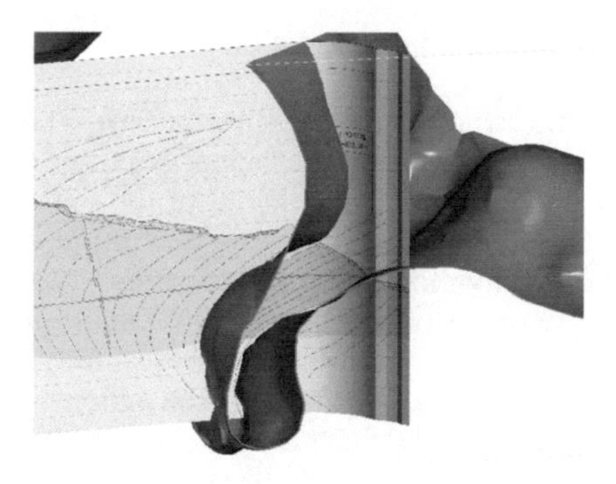

Figure 2. modelling folded Cambrian Units (purple surface).

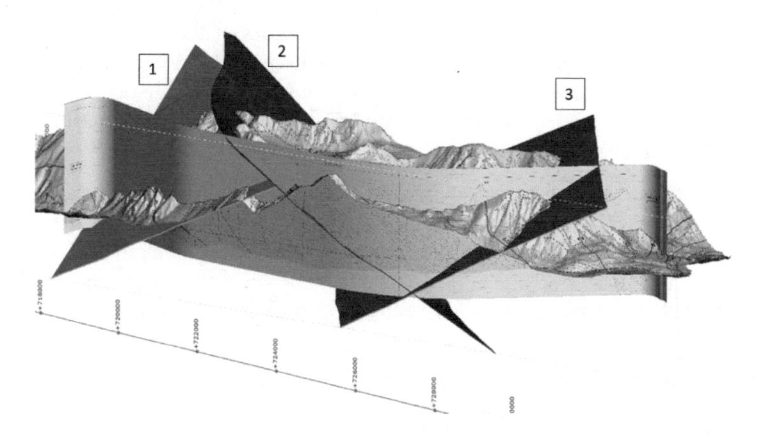

Figure 3. the modelled frontal thrust with a double possible interpretation: (1) the pink plane with southward trend and (2) the red plane south verging; the back thrust is showed by the purple plane (3), exposed without the DTM.

4.2 *Example B – 3D Geological Modelling - Topo River Valley (Ecuador)*

The geological 3D model presented in this section summarized the results of the geological and structural studies carried out along the Topo River Valley (Ecuador) to assess the suitability of the valley for hydroelectric power plant design and construction. Because of the lack of a reliable local geological model, the studies were performed to enquire and represent the geological layout of one of the most complex area of the country. This location shows brittle and ductile deformation with scarce outcrops and a widespread quaternary cover.

The geological setting consists of different tectonic units including metamorphic and Mesozoic to Tertiary sedimentary geological formations (Hollin, Napo, Tena Formations). Most common rock types include fine grained phyllite, slate, shale with interbedded meta arenite and quartz-arenite. Granite intrusions locally cut the whole sedimentary sequence. Rock mass are strongly deformed, faulted and folded. Regional cleavages, tight to isoclinal folds, major to local fault planes both intersecting and affecting rock masses formed a composite geological scenario to investigate and assess the suitability of the area for the project.

Due to the difficulties to get access to the crucial sectors of the valleys (caused by steep morphology and vegetation cover, presence of private proprieties, distance from the roads, etc.), the geotechnical investigations were limited to very small areas. For this reason, extensive geological surveys produced detailed geological maps, regularly spaced cross sections and a longitudinal profile along the Valley axis where the main underground works were foreseen.

Based on the outcomes of the performed ground investigations, the model was built according to the following steps:

1. Uploading all the available input data; elevation surface data, geological map and cross sections images (Figure 4);
2. Digitizing the geological and structural contacts was a crucial step in this workflow because of the complexity of the local geology and the lack of borehole outcomes in several relevant areas. The digitized 3D polylines represent therefore the only constraint to build the 3D surfaces and, consequently, lithology volumes are given in Figure 5;
3. Once the contacts were correctly modelled on the assumed geometries and reciprocal relationships (stratigraphic pile, deposition, erosion, vein, intrusion) it was possible to get into the next step and deal with tectonic elements;
4. Faults could therefore be modelled as surfaces whose action will split the volume in fault blocks that can be managed separately. Each fault block was checked independently, making sure the correct displacements were shown as well as stratigraphy elements activated;
5. This was followed by modelling the faults as 3D elements to show the expected fault affected zone and to quantify its volume (Figure 6);
6. All the geological elements were included in the final block model (Figure 7) that shows the interpolated geology (surface, volume) in different sectors of the valley. Furthermore, it has been possible to extrude the local geology (3D block volume) along design structural elements with the aim to assess the expected lithologies and the possible occurring hazards to be considered. It was therefore possible to give a relatively reliable preliminary estimation of the volumes by exporting the properties table.

In this case, 3D modelling allowed designers to define the most credible geological scenario of the area. The model included different geological elements (stratigraphic units, brittle and ductile tectonics structures, geomorphologic elements, etc.) and their mutual interactions with internal and external geodynamic processes. Ultimately, the model highlighted the principal hazards related to unfavorable geological and hydrogeological conditions.

4.3 *Example C – Geotechnical modelling applied to cavern design*

This example deals with the design of a hydroelectric power plant in a complex sedimentary sequence of Carboniferous age. The structural framework shows a complex brittle deformation process where joints and faults associations from a compressive regime have been found

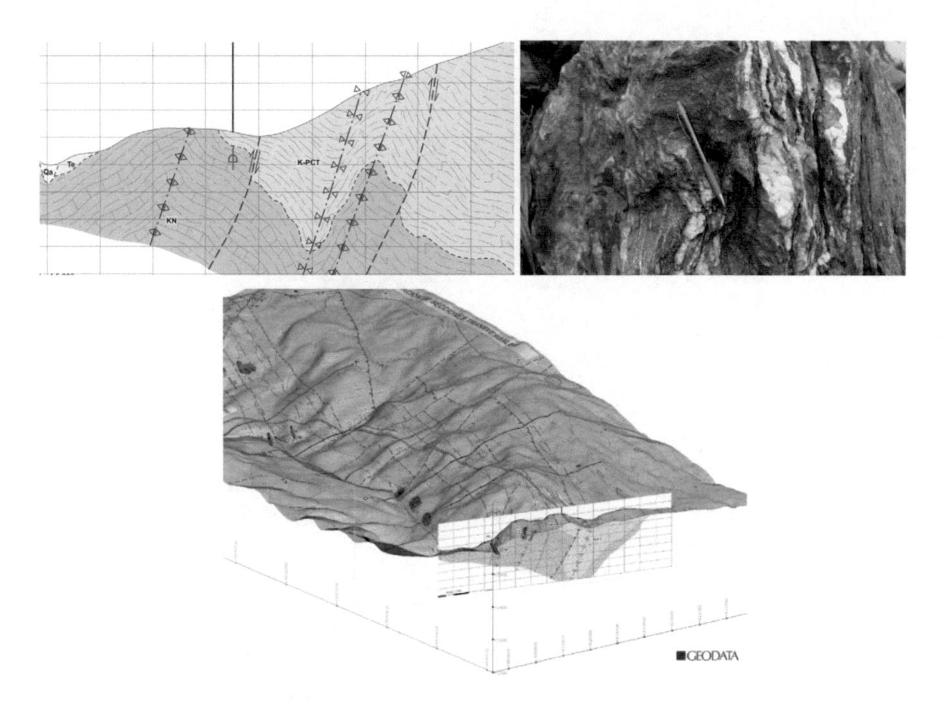

Figure 4. Input data for the 3D geological model represented by a geological map, regularly spaced cross sections and structural measurements. The accuracy of the model depends on the quantity, quality and accuracy of the input data. A detailed geological survey is required for a reliable 2D Baseline model (maps and sections).

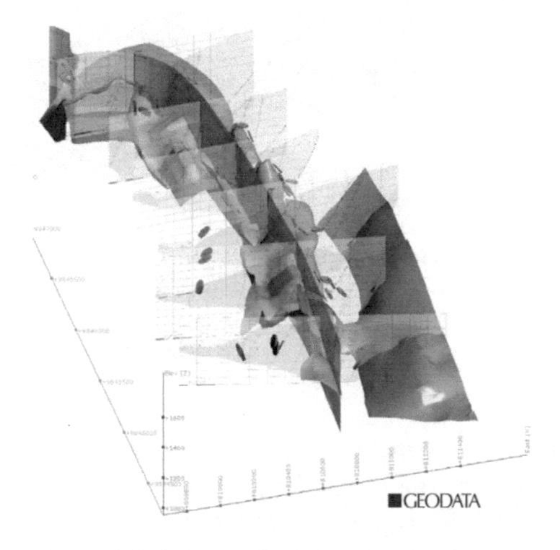

Figure 5. Geological contacts modelled onto a folded structure that shows the effect of deformation (fold and fault displacement). The definition of complex tectonic structures needs to rely on an adequate amount of field data to conceive the most likely geometry and rock types distribution especially at depth. Each measurement represents a "natural" constrain for surfaces (geological contacts) to be modelled.

in the rock mass. The focus is on the caverns design and construction when specific requests are formulated by the engineering teams to obtain input for their numerical simulations. The main target of the model was to provide a 3D geotechnical characterization of the caverns location.

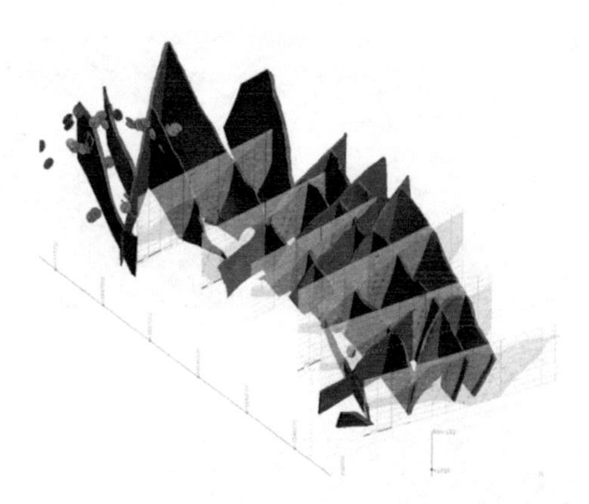

Figure 6. Volumetric model of fault shear zones surveyed in the areas. For each fault system, the main features such as orientation, type of movement and expected thickness were defined. Fault zone excavation is one of the most critical hazards to be faced in tunnel construction; fault zones were modelled as discrete rock mass volumes displacing rock units.

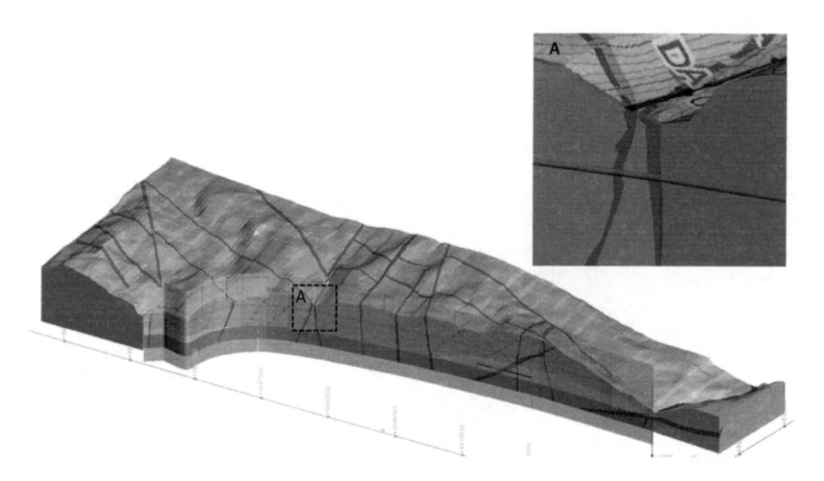

Figure 7. Final geological block model. It includes different volumes displaced by tectonic structures. Geological sections and 3D orientation of tectonic structures can be easily visualized within the study area. On the right-hand corner, a detail showing the expected geology and fault thickness along the tunnel alignment. (A) Fault shear zones are displayed in red, the rock formation in brown.

The 3D numeric model of this area was based on RMR (Rock Mass Rating, Bieniawski, 1989) index data calculated over the boreholes and excavated tunnels. The workflow in this instance included the study of different aspects within the area:

1. A global model of area was made by imposing the effect of anisotropies (stratigraphy, fault zones, master joints, etc.) on the RMR numeric model (Figure 8) that generate the RMR classes distribution. It is important to highlight how the envelope surfaces are considered reliable only in the proximity of the factual data (boreholes, face mapping).
2. A discrete model of fault planes (Figure 9) was made and then refined over the cavern volume, this allowed to study their geometries and used as an input for wedge stability analysis.
3. Faults zones were also modelled as 3D elements (volumes) defining their influence zone (core zone and damage zones). To these horizons RMR class IV was assigned.
4. The global RMR model was refined over the cavern volumes taking into the effect of the fault zones (Figure 10).

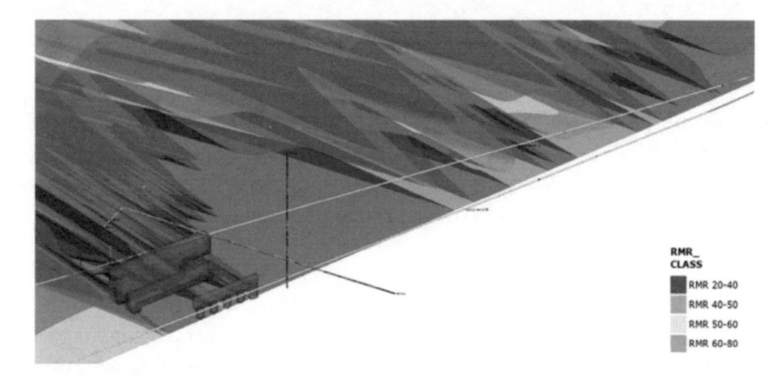

Figure 8. RMR Global model showing the effect of anisotropies in Bieniawsky's classes distribution. In the left-hand corner, it is shown the caverns location.

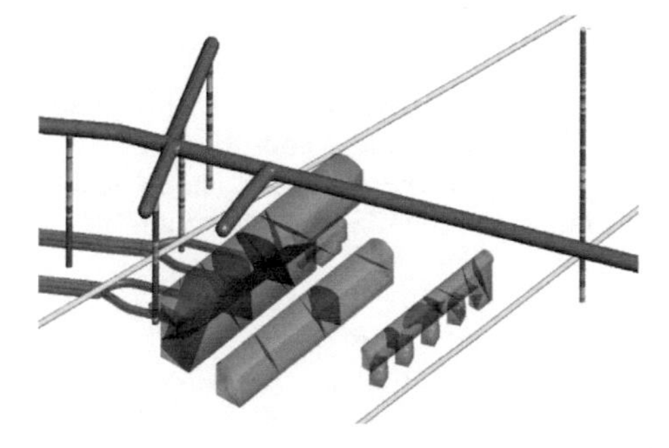

Figure 9. Discrete model of fault planes affecting the caverns excavation. Fault planes were modelled according to face mapping outcomes.

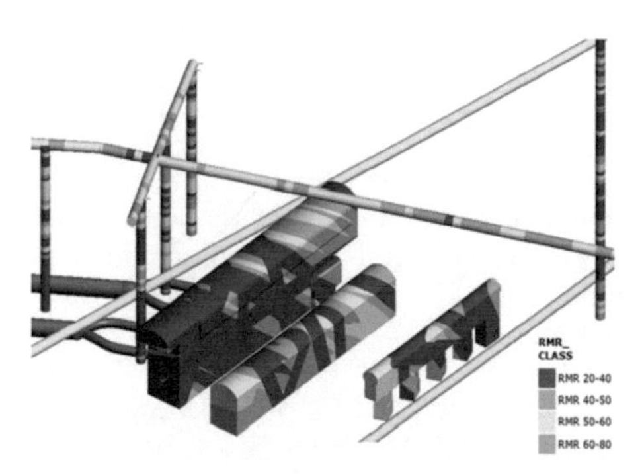

Figure 10. Final RMR local model (caverns) that account for the fault influence volumes.

5 ADVANTAGES OF 3D GEOLOGICAL MODELLING IN UNDERGROUND DESIGN

Geodata Engineering has undertaken a systematic approach toward geological 3D modelling which is now deployed in all the new ongoing projects. Since the earliest projects in which it was applied, the modelling has quickly shown several advantages, listed in the below paragraphs.

5.1 *Iterative process*

The design of a geological conceptual model is inherently an iterative process as the model is built with a little amount of data at the start and as soon as new data comes in it is improved in time. It is also a dynamic object because the design of the infrastructure is modified due to multiple factors as geological conditions and other issues. It is therefore fundamental to be able to update the model in a fast and efficient way to provide a prompt response to the designer's needs.

5.2 *Multiple versions*

The ability to create and process a 3D geological model in a time-efficient manner helps to make different scenarios and test them in different model versions. The capability to calibrate quickly and in progress the most credible geological layout improves the efficiency of the whole design process. This lends to a robust geological reference model that can be used to take decision and find design solutions.

5.3 *Integration with BIM*

The BIM elements can be managed by Leapfrog Works, this significantly helps to investigate the impact of site geological and geotechnical conditions on the design. Infrastructure 3D models from BIM can be imported into the 3D geological Model with a limited post processing effort that consists into simplifying the file to make it suitable for Leapfrog Works.

5.4 *3D Model as a Communication tool*

3D geological models have a great potential as a communication tool. If displayed correctly they help to visualize complex geological geometries that the human brain cannot easily process from a 2D geological map. This means geological reasonings are much more accessible to nonprofessional figures. Moreover, they allow a much better understanding of the geological reference model and its associated risks. A common ground is created, and this improves the collaboration side of the decision-making process.

6 CONCLUSION

The use of geological 3D models is paramount in the approach to Geodata Risk Analysis Based Design (GRBD) since it has a capability to simultaneously store, elaborate and interrogate multiple sources of input data. It can efficiently support designers in the crucial phases both of design and construction process.

Notwithstanding the initial effort in collecting, processing and storing the input data, the consequent advantages of getting a reliable 3D model become consistent during the design process when the interactions among different professionals need a clear and common base of discussion.

he chosen software is a flexible tool because it is useful not only for preliminary and quick considerations, but it allows a faster updating process and the construction of multiple versions to test several interpretations. This can be considered an added value especially in tunnel or underground works design because the construction of a Reliable Reference Model is characterized by an intrinsic uncertainty level.

In these contexts, especially for deep tunnel design the introduction of new data (borehole outcomes exploratory galleries and face mapping) in the model, can deeply affect the interaction among the different elements changing the final interpretation. Extensive sensitivity analysis of the different scenarios are often carried out to assess and mitigate the potential impacts of the hazards. Most of them can be proper identified (distribution of the hazards along the alignment), quantified ("volumes" definition for quantities estimation) and visualized to define the design solutions.

REFERENCES

Bieniawski, Z.T., 1989. *Engineering Rock Mass Classification*. John Wiley & Son.

Cowan, J., Beatson, R., Fright, W.R., McLennan, T.J., Mitchell, T.J., 2002. *Rapid geological modelling*. International Symposium, Kalgoorlie 23–25.

Cowan, E.J., Lane, R.G., Ross, H.J., 2004. *Leapfrog's implicit drawing tool: a new way of drawing geological objects of any shape rapidly in 3D 3*.

IAEG commission 25 S. Parry, F. J. Baynes, M. G. Culshaw, M. Eggers, J. F. Keaton, K. Lentfer, J. Novotny, D. Paul., 2014. *Engineering geological models: an introduction*. Bulletin of Engineering Geology and the Environment, Volume 73, Issue 3, pp 689–706.

Knill, J. (2002). *Core Values: The First Hans Cloos Lecture'*, 9th Congress of the International Association for Engineering Geology and the Environment, Durban S. Africa, 45p.

Riella A, Vendramini, M., Eusebio A, Soldo L., 2015. *The Design Geological and Geotechnical Model (DGGM) for Long and Deep Tunnels*. Springer International Publishing - Engineering Geology for Society and Territory - Volume 6, pp. 991-994.

Soldo L., Vendramini M., Eusebio A. 2014. *The Design Geological and Geotechnical Model (DGGM) for major infrastruc*tures. The contribution of Tectonics and Structural Geology. Lyon, AFTES Congress 2014.

Venturini, G., Damiano, A., Dematteis, A., Delle Piane, L., Fontan, D., Martinotti, G., Perello, P., 2001. *"L'importanza dell'affidabilità del Modello Geologico di Riferimento negli studi per il tunneling"*. Atti del congresso "Geoitalia 2001 3°Forum Italiano di Scienze della Terra" – FIST, Chieti, 5-8 settembre, pp. 426-427.

Tunnels and Underground Cities: Engineering and Innovation meet Archaeology, Architecture and Art, Volume 3: Geological and geotechnical knowledge and requirements for project implementation – Peila, Viggiani & Celestino (Eds)
 ISBN 978-0-367-46583-4

Risk maps for cutter tool wear assessment and intervention planning

J.G. Grasmick & M.A. Mooney
Colorado School of Mines, Golden, CO, USA

ABSTRACT: The risk assessment relating to geological and geotechnical conditions depends on the uncertainty in key parameters. However, understanding and communicating the spatial distribution of the uncertainty and corresponding risk is often a challenge. This paper presents a case study where three-dimensional risk maps were generated using site investigation data and spatial interpolation tools for the interpretation and communication of cutter tool wear risk. In addition, the maps are used to forecast the number of cutter tools requiring replacement during construction, which can serve as a decision aid for intervention planning. Construction data and observation logs are used to validate these maps and demonstrate how they can provide a more comprehensive assessment of tunneling risks.

1 INTRODUCTION

Soft ground tunneling projects often encounter rapid changing ground conditions that pose significant risks. However, deterministic geological profiles fall short of conveying the uncertainty in ground conditions along with the spatial extent of tunneling risks. The ability to model the spatial variability and uncertainty in geological/geotechnical conditions would improve the risk assessment, taking into consideration the spatial distribution of such risks and incorporating uncertainty in the geological profile. Geostatistical (or random field) methods can be employed to assess the spatial characteristics of the geological/geotechnical conditions and develop 3D models of variability and uncertainty (Chiles and Delfiner 2012; Pyrcz and Deutsch 2014).

In geotechnical engineering, spatial modeling of risks (i.e., risk maps) on both the project and city scale have increasingly gained attention in the literature. Huber et al. (2015) presented a spatial modeling approach for the risk-based characterization of an urban building site using geostatistical simulation methods. These maps aid in the communication of the allowable ultimate loads on the soil over the building site. Wang et al. (2017) used CPT data to map the spatial variability in liquefaction potential of soils on a city-wide scale for Christchurch, New Zealand. These risk maps have demonstrated to be effective in the communication of risks and serve as a decision tool for mitigation and planning.

This paper presents an example of employing geostatistical methods to develop 3D risk maps (or models). Site investigation data from the Northlink Tunnel project in Seattle, WA is used to map the risk for cutter tool wear. These risk maps allow for a more rigorous and comprehensive assessment of ground conditions. These maps also serve as a tool for improved communication between all parties on the risks faced in a particular project.

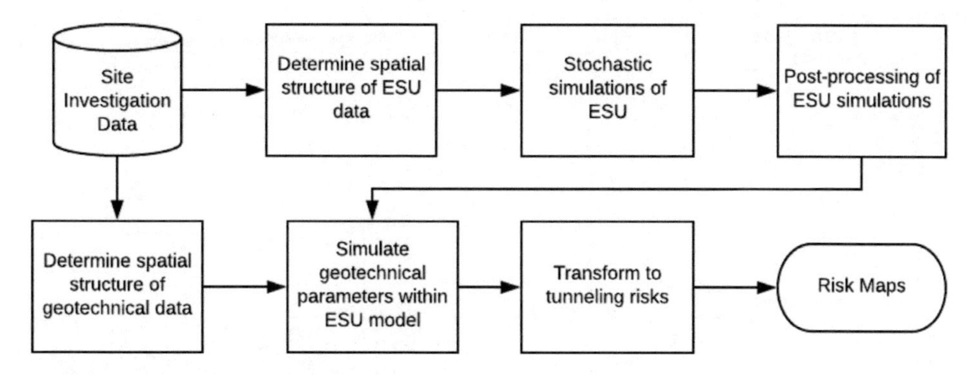

Figure 1. Flow chart of the procedure for developing risk maps.

2 METHODOLOGY

Geostatistical methods are powerful tools used for spatial interpolation of geological features/ geotechnical parameters. The spatial correlation/variability structure of these parameters can be assessed and modeled using functions that quantify the correlation/variability as a function of distance and direction (e.g. auto-correlation functions, variograms, and Markov chains) (Chiles and Delfiner 2012; Carle and Fogg 1996). Both categorical (e.g., soil type) and continuous (e.g., geotechnical parameters) variables can be simulated using such models. Stochastic interpolation tools allow one to generate many statistically equally probable realization of the subsurface ground conditions. Post-processing these realizations derives average, most probable, uncertainty and other statistical measures spatially. The reader is referred to (Chiles and Delfiner 2012; Pyrcz and Deutsch 2014) for a more theoretical explanation of these methods.

In this paper, a transitional probability geostatistical method is used to simulate geology in terms of engineering soil units (ESU) in 3D. This method incorporates the conditional probabilities of a certain category (e.g., ESU) to occur adjacent to the others along particular directions. Proportions of each ESU, average lateral extend and thickness, connectivity and juxtaposition tendencies are also honored. In a sequential framework, conditional simulations are performed to generate many equally probable realizations of the ESU in the project extent. These simulations are conditioned to the site investigation data such that simulated values at locations where ESU samples were reported are equal to the observed ESU. The ESU simulation results (100 total) are post-processed to determine the most probable ESU profile, probabilities of particular ESUs to exist at any location, and uncertainty in the predicted (most probable) ESU based on the variability of simulated ESUs at each location.

Within the ESU model, geotechnical parameters are simulated using the sequential Gaussian simulation approach. In this approach, the spatial structure of relevant geotechnical data is inferred from the borehole data. The simulations are conditioned to the site investigation data as well as the ESU model. This allows one to take into account the different spatial characteristics and distributions of the geotechnical data within each ESU type. These simulation results are post-processed to determine the average and uncertainty in the geotechnical parameter of interest at any location. Figure 1 presents a flow chart of the general procedure for the geostatistical analysis described here.

3 PROJECT OVERVIEW

The Northgate Link tunnels project in Seattle, WA consisted of two (twin) tunnels, each approximately 5.6 km in length, excavated by 6.6 m diameter earth pressure balance tunnel

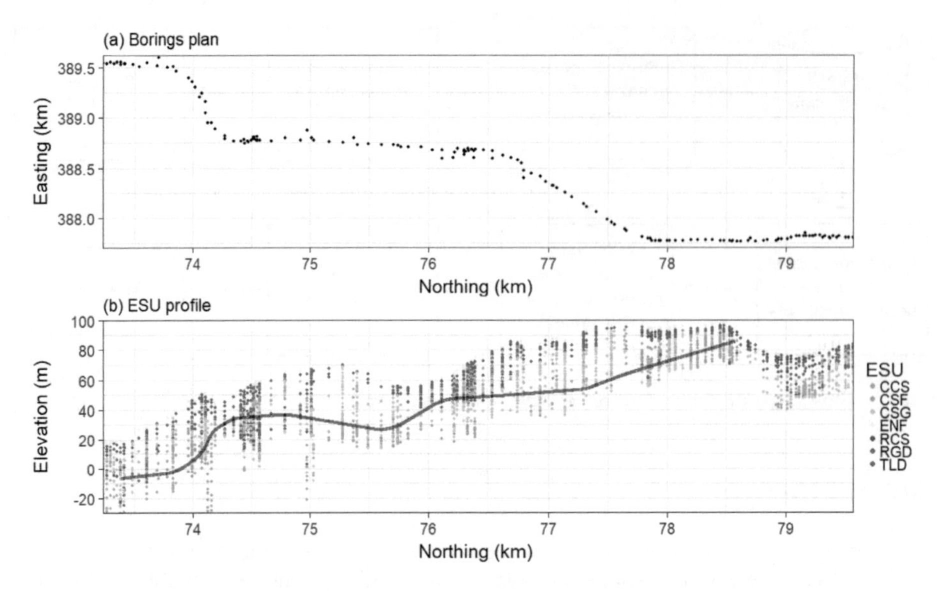

Figure 2. Overview of tunnel site and site investigation data. a) Plan view with borehole locations; b) Profile view of reported ESU samples.

boring machines (EPB TBM). The site investigation consisted of 158 boreholes with varying sampling methods for determining parameters including ESU classification, water content, Atterburg limits, grain size distribution, etc. The majority (58%) of the boreholes are within 50 m of another borehole. These boreholes are typically located at tunnel stations, where a higher density of boreholes is generally present. Along the tunnel drive (i.e., between stations), borehole spacing generally exceeds 50 m, and even 100 m in some cases. Figure 2 presents the

Table 1. Description of Engineering Soil Units encountered in project area.

Engineering Soil Unit	USCS	Description
Cohesive Clay and Silt (CCS)	CH, CL, MH, ML, OH, OL, PT, SC, SM	Hard, interbedded silt and clay. Includes multiple layers and lenses of cohesionless silt, sand, and gravel, with varying lateral extent and thickness.
Cohesionless Silt and Fine Sand (CSF)	ML, SM, SP	Fine-grained granular soil consisting of very dense silt, fine sandy silt, and silty fine sand.
Cohesionless Sand and Gravel (CSG)	GM, GP, SM, SP, SW	Dense to very dense silty sand to sandy gravel. May contain lenses of clay and clayey silt.
Engineered and Non-Engineered Fill (ENF)	CL, GM, GP, ML, OH, OL, SC, SM, SP, SW	Very loose to very dense sand with varying amounts of silt and gravel. Also includes wood, concrete, metal, brick, and other debris.
Recent Clay and Silt (RCS)	CH, CL, ML, OL, PT, SC, SM	Soft to stiff silty clay and clayey silt with variable amounts of sand and gravel in localized zones.
Recent Granular Deposits (RGD)	GC, GM, GP, ML, SM, SP	Loose to dense or locally very dense silty sand, medium stiff to hard sandy silt, and silt. Contains localized lenses of sandy gravel and gravelly sand with varying lateral extent and thickness.
Till and Till-Like Deposits (TLD)	CL, GC, GM, GP, GW, ML, SC, SM, SP, SW	Has a high spatial variability and will grade over short distances from an unsorted mixture of gravel, sand, silt, and clay, to an unsorted mixture of silt, sand, and gravel to clean or relatively clean sand and gravel.

spatial layout of the site investigation and the sampling frequency for ESU classification. The regional geology consists of highly variable glacial and non-glacial sediments and according to the geotechnical baseline report, seven major ESUs are identified and summarized in Table 1 (Jacobs 2013).

4 CUTTER TOOL WEAR RISK MAP

To illustrate risk map development, an example for cutter tool wear is presented. We rely upon tool wear prediction models from Köppl et al. (2015) in this analysis. However, the approach can be applied using a variety of different tool wear prediction models. According to Köppl et al., the rate of cutter tool wear in soft ground tunneling is a function of the soil abrasivity index (*SAI*):

$$SAI = \frac{EQC}{100} * \tau_c * D_{60} \tag{1}$$

where EQC is the equivalent quartz content (%), τ_c is the Mohr-Coulomb shear stress (kN/m^2) and D_{60} is the grain size where 60% of all grains are finer (mm). From the site investigation data relating to relevant geotechnical parameters EQC,τ_c and D_{60}, the distributions of SAI for each ESU can be determined. Boxplots representing the distribution of SAI for each ESU is presented in Figure 3. As expected, CSG and TLD exhibit higher SAI compared to other ESUs due to the presence of gravel and cobbles.

In the geostatistical simulation, 10,000 individual, statistically possible realizations of the ground conditions were generated. The results of the geostatistical simulation is summarized in Figure 4. Here, the average predicted *SAI* and corresponding uncertainty (in terms of standard deviation σ_{SAI}) along the tunnel axis is presented. Examination of the profiles reveals that *SAI* can vary significantly over the alignment, including drastic changes in *SAI* over short (< 100 m) distances. In addition, the uncertainty in *SAI* is quite variable along the alignment. Regions of high uncertainty imply that actual *SAI* is not well understood locally, and is a result of a combination of limited sampling and local variability in the geotechnical data.

The range and average predicted *SAI* within the tunneling horizon (excavated ground) from the risk map is presented in Figure 5. The deterministic estimate of SAI is also plotted for comparison. It can be observed that the change in *SAI* can vary significantly

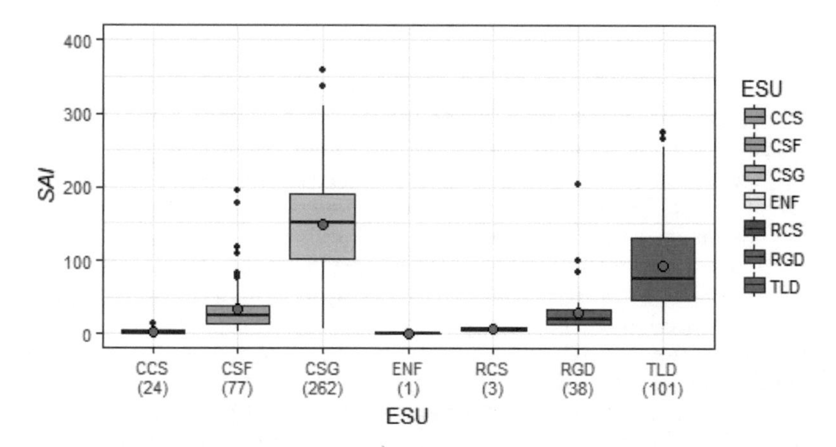

Figure 3. Boxplot of *SAI* parameter for each ESU. The number in parentheses corresponds to the number of samples available from the site investigation.

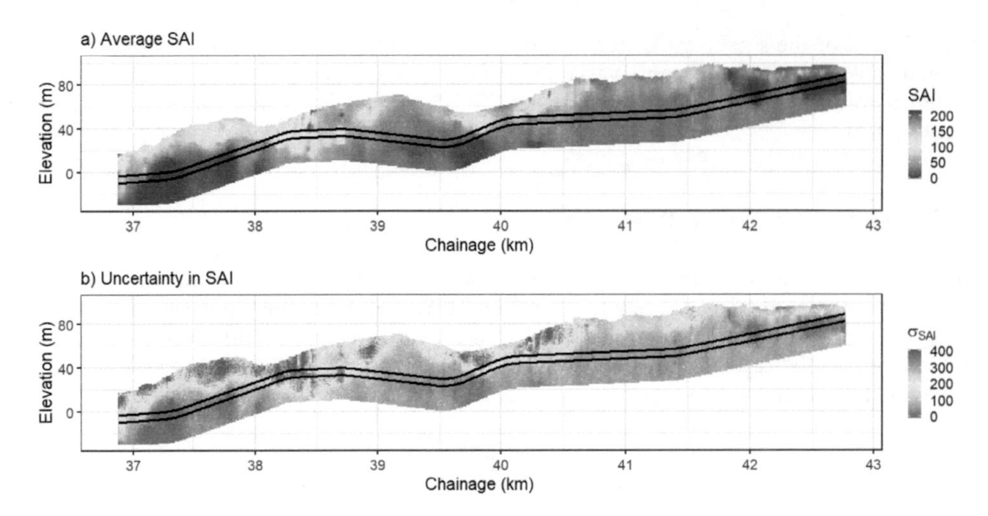

Figure 4. Results of the geostatistical simulation for *SAI*. a) Average *SAI*; b) Uncertainty in *SAI*.

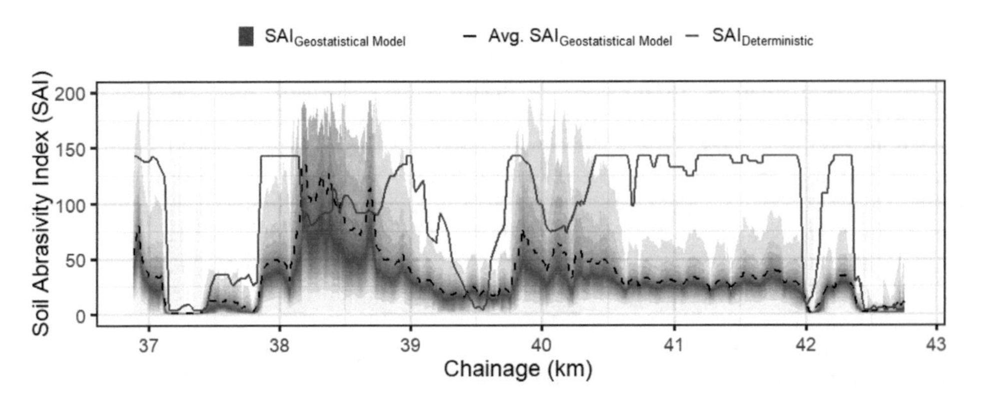

Figure 5. Range and average predicted SAI within the tunnel horizon (crown/invert). The estimated SAI using a deterministic approach based on the geotechnical baseline report (GBR) geological model is presented for comparison.

over ~250 m of excavation. In general, the geostatistical model predicts lower *SAI* compared to the deterministic estimate, with the exception of chainage 38.2–38.7 km, 39.5 km and, to some degree, 39.9–40.2 km. At chainage 40.5–42 km, the deterministic approach predicts significantly higher SAI than the geostatistical model. This is a section of the tunnel alignment where predominately CSG soils are encountered. The deterministic approach utilized the deterministic geological profile in the GBR and takes the average *SAI* of all samples for the respective ESU, rather than considering the regional distribution at any section of the tunnel alignment. The geostatistical model reveals that the average *SAI* in the 40.5–42 km region within the tunnel horizon is actually significantly lower than the average of all *SAI* samples for CSG.

To estimate the number of cutter tool replacements during TBM excavation, we adopt the equations proposed in Köppl et al. (2015). To estimate the expected cutting distance s_c for each of the tools based on the geostatistical model, the predicted *SAI* for each realization of the simulation is used. For each ring, the average s_c within the tunnel horizon for each

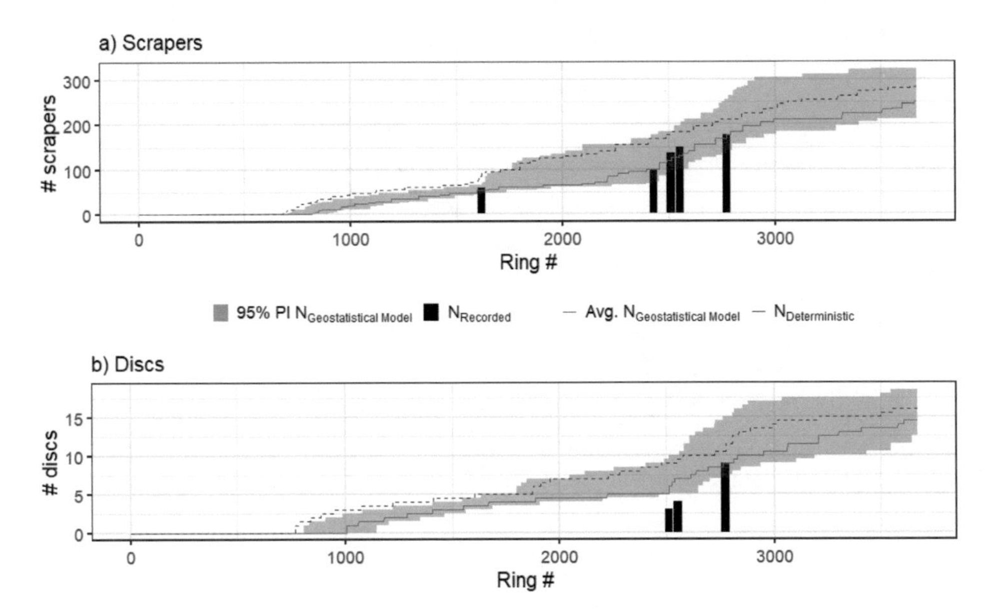

Figure 6. Estimated cumulative tool replacement for (a) scrapers and (b) discs compared with the deterministic estimate and the recorded tool changes during the project.

realization closest to the respective ring is taken and divided by the actual cutting distance s_{cd} for each of the cutting tools (using the average penetration rate p_e for the respective ring) to get the partial utilization factor e_{cd}. Figure 6 presents the average and 95% probability interval (PI) of the cumulative number of tool replacements N vs. ring number according to the geostatistical model. The deterministic estimation according to the GBR geological model is also plotted for comparison. For both cutting tool types, the geostatistical model on average predicts fewer tool replacements than the deterministic model. Furthermore, comparison with recorded tool replacements from the project validates the geostatistical model estimates and demonstrates the approach presented in this paper performs better than the deterministic estimate.

5 CONCLUSIONS

Geostatistical methods have been widely adopted in industries such as mining, petroleum/gas, hydrology and environmental engineering. However, few efforts have been made to extend these tools to heavy civil applications such as tunneling (particularly in soft ground). The example presented in this paper demonstrates the added value of performing geostatistical analysis on site investigation data for a more rigorous/comprehensive risk assessment. The key advantages of applying geostatistical methods are 1) to determine spatial characteristics/trends in the data and 2) quantify uncertainty in ground conditions.

Through validation with field records, the results of the risk mapping presented here demonstrate to provide more accurate estimates of the ground conditions. As a result, estimates of cutter tool replacements are significantly improved. Furthermore, quantifying spatial uncertainty in the ground conditions provides a direct means for quantifying the uncertainty in the required mitigation to account for risks. This information can serve as an aid for improved intervention planning.

The outcomes of the geostatistical analysis and the development of the risk maps can serve as communication tools to key stakeholders of a project. This approach can be applied to a range of other tunneling risks including clogging, ground deformation, and mixed face conditions, among others. 3D models such as those presented in Figures 7 and 8 can aid in

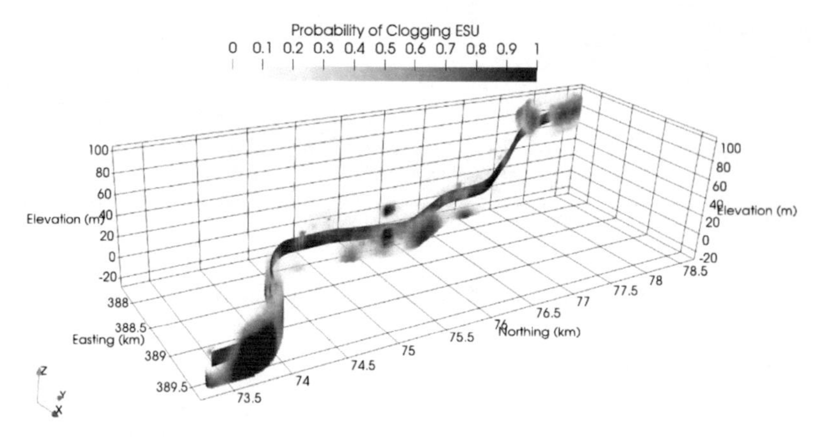

Figure 7. 3D risk map of the probability for clogging ESUs.

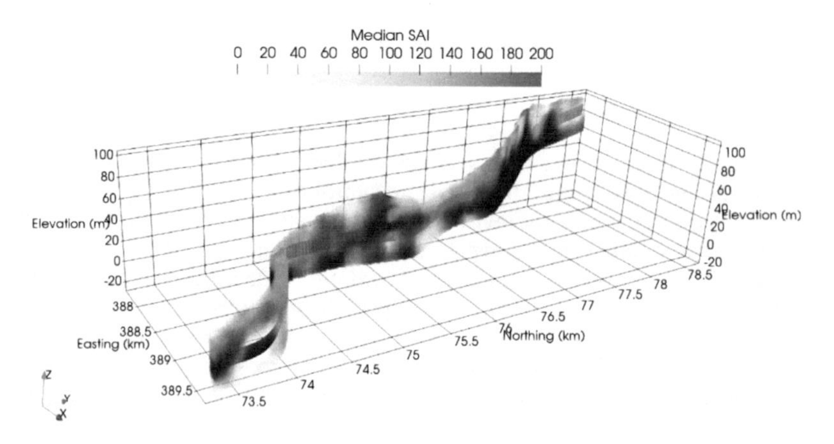

Figure 8. 3D risk map of soil abrasivity index.

procurement, establishing baselines, design and risk mitigation planning. Site investigation data is often under-utilized. Applying geostatistical methods enables us to extract the most information out of the data, and better convey areas where further site investigations and/or mitigation measures may be warranted.

REFERENCES

Carle, S. F., & Fogg, G. E. (1996). Transition probability-based indicator geostatistics. Mathematical Geology, 28(4), 453–476.

Chiles, Jean-Paul, and Pierre Delfiner. Geostatistics: Modeling Spatial Uncertainty. Wiley, 2012.

Huber, M., Marconi, F., & Moscatelli, M. (2015). Risk-based characterisation of an urban building site. *Georisk: Assessment and Management of Risk for Engineered Systems and Geohazards*, *9*(1), 49–56.

Jacobs, A. (2013). "Northgate Link Extension Light Rail Project Contract N125: TBM Tunnels (UW to Maple Leaf Portal) Geotechnical Baseline Report."

Köppl, F., Thuro, K., & Thewes, M. (2015). Suggestion of an empirical prognosis model for cutting tool wear of Hydroshield TBM. *Tunnelling and Underground Space Technology*, *49*, 287–294.

Maidl, B., Herrenknecht, M., Maidl, U., & Wehrmeyer, G. (2013). *Mechanised shield tunnelling*. John Wiley & Sons.
Pyrcz, M. J., & Deutsch, C. V. (2014). *Geostatistical reservoir modeling*. Oxford university press.
Wang, C., Chen, Q., Shen, M., & Juang, C. H. (2017). On the spatial variability of CPT-based geotechnical parameters for regional liquefaction evaluation. *Soil Dynamics and Earthquake Engineering*, *95*, 153–166.

Tunnels and Underground Cities: Engineering and Innovation meet Archaeology, Architecture and Art, Volume 3: Geological and geotechnical knowledge and requirements for project implementation – Peila, Viggiani & Celestino (Eds)

Geotechnical challenges in design and construction of tunnels for Mumbai metro line – 3

S.K. Gupta
Mumbai Metro Rail Corporation Ltd., India,

M.G. Khare & G.V.R. Raju
AECOM, India

ABSTRACT: Mumbai Metro Line – 3 (MML-3) is a 33.5 kilometer long underground corridor presently under construction. The project scope includes twin tube tunneling by Tunnel Boring Machines (TBM), 19 Cut and Cover Stations, 7 Stations, cross passages and cross overs using New Austrian Tunnelling Method. The city of Mumbai was formed by merging of seven islands through multiple land reclamations. Mumbai geology consists of soft ground followed by volcanic rocks such as Basalt, Breccia, Tuff and intertrappean sedimentary rocks such as Shale. The ground water table is shallow. The paper describes challenges including tunneling under mixed ground conditions, water drawdown and associated settlements. The observed settlements during TBM tunneling are analyzed to estimate a range of volume loss and trough width parameters for Mumbai geology. The paper illustrates the benefits of instrumentation and monitoring scheme to understand the ground and groundwater response in urban tunneling.

1 INTRODUCTION

India's urban population is projected to be about 820 million by 2051. The rapid urbanization has led to a high growth in the number of vehicles leading to traffic congestions, higher air pollution, associated health risks, and higher number of road accidents leading to loss of life. Therefore it is necessary to plan public urban transport projects in a sustainable way to support the desired economic growth, protect the environment and to improve the quality of life. The city of Mumbai is the financial capital of India and home to about 22 million people. Mumbai Metro Rail Corporation (MMRC) is constructing the Mumbai Metro Line 3 (MML-3) which is a 33.5 kilometer fully underground corridor through the city center integrating the existing modes of transportation. The Figure 1 shows MML-3 alignment plan. The project scope includes 33.5 kilometer of twin tube tunnels by TBM, 19 Cut and Cover Stations, 7 Stations including cross passages and cross overs to be constructed using NATM. The shallow urban tunneling is being undertaken for the first time in Mumbai geology. The objective of the paper is to present the ground and groundwater response to TBM tunneling which would be of value to the ongoing and future projects of similar nature in similar hydrogeological conditions.

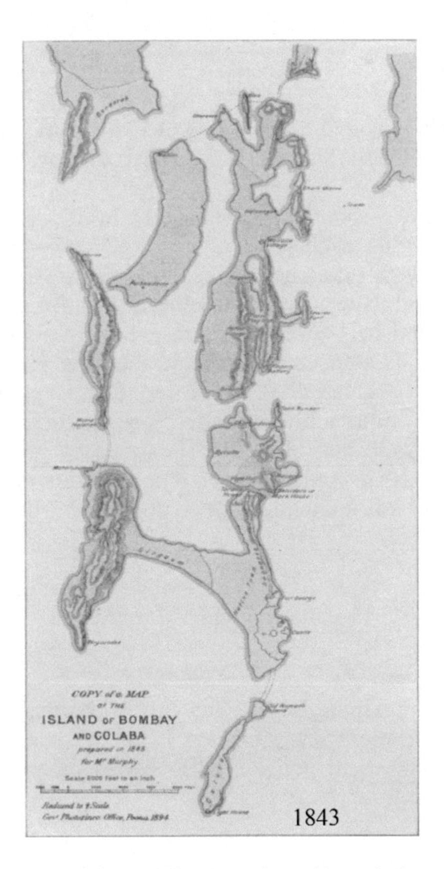

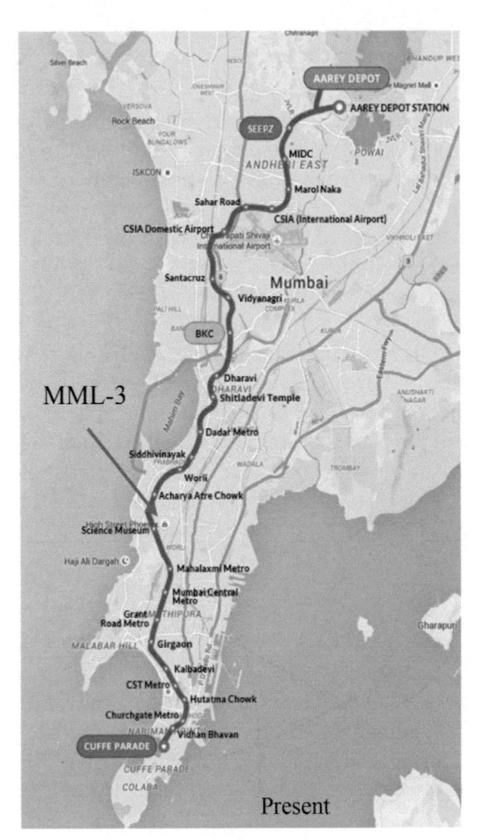

Figure 1. History of reclamation of Mumbai.

2 MUMBAI GEOLOGY

Seven islands of Mumbai off the West Coast of India were merged into one landmass by means of multiple land reclamation projects in 19th and 20th century as illustrated in Figure 1.

The ground investigations show strata of soft ground (predominantly silty, clayey sand, clay of medium to high plasticity and boulders) followed by Deccan trap formation. The thickness of soft ground varies from about 1m to 14m. The Deccan traps were formed about 65 million years ago as a result of various subaqueous lava flows showing a distinct westerly dips. The major volcanic rocks are basalt, breccia, and tuff. Intertrappean sedimentary rocks such as shale are also found which indicate a long period of quiescence in volcanic activity (Sethna, 1999). The ground water table is shallow and rises close to ground level during monsoon.

The ground and groundwater response to TBM tunneling in varying geological conditions of Mumbai is presented in the following sections.

3 SETTLEMENT DATA FROM TBM TUNNELS

The settlement data from TBM tunneling in varying ground conditions which are representative of the Mumbai geology are analyzed and presented in this section. The basis for back analysis of the settlement data is explained here.

The shape of a settlement trough above mining excavations was examined by Martos (1958) and he represented it by a Gaussian or Normal distribution curve.

Later, Schmidt (1969) and Peck (1969) showed that the surface settlement trough above tunnels took a similar form. O'Reilly and New (1982) developed the Gaussian model by making the assumptions that the ground loss could be represented by a radial flow of material toward the tunnel and that the trough could be related to the ground conditions through an empirical "trough width parameter" (k). The model was guided by an analysis of case history data. These assumptions allowed to develop equations for vertical and horizontal ground movements that were also presented in terms of ground strain, slope and curvature (both at, and below, the ground surface). The equations have since widely used to assess the potential impact of tunneling works during the design stage. The base equation is given as:

$$S = S_{\max} \exp\left(\frac{-x^2}{2(kz)^2}\right) = \frac{AV}{(kz)\sqrt{2\Pi}} \exp\left(\frac{-x^2}{2(kz)^2}\right) \tag{1}$$

where S = ground settlement at a point; S_{max} = maximum settlement; A = cross-sectional area of tunnel; V = percentage of ground loss assuming the ground is incompressible i.e. V = V_s/A, where V_s is the volume loss; k = empirical constant also called as trough width parameter; and z = depth of tunnel axis.

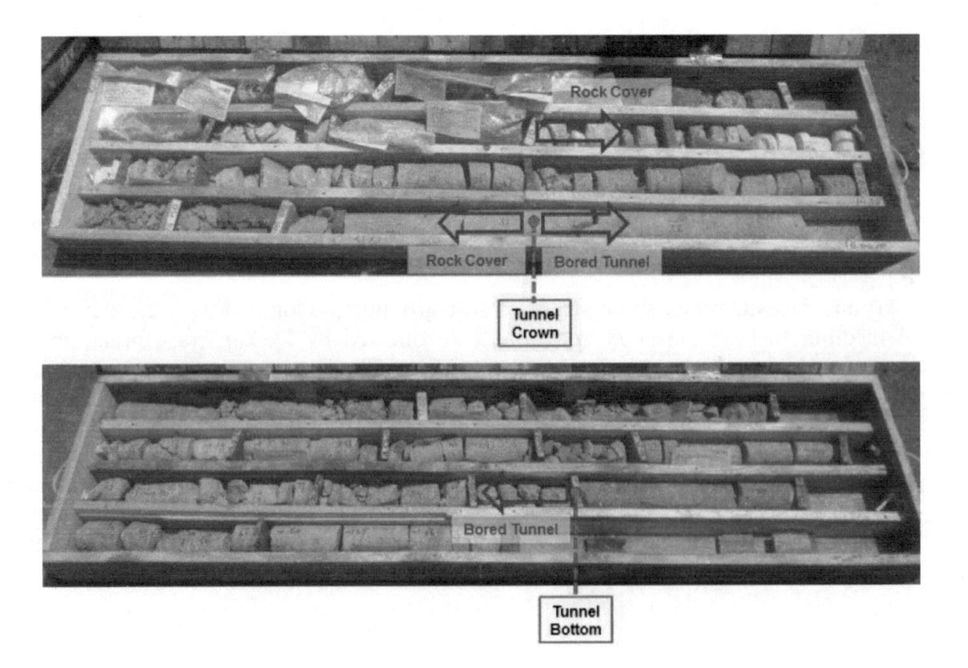

Figure 2. Representative corebox (BH-1) – Site 1.

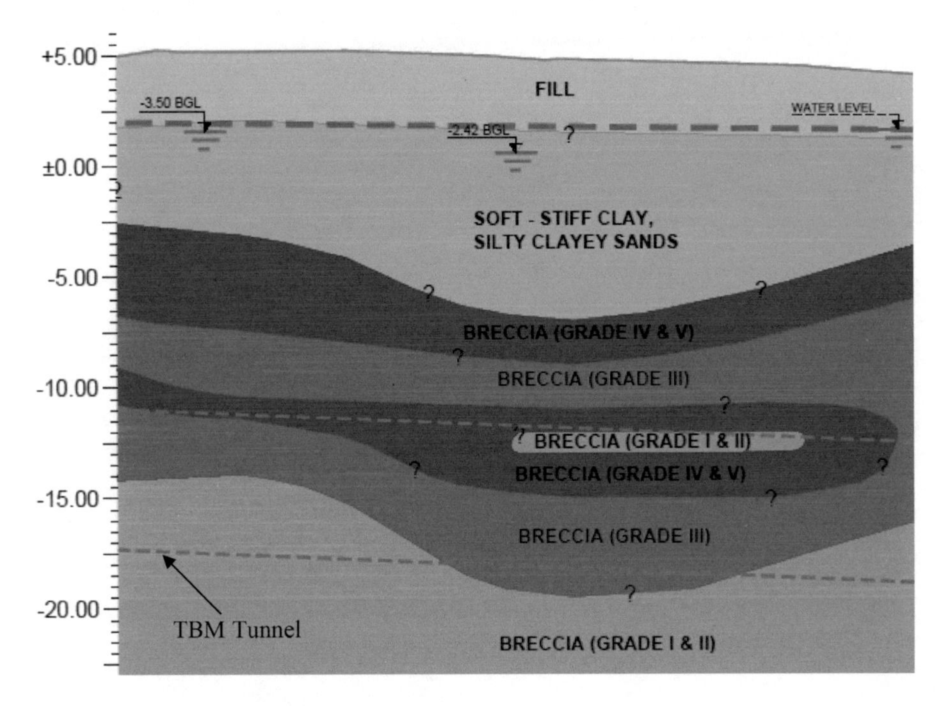

Figure 3. Interpreted Geological Section – Site 1.

The ground settlement induced by twin tunneling is calculated by superposition of two single tunnels. The back analysis of settlement data is carried out at Site 1 and Site 2.

3.1 *Site 1*

The TBM tunneling was completed through a mixed ground comprising Volcanic Breccia rock of Grade III with Rock Quality Designation (RQD) of 40% - 60% and Grade IV with RQD of 10% - 40%, with a small rock cover of about only 0.5 tunnel diameter (minimum value), overlain by layer of alluvial soils. The tunnel depth (measured from tunnel crown) ranges from 15.5m to 16.5m. The groundwater is close to the ground surface. A representative corebox BH-1 from this section is shown in Figure 2. The interpreted geological section is illustrated in Figure 3.

The rock earth pressure balance (EPB) machines were chosen along this section of the alignment primarily because of largely varying geological conditions, surface features such as rivers and the reduced cover to sensitive structures. The design features of the EPB TBM are summarized below:

- Large opening ratio of between 28 and 35%,
- Additional injection ports in the cutter head to enhance soil conditioning,
- Closer spacing than normal for forward probing and ground enhancements,
- Bespoke cutter head design with closely spaced disk cutters, scrapers and peripheral cutters to accommodate the large opening ratio.

The TBM was operated in open mode at the section presented here. The ground movements were monitored using arrays of ground settlement markers at regular intervals. The observed maximum ground settlement after completion twin tube tunnels is only 7.6 mm. The Figure 4 shows ground settlements measured using 4 instrumentation

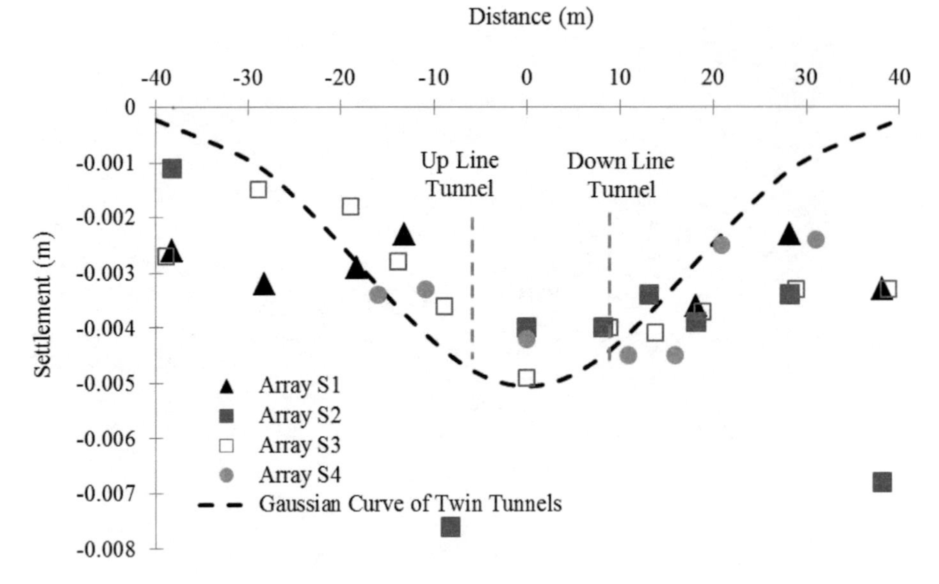

Figure 4. Ground Settlement Profile – Site 1.

arrays with settlement markers installed transverse to the direction of twin tunnel drives.

Using equation (1) and the principle of superposition, the Gaussian distribution curve is fitted for the observed ground settlement data from Site 1 and is shown in Figure 4. The back analyzed value of trough width parameter (k) is 0.7 and the volume loss is 0.3%. These values are similar to the range of k values of 0.5 to 0.7 and volume losses of 0.4 to 0.6 % for tunnels in weak rock, mixed ground condition as reported by Guglielmetti et al. (2008).

3.2 *Site 2*

The site 2 represents mixed ground conditions comprising Basalt rock of Grades I (with RQD 90%-100%) to V (with RQD of 0% to 10%), overlain by loose to medium dense silty sands. Intertrappean Shale of Grades IV (10%-40%) to Grade III (40% to 75%) is observed below tunnel axis. Tunnel depth (measured from tunnel crown) is about 15m below ground level. The weak Shale having compressive strength of 5 MPa to 15 MPa together with high strength Basalt of about 50 MPa compressive strength presents difficult ground conditions for tunnelling. A representative corebox of borehole BH-2 for this section is shown in Figure 5. The interpreted geological section is illustrated in Figure 6. The geological section is further divided into favourable ground (titled as stretch A) and unfavourable ground (titled as stretch B) conditions. The unfavourable ground comprising mixed of weak Shale and strong Basalt is found in a very localised area beyond which the geology again changes similar to that of Stretch A comprising Grade III or better Basalt.

The tunneling in this geology was carried out by dual mode TBM. The machine features are summarized below:

- Heavy duty main bearing and modified cutter head to cater for predominately hard rock,
- Change over from “Open” to “Closed” mode for mixed face conditions,
- Capable of high excavation rates in the predominantly hard rock conditions,
- Ability to “close off” the machine should soft flowing ground, or high water inflows are encountered.

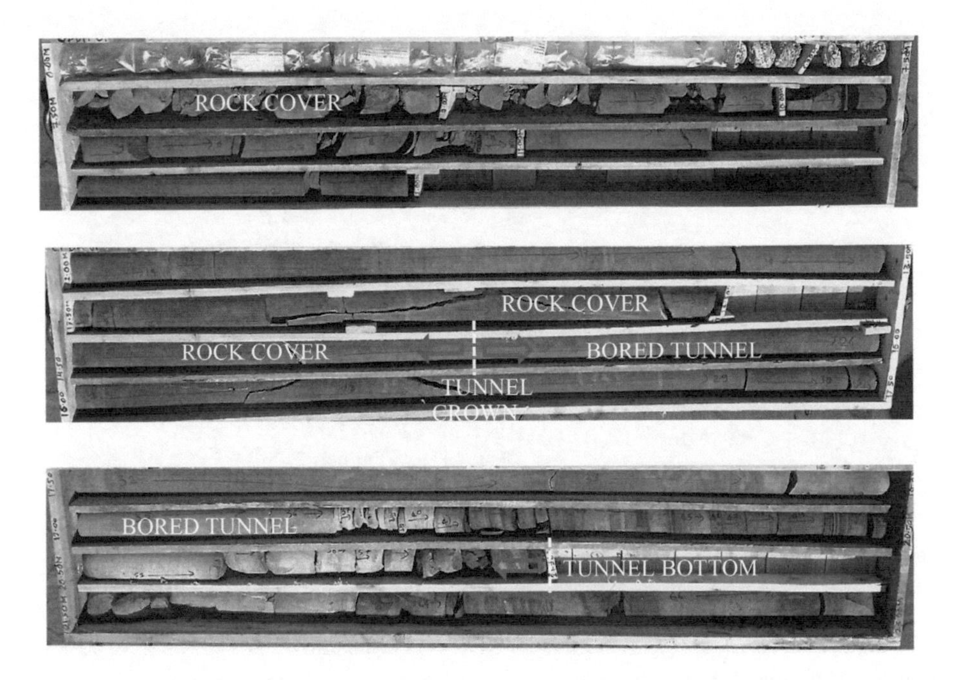

Figure 5. Representative corebox (BH-2) – Site 2.

The tunnel drive in the section was carried out with open mode of operation. The open mode drive was chosen because the tunnels are under open ground and there are no nearby sensitive structures. The tunnel stretch discussed here was initial drive when the TBM operating parameters were yet to be refined and considered as a learning curve for tunneling through such geological conditions. The settlement data for stretch A and stretch B is presented in Figures 7 and 8 respectively. It is observed that the settlement data over the downline tunnel is limited due to the restricted

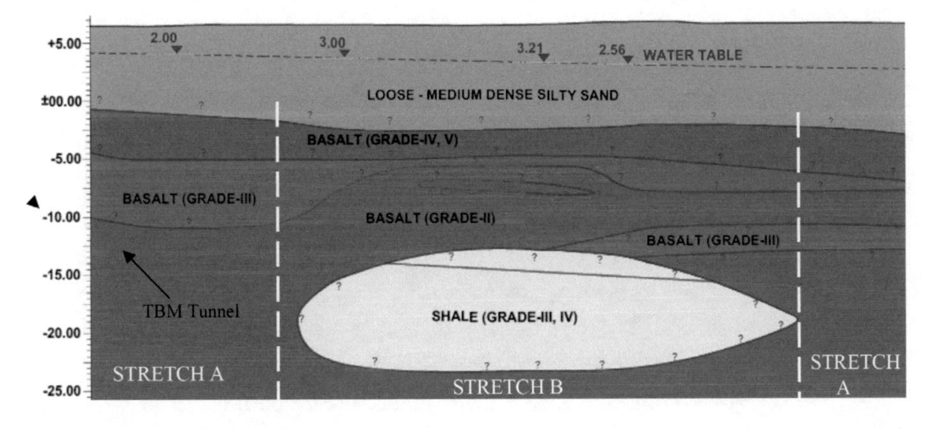

Figure 6. Interpreted Geological Section – Site 2.

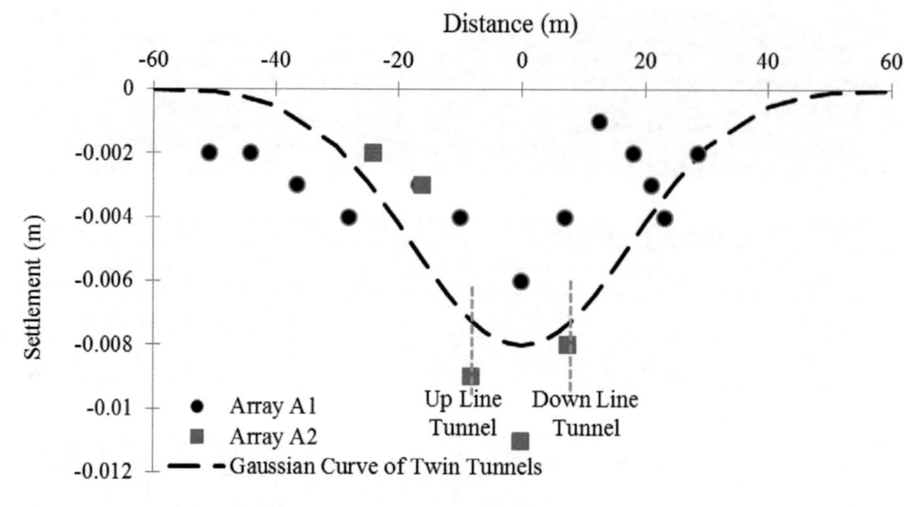

Figure 7. Ground Settlement Profile – Site 2(STRETCH A).

access. It is assumed that the ground response for the both tunnels is likely to be similar.

For the favorable ground conditions referred as the "Stretch A", the maximum observed settlement is 11 mm. The back analyzed value of trough width parameter (k) is 0.5 and the volume loss is 0.5%.

For the mixed ground conditions observed during the stretch B, the maximum observed settlement is 25mm. The back analyzed value of trough width parameter (k) is 0.5 and the volume loss is 1.2 %. The high magnitude of settlements in the Stretch B are attributed to the mixed ground conditions comprising weak sedimentary Shale and strong volcanic Basalt rock observed at the tunnel face. The open mode TBM operation in this geology resulted in

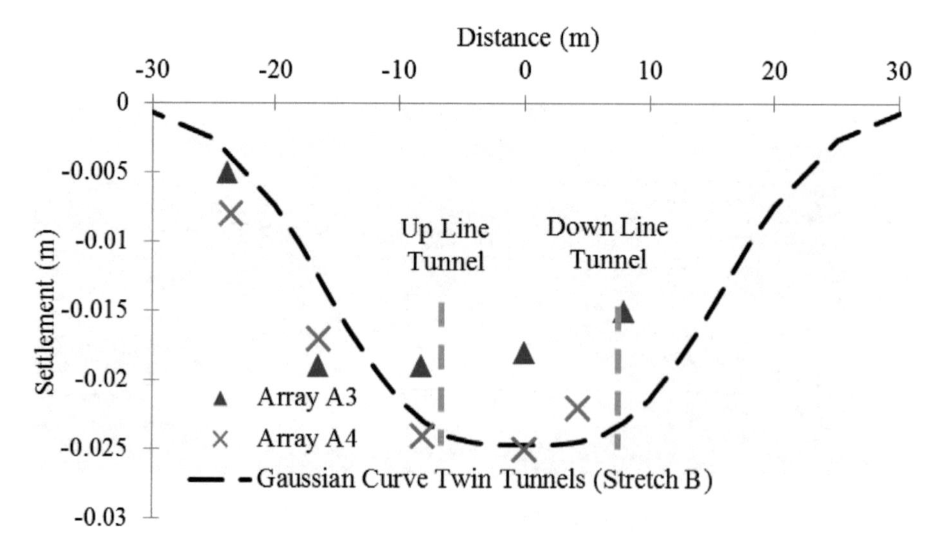

Figure 8. Ground Settlement Profile – Site 2(STRETCH B).

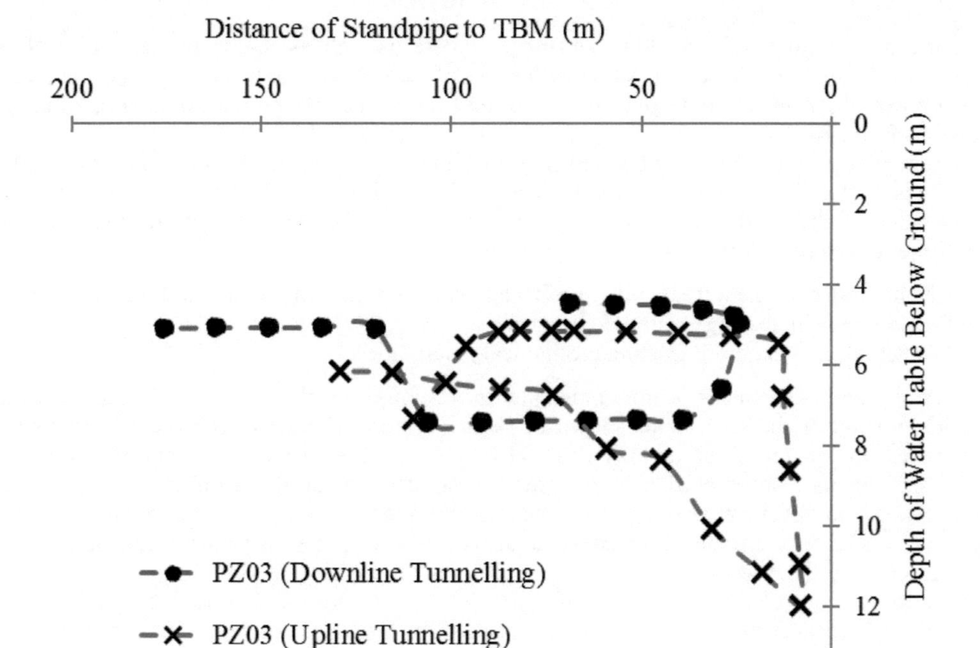

Figure 9. Standpipe PZ 03 Data –Site 2 (STRETCH B).

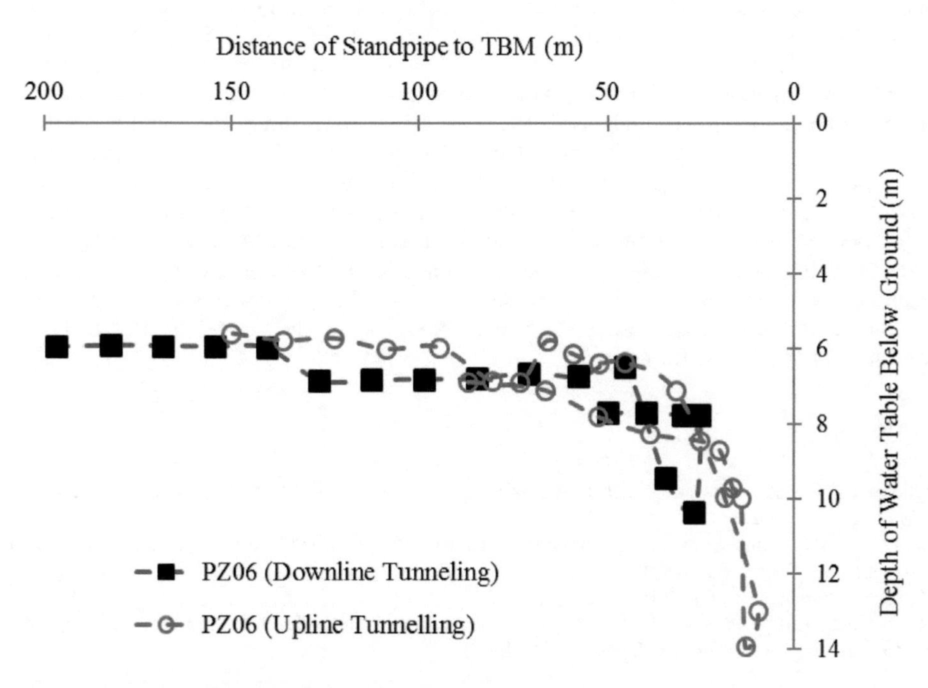

Figure 10. Standpipe PZ 06 Data –Site 2 (STRETCH B).

significant groundwater drawdown during downline as well as upline tunnel drives. The groundwater drawdown was measured with series of standpipes installed along the drives. The relationship between groundwater drawdown and the tunnel boring machine is presented in Figures 9 and 10.

From Figures 9 and 10, it is observed that open mode tunneling in the mixed ground conditions resulted in drawdown of 6m to 8m. The drawdown is observed when tunneling is within a distance of about 50m from the piezometer monitoring location. The drawdown induces settlements as a result of:

- The hydraulic gradient weakening the mechanical conditions at the face and on the tunnel walls thereby increasing ground deformations;
- Worsening effects on pre-existing mechanical instabilities.

The ground settlements were immediate in response to the drawdown. Therefore, a major portion of the settlements shown in Figure 8 are attributed to the water drawdown. However, the groundwater table rose as TBM moved away from the monitored location. The settlements also showed a stable trend post recovery of the groundwater table. As there were no sensitive structures in the vicinity there was no adverse impact on the ground surface due to the observed settlements. However, the instrumentation data demonstrates that the successful tunneling in rock under high water table requires effective groundwater control measures to be in place. The mixed ground features observed in Site 2 are localized and the rest of the tunnel drive is in Grade III or better Basalt very with negligible settlements.

4 CONCLUSIONS

Instrumentation data from shallow urban tunneling by TBM in Mumbai geology is presented. The settlement data in varying geological conditions is analyzed. The data is fitted with Gaussian model to estimate the volume loss and trough width parameters. The study indicates that volume loss in the range of 0.3% to 0.5% and trough width parameter range from 0.5 to 0.7. However the absence of groundwater drawdown control measures together with unfavorable geological condition could increase the magnitude of volume loss to as high as 1.2%. The initial data indicates that the settlements for tunnel drives in Breccia with mixed weathering grades and even with open mode of TBM operation are generally less than 10mm. The contact boundaries of volcanic (e.g. Basalt) and sedimentary (e.g. Shale) rocks presents possibility of groundwater drawdown and associated ground settlements of higher magnitude. Appropriate measures such as TBM operating in closed mode, forward probing and grouting should be implemented to control groundwater drawdown and settlements. The instrumentation should be used to understand the ground and groundwater response to the tunneling and is an indispensable tool for urban tunneling.

REFERENCES

Guglielmetti, V. (Ed.), Grasso, P. (Ed.), Mahtab, A. (Ed.), Xu, S. (Ed.). 2008. *Mechanized Tunnelling in Urban Areas*. London: CRC Press.

Martos, F. 1958. Concerning an approximate equation of the subsidence trough and its time factors. *International Strata Control Congress, Leipzig, (Berlin: Deutsche Akademie der Wissenschaften zu Berlin, Section fur Bergbau)*, 191–205.

O'Reilly, M.P. and New, B.M. 1982. Settlements above tunnels in the United Kingdom – their magnitude and prediction. *Tunnelling'82, London*, 173–181.

Peck, R.B. 1969. Deep excavations and tunneling in soft ground. *7th International Conference on Soil Mechanics and Foundation Engineering, Mexico City State-of-the-Art volume*, 225–290.
Schmidt, B. 1969. Settlements and ground movements associated with tunneling in soils. *Ph.D. Thesis, University of Illinois, Urbana.*
Sethna, S.F. 1999. Geology of Mumbai and surrounding areas and its position in the Deccan volcanic stratigraphy. *Journal of the Geological Society of India*, 53, 359–365.

Tunnels and Underground Cities: Engineering and Innovation meet Archaeology, Architecture and Art, Volume 3: Geological and geotechnical knowledge and requirements for project implementation – Peila, Viggiani & Celestino (Eds)

Grand Paris: The exceptional line 16 Lot 1

T. Hugues Beaufond, A. Truphemus & R. Ndouop Molu
Société du Grand Paris, Paris, France

L. Samama, S. Ghozayel, T. Couttet & M. Ferrari
Egis, Paris, France

P. Hamet
Eiffage, Paris, France

ABSTRACT: Line 16, a 30 km long underground metro line, is a part of a huge construction programme of over 200 km of automatic metro line with 68 stations in the Paris region.

The civil works of contract Lot 1 are for 20 km of tunnel. This contract concerns four new lines: 14, 15, 16 and 17. It constitutes a challenge for every stakeholder, with a common goal: fulfil the commitment to commissioning in time for the 2024 Olympic Games.

The simultaneous construction of all the underground works at a depth of 30 m in sands and soft rocks under high ground water pressure (5 stations, about 20 shafts and 20 km of tunnels simultaneously excavated by 6 confined type shield TBMs) makes contract lot n°1 a quite exceptional project.

1 PROJECT DESCRIPTION

1.1 *The Grand Paris Express project*

The *Société du Grand Paris* (SGP) was established pursuant to the French Act of Parliament n °2010–597 dated 3 June 2010 on the design and construction of the Greater Paris (or "*Grand Paris*") public transport network.

One year later, the master plan of the Grand Paris public transport network was approved by the government through the issuance of decree n° 2011–1011 dated 24 August 2011.

The Grand Paris public transport network (RTPGP) is a programme of 200 km of automatic metro lines, as extensive as the current Paris metro, and with 68 stations. It comprises three separate lines:

A red line (100km), providing services to the departments of Seine-Saint-Denis, Val-de-Marne, Hauts-de-Seine, Seine-et-Marne and Val-d'Oise. The red line comprises lines 15, 16 and 17;

A blue line (30km) connecting Paris with the hubs of Saint-Denis Pleyel and Orly (extension of line 14 north and south);

A green line (21km) providing services to a scientific and technological cluster and the main residential and employment basins of the departments of Yvelines and Essonne, connecting with the main transportation hubs in the west and south of Paris (line 18).

The four new lines of Grand Paris Express (15, 16, 17 et 18), and line 14 extended north and south will all connect with the existing transport network.

This project will provide transportation from one point of the Paris region to another without passing through the centre of Paris, while also providing faster connections to the centre of the capital from its outskirts. As a new alternative to the personal car, Grand Paris Express will reduce pollution, congestion and will contribute to creating a more environmentally friendly metropolitan region.

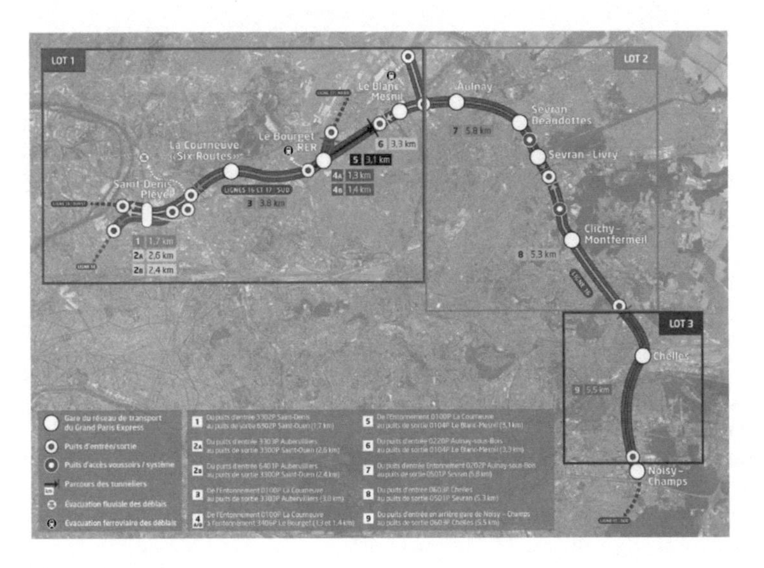

Figure 1. Metro L16 Paris.

1.2 *Line 16, Lot 1*

The project ownership of this line (as for the entirety of the Grand Paris Public Transport network) falls under the responsibility of SOCIÉTÉ DU GRAND PARIS (SGP).

The engineering, procurement and construction management is the task of the EGIS-TRACTEBEL consortium. It relates to the civil engineering works and all other structures required for the delivery of the transport infrastructure for the following sections:

- Le Bourget RER (including station) – Noisy-Champs (excluding station) on line 16,
- Saint-Denis Pleyel (including station) – Le Bourget RER (excluding station) common to lines 16 and 17,
- Mairie de Saint-Ouen – Saint-Denis Pleyel on line 14
- And the connection of line 16 to the maintenance depot at Aulnay-sous-Bois.

The contractor consortium coordinated by EIFFAGE holds the contract for civil works and system equipment for a section of lines 14, 15, 16 and 17 of Grand Paris Express for the value of €1.84 billion, which was officially notified on 20/02/2018.

The works consist mainly of building:

- Five underground stations (including two flagship stations): Saint-Denis-Pleyel, Stade de France, La Courneuve, Le Bourget, Le Blanc-Mesnil, including the structural works of the above-ground entrances to the stations of Saint-Denis-Pleyel and La Courneuve;
- 19.3 km of tunnel bored using TBMs;
- 18 related structures (including 5 TBM launch shafts).

This lot, featuring in Paris's application to hold the 2024 Olympic and Paralympic Games, will provide transport services to several major sporting venues. These sections are therefore due to be commissioned at the end of 2023.

At time of writing this article, the TBM launch shafts are currently being excavated, and the station diaphragm walls are being constructed.

The TBMs are due to start boring in summer 2019. In order to make progress in accordance with a very tight schedule, six TBMs will be required.

2 CRITICAL ISSUES

Lot 1 of line 16 covers a perimeter stretching 20 km in length. This constitutes a major challenge for all stakeholders contributing to the project, with a common goal: fulfil the commitment to the commissioning of this section before the beginning of the Paris 2024 Olympics.

Lot 1 is made up of a number of structures for which a considerable quantity of technical challenges lies ahead. We could give the following examples:

- TBMs will be required to bore several times beneath railway lines, on HSL and regional rail routes heading to countries and regions to the north of France, and on which railway services may under no circumstances be interrupted.
- The flagship station of Saint Denis with its exceptional size, the construction of its concourse and its connection to the existing network. This will also be the gateway to the future Olympic Park.
- The flagship station of Le Bourget, built in a confined environment. It will be hemmed in by RER line B tracks, motorways, etc.
- The management and logistics of the 2.5 million m^3 of materials to be excavated and disposed of in an already saturated urban network.
- The criss-crossing of the tunnels to be built for lines 14, 15, 16 and passing under the existing line 13, in a dense and sensitive urban environment.

The simultaneous construction of all these works (five stations, six TBM's and many associated structures) combine to make this Lot 1 an exceptional project.

2.1 *BIM, from design to construction.*

BIM was introduced on Line 16 in the design phase for the 3-D modelling of stations, additional structures, tunnels and their fixtures and equipment, including that of the transport system.

Each discipline (architecture, civil engineering of stations and additional structures, civil engineering of the tunnel, ventilation, drainage, electrical equipment, plumbing, traction power, rail track, catenary, etc.) produces its own digital model. These are each updated throughout the design process and shared with the other contributors according to procedures implemented in an information system (Projectwise ®, BIMsync®ou©).

In all, this constitutes more than 200 discipline, or "speciality template" digital models to be coordinated by a team made up of a BIM manager and several BIM coordinators.

This modelling process was used to optimise the design, coordinate design studies, facilitate dialogue between discipline specialists to manage the complex interfaces between infrastructure and systems, and manage interfaces at connection points between the tunnel, stations and additional structures.

The digital models also enabled the delivery of combined drawings and project reviews.

The bored tunnels, the services and the equipment inside the tunnel were modelled in the design phase using software solutions Civil 3D, REVIT and Dynamo by Autodesk and Geomensura for the drainage.

But it is in the works phase that the application of BIM in the tunnel proves to be the most valuable.

This is because the 3D verification of the tunnels corresponding to the definitive geometry (post-construction) down to the detail of each tunnel wall segment will help to finalise the shop drawings for the fixings or even for the connections with stations, with the cross-passage tunnels to the offset shafts and with special structures (tunnel turnouts, centred shafts) whilst abiding by functional restrictions and in accordance with the rolling stock gauge.

Ultimately, the Employer requires of the works contractors, in every discipline, that they produce a digital As-Built Drawings (ABD) file, with the aim of archiving and passing on to

Figure 2. Saint Denis Pleyel station.

the operator and maintainer a digital duplicate of the structures built, comprising digital models in an open format (IFC) and compatible with a range of software solutions, containing data attached to the objects (product characteristics, construction data) and links between the ABD documentation and the objects in the model to which they apply.

The operator and maintainer will thus have easy access to reliable (and enduring) data on installations, including with regard to construction.

To meet these goals, a BIM execution plan is included in the works contractors' contracts for line 16.

As the production of digital models for the tunnel and its fixtures is a recent request of Employers, but also an opportunity to exercise better control over certain risks, the industry's stakeholders have engaged in an approach to develop modelling and simulation methodology, as presented at the AFTES colloquium (Bron, France) on 25 May 2017.

Finally, in order to make the BIM databases operational over the long term, international data standards (IFC/OGC) must be developed for tunnels, underground structures and their fixtures and for geotechnics.

This is what has prompted several French firms and bodies including ANDRA, CETU and Egis, within the national research project MINnD, to conduct pre-standardisation works on the tunnel IFC standard.

This project resulted in Egis winning the BIM d'Or award in 2017, the most highly-acclaimed award in France in the BIM field.

2.2 *The six TBMs used for Lot 1*

Given the geology of the Paris region and the results of approximately 1000 boreholes carried out for the construction of line 16, and more particularly those of around 450 boreholes conducted solely for the purposes of lot 1, a multi-criteria analysis carried out by the contractor EIFFAGE and its subsequent consultation of several manufacturers led to the selection of six Herrenknecht earth pressure balance tunnel boring machines (EPB TBM).

These tunnels will be bored in a highly built-up environment. In some places, they will pass beneath private buildings, transport infrastructure (RER line B, metro line 13; national rail network, A1 motorway) or mains sewer infrastructure.

The main data relating to the different tunnels are given in the table 1:

Table 1. Main data relating to the different tunnels.

Tunnels	Tunnel type	Interior diameter	Depth underground	Bore distance
Line 14	Single tube	7.75	20 - 50 m	1,613 m
Line 15	Single tube	8.70	15 – 40 m	2,351 m
Line 16	Single tube	8.70	10 – 40 m	16,000 m
Line 17	Dual tube	6.70	15 – 40 m	2 x 1,150 m

The TBM-bored tunnel alignment will cross through the following geological formations:

- Coarse limestone (CG)
- Marl and loose stone (MC)
- Beauchamp Sands (SB)
- Saint-Ouen limestone (SO)

The hydraulic load can reach 30m at the tunnel axis.

The TBMs are all fitted with an Additional Active Confinement System. This helps to prevent sudden fluctuations in the confinement pressure in the excavation chamber when boring is at a halt (to install a ring, conduct maintenance, etc.).

If the confinement pressure drops below a predetermined threshold, the system is automatically activated by sensors. Bentonite is injected in the upper section of the excavation chamber to counterbalance the measured drop in pressure.

Additionally, in consideration of the risks associated with this project (decompressed zones, gypsum dissolution zones and/or voids, etc.); the TBM's are equipped with a forward recognition system.

Lot 1 of line 16 represents the largest civil works lot for an urban transport project (in terms of contract value) ever to have existed in France, in terms of both contract value and the number of TBMs used simultaneously.

Another specificity of this "civil engineering" lot is its incorporation of the equipment relating to the "line" subsystems; in particular: rail track, the rigid overhead contact rail, and fire protection.

3 STATION CONSTRUCTION METHODS AND TBM PROGRESS THROUGH THEM

3.1 *Stations*

Four underground stations (including two "flagship" stations) and the underground "box" of a fifth are included within this Lot 1. The methodology of their construction is presented below.

Considering the dense urban environment, the many interfaces with nearby infrastructure and the reduced footprint of work sites, the project necessitated specific attention to construction sequencing (switching footprint to divert traffic, etc.), "top down" works, boring the tunnel through the station before its excavation, etc.

3.1.1 *Saint Denis Pleyel station: Architectural design: KENGO KUMA*

This flagship Line 16 station displays the following features:

- A sensitive neighbouring environment, such as the Landy railway sidings,
- Exceptional dimensions: 120 m long, 80 m wide and 33 m deep,
- Three TBMs passing through (lines 14 and 16 before excavation, line 15 after the completion of the base slab),
- Top and down construction,
- The construction of an above-ground structure made from a metal framework.

The structure of this building is made up of diaphragm walls 1.50 m thick, floated into the Coarse Limestone to a depth of 54 m, and 36 pre-founded columns.

Work on the station will be sequenced as follows:

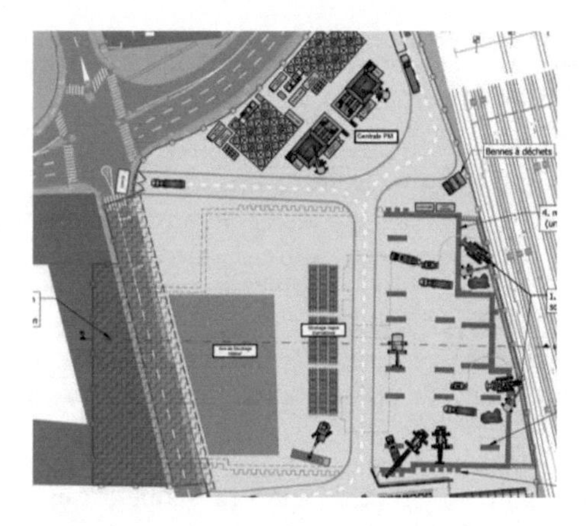

Figure 3. phase one: diaphragm walls and pre founded columns Saint Denis Playel Station.

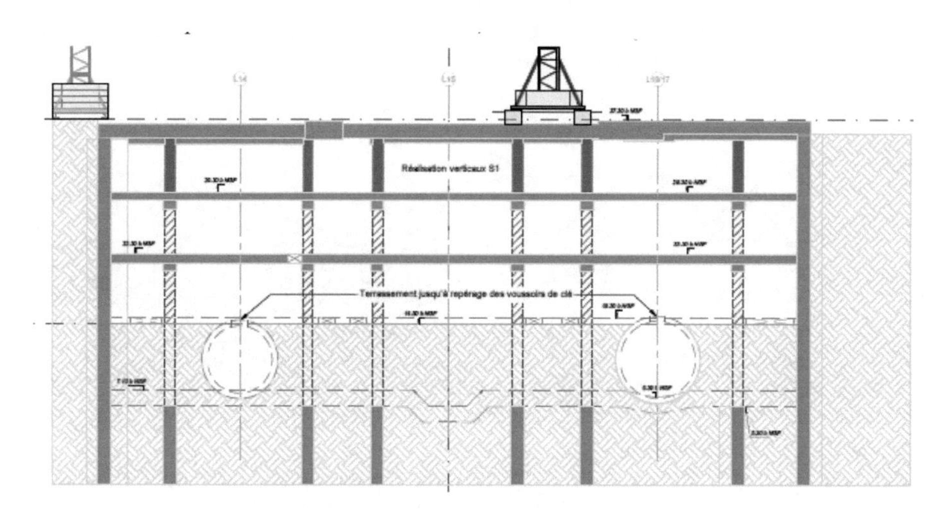

Figure 4. Cross section Saint Denis Playel Station.

Construction of diaphragm walls and pre founded columns,-
Staggered construction of the covering slab, according to the necessary road diversions,
Excavation and construction of slab S1,
Tunnel boring by TBMs on lines 14 and 16,
Excavation and construction of slab S2,
Steel reinforcement of line 14 and 16 tunnels,
Excavation and partial construction of slab S3,
Excavation up to the level of the apron with demolition of line 14 and 16 tunnel liners,
Construction of the base slab and of remainder of slab S3 and platforms.

3.1.2 *Le Bourget station: architectural design: Elisabeth and Christian DE PORTZAMPARC*
Another flagship station on line 16, the Le Bourget station has the following distinctive features:

- A sensitive neighbouring environment, in particular with a mains sewer and the proximity of major railway lines to the north east and to the south, requiring some temporary service interruptions, and temporary train speed reductions,
- Exceptional dimensions: constructed within a 32 m diameter circular shaft, 33 m deep

- Three TBMs pass through the station (line 16 and line 17 North before station excavation, line 17 South after base slab construction),
- Top down construction.

The structure is made up of diaphragm walls measuring 1.50 m thick, floated into coarse limestone at a depth of 47 m, and 32 pre-founded columns, sunk up to 55 m in depth.

The TBM entrance and exit arrangements are defined according to the kinematics of the TBMs (passing through the station before or after earthmoving), the geology and the interface with the neighbouring features described above.

On the line 17 North and South tunnels, as the station is a cylindrical shape, the tunnel enters the station at an angle, and this requires the tunnel liner segments to be cut in a complex configuration.

To the north-west, the TBM used for line 17 North enters the station before its excavation. The major interface is the presence of a 2.5 m diameter sewerage network lying 15 m underground. Preparatory works have taken place on this infrastructure (lining and reinforcement grouting) to mitigate the risks relating to the construction of the station and the transit of the TBM in its vicinity. The rock depth between the tunnel and the sewer located in Saint Ouen limestone is less than 4 m. The ground located beneath this sewer will be treated with jet grouting to reduce settlement and ensure the TBM can enter the station without any problems.

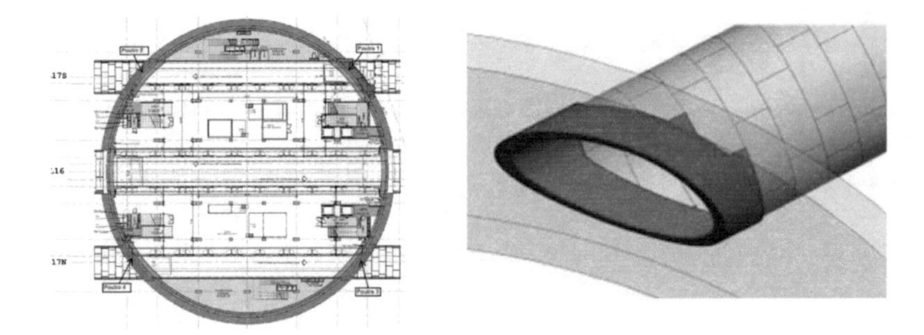

Figure 5. plan view of TBM entrances/exits and 3-D detail of angled entrance.

Figure 6. DEA/SIAAP sewer following reinforcement works and installation of monitoring optic fibres - Gauss/DEA.

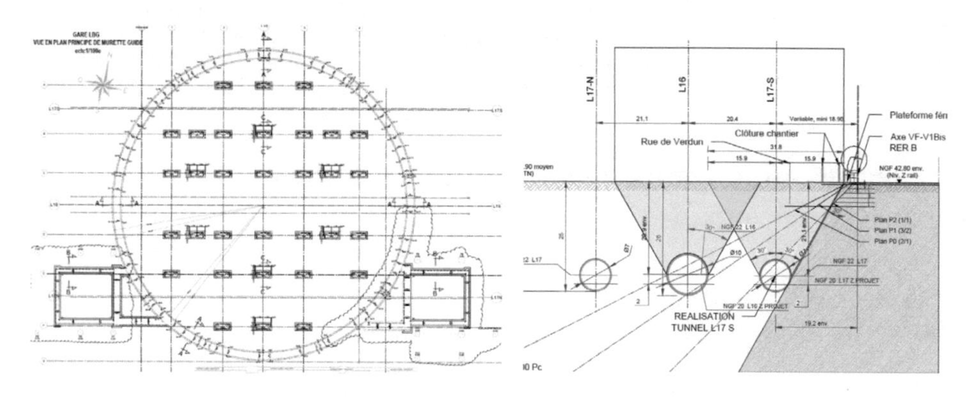

Figure 7. Cross-section of tunnels L17N, L17S and L16 near to railway tracks– installation of diaphragm walls in LBG station.

Optic fibres embedded into the concrete were installed inside the sewer during reinforcement works so as to monitor this structure throughout construction work. To fully analyse any distortion caused to this sewer, manual convergence measures will be carried out with the sewer closed off.

To the south of the station, the Line 17 South TBM will enter the station following its excavation. The railway lines lie approximately 15 m from the tunnel. Two chambers made from diaphragm walls will be used for the entrance and exit of the TBM. These chambers will help to lower the groundwater table locally and gradually reduce the confinement pressure. The diaphragm walls of the chamber will isolate the railway lines from any ground deformation.

At the exit, the same type of chamber is used. In addition, the railway line structure will be stabilised using hydraulic jacks to prevent any deformation during TBM boring, and also during the excavation of the station.

In addition to this system, the railway lines will be subject to automated monitoring which will begin on commencement of construction work.

Work on the station will be sequenced as follows:

- Creation of diaphragm walls and pre-founded columns,
- Excavation and laying of level S1, boring by TBMs on line 16 and line 17 North, reinforcement of rings,
- Excavation and laying of level S2,
- Excavation as far as the base slab with demolition of liner segments, construction of base slab,
- Passage of TBM line 17 South, and end of interior civil works.

4 CONCLUSION

The line 16 Lot 1 project is a monumental task due to its critical importance, the magnitude of its structures built at the same time, and its complexity which is chiefly in evidence beneath buildings and below sensitive surrounding structures. It is evident that the successful delivery of the project relies heavily on an ability to continuously adapt during the works, once all the foreseeable risks have been anticipated in design phase. To achieve its completion, contributors must work in close coordination in a spirit of partnership on this project. And, in view of this complexity, the contractor consortium made up of EIFFAGE GENIE CIVIL SAS, RAZEL BEC, EIFFAGE RAIL, SNC, TSO and TSO-CATENAIRES, and the Engineers EGIS and TRACTEBEL will continue to support the project owner, SOCIETE DU GRAND PARIS, to fulfil the challenge of this major project with a shared objective: succeed together.

Tunnels and Underground Cities: Engineering and Innovation meet Archaeology, Architecture and Art, Volume 3: Geological and geotechnical knowledge and requirements for project implementation – Peila, Viggiani & Celestino (Eds)
 ISBN 978-0-367-46583-4

Countermeasures against the deformation of the pilot tunnel of the Seikan Tunnel, Japan

H. Kakinuma & T. Okada
Japan Railway Construction, Transport and Technology Agency, Sapporo, Japan

K. Yashiro
Railway Technical Research Institute, Kokubunji, Japan

Y. Kobara
A-Tic Co. LTD, Sapporo, Japan

ABSTRACT: The Seikan Tunnel is a 53.85-kilometer-long undersea railway tunnel in Japan. This tunnel is 240 m deep below sea level at the maximum. In 2013, heaving of the roadbed and tunnel convergence were observed at the Yoshioka pilot tunnel, which is used for the ventilation and the drainage of the ground water. As a result of geological surveys, it was found that the strength of the ground was low, the ground slaked easily and a large amount of ground water flowed through the central drainage. Therefore, the deformation of the pilot tunnel was considered to be attributable to a gradual decrease in the strength of the ground surrounding the tunnel and plastic earth pressure thereby acting on the tunnel. In 2017, struts and rock bolts were applied, and deformation was suppressed. This paper describes the deformation of the pilot tunnel, its causes, countermeasures and their effects.

1 INTRODUCTION

The Seikan Tunnel is a 53.85-kilometer-long railway tunnel linking Honsyu Island and Hokkaido Island of Japan. Figure 1 shows the outline of the Seikan Tunnel. There are three types of tunnel (main tunnel, service tunnel and pilot tunnel) in the undersea section of the Seikan Tunnel.They are 240 m deep below sea level at their maximum and receive a water pressure of a maximum of 2.4 MPa. Pilot tunnels are used for the ventilation of the main tunnel and the drainage of the ground water, and they are very important structures in service of the Seikan Tunnel.

In 2013, 25 years after the start of service, heaving of the roadbed and tunnel convergence were observed in an 80m-long section of the Yoshioka pilot tunnel at Hokkaido Island side. As a result of geological surveys, it was found that the strength of the ground was low, the ground slaked easily and a large amount of ground water flowed through the drainage. Therefore, the deformation of the pilot tunnel was considered to be attributable to a gradual decrease in the strength of the ground surrounding the tunnel and plastic earth pressure thereby acting on the tunnel. In 2017, struts and rock bolts were applied, and this deformation was suppressed. In addition, a numerical analysis was carried out to simulate the deformation of the pilot tunnel and to evaluate the effect of the countermeasures, using an analytical method with due consideration for a decrease in the strength of the ground.

This paper describes the deformation of the pilot tunnel, its causes, countermeasures and their effects.

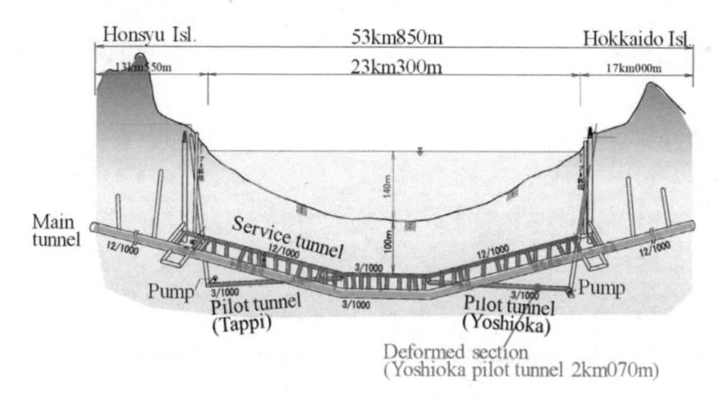

Figure 1. Outline of the Seikan Tunnel.

2 DEFORMATION OF THE PILOT TUNNEL

In the regular inspection conducted in December 2013, deformation was confirmed in the Yoshioka pilot tunnel at Hokkaido Island side. Figures 2 - 3 show the Yoshioka pilot tunnel in the deformed section. Specifically, they show heaving of the roadbed and tunnel convergence, and spalling of sprayed concrete caused by these kinds of deformation. This deformed section was constructed between 1971 and 1972, and 46 years have passed.

Figure 4 shows the level of the roadbed. The heaving of the roadbed was 50 mm high in 20 months (heaving rate 2.5 mm/month). Figure 5 shows the horizontal convergence. The horizontal convergence was 25 mm contraction in 16 months (convergence rate 1.7 mm/month). From these results, it was found that the deformation was progressing rapidly. There was no deformation in the main tunnel which was distant from this section.

Figure 6 shows the geological profile of the Seikan Tunnel. The deformed section is located in the Cenozoic (Neogene) sedimentary rock. A fracture zone called F10 disturbance zone is distributed over about 1 km-long around the deformed section. At the time of construction, it

Figure 2. Heaving and convergence.

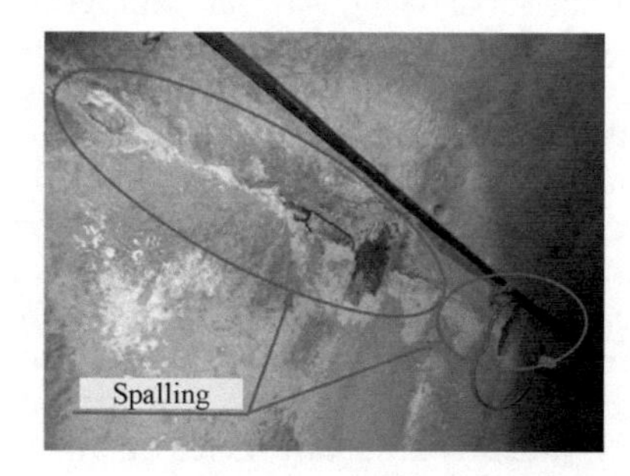

Figure 3. Spalling of shotcrete.

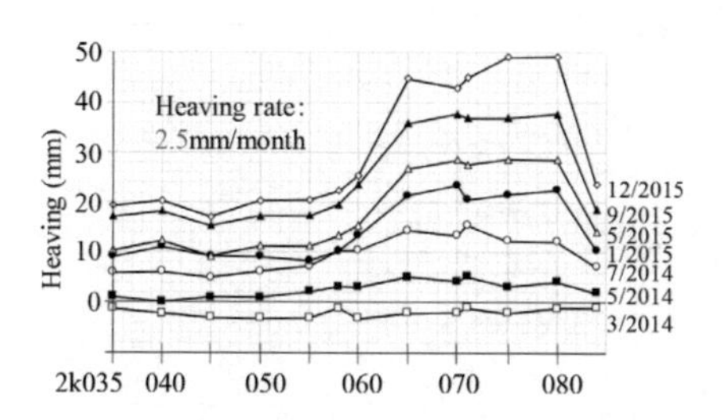

Figure 4. Level of the roadbed.

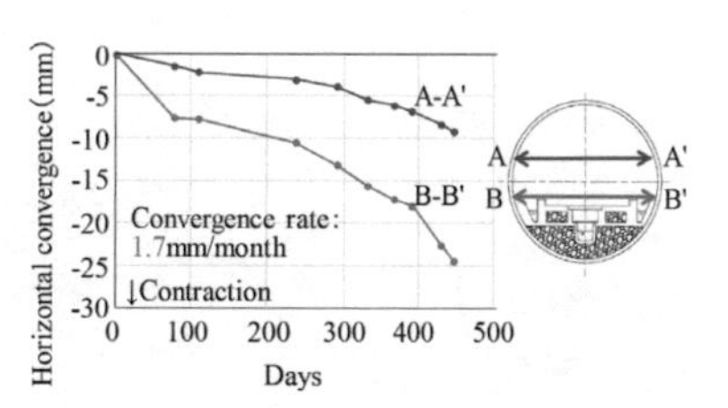

Figure 5. Horizontal convergence.

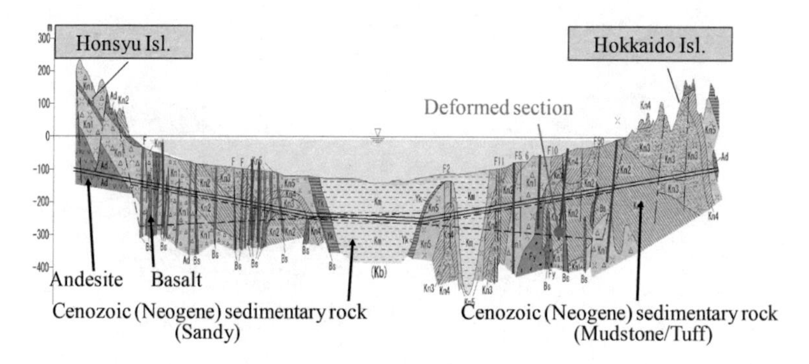

Figure 6. Geological profile of the Seikan Tunnel.

was recorded that there were a lot of difficulties in excavating the deformed section, such as the convergence of the tunnel due to squeezing earth pressure of the ground (Inoue, 1986.)

Figure 7 shows the cross section of the Yoshioka pilot tunnel around 2 km 070m. In the deformed section, the tunnel was excavated originally in a horseshoe shaped cross section. Because large heaving and convergence occurred, the tunnel was re-excavated in a circular cross section and steel supports were replaced. In the literature, it is written that the horizontal convergence was approximately 20 to 30 cm contraction in 2 weeks and the heaving was about 40 to 80 cm high in 3 months at the time of excavation. In addition, grout injection was carried out to stop the ground water at the time of construction at the Seikan Tunnel, but it was not carried out in the deformed section. The deformed section is located at a water depth of about 50 m and an earth covering of about 240 m, and it is thought that the competence factor is small due to the large vertical earth pressure.

Pilot tunnels are used for the ventilation of the main tunnel and the drainage of ground water. Figure 7 also shows the drainage pathway of the Seikan Tunnel. The deformed section is almost the deepest point in the Seikan Tunnel, and a large amount of ground water is flowing through the drainage.

Boring surveys and various geological tests were conducted to clarify the cause of the deformation and to design countermeasures. Figure 8 shows the deformation coefficient of the ground derived from borehole loading tests (pressuremeter tests). The loosening depth is about 3.5 m, and based on the judgement on the safe side, the maximum loosening depth is about 5m. Figure 9 shows an example of the results of various geological tests conducted using boring cores, and Table 1 shows a summary of the geological tests. The slaking durability is roughly 4, the competence factor is 2 or less, the natural water content ratio is 20% or more, the smectite content is 20% or more, and it is found that the conditions are satisfied under which the plastic earth pressure is likely to act on the tunnel.

From these geological tests, it is estimated that the plastic earth pressure acts on the tunnel with its intensity gradually increased and deformation occurs because the competence factor is small, the ground around the tunnel slakes easily and a large amount of ground water is flowing through the drainage.

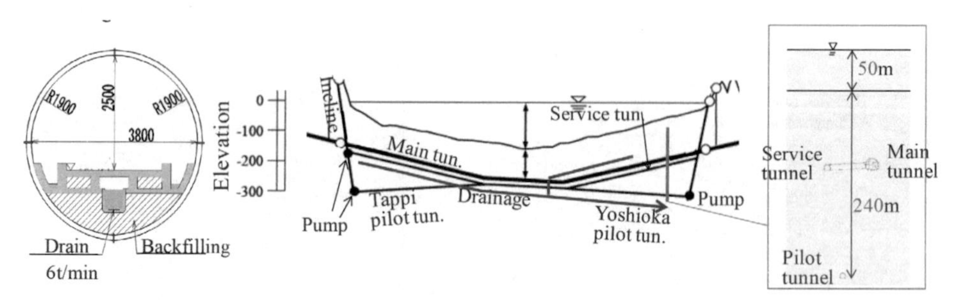

Figure 7. Cross section of the Yoshioka pilot tunnel around 2 km 070 m.

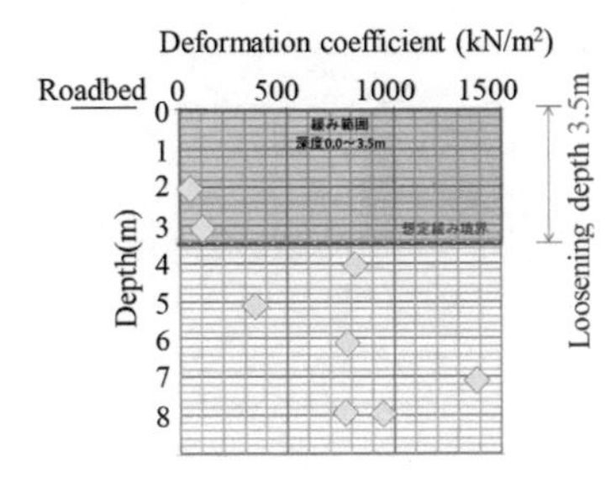

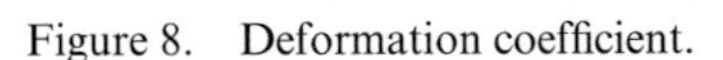

Figure 8. Deformation coefficient.

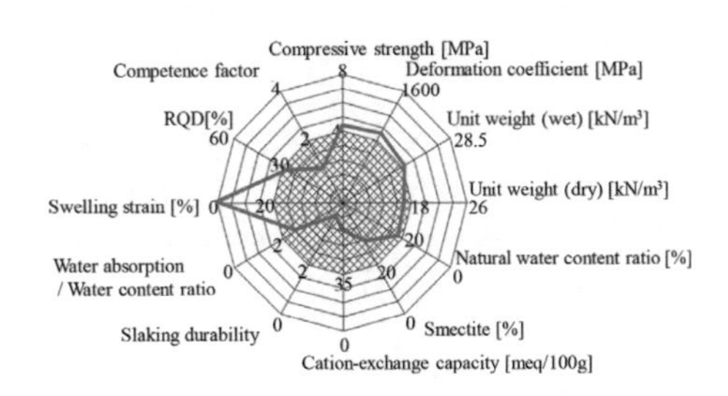

Figure 9. An example of geological test results.

Table 1. Geological survey results.

Item	Unit	Range	Ave.
Unit weight (wet)	kN/m^3	1.96 ~ 2.22	2.08
Unit weight (dry)	kN/m^3	1.52 ~ 1.90	1.69
Natural water content ratio	%	16.8 ~ 29.1	23.2
Compressive strength	MPa	1.90 ~ 9.20	3.83
Deformation coefficient	MPa	369 ~ 1305	735
Swelling strain	%	0 ~ 1.68	0.237
Swelling pressure	kPa	0 ~ 23.1	3.87
Slaking durability		3 ~ 4 (Flakes - Mud)	4
Smectite content	wt g	23 ~ 49	34.5
X-ray diffraction type		(Various)	
Cation exchange capacity	meq/100g	37.3 ~ 104.6	73.2
Competence factor		0.21 ~ 1.67	0.69

3 COUNTERMEASURES BY STEEL STRUTS

After confirming the deformation, the heaving of the roadbed was monitored by the level measurement and the horizontal convergence was measured by a light wave range finder. Since these kinds of deformation continued, JR Hokkaido implemented countermeasures by steel struts as an emergency measure when the amount of the roadbed heaving reached 53 mm and horizontal convergence reached 25 mm. Figure 10 shows the outline of the steel strut. In the deformed section of Yoshioka pilot tunnel, rectangular steel pipes of 150mm in width, 150mm in height and 12mm in thickness were placed at intervals of 1 m at the roadbed.

Figure 11 shows the horizontal convergence and axial force. Although the convergence rate temporary increased during the installation of the struts, after completion of the countermeasure work, the convergence rate reduced to about 1/10 of what it was before, and a large effect of the countermeasure was recognized. Here, since the total amount of the deformation so far and the convergence rate before the implementation of countermeasures were considerably large, axial force transducers were attached to the strut and their behaviors were monitored. According to Figure 11, although the axial force had sufficient margin for the strength of the strut, it can be seen that the value increased immediately after the installation of the struts.

Figure 12 shows the steel support. After the installation of the struts, deformation and buckling were confirmed in the steel supports. For this reason, strain gauges were also attached to the steel supports, and monitoring was carried out. Figure 13 shows the strains of the steel support. From this figure, it is found that strains are also increasing.

Figure 14 schematically shows the status of axial force and strain in the tunnel. It is thought that the steel supports are buckling near the shoulders of the arch and the load bearing capacity of the support is reduced in the upper half. On the other hand, the reinforcement of the lower half of the structure by struts, that is, a partly increase in the rigidity of the tunnel arouse a concern that the axial force and strain would continue to increase in the tunnels in the future.

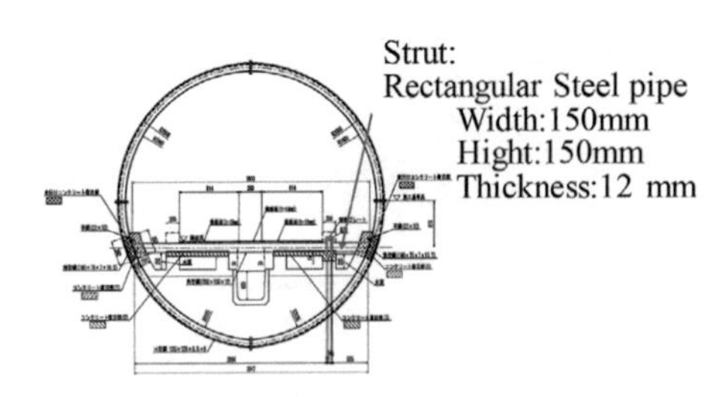

Figure 10. Strut.

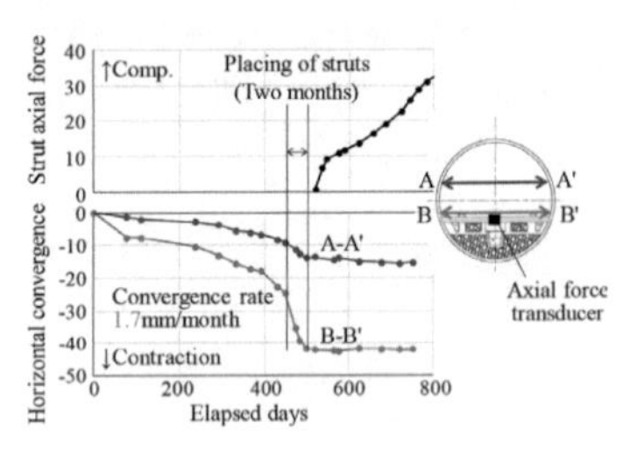

Figure 11. Axial force of the strut and horizontal convergence.

Figure 12. Buckling of the steel support.

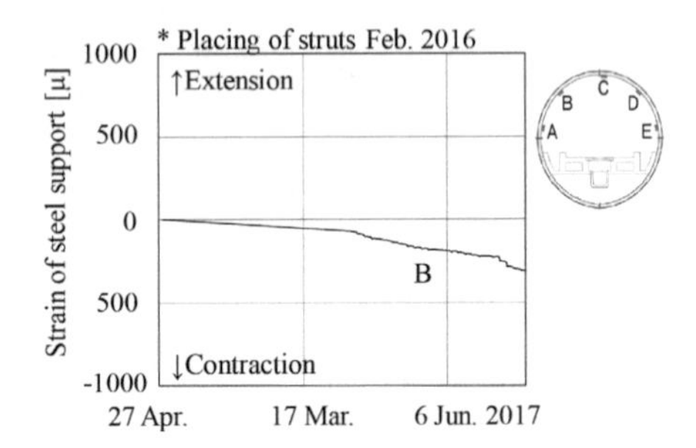

Figure 13. Strains of the steel support.

In the past, there was a tunnel where deformation and swelling occurred due to the plastic earth pressure. In this tunnel, the deformation occurred in the upper part of the tunnel because the structural rigidity was increased by closing the tunnel by adding an invert. It was thought that the Seikan Tunnel would also follow a similar process of the deformation of this tunnel.

In addition to directly strengthening the tunnel structure by the struts, we have carried out rock bolting, expecting the ground reinforcement effect of the rock bolts by tensioning stress.

4 COUNTERMEASURES BY ROCK BOLTS

In designing rock bolts, it is necessary to decide the length and number of rock bolts. Figure 15 shows the arrangement of rock bolts. Referring to the past cases, 12 rock bolts per section at regular intervals were arranged around the tunnel. According to the results of the boring survey shown in Figure 8, the loosening depth is thought about 3.5 m and the maximum is about 5 m based on the judgement on the safe side.

Regarding the length of the rock bolts, based on that result, the fixed length of the bolts to the ground was set to 5 m. The diameter of the rock bolts was set to 25 mm.

As the deformation of the steel supports was progressing rapidly, which indicated a situation where even a momentary delay was unacceptable, it was decided to start the installation

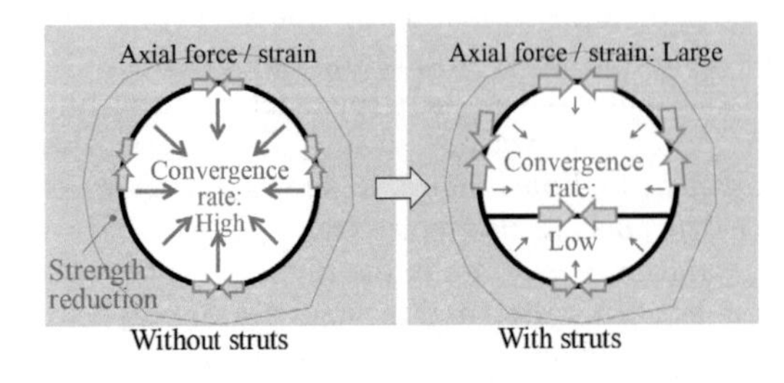

Figure 14. Status of axial force and strain in the tunnel.

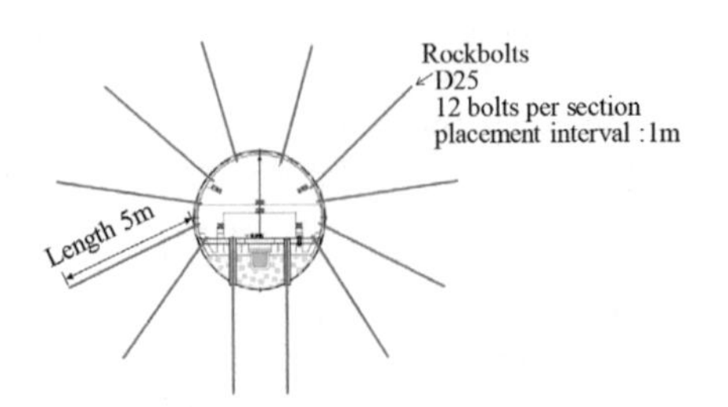

Figure 15. Arrangement of the rock bolts.

of rock bolts provided that the specifications of the rock bolts would be modified based on the results of the measurement.

Since the total number of rock bolts was approximately 1000, and it would be installed in a narrow tunnel, we carried out installation testing and determined the specifications of the rock bolt and the construction machine. It was possible to encounter a sudden ground water and the collapse of the borehole during drilling. So, various types of rock bolts were prepared in advance. As a result of the installation testing, the ground around the construction site consists of relatively uniform mudstones and fortunately neither the sudden ground water nor the collapse of the borehole wall was seen. So, a usual rock bolt (bar steel D25) was selected.

Figures 16 - 19 show the bolting machines. Four downward bolts of the total of 12 bolts were installed with three rotary boring machines and one rotary percussion drill which were relatively suited to muck or deal with embedded objects such as piles and steel materials. Remaining eight upward bolts were installed with one drill jumbo with fast drilling speed. The smallest one was chosen for the drill jumbo and it was loaded on the carrier of the cable car of the incline and transported to the construction site through the service tunnel.

As the fixing material, cement milk was chosen for the rotary type boring machines and the rotary percussion drill, and dry mortar for the drill jumbo respectably.

From the installation test, it was found that the rotary boring machine took nearly one hour to drill up to a depth of 1 m, and the drilling speed was extremely slow. Therefore, the bits were improved by attaching a partition plate to the central part of them so that non-core boring can be performed instead of the core boring and the machine was loaded on the roller mount to shorten the travel time of the machine. The drill jumbo has a high drilling speed nearly three times as high as a rotary boring machine but the machine is large. Consequently there was a problem in that the moving speed was slow. So, we reduced the weight of the equipment and shorten the travel time. There were steel sheet piles between the sprayed concrete and the steel supports, and it took a long time to pierce them with an ordinary bit, so we improved the material and the shape of the bit so that it can drill the sheet pile easily (Fig. 19).

The construction was carried out from spring to autumn of 2017, and the installation of 1000 rock bolts was completed in about 5 months.

Figure 16. Rotary type boring machine.

Figure 17. Drill jumbo.

Figure 18. Drilling by a drill jumbo.

Figure 19. Improved bits.

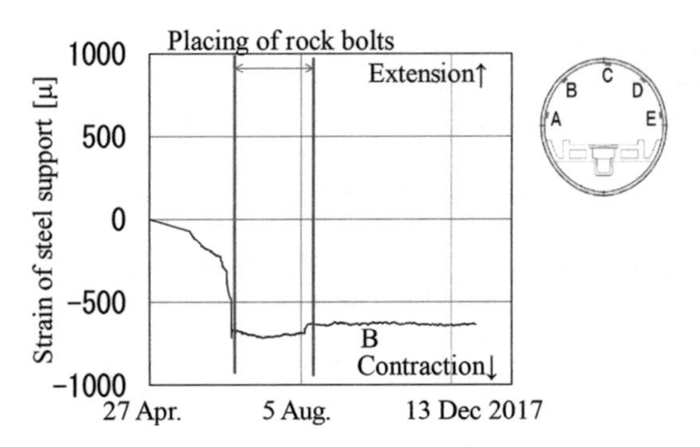

Figure 20. Strains of the steel support.

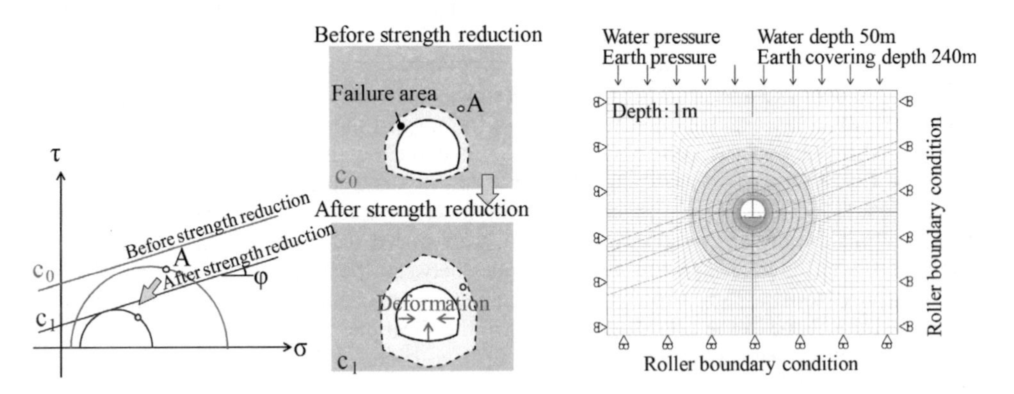

Figure 21. Concept of the ground degradation model. Figure 22. Analysis model.

Figure 20 shows the strains of the steel support. From this figure, it is found that the increase in the strain of the steel support is suppressed by the rock bolts.

5 EVALUATION OF COUNTERMEASURES BY NUMERICAL ANALYSIS

In order to evaluate the effectiveness of countermeasures, numerical analysis was conducted in parallel with the design and installation test of rock bolts.

For the analysis, a method under which the tunnel deformation and heaving of the ground can be simulated by the reduction of the ground strength was adopted (Shimamoto et al., 2009.) In the analysis the finite difference code FLAC3D (Itasca., 2013) was used.

Figure 21 shows the conceptual diagram of this method. The point A is initially stable, whereas when the strength of the ground (the cohesion c in this case) is lowered, the point A becomes unstable. The stress of the ground is redistributed by the failure of the ground, and the deformation of the tunnel occurs. In addition, we applied a method for consistently expressing both the tunnel excavation stage and tunnel deformation stage.

The sequential excavation of the tunnel is expressed by deleting ground elements, and the deformation of the tunnel after the completion of the tunnel are modeled by lowering the shear strength of the ground (the cohesion c) according to the distance between the stress status and the failure envelope of the ground. This makes it possible to express the deformation after the completion of the tunnel considering the stress status of the ground.

Figure 22 shows the analysis model. The tunnel and the ground are modeled by a 2D model. A roller boundary condition is used at both the sides and the bottom of the model, and the free

Table 2. Analysis input value of the ground.

	Item	Input value
Unit weight	20	kN/m^3
Uniaxial compressive strength	2.15	MPa
Deformation coefficient	50	MPa
Initial cohesion	0.62	MPa
Initial internal friction angle	30	deg
Poisson's ratio	0.3	

Table 3. Analysis input value of the tunnel.

	Item	Input value	
Shotcrete and invert	Unit weight	25	kN/m^3
	Deformation coefficient	22,000	N/mm^2
	Initial cohesion	4.90	N/mm^2
	Internal friction angle	40	deg
	Poisson's ratio	0.2	
	Initial tensile strength	1.75	MPa
Steel support and strut	Elastic coefficient	210,000	MPa
	Poisson's ratio	0.2	
Rockbolt and ground	Cohesion	20	N/m
	Shear rigidity	10	MN/m^2

boundary condition is used at the upper surface of the model and the remaining load equivalent to the vertical earth pressure is applied. The stress release rate by excavation is set at 100%.

Table 2 shows analysis input values of the ground. Referring to the results of the uniaxial test of the boring core, the uniaxial compressive strength of the ground is set to 2.15 MPa, so as to make the competence factor 0.5, and the geophysical property values adopted for the analyses are set referring to the relationship between uniaxial compressive strength and them (Aydan et. al., 2000; Jiang et. al., 1994.) The decrease in the ground strength is represented by an exponential function, and the strength is lowered according to the time elapsed so as to match the actual measurement.

Table 3 shows the analysis input value of the tunnel. A strain softening model is used for the shotcrete and inverted concrete, where the tensile strength and the shear strength are reduced according to the plastic tensile strain and the plastic shear strain which occurred after plasticization.

The steel supports and the struts were modeled with beam elements, and nonlinearity according to strength is considered. The rock bolts are modeled with cable elements and the cohesion between the rock bolts and the ground was considered. The cohesion between the rock bolts and the ground is modeled by sliders, whereby a slip is supposed to occur between the rock bolts and the ground elements when a shearing force exceeds a certain value (cohesion cut). Table 3 shows the values adopted for the rock bolts and the ground.

The analysis is carried out by the following steps [1] to [4].

[1] Initial stress analysis is carried out by applying self-weight.

[2] Tunnel excavation analysis is carried out, where the stress release rate is set as 100%. After the displacement due to excavation converges, the shotcrete and the steel supports are installed and the self-weight is applied.

[3] The strength of the ground is decreased. Specifically, the cohesion is lowered according to the distance between the stress status and the failure envelope of the ground.

[4] The strut and rock bolt elements are added while the strength being reduced.

Figure 23 compares the convergence between the actual measurement and the analysis. From this figure, it is found that the convergence becomes large according to the number of elapsed days, and it is appreciated that the convergence is suppressed appropriately by the installations of the rock bolts and the struts.

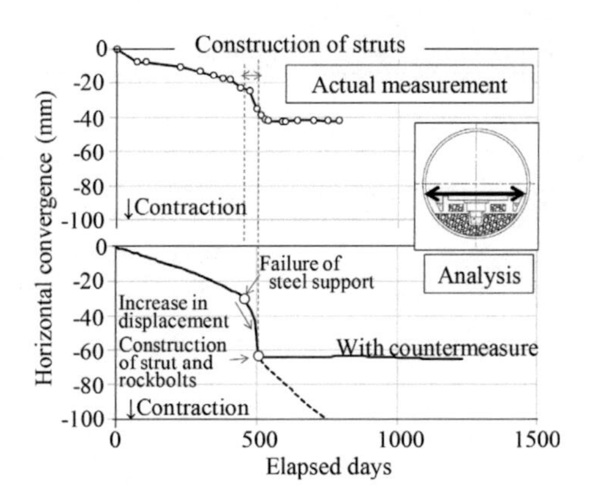

Figure 23. Horizontal convergence.

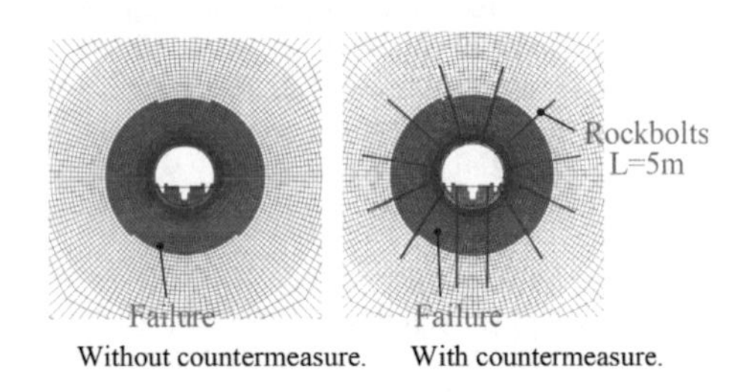

Figure 24. The failure area of the ground at the time the ground is degraded sufficiently.

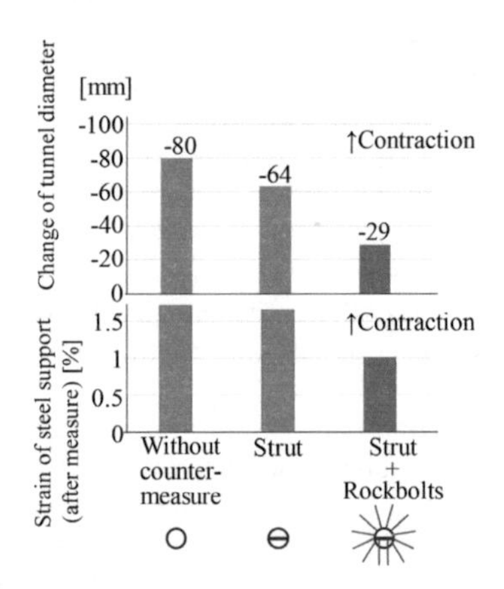

Figure 25. Horizontal convergence and the increment of the strain of the steel support after the countermeasure work.

Figure 24 shows the failure of the ground at the time when the strength of the ground decreased sufficiently. From the figure, the depth of the failure area is about 3.5 - 4 m, which is almost the same as the result of the boring survey shown in Figure 8. The length of 5 m of the rock bolt can be judged to be appropriate.

Figure 25 shows the horizontal convergence and the increment of the strain of the steel support after the countermeasure work at the time when the strength of the ground has decreased sufficiently. It is confirmed that by installing rock bolts in addition to the struts, it is possible to further suppress the increment of the convergence and the strain. From the above, it is confirmed that the specifications of the countermeasure works at this time are appropriate.

6 CONCLUSION

From 2013, 25 years after the start of service, heaving of the roadbed and tunnel convergence were observed in an 80m-long section of the Yoshioka pilot tunnel at Hokkaido Island side, and various geological surveys, designs, countermeasure were carried out from then through 2017.

The cause of the deformation was assumed to be a plastic earth pressure due to a decrease in the strength of the ground, and the struts suppress the convergence. On the other hand, the progressive deformation and buckling of the steel supports were observed. It was thought that this was because the strengthening of the tunnel structure by the struts, and there is a risk of the axial force and strain continuingly to increase in the arch portion. For this reason, additional rock bolts were installed. The installation of the rock bolts was also able to suppress the deformation of the support, and the effect of the countermeasure works was confirmed. We also evaluated the effect of these countermeasures by numerical analysis.

Currently, we are conducting follow-up on the effect of a series of countermeasures.

Although mountain tunnels may be subjected to earth pressure due to poor geology, it can be said that the mountain tunnel is a structure that can be used for a long time if proper maintenance, repair and reinforcement are carried out. It is thought that this also applies to the Seikan Tunnel which is an undersea tunnel. The Seikan Tunnel used by the Hokkaido Shinkansen trains and freight trains is located with the key route to the movement and logistics between Honshu Island and Hokkaido Island. Now that the speedup of the Shinkansen trains in the Seikan Tunnel and the extension of the Hokkaido Shinkansen to Sapporo are planned, the importance of the Seikan Tunnel will increase more and more in the future. We will reward the hard work of the forerunners who constructed the Seikan Tunnel and we will continue to maintain and manage it properly so that everyone can use the tunnel safely for a long time.

7 ACKNOWLEDGEMENT

The planning and design of this construction were carried out under the guidance by the Seikan Tunnel Technical Committee (Chairperson: Yoshiyuki Kojima (Railway Technical Research Institute), Advisor: Toshihiro Asakura (Kyoto University Emeritus Professor.)) We express our deep appreciation to everyone concerned.

REFERENCES

Aydan, Ö., Dalgıç, S. and Kawamoto, T., 2000. Prediction of squeezing potential of rocks in tunnelling through a combination of an analytical method and rock mass classifications, *Italian Geotechnical Journal*, 34: 41–45.

Inoue, T. 1986. Survey of the Seikan Tunnel, *Tunnelling and Underground Space Technology*, 1

ITASCA Consulting Group, Inc. 2013. FLAC3D Version 5.01 Theory and Background

Jiang, Y., Ezaki, T., Yokota, Y. and Kamuro, K. 1994. Quantitative Analysis of the Ground Characteristic Curve in Tunneling, *Proceedings of the 9th Japan Symposium on Rock Mechanics*: 767–772. (in Japanese)

Shimamoto, K., Yashiro, K., Kojima Y. and Asakura, T. 2009. Prediction Method of Tunnel Deformation Using Time-dependent Ground Deterioration Model, *Quarterly Report of RTRI*, 50(2): 81–88

Tunnels and Underground Cities: Engineering and Innovation meet Archaeology, Architecture and Art, Volume 3: Geological and geotechnical knowledge and requirements for project implementation – Peila, Viggiani & Celestino (Eds)
 ISBN 978-0-367-46583-4

A numerical study on effect of wall penetration on basal heave stability during vertical shaft excavation in soft ground

S.J. Kang, E.S. Hong & G.C. Cho
Korea Institute of Science and Technology, Daejeon, South Korea

ABSTRACT: The Previous study on the problems that occur during vertical shaft excavation is relatively insufficient although it is more directly related to the excavation. In particular, basal heave, a phenomenon that the excavation surface rises up, were studied little although the basal heave causes equipment damage or ground settlement. In this study, numerical analysis and experimental verification were performed to understand the effect of wall penetration, which is typically used to prevent the occurrence of the basal heave. The experimental study was conducted using a centrifugal model test for two cases depending on the presence of the wall penetration. The numerical analysis results were verified by comparing with the results from the centrifugal model test. The results show that the wall penetration suppresses the occurrence of basal heave and reduces the deformation even after the occurrence of the basal heave.

1 INTRODUCTION

1.1 *Research background*

As the population and facilities are concentrated in urban areas, the demand for creating new area such as underground space is increasing. In order to develop the underground space, it is necessary to consider the geotechnical engineering for underground tunneling and reinforcement works. Construction of underground space needs vertical shaft for ventilation and passage of excavating equipment. Since the vertical excavation proceeds with increasing depth, it is accompanied with the risk of ground failure. Therefore, the excavation method and the soil stiffness should be taken into consideration before the vertical shaft excavation. During vertical shaft excavation, piping, boiling, and basal heave phenomena occur representatively. The phenomena called boiling and piping occur due to the inflow of groundwater. The problems happen in the sandy soil with high permeability. On the other hand, the basal heave occurs in clayey soil with low permeability, making the excavation surface rise up vertically due to the weight of surrounding soil. The phenomena may cause serious accidents during the excavation, but the methods for consideration and prevention are different because the causes are not same each other. In this study, only basal heave was considered as the problem occurs during vertical shaft excavation in soft clay soil. Previous methods for preventing the basal heave have been ground reinforcement and wall penetration. In particular, the method penetrating outer wall of the vertical shaft below the excavation surface can be applied during excavation, and the penetrating depth can be controlled according to the characteristics of soil or excavation geometry. It is expected that application of wall penetration prevents the basal heave effectively with less cost than the soil reinforcement. So, the effect of wall penetration needs to be reflected properly in the consideration of the safety against basal heave for the optimal design of vertical shaft excavation.

1.2 *Previous studies related to the safety of vertical shaft excavation*

Previous studies related to the vertical shaft excavation have been conducted the researches for securing safety of vertical shaft after completion of excavation. The studies mainly deal with the earth pressure distribution acting on the outer wall of vertical shaft, considering the geometrical characteristics such as three-dimensional arching effect (Kim et al., 2012; Shin et al., 2005). Also, the deformation of vertical shaft caused by earth pressure was analyzed through various methods including actual experimental and field data (Shin et al., 2007). However, previous study about the safety against problems that can occur during excavation such as boiling, piping, and basal heave are relatively minor. The previous studies on the stability against basal heave are lacking also. Most studies are conducted research based on theoretically derived equations for the factor of safety against basal heave from Terzaghi and Peck (1967), and have conducted numerical analysis (Goh, 1994; Goh, 2017; Khatri and Kumar, 2010). The studies conducted with numerical analysis have analyzed the effect of each factors included in existing equation through parametric study and complemented the theoretical study by proposing additional factors for considering three-dimensional excavation and the effect of wall penetration. However, there is no verification of the numerical method using experimental or field data although the numerical results may show different trends according to used software. In this study, parametric study with FDM-based numerical software was conducted to complement theoretically derived equation considering the effect of three-dimensional geometry and wall penetration. Also, the numerical software used in this study was verified using the data from centrifugal model test which is conducted in previous study (Kang et al., 2017). As the result of the parametric study, additional factor reflecting the effect of three-dimensional geometry and wall penetration was proposed, and the existing equation was complemented.

2 NUMERICAL METHOD FOR CALCULATING FACTOR OF SAFETY

2.1 *Introduction of numerical software*

The numerical software used in this study is FLAC 3D (ver. 5.00), Fast Lagrangian Analysis of Continua in Three-dimensions. It solves differential equations with finite-difference methods (FDM), and this is designed for geotechnical analysis targeting soil, rock, constructs, supports, and so on. The main features of this software are that it is easy to analyze the mechanism of continuum in the application of civil or geotechnical situations and that it can simulate the behavior of large deformation, non-linear materials, or unstable condition. The steps for numerical analysis were designed as Table 1 in this study.

2.2 *Strength reduction method*

The factor of safety was calculated in FLAC 3D based on Strength reduction method, one of the representative method for obtaining factor of safety in numerical analysis. The governing equation of the method is expressed in Eq. (1) and Eq. (2). The representative factors reflecting soil stiffness, cohesion and friction angle, decrease until the numerical model gets failure. The decreased rate of factors when the failure occurs determined as the factor of safety of the numerical model.

Table 1. Steps for numerical analysis.

Steps for numerical analysis
Formation of zones for initial ground/Entering properties
Application of gravity/Equilibrium reached
Simulation of vertical excavation
Installation of the lining as supports
Calculation of differential equations to find the factor of safety

$$c^{trial} = 1/F.S.^{trial}c_{initial} \tag{1}$$

$$\emptyset^{trial} = \arctan\left(1/F.S.^{trial}tan\emptyset_{initial}\right) \tag{2}$$

2.3 *Verification of numerical software*

The result of numerical analysis was verified using the data from centrifugal model test. The centrifugal model test was conducted with two cases with and without wall penetration. The effect of the wall penetration was analyzed in the actual test and compared with the result from numerical analysis. Figure 1 shows the overall setup of centrifugal model test, and Figure 2 shows the test cases composed with and without wall penetration. The effect of wall penetration was found that it suppresses the occurrence of basal heave until high gravitational acceleration and that it also reduces the deformation through the basal heave throughout the test. Figure 3 shows the results of centrifugal model test, deformed soil at 50 g of gravitational acceleration. The installation of wall penetration reduced the vertical displacement of excavation surface by 78%. The test procedure was simulated identically using FLAC 3D (Figure 4), and the results from two methods were compared. In the results of numerical simulation, the displacement of excavation surface was also decreased due to wall penetration, and the reduction rate was 75%. Since the reduction of displacement from two methods showed similar trends, it was found that the numerical software has well reflected the actual condition.

Figure 1. Test equipment.

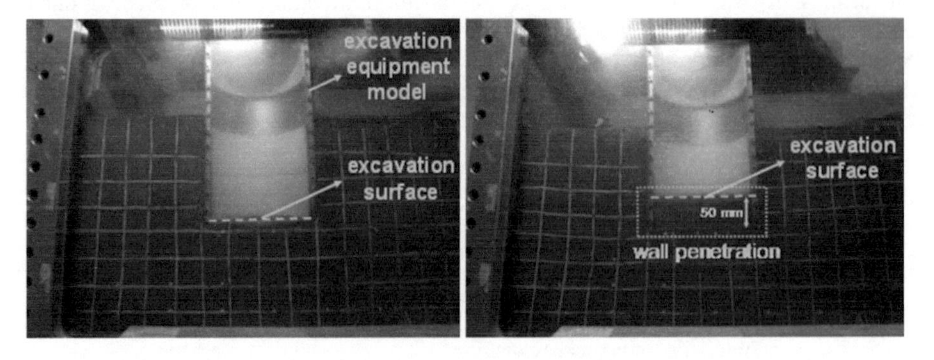

Figure 2. Test cases with and without wall penetration.

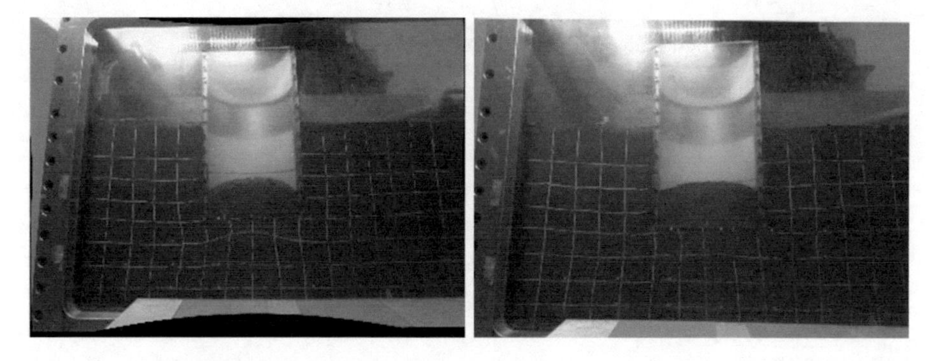

Figure 3. Results of centrifugal model test.

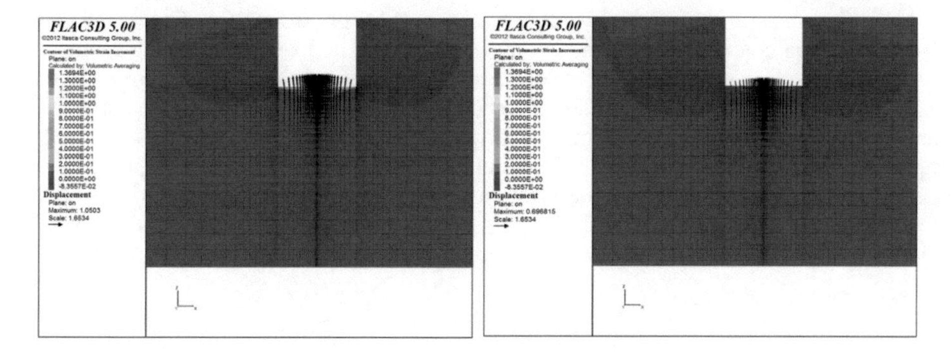

Figure 4. Results of numerical analysis.

3 PARAMETRIC STUDY

3.1 *Equation for factor of safety as a standard*

As a standard of numerical study, the existing equation from Terzaghi and Peck (1967) was considered. The equation was theoretically derived with assumption of two-dimensional shape and is expressed as Eq. (3). In the equation, c is cohesion, γ means unit weight, H means excavation depth, B means excavation diameter, and q is external surcharge. The main variables are cohesion, unit weight, excavation depth and excavation diameter, so the parametric study was conducted on those variables included in the equation.

$$\text{F.S.} = 5.7c/(\gamma H - \sqrt{2}cH/B + q) \tag{3}$$

3.2 *Input parameters*

For the parametric study, cohesion, unit weight, excavation depth, and excavation diameter were determined as input parameters, and they were analyzed in the process of deriving factor of safety to obtain the effect of each parameter on the factor of safety. Input values were set as shown in Table 2. The range of input values includes the lower and upper part of the value which is composed in the centrifugal model test without wall penetration. As one of the input values varies, other parameters were fixed as the value of centrifugal model test for distinguishing the effect of each parameter. For consideration of wall penetration, the depth of penetrated wall was controlled also. The depth of wall penetration is usually determined according to the excavation diameter, and the range of the input variable was set as from 0 % to 45% of standard excavation diameter, 10 m.

Table 2. Input parameters.

Input parameter	Input values					
Cohesion [kPa]	18.9	25.3	31.6	37.9	44.2	50.5
Unit weight [kN/m^3]	13.7	15.7	17.6	19.6	21.6	23.5
Excavation diameter [m]	2.5	5	10	15	20	30
Excavation depth [m]	5	10	15	20	30	50
Wall penetration depth [m]	0	0.5	1.5	2.5	3.5	4.5

3.3 *Result of parametric study*

As the result of the parametric study, the influences of each parameter on the factor of safety were found as shown in Figure 5 to Figure 8. Since the methods for deriving factor of safety are different, the exact values of factor of safety derived from two methods may not be same, but there found specific trends of variation according to the changes of parameters. The factor of safety with changing soil characteristics, cohesion and unit weight, had similar changing trends in two deriving methods. However, the results with changing parameters related to the geometry, excavation depth and diameter, showed different trends. It seems that the effect of three-dimensional geometry was reflected in FLAC 3D, but the theoretical equation could not reflect it properly. For comparing the effect of wall penetration from numerical analysis and theoretical equation, the typical assumption of considering wall penetration in theoretical equation was used, and the wall penetration was applied as the increased weight of soil inside the vertical shaft. There were also similar trends though the result of a case with zero wall penetration had inconsistent value.

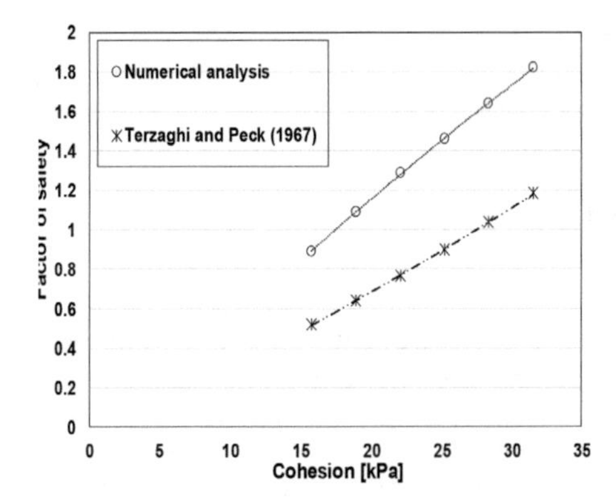

Figure 5. Effect of cohesion on factor of safety.

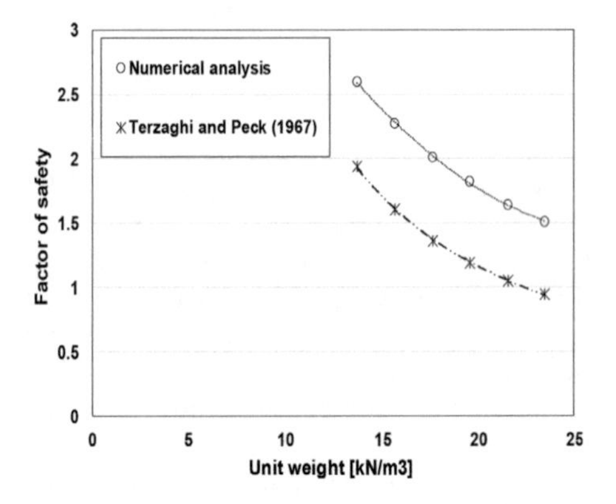

Figure 6. Effect of unit weight on factor of safety.

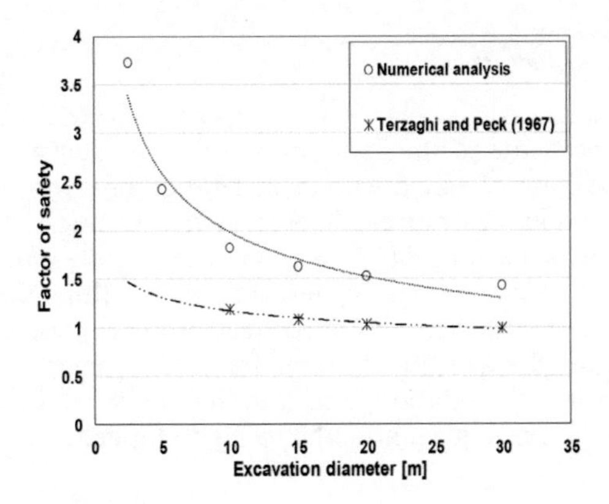

Figure 7. Effect of excavation diameter.

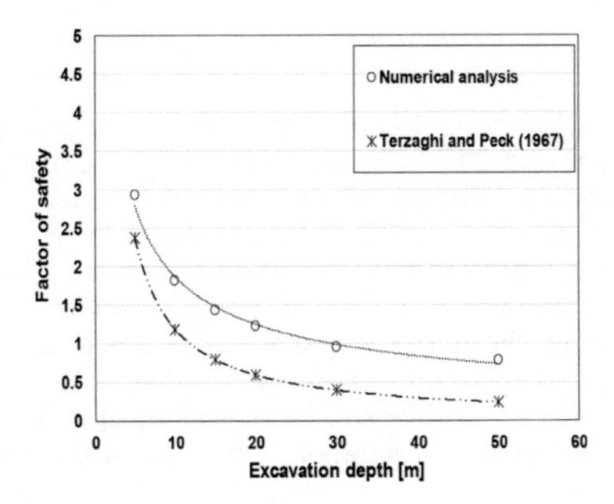

Figure 8. Effect of excavation depth.

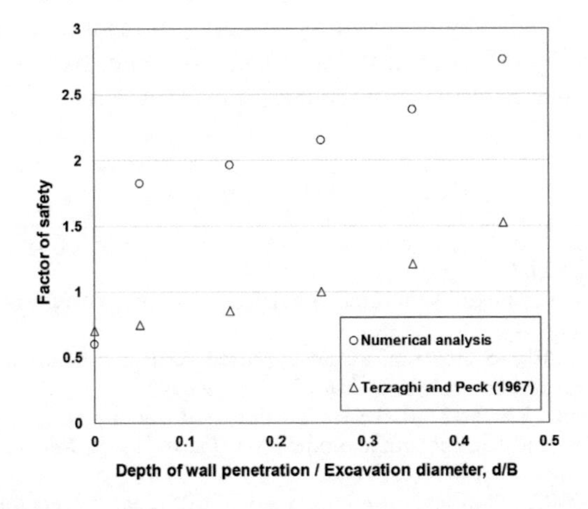

Figure 9. Effect of wall penetration.

3.4 *Proposal of additional factors*

Since the existing theoretical equation has not properly reflected the actual condition with three-dimensional geometry due to the assumption of two-dimensional shape, the factor of safety equation has to be complemented with consideration of three-dimensional shape. Also, the effect of wall penetration was not considered in the equation, so the additional factor including the effect of wall penetration is needed. Two additional factors were proposed according to the result of parametric study, and they each reflects the effect of three-dimensional shape and wall penetration. The additional factors are expressed as Eq. (4) and Eq. (5). In the equations, α_s means the additional factor for three-dimensional shape, α_d means the additional factor for the wall penetration, and d means the depth of wall penetration. These factors can complement the existing equation by being multiplied

$$\alpha_s = 1.05B^{-0.06}D^{0.42} \tag{4}$$

$$\alpha_d = 1 + 0.585(d/B - 0.5) \tag{5}$$

4 CONCLUSIONS

In this study, numerical analysis using FLAC 3D was conducted to complement existing theoretical equation for basal heave stability. The used software was verified with data from centrifugal model test by comparing the displacement of excavation surface after occurrence of basal heave. Parametric study with main factors in existing equation showed the similar trends of factor of safety obtained from two methods, numerical analysis and theoretical equation. However, it was found that the theoretical equation assuming two-dimensional shape has not properly reflected the actual three-dimensional condition, so two additional factors were proposed by the results of parametric study to complement the two-dimensional equation. Two additional factors may act as a subsidiary factor for considering the basal heave stability in three-dimensional shape with installation of wall penetration. It is expected that using the factors will help in optimal design of vertical shaft excavation ensuring the safety against basal heave.

ACKNOWLEDGEMENTS

This research was supported by a grant (18SCIP-B105148-04) from the Construction Technology Research Program funded by the Ministry of Land, Infrastructure, and Transport of the Korea government and by Korea Minister of Ministry of Land, Infrastructure and Transport (MOLIT) as 「U-city master and Doctor Course Grant Program」

REFERENCES

Goh, A. 1994. Estimating basal-heave stability for braced excavations in soft clay. *Journal of geotechnical engineering* 120(8): 1430-1436.

Goh, A. 2017. Basal heave stability of supported circular excavations in clay. *Tunneling and Underground Space Technology* 64: 145-149.

Khatri, V. N., Kumar, J. 2010. Stability of an unsupported vertical circular excavation in clays under undrained condition. *Computers and Geotechnics* 37(3): 419-424.

Kim, K.Y., Lee, D.S., Jeong, S.S. 2012. Analysis of earth pressure acting on vertical circular shaft considering aching effect (I): a study on centrifuge model tests. *Journal of the Korean Geotechnical Society* 28 (2): 23-31.

Shin, Y.W., Sagong, M. 2007. A rational estimating method of the earth pressure on a shaft wall considering the shape ratio. *Journal of Korean Tunnelling and Underground Space Association* 9(2): 143-155.

Terzaghi, K., Peck, R. 1967. *Soil mechanics in engineering practice.* New York: John Wiley and Sons.

Tunnels and Underground Cities: Engineering and Innovation meet Archaeology, Architecture and Art, Volume 3: Geological and geotechnical knowledge and requirements for project implementation – Peila, Viggiani & Celestino (Eds)
 ISBN 978-0-367-46583-4

Numerical study of portal displacement and stress-strain distribution mechanism at a tunnel portal

B. Khadvi Borujeni
Amirkabir University of Technology, Tehran, Iran

ABSTRACT: Failures commonly occur in high angle approach cuts and initial subsurface of tunnel portal. Because the portal comprises an interface between the surfaces and subsurface, both rock slope and tunnel engineering approaches are needed for portal design. Also, specific information in portal related failures do not exist and traditional design methods are inadequate for utilization in complex three dimensional area. In this paper, portal of a tunnel is simulated by FDM method. Among many factors which are affected the portal stability, trench angle, cohesion and friction angle has been studied. The friction angle in the main model is 36 degree and cohesion is 8 MPa. Also, trench angle has decreased and increased 15 degree, cohesion and friction angle increased and decreased 15 percent and their effect on portal stability and displacement studied separately. Finally, based on tunnel span and plastic zones, 2D and 3D stress-strain field has distinguished.

1 INTRODUCTION

In order to analyze and simulate a model, first of all, the critical area have been selected based on the geotechnical reports. In this research, portal is the critical area and a portal of a road tunnel in Chaharmahal and Bakhtiyari province in Iran selected. After portal simulation, based on portal definition, trenches have excavated through three stages and the exposure of external part of the portal has showed. After that, inner part of portal has excavated through 4 stages that each of them has 3 meter in length. In the previous case studies which was held by Roger in 1989, the crown face of tunnels collapse more than other part of portal's part (Rogers,1989). Therefore, the effect of upper trench angle have been studied. Finally, distribution of plastic zones and strain-stress field in portal have been studied.

2 DEFINITION OF PORTAL

On the basis of a good work by Rogers about the portal of tunnel, the portal definition is: "The approximately horizontal entrance (e.g. i 35° from horizontal) to an underground excavation, which typically consists of both an approach cut and a section of underground entry. The minimum length of the portal zone is considered to be approximately two spans, one span out by (e.g. outward) the portal interface, and one diameter in by (e.g. inward) the portal interface" (Rogers, 1989). Figure 1 presents a portal zone and its features.

3 NUMERICAL MODEL AND PARAMETERS

In this study, a three dimensional model is employed for portal simulation. This model is developed by using the FDM method. Figure 2 presents dimensions and volume of the model which is 280×170×130 m^3 (about 1million zones). The tunnel centerline has a depth of about

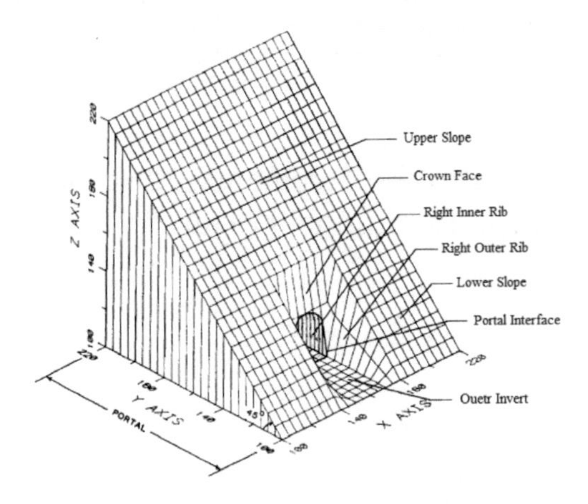

Figure 1. External and internal views of a Portal zone (Rogers, 1989).

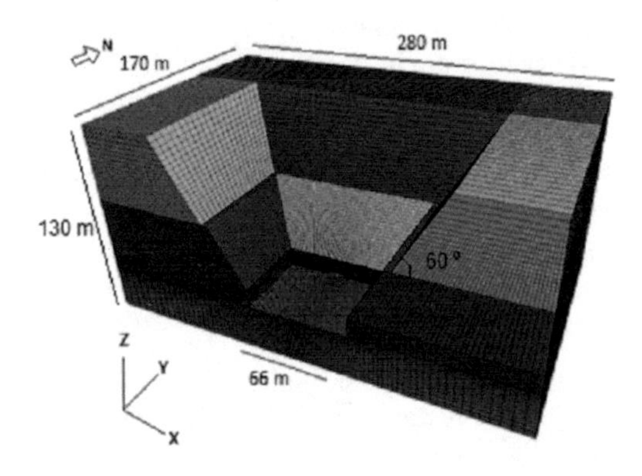

Figure 2. Dimension of simulated model.

Table 1. Mechanical properties of the rock according to Mohr-Coulomb model.

Friction angle (degree)	Cohesion (MPa)	Density (kg/cm^3)	Young Modulus	Poisson's Ratio (GPa)
36	8	2500	11	0.27

100–115 meter. The outer part of the portal is divided into 3 separately trenches. In the base model, all of the trench angles are 60 degrees. The height of trenches are 10, 50 and 50 meter respectively. Based on the case study, the portal designed in a typical horse shoe. The tunnel span is about 11 meter and its area is 100 m^2.

Some rock samples were obtained from exploratory borehole and prepared for tests by the proposed ISRM method (Brown, 1981). As noted in the ISRM, the ratio of length to diameter in the rock samples should be in the range of 2.5 to 3. Since the exploratory borehole diameter was 60 mm, samples were about 20 mm in diameter. The mechanical property of material used in the numerical simulation are listed in Table 1. bounderies are placed far enough from the tunnel center and trenches to avoid of boundary effect on calculation.

Based on the Equation 1 and Equation 2, bulk and shear modulus have been determined.

$$K = E/3(1 - 2v) \quad (1)$$

$$G = E/2(1 + v) \quad (2)$$

Where; K = bulk modulus; G = shear modulus; and v = Poisson's ratio.

It is very important to determine the rock mass properties for numerical simulation in every underground excavation for which geological strength method is widely used (Mahanta et al, 2016). The design and successful execution of any rock engineering projects require careful determination of deformation modulus along with other rock mass properties and choose a practical consecutive model like MC model (Panthee et al, 2016). Bulk and shear modulus are 4.33 GPa and 7.97 GPa respectively. Also tensile strength is between 0.5–0.7 MPa.

4 NUMERICAL ANALYSIS

After simulating the model and assign material properties, trenches are excavating thorough 3 stages. Then, tunnel is excavating in portal zone. Normally, after each step, the tunnel's advance is approximately 3.5 meter. To a large extent, the tunnel's advance rate depends on rock mass and other geomechanical properties. After removing the muck, with regards to the estimated displacement and in order to control the tunnel convergence, a proper support system has been installed after each excavation phases. There are some methods to accommodate these loads during simulation, such as "internal pressure reduction method". These techniques allow a certain amount of displacement at the tunnel face and the unsupported upper section of the tunnel. Due to this geomechanical properties in this case study, installing support system is not needed.

There are a number of practices available to examine the stability of a tunnel, such as empirical method, physical method, mathematical method and numerical method (Verma et al, 2010; Kainthola et al, 2012; Dindarloo et al, 2015). Physical models have many limitations, such as proper selection of structural models, nonlinearity of rocks, and high cost of experimental setup (Yang et al, 2010). Recent development in numerical methods delivers strong supports for the behavioral study of underground constructions under different geoenvironmental conditions. Numerical models are computer programs which can simulate the mechanical behavior of a rock unit subjected to a set of predefined initial environments like boundary conditions, in-situ stresses, and geometry (Kainthola et al, 2012).

There are numerous of numerical methods available today. This study concentrates on the ground displacement due to excavation of portal of a tunnel. For this purpose, 3D finite difference numerical analysis software engaged. The simulated model in Figure 1 is showing dimensions of tunnel, topographic undulation, boundary conditions.

In situ measurements and analytical modelling of excavations show that an area of 2d (d is the span of tunnel) is mostly affected in terms of stress redistribution and resulting strain (Brown, 1981; Kontogianni et al, 2008). Also, due to the importance of trenches' displacement in this research, in order to minimize the effect of boundary condition on the deformational behavior of portal zone, the outer boundary was constructed twenty times the diameter of the tunnel's diameter (four times of trench width). Fixed restraints is applied to the bottom and sidewalls of the model (i.e. no movement is in X and Y directions), while top ground surface is left free (Fig 1).

The geomechanical properties of rock mass which used in numerical simulation describe in Table 1. For complex in topographic conditions such as hills or valleys, rock mass is assumed to be under gravitational load (Zhang, 2013). Hence, gravitational load applied to the model and actual undulating ground surface selected. One case have been compared in this study: unsupported tunnel models.

Unfortunately, a few publication is found related to the portal zone. There is just a comprehensive field investigation on portal failure types which was published by Rogers's in 1988. The vulnerable zones were chosen in portal of the tunnels to monitor the deformation pattern under unsupported conditions. These locations were placed at crown face of tunnel, right wall of tunnel and invert level of the tunnel. The displacement contours are shown in Figures 3–4.

5 RESULTS AND DISCUSSION

Figure 3 shows a plan of ground displacement which is parallel to XY plane when trench excavated (before tunnel excavation). The maximum displacement occurred in trenches wall and crown face of tunnel. Also, Figure 4 indicates ground displacement that is parallel to the tunnel span. Maximum displacement occurred in the corner of trenches in portal area.

5.1 *Upper trench angle*

After trench excavation, tunnel excavated through four stages. Three first stages have 3.5 meter length. As expected, maximum displacement occurred in crown face that confirmed field observation. Also, based on case studies which were developed by Roger, most of collapse in portal area divided into crown face over break and upper slope collapse. Figure 5 shows displacement vectors in a plan parallel to tunnel axis. It presents maximum displacement vectors at crown face part. Hence, upper trench angle selected as a variable to study its effect on amount of displacement. Figure 6 presents crown face displacement in three modes and in all steps of portal excavation. In other words, due to the fact that in this case study failure occurred, sensitivity analysis of trench angle is given as a factor to evaluate its impact on potential failure.The upper trench angle in the base model is 60 degree. By increasing and decreasing 15 degree, the displacement increase 85 and decrease 47 percent in average respectively. The increasing rate is more than decreasing rate because the mass of upper slope tend to move toward tunnel span when slope angle increased.

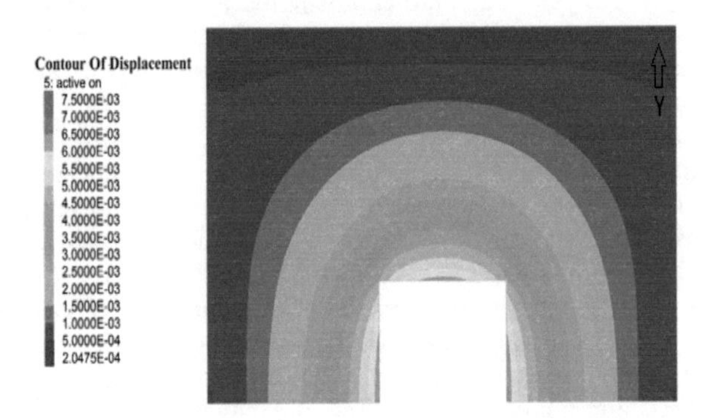

Figure 3. Ground displacement in a parallel plan to XY plane that crossed from tunnel span.

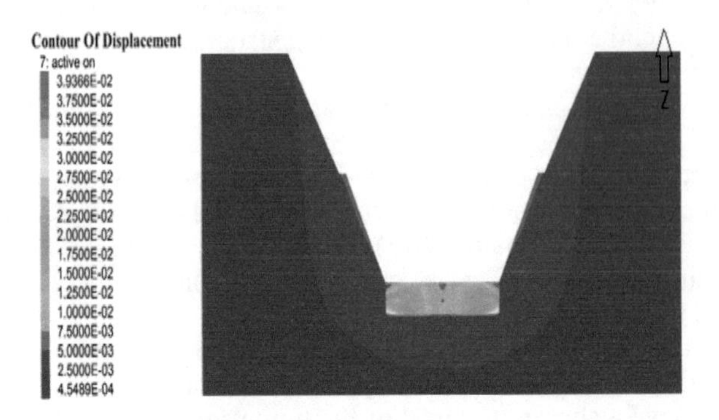

Figure 4. Ground displacement in contact with tunnel span (Y=0, XZ plane).

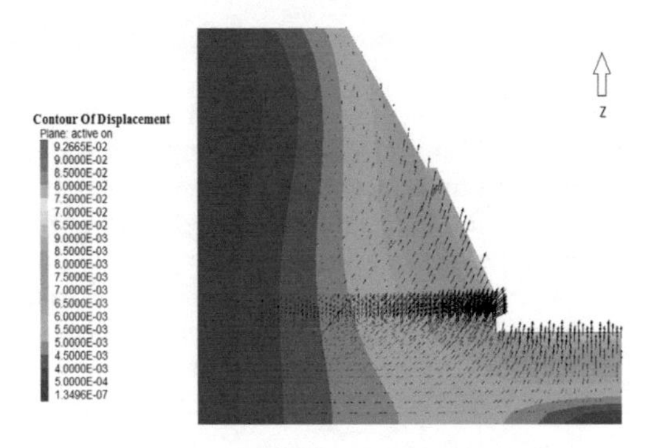

Figure 5. Displacement vector and contour in a plan parallel to tunnel axis.

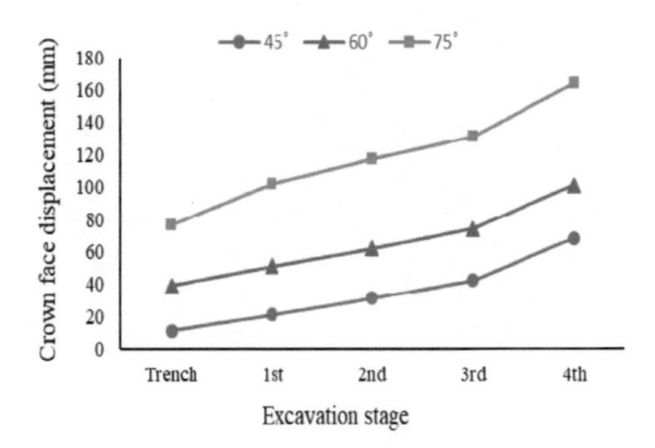

Figure 6. Displacement comparison in different upper trench angle.

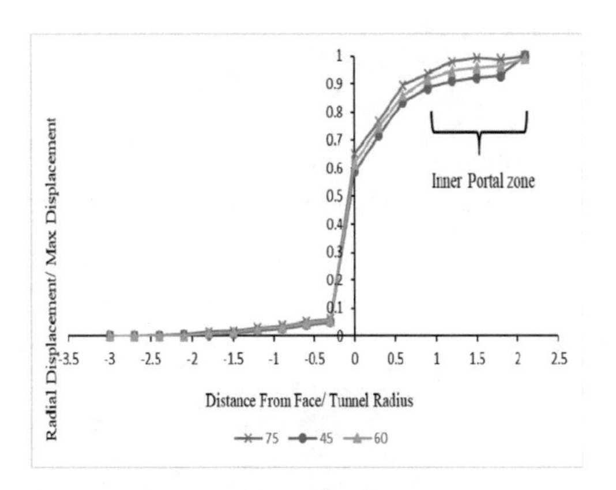

Figure 7. LDP curve.

5.1.1 Longitudinal Deformation Profile (LDP)

In this study, LDP curve drawn from nodes that placed on the crown of tunnel where maximum displacement occurred. Figure 7 presents the LDP curve. Maximum radial displacement occurred in the portal area and by increasing upper trench angle, maximum displacement increased. The amount of maximum radial displacement in this zone were 1.5, 3 and 7.5 centimeter respectively.

5.2 *Zone Status*

Figures 8–9 present zone status when trenches excavated completely. It is interesting to note that distribution of plastic zones did not extended over inner portal zone. In the green zones, plastic displacement and failure did not occur and these zones are in elastic state. During the model running in Mohr-Coulomb failure criteria, displacement occurred in red zones but they are without any displacement now. During the trench excavation, maximum tensile stress is more than tensile strength. Therefore, displacement occurred and based on the amount of displacement failure could be happen. By comparison amount of displacement, suitable support system should be install. Blue zones are important compared to the other zones, because they placed in portal zone and failure could be occurred in these zones.

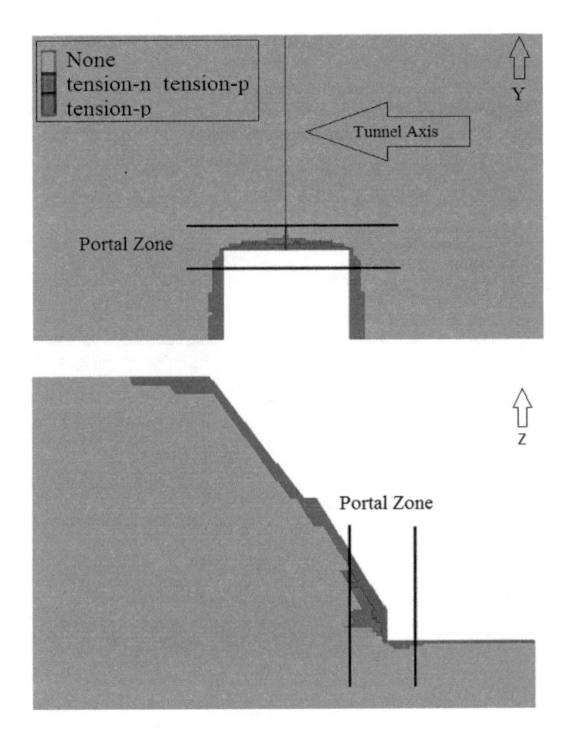

Figure 8. Zone status in portal area (up: plane parallel to tunnel axis and crossed from middle of tunnel. down: parallel to tunnel axis and ZY plane and crossed from middle of tunnel).

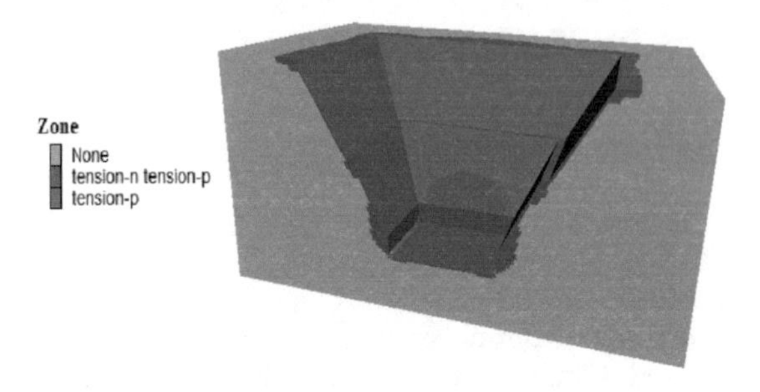

Figure 9. Zones status in outer portal area.

5.3 *Cohesion and friction angle*

Another important factors in this research is cohesion and friction angle of material in portal zone. Trench angle is 60 degree and cohesion is 8 MPa at the base model. It increased and decreased 15 percent respectively while other parameters remain constant. In addition, the effect of friction angle on portal displacement studied. Friction angle of model is 36 degree and increased and decreased 15% respectively. Figure 10 presents the effect of cohesion and friction on portal excavations while all other parameters were constant.

5.4 *Stress-Strain measurement*

Figure 5 presented displacement when trench excavated (outer part of portal). Figure 11 presents failure at crown face in portal zone. Figures 12–14 present the principle strains in a longitudinal profile along with the axis of tunnel. This profile selected based on the maximum displacement which occurred in the simulation. With the aid of these profiles, the maximum area of three dimensional strain can be determined. The maximum principle strain begin to reduce and in the middle of tunnel (60 meter from tunnel opening) reached to zero. Hence from this part of tunnel to the end, strain rate will be two dimensional.

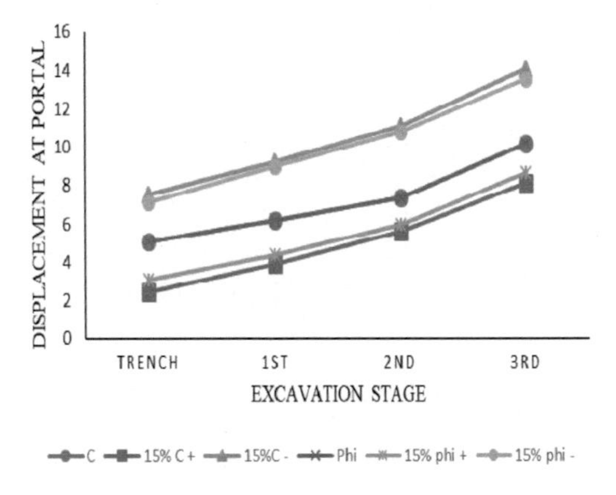

Figure 10. Displacements in different cohesion and friction amount in different step of portal excavation.

Figure 11. Crown face over break and upper trench collapse.

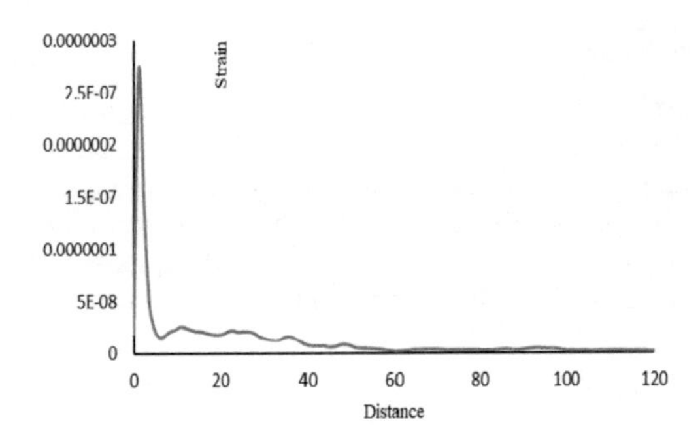

Figure 12. Minimum principle strain rate in a longitudinal profile at roof of tunnel.

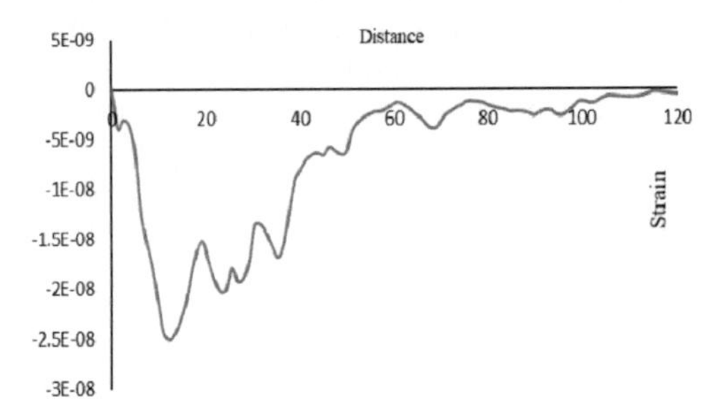

Figure 14. Intermediate principle strain rate in a longitudinal profile at roof of tunnel.

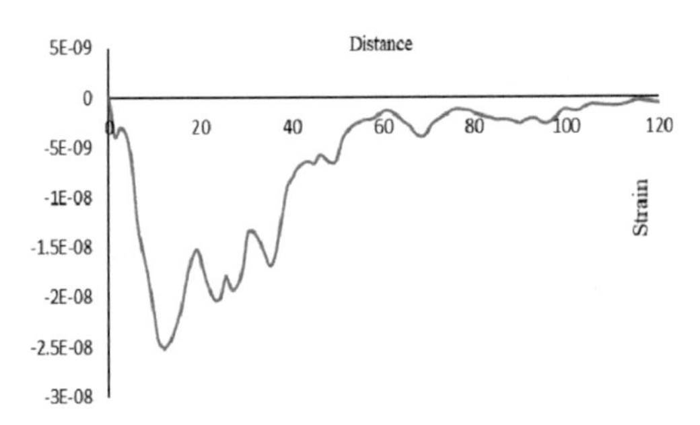

Figure 13. Maximum principle strain rate in a longitudinal profile at roof of tunnel.

5.4.1 Stress measurement

Figure 15 shows a lateral profile in portal zone where the maximum displacement occurred. Figure 16 presents the principle stresses in a lateral profile above the tunnel span when outer part of portal zone excavated. The shear stresses are zero and these are principle stresses in x, y and z direction. At corners, stress concentration is more than other places. Minimum principle stress is zero because trench excavated. Figure 17 presents the principle stresses in a lateral profile which is parallel to the previous profile and in the end of portal zone in tunnel (Y=10 m). At corners, stress concentration is more than other places. Minimum principle stress is a bit more than zero in this profile. Also, Figure 18 presents the principle stresses in the lateral profile in the middle of tunnel. In this area, field stress distribution mechanism is three dimensional.

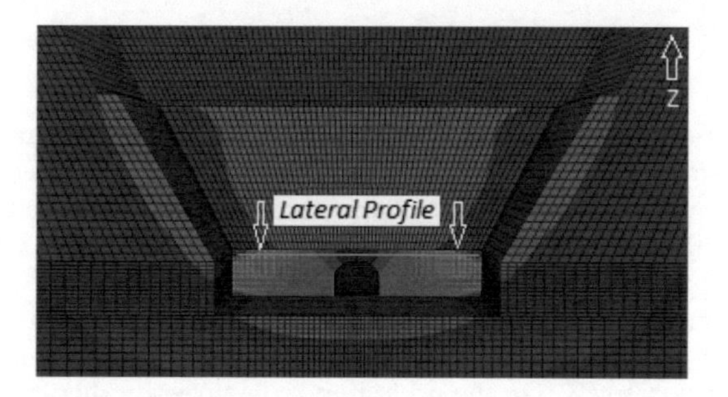

Figure 15. Lateral Profile.

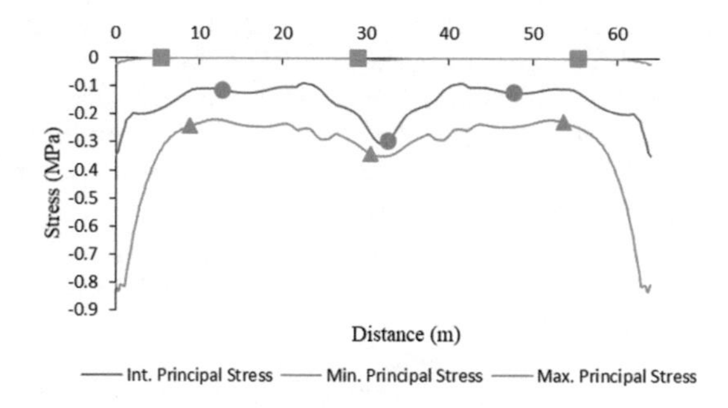

Figure 16. Principle stresses in a lateral profile in portal zone (Y=0.1 m).

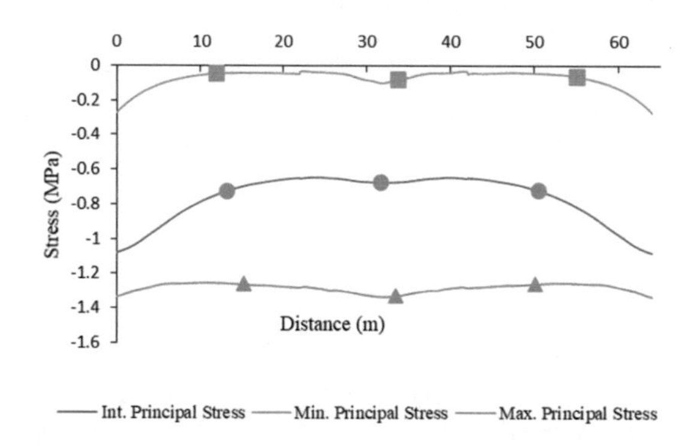

Figure 17. Principle stresses in a lateral profile in portal zone (Y=10 m).

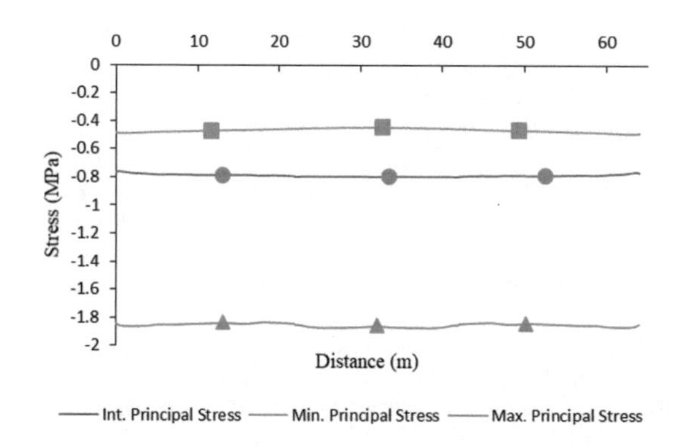

Figure 18. Principle stresses in a lateral profile in middle of tunnel (Y=60 m).

6 CONCLUSION

In this paper, the effect of three factors including trench slope, internal friction angle and cohesion on portal stability studied. Each of the above factors considered with regard to the constancy of other factors and the effect of increasing and decreasing on the tunnel portal stability and displacement rate have been studied separately. Also stress-strain distribution mechanism in portal area have studied by profile and histories.

The trench slope angle is one of the factors affecting the stability of the portal. By Increasing and decreasing 15 percent, the displacement in the portal of tunnel increase 80 percent and decrease 47 percent respectively. On the other hand, the growth rate of displacement is higher than its lowering rate, which may be due to the location of the center of the trench mass toward the portal. This causes the material to move towards the portal due to its weight.

Another effective factor is cohesion which has been studied. As cohesion increases, the displacement rate in the portal decreases and vice versa. The rate of increasing and decreasing of displacement have measured on average by 46% and 30% respectively.

Another factor that has been studied is the angle of friction. By increasing the amount of friction angle, the displacement in the portal decreases and vice versa. The average increasing and decreasing were in average 41 and 23 percent respectively.

The plastic zones extend over the surface of the tunnel and upper trench portal and their expansion is limited by portal zone. Most of the failures happened within the portal range in these zones and case study confirmed that.

Due to the fact of portal definition, three-dimensional strain range is 2.5 times of tunnel area from the beginning of the tunnel. After that, two-dimensional strain dominates in the tunnel.

The main components of stress were studied by profiles and histories. In the initial part of the portal range, the shear stress components are almost zero and the value of σ_{yy} is also near zero. Therefore, the stress distribution mechanism in this range is two dimensional. Thereafter, the stress distribution is three-dimensional by moving towards the center of the tunnel. By analyzing stress mechanism at the portal of tunnel as well as the slope stability, consideration of three-dimensional mechanism for the entire portal zone considered to be quite favorable and conservative.

REFERENCES

Brown ET. 1981. Rock characterization, testing and monitoring: *ISRM suggested methods*. Published by Pergamon Press.

Dindarloo SR, Siami-Irdemoosa E. 2015. Maximum surface settlement based classification of shallow Tunnels in soft ground. *Tunnelling and Underground Space Technology*; *49:320e7*.

Kainthola A, Singh PK, Wasnik AB, Sazid M, Singh T. Finite element analysis of road cut slopes using Hoek & Brown failure criterion. International Journal of Earth Sciences and Engineering 2012; 5 (5):1100e9.

Kontogianni V, Papantonopoulos, Stiros S. 2008. Delayed failure at the Messochora tunnel, Greece. *Tunnelling and Underground Space Technology*; 23:232e40.

Mahanta B, Singh HO, Singh PK, Kainthola A, Singh TN. 2016. Stability analysis of potential failure zones along NH-305, India. *Natural Hazards*; 83(3):1341e57.

Panthee S, Singh PK, Kainthola A, Das R, Singh TN. 2016. Comparative study of the deformation modulus of rock mass. *Bulletin of Engineering Geology and the Environment*.

Rogers, G. K. 1989. "The Stability of Portals in Rock", Thesis (PhD), *Virginia Polytechnic Institute and State University*, 334 pages.

Verma AK, Singh TN. 2010. Assessment of tunnel instability e a numerical approach. *Arabian Journal of Geosciences*; 3(2):181e92.

Yang Y, Xie X, Wang R. 2010. Numerical simulation of dynamic response of operating metro tunnel induced by ground explosion. *Journal of Rock Mechanics and Geotechnical Engineering*; 2(4):373e84.

Zhang L. 2013. Engineering properties of rocks. *Journal of Chemical Information and Modeling*; 53 (9):1689e99

Tunnels and Underground Cities: Engineering and Innovation meet Archaeology, Architecture and Art, Volume 3: Geological and geotechnical knowledge and requirements for project implementation – Peila, Viggiani & Celestino (Eds)
 ISBN 978-0-367-46583-4

Forensic study on tunnel collapse in deep and fractured rock

K.H. Kim, J.H. Jeong & J.H. Shin
Department of Civil Engineering, Konkuk University, Republic of Korea

S.H. Kim
Department of Civil Engineering, Hoseo University, Republic of Korea

S.S. Jeong
Civil and Environmental Engineering, Yonsei University, Republic of Korea

ABSTRACT: A tunnel collapse occurred in a deep fractured rock. The cover depth of the tunnel was over 70m, and its height and width were 7.7m and 11.6m, respectively. The length of tunnel collapse was 14m and total volume of rock fragments flowed into the tunnel was about 600m^3. To identify the failure mechanism, detailed investigation including additional rock drilling, theoretical review and numerical back analysis were performed. Vertical and horizontal drilling revealed that the rock was severely fractured with faults, and the area of rock fall developed to the distance of 14.4m from the tunnel crown. It was revealed that the blast excavation disturbed the rock mass and triggered raveling of rock fragments, and finally caused mass rock fall. Measures reinforcing the ground around the tunnel and further excavation method and procedure were proposed.

1 INTRODUCTION

Many difficulties are faced when tunnel excavation in mountainous terrain with faults and fracture zones. Although, the technology of tunnel construction has been greatly improved due to the development of various auxiliary methods and reinforcing materials (Erdem & Solak, 2005; Ochi et al., 2007), but collapses still occur frequently due to the limit of investigation and uncertainty involved in design and construction of tunnels. This paper investigated the collapse of a mountainous terrain tunnel with depth of 70m.

In-situ drilling investigation for tunnel design is only a line survey, which is a very small part of the entire tunnel path. Moreover, there is a problem that the precision of the geophysical exploration is doubtful. Unexpected boulder and core stones in the area of tunnels often cause collapses by the gravity action (Shirlaw et al., 1988). The major factors governing the tunnel collapse are the ground relaxation and joint development at the tunnel face (Barton & Abrieu, 2007), poor contact of shotcrete and steel rib, and ground disturbance due to blasting (Shin et al., 2011).

At the collapse site, it was found that the ground condition during the tunnel excavation was significantly different from the ground profile evaluated at the design stage. In order to analyze the cause of the collapse, additional ground investigation was carried out to identify the rock profile. Drilling results showed the extremely fractured zone which was not predicted at the design stage. In order to investigate the failure mechanism, failure mode of the collapse site was estimated and numerical analysis was conducted. The recovery method of the collapse tunnel was proposed and the modification of support design was suggested.

2 TUNNEL DESIGN AND CONSTRUCTION

2.1 *Ground condition*

The tunnel where the collapse occurred was a road tunnel planned as a twin tunnel. The preliminary geological survey indicated that the amphibole granite was a main rock type. The boring result for design was shown in Figure 1(a). TCR and RQD were measured through drilling, and unconfined compression tests and triaxial compression tests of drilling sample were performed to determine compressive strength. Base on this, RMR was determined using field tests and laboratory tests. The strata covering the tunnel were decomposed rock, soft rock and weathered rock from the ground surface. Electrical resistivity survey and seismic survey were also performed. The rock mass classification of collapse area was grade III at design stage.

During construction, face mapping for each excavation round was conducted. Figure 1(b) shows the result of face mapping just before the collapse. Weathering has severely developed, and consequent excavation surface showed mixed ground condition between decomposed rock and weathered rock. A fractured zone was found at the left side of the face. Joints were generally wide and filled with cohesive gouges. RMR score was 7 ~ 14, and the rock classification grade was V.

2.2 *Tunnel design and construction*

The tunnel is 700m long, 11.6m wide and 7.7m high. When designing, the rock mass classification grade was III, and it was expected to be constructed by the full face cut. However, during excavation, significant site differing condition was found. The rock mass classification was revaluated as the grade of V based on the results of face mapping and instrumentation. The support and excavation method have been modified. To reinforce the support additional, multi-level pipe roof method was adopted at the crown, and a rock bolt was added to the tunnel sidewall. The final lining was reinforced with rebar. The tunnel section was divided into upper and lower half for excavation and ripper was used to excavate the rock.

Figure 2(a) shows the excavation profile just before the tunnel collapse. The inbound tunnel was excavated to 246m from the portal in the upper half and 157m in the lower half.

The accumulated convergence and crown displacement were shown in Figure 2(b). Displacement at the point No. 1 ~ 3 were generally safely converged, but at the point No. 4 ~ 8,

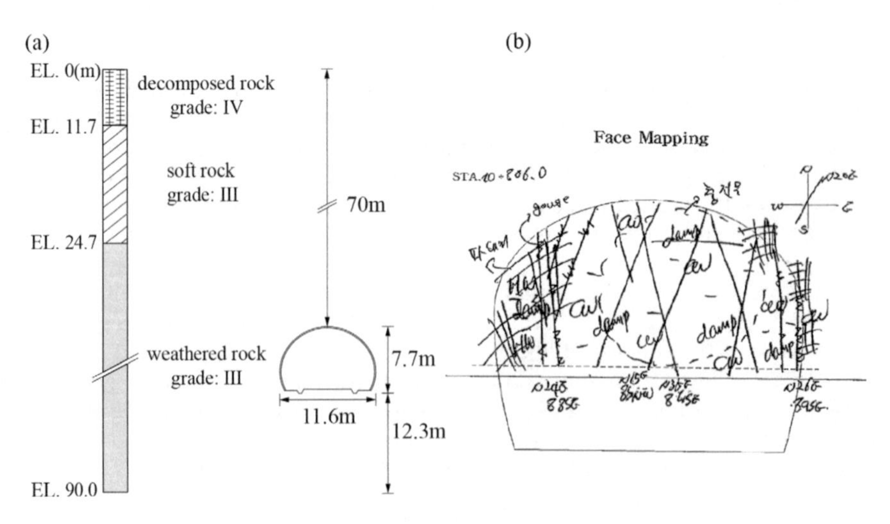

Figure 1. Ground profile and face mapping, (a) initial ground profile and (b) face mapping before tunnel collapse.

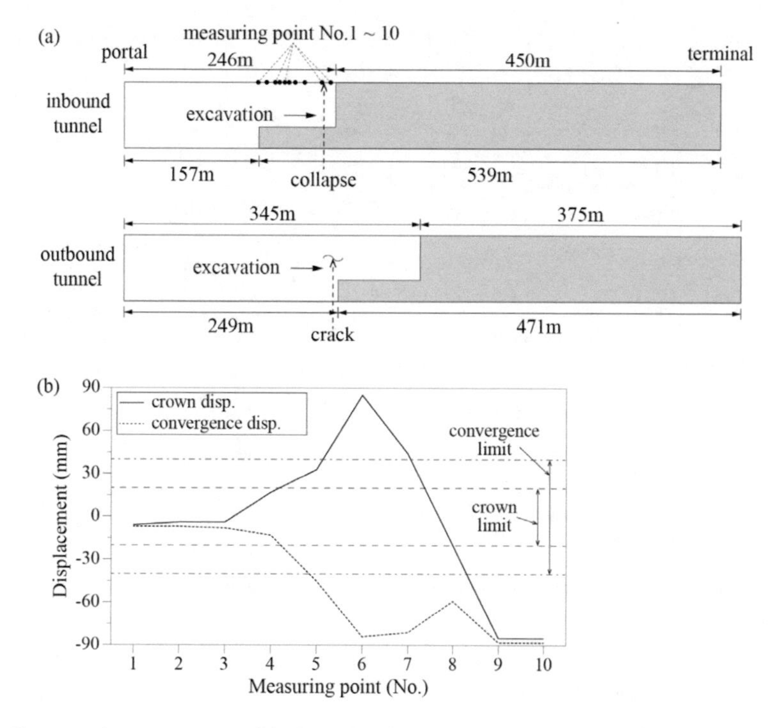

Figure 2. Construction progress and instrumentation results, (a) excavation progress and (b) accumulated displacement.

large displacement exceeding the management limit occurred. This implied the possibility of tunnel collapse. The collapse occurred at the location of point No. 9 and 10.

3 CAUSE OF TUNNEL COLLAPSE

3.1 *Collapse status*

Figure 3(a) shows the inflow of rock falls due to collapse. The location of collapse occurred at 232 m from the portal of inbound tunnel. The collapse of the inbound tunnel was started at the crown, and the rock fragment flow in the tunnel. The length of collapse was 14m and the amount of collapsed rock fragment was about 600m^3. A 280m^3 of counterweight fill was surcharged as an emergency measure, but, one day after the collapse of the inbound tunnel, the side wall of outbound tunnel cracked at 234m from the portal. The length of the crack was 16m. Additional counterweight fill of 1,300m^3 surcharged to prevent from further deformation and crack. Figure 3(b) shows the sidewall cracks of the outbound tunnel.

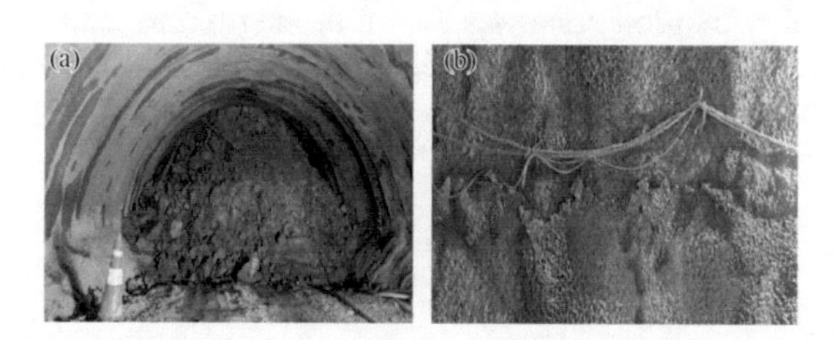

Figure 3. Collapse and crack, (a) collapse of inbound tunnel and (b) crack (sidewall) in outbound tunnel.

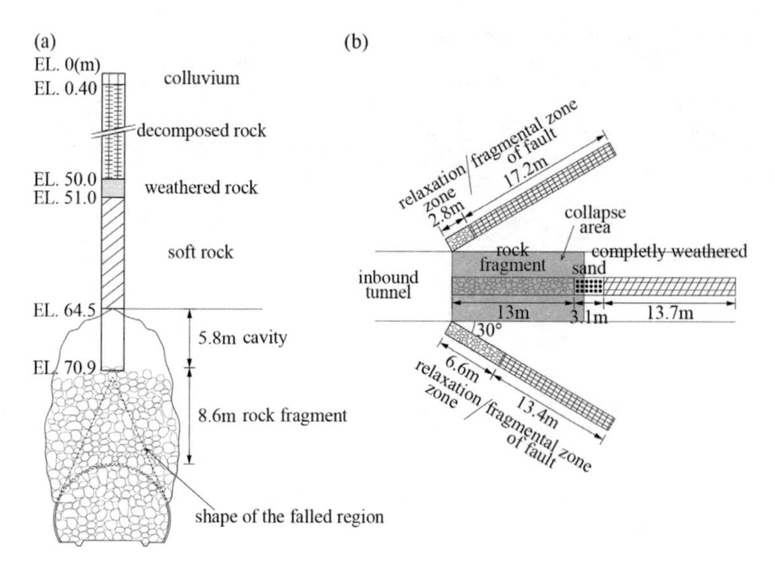

Figure 4. Additional drilling, (a) vertical and (b) horizontal of tunnel face.

3.2 *Additional ground investigation*

In order to investigate of the collapsed area, the drilling investigation was performed. Four boring logs, including vertical, horizontal and 30° left and right to the tunnel face were carried out and the results were shown in Figure 4. Additional geological investigation for the collapse area had revealed that the andesite is dominant and the severe fracture zone exists. The vertical drilling investigation identified the ground profiles from the surface: colluvium (silt sand mixed with gravel), decomposed rock (silt sand mixed with coarse fragment), weathered rock, and soft rock (sand stone). The soft rock layer is appeared at the elevation of 51.0 ~ 70.9m. The fracture zone is located at 55.9m ~ 56.1m, 60.5m ~ 60.7m and 63.3m ~ 63.5m directly above the collapse boundary. The height of collapse from the tunnel crown was determined to be 14.4m. The RQD ranged from 0 to 53%, with an average of 12%.

To analyze geological profile of the tunnel face, the horizontal drilling investigation was performed. The coarse rock fragment was found at the distance of 0.0m ~ 13.0m from the face and the decomposed rock of which RQD was 0 ~ 10% was observed at the distance of 13.0 ~ 16.1m from the face. It was found that the whole area in front of the face was fractured and the self-support capacity is almost negligible.

Relaxation zone due to collapse was evaluated from the sidewall drilling. It was estimated to be 6.6m on the right side and 2.8m on the left side.

3.3 *Evaluation of relaxation zone*

The evaluated range of relaxation zone was 14.4m in length and 2.8 ~ 6.6m laterally wide at the tunnel side. The theory of Terzaghi modified by Deere (1970) was considered to estimate the relaxation zone. According to the rock load classification, when the RQD is 2 ~ 30%, the rock load (H_p) and width (B) are calculated as follows:

$$H_p = 1.1C, \quad B = 2\left[\frac{b}{2} + m \times \tan\left(45 - \frac{\phi}{2}\right)\right] \tag{1}$$

where $C = m+b;$ m = excavated height of the tunnel; b = excavated with of the tunnel; ϕ = friction angle.

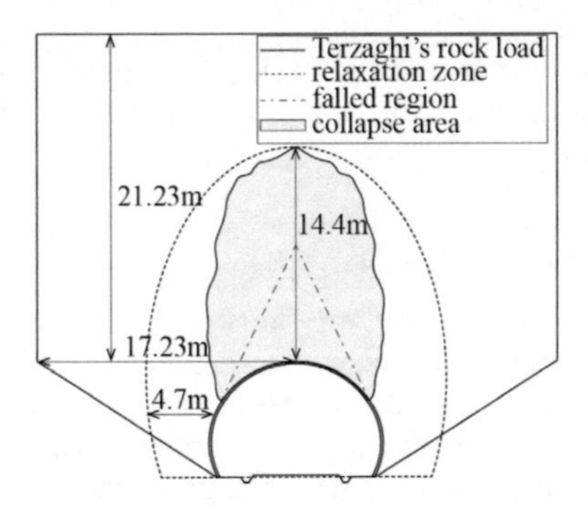

Figure 5. Zone of relaxation load.

The average RQD is 12% and the cross section of the tunnel is 11.7m in width and 7.7m in height. Therefore, Equation 1 produced the rock load (H_p) and width (B) as 21.23m and 17.23m, respectively. The calculated results are presented in Figure 5 and compared with the measured profile. The collapse zone was much smaller than the Terzaghi profile.

Inversely, estimating the ground condition based on the height of the relaxation zone determined by the drilling investigation, the coefficient of was C is 0.75, which corresponds to "very blocky, seamy and shattered" with RQD of 30% ~ 75% in the Terzaghi's Rock Load Classification. In view of relaxation zone the measured zone was significantly different from the general tendency of relaxation zone due to tunnel excavation. More detailed analysis considering the initial stress was carried out.

3.4 *Analysis of initial stress and failure mode*

When the tangential stress exceeds the yield strength collapse may occur at the excavation boundary. Figure 6(a) shows the plastic zone and the slip surface from the stress condition of $K_o > 1.0$. In the plastic zone, logarithmic spiral shear failure surfaces can be formed at an angle with respect to the radial direction.

In this accident, the tunnel crown was collapsed and the rock fragments at the upper ground of the tunnel were fallen. The failure pattern was similar to the shear extrusion failure. K_o was estimated by using the failure mode superposition, and $K_o \approx 1.5 \sim 2.0$ (Martin et al., 1999, Figure 6(b)). However, the overall behavior of the tunnel was different from the case of $K_o > 1.0$, and the collapse mode such as side crack occurred. Therefore, the probability that the initial stress condition of $K_o > 1.0$ was not great.

Therefore, although it was $K_o < 1.0$ condition, it was more potentially possible that the discontinuous joint was developed and the rock fragment at the fractured zone was fallen by

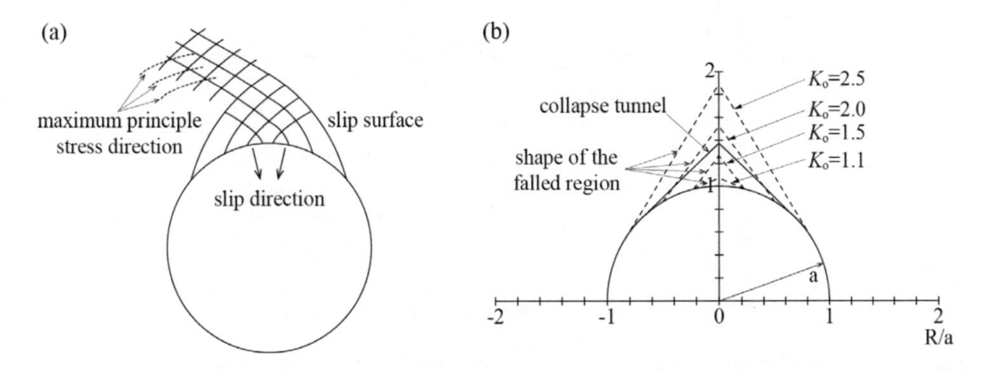

Figure 6. Case study on collapse mode, (a) shear failure mode and (b) effect of K_o.

gravity. The initial stress condition and the rock joints discovered using the ground drilling has shown the geological weakness, and the action of gravity threatened the stability of rock fragment at the tunnel crown.

3.5 *Numerical analysis: tunnel stability analysis*

The mode of collapse in this case was not very mush similar to the general shear failure mode of a typical deep tunnel. Attempts to review the adequacy of the support were made by performing numerical analysis. Moreover, the influence of K_o was also investigated.

To represent ground behavior, the elasto-plastic model based on the Mohr-Coulomb criterion was adopted. The model used and the material properties are shown in Figure 7 (a). The construction stage was set up by simulating the actual construction process and considered the upper and lower excavation for both the inbound and outbound tunnels. The finite element based MDIAS NX (Shin et al., 2011; Midas Information Technology Co., 2013) was used.

Bending compressive stress exceeding 8.4 MPa which is the allowable compressive stress (σ_a) of shotcrete, occurred under the condition of initial stress condition of K_o <1.0. The bending compressive stress of shotcrete in the inbound tunnel was 8.83 MPa, meanwhile it was 8.2 MPa in the outbound tunnel. It can be inferred that cracks of the outbound tunnel occurred due to the load transfer from the collapse of inbound tunnel.

Combining the theoretical and the numerical analysis results, it can be concluded that the gravity action caused the collapse of rock fragments. The excavation disturbance under K_o <1.0 condition would make it worse. The rock mass originally fractured must be strongly disturbed by the ripping excavation, therefore the gravity mobilized full weight exceeding the shear resistance and consequently caused rock fall collapse.

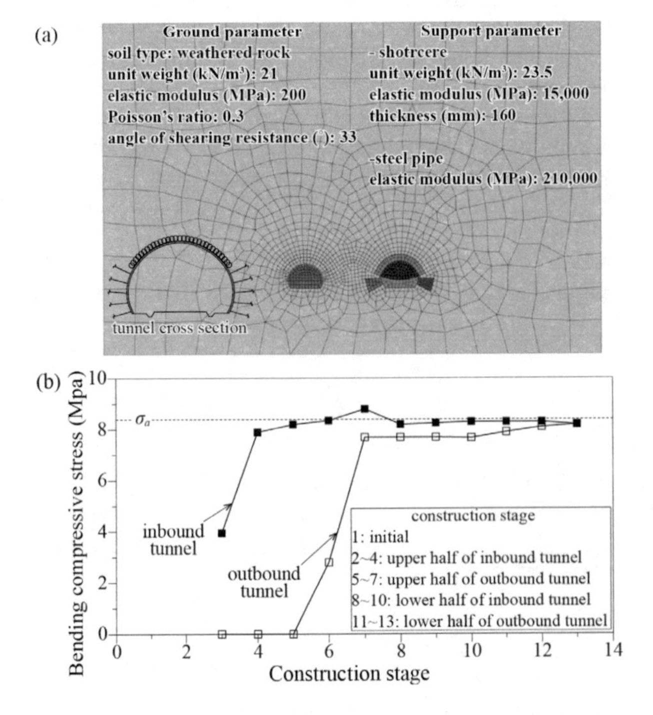

Figure 7. Numerical analysis, (a) mesh and material properties and (b) bending compressive stress of shotcrete.

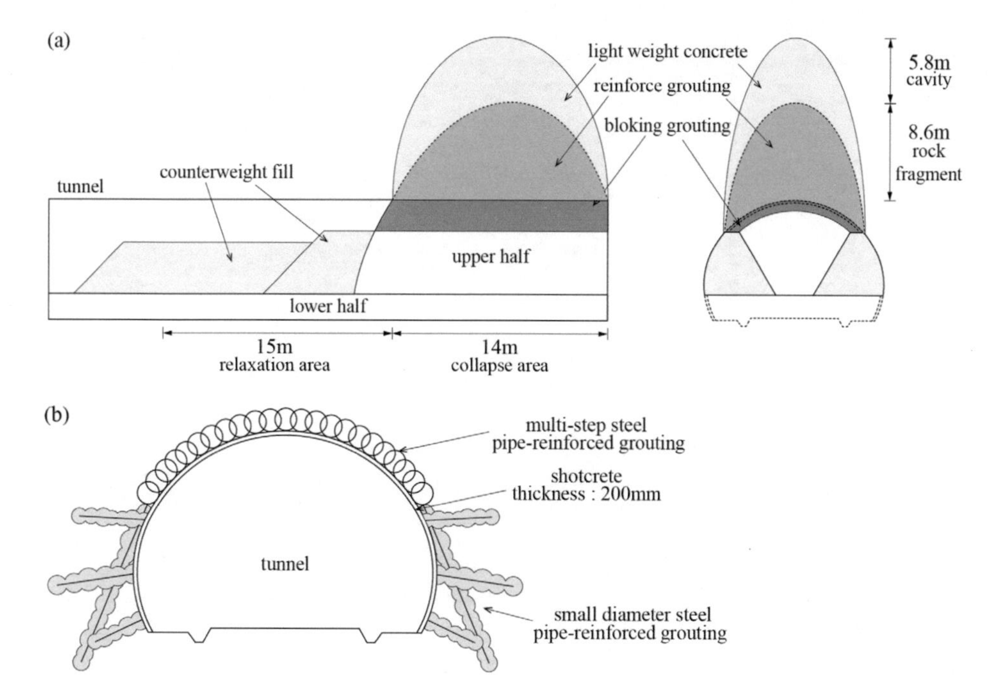

Figure 8. Recovery Plan of tunnel collapse, (a) recovery plan and (b) new support pattern.

4 RECOVERY OF COLLAPSE AND REINFORCEMENT OF SUPPORT

4.1 *Recovery and support pattern change*

Recovery method of tunnel collapse and new support design were proposed. To restore the collapse, the cavity above the tunnel was first filled. Blocking grouting to the tunnel crown was followed to prevent the filling material from moving downward to the cavity, and then filling the cavity with light weight concrete was conducted. Finally reinforce grouting was added to the height of 8.6m from the tunnel crown to improve the strength and secure the re-excavation. The recovery details are presented in Figure 8(a).

It was also necessary to improve the support capability for the weak ground. Multi-step steel pipe-reinforced grouting at the tunnel crown and small diameter steel pipe-reinforced grouting at tunnel sidewall were applied. Shotcrete thickness increased from 160mm to 200mm, and the proposed support pattern is shown in Figure 8(b).

4.2 *Stability analysis of reinforcement*

Numerical analysis was performed to investigate the stability of recovery plan and the proposed support. The equivalent material properties of the reinforced ground were calculated by considering the spacing and materials involved:

$$E_{eq} = \frac{V_{bulb}E_{bulb} + V_{ground}E_{ground} + V_{steel}E_{steel}}{V_{bulb} + V_{ground} + V_{steel}} \tag{2}$$

$$c_{eq} = \frac{V_{bulb}c_{bulb} + V_{ground}c_{ground} + V_{steel}c_{steel}}{V_{bulb} + V_{ground} + V_{steel}} \tag{3}$$

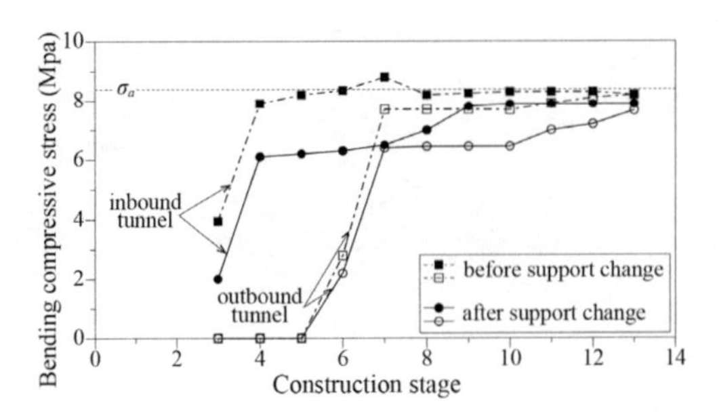

Figure 9. Results of proposed support modification.

where E_{eq}, c_{eq} = equivalent Young's modulus and cohesion; V_{bulb}, V_{ground}, V_{steel} = Volume of the bulb, ground and steel, respectively; E_{bulb}, E_{ground}, E_{steel} = Young's modulus of the bulb, ground and steel, respectively; c_{bulb}, c_{ground}, c_{steel} = cohesion of the bulb, ground and steel, respectively.

The results are summarized in Figure 9, and the maximum shotcrete bending compressive stress calculated is 7.88 MPa (inbound tunnel) and 7.66 MPa (outbound tunnel), which is below the allowable stress (8.4 MPa).

5 CONCLUSIONS

In this study, the collapse mechanism and the recovery method of the collapse occurred in a deep tunnel were investigated. The observed failure mode is like an extrusion of the tunnel crown under conditions where the horizontal pressure is greater than the vertical pressure. However, the actual failure was found to be gravity mobilized rock fall and shotcrete failure due to insufficient shear resistance of fractured rock mass. The stability analysis also showed that the bending compressive stress of shotcrete exceeds the allowable stress model K_o <1.0 condition. In addition the sidewall crack of the outbound tunnel confirmed the results.

Recovery of collapse area and support modification were proposed. The cavity above the tunnel was filled with light weight concrete. And reinforce grouting was conducted in the tunnel crown to improve the strength. To secure the re-excavation, new support pattern was proposed. Numerical analysis was performed to check the stability of the recovery measures and the proposed support. It was confirmed that the reinforcing measures are appropriate and the modified support design is structurally safe.

ACKNOWLEDGEMENT

This research was supported by Development of Design and Construction Technology for Double Deck Tunnel in Great Depth Underground Space (18SCIP-B089420-05) from Construction Technology Research Program funded by Ministry of Land, Infrastructure and Transport of Korean government.

REFERENCES

Barton, N. & Abrieu, M.2007. City metro tunnels and stations that should have been deeper. *In Proc. of Int. Workshop, Underground Works under Special Conditions*, Madrid.

Deere D.U. & Deere, D.W. 1988. The RQD index in practice. In: Proceedings of the Symposium of Rock Classification for Engineering Purposes. Philadelphi. ASTM Special Technical Publication: 91–10.

Erdem, Y. & Solak, T. 2005. *Underground Space Use. Analysis of the Past and Lessons for the Future.* CRC Press.
Martin, C.D., Kaiser, P.K. & McCreath, D.R. 1999. Hoek-Brown parameters for predicting the depth of brittle failure around tunnels. *Canadian Geotechnical Journal*, 36(1): 136–151.
Midas Information Technology Co. 2013. Geotechnical and tunneling analysis system and User's manual.
Ochi, T., Okubo, S. & Fukui, K. 2007. Development of recycled PET fiber and its application as concrete-reinforcing fiber. *Cement and Concrete Composites*, 29(6): 448–455.
Shin, J.H., Choi, K.C., Yoon, J.U., & Shin, Y.J. 2011a. Hydraulic significance of fractured zones in subsea tunnels. *Marine Georesources & Geotechnology*, 29(3): 230–247.
Shin, J.H., Moon, H.G. & Chae, S.E. 2011b. Effect of blast-induced vibration on existing tunnels in soft rocks. *Tunnelling and Underground Space Technology*, 26(1); 51–61.
Shirlaw, J.N., Doran, S. & Benjamin, B. 1988. A case study of two tunnels driven in the Singapore 'Boulder Bed'and in grouted coral sands. *Geological Society, London, Engineering Geology Special Publications*, 5(1), 93–103.

Tunnels and Underground Cities: Engineering and Innovation meet Archaeology, Architecture and Art, Volume 3: Geological and geotechnical knowledge and requirements for project implementation – Peila, Viggiani & Celestino (Eds)

Comparative analysis of seismic refraction profiles with Drilled Borehole Data – Delhi Metro Rail Corporation – Phase-IV Corridor

U. Kumar
Delhi Metro Rail Corporation Limited, New Delhi, India

S. Pateriya
Cengrs Geotechnica, Noida, India

ABSTRACT: Drilling operation for the geotechnical survey is a challenge in an urban area. Seismic Refraction Test is an effective way to map overburden soil & depth of rock without carrying out the drilling operation. This case study presents the comparative analysis between Seismic Refraction profiles of soil/rock with drilled borehole data along the proposed alignment. From the available seismic refraction data, detailed velocity-depth model of the soil and rock was derived and was utilized to map geotechnical profile of the stretch. Seismic refraction is geophysical principle governed by Snell's law. It depends on the fact that seismic waves have different properties in the different type of soil or rock. Also, the waves are refracted when they cross the boundary between different types of soil or rock. This method of geotechnical analysis can be helpful to provide geotechnical data for Tunneling under dense urban agglomerations where drilling operation is not feasible.

1 INTRODUCTION

The metro rail services in Delhi have changed the face of transportation in the city and have been appreciated all over the world. The metro railway network has almost covered the entire city and with continuous endeavors of expansion, most of the neighboring areas and suburbs will be covered by the metro railway by the end of 2022. This study will concentrate on the geotechnical investigation for Delhi Metro Rail Corporation Limited (DMRC) Phase-IV Lajpat Nagar – Chattarpur Corridor and present a comparative analysis between the field investigation and geo physical investigation.

The scope of work for the corridor included drilling of two hundred and eighty five (285) boreholes along the proposed project alignment, as well as carrying out seismic refraction testing (SRT) at selected locations where the bore holes were not physically possible.

The corridor passes in the dense urban agglomeration and area of historical importance. The dense urban area which comprised of primarily non engineered buildings offered very limited space for set up of conventional bore equipment's. Also drilling operation in such areas posed threat to property. So we had to adopt the methodologies which were safe to carry out in such situation and carry out bore-holes in the feasible locations.

2 SEISMIC REFRACTION TEST

In order to carry out geotechnical investigation in the non-feasible stretches, it was planned to carry out seismic refraction test for such stretches and co-relate the data with the near most available bore-hole data. Accordingly if the comparative shows some distinct pattern, the details were used to draw the geotechnical mapping along the alignment.

Figure 1. Alignment with Boreholes location which fall in Mehrauli –DMRC.

2.1 *Concept of SRT:*

A stress applied at the surface of an elastic medium creates conditions for the associated strains to propagate as elastic waves (P&S) in the subsurface material (Figure 2). The waves travel as a pattern of particle deformation, with velocities that are dependent on the elastic properties and densities of the media through which they travel.

The seismic refraction test involves the measurement of travel times of p-waves from an impulse source to a linear array of receiver geophones spread along the ground surface. The geophones are placed in line and connected to the seismograph via geophone cables. The seismograph registers these times and displays them as traces of time for individual geophones. Both normal and reverse profiles were considered during the field execution.

The test uses the travel times of the first arriving wave component only regardless of its travel path. Depending on the source-receiver distance, the first arriving component is either a direct wave or a refracted wave (refracted at the interface of two soil layers). Based on the position of the point where this qualitative change of the signal occurs the layer thickness and velocity can be calculated.

The success of the refraction test depends highly on the properties of the layer interfaces (refractors). An interface can only be found if the layer below the interface shows a significantly higher wave speed than the layer above the boundary.

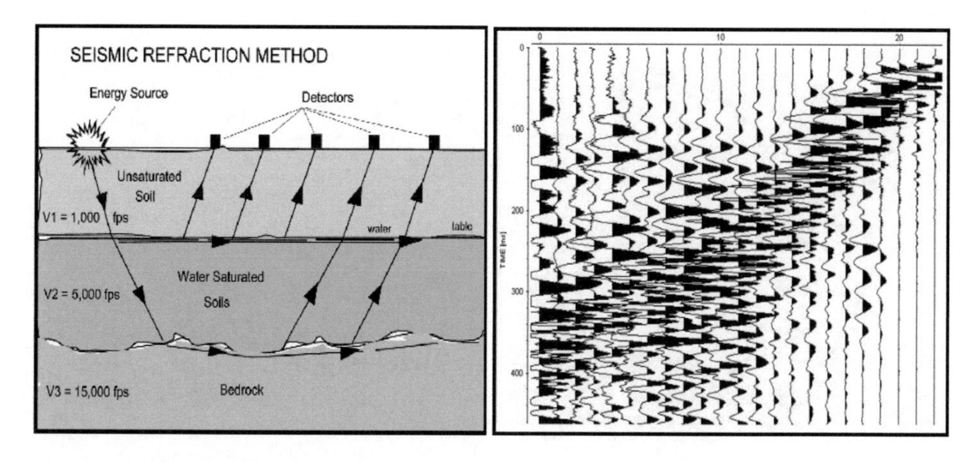

Figure 2. SRT – Pictorial Representation.

2.2 *Equipment Used:*

The seismic refraction test equipment used on site consists of the following basic components:

i. Energy Source: Sledge hammer with steel impact plate.
ii. Geophones: Twenty-four (24) moving-coil type digital grade vertical geophones [natural frequency = 10 Hertz] and one (1) starter geophone
iii. Cables: Geophone spread cables with NK2721C connectors and single takeouts
iv. Data Acquisition System: 24-channel Seismograph: Model PASI (GEA24), Italy
v. Processing Software: Rayfract

2.3 *Data Acquisition:*

The equipment installation and data acquisition involves the following stages:

1. The seismic line to be surveyed is marked on the ground.
2. Low frequency (10Hz) geophones are used to record seismic signals. Depending on the length available along seismic lines, 12 channels were used for each spread. Care was taken to ensure that the pointed ends (spikes) of the geophones are fully embedded in the topsoil.
3. Each seismic spread typically consists of 7 shots. Line length is computed from first End-of-Line shot to the last End-of-Line shot with two far shots.
4. For the seismic refraction survey work, a sledge hammer struck vertically on a steel plate was used as the energy source at each shot point to generate P-waves. A starter geophone is installed next to the shot location as "trigger".
5. A 24-channel seismograph (PASI GEA24) was used to record the field data (Figure 3). The seismograph records the arrival of seismic waves through 12 channels, which were used for each spread. The seismic waves detected by each geophone are displayed simultaneously on the screen. The seismograph has the signal enhancement or stacking capability.
6. The field data is stored in a laptop attached to the seismograph in SEG-2 format data files. These files are later processed using commercial software's, as described in the following section.

2.4 *Analysis of Data and Interpretation:*

The seismic refraction processing software used (Rayfract), is an automatic refraction interpretation package that contains modules for performing velocity optimization and visualization. Rayfract is the Graphical User Interface to the suite of modules for performing the

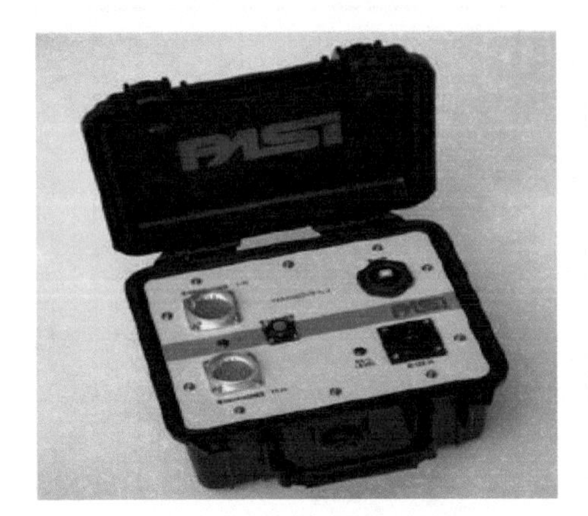

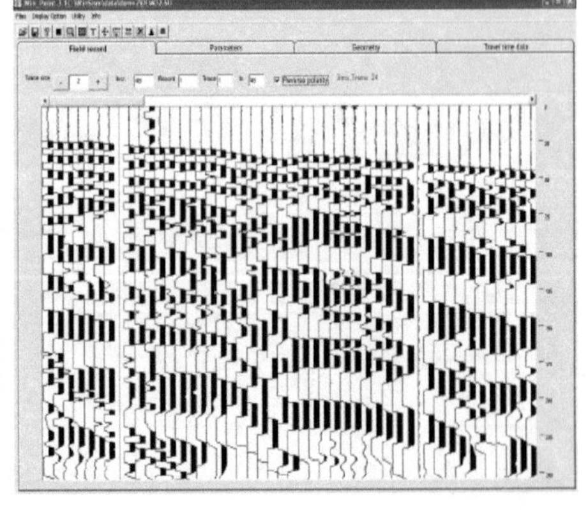

Figure 3. Seismograph & Typical Recorded Signal.

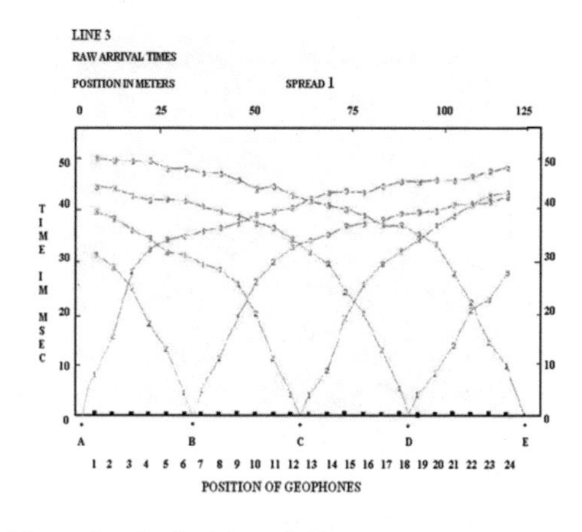

Figure 4. Typical Travel Time Curve.

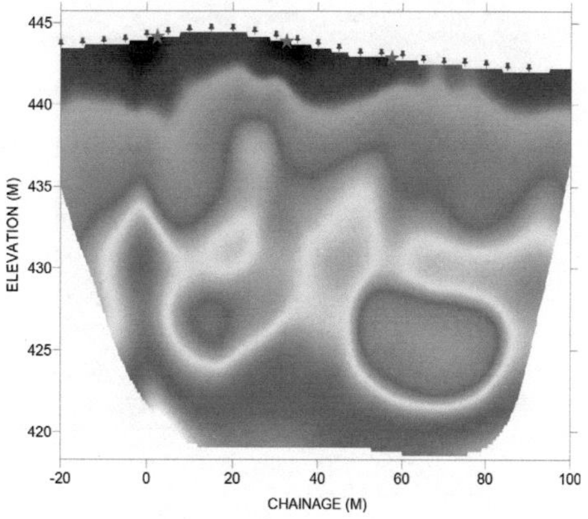

Figure 5. Velocity Model.

velocity model optimization and visualization, interactive seismic survey design, and outputting postscript images of the results for printing.

Interpretation of the seismic data involves resolving the number of velocity layers, the velocity of each layer, and the travel-time (Figure 4) taken to travel from a given refractor up to ground surface. This time is then multiplied by the velocity (Figure 5) of each overburden layer to obtain the thickness of each layer at that point. The processed data is normally presented as a series of plots: a true depth profile for the identified refractors and a velocity profile for the overburden and refractor.

For preparation of the seismic sections; GRM (Generalized Reciprocal Method) has been used which is further refined by using WET (Wavepath Eikonal travel time tomography inversion).

Results of all shots positions in a seismic spread have been integrated into a single section. The seismic section gives the information about the subsurface stratigraphy in terms of their seismic velocities, which are directly related to the quality and strength of the medium.

3 STUDY AREA:

Table 1. SRT Array.

Location	Line Designation	UTM Coordinates, m (Zone-43 R)				Line Length, m
		Start		End		
		Easting	Northing	Easting	Northing	
Azim Khan Tomb	Line-11	714452	3157032	714302	3156956	180
	Line-12	714219	3156892	714090	3156773	180
ASI Park	Line-13	714077	3156781	713925	3156865	180
	Line-14	713916	3156875	713701	3156896	240
MCD Park	Line-15	713603	3156946	713582	3157002	60
Bhool Bhulaiya Parking	Line-16	713458	3157047	713488	3157045	30
School Playground	Line-17	713315	3157094	713254	3157095	60

Figure 6. SRT Line-11.

Figure 7. SRT Line 12, 13 & 14.

Figure 8. SRT Line-15.

Figure 9. SRT Line 16.

Figure 10. SRT Line-17.

4 SRT PROFILES

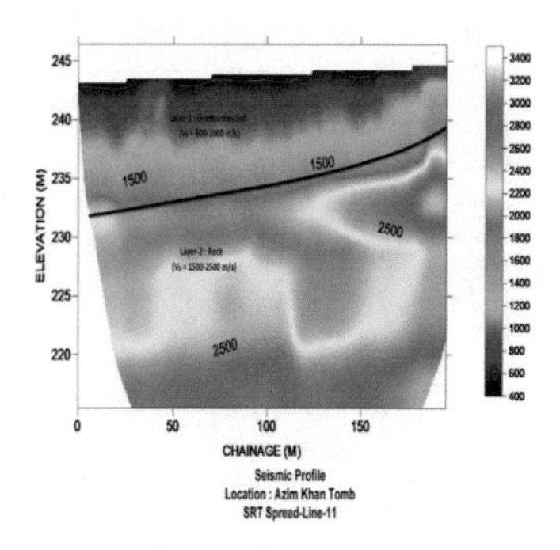

Figure 11. Seismic Profile-Line-11.

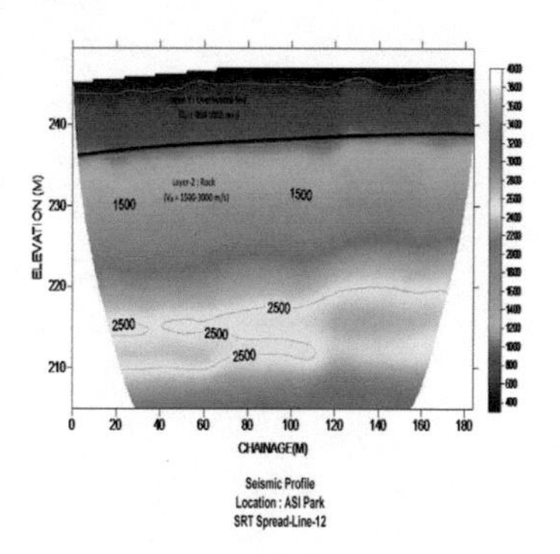

Figure 12. Seismic Profile-Line-12.

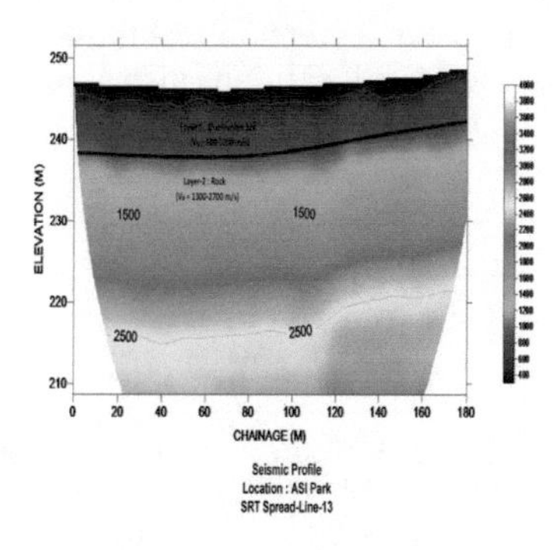

Figure 13. Seismic Profile-Line-13.

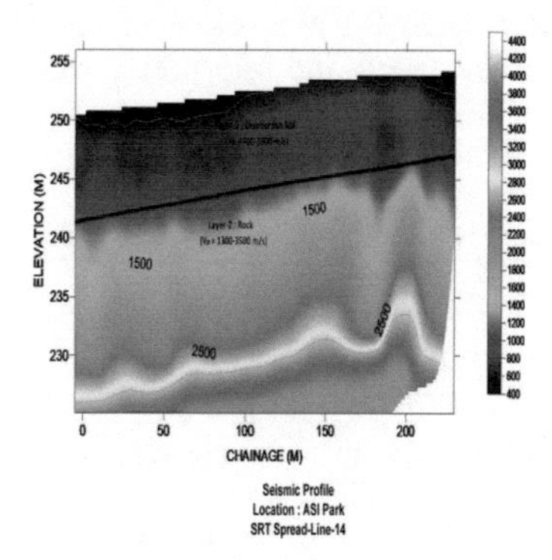

Figure 14. Seismic Profile-Line-14.

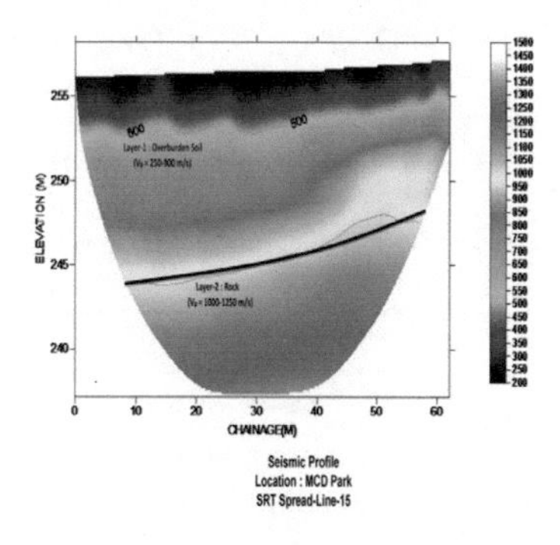

Figure 15. Seismic Profile-Line-15.

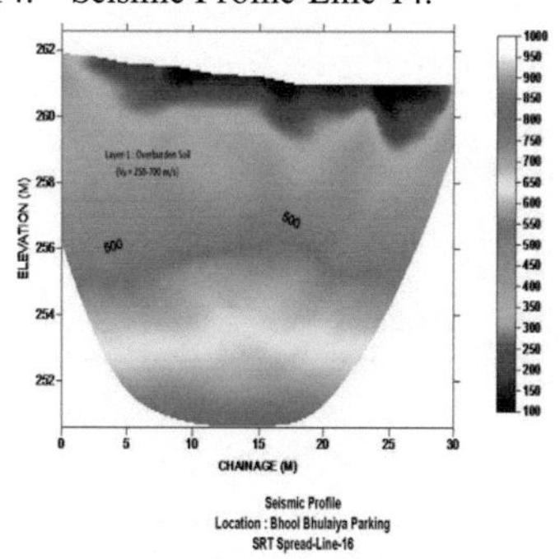

Figure 16. Seismic Profile-Line-16.

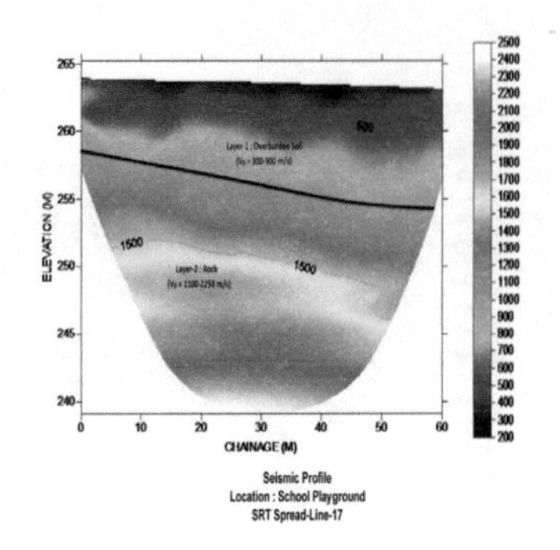

Figure 17. Seismic Profile-Line-17.

Seismic Profile of SRT line from 11 to 17 showing the clear cut two stratum, soil and rock. The interpreted P-wave velocities (computed by applying linear regression techniques on the travel time curves) and interpreted strata information at each test location is also summarized in the table below:

Table 2. SRT Interpretation.

Location	SRT Spread	Layer	Depth, m		Interpreted V_P, m/s	Interpreted Strata
			From	To		
Azim Khan Tomb	Line-11	1	0	5~12	600~1000	Overburden Soil
		2	5~12	30	1500~2500	Rock
ASI Park	Line-12	1	0	7~8	400~1000	Overburden Soil
		2	7~8	30	1500~3000	Rock
	Line-13	1	0	8~9	600~1000	Overburden Soil
		2	8~9	37	1300~2700	Rock
ASI Park	Line-14	1	0	4~8	400~1000	Overburden Soil
		2	4~8	25	1300~3500	Rock
MCD Park	Line-15	1	0	8~10	250~900	Overburden Soil
		2	8~10	20	1000~1250	Rock
Bhool Bhulaiya Parking	Line-16	1	0	12	250~700	Overburden Soil
School Playground	Line-17	1	0	6~9	300~900	Overburden Soil
		2	6~9	25	1100~2200	Rock

Please note that the strata interpretation presented in the table above is approximate and based on limited borehole and seismic refraction test data along the alignment.

5 GEOLOGY

The deposits in the area belong to the "Indo Gangetic Alluvium" and are river deposits of the Yamuna, and its tributaries. The alluvial tract is in the nature of a synclinal basin formed concomitantly with the elevation of the Himalayas to its north. It was formed during the later stages of the Himalayan Orogeny by the buckling down of the northern border of the peninsular shield beneath the sediments thrust over it from the north.

The Pleistocene and Recent Deposits of the Indo-GangeticBasin are composed of gravels, sands, silts and clays with remains of animal and plants. A generalized description of geological formations encountered in Gurgaon and Delhi is as follows:

Table 3. A generalized description of geological formations encountered in Gurgaon and Delhi.

Period	Formation	Description
Recent	Newer Alluvium (Younger alluvium)	Unconsolidated, inter-bedded lenses of sand, silt gravel and clay confined to flood plains of Yamuna river.
Quaternary	Older Alluvium	Unconsolidated inter-bedded, inter-fingering deposit sand, clay and kankar, moderately sorted, thickness variable, at places more than 300 m.
Pre-Cambrian	Pegmatite and Quartz Veins Quartzites and minor Schist Bands	Well stratified, thick-bedded brown to buff colour, hard and compact, intruded locally by pegmatite and quartz veins inter-bedded with mica schists.

The older alluvium is rather dark colored (locally called "Bhanger") and is generally, rich in concretions or nodules of impure calcium carbonate (kankars). The kankars are of all shapes and sizes, varying from small sand sized grains to big grains and big lumps. The age of the "Bhanger" alluvium is Middle to Upper Pleistocene.

The newer alluvium (locally called "Khadar") is light colored and poor in concretions. It contains lenticular beds of sand and gravel as well as peat beds. It is merged by insensible gradations into the Recent or deltaic alluvia and its age is Upper Pleistocene to Recent.

6 COMPARISION WITH BOREHOLES

When we compare the SRT data with boreholes, we can see that there is similarity and some deviations in strata. BH-A3 soil is encountered through the boreholes (Figure 18). However in

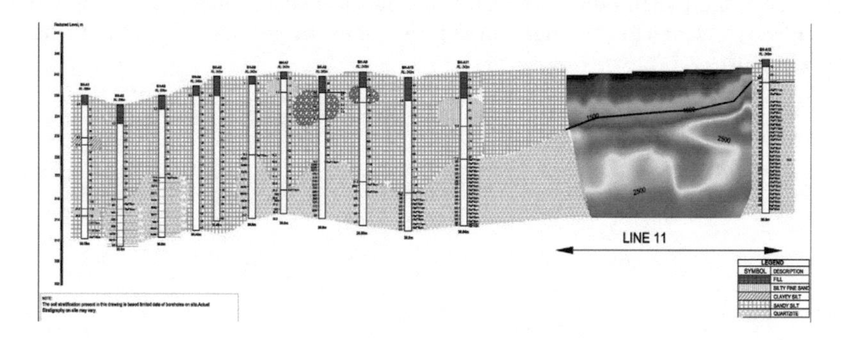

Figure 18. Seismic Profile-Line-11 with bore well data.

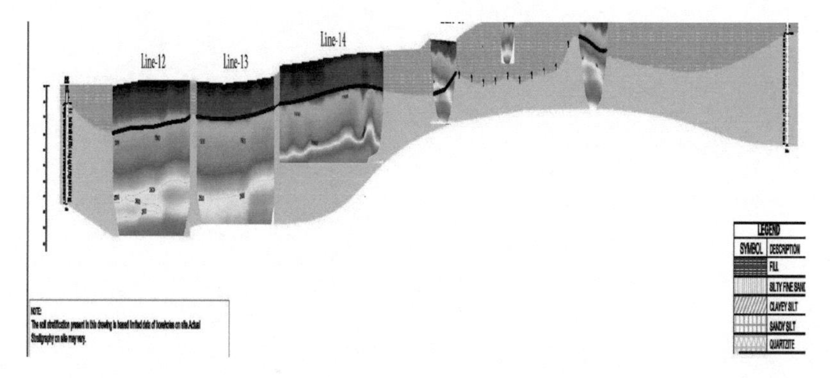

Figure 19. Seismic Profile-Line-12-17 with bore well data.

BH-11 rock is encountered at about 17 m depth. And in the continuity with SRT profile rock level is decreasing towards the BH-11. In BH- 13 rock is encountered at about 4.5 m depth. Thus we can see that the SRT profile is following the trend of boreholes data & it can be used to interpret the data for the sections where physical bore are not feasible.

There is no space for drilling operation between BH-A13 to BH-A30 (Figure 19). We can see that in BH- A30 rock is encountered at about 6.0 m depth. There is not much variation between BH-A13 to BH-A30. Seismic profile also predicts that the rock will encounter in that area below 10 m depth.

7 CONCLUSION & APPLICATION

- Methods like SRT can be very helpful in gathering geotechnical and strata information in areas where physical bore well is not feasible to carry out.
- P waves have a higher velocity than S waves when travelling through several mineral types. The speed at which seismic waves travel depends on the properties of the material that they are passing through. For example, the denser a material is, the faster a seismic wave travels. P waves can travel through liquid and solids and gases, while S waves only travel through solids. Therefore we only use P waves in seismic refraction tests (SRT) instead of S waves.
- SRT since provides instant data and result for interpretation, it is a fast method and can be implemented on a large scale.
- SRT can be useful in dense urban agglomeration in congested cities without posing any threat to the existing structures. Also it can be used to carry out stratigraphic investigation in areas with heritage monuments without disturbing the surrounding.
- SRT is free from any demands of post investigation restoration.
- The result from the geo-physical investigation reflects close relation with the bore holes in vicinity. Thus these methods can be used for indicative geotechnical mapping, however for the detailed geotechnical investigation lab tests shall be conducted for finalizing the Geotechnical reports.

Tunnels and Underground Cities: Engineering and Innovation meet Archaeology, Architecture and Art, Volume 3: Geological and geotechnical knowledge and requirements for project implementation – Peila, Viggiani & Celestino (Eds)
 ISBN 978-0-367-46583-4

Consideration of heaving and deformation behavior of mountain tunnel in service

N. Kuramochi
East Nippon Expressway Co., Ltd, Ohmiya, Japan

S. Yamashita & K. Nakano
NEXCO-East Engineering Co., Ltd, Tokyo, Japan

S. Kunimura
Oyo Corporation, Saitama, Japan

H. Watanabe
Nippon Civic Consulting Engineers Co., Ltd, Tokyo, Japan

K. Nishimura
Tokyo Metropolitan University, Tokyo, Japan

ABSTRACT: Recently, in mountain tunnels in service on expressway, heaving has occurred in specific areas, and various investigations for deformation and countermeasures by invert have been performed. Several past countermeasures for heaving have been reported on mountain tunnel, but from investigations and the long-term monitoring results, quantitatively research are few that analyzed behavior at the time of transformation and the cause of transformation and effect of countermeasures. The authors clarified by long-term monitoring the heaving behavior of the mountain tunnel of Expressway and the effect of countermeasures, and have reflected them in the design of countermeasures.

1 INTRODUCTION

Recently, there have been increasing cases of heaving of Expressway constructed through mountains happening in service (Figure 1), and renewal works are being done at various locations to set up invert against heaving phenomenon in Japan. Mechanism of heaving phenomenon is typically mentioned that road upheaval occurs in tunnels without invert. It causes by the reduction of ground strength due to stress release by excavation in construction and swelling by underground water in service in spite of relatively stable ground condition at that time of construction.

Over the course of several years, authors have set up countermeasure invert at various locations, as well as measuring the heaving and reviewing the design to collect enough data to analyze and evaluate the effectiveness of the invert.

In this thesis, we will clarify the trends of the heaving and deformation behavior of the tunnels based on the measurements done at locations and viewed the major trends of causes and mechanisms of deformation by analyzing the results of natural ground tests.

1) Tunnel No.1

2) Tunnel No.2

Figure 1. Heaving Examples on Mountain Tunnel.

2 HEAVING AND DEFORMATION BEHAVIOR

We selected two tunnels in which relatively large heaving occurred to analyze the deformation behavior by comparing the rising of the road surfaces, convergence of displacement in the tunnels, and ground displacement.

2.1 *Conditions of Natural Ground*

Tunnel A was constructed through the natural ground that consists of a conglomerate formed in the third Miocene epoch, tuff, and rocks mainly made from tuff formed in the fourth Pleistocene epoch. The maximum overburden is approximately 300m, and the length is 3,200m. Tunnel A was constructed mainly using Standard Supporting "CII". Sinking of the crown and convergence of displacement were relatively a small value (approximately 15mm). Tunnel B was constructed in natural ground that mainly consists of mudstone and tuff breccia. The maximum overburden is approximately 150m, and the length is 1,019m. Tunnel B was constructed using Standard Supporting "DI". Shrinkage of the crown was uncertain. However, it is presumed that the shrinkage converges to a small value.

They were judged that the natural ground in which the tunnels were constructed relatively stable condition. Therefore, the support structure without invert was adopted. As a result, the road surfaces rose several years after the use of the tunnels started. Setting up invert are considered against heaving.

2.2 *Rising of Road Surface*

The Rising of road surface has been measured with leveling due conditions of lane regulations. Figure 2 shows the rising road surface measured in the vertical direction of the tunnels. For Tunnel A, the road surface rises continuously in a section where an invert was not constructed. The length of each rising varies from 30m to 200m. For Tunnel B, the road surface rises in one place in a section where an invert was not constructed. The length of the rising is approximately 150m. In the both tunnels, the degree of road surface rising is greater than that of the convergence of displacement.

Figure 3 shows the transition of the accumulated road surface rising with time in a place where the largest rising occurs in chronological order. Tunnel A rose 41mm in the past 10 years from 2007. The rising speed is 4mm/year. For Tunnel B, the road surface was repaired by the cutting-and-overlay method. The measurements saw years where rising reduced. Overall, the surface of Tunnel B rose 76mm in the past 17 years from 1999. The rising speed is approximately 4.5mm/year. The maximum rising of both tunnels is relatively small at 10mm/year. However, both roads continue to rise even 24 years after the use of the tunnels started.

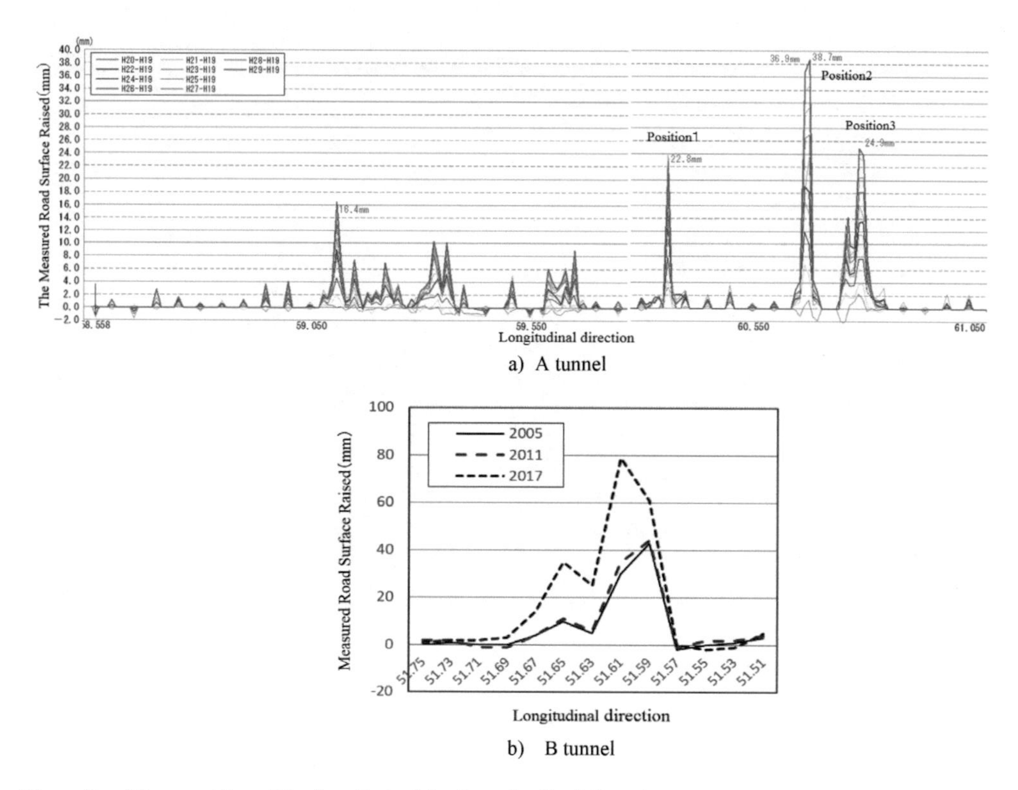

Figure 2. Measured Road Surface Raised for Longitudinal direction.

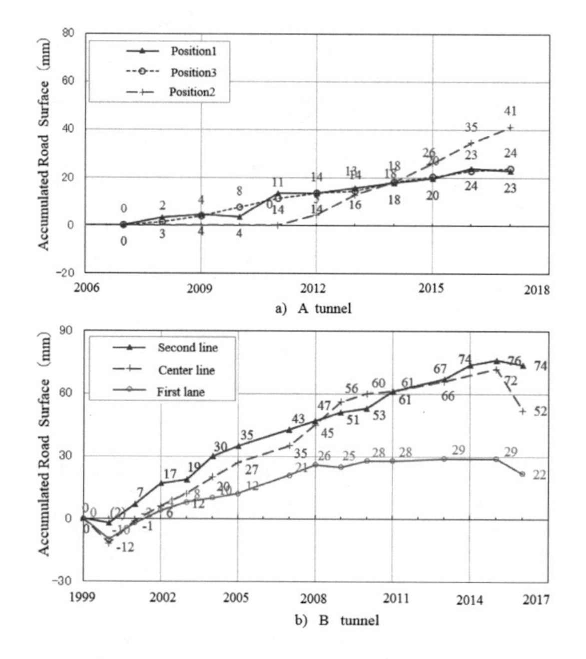

Figure 3. Accumulated Road Surface.

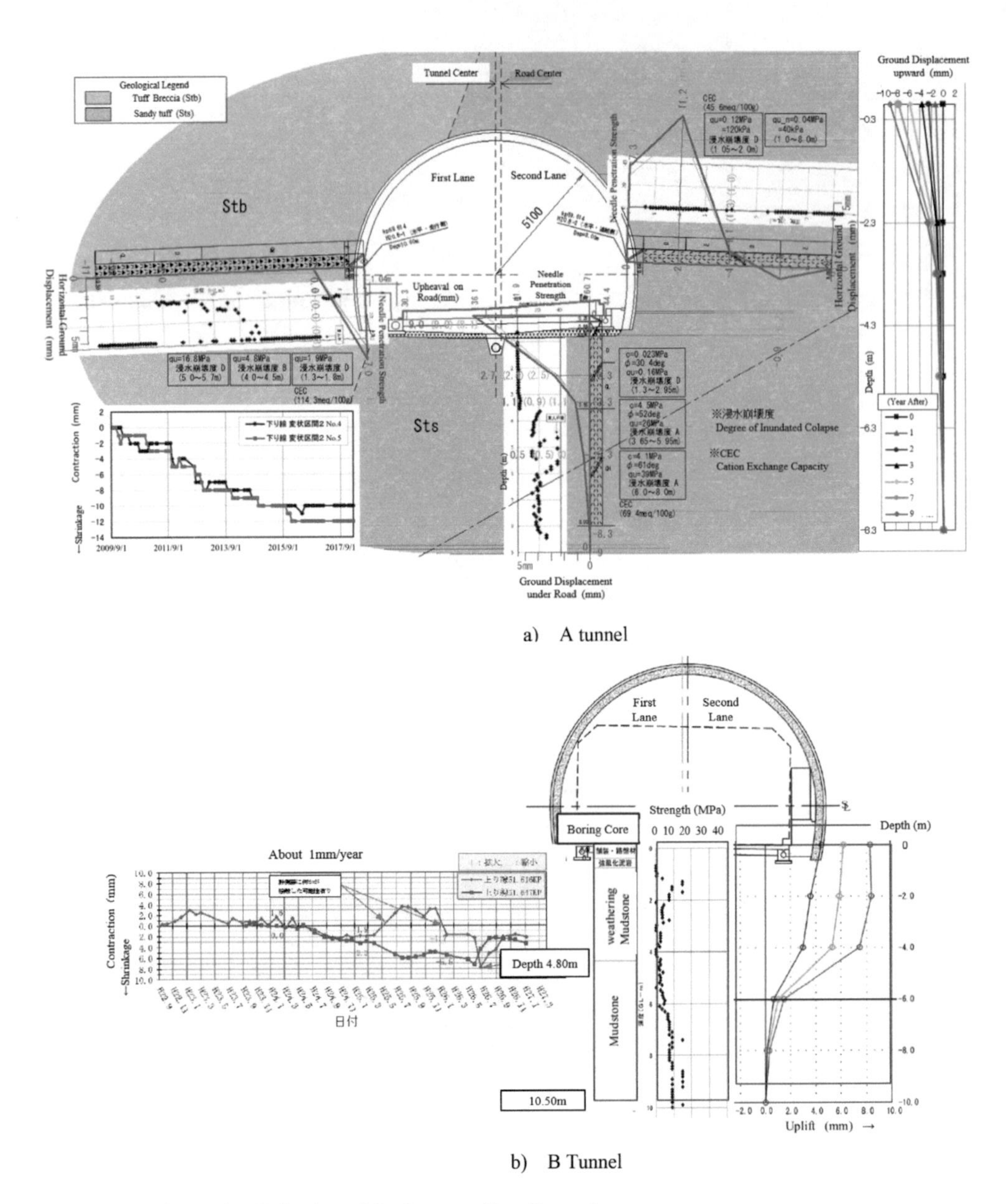

Figure 4. Deformation Behavior of the Surrounding Ground.

2.3 *Correlation between Ground Displacement and Strength of Natural Ground*

Figure 4 shows the ground displacement, shrinkage displacement of convergence measurement and road surface raised, and physical property value of the natural ground, in a place where relatively large deformation occurs. The following observations can be deducted from these data:

1. Tunnel A
   A ground displacement meter was installed after the deformation occurred under the road surface in vertical (L=10m) and horizontal directions (L=10 m) for left and right at the side wall part to measure underground displacement at various depths. A laser convergence measurement meter was used to measure shrinkage between the side walls. The physical property value of the natural ground was tested from the boring core sample obtained by the installment of the ground displacement meter. Tests such as needle penetration test,

traxial compression test, inundated colapse test, cation exchange capacity (CEC) test, and plasticity index test were done. The results including the strength from needle penetration, c , φ , the degree of inundated colapse, and CEC test figures are plotted in the figure.
In Figure 4, a cross section is the place where the largest shrinkage displacement by convergence measurement occurs. The side wall on the first lane side is displaced inward (horizontally), and the side wall on the second lane is displaced inward 4 meters from the natural ground at the back of tunnel due to ground displacement in the vertical direction. Natural ground consists of sandy tuff and tuff breccia, in which sandy tuff is weaker, so the displacement on the right side is larger compared to the left. In the vertical direction, ground displacement occurs upward at a depth of 5.3m, and the depth of displacement continues to become greater, according to measurements. The speed of road rising is 1-3 mm/year, and the total degree of rising is approximately 60mm. The shrinkage speed of convergence measurement is approximately 1.5 mm/year, and the total degree of shrinkage is approximately 50mm. The natural ground in which large ground displacement occurs clearly has low strength (with small figures measured by penetration tests), both horizontally or vertically, and exhibits characteristics of the natural ground whose degree of inundated colapse caused by water permeation is about "D." (The natural ground collapses without a trace of the original form after 24 hours from water permeation.) This shows this natural ground is fragile to water.

2. Tunnel B
   As with Tunnel A, the ground displacement meter measures the displacement horizontally (L=10m) for every 2.0m in depth. The setups for convergence measurement meter and physical property value of the natural ground are as the same as Tunnel A. Needle penetration and other tests were tested to obtain the physical property values of the natural ground. The following figure shows the core properties and ground strength from needle penetration.

In Figure 4, a cross section is the place where the largest surface raised occurs. This surface sample had the largest upheaval. Displacement of natural ground occurred up to 6m in depth, and the displacement area tended to expand. For Tunnel B, we confirmed that the core was weathered mudstone, which is different from the samples obtained from under it. As with Tunnel A, weathered mudstone's degree of inundated colapse by water permeation is "D", having it fragile to spring water. The amount of surface upheaval is as Figure 3b, and the shrinking speed by the convergence measurement is approximately 1 mm/year. The speed of road rising was less than 10 mm/year, which implies a trend of deformation behavior where surface upheaval is larger.

2.4 *Conclusion*

From the measurements of deformation and testing of physical property value of the natural ground, we confirmed that the area where the strength of natural ground reduced expanded year after year. This ground was relatively stable condition without invert when the tunnel was constructed, and the upheaval of the base of the natural ground did not occur. We observe that the natural ground surrounding the tunnel became fragile due to spring water and other influences, which caused ground displacement. When this continued, it resulted in the upheaval of the roadbed and road surface.

3 ESTIMATED CAUSES OF DEFORMATION

The following are estimates on the causes of such heaving, based on analysis of past heaving cases of tunnels without invert compared to Tunnels A and B.

3.1 *Countermeasures (past tunnels)*

Figure 5 shows the countermeasures to heaving by roadbed excavation of the three tunnels. All had spring water seeping all time while excavating, or the roadbed was wet. Also, the degree of natural ground collapse of the three tunnels were mainly "D". The comparison of

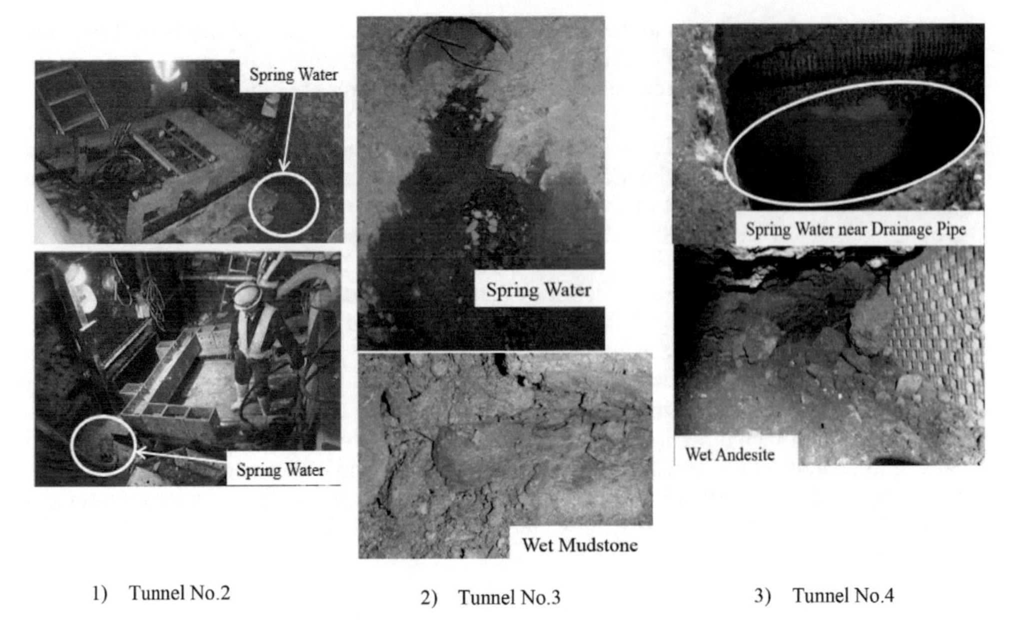

Figure 5. Ground Condition under Construction of Countermeasure.

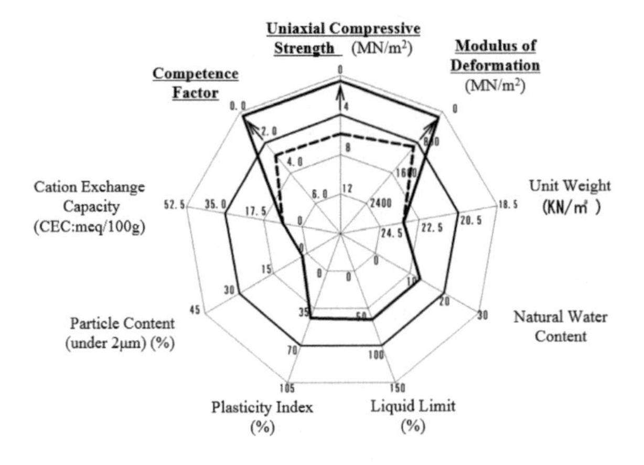

Figure 6. Reduction of Ground Strength due to Swelling by Spring Water.

the physical property values of natural ground at construction and heaving of the tunnel (as in Figure 5a) can be observed in Figure 6. The thick line shows the threshold of expansion. One can see that the items of strength, such as uniaxial compressive strength, competence factor, modulus of deformation, are under threshold at the time of construction but overwhelms the threshold extremely at the time of deformation. This shows that the influence of spring water and the high groundwater level reduced the strength of the roadbed.

3.2 *Tunnel A*

Figure 7 shows the physical property values of natural ground. The above contains swelling clay mineral (smectite). Items such as strength, degree of inundated colapse and cation exchange capacity (CEC) are all above threshold. As the degree of inundated colapse of the tunnel being "D", which indicates significant strength deterioration, and in consideration of the above (1), we assume that the heaving is caused by the combination of deterioration by water infiltration and swelling by absorption of water.

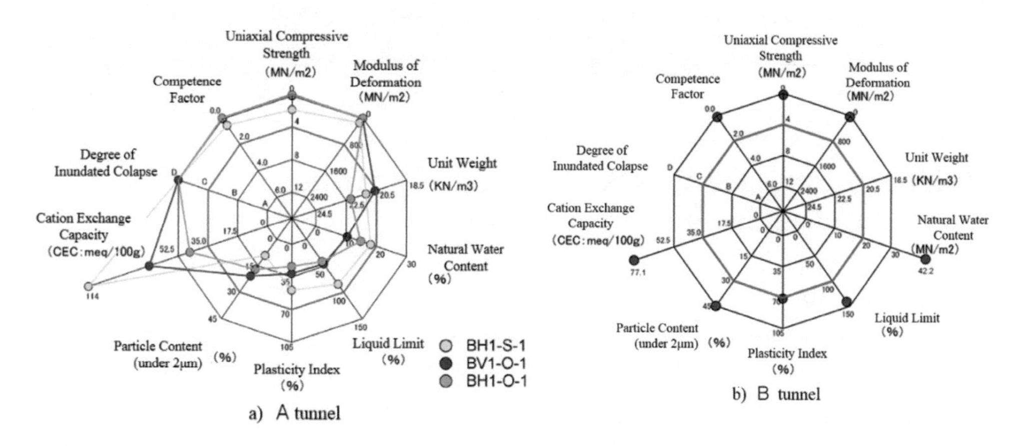

Figure 7. Ground Physical Properties and Cause for Deformation.

3.3 *Tunnel B*

Swelling clay mineral (montmorillonite) was found from the mudstone and weathered mudstone. Items such as strength, cation exchange capacity (CEC), particle content (under 2μm), liquid limit and natural water content were both above threshold. The degree of inundated colapse was mainly "A: almost no change exhibited", with the shallow part of the weathered mudstone being "D", and with the high natural water content, together in consideration of above (1), we assume the strength deteriorated by the absorption of water.

3.4 *Conclusion*

The roadbeds of both tunnels exhibited the degree of inundated collapse "D", and as it was obvious that strength deterioration was caused by spring water, we observe that the deterioration by swelling is the main cause of deformation. This is clearly different from squeezing ground, where the stress on the natural ground exceeds its strength from the time of construction, causing a significant deformation. Thus, it can be considered that the heaving was caused by the deterioration of the strength of natural ground by a long-term stress released during excavation along with exposure to spring water.

4 FINAL CONCLUSION

Through the consideration described above, we obtained the following new findings about heaving behavior and the cause of the deformation:

1. Where the natural ground was relatively stable condition, strong state at the time of construction and invert was not set up, the degree of road surface rising is greater than that of the displacement by convergence measurement. In the places where maximum road surface rising occurs, deformation speed is relatively low at 10 mm/year. However, the rising continues even 24 years after the start of the service.
2. The strength of the risen roadbed has been seen to decrease, and there is a tendency for the deteriorated area to expand with time.
3. The deteriorated roadbeds have high water content and characteristics that they are easily deteriorated by water, which can be assumed that the area was exposed to spring water. In the places where the roadbed is deteriorated, water seeps out.

We elucidated the mechanism of the continuation of road surface rising such that the roadbeds deteriorate from the release of long-term stress and the effect of spring water, and such deterioration leads to the continuous rising of the road surfaces.

Tunnels and Underground Cities: Engineering and Innovation meet Archaeology, Architecture and Art, Volume 3: Geological and geotechnical knowledge and requirements for project implementation – Peila, Viggiani & Celestino (Eds)

Loading magnitude for rock tunnel during earthquake estimated by dynamic measurement at an actual tunnel

A. Kusaka & S. Hara
Public Works Research Institute, Japan

N. Isago
Tokyo Metropolitan University, Japan (Formerly, Public Works Research Institute)

ABSTRACT: Rock tunnels suffered severe damage from recent large earthquakes in Japan, despite the empirical knowledge that tunnels tolerate earthquakes better than surface structures. Seismic design for rock tunnels have therefore become a fascinating subject for many tunnel engineers. However, even loading magnitude during earthquake is not fully understood. In a recent study, authors have installed dynamic measurement instruments in some actual rock tunnels and then obtained dynamic data which was measured during some large earthquakes. In this paper, outline of the results from the measurement is described and loading magnitude during earthquake for the tunnel is discussed through numerical analysis. The main discussion is based on the strongest aftershock of "The 2011 off the Pacific coast of Tohoku Earthquake", occurred on April 7th, 2011. Major conclusions include: Strain of the tunnel lining is much smaller than compressive failure strain in a normal ground condition for rock tunnels according to the measurement results. Loading magnitude is approximately 26 kPa with k-ratio of 0.5 to simulate the strain of the lining in a simple static numerical analysis.

1 INTRODUCTION

Large earthquakes can pose serious risks even for rock tunnels, as evidenced by large earthquakes in the past. In the Niigataken Chuetsu Earthquake 2004 in Japan (Mw = 6.6), some tunnels suffered severe damage including collapse of their permanent lining (Mashimo, 2005). In the Kumamoto Earthquake 2016 in Japan also brought a similar damage (Isago *et al.*, 2018). In addition, numerous damage modes were actually reported due to poor ground conditions (eg. Wang *et al.*, 2001).

The design criteria against earthquake is not fully established in practical tunnel engineering area, partly due to lack of knowledge as to loading-related matters during earthquake. In Japan, a single reinforcing steel bar is used for tunnel permanent lining for portal areas and other poor ground conditions as a countermeasure against earthquake and other external forces. But this is only an empirical measure and even basic mechanical behavior of tunnel during earthquakes have not been fully understood.

A recent study conducted a dynamic measurement at an actual tunnel, and dynamic data obtained during the strongest aftershock of "The 2011 off the Pacific coast of Tohoku Earthquake" occurred on April 7th (Kusaka *et al.*, 2016). This measurement showed that only 20×10^{-6} strain on the lining occurred during strong motion where 200 gal acceleration observed in the tunnel. However, loading magnitude to tunnel lining during such large earthquake were not calculated.

In this study, a series of static numerical analysis using a frame model is conducted to estimate rough loading magnitude during such large earthquakes, referring to a result from measurement in an actual tunnel during a large earthquake.

2 ROUGH DESCRIPTION OF MEASUREMENT RESULTS AT SANT JUAN TUNNEL

2.1 *Outline of the measurement*

The measurement (Kusaka *et al.*, 2016) were conducted at the Sant Juan Tunnel, which is located in Ishinomaki City on the Pacific coast side of the Tohoku Region of Japan. Topography around the tunnel and an outline of the tunnel features are shown in Figure 1 and Table 1 respectively. The tunnel is a two-lane road tunnel with approximately 700m in length constructed by the conventional tunneling method (NATM), and had been in service for about 15 years at the time of the measurement since being completed in 1996. The bedrock around the tunnel consists mainly of sandstone and shale. Obvious fault-fracture zones were not observed during construction. In addition, no major defects around the measuring area in the tunnel were observed before the measuring equipment was installed or after the earthquake.

A dynamic measurement system was installed in a section which was located approximately 100m from the tunnel portal under about 40m of overburden. Ground classification, commonly used for road tunnels in Japan (Japan Road Association, 2003), was DI, in which the invert was installed. As shown in Figure 2, acceleration was measured at one point for three components, and uniaxial strains in the circumferential direction were measured at five points on the inner surface of the lining.

Measured data were obtained for an earthquake which occurred at 11:32pm (Japan standard time) on April 7th, 2011. It was a strong aftershock of "The 2011 off the Pacific coast of Tohoku Earthquake", with JMA seismic scale of "6 Upper" around the tunnel. The magnitude of the earthquake was 7.1 (Mw) and the tunnel was about 60 km from the epicenter of the earthquake.

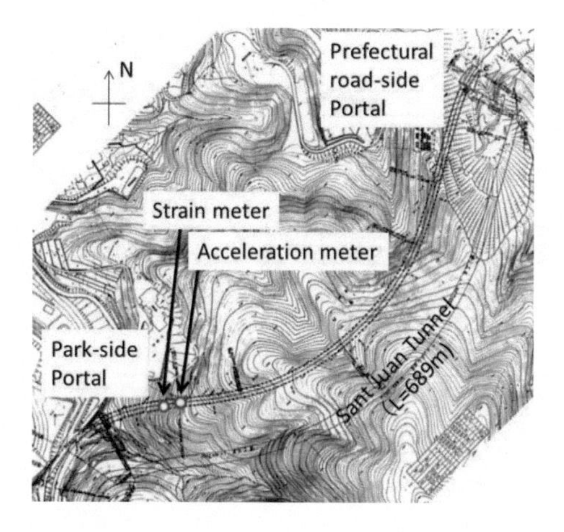

Figure 1. Topography around San Juan tunnel.

Table 1. Outline of the tunnel.

Name	Sant Juan Tunnel
Administrator	Ishinomaki City
Purpose	Two lane road tunnel
Length	689 m
Completion year	1996
Construction method	Conventional tunneling method
Bedrock	Mainly sandrock and shale

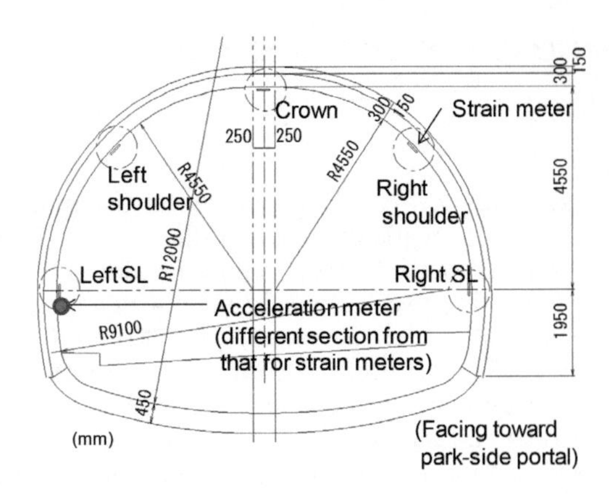

Figure 2. Measuring points of acceleration and strain of the lining.

2.2 *Obtained strain data of the lining from the measurement*

The peak acceleration at ground surface (K-net MYG010, by NIED) was 300 gal. On the other hand, the maximum acceleration at the tunnel sidewall was approximately 200 gal, and there was almost no difference between the three components (Kusaka *et al.*, 2012). These data cannot be directly compared since the observation points are different, but the peak acceleration of the tunnel was smaller than that of the ground surface.

Figure 3 indicates the time history of the strain on the lining, where positive value indicates tension. Only major period of time is extracted in the figure despite the longer duration time

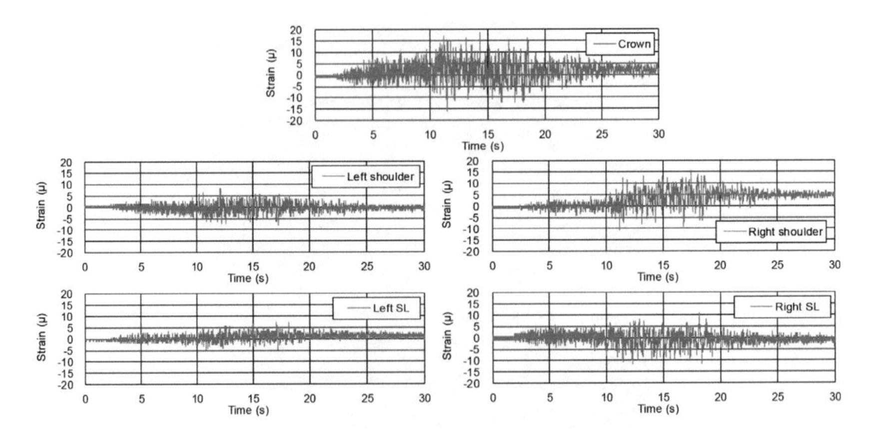

Figure 3. Time history of strain on the lining.

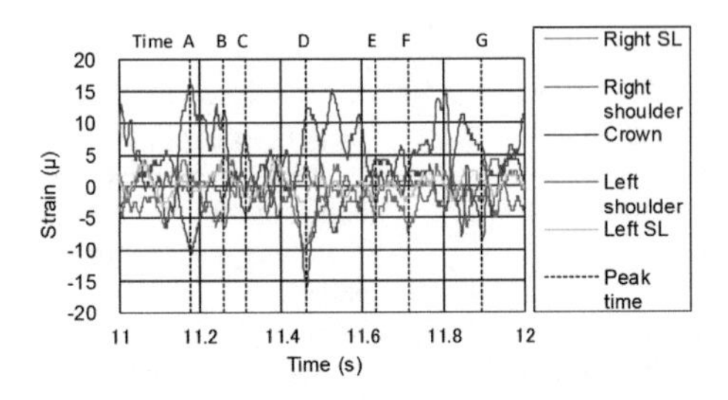

Figure 4. Time history of strain on the lining (11-12 sec).

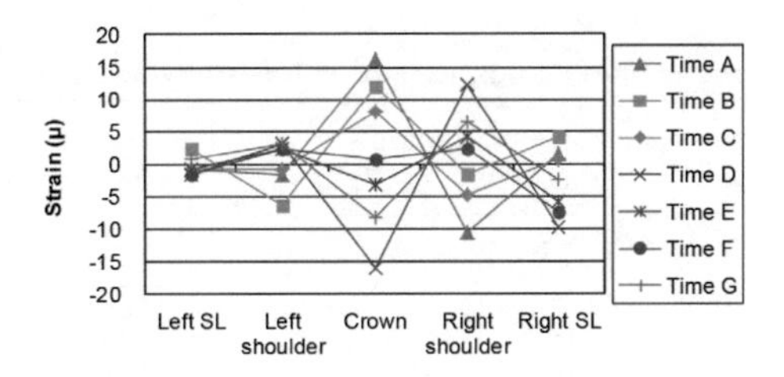

Figure 5. Strain distributions at the peak time.

of the earthquake. Although the largest strain occurred at the right shoulder and crown, they were 20×10^{-6} at most for both compression and tension, which can be said to be in the elastic range of concrete. Note, however, that residual strain was observed despite its smallness, so that the influence of repeated earthquakes should be a problem to be examined.

Figure 4 shows the detailed strain for 11 to 12 seconds, during which maximum strain occurred. Strain distribution at peak times in the time range, indicated by time A through G in the figure, is shown in Figure 5.

3 OUTLINE OF NUMERICAL ANALYSIS

3.1 *Purpose of the analysis*

Most major seismic damage mode of tunnel should be generally caused by shear deformation of surrounding ground and shoulders of tunnel are most likely to be damaged by earthquake (eg. Asakura *et al.*, 2001, Hashash *et al.*, 2001, Wang *et al.*, 2001). However, actually-measured strain mode, mentioned above, showed that the largest strain occurred at crown. Other seismic damages in actual tunnels (eg. Mashimo, 2005) also support that major damage may occur at crown. Based on these facts, the authors have recognized that vertical and horizontal principal loading modes should also be considered in order to evaluate the seismic actions.

As mentioned above, actually-measured strain of the lining was only 20×10^{-6} at most during the large earthquake. Although this amount of strain was much smaller than compressive yield strain of concrete, generally mentioned as $2{,}000 \times 10^{-6}$, loading magnitude is examined to grasp rough importance of the influence of earthquake in design stage in future.

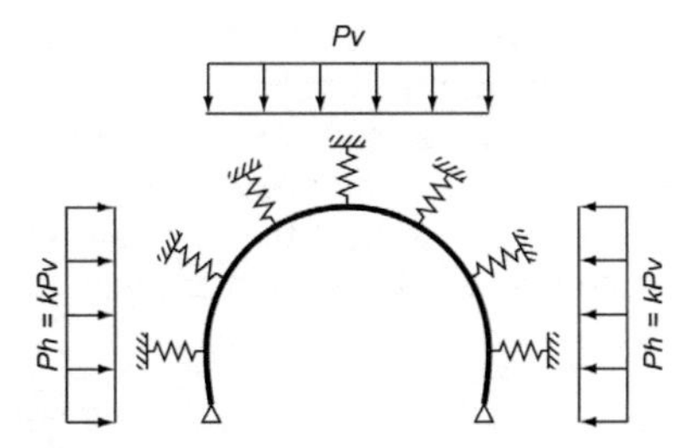

Figure 6. Outline of the beam-spring model.

Table 2. Physical properties of the model of lining.

item	Value	Unit
Thickness	0.3	m
Young's modulus	22,000	MPa
Poisson's ratio	0.2	–

Table 3. Ground reaction modulus used in this study

Assumed ground class	Ground Young's modulus E (MPa)	Ground reaction modulus Kn (MN/m3)
DII	150	37
DI	500	123
CII	1,000	246
CI	2,000	491

3.2 *Outline of the analysis*

The numerical simulation was upon a simple beam-spring model as illustrated in Figure 6.

The lining was modeled as a linear elastic beam with physical properties shown in Table 2. The thickness of the lining was 30 cm with depth of 1 m in the model. The Young's model of the lining was 22 GPa with Poisson's ratio of 0.2, assuming a plain concrete with uniaxial compressive strength of 18 MPa.

The ground was modeled as no-tension ground spring. The ground reaction modulus used in this model was listed in Table 3, assuming Young's modulus of 150, 500, 1,000, 2,000 MPa which are commonly used for Japanese ground classes DII (poor), DI, CII and CI (fair).

The load was assumed as vertical and horizontal pressure with k-ratio of 0.5 and 2. This mode of the loading was based on the measured strain mode which implies a presence of vertical and horizontal pressure (Kusaka *et al.*, 2011). Unit load was applied as Pv = 100 kPa or Ph = 100 kPa.

Both foots of the lining were supported as pins, since the tunnel poses enough ground bearing force by installing the invert concrete.

4 RESULTS AND DISCUSSIONS ABOUT LOADING MAGNITUDE

4.1 *Stress distribution*

Figures 7, 8 and 9 show the deformation, axial force and bending moment of the lining, respectively. In the case of k = 0.5, positive bending moment occurs around the crown with inwardly-convex settlement. The axial force on the sidewall is larger than that on the crown. On the other hand in the case of k = 2, positive bending moment occurs around the SL with inwardly-convex convergence. The axial force on the crown is larger than that of the sidewall. The positive moment around the crown despite dominance of horizontal pressure is due to the straight part of the lining at the crown in this tunnel.

Figure 10 illustrates the strain of inner surface of the lining calculated from the axial force and bending moment. In the case of k = 0.5, the opposite sign of the strain occurs between the crown and the shoulder. Also, the strain around the sidewall is relatively small. This trend is in good agreement with the actually measured data in the San Juan tunnel shown in Figure 5.

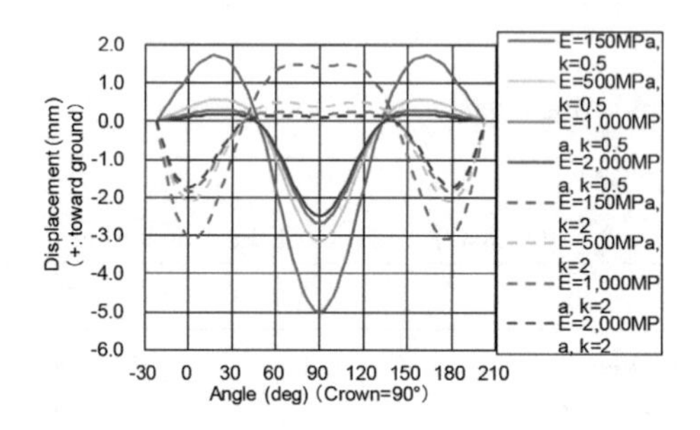

Figure 7. Displacement diagram.

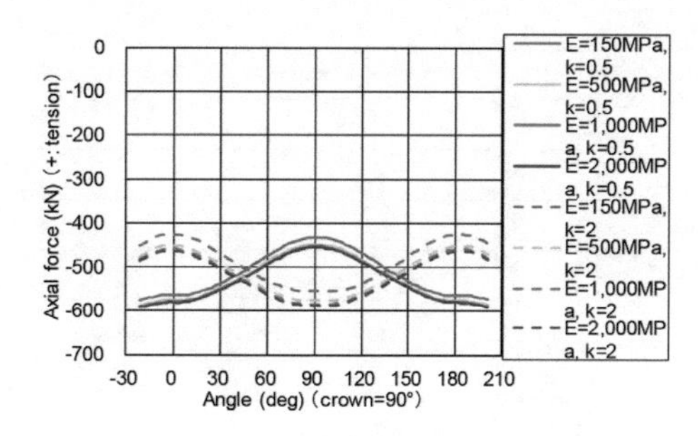

Figure 8. Axial force diagram.

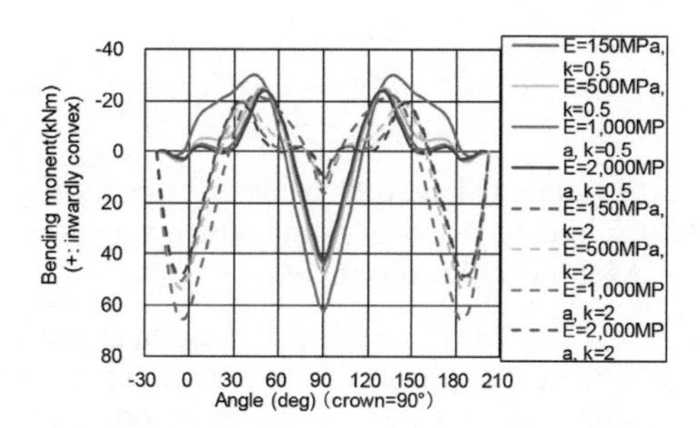

Figure 9. Bending moment diagram.

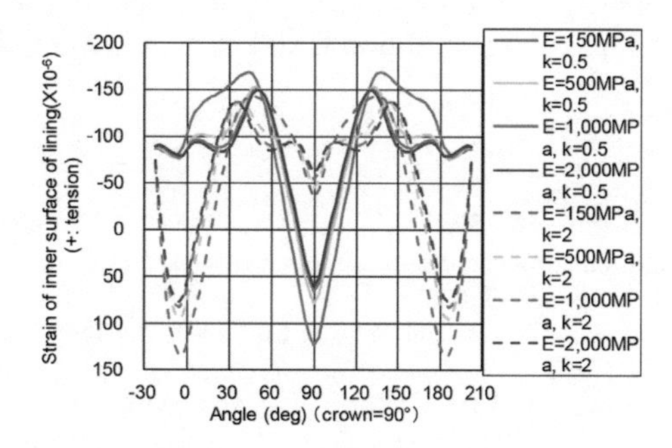

Figure 10. Strain distribution of inner surface of lining.

Judging from these results, the loading mode of San Juan tunnel during the earthquake described above can be closed to k = 0.5.

4.2 *Loading magnitude estimated from actual measurement*

The unit load was applied in the analysis, therefore the loading magnitude to simulate the strain of 20×10^{-6} at the crown can be calculated by a simple multiplication. Figure 11 shows the loading magnitude that represents 20×10^{-6} at the crown. To focus on the case of k = 0.5, which can simulate the strain mode of San Juan tunnel most fairly, the loading magnitude would be in the range between Pv = 16 kPa for poor ground (DII) and Pv = 33 kPa for fair ground (CI). As the ground class at the measuring point of San Juan tunnel was judged as DI at the time of construction, most possible loading magnitude is Pv = 26. This is relatively small compared with classical Terzaghi's rock load or full overburden pressure for a generally-considered scale of road tunnels.

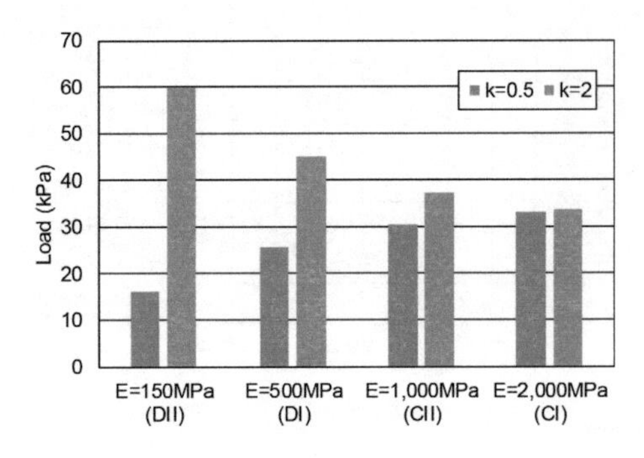

Figure 11. Loading magnitude to simulate the strain of 20×10^{-6} at the crown.

5 CONCLUSIONS

In this study, a series of static numerical analysis using a frame model is conducted to estimate rough loading magnitude to simulate strain of lining which was measured in an actual tunnel during a large earth quake. Major conclusions include:

- In order to simulate the strain mode of the tunnel lining measured during the earthquake, a static load of vertical and horizontal pressure with k-ratio 0.5 can be applied.
- According to the simulation, Pv = 26 kPa is the most possible maximum magnitude of loading during the earthquake.

Note, however, that the conclusions described above is a result from the measurement in a tunnel in an ordinal ground condition as a rock tunnel. The mechanism of actual tunnel damage in an extremely poor ground condition is still remained to be clarified.

REFERENCES

Asakura T., Matsuoka S., Yashiro K., Shiba Y., and Oya T. 2001. Damage to mountain tunnels by earthquake and its mechanism, *Proceedings of the International Symposium on Modern Tunneling Science and Technology*, Vol. 1, pp. 351–356.

Hashash, Y.M.A., Hook, J.J., Schmidt, B., and Yao, J.I.C. 2001. Seismic design and analysis of underground structures, *Tunnelling and Underground Space Technology*, Vol. 16, pp. 247–293.

Isago, N., Kusaka, A., and Koide, T. 2018. Deformation and reinforcement mechanism for mountain tunnel in severe earthquake, *ITA-AITES World Tunnel Congress 2018 Proceedings*, No.1023, pp.2–15.

Japan Road Association. 2003. *Technical standard for road tunnels (structures) and its commentary.* Tokyo: Maruzen. (in Japanese)

Kusaka, A., Mashimo, H., Isago, N., and Kadoyu, K. 2011. Seismic behavior of mountain tunnel affected by difference of lining structure with numerical analysis, *ITA-AITES World Tunnel Congress 2011 Proceedings*, pp.450–458.

Kusaka, A., Isago, N., Mashimo, H., and Kadoyu, K. 2012. Dynamic Measurement of an Actual Mountain Tunnel during a Large Earthquake, *Proceedings of ITA-AITES World Tunnel Congress 2012*, pp.1–6.

Kusaka, A., Kawata, K., and Isago, N. 2016, Tunnel Deformation Mode and Loading Magnitude during Large Earthquake, *Proceedings of ITA-AITES World Tunnel Congress 2016, No. 605*, pp.1–9.

Mashimo, H. 2005. Damage of road tunnels caused by 2004 Mid Niigata Prefecture Earthquake. *Journal of JTA-Tunnel and Underground*, Vol. 11, No. 35, pp. 55–63. (in Japanese)

Wang, W.L., Wang, T.T., Su, J.J., Lin, C.H., Seng, C.R., and Huang, T.H. 2001. Assessment of damage in mountain tunnels due to the Taiwan Chi-Chi Earthquake. *Tunnelling and Underground Space Technology*, Vol. 16, pp.133–150.

Tunnels and Underground Cities: Engineering and Innovation meet Archaeology, Architecture and Art, Volume 3: Geological and geotechnical knowledge and requirements for project implementation – Peila, Viggiani & Celestino (Eds)
 ISBN 978-0-367-46583-4

Ensuring safety during construction of double-track subway tunnels in quaternary deposits

M.O. Lebedev, K.P. Bezrodny & R.I. Larionov
OJSC "NIPII "Lenmetrogiprotrans", Saint-Petersburg, Russia

ABSTRACT: For the first time in the subways construction practices in the territory of the Russian Federation, a double-track running tunnel was built on two sections of the Saint-Petersburg subway. The drivage carried out by a Tunnel-Boring Machine (TBM) with Earth Pressure Balance (EPB). The tunnels' route crosses road junctions and railway; buildings and structures are also located in the construction influence zone.

Ensuring the construction safety requires geotechnical monitoring, and one of its goals was adjusting the process parameters of the shield. Adjustment of TBM parameters was made according to the data from borehole extensometers on the tunnel axis at different depths.

The result of geotechnical monitoring, and also the performed analytical and numerical simulations, is conclusion that a major influence on the development of the stress-and-strain state for the tunnels sections passed within the quaternary deposits has the depth of the tunnel and, to a lesser extent, the mechanical characteristics of the soil massif.

1 INTRODUCTION

The subway construction in the city of Leningrad started at significant depths, approximately 60 metres under the ground, on average. The reason for such deep placement was the presence of a thick layer of quaternary deposits in the form of water-saturated unstable soils. Since the pick hammer served as the main tool for tunnelling up to the middle of the 20th century, the water tightness of the subway facilities could only be achieved within the aquilide of dense claystone-like Kotlin underclays of the quaternary deposits.

The development of the technologies and the international experience of their application for the construction of wide diameter subway tunnels (Maslak et al. 2014, Merkin & Khokhlov 2016) allowed constructing a two-track tunnel in Saint-Petersburg. This, in turn, made it possible to apply modern engineering solutions when constructing subsurface stations (Boytsov & Evstifeeva 2016).

The world experience of tunnels construction with a large cross-section by EPB-shield has a broad geographic spread. Tunnels of this type are built in Spain, Italy, Portugal, Germany, America, China, Brazil, Egypt, Greece and other countries. The construction of the first double-track running tunnel completed by EPB-shield on the section of Frunzensky radius of Saint-Petersburg subway.

A major part of the 3800 metres long running tunnel is in the completely unstable soils at the depth of 10.0 to 13.6 metres (measured at the height of the vault crown). The TBM reaches the dismantle chamber at the depth of more than 50 metres.

A tunnel with inner and outer diametrs of 9.4 m and 10.3 m respectively excavated by "Herrenknecht" shield with EPB in the face. The cross-section of the tunnel is presented in Figure 1.

The influence zone of the tunnel includes the Ring Road, railroad tracks, buildings and structures, tramway tracks and motor roads.

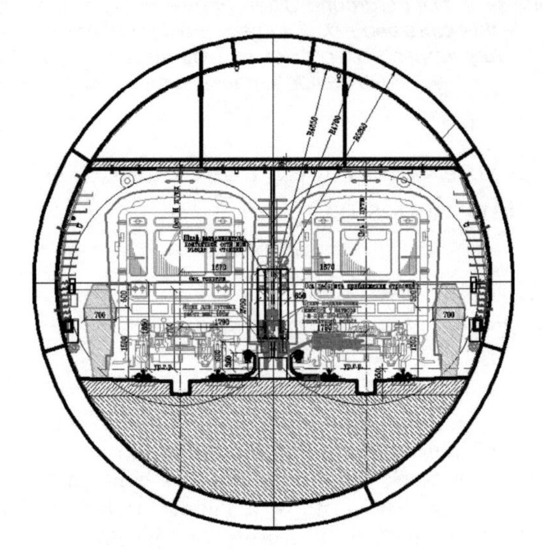

Figure 1. Cross section of a double-track tunnel.

The construction of the second double-track running tunnel completed by EPB-shield on the section of Nevsko-Vasileostrovskaya line of Saint-Petersburg subway with similar engineering and geological conditions, with the exception of tunnel section passed under the Gulf of Finland.

During excavation the geotechnical monitoring was set up, aimed at reducing the negative anthropogenic impact on the environment and at the safety of tunnel excavation. The geotechnical monitoring served the following purposes:

- making engineering and geological, as well as hydrogeological forecast beyond the face;
- identification of the surface and deep level movements of soil massif;
- visual and instrumental monitoring of the structures in the area of tunnel construction;
- backfilling quality control of the area outside of the tunnel lining;
- assessing the stress-and-strain state of the tunnel lining.

The safety of tunnelling works depends on a large extent on the availability of ultimately reliable and up-to-date data on the stress-and-strain state of the "tunnel – soil massif" system, as well as of the ground surface. This data can be obtained through geotechnical monitoring with the use of automated systems that assure the availability of up-to-date information received from all fixed control and measuring devices (Lebedev 2016).

2 GEOTECHNICAL MONITORING

2.1 *Adjustment of shield TBM process parameters*

The main problem of soft grounds tunnel construction at shallow depths is support of ground surface. The presence of geodetic control of deformations doesn't allow determining the specific TBM process parameters that affect the development of deformations in the enclosing solid mass. To this end, boreholes with extensometers were placed along the entire length of the running tunnel, both in the shallow (Figure 2a) and deep level (Figure 2b). Lower extensometers are located 1.5 meters above the tunnel lining with diameter of 10.3 meters. Automated measurement system located in a protective cabinet took the gauge every 2 hours with data transfer to a remote Internet portal. The measurement results at one of the tunnel sections are shown in Figure 3.

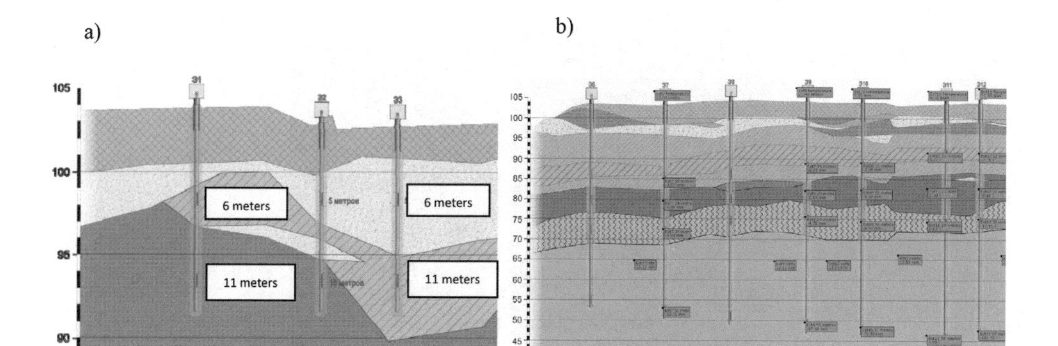

Figure 2. Location of wells with extensometers over a double-track running tunnel: a) in shallow level; b) in deep level.

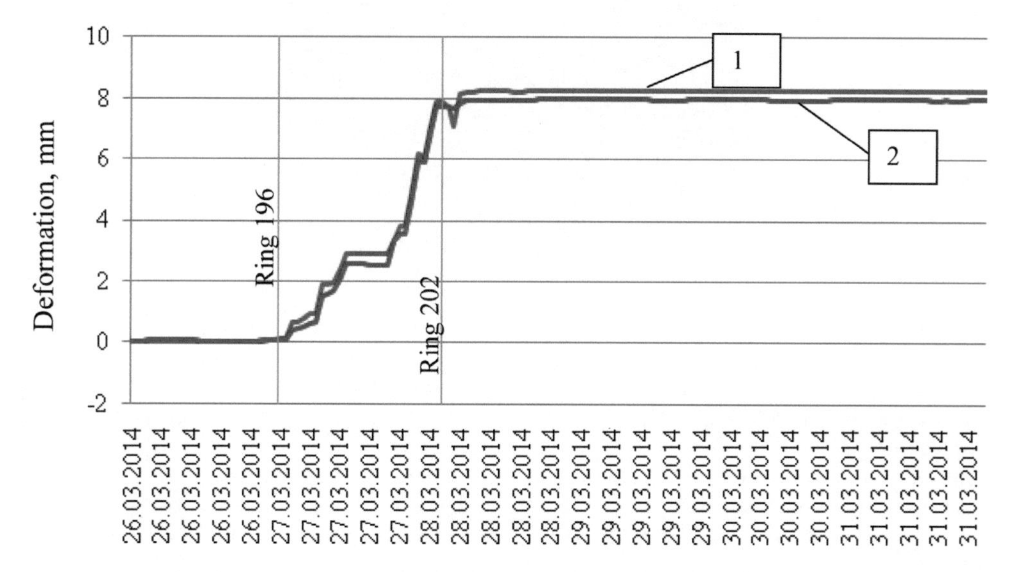

Figure 3. Development of deformations in the depth of solid mass for two extensometers located in one well: 1 - at a depth of 5 meters; 2 - at a depth of 10 meters.

Following the results obtained, a timeframe has been set up to establish correlation between strains and technological parameters of the TBM guiding, and also the strain spread velocity from the lining contour to the surface was calculated. The optimal pressure of the face's EPB did not incur any strain of the soil massif either beyond the face or during the movement of the shield under the borehole. Ascent strain occurred only when delivering grout beyond the lining. Given the absence of the engineering communications in this section of the tunnel, it became possible to exceed the pressure of delivery of backfill compared to the estimated threshold of the surface rising (Figure 4). Besides, it was discovered that the 12 metres thick layer of soil above the tunnel almost does not get compressed, that is why the excessive pressure produced when backfilling the space beyond the tunnel lining was immediately transferred to the ground surface. Within 2 days from 26 March, 2014 till 28 March, 2014, the ground level ascended by 45 mm, though its subsidence had been estimated by 20 mm.

Another result obtained during the study of the soil massif strain was indirect limitation of the "survival time" of the grout. The extensometer ascent strain lasts for more

Figure 4. Crack on the earth surface when inflating the grouting mortar beyond lining.

than 24 hours. Within this period, 6 ring sections of the lining were erected; the distance from the shield was about 10.8 metres. This is only possible when the pressure is applied from the contour of the tunnel, that is, through the pressure of the grout. The setting time of the grout determine not only the surface strain, but also the stress-and-strain state of the lining.

The adjustment of the TBM technological parameters at the subject section allowed the excavation of the rest subsurface tunnel route with minimal surface strain, ranging from -5 mm (sinking) to + 15 mm. The areas with an ascent of the surface commonly did not sink later on.

Altogether, 6 extensometer boreholes have been placed along the tunnel route in the subsurface area, and 2 ones in the deep level area.

In the deep level area the extensometers registered sinking strain of the near-lining zone by 0.5 mm. No surface strain has been observed following these sinking strains of the near-lining zone.

In the construction of the second double-track running tunnel, 5 boreholes with extensometers were placed in the shallow and 7 ones in the deep level. The measurement results at one of tunnel sections in the shallow level are shown in Figure 5 and show the emergency mode of TBM.

The deformations' development begins when the face is located at a distance of 7.5 from the well. Having reached the value of 18 mm on the lower extensometer, the deformations are stabilized. By this time was installed ring 608, the surface deformations were 11 mm.

A rapid increase of subsidence begins when the lining ring is mounted under the borehole and the next two rings reaching 65 mm along the lower extensometer and 13 mm along the upper extensometer. The difference in the readings between the upper and lower extensometers shows the value of "segregation" of the solid mass over the tunnel.

Subsequent extensometer readings show more than 100 mm subsidence of the ground surface.

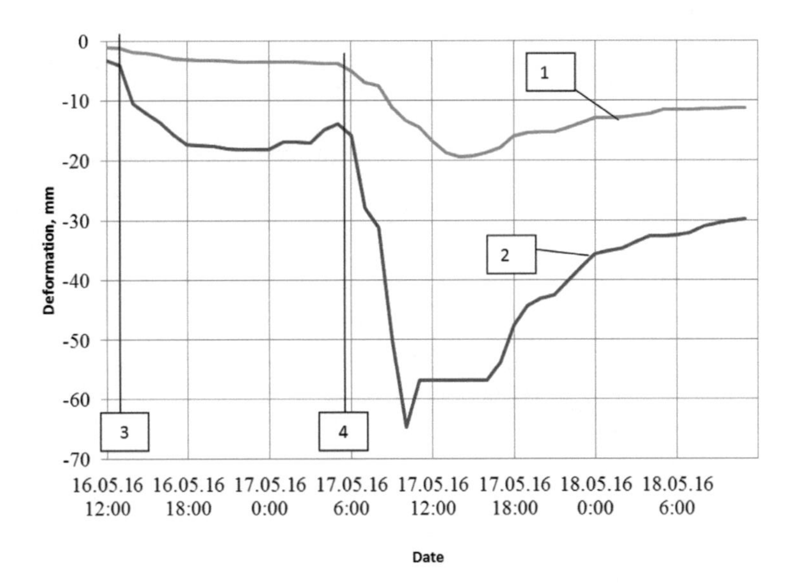

Figure 5. Development of deformations in the depth of solid mass for two extensometers located in one well: 1 - at a depth of 4 meters; 2 - at a depth of 8 meters; 3 – face at a distance of 7.5 from the well; 4 – erection of a lining ring under the well.

2.2 *Monitoring of buildings and structures*

The visual and instrumental monitoring of buildings plays a significant role in assessing the impact of the construction process on the pre-existing buildings and structures or the absence of such impact.

Along the route of both tunnels, residential and public buildings fell into the construction influence zone in the section of shallow and deep levels.

In order to assess the stress-and-strain state of the buildings, a fixed system for monitoring architectural structures was used. The measuring devices regularly registered the change in the crack openings, in the vibration intensity, in the tilt angle and the temperature in the points where the devices were installed; the collected data was further transferred through radio channel to a remote Internet portal.

The 2-year-long monitoring with the help of the control and measurement devices installed determined that the crack openings mostly depended on the changes in the temperature alone. The range of the crack opening/closing reached 2 mm. The cracks did not return to their previous state. Neglecting the temperature, the accumulated crack opening amounted to 0.5 mm. At the same time, no crack opening resulting from the tunnel construction was observed on the façades of the buildings or on the inner structure.

The use of an automated geodetic monitoring system for the construction of the Nevsko-Vasileostrovskaya line showed its effectiveness in controlling the deformations of WHSD (Western High-Speed Diameter) supports (Figure 6a) in line of the small Neva river and the Gulf of Finland.

The supports receiving loads from the cable bridge and the support of exit from the highway are in the zone of construction influence of a double-track running tunnel. By means of a robotic tachymeter located outside the zone of construction influence the deformations of supports were monitored with an interval of one measurement per hour. The measurement results (Figure 6b) were transmitted in real time to a remote Internet portal.

The largest influence from piercing had the support with prism No.12 which is also the most loaded. The vertical deformations along this support were 16 mm. Actual values of vertical deformations along all monitored supports coincided with the calculated values or were less than calculated values.

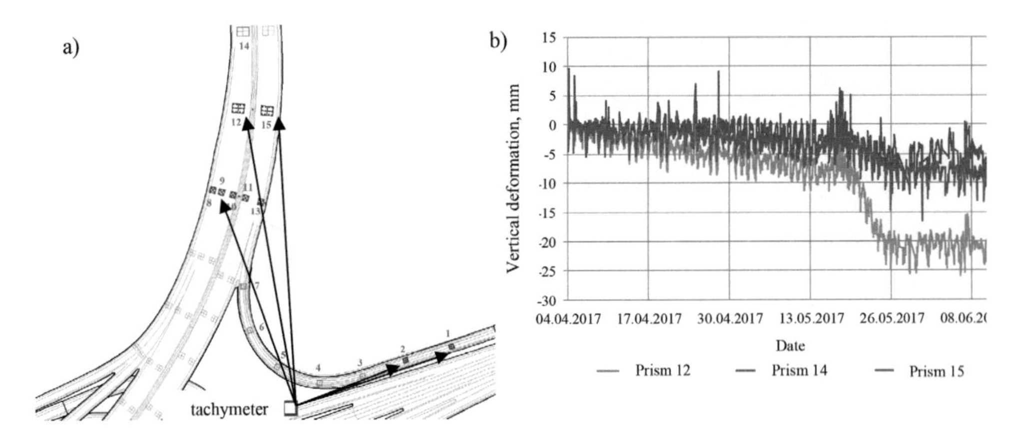

Figure 6. Layout of the WHSD supports equipped with mirror prisms a) and measurement results for vertical deformations b).

2.3 *Stress-and-strain condition of the lining*

The stress-and-strain state of the tunnel lining was assessed with the use of control and measurement devices installed in the blocks during their production. The lining was equipped with string strain meters – tensometers with 200 mm spacing. The registered local relative strains were used to calculate the regular tangential stresses. In concrete and reinforced concrete structures the stresses are calculated using a special method that takes into account the loading of the concrete at an early stage and concrete crawling.

The most extensive data on the formation of the stress-and-strain state of supports and linings is provided by a combination of sensors (strain meters) inside the structure (Figure 7*a*) and measurement of the inner contour strain starting from the moment of construction.

Considering the construction technology, it is possible to control the qualitative and the quantitative changes in the stress-and-strain behaviour of the lining (Figure 7*b*) from the moment of the ring assembly to under the protection of the TBM. The comparison of the stress in the lining with the strains in the inner lining for a specific cross-section allows assessing the load-bearing capacity of the other sections of the tunnels with fewer costs, limiting it to inner lining strain control only. For reliable and sufficient determining of the load-bearing capacity of the lining along the tunnel route using this method, the lining is equipped with sensors within all lithological varieties that the tunnel crosses. Along the route of the tunnel analysed, 18 rings were equipped with sensors.

The control of the stress-and-strain state of the lining as concerns the duration and the change speed of the observed values allows assessing not only the absolute values determined through calculation, but also assessing indirectly the quality of backfill of the space beyond the tunnel lining and the time of setting of the grout.

3 SIMULATION OF STRESS-AND-STRAIN STATE OF THE SYSTEM "LINING - ENCLOSING SOLID MASS"

The prediction of stress-and-strain state of the lining for double-track tunnel and enclosing soils was carried out using 3D-finite element modeling.

The basic scheme for simulation of construction for double-track tunnel includes the deformation area of ground in front of the face crown, the deformation of ground along the shell length of the shield (due to their joint deformation and the loss of tunnel volume due to the conical shape of shield shell), and the ground softening in the grouting zone behind the lining.

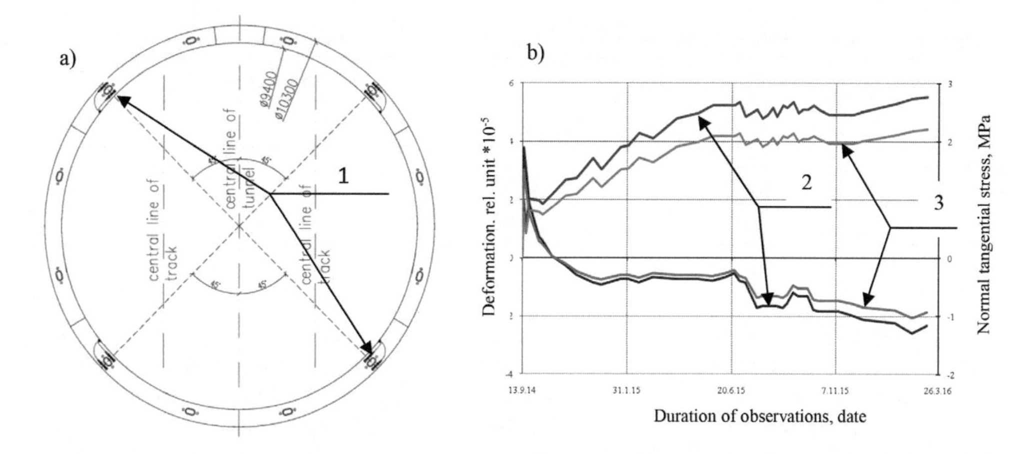

Figure 7. Layout of sensors in the lining of a double-track running tunnel a) and indicative schedule for development of forces in the lining b): 1 - stringed tensometers; 2 - relative deformations on inner and outer contours*10^{-5} relative units; 3 - normal tangential stresses in concrete on inner and outer contours, MPa.

The simulation of construction for double-track tunnel was carried out using the software complex Plaxis 3D. The basic design scheme embedded in the model is presented in Figure 8.

The model geometry was divided into 250,000 spatial solid-state elements of tetrahedral shape. The finite element grid was created with differentiation in size: it thickened the tunnel contour and thinned out near the model faces.

The elasto-plastic model Hardening Soil was used in the framework of studies to describe the behavior of solid mass. The linear solid model was used to describe the mechanical behavior of lining for double-track tunnel.

The tunnel construction was simulated by sequential execution of calculated stages, during which the drivage for the next pass and the erection of lining were made. In total, the model included 23 calculation stages, the first of which assumed the formation of natural stress in solid mass. The remaining stages enabled to simulate the drivage for a section of double-track tunnel with a length of about 40 m.

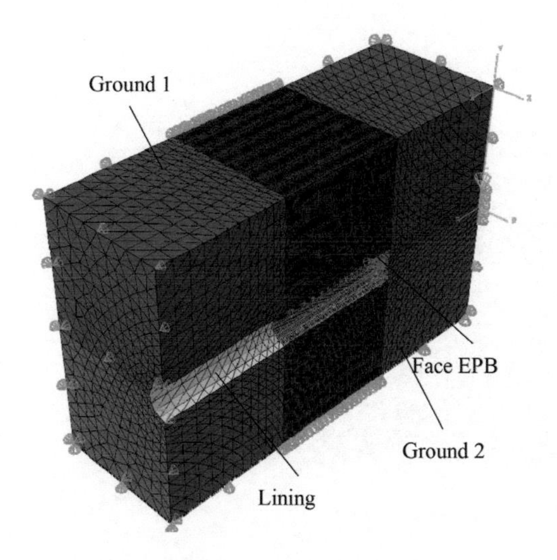

Figure 8. Basic design scheme of finite element model for construction of double-track tunnel.

Table 1. Process parameters for tunnel construction.

Plot	Pressure of face crown EPB, kPa	Pressure of grouting mortar injection, kPa	Force from shield jacks, kN	Solid mass properties			
				E_0, MPa	ν	ϕ, deg.	C, kPa
1	300	500	41000	14.5	0.35	22	27
2	250	700	39000	14.5	0.35	22	27
3	250	300	31500	14.5	0.35	22	27

The calculations are performed for three plots which are practically in the same geological engineering conditions, the tunnel face is practically homogeneous in the geological plan and is represented by claying soils, the laying depth is in the range of 12–13 m (above the tunnel).

The process parameters of tunnel construction were taken according to the EPB-shield protocols created by the results of drivage for each ring. The process parameters of construction of double-track tunnel by means of tunnel-boring machine are summarized in Table 1.

The largest predicted subsidence of the ground surface was obtained for the plot No.2 where there is the largest spread of EPB pressure and pressure of grout injection.

Absolute values of vertical subsidence at the final design stage are: for the plot No.1 — 2 mm, for the plot No.2 — 3 mm, for the plot No.3 — 1 mm.

The major principal stresses in the lining side of double-track tunnel have only one growth stage — immediately after insertion into the work (Figures 9, 10). The major principal stresses in the lining side stabilize fairly quickly. Absolute values of the major principal stresses in the lining side at the final design stage are: for the plot No.1 — 4.5 MPa, for the plot No.2 — 4 MPa, for the plot No.3 — 5 MPa. The calculations conform to the results of field studies.

Calculations of the stress-and-strain state of a system "lining – soil" show that the process parameters of TBM influence the uniformity of load transfer from the solid mass to the tunnel lining, and as a result, the qualitative and quantitative force distribution.

To predict the stressed state of the lining for double-track running tunnels in the conditions of Saint-Petersburg, multivariate calculations were performed by the analytical solution and the numerical method in a plane-deformation setting.

As a basis for analytical method, a decision was made in (Einstein & Schwartz 1979) which corrections were made to take into account the change in deformation characteristics of the ground from the value of acting stresses of the rock mass, as well as the presence of groundwater. The problem is considered in effective stresses.

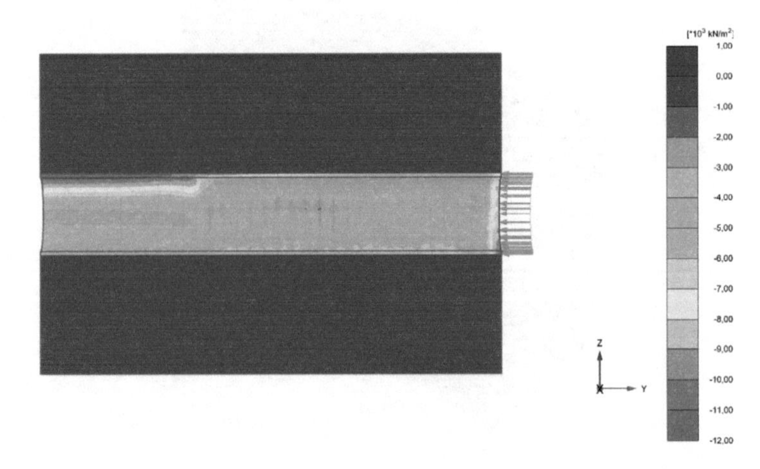

Figure 9. Isochromatic curve of distribution of the major principal stresses in the lining.

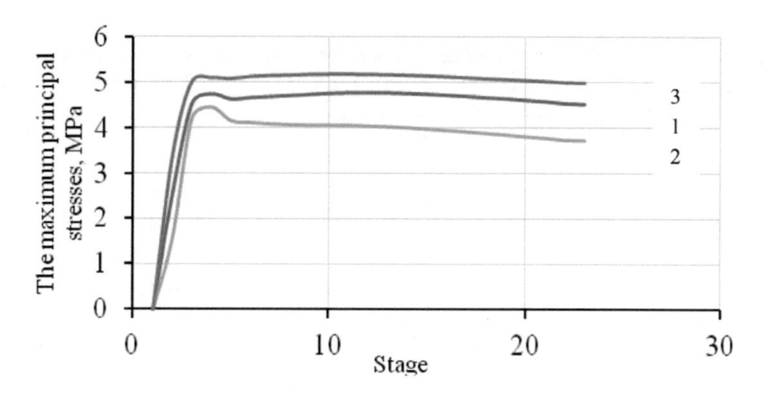

Figure 10. The curves of development of the major principal stresses in the lining side for the plots 1–3.

To perform a comprehensive analysis of the stressed lining state, the following geological engineering conditions are adopted: a) dry rock mass formed by sandy grounds; b) dry rock mass formed by clay grounds; c) watered rock mass formed by sandy grounds; d) watered rock mass formed by clay grounds.

The calculation results are presented in the form of change dependencies of integral indicators of stressed state of the tunnel lining section and the maximum values of tangential normal stresses on the inner and outer lining contours (Figure 11a).

The tangential lining stresses are predominantly compressive, and starting from a depth of 12 m, the normal tangential stresses go completely into compressive stresses.

In general, it can be noted that from the point of view of lining behavior under the conditions in question, the clay watered grounds are more favorable, while the lining arrangement in sandy dry grounds leads to the appearance of significant bending moments forming an uneven distribution of tangential stresses in the lining section.

The comparison of calculation results for the stress lining state performed on the basis of analytical calculations (Figure 11a) and numerical modeling (Figure 11b) allows us to draw the following conclusions. The values of longitudinal forces are insignificantly dependent on the chosen model of solid mass deformation and calculation method. The change in bending moment at the tunnel depth of 10 m from the ground surface on the basis of numerical simulation and the application of elasto-plastic model with double hardening varies from 21 to 50 kN*m, then on the basis of analytical solution the bending moment in such grounds varies from 54 to 89 kN*m. Thus, the value of bending moment on the basis of analytical solution is overestimated 1.5–2 times. By increasing the tunnel depth to 40 m, the value of bending moment obtained on the basis of numerical simulation varies from 61 to 116 kN*m, while on

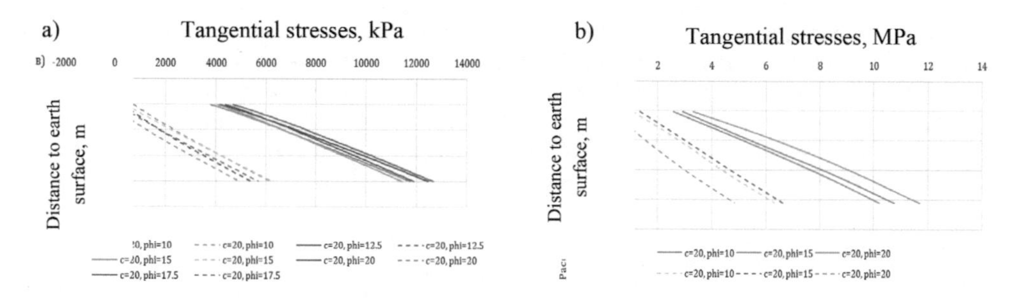

Figure 11. The calculation results for stressed lining state located in the clay watered grounds by analytical method a) and numerical modeling b): c – rock coupling value (kPa); phi – internal friction angle value (deg.); the solid line represents the stresses on inner lining contour, the dashed line — on outer lining contour.

the basis of analytical solution from 87 to 130 kN*m. Thus, the value of bending moment on the basis of analytical solution is overestimated 1.15–1.4 times. The difference in calculation of bending moment in the tunnel lining leads to the change in predicted values of tangential stresses.

4 CONCLUSION

The construction of a large cross-section running tunnel in the complex engineering and geological conditions of Saint-Petersburg confirms that the safety during construction is assured through the implementation of the modern technologies to complement the geotechnical monitoring.

The results of the studies allowed determining the main factors that influence the movement of the soil massif across the entire soil layer and working out recommendations for adjustment of the construction technology parameters, for reduction of the negative impact of the ground sinking and the related process of building deterioration within the shift trough, as well as for the reduction of the vibration impact on such building to the admissible level.

The use of modern automated geotechnical monitoring systems during tunnel excavation for different purposes, particularly in urban construction areas, constitutes an efficient element of the technological process allowing to mitigate considerably the risks of emergencies and to increase the efficiency of the measures adopted following the monitoring results.

The placement of the control and measuring devices within the lining for assessment of the stress-and-strain state helps not only to control the pressure formation during the tunnel construction, but also to estimate the load-bearing capacity of the lining in order to assure safe operation of the transportation tunnels.

Generalizing the accumulated experience of geotechnical monitoring, it is important to recognize its role in the process of subterranean construction and to point out the following advantages of such monitoring:

- the research tasks allow formulating an overall understanding of the cooperation and joint functioning of the "tunnel – soil massif" system to be used for rapid decision-making related to the need of adjustment of the construction technical parameters;
- the aims of the monitoring complement each other and help avoiding possible mistakes in the interpretation of the data collected;
- the monitoring assists in reducing the negative impact of the tunnelling on the environment; it assures industrial safety during the construction phase.

If the tunnel depth is constant, the qualitative and quantitative distribution of stress-and-strain lining state is affected by process parameters of TBM-shield, such as the pressure of face EPB, the pressure of grout injection behind the lining, and the pressure of shield jack.

Taking into account the results of field researches, the numerical simulation of the stress-and-strain state of the double-track tunnel lining and the soil mass is performed. The obtained regularities after verification of the design scheme show that the main influence on the development of the stress-and-strain state of the lining has a depth of the tunnel and to a lesser degree the mechanical properties of the soil mass. The calculated values of the stresses in the lining are close to the actual stresses, obtained by field researches.

REFERENCES

Boytsov D.A., Evstifeeva O.V. 2016. Present-day developments in the design of subway stations, *Metro and tunnels*, 6, Moscow: 47–52.

Einstein, H.H., Schwartz C.W. 1979. Simplified Analysis for Tunnel Supports, *J. Geotech. Engineering Division*, 105, GT4, 499–518.

Maslak, V.A., Bezrodny, K.P., Lebedev, M.O., Gendler, S.G. 2014. New technical and technological solutions for the construction of subway tunnels in a megacity, *Mining Journal*, 5, Moscow: 57–60.
Merkin, V.E., Khokhlov, I.N. 2016. On the conditions for the effective use of double-track tunnels in the construction of a subway in Moscow. *Metro and tunnels*, 6, Moscow: 83–86.
Lebedev, M.O. 2016. Automated systems as a part of geotechnical monitoring in construction and operation of transport tunnels. in The 2016 15th World Conference of Associated Research Centers for the Urban Underground Space (ACUUS-2016). Procedia Engineering, 448–454. DOI: 10.1016/j.proeng.2016.11.719.

Tunnels and Underground Cities: Engineering and Innovation meet Archaeology, Architecture and Art, Volume 3: Geological and geotechnical knowledge and requirements for project implementation – Peila, Viggiani & Celestino (Eds)
 ISBN 978-0-367-46583-4

Response of tunnel surrounding rocks in heavily deformed ground: Case study of Dongao Tunnel, Taiwan

C.-S. Lin
Fortune Construction Co, Ltd, Taipei, Taiwan

H.-J. Shau
Suhua Improvement Engineering Office, Directorate General of Highways, Ilan, Taiwan

Y.-P. Chen
Directorate General of Highways, Taipei, Taiwan

T.-T. Wang
National Taiwan University, Taipei, Taiwan

ABSTRACT: Dongao Tunnel underwent the strata near the collision belt of the Eurasian Plate and the Philippine Sea Plate. The excavated geological formations suffered from deformation caused not only by the global tectonics, but also by local deep-seated gravitational slope movements (DSGS), resulting in different tunneling features. The ground deformed by DSGS exhibits open fractures between rock blocks, with infilling gouge. During initial excavation stage, rock wedges were displaced with various magnitudes around tunnel perimeter. Although cave-in due to falling of rock wedges was considered a possibility, the excavated face was generally stable. Tunneling encountered also a tectonically squeezing ground. The crew suffered from multiple severe face instabilities and squeezing problems. The responses of tunnel surrounding rock after excavation are investigated. Some innovative approaches for construction management are also introduced, including bidding strategy for contractor selection, integrated predictions on multi-temporal geological data at different scales, risk management and timely notification system.

1 INTRODUCTION

The stability of excavated face and associated self-standing time are key factors for tunneling in rocks. A stable excavated face with sufficient self-standing time allows large excavating distance for each rounding, moderate or even no rock support needed after excavation, resulting in high advancing rate of tunneling. In contrast, an excavated face with limited self-standing time, or even unstable immediately after excavation, always jeopardizes rock tunneling. Such a case needs ground reinforcement or improvement for the completion of tunnel excavation and installation of necessary supports typically. However, ground reinforcement or/and improvement is costly and time consuming.

Rock mass classification is typically adopted for the selection of excavation and support type in rock tunneling. Engineering characteristics of rock masses are generally classified into five types, sometimes slightly more. Amongst which, auxiliary countermeasures are typically prepared for the worse type. Nevertheless, a large number of tunnel cases indicated that the value of rock mass rating has low correlation with the stability of excavated face. As such, identifying the stability of excavated face of a rock tunnel and factors affecting tunnel instability are still challenging tasks for rock tunneling.

Dongao Tunnel underwent the strata near the boundary of the collision belt of the Eurasian Plate and the Philippine Sea Plate. The excavated geological formations suffered from deformation caused not only by the global and regional tectonics, but also by local deep-seated gravitational slope movements (DSGS), resulting in different responses of surrounding rock and thus tunneling features. Taking the Dongao Tunnel as an example, this manuscript investigates the response of surrounding rocks in heavy deformed ground after excavation, related challenges met in terms of contractual practice, and introduces some innovative approaches for construction management.

2 PROJECT BACKGROUND

The Dongao Tunnel is a twin-bore highway tunnel located between Suao and Nanao, Ilan (Yilan) County, Taiwan. The tunnel, with 3.32 km in length and 11.52 m in net width, passes the Mt. Houi and Dongao Ridge in northeast Taiwan. The Dongao Tunnel is the critical-path engineering of the Suhua Highway Improvement Project that is going to reconstruct and bypass the dangerous section of existing Suhua Highway to meet the safety expectation of eastern Taiwan populations. Figure 1 shows the location of the Dongao Tunnel.

2.1 *Regional geology*

Figure 2 shows the regional geology in the vicinity of the Dongao Tunnel. The tunnel passes through the Suao Formation, Xiaomaoshan Fault, Nansuao Formation, Houishan Fault and Dongao Schist. The Miocene Suao Formation is mainly composed of slate, with occasionally interbedded thin meta-sandstone. The Eocene Nansuao Formation consists of arkose (feldspar sandstone), diabase, slate, phyllitic slate and phyllite. The Dongao Schist is equivalent to the late Paleozoic to early Mesozoic Dananao Schist near the tunnel, and comprised of quartz mica schist, chlorite schist, meta-chert. In addition to these major formations, amphibolite and gneiss can be also founded in the vicinity of the tunnel (Lin and Kao, 2009).

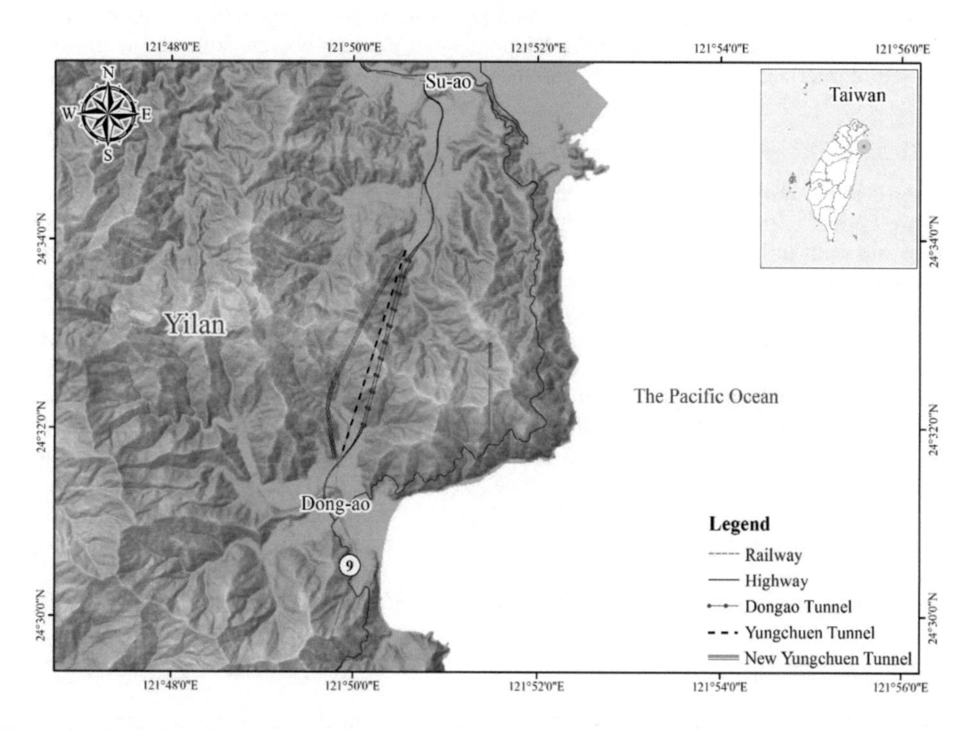

Figure 1. Location of Dongao Tunnel.

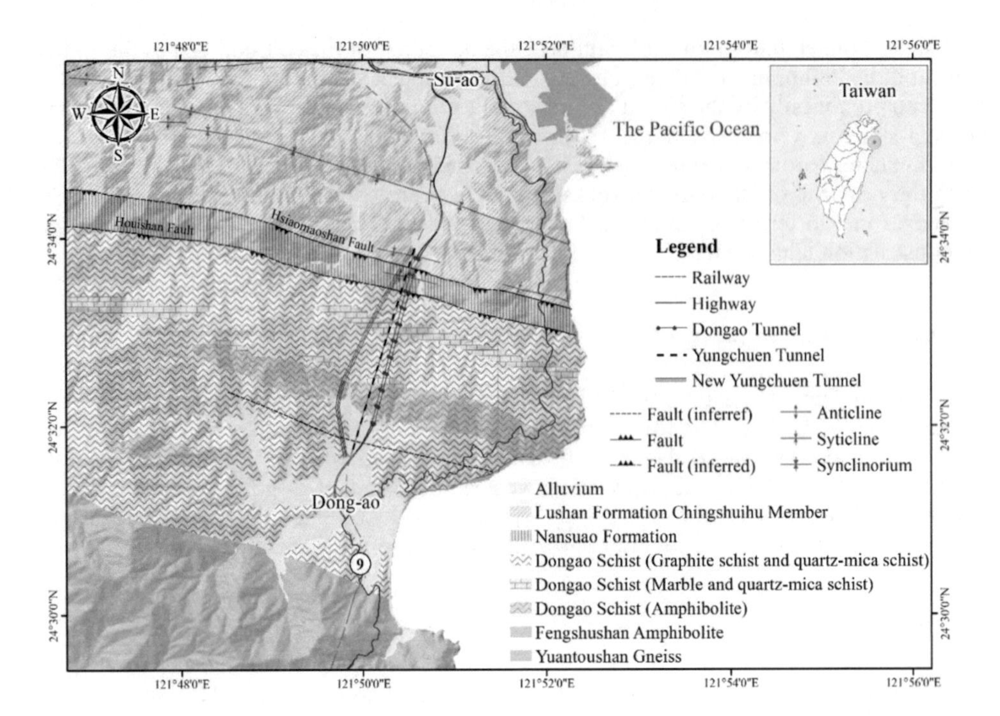

Figure 2. Regional geology near Dongao Tunnel.

The Xiaomaoshan Fault and the Houishan Fault are respectively the boundaries of the Suao Formation and Nansuao Formation, and the Nansuao Formation and Dongao Schist. The Xiaomaoshan Fault has a WNW strike and thrust to north. The fault zone, with a thickness of 30 m approximately, is mainly composed of fractured slate. Both the attitude and thickness of the Houishan Fault are similar to that of the Xiaomaoshan Fault, however, fault gauge and breccia composed of quartz and meta-chert are common seen in the Houishan Fault (Lin et al., 2017).

2.2 *Tunneling experience nearby Dongao Tunnel*

There are two tunnels, the Yungchuen Tunnel and New Yungchuen Tunnel close to the Dongao Tunnel with distance less than 1 km. These tunnels were both single track railway tunnels and built in late 1970s and late 1990s, respectively. The mechanical excavator (known colloquially as Big John) used for the construction of Yungchuen Tunnel was abandoned after 656 m excavation due to its poor advance rate and several collapses (NLREB, 1980). The drilling and blasting method was used for the subsequent excavation. However, a number of construction problems were still encountered. In the northern part of tunnels, tunneling suffered from huge deformation (> 200 mm) caused by squeezing ground. And catastrophic groundwater inflows jeopardized construction of these two tunnels (Wang et al., 2011).

The experience learned from the Yungchuen Tunnels provides valuable information for the design and construction of the Dongao Tunnel. A three dimensional geological map was proposed in detailed design stage, based on the geological information revealed from the Yungchuen Tunnels and results of supplement geological investigation (Fig. 3). To avoid the catastrophic groundwater inflow, the route of the Dongao Tunnel is shifted 63 m to the east of the Yungchuen Tunnel, with a raise of tunnel elevation about 20–35 m above the Yungchuen Tunnels.

2.3 *Tunneling design*

The complicated geological conditions of the Dongao Tunnel and associated possible influence on tunnel construction were realized during planning and design stages. The New

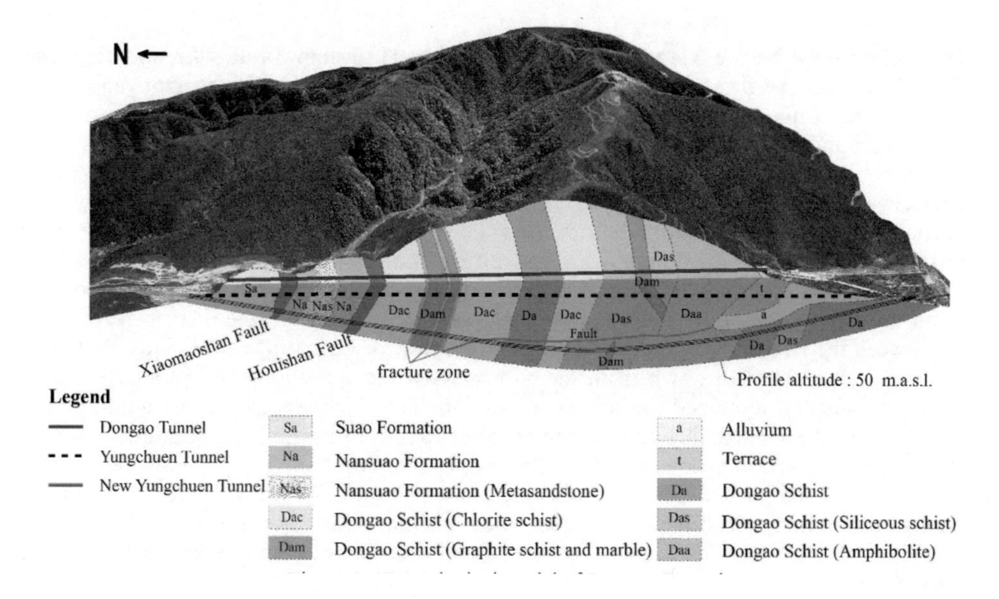

Figure 3. 3D geological model of Dongao Tunnel.

Austrian Tunneling Method (NATM) was adopted for the design and construction of the tunnel, associated with the use of the rock mass rating (RMR). The top heading and bench excavation using the drilling and blasting method was designed, and could be separated into three to five steps to mitigate the disturbance to the surrounding ground. Steel fiber shotcrete, rock bolt, sectional steel and lattice girder were major support elements.

A series of auxiliary methods were also prepared for the mitigation of possible hazards caused by unfavorable geological situation. First, geological investigation during tunnel construction to grasp and confirm the geological information from Fig. 3, including the use of Resistivity Image Profiling (RIP) technique on surface above the tunnel, the Tunnel Seismic Prospecting (TSP) in the working area inside the tunnel, and advancing drilling with and without core recovery. Second, forepoling using spiles and the roof pipe method. Third, consolidation grouting in weak zone using cement or chemical materials. Finally, surrounding rock deformation and variation of groundwater level are close monitored to grasp the influence of groundwater inflow in hydrogeologically sensitive sections (Shau et al., 2018).

3 RESPONSE OF SURROUNDING ROCK AFTER EXCAVATION

The construction of the Dongao Tunnel was excavated from the southern and northern portals of both lines with four working areas totally. The response of surrounding rock after tunnel excavation exhibited huge differences at different rock types and relative distances to the Xiaomaoshan Fault and the Houishan Fault.

3.1 *Stability of excavated face in various lithology*

The overburden of the tunnel north of the Xiaomaoshan Fault ranges 30–50 m. Lithology mainly consist of slate, phyllitic slate, phyllite, and meta-sandstone. Foliation is well developed except for meta-sandstone. The excavated face exhibited displaced rock masses with wet groundwater condition. After tunnel excavation, the excavated faces were typically stable, except that minor cave-ins occasionally occurred in the phyllitic slate and slate sections (Lin et al., 2017).

The section between the Xiaomaoshan Fault and the Houishan Fault, with an overburden of 50–140 m, was the most difficult section during the construction of the Dongao Tunnel. The forepoling using spiles and pipe roof method associated with grouting for the improvement of surrounding rocks, and ring excavation with reserved core part and heavily support was commonly adopted to protect the working area. The stability of excavated faces can be further classified into three subsections, i.e., the northern Xiaomaoshan Fault, the southern Houishan Fault, and the disturbed transition zone between these two faults. In the Xiaomaoshan Fault zone, the maximum discharge during tunnel excavation was 1,140 l/min. Rock blocks and debris commonly fell from face and tunnel wall after excavation, with limited influence on the adjacent support elements that have been completed. In the disturbed transition zone between the two faults, the ground water might flow or gush out with various discharges. Fractured rock blocks fall down from the gap between the spiles and roof pipes, broadened the cave-in zone and damaged the supported section. The thickness of sheared mud and pulverized rock in the Houishan Fault is much greater than that in the Xiaomaoshan Fault and in the transition zone, with the maximum thickness exceeding 5 m. The groundwater consistently flowed or even gushed out from excavated face and tunnel wall, with the maximum discharge of 3,600 l/min. Although the auxiliary methods mentioned above were used to enhance the stability of excavated face, several collapses and cave-ins successively jeopardized tunnel construction.

The overburden of the Dongao Tunnel south of the Houishan Fault reaches 515 m. The rock mass mainly consists of schist, marble and amphibolite. The engineering characteristics of rock masses in the southern part of the tunnel are better than that in the northern part. Groundwater also impacted tunnel construction, the excavated face during tunneling were generally stable, with minor cave-ins caused by falling rock wedge and collapses occurred in local sheared zones.

3.2 *Tunnel deformation in heavy deformed ground*

Due to the global tectonic collision of Eurasia Plate and Philippine Sea Plate, the metamorphic rock formation in Taiwan are typically folded with various wavelengths and developed foliation. Faults and shear zones can be found everywhere. The Ryukyu Trench causes another regionally tectonic activity, resulting that the rock mass in northeastern Taiwan exhibits tensile trend. The strike of Central Ridge Mountains shifts from the northeast-southwest direction to the east-west, or even southeast east direction near Dongao. Rock masses in this area have anisotropic engineering characteristics because of inherent foliation and secondary discontinuities such as joint sets and fractures. Gravitation-induced deep-seated deformations with various depths are also common near slopes, leading to the break of rock formations and open fractures. During tunneling the surrounding rocks exhibited diverse responses in DSGS deformed zones, faults and sheared zones.

Figure 4 shows the images taken in the location north of Xiaomaoshan Fault after tunnel excavation. The foliation in slate (cleavage) is clearly observed and dip to right hand side steeply in the image. Sheared gouge fills in some fractures. The rock separated by an arc (dotted curves in the Fig. 4) in the middle part of these images are obviously displaced with open fractures, being in line with the representative appearances of DSGS formation. Along with the tunnel excavated round by round, the variation of fracture opening and rock wedges rotation can be observed. Tunneling in this section has limited groundwater discharge and pressure, minor surrounding rock deformation. The excavated faces were typically stable with occasionally local cave-ins and no support failure was reported.

Figure 5 shows the images taken in the Xiaomaoshan Fault after tunnel excavation. The rock formation has been disturbed by the fault that is heavily fractured. Fractured rock block and debris always fell out from excavated face. The groundwater flow might increase the volume of fallen rock block and debris. The designed support sustained the tunnel well with limited surrounding rock deformation, perhaps due to the limited overburden stress.

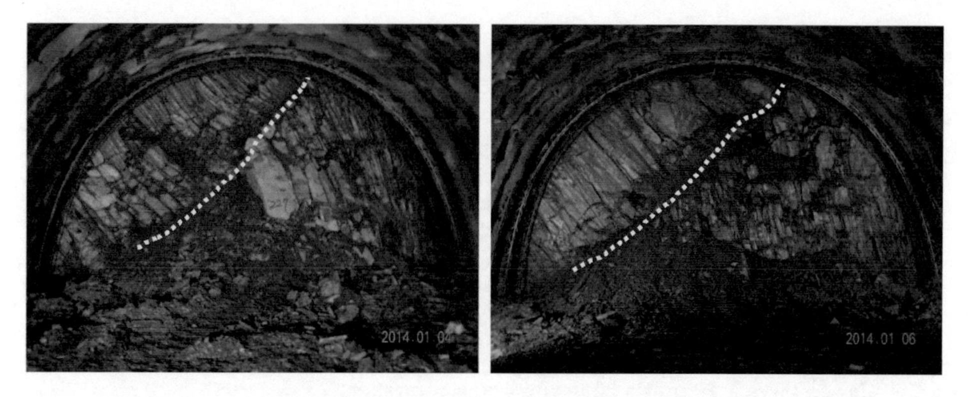

Figure 4. Images taken in the location north of Xiaomaoshan Fault after tunnel excavation.

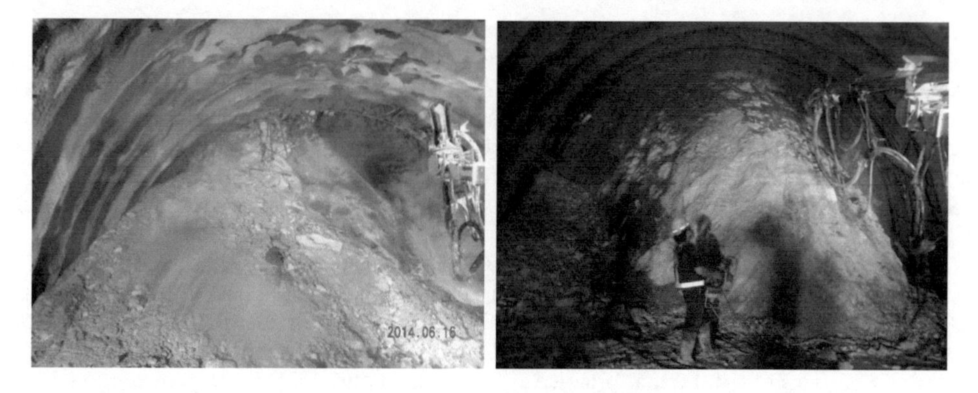

Figure 5. Images taken in Xiaomaoshan Fault after tunnel excavation.

Figure 6 illustrates the images taken in the disturbed transition zone between the Xiaomaoshan Fault and the Houishan Fault. The rocks exposed in this section, including slate, meta-sandstone, diabase, and phyllite are all disturbed by these two faults and disseminated with dense folds and sheared zones. The groundwater discharge increases, and the excavated faces are rarely self-standing. Collapses occurred when groundwater gushed out and damaged the supports that have been completed.

Figure 7 illustrates the images taken in the Houishan Fault after tunnel excavation. The excavated face lost self-standing time totally, pipe roof associated with grouting were necessary for

Figure 6. Images taken in the disturbed transition zone between Xiaomaoshan Fault and Houishan Fault after tunnel excavation.

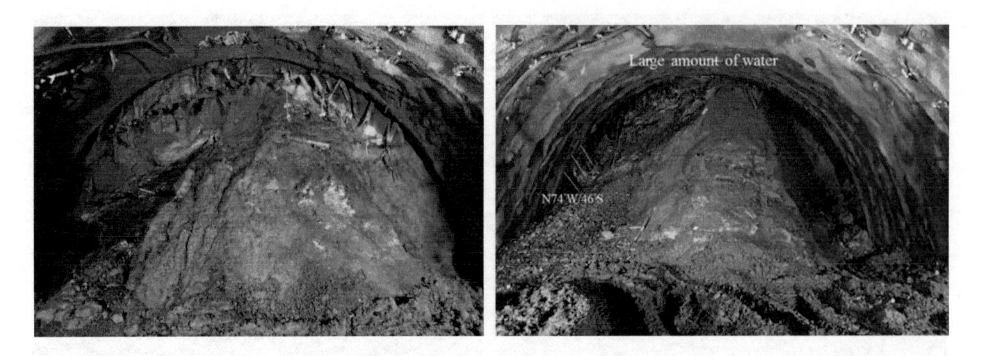

Figure 7. Images taken in Houishan Fault after tunnel excavation.

the strengthening of surrounding rock before excavation. Ring cut with reserved core part of excavated face are always needed to protect the working area. Rock masses in this section are disturbed by the fault that is heavily fractured, mixed with sheared mud with various thickness. Fractured rock block and sheared mud were squeezed out from the gap between roof pipes, broadened the cave-in scope in the heading area behind excavated face and increased the stress on the support. Deformation of the surrounding rock was huge and resulted in insufficient clearance. Excavated faces became instability as soon as groundwater gushed out or gradually deteriorated by the increases of groundwater discharge, resulting in severe collapses.

There were more than 200 cave-ins, or even collapses occurred during the construction of the Dongao Tunnel, among which, more than 2/3 happened in the heavily deformed ground that is studied in this manuscript. These cave-ins and collapse increase the volume of debris with a magnitude exceeding 280,000 m^3 and more than 24,000 m^3 lining concrete. The challenges met in terms of contractual practice are under mediation. Obviously, conventional rock mass rating method was not able to adequately distinguish the engineering characteristics of ground with pre-existed deformations, therefore its benefit in the section of heavy deformed ground of the Dongao Tunnel is limited.

4 PROJECT MANAGEMENT

Some innovative management approaches were applied to smooth tunnel construction (Wang et al., 2018).

4.1 *Entity's different quality procurement with lowest tender*

After completing the design of the Dongao Tunnel project, the conventional design and build (DB) contract was suggested for the contractor selection of the tunnel, and the procurement considers both qualities of the candidates and the tender amount. A special proposal and specification review stage is added between the stages for qualification examination of contractor and the award of bid by minimum total prize. The review contents include the composition of construction teams and subcontracting proportion, experience and credibility of the bidders, other engineering quantities and execution statue at the time of bidding, and the understanding regarding the engineering geology of tunnel and related construction plan and process arrangements, etc.

The bidding policy aims to guide the potential contractor to understand the engineering characteristics and tunneling risks in the bidding stage, enhance the construction concept and organize construction team and required machinery and materials. In addition, the client benefits from the review process by understanding different views of bidders on special and difficult construction topics in advance, facilitating the whole construction team later to overcome foreseen tunneling difficulty in unfavorable geological conditions.

4.2 *Integrated prediction of geological characteristics at different scales obtained in multiple periods*

The Suhua highway improvement project has extensively collected geotechnical and geological data near the route of the project, including the early geological investigation reports and the construction record of the North-link railway in 1970's and their improvement projects in 1990's, and the previous investigation results of the suspended Suhua expressway project in 2000's. Descriptive model for engineering geology along the alignment of the Suhua highway was proposed based on integrated interpretation on these data. A geological model was then confirmed by further detailed investigation that was implemented in detail design stage. Accordingly, a large-scale geological profile, the geological map and the distribution of rock masses rating results along the route of Dongao Tunnel were prepared for tunnel excavation and support design, so as the reference for construction. To mitigate possible tunneling difficulty caused by the variation of geological conditions, a considerable number of geological investigations were mandatory preformed in construction stage, such as the RIP in the surface above the tunnel, TSP inside the tunnel, drilling investigation with and without core recovery at the face, etc. Monitoring boreholes were also installed in some hydrogeological sensitive areas for the observation of variations of groundwater level while the excavating face was close to and far away these areas, providing reference for the mitigation on catastrophic groundwater inflow.

4.3 *Risk management*

To avoid or mitigate construction accidents caused by the predicted unfavorable geological conditions and potential uncertain factors, the administration of the Dongao Tunnel – the Suhua Improvement Engineering Office, Directorate General of Highway, MOTC, Taiwan (SIEO), carried out risk management for some working items that might jeopardize tunnel construction, the high risk working items. High risk working items were identified in detailed design stage first, and then reviewed in the preparation stage of supervision plan and construction plan. Risk management strategies associated with prevention and/or mitigation countermeasures were formulated with these items for the establishment of operation plan and detailed procedure. The work section of the SIEO supervised the operation of risk management before, during and after the construction of high risk working items. Some intense countermeasures could be applied in case of the occurrence of critical response and a signal of accident event. The supervision unit identified the high risk working items and possible hazard factors during construction, developed the hazard prevention measures and scheduled control period, and implemented the control procedure of high risk management plan. The construction unit established a standard flow chart of control operation to each high risk working item, including the emergency notification and processing mechanism, and implemented situational simulation to familiarize the relevant personnel with the operation process. Figure 8 shows the operational procedure of risk management to high risk working item.

4.4 *Timely notification*

Another innovation of the Suhua Highway Improvement Project is the use of instant notification systems. Traditional construction notifications need to be reported on a layer-by-layer basis, sometimes missing opportunities for timely processing or emergency response. Benefiting from the rapid advancement of wireless networks and communication software, hand-held mobile de-vices such as smart phones and tablet computers can easily transfer files and images. The SIEO makes good use of these technologies, organizes various timely notification groups according to the engineering characteristics and the responsibility of personnel in the construction team. Such a timely notification system contributes the construction of the Dongao Tunnel significantly since the construction team, the site supervisor, the engineers of the work section of SIEO and even its director received construction information in close time. The combination of timely no-tification system, integrated prediction of geological

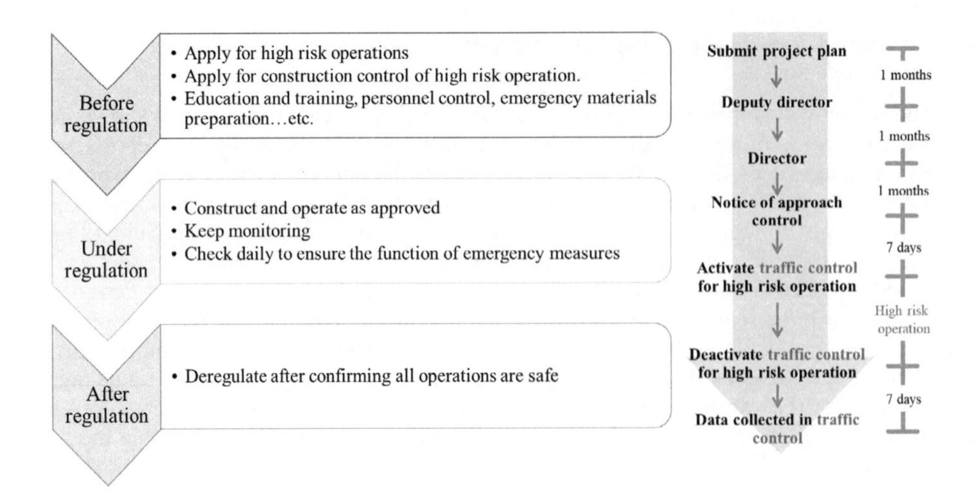

Figure 8. Operating procedure of high risk management

characteristics at different scales obtained in multiple periods and risk management operation delivery tunneling information directly, after quick discussion the decision-making can be more rapid and accurate, and the effect of applied countermeasures can be examined in short time. In addition, team members can visualize the changes of site condition, learn and accumulate experience together. In a word, these innovative management approaches play a huge role in the success of the construction of Dongao Tunnel.

5 CONTRAUCTUAL PRACTICE CHALLENGES

The contractor, being at the front line, has the most immediate feedback from the ground conditions and excavated faces. With years of experiences, the contractor is in the best position to suggest modifications to designed tunneling techniques to improve safety as well as efficiency; this is also the spirit of NATM tunneling method. On the other hand, the client and/or consult-ants often see project completion dates and budget control as priorities; if project schedule is compressed without proper support in funding, flexibilities in construction methods, and communication channels for all levels of personnel, risk of delaying or worst case collapsing of project grounding, both figuratively and physically speaking, increases drastically.

For contractual practices of this scale, it is important to realize two factors are in play: 1. de-signs can NOT be perfect, even with unlimited budget; and 2. contracts can NOT be perfect (Hart, 1995). Disagreements on how the project should be executed therefore should not be a surprise given the high variability nature of the work, and all parties involved should take the responsibility of resolving the differences positively, proactively, and responsively. Finding suitable mechanisms that can encourage collaboration between parties and increase transparency of projects such as this should be of high importance and based on experience and data, so the potential of public resources can be maximized.

6 CONCLUSION

Dongao Tunnel underwent the rock formations deformed by the global and regional tectonic activities and local DSGS deformation, resulting in diverse deformed ground characteristics and tunneling features. The ground deformed by DSGS exhibits open fractures between rock blocks, with gouge fills in some fractures. During initial excavation stage rock wedge was displaced with various magnitudes around tunnel perimeter. Although cave-in due to falling wedge was considered a possibility, but the working face was generally stable. Perhaps the interlocking of rock wedge contributes to tunnel stability. The tectonically squeezed ground commonly involves highly sheared rocks and heavily fractures, with faulty gouges filled in-betweens. Tunnel construction suffered from multiple severe face instabilities and squeezing problems, even with forepoling and ring excavation together with pre-grouting and reinforced shotcrete in excavated face.

The construction of the Dongao Tunnel adopted the NATM and some innovative management approaches. First, both the most advantage tender and the entity's different quality procurement with lowest tender have been applied to select excellent consultant and construction teams. Multi-temporal geological data and tunneling records which obtained in previous stages with various precisions have been integrated to predict geotechnical characteristics and tunnel behavior in advance, associated with preparing necessary construction resources for risk management. Modern communicative software is adopted for timely notification to related personnel, and engineering information is disclosed typically; not only does it benefits timely decision-making, but also make the team members keep abreast of changes in construction information. Thus, participants in a project can learn and accumulate experience together. These innovative management approaches play a huge role in the success of the construction of Dongao Tunnel, especially in the sections of heavy deformed ground, making the tunnel was opened to traffic in early 2018 in the light of its project plan.

REFERENCES

Hart, O. 1995. *Firms, Contracts, and Financial Structure*. Clarendon Press, Oxford.

Lin and Kao, 2009. *Explanatory text of the Geological Map of Taiwan- Suao Sheet*, 2ed Edition, Central Geological Survey

Lin, Y.C., Shen, T.Y., Hsian, S.C. & Shau, H.J. 2017. Case study on the geological characteristics and collapse of the weak intercalation of schist at the Dongao Tunnel in the Suhua Improvement Engineering Project, *Sino-Geotechnics*, 151, 35–44.

NLREB (North-Link Railway Engineering Bureau). 1980. *Report on the Construction of North-Link Railway*, Ilan, Taiwan (in Chinese).

Shau, H.J., Lee, M.C., Lee, Y.T. & Lin, C.C. 2018. Route selection and countermeasures of difficult geology of Dongao Tunnel, *Sino-Geotechnics*, 157, 47–56.

Wang, T.T., Chen, C.H. & Shau, H.J. 2018. Review on application of NATM in rock tunneling in Taiwan and associated advancement in Suhua Highway Improvement Project, *Sino-Geotechnics*, 157, 7-16.

Wang, T.T., Jeng, F.S., & Lo, W. 2011. Mitigating large water inrushes into the New Yungchuen Tunnel, Taiwan, *Bulletin of Engineering Geology and the Environment*, 70(2),173–186.

Tunnels and Underground Cities: Engineering and Innovation meet Archaeology, Architecture and Art, Volume 3: Geological and geotechnical knowledge and requirements for project implementation – Peila, Viggiani & Celestino (Eds)
 ISBN 978-0-367-46583-4

A 3D numerical model for the evaluation of the tunneling water inflow: The Fortezza – Ponte Gardena case study

F. Marchese, S. Rodani, A. Sciotti & S. Vagnozzi
Italferr S.p.A., Rome, Italy

G. Bernagozzi, F. Sciascia & A. Scuri
Enser s.r.l., Engineering Company, Faenza (RA), Italy and DICAM, University of Bologna, Bologna, Italy

L. Piccinini
Geoscience Department, University of Padova, Padova, Italy

L. Borgatti
DICAM, University of Bologna, Bologna, Italy

G. Benedetti
Enser s.r.l., Engineering Company, Faenza (RA), Italy

ABSTRACT: The Brenner Base Tunnel (BBT) is being built to connect Italy with Austria. The southern access connects Ponte Gardena with Fortezza by two railway tunnels named Gardena and Scaleres. A 3D groundwater flow model, extended through an area of about 200 km^2, has been built using MODFLOW code to evaluate the water inflow. The model thickness is 1200 m, divided into 8 layers varying from 100 m to 400 m. The model has been solved in transient conditions, allowing the water inflow both during 6 excavation stages reproducing the design excavation time schedule and during the subsequent equilibrium phase to be evaluated. Based on the geological and geotechnical model, the excavation methods have been defined; the results of the evaluation of water inflow and of the possible impact on the hydrogeological resources have been taken into account by defining risk mitigation measures to ensure hydraulic ante-operam conditions.

1 INTRODUCTION

The construction of a tunnel can in principle determine the alteration of the natural hydrogeological regime and the drainage of groundwater. In the design phase it is therefore important to estimate the total water inflow in order to assess the potential impacts on the hydrogeological system, with the purpose to design proper mitigation measures.

In Italy, one of the most recent case study is the Bologna-Firenze segment of the high-speed railway line, which represents a detailed hydrogeological investigation regarding the interactions between underground constructions and aquifers (Gargini et al., 2006; Gargini et al., 2008; Vincenzi et al., 2009; Piccinini & Vincenzi, 2010; Vincenzi et al., 2014).

Two tunnels, named Scaleres and Gardena represent the southern access to the Brenner Base Tunnel (BBT), the longest railway tunnel in the world. These two tunnels, respectively 15 km and 6 km long, have a typical overburden up to 800 m and a minimum between 100 and 45 m.

This context has been reproduced in a 3D numerical groundwater flow model adopting the equivalent porous medium (EPM) approach with the finite difference code MODFLOW, that has allowed the water inflows drained during the excavation phase and during the 6 stages of construction phases to be estimated.

The results of these studies have been used in the design of the tunnels as far as the hydraulic conditions are concerned.

2 REFERENCE GEOLOGICAL MODEL

The geological layout of the area is characterized by the presence of a metamorphic basement prevalently formed by quartz-rich phyllites (Arboit et al., 2018), with the occurrence of localized Permian granitic and dioritic intrusions. Thick Quaternary deposits (fluvio-glacial and alluvial deposits) usually cover the slopes and the valley floor of the Isarco River (Rodani et al., 2019).

The reference geological model has been built using 18 deep boreholes reaching the maximum depth of 615 m below ground level and 76 boreholes reaching 150 m depth maximum. A total of 133 Lefranc tests on Quaternary deposits, 215 Lugeon packer tests on bedrock, 176 dilatometric tests, 69 hydraulic fracturing tests, 385 Standard Penetration Tests, 97 soil pressuremeter – rock dilatometer tests were carried out during the boreholes drilling. A total of 823 samples of rocks and soils were collected to perform laboratory tests.

Two long-term pumping tests were also performed using an extraction well and 3 control piezometers to quantify the hydraulic conductivity (K) of the two main tectonic features, i.e. the so-called Funes and Scaleres fault zones.

Geophysical surveys were also carried out, for a total of 50 seismic refraction profiles acquiring P and S-wave velocity, 29 geoelectrical tomography profiles, 63 HVSR, 33 seismic passive and active arrays (Re.Mi. and MASW), 4 down-hole and 5 cross-hole P and S-wave velocity profiles and 33 magnetotelluric stations.

The geological field survey counts up to 575 observational points and 91 geomechanical stations for rock mass characterization.

The monitoring data acquired from a total of 96 water points were used for the definition of the hydrogeological conceptual model: 667 springs, 14 wells, 15 rivers and 67 piezometers (41 of them used for the calibration of the numerical model).

3 HYDROGEOLOGICAL CHARACTERIZATION

The Quaternary deposits are permeable by porosity: they are mostly composed by loose or slightly cemented fluvio-glacial and alluvial deposits, debris flow and colluvial deposits. The hydraulic conductivity is extremely variable, mainly as a function of the amount of the fine fraction (silt and fine sand). Their thickness can reach tens of meters such as in the Isarco river valley. The Quaternary deposits can host groundwater and, in some cases, they may constitute a permanent supply for the underlying rock mass.

The bedrock is characterized by the presence of a metamorphic crystalline basement prevalently formed by quartz-rich phyllites with the occurrence of localized Permian granitic and dioritic intrusions.

The superficial portion of the crystalline bedrock is partially weathered with a consequent increase of the hydraulic conductivity, while at depth fractures and discontinuities tend to close progressively. This phenomenon can be observed by plotting the K values obtained from the Lugeon tests as a function of the depth from ground level. The graph of Figure 1 shows the distribution of the hydraulic conductivity of the phyllites, expressed as Log10 K, with K in m/s, as a function of the depth. The tendency of K to decrease with depth can be expressed by the following Equation 1:

$$K = 10^{(-0.002 \cdot Z) - 6.87} \tag{1}$$

Where K = hydraulic conductivity (m/s); and Z = depth (m).

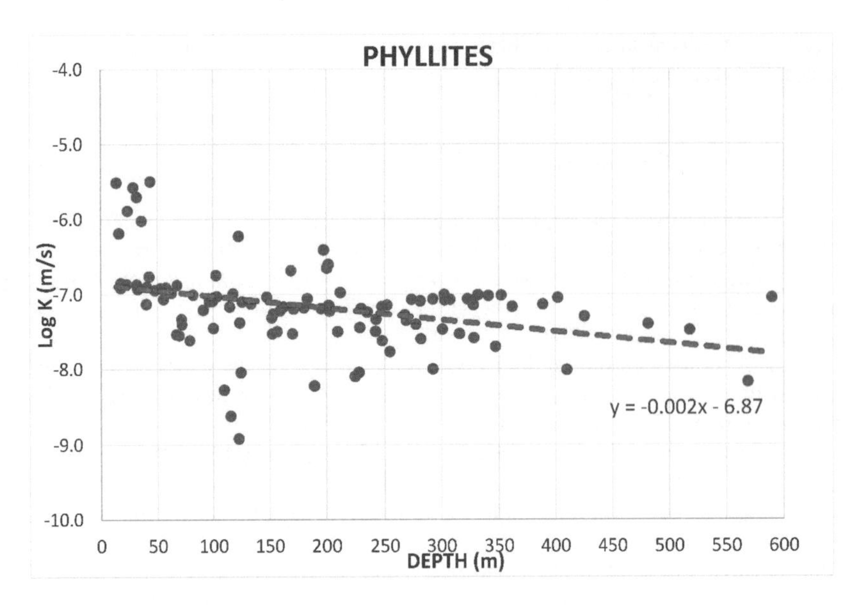

Figure 1. Lugeon tests on phyllites: hydraulic conductivity (K) vs. depth (Z).

In a similar way, the calculation was done for all the other lithotypes such as granites, diorites, and paragneisses.

In the investigated area some regional fault zones can be found: the Funes fault, the Scaleres fault and the Rio Bianco fault. The first two fault zones will be intercepted during the tunnels' excavation. To identify a reference hydraulic conductivity value to be associated with these two tectonic structures, two long-term pumping tests were carried out, one in the Funes valley, the other in the Scaleres valley. The obtained values, however, are to be considered as site specific, since the hydraulic conductivity along the fault zones can vary considerably depending on local rocks mass conditions (see for example the Damage Index classification in Bistacchi et al., 2010).

4 HYDROLOGICAL WATER BALANCE

The Scaleres and Gardena tunnel alignments cross 7 hydrological river basins: Varna Nord, Varna Sud, Bressanone, Snodres, Chiusa, Funes and Laion (Figure 2). Before the implementation of the numerical model, the hydrological balance of these basins was calculated using the following water balance (Equation 2) in a GIS environment:

$$P = ET + I + D \tag{2}$$

where P = average annual rainfall (or equivalent water in case of snowfall) in mm/year; ET = average annual evapotranspiration in mm/year; I = average annual infiltration in mm year; and D = average annual surface runoff in mm/year.

P was calculated using the rainfall data available for the area. Data from 2008 to 2016 were selected from 4 meteorological stations located along the bottom of the Isarco valley and from 3 stations located at high altitude to calculate the average annual rainfall and the temperature.

Figure 3 shows the altitude in m above the sea level (horizontal axis), the average annual rainfall in mm/year (vertical axis on the left) and the average annual temperature in °C (vertical axis on the right) for the selected stations.

The evapotranspiration ET was calculated through the formula of Turc (1954), while the average annual infiltration (I) was calculated by the inverse balance method (Civita, 2005).

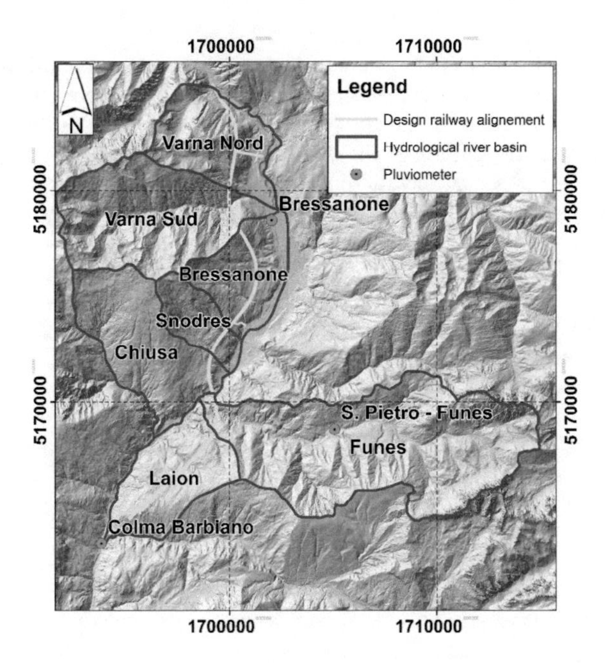

Figure 2. Map of the 7 hydrological river basins and three of the seven meteorological stations.

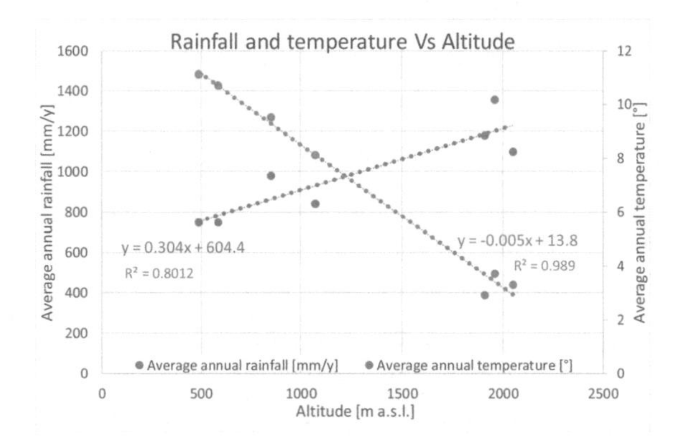

Figure 3. Correlation between average annual rainfall vs altitude and average annual temperature vs altitude.

In the 7 hydrological river basins the infiltration ranges from 99 to 160 mm/year. The I values have been assumed as the vertical recharge of the deep groundwater in each hydrological river basin included in the model domain. In correspondence of local relatively large loose deposits, the infiltration values were increased in order to account for the specific local conditions.

5 MODEL SET UP

The assessment of the inflow rates due to the excavation of the tunnels in transient conditions and at the steady state was performed with MODFLOW 2005 (Harbaugh, 2005) using the graphical user interface Model Muse (Winston, 2009; Winston, 2014).

The model refers to 3 phases: (1) an undisturbed condition, corresponding to the initial condition with no excavation; (2) the excavation phase, referring to the period from the starting of the works to the completion of the tunnels; (3) the stabilized condition, which refers to a long-term condition after the ending of the works.

The model grid is composed of rectangular cells with variable dimensions and in particular: the grid rows (x axis), following the average direction of the railway alignment, generally have a dimension of 100 m which is progressively reduced to 50 m where it intersects the main tectonic discontinuities. The grid columns (y axis) have a width of 200 m which is progressively reduced to 50 m near the design path.

In order to take into account the variation of the hydraulic conductivity with depth, 8 layers of constant thickness were created being the top and bottom of the model domain always parallel to the topographic surface, as shown in Table 1.

In the model it was assumed that the water inflow, i.e. the model recharge, originates exclusively from the rainfall and that the outflows correspond to the discharges due to rivers and springs and, both during the excavation phase and at the completion of the works, from the drainage of the tunnels. The water exchanges with adjacent hydrogeological river basins were considered negligible.

A drain package that matches with the topographic surface was inserted at the top of the domain. It was thus assumed that the topographic surface acts as a draining surface. The drains package discharges water from the model when the groundwater level reaches the topographic surface. The Figure 6 summarizes the applied boundary conditions.

The EPM approach assumes that the rock matrix including fractures and conduit networks can be represented by an equivalent porous medium with equivalent hydraulic conductivity in a certain area (Long et al., 1982), see Figure 4 on the left.

The main water inflows in the tunnel can be attributed to the interception of the main regional tectonic discontinuities, while the common fracture networks in the rock mass tend to cause diffuse drippings.

Table 1. Depth of the top and the bottom of the different layers with respect to the ground level.

Layer	1	2	3	4	5	6	7	8
from m	0	100	200	300	400	500	600	800
to m	100	200	300	400	500	600	800	1200

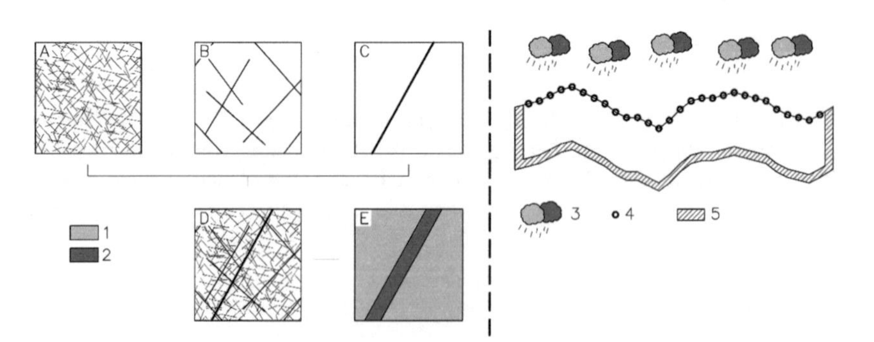

Figure 4. On the left: conceptual diagram for the implementation of the tectonic discontinuities using the EPM approach. A: joint and fractures at the outcrop scale. B: faults and discontinuity interpreted from remote sensing data. C: faults of regional importance. D: real rock mass. E: modelled rock mass. 1: fractured rock mass. 2: regional discontinuity. On the right: graphic summary of the model boundary conditions. 3: infiltration I. 4: drains package. 5: boundary conditions.

The geological units outcropping in the investigated basin have been grouped into 7 homogeneous hydrogeological units: the most important for the parameterization are the units including all phyllites lithotypes (BSS) and granitic lithotypes (GRA).

5.1 *Calibration*

A total of 41 piezometers are located on the design alignment and each point has been monitored in the period 2013 – 2017. The hydraulic head variations detected in each point are generally lower than few meters. For this reason, the average groundwater level was considered as the value to be used for calibration purposes. The calibration was performed using the MODFLOW functions HOB (Head Observation package), which basically allows a reference piezometer to be introduced into the model grid.

The calibration was carried out manually through a trial and error procedure (Anderson & Woessner, 1992). The hydraulic conductivity associated to the 7 hydrogeological units was updated several times to obtain the best fit between the calculated and observed hydraulic head data. The Table 2 shows the hydraulic conductivity values assigned to the phyllites (BSS) and granitic (GRA) hydrogeological units at the end of the calibration process.

The graph in Figure 5 shows the relationships between the observed and the simulated values. The two dashed lines represent an ideal buffer of 25 m around the ideal calibration line.

This buffer can be considered as an error associated with the reconstruction of the topographic surface, created starting from a map having a distance of 25 m between the elevation contour lines.

At the end of the calibration process the Root Mean Square RMS is 58.66 m. Considering a maximum observed value of 1083 m and a minimum observed value of 464.1 m, it is possible to calculate a normalized RMS (nRMS) of 9.5%. This value is less than 10% and therefore falls within the limits of acceptability commonly used (Anderson & Woessner, 1992).

Table 2. Hydraulic conductivity values assigned to BSS and GRA hydrogeological units after the calibration process.

Layer	BSS	GRA	Layer	BSS	GRA
1	$1.0 \cdot 10^{-7}$	$2.0 \cdot 10^{-7}$	5	$1.1 \cdot 10^{-8}$	$1.0 \cdot 10^{-9}$
2	$6.0 \cdot 10^{-8}$	$2.0 \cdot 10^{-8}$	6	$6.6 \cdot 10^{-9}$	$1.0 \cdot 10^{-9}$
3	$3.4 \cdot 10^{-8}$	$8.0 \cdot 10^{-9}$	7	$3.8 \cdot 10^{-9}$	$1.0 \cdot 10^{-9}$
4	$2.0 \cdot 10^{-8}$	$5.0 \cdot 10^{-9}$	8	$2.2 \cdot 10^{-9}$	$1.0 \cdot 10^{-9}$

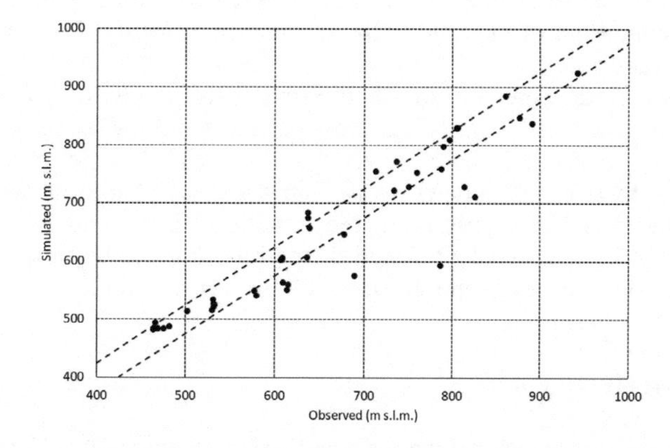

Figure 5. Observed vs. simulated hydraulic head values at the end of the calibration process.

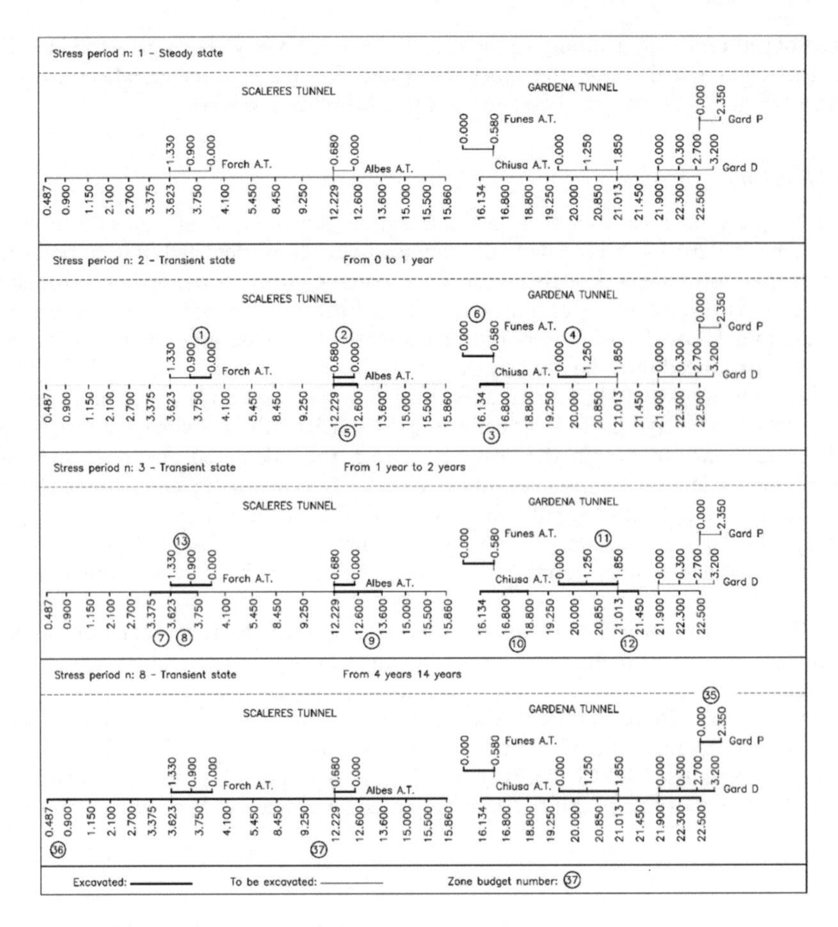

Figure 6. Construction time schedule: conceptual diagram of the tunnels' discretization.

5.2 *Simulation*

The excavation time schedule was considered both for the simulation in transient and in steady-state conditions. All the design tunnels (the main railway tunnels, the interconnection, the service and construction tunnels) were discretized into a series of segments and each segment was activated at a certain time of the simulation in accordance with the excavation time schedule. A third type boundary condition (BC), i.e. the drain package of MODFLOW (DRN function) was used to simulate the drainage operated by the tunnels and the rivers or streams, while a second type BC, i.e. the recharge package of MODFLOW was used to reproduce the infiltration process.

In total, 8 stress periods were considered: the first corresponds to the undisturbed conditions and it was solved at steady-state. The stress periods from the second to the seventh represent 6 excavation stages and therefore simulate the situation in progress in transient conditions. The eighth stress period was solved in transient conditions as well, using a 10 years period as the reference representing the situation at the end of excavation works. Figure 6 shows the stress periods 1, 2, 3 and 8: the bold lines indicate the excavated tunnel segments, while thin ones represent the segments still to be excavated.

6 DESIGN OF MITIGATION MEASURES

As mentioned above, the work includes the Scaleres Tunnel (length about 15 km) and the Gardena Tunnel (length of about 6 km), separated by a short stretch crossing the Isarco River on a viaduct.

To ensure appropriate safety standards in operation, the configuration planned for all the running tunnels (inner radius of 4.2 m) is that of double-tube single-track rail tunnels, with a horizontal axis distance of 40 m and cross passages every 500 m.

The total length of all the underground works is approximately 62 km, including complementary works. In the Scaleres Tunnel (maximum overburden of 800 m) the rock mass is mainly composed of Bressanone Granite and Quartz Phyllites. In the Gardena Tunnel (maximum overburden of 580 m) the rock mass is mainly made up of Quartz Phyllites. Both rock masses generally display good geomechanical properties, although some fault zones can be found, with a distribution that is sometime dispersed or concentrated in some sections. For the line and interconnection tunnels both the conventional excavation method and the mechanized excavation are foreseen. In particular, the Scaleres Tunnel will be excavated using both conventional method and mechanized method with two Hard Rock TBM-S, while the Gardena Tunnel will be built using the conventional excavation method. Finally, the two interconnection tunnels of Ponte Gardena (about 5 km in length overall) will be built using a TBM-EPB; this last choice is due to the need to safely underpass an embankment and some foundations of a Brenner motorway viaduct in an area where the cover is shallow.

Any hydrogeological impact and the possible risk of water resources depletion as a result of tunnel excavation can be addressed and mitigated during the excavation with design and construction solutions, such as the use of drainages in advance for the temporary and localized reduction of hydraulic loads and the control of the water inflows to the tunnel and, in case of high water discharge, the eventual use of interventions to reduce rock mass permeability. In this case, in addition to the use of cement injections, the use of bicomponent organo-mineral resins is also envisaged. The application of the interventions is based on the results of the MODFLOW model and of the reference geological model which are aimed at ascertaining the presence of the fault zones at the tunnel depth, examining the degree of fracturing and relative permeability and verifying the presence of groundwater eventually connected to springs or sensitive watercourses.

During the railway operation phase, for the sections excavated by means of TBM, the installation of the lining made of precast concrete segments with hydraulic sealing gaskets ensures the impermeability of the tunnel and the hydraulic ante-operam conditions up to hydraulic load values compatible with the structural characteristics of the final lining. For sections excavated by conventional method, in the undercrossing of hydraulically sensitive zones, the tunnel will be waterproofed over the entire perimeter.

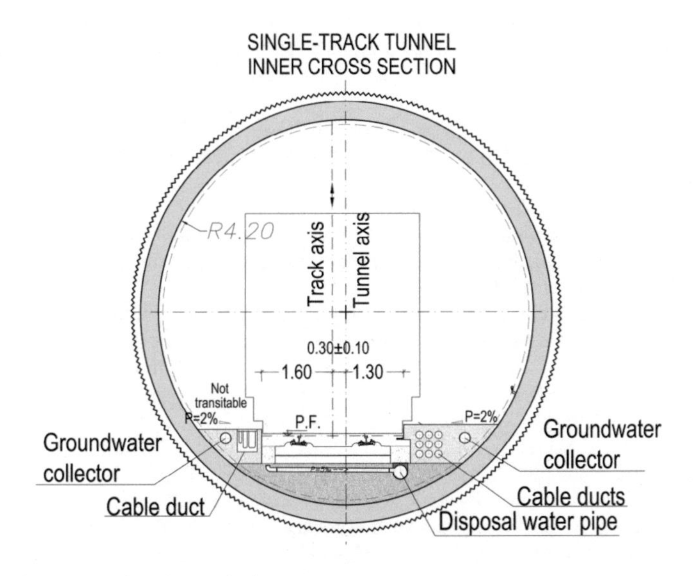

Figure 7. Tunnel cross section: inner drainage system.

As far as the the tunnel water management, during its operation the disposal system provides for the separation of any water coming from rock mass and those coming from the train platform. For platform waters, a system of collection and disposal of hazardous liquids that may accidentally spill from freight wagons on the railway platform is available (Figure 7).

7 CONCLUSION

This case study has required a series of specifically designed monitoring and modelling solutions, in relation to the dimension of the investigated area and to its geological and geomorphological settings.

The Lugeon tests carried out on site effectively demonstrated that the hydraulic conductivity of the crystalline rock masses decreases with the increasing of the depth.

The EPM approach, integrated with the distinct introduction of regional tectonic discontinuities, allowed MODFLOW to be adopted in a complex tectonic context with fractured rock masses.

The upper boundary of the model was ideally covered by a draining surface. This surface drains package prevented the groundwater from rising above the ground level even in the presence of strong topographic variations in altitude. At the bottom of the valley, the drains package allowed the water flow in the rivers and in the streams to be simulated.

The simple modelling approach that divides the entire tunnels path into several segments activated at different time intervals to simulate the designed construction stages proved to be reliable. This approach also allowed not only to perform a long-term forecast but also to estimate the water inflows during the excavation. The results of the monitoring and of the modelling activities have been incorporated in the design of mitigation measures that ensure ante-operam hydraulic conditions and no impacts on groundwater resources.

This work can represent a simple but rigorous methodological approach that can also be applied to other mountain contexts where tunneling is been designed.

REFERENCES

Anderson, M.R. & Woessner, W.W. 1992. *Applied groundwater modeling simulation of flow and advective transport*. I. Academic Press, Ed. San Diego/New York/Boston/London/Sydney/Tokyo/Toronto: Academic Press, Inc.

Arboit F., Chew D., Visoná D., Massironi M., Sciascia F., Benedetti G. & Rodani S., 2018. The geodynamic evolution of the Italian South Alpine basement from the Ediacaran to the Carboniferous: was the South Alpine terrane part of the Peri-Gondwana arc-forming terranes? *Gondwana Research* 65: 17–30.

Bistacchi, A., Massironi, M. & Menegon, L. 2010. Three-dimensional characterization of the crustal-scale fault zone: the Pusteria and Sprechenstein fault system (Eastern Alps). *Journal of structural geology* 32: 2022–2041.

Civita, M. 2005. *Idrogeologia applicata e ambientale*. Milano: CEA.

Gargini, A., Vincenzi, V., Piccinini, L., Zuppi, G.M. & Canuti, P. 2008. Groundwater flow systems in turbidites of the Northern Apennines (Italy): natural discharge and high-speed railway tunnel drainage. *Hydrogeology Journal* 16(8): 1577–1599.

Gargini, A., Piccinini, L., Martelli, L., Rosselli, S., Bencini, A., Messina, A., & Canuti, P. 2006. Idrogeologia delle unità torbiditiche: un modello concettuale derivato dal rilevamento geologico dell'Appennino Tosco-Emiliano e dal monitoraggio ambientale per il tunnel alta velocità ferroviaria Firenze-Bologna. *Bollettino della Società Geologica Italiana* 125(3): 293–327.

Harbaugh, A.W. 2005. MODFLOW 2005: the U.S. Geological Survey modular ground-water model. The ground-water flow process.

Long J.C.S., Remer J.S., Wilson C.R., & Witherspoon P.A. 1982. Porous Media Equivalents for Networks of Discontinuous Fractures. *Water Resources Research* 18(3): 645–658.

Piccinini, L., & Vincenzi, V. 2010. Impacts of a Railway Tunnel on the streams baseflow verified by Means of numerical modelling. *Aqua Mundi* 1: 123–134.

Rodani S., Scuri A., Piccinini L., Cervi F., Benedetti G., Sciascia F. & Castioni D. 2018. Monitoring the water resources in the large railway projects: the Fortezza – Ponte Gardena case study, Southern BBT access tunnel, WTC 2019.

Turc, L. 1954. Calcul du bilan de l'eau: evaluation en function des precipitation et des temperatures. IAHS. 37.

Vincenzi, V., Gargini, A., Goldscheider, N., & Piccinini, L. 2014. Differential Hydrogeological Effects of Draining Tunnels Through the Northern Apennines, Italy. *Rock Mechanics and Rock Engineering* 47 (3): 947–965.

Vincenzi V, Gargini A, & Goldscheider N. 2009. Using tracer tests and hydrological observations to evaluate effects of tunnel drainage on groundwater and surface waters in the Northern Apennines (Italy). *Hydrogeology Journal* 17(1):135–150.

Winston, R.B. 2009. ModelMuse: A Graphical User Interface for MODFLOW-2005 and PHAST: U.S. Geological Survey Techniques and Methods 6–A29. U.S. Geological Survey.

Winston, R.B. 2014. Modifications made to ModelMuse to add support for the Saturated-Unsaturated Transport model (SUTRA): U.S. Geological Survey Techniques and Methods, book 6, chap. A49, 6 p. U.S.G. Survey.

Tunnels and Underground Cities: Engineering and Innovation meet Archaeology, Architecture and Art, Volume 3: Geological and geotechnical knowledge and requirements for project implementation – Peila, Viggiani & Celestino (Eds)
© 2020 Taylor & Francis Group, London, ISBN 978-0-367-46583-4

Requirements for geo-locating transnational infrastructure BIM models

Š. Markič & A. Borrmann
Chair of Computational Modeling and Simulation, Technical University of Munich, Munich, Germany

G. Windischer, R.W. Glatzl, M. Hofmann & K. Bergmeister
Brenner Base Tunnel BBT SE, Innsbruck, Austria

ABSTRACT: The transfer of design data into nature is a necessary task during the construction of a building. For this, the geodetic Coordinate Reference System (CRS) used during the design process needs to be accounted for and the induced distortions handled appropriately. In the context of Building Information Modelling (BIM), the CRS represents the metadata of the initial model which needs to be included and maintained throughout project's lifetime. When a project's construction site spans multiple countries, the initial data is available in different national CRSs and transforming them completely free of residuals is impossible. For the Brenner Base Tunnel between Austria and Italy, a new compound CRS was designed, representing a homogeneous reference system for all surveying and construction work. We present its definition and the rationale behind it. We highlight the requirements for BIM models and design systems and present current deficiencies.

1 INTRODUCTION

1.1 *Geolocation*

The transfer of design data into nature is a necessary task during the construction of a building. The structure must have correct dimensions and especially be placed on the correct land slot at the correct elevation. This task, also known as setting out, is the principal responsibility of surveyors. Setting out can be done within a local or a global context, depending on the building at hand, and generally counts as a solved and manageable problem.

For infrastructure objects (e.g. roads or railways) the setting out is primarily done in the global context. Shared elements with neighboring objects (like junctions or railway stations) represent compulsory points (i.e. geometric boundary conditions) which need to be considered and met. All objects that influence the design can be included within a large-scale geodetic Coordinate Reference System (CRS). These are generally provided by the surveying agencies of individual countries.

CRSs are split in two independent systems: the location and the height reference, named the geodetic and the vertical datum, respectively. A geometric projection of the geodetic datum flattens the Earth's curvature and together with the vertical datum establishes a well-defined CRS (Kaden & Clemen 2017, ISO 19111:2007, 2007). A comprehensive collection of these systems and their combinations is the database of the European Petroleum Survey Group (EPSG), where nearly 6000 CRSs from around the world are listed together with datum definitions and transformations (EPSG 2018).

The CRSs usually do not provide sufficiently precise data for tunnel construction. Therefore, it is common to define a new, local CRS on the tunnel construction site, which provides a common ground for all documentation and construction. The geospatial data is then transformed from existing databases or measured anew ensuring high precision and quality.

When a construction project of a tunnel spans multiple countries, the initial data is available in different national CRSs. Consequently, problems arise already at the beginning of the design stage as transforming between the CRSs completely free of residuals is a demanding and peculiar task. Using project's own CRS, the need for handling the coordinates of each and every object in two different CRSs is avoided.

1.2 *Design*

The engineering design takes place in a right-handed three-dimensional (3D) Cartesian coordinate system – the Project Coordinate System (PCS). This is beneficial for visualization, perception, and performing calculations and hence less error-prone. Therefore, the definition of a suitable PCS is the first task in the engineering design. Knowing the axes, their point of origin, and their scale provides a common basis for all actors participating in a project.

For a long time, the design was done by hand and has been digitalized during the last century with the introduction of Computer-Aided Design (CAD). In the last decades, the Architecture, Engineering and Construction (AEC) industry has undergone an evolution from two-dimensional (2D) CAD drawings to three-dimensional (3D) object-oriented models. This new paradigm was branded Building Information Modelling (BIM) and has found quick adoption in the building sector (Borrmann et al. 2015).

buildingSMART International (bSI), formerly known as the International Alliance for Interoperability (IAI), is the international community behind the BIM standards. These range from the vendor-neutral BIM data formats *Industry Foundation Classes* (IFC) (ISO 16739:2013, 2013) and *BIM Collaboration Format* (BCF) to the *Information Delivery Manuals* (IDM) (ISO 29481) and *International Framework for Dictionaries* (IFD) (ISO 12006-3).

In the last years, the infrastructure sector has shown increased interest for the usage of BIM methods and the benefits they bring (Barazzetti & Banfi 2017). When considering BIM models of buildings with comparatively limited extends, the question about the underlying PCS has been handled by surveyors separately from the designers. However, when large linear objects are being modelled, aspects of geodetic CRS and the distortions they imply have a much more significant impact. These aspects need to be adequately addressed when expanding the established work flows and data formats (Markič et al. 2018).

1.3 *Goal*

The often-forgotten fact is that the well-defined CRS from Section 1.1 and the PCS from Section 1.2 are two sides of the same coin: the geodetic and the engineering perspective of the truth, respectively. This has significant implications as correcting such a mistake can be demanding and challenging. Its consequences are at best only shameful (Conchúir 2016) and at worst really expensive (Spiegel 2015).

This paper provides an insight at geo-location in general and geo-referencing BIM models in particular. The correct handling of transnational geospatial data is exemplary presented on the Brenner Base Tunnel (BBT) project. Following questions are in focus:

- How is a CRS defined?
- How do BIM systems handle geospatial data?
- How to exchange geo-referencing metadata in the BIM model?

The paper is structured as follows. This section provides a short introduction with our motivation. Geodetic background is briefly explained in Section 2. Section 3 provides a short overview of related recent works in the field of geo-referencing BIM models. The BBT project is presented in Section 4 followed by handling of geo-referencing data in BIM design systems in Section 5. The paper concludes with discussion in Section 6.

2 THEORETICAL BACKGROUND

In civil engineering, the PCS is a depiction of the real world by the chosen geodetic CRS. Having the underlying CRS and thus the PCS well-defined, geospatial data from different sources can be incorporated in the project by according transformations.

2.1 *Geodetic Datum*

The Earth is roughly a sphere and as such, the use of spherical coordinates offers itself as a way of referencing points on Earth surface. Since the Earth is a sphere squished at the poles (due to the rotational forces), a really good approximation is an oblate ellipsoid, which is an ellipse rotated around the minor axis (ISO 19111:2007, 2007). The longitude Λ and latitude Φ denote the angles from the reference lines, e.g. the Greenwich meridian and the Earth's Equator, respectively. A pair of angles (Λ, Φ) defines a unique location on the ellipsoid.

Through the history, many ellipsoids have been defined and used with different areas of best fit. The "best fit" objective is to minimize the differences between the Earth's equipotential surface and ellipsoid in a specific area or globally. For example, the ellipsoid *WGS84* used in World Geodetic System 1984 geodetic datum is an Earth-centered ellipsoid and has a global best fit. It is widely used, e.g. by the Global Navigation Satellite System (GNSS).

It is common to describe ellipsoid shape in geodetic context by providing its major axis R_{major} and instead of its minor axis R_{minor} its inverse flattening f^{-1}, which is defined as (ISO 19111:2007, 2007, EPSG 2018):

$$f^{-1} = \frac{R_{major}}{R_{major} - R_{minor}} \tag{1}$$

2.2 *Coordinate System*

The Cartesian coordinates (X, Y) of the PCS are obtained by projecting the ellipsoidal coordinates (Λ, Φ) onto a plane using some sort of map projection. Since projecting the curved surface of an ellipsoid onto a plane without any deformation is not possible, a map projection can only preserve one of: either angles, distances or surface areas. The compromise most frequently chosen is to preserve angles by using the so-called conformal map projections, such as the Transverse Mercator (TM) or Universal Transverse Mercator (UTM) projections.

To keep the distortions of distances and surface areas in an acceptable range for applications like large-scale topographic mapping or cadastral surveying, strips of the ellipsoid are defined and projected onto a cylinder's surface. Figure 1 shows two conformal map projections that differ from each other in the radius of the cylinder and the width of the strips. The Gauss-Kruger projection (GK) (a type of TM) uses a cylinder that touches the ellipsoid at a meridian (see Figure 1, left). Therefore, only the distances along the meridian are not distorted and get increasingly more distorted the further away from meridian the location is. This is why, the strips of the projection

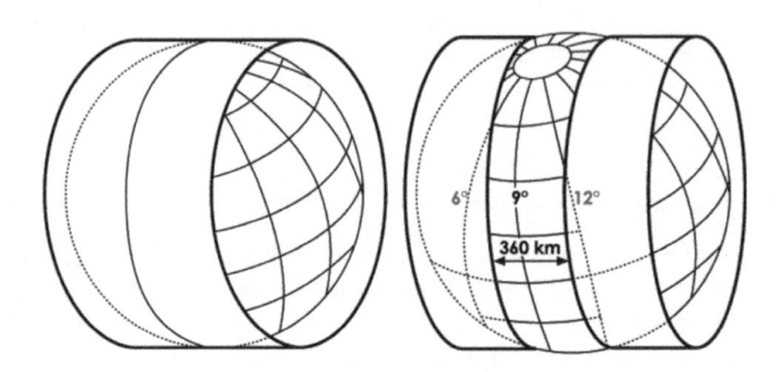

Figure 1. Different ways of projecting the ellipsoid on a cylinder: the Gauss-Kruger projection (GK) (left) and the Universal Transverse Mercator projection (UTM) (right) (Markič et al. 2018).

usually have a width of 3 degrees. In the UTM projection, the cylinder intersects with the ellipsoid 180 km east and west of the central meridian of a specific strip, which has a width of 6 degrees (see Figure 1, right). To keep the distance distortions in an acceptable range, even at the borders of the strip, the central meridian is shortened with a scale of $m = 0.9996$.

2.3 *Vertical Datum*

There are several possible definitions of elevation on Earth. One of them is to define the verticality on the Earth's surface as the (opposite) direction of the Earth's gravity pull. In this way, the water does not flow between two points with the same elevation which is very practical in construction. The vertical axis (H) follows the so-called plumb line and for easier notation, the coordinate value is usually given as a distance to some reference plane and not to the point of origin. This reference plane – the zero orthogonal height $H = 0$ – is the Earth's equipotential gravity field and defines the geoid form (see Figure 2). The most common plane is the mean sea level (ISO 19111:2007, 2007).

The geoid and the ellipsoid forms disagree to a certain extent. This so-called undulation N can be determined with measurements and can amount to up to 100 m, which induces additional distortions in dimensions. Additionally, the plumb line from a point on the Earth's surface to the geoid (which follows the direction of gravity by its definition) and the perpendicular line from that same point to the chosen geodetic datum may not coincide. The reason for this is the gravity anomaly caused by variations in the density distributions within the Earth (Kaden & Clemen 2017). The ellipsoid height h is defined:

$$h = H + N + \Delta \tag{2}$$

where the error Δ comes from the influence of gravity anomaly and is determined from the deflection of the vertical direction (see Figure 2). For small areas the gravity anomaly is constant $\Delta = const.$ and can be added to coordinates of the whole project. However, elongated infrastructure objects like tunnels need to appropriately account for the changing environment by conducting additional measurements and adjusting their models accordingly.

2.4 *Projected and Compuond CRS*

To summarize, a CRS is composed of multiple parts. The choice of ellipsoid's size, position and orientation together with the height reference define the geodetic and vertical datums, respectively. The chosen projection defines the transformation from a double-curved surface of ellipsoid to a Cartesian CS. The map projection together with a geodetic datum is called a projected CRS, which uniquely defines the transformation of the Cartesian CS to the ellipsoid

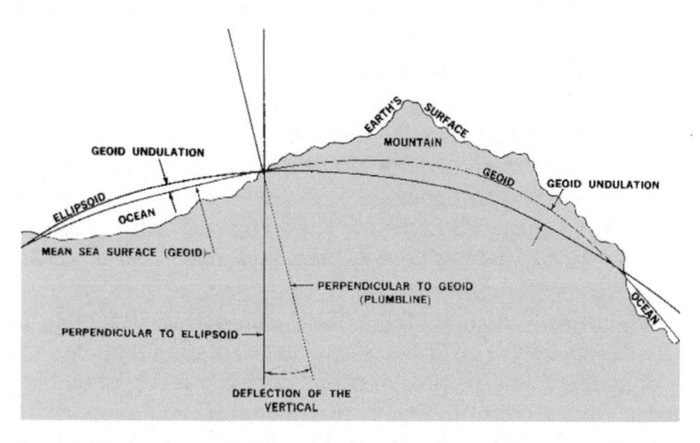

Figure 2. Depiction of the geoid, ellipsoid and Earth's surface as well as the undulation of the geoid and the deflection of the vertical (National Imagery and Mapping Agency 2002).

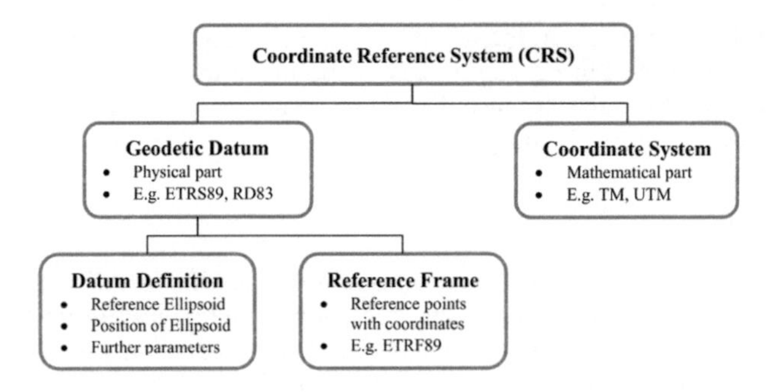

Figure 3. Components of a locational Coordinate Reference System (CRS) in the geospatial domain (ISO 19111:2007, 2007).

surface as presented in Figure 3. In combination with a vertical CRS, the reference system is called a compound CRS (ISO 19111:2007, 2007).

3 RELATED WORK

The geo-referencing of BIM models has become a topic of interest with the introduction of elongated objects from the infrastructure to the BIM context. bSI members have addressed this issue in one of their latest projects *Model Setup IDM*. The main focus of the project was the use case of geo-referencing. IFC versions 2x3 and 4 have been looked at in detail and a guideline for implementers has been published. As a general remark, they call for increased communication between all participants and free sharing of information (buildingSMART International 2018).

Barazzetti & Banfi (2017) list multiple examples from all areas of AEC industry and evaluate the applicability of a bridge BIM model for traffic simulation analysis in detail. They conclude that an integration of geospatial data with parametric BIM modelling could simplify and speed up the design processes in the AEC industry.

Kaden & Clemen (2017) walk through an example study on the coordinate systems from the geodetic perspective. They noted that a correct understanding of geodetic CRSs is crucial for the success of BIM projects, especially in the infrastructure sector. However, in their words, *most CAD data is created without this consideration*. They go in great depth in linking the geodetic CRS and the BIM model using an intermediate CRS of a so-called local surveying coordinate system that represents a PCS.

Heunecke (2017) provides the equations and the reasoning behind the geodetic distortions of projections and even exemplarily calculates their exact values. For example, a curve's radius of $R_{BIM} = 1000\ m$ in a BIM system changes to $R_{UTM} = 999.46\ m$ when projected by a map prijection (here, to UTM) or transformed to $R_{real} = 1000.54\ m$ when calculating the set-out values (here, from UTM). This distortion is location-dependent (!) and varies in different CRSs. Such differences influence the drive dynamics insignificantly; however, they can be the reason to violate a compulsory point's margin like a railway platform's edge.

Uggla & Horemuz (2018) argue that a BIM model is to be *viewed as a 1:1 representation of the terrain at the construction site* and that it is not described in a geodetic CRS. As such, the engineering system's agreement with the real world stays within given tolerances only within a small area (1 km) around the project base point. They conclude that the current implementation in the IFC schema is not usable and wishes for *addition of support for object specific map projections and separate scale factors for different axes*.

Markič et al. (2018) critically evaluate the IFC schema and its capability to store geospatial metadata. With the introduction of BIM-based exchanges of digital models the handling of this metadata needs to be addressed in the processes and correctly incorporated in the models. They argue that the current official version (IFC4) provides sufficient support for the typical case occurring in the majority of projects. However, based on two recent real-world infrastructure projects the implementation is rendered insufficient. They propose a solution by extending the IFC schema to support such peculiarities with the inclusion of grid shift parameter sets.

4 BRENNER BASE TUNNEL

The Brenner Base Tunnel (BBT) is a major European infrastructure project of the Helsinki (Finland) – La Valletta (Malta) North-South Trans-European Network (TEN) corridor. The BBT's two 55 km long, parallel, single-track railway tunnels enable freight and passenger trains to cross the Alps between Innsbruck (Austria) and Franzensfeste (Italy). At its completion, the 1371 m high Brenner Pass will be avoided, and the travel time reduced significantly while increasing the safety and throughput (Bergmeister 2011).

For historical reasons, most of the European major countries base their geospatial data in their own national CRSs. In the recent years, the European Union (EU) pushed for a common system and the member countries have started updating and transforming their systems accordingly. This is a long process that is nowhere near its end; the companies and public authorities have not yet adapted their processes and thus still provide geospatial data in the established data formats and (old) CRSs (Donaubauer & Kolbe 2017).

At the beginning of the project in 2001 the geospatial data of the construction site from the two countries participating in the project needed to be merged. This ensures a clear planning process and avoids mistakes during the underground construction. However, both countries used a completely different CRS as presented in Table 1 and as such three options were available.

1. Convert the Austrian geospatial data into Italian CRS and work in Italian CRS.
2. Convert the Italian geospatial data into Austrian CRS and work in Austrian CRS.

Table 1. The properties of the geodetic and vertical datums and the projected CRSs used by the countries participating in and by the Brenner Base Tunnel (BBT) project itself (Mugnier 2005, Macheiner 2015). For each element its code and name from the European Petroleum Survey Group (EPSG) database as well as additional parameters are provided (EPSG 2018).

Property	Austria	Italy	BBT
Responsible authority	Bundesamt für Eich und Vermessungswesen (BEV)	Instituto Geografico Militare (IGM)	Prof. Ing. Franco Guzzeti*
Geodetic datum	MGI	Monte Mario	WGS84**
– EPSG	4312	4265	4326
– Ellipsoid	Bessel 1841	International 1924	WGS84**
○ EPSG	7004	7022	7030
○ R_{major}	6,377,397.155 m	6,377,388 m	6,378,137.0 m
○ f^{1}	299.1528128	297	298.257223563
Projected CRS	Austria M28, M31 & M34	Italy zone 1 & 2	BBT_TM-WGS84
– EPSG	31,257, 31,258 & 31,259	3003 & 3004	not set
– Scale factor	1.0000	0.9996	1.000121
– False easting	150 km	1500 & 2520 km	20 km
– False northing	-5000 km	0 km	-5,105.739717 km
– Projection	Gauss-Kruger	Gauss-Boaga	TM
○ EPSG	9807	9807	9807
– Central meridian	10°20'E, 13°20'E & 16° 20'E	9°0'E & 15°0'E	11°30'42.5775"E
– CS Origin***	48°16'15.29"N 16°17'41.06"E	41°55'25.51"N 12°27'08.40"E	46°58'50.7947"N 11°31'42.5775"E
Vertical datum	Trieste datum	Genova datum	EVRF2007****
– EPSG	1050	1051	5215

* Prof. Ing. Franco Guzzetti is associate professor at the Polytechnic University of Milan.
** WGS84 stands for World Geodetic System 1984 and is the name of the geodetic datum as well as its underlying ellipsoid.
*** The reference lines are the Equator and the Greenwich meridian.
**** EVRF2007 is the European Vertical Reference Frame 2007 realized by geopotential numbers and normal heights of the United European Leveling Network (UELN).

3. Plan in a third CRS and convert both Austrian and Italian data into it. This system may be an existing CRS or defined completely anew.

The project team decided for the third option and defined a completely new CRS named "*BBT_TM-WGS84"*. A short overview of the properties is shown in Table 1, right column. The rationale behind the decision, as well as the chosen values, is explained in the following paragraphs.

4.1 *Geodetic Datum*

Austria still uses a local best-fit ellipsoid *Bessel 1841*, while Italy uses one of the first global best-fit ellipsoids *International 1924*. These differ in all parameters (see Table 1) and a transformation between them – although possible – is computationally very demanding. With the introduction of GNSS the ellipsoid *WGS84* has become a global reference and the transformations from all other ellipsoids have been precisely calculated (EPSG 2018). The project team decided to use the *WGS84* as this simplifies the transformations for obvious reasons. Additionally, GNSS-based tachymetry results are more easily incorporated within the project as only the special projection needs to be applied. It also provides good basis for future maintenance works and enables portability in future scenarios.

4.2 *Vertical Datum*

The chosen vertical datum is the European Vertical Reference Frame 2007 (EVRF2007), realized through the United European Leveling Network (UELN). This again allows for easier future maintenance and redesign works.

4.3 *Reference Plane*

The undulation of the geoid to the ellipsoid *WGS84* spans from $N = 49$ *to* $51\ m$ in the area around the tunnel. To achieve better agreement between the nature and the geospatial data and to lessen the computational burden, the project's reference plane has been defined anew, which lies $H = 770\ m$ $(H_{ell} = 720\ m)$ above the ellipsoid (see Figure 4). This scales the data from the ellipsoid by $m_{\mathrm{ref.plane}} = 1.000121$, which was done at the beginning of the project.

4.4 *Coordinate Reference System*

There are many already defined projected CRSs based on *WGS84*, like the *WGS84/World Mercator* (Eurocentric view of the world excluding polar areas, EPSG code 3395) and

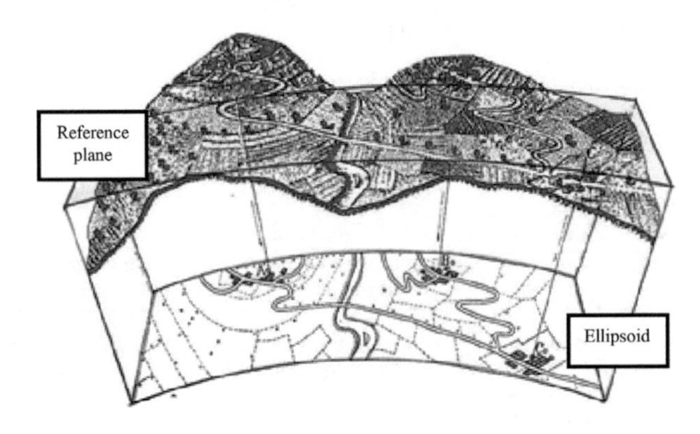

Figure 4. The *WGS84* ellipsoid in the project scope and the reference plane for the elevation, which lies $H = 770\ m$ above the ellipsoid.

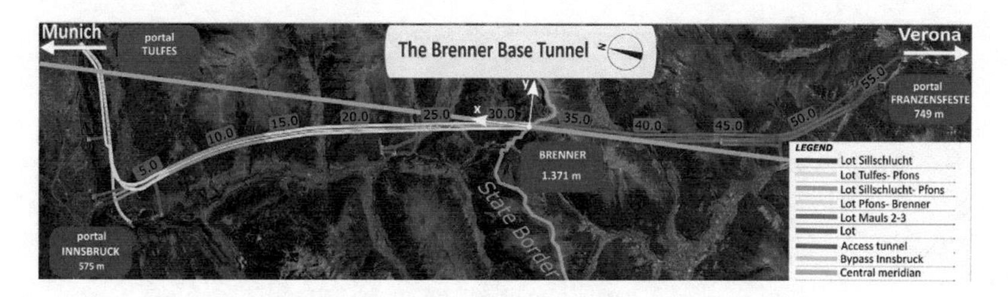

Figure 5. Plan of the BBT project site, where the topography, the state border Austria-Italy, the axes of the tunnel together with the central meridian, project's origin, and coordinate axes are marked.

WGS84/UTM grid system (EPSG codes 32600-32660). As the geospatial data was going to be transformed without regard for the chosen projection, it was optimal to choose the best suitable one. However, none of the available options was providing an extra edge over the others.

As presented by Markič et al. (2018), infrastructure projects sometimes opt for custom orientation and/or relative position of the projection plane. It is beneficial for the project area to be distorted as little as possible through projection operations. Because the project site extends primarily in the North-South direction, it was optimal to define a TM projection in such a way, that its meridian runs as close to the tunnel axis as possible to ensure a constant scale across the whole project area (see also Figure 1, left). The chosen meridian was 11°31'42.5775"E from Greenwich as presented in Figure 5 which ensures the whole project lies within +/- 10 km of the meridian. Therefore, the distortions of the projection are neglectable.

4.5 *Deflection of the Vertical*

Because of the big variance in the direction of the gravity pull along the tunnel, these were separately recorded along the axis of the tunnel by Technical University Graz as part of the definition of the vertical datum. The corrections of the project height due to gravity anomalies lie between 5,5 cm and 17 cm and their detailed progression is shown in Figure 6.

5 DESIGN WITH BIM TOOLS

The design data needs to be exchanged multiple times during the project and later archived for maintenance purposes. Currently, this is done by exchanging paper-based and digital blueprints together with additional supportive project information. Geo-referencing is one of

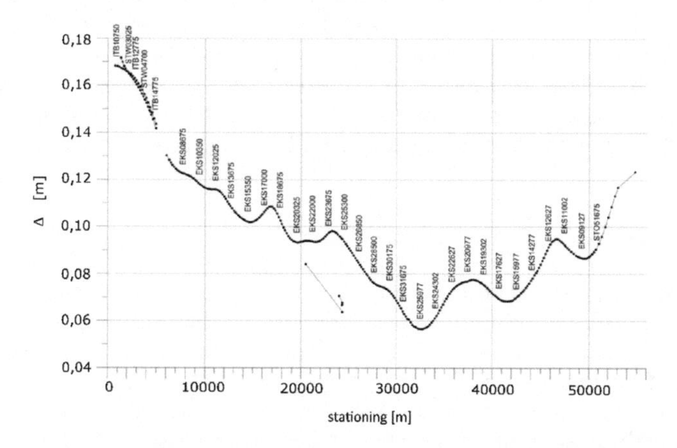

Figure 6. Detailed measurements of the deflection of the vertical resulted in precise height corrections along the tunnel axis which are presented together with identifiers of the individual points.

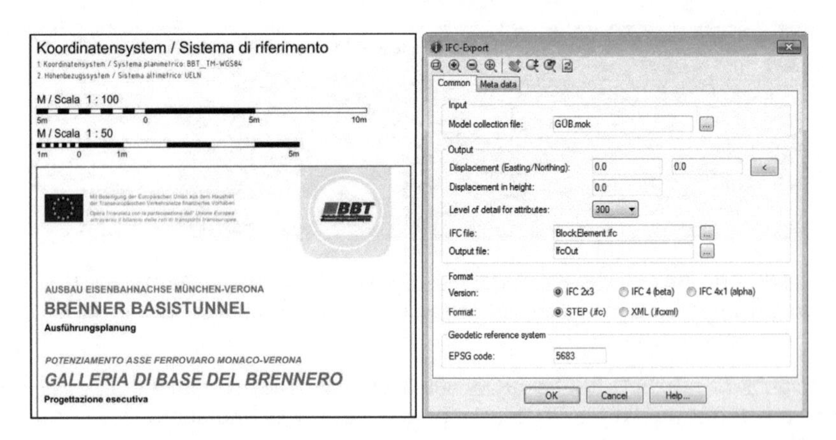

Figure 7. The geo-referencing metadata provided on a blueprint (left) and the parameters of IFC export in the infrastructure design software ProVI 6.0 (right). The metadata about the underlying compound CRS is clearly stated on the blueprint (top). Similarly, the software user can specify the translation vector as well as set the EPSG code of the underlying geodetic CRS used in the project by filling out the export mask.

the supporting elements set at the beginning and looked at when setting-out the design. In the BBT project, every blueprint is labeled with the information about the underlying CRS (see Figure 7, left).

In the long term, the blueprints will be succeeded by digital models (Borrmann et al. 2015). The goal is to make machine guidance and prefabrication possible directly from the model itself. Therefore, this supportive information needs to be addressed and considered properly (Markič et al. 2018). As such, the BIM design programs need to provide functionality to include and correctly handle the model's metadata. As the design takes place in the PCS (X, Y, Z), the designers usually do not know about the discrepancies between their geospatial data. It is first recognized when including freely accessible ortho-photo images and finding misalignment between objects of interest. Additionally, some programs have a problem handling bigger coordinates. This can be fairly easily solved with introducing a new, local PCS, whose origin and rotation is known within the geodetic CRS (Kaden & Clemen 2017). For example, the user of software ProVI 6.0 can specify a displacement in location as well as elevation (see Figure 7, right).

The IFC schema enables the exchange of geo-referencing metadata in its entities *IfcMapProjection* and *IfcProjectedCRS*. The former allows for diminishing of coordinates' values and the latter saves the information about the underlying geodetic CRS by providing its EPSG code (see Figure 7, right) (ISO 16739:2013, 2013, buildingSMART International 2018). Although this covers most cases, this is not sufficient for the BBT project. Because the BBT project defined new projected and compound CRSs, that do not have an EPSG code (see Table 1) and thus cannot be exchanged using the popular IFC format. This inadequacy has been addressed by Markič et al. (2018) which have proposed a solution by extending the IFC schema accordingly. Currently, such information needs to be exchanged in an additional file attached to the project documentation.

6 CONCLUSIONS

Exchange of design data and its transfer to the construction site is currently done with paper-based and digital blueprints. With the introduction of Building Information Modelling (BIM) methods, the AEC industry will switch to model-based exchanges (Borrmann et al. 2015). This demands a careful definition of the structure of the BIM model in order to be clear about the data and its context. The focus of this paper is the geo-referencing of BIM models and its evaluation on a real-world example, the Brenner Base Tunnel (BBT).

The BBT project's surveyors decided to define a new, compound geodetic Coordinate Reference System (CRS) because of the benefits it brings to the accuracy and clearness of the data. The CRS "*BBT_TM-WGS84*" was based on the *WGS84* ellipsoid (used by global navigation satellite systems) and projected using a Transverse Mercator (TM) projection. The vertical datum was based on the United European Leveling Network (UELN) and its corrections measured in detail along the axis of the tunnel. Each blueprint has been foreseen with the metadata about the chosen geodetic CRS (see Figure 7, left).

A BIM model provided in the vendor-neutral format IFC can include data about the underlying geo-referencing context. However, there are still some missing functionalities if we wish to use IFC format in projects like the BBT. Since the defined CRS does not have an EPSG code foreseen to be included in the IFC file, this metadata needs to be exchanged in another way which is suboptimal. The software implementers should sign an agreement of how to handle such peculiarities in order to allow for clear and concise design processes.

REFERENCES

Barazzetti L. & Banfi F. (2017). BIM AND GIS: WHEN PARAMETRIC MODELING MEETS GEOSPATIAL DATA. *ISPRS Annals of Photogrammetry, Remote Sensing & Spatial Information Sci. 4.*

Bergmeister K. (2011). Brenner Basistunnel. Brenner Base Tunnel. Galleria di Base del Brennero. „Der Tunnel kommt", „The Tunnel Will Be Built"„La Galleria diventa realtà". Tappeinerverlag Lana. 263p, ISBN 978-88-7073-587-1.

Borrmann A., König M., Koch C. & Beetz, J. (2015). Building Information Modeling: Technologische Grundlagen und industrielle Praxis. [Building Information Modeling - Technological foundations and industrial practice]. VDI-Buch, Springer Fachmedien, Wiesbaden.

buildingSMART International (2018). *Model Setup IDM, Vol I: Geo-referencing BIM.* Retrieved from buildingSMART website: https://bsi-intranet.org/kos/WNetz?art=Folder.show&id=5568, accessed on August 31st, 2018.

Conchúir D.Ó. (2016). Great miscalculations: the importance of validating project management assumptions. *Engineersjournal, Engineers Ireland.* Retrieved from Engineers Ireland website: http://www.engineersjournal.ie/2016/06/28/gmiscalculation-laufenburg-bridge/, accessed on May 5th, 2018.

Donaubauer A. & Kolbe, T.H.E. (2017). Leitfaden Bezugsystemwechsel auf ETRS89/UTM: Grundlagen, Erfahrungen und Empfehlungen. [Guide for reference system change to ETRS89/UTM: Fundamentals, experiences and recommendations] Retrieved from Rundertisch-GIS website: https://rundertischgis.de/publikationen/leitfaeden.html, accessed on August 31st, 2018.

EPSG (2018) The European Petroleum Survey Group homepage: http://www.epsg.org/, accessed on August 31st, 2018.

Heunecke O. (2017). Planung und Umsetzung von Bauvorhaben mit amtlichen Lage- und Höhenkoordinaten. [Planning and implementation of construction projects with geodetic coordinates] In: *zfv* 142, pp. 180–187.

ISO 16739:2013 (2013). *Industry Foundation Classes (IFC) for data sharing in the construction and facility management industries.* Standard International Organization for Standardization Geneva, CH.

ISO 19111:2007 (2007). *Geographic information – Spatial referencing by coordinates.* Standard International Organization for Standardization Geneva, CH.

Markič Š., Donaubauer A. & Borrmann A. (2018). Enabling Geodetic Coordinate Reference Systems in Building Information Modelling for Infrastructure. In: *Proceedings of the 17th International Conference on Computing in Civil and Building Engineering.* Tampere, Finland.

Kaden R. & Clemen C. (2017). Applying Geodetic Coordinate Reference Systems within Building Information Modeling (BIM). In: *Technical Programme and Proceedings of the FIG Working Week 2017,* Helsinki, Finland. ISBN: 978-87-92853-61-5.

Spiegel (2015). Behörde baut Autobahnbrücke an falsche Stelle. *Spiegel Online.* Retrieved from http://www.spiegel.de/wirtschaft/soziales/steuerverschwendung-behoerde-laesst-autobahn-bruecke-an-falsche-stelle-bauen-a-1064758.html, accessed on May 5th, 2018.

Macheiner K. (2015). Drei große Eisenbahn-Tunnelprojekte in Österreich – ein Vergleich ausgewählter Aspekte aus der Sicht der ingenieurgeodätischen Praxis [Three big railway projects in Austria – a comparisson of chosen aspects from the view of the engineering survey]. *Vermessung & Geoinformation* (4), pp. 221–234.

Mugnier, C. (2005). Italian Republic. *Grid&Datums* (71), No. 8, pp. 889–890.

National Imagery and Mapping Agency (2002) *The American Practical Navigator: An epitome of Navigation. 2002 Bicentennial Edition.* Bethesda, Maryland, USA. NSN 7642014014652

Tunnels and Underground Cities: Engineering and Innovation meet Archaeology, Architecture and Art, Volume 3: Geological and geotechnical knowledge and requirements for project implementation – Peila, Viggiani & Celestino (Eds)
 ISBN 978-0-367-46583-4

Analysis of critical condition during the advancement into the "Argille a Palombini" formation

N. Meistro, A. Mancarella, A. Di Salvo, C. Iemmi & I. Larosa
COCIV Consorzio Collegamenti Integrati Veloci, Genova, Italy

ABSTRACT: The Milan Genoa high speed/high capacity line is expected to realize 90 km length tunnel; about 60% of them should be excavated in a shale formation called "Argille a Palombini"This shale formation has characterized to have a low metamorphic range and, moreover, inside the rock mass sudden mutations of the tectonics stress are observed.The monitoring of these stress and daily viewing of excavations has underlined how this typical geotechnical behavior didn't influenced the excavation features, but it has modified the tunnel profile deformation, in some case intensely.Inside the paper, this particular behavior and the resulting technical difficulties are therefore discussed and analyzed. The design tools and solution, like geological and geo-technical profiles and ground excavation consolidation, were demonstrated a good means to understand and solve the advancement in critical conditions.

1 PROJECT DESCRIPTION

The railway Milan–Genoa, part of the High Speed/High Capacity Italian system (Figure 1), is one of the 30 European priority projects approved by the European Union on April 29th 2014 (No. 24 "Railway axis between Lyon/Genoa – Basel – Duisburg – Rotterdam/Antwerp) as a new European project, so-called "Bridge between two Seas" Genoa – Rotterdam. The new line will improve the connection from the port of Genoa with the hinterland of the Po Valley and northern Europe, with a significant increase in transport capacity, particularly cargo, to meet growing traffic demand.

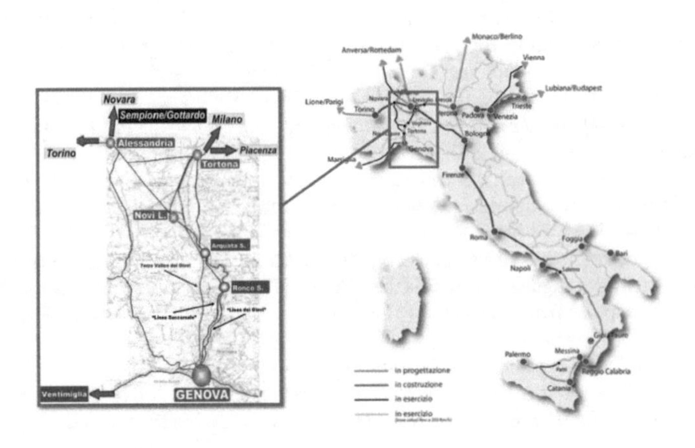

Figure 1. Italian railway system.

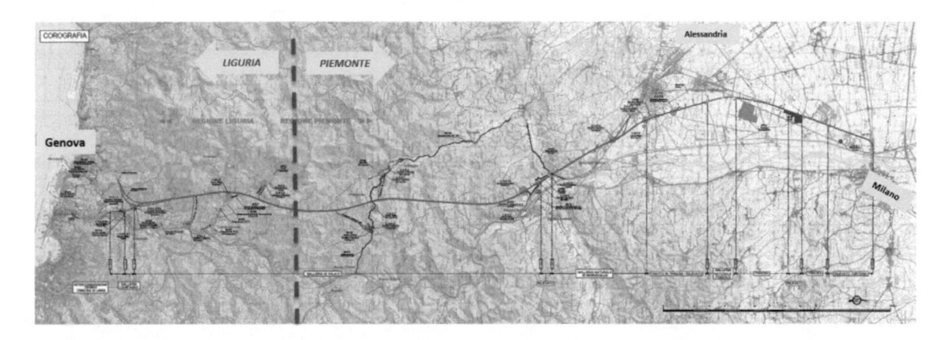

Figure 2. Terzo valico project.

The “Terzo Valico” project is 53 Km long and is challenging due to the presence of about 36 km of underground works in the complex chain of Appennini located between Piedmont and Liguria. In accordance with the most recent safety standards, the underground layout is formed by two single-track tunnels side by side with by-pass every 500 m, safer than one double-track tunnel in the remote event of an accident.

The layout crosses the provinces of Genoa and Alessandria, through the territory of 12 municipalities.

To the South, the new railway will be connected with the Genoa railway junction and the harbor basins of Voltri and the Historic Port by the “Voltri Interconnection” and the “Fegino Interconnection”. To the North, in the Novi Ligure plain, the project connects existing Genoa- Turin rail line (for the traffic flows in the direction of Turin and Novara – Sempione) and Tortona – Piacenza –Milan rail line (for the traffic flows in the direction of Milan- Gotthard).

The project crosses Ligure Apennines with Valico tunnel, which is 27 km long, and exits outside in the municipality of Arquata Scrivia continuing towards the plain of Novi Ligure under passing, with the 7 km long Serravalle Tunnel, the territory of Serravalle Scrivia (Figure 2). The underground part then includes Campasso tunnel, approximately 700 m long and the two “Voltri interconnection” twin tunnels, with a total length of approximately 6 km.

Valico tunnel includes four intermediate adits, both for constructive and safety reasons (Polcevera, Cravasco, Castagnola and Vallemme). After tunnel of Serravalle the main line runs outdoor in cut and cover tunnel, up to the junction to the existing line in Tortona (route to Milan); while a diverging branch line establishes the underground connection to and from Turin on the existing Genoa-Turin line.

From a construction point of view, the most significant works of the Terzo Valico are represented by the following tunnels:

- Campasso tunnel 716 m in length (single-tube double tracks)
- Voltri interconnection even tunnels 2021 m in length (single-tube single track)
- Voltri interconnection odd tunnels 3926 m in length (single-tube single track)
- Valico tunnel 27250 m in length (double tube single track)
- Serravalle tunnel 7094 m in length (double tube single track)
- Adits to the Valico tunnel 7200 m in length
- Artificial tunnels 2684 m in length
- Novi interconnection even tunnels 1206 m in length (single-tube single track)
- Novi interconnection odd tunnels 958 m in lenght (single-tube single track)

The project standards are: maximum speed on the main line of 250 km/h, a maximum gradient 12,5 ‰, track wheelbase 4,0 – 4,5 m, 3 kV DC power supply and a Type 2 ERTMS signalling system.

2 THE "ARGILLE A PALOMBINI" FORMATION: MINERAL-PETROGRAPHIC CHARACTERS AND GEODYNAMIC CONTEXT OF ORIGIN

The formation of Argille a Palombini, from the minero-petrographic point of view, can be classified as a shale, a metamorphic rock deriving from the transformation of clayey rocks by a slight regional metamorphism.

The descriptive detail of the formation sees the presence of two distinct members: the "Argilliti di Costagiutta" and "Argilliti di Murta" (respectively identified with the abbreviations AGI and AGF). The first ones are composed of alternations of shale, by far prevailing, and crystalline limestone (the palombini in fact) with veins of quartz and albite. The latter are composed of black Filladic shale, always with quartz and albite veins and differ from the former due to the absence of calcareous banks.

In the excavation phase, this distinction is not always clearly evident anyway. For both the members described, the metamorphic rock formation range foresees the presence of temperatures between 200 ° C and 300 ° C and pressures not higher than 4 kbar (Figure 2). It is therefore a lithology characterized by a low metamorphic degree, in which the mineralogical composition does not undergoes modifications, on the whole, compared to clayey original rock from which it comes. This aspect, as we will see later in the discussion, assumes a significant importance in the geomechanical behavior of the rock mass, both during the excavation phase and in the period immediately following the advance of the face.

The action of metamorphism (by temperature and pressure changes), although not able to generate new mineralogical structures, is however clearly visible in the rock mass, as resulting in a recrystallization and mechanical reorientation of the minerals of the clays, subject to evident oriented pressures, due to strong tectonic stresses in the past, during the Alpine orogeny.

In particular, the result of these metamorphic processes is represented by the presence of a marked schistosity in the final product, a distinctive element of the formation under examination and very visible, practically, in all the observations carried out on the sections until now excavated (Figure 3).

On the basis of foregoing discussion, in providing a characterization of the geomechanical behavior of the formation, for the purposes of excavation behavior, two main aspects must be considered:

- Presence of a low metamorphic degree and consequent maintenance of the mineralogical structure of the material, even after the metamorphic phases have been explicated
- Presence in the rock mass, in the investigated sector, of strong oriented tectonic thrusts, responsible for the formation of the schistosity, clearly visible in the outcropping rock, whose action strongly conditions the excavation modalities and, above all, the recorded deformative trend of the rock mass, as we will see in the continuation of the discussion.

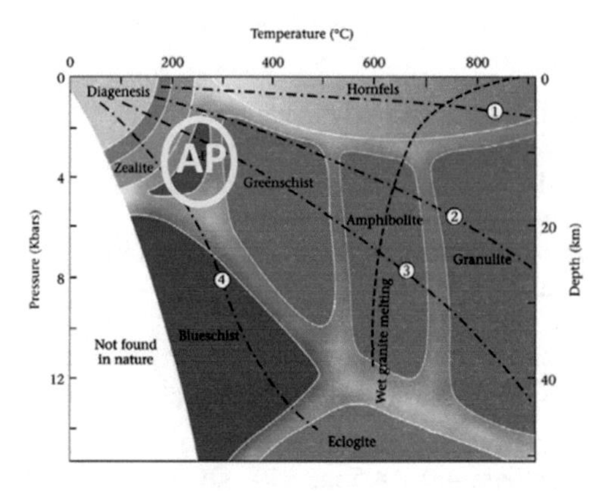

Figure 3. Diagram of phases of the rock mass metamorphism, in which the temperature and pressure range of Argille a Palombini formation is highlighted.

Figure 4. Excavation face in Argille a Palombini, with evident schistosity, associated with fold structures and calcitic quartz veins.

3 THE EXCAVATION OF THE ARGILLE A PALOMBINI: THE PROJECT, PREDICTED GEOMECHANICAL CONDITION AND RELATED TYPE ADVANCEMENT SECTIONS

Due to the considerable extension of rock volumes crossed in Argille a Palombini along the route, the lithomechanical characteristics of the rock mass (composition, GSI, characteristic modules, etc.) cannot be defined in a univocal way, but will vary, even significantly, depending on numerous factors, such as overburden or tectonic stress concentration in fault zones.

For this reason, during the design, five distinct geomechanical reference groups were associated with this formation, based on the presence of diagnostic elements such as the extension of calcareous interlayers, the foliation structure, more or less frequent presence of mesoscale and microscale folds, estimated GSI values, uniaxial compression strength and elastic modulus.

On the basis of the data cataloged in the cognitive phase, the type of advancement sections evaluated most appropriate to proceed, during the excavation, have been defined in the different geomechanical contexts envisaged for the formation of Argille a Palombini (Figure 5).

In Figure 5, the alphanumeric progression of the abbreviations corresponds to a progressive weighting of the stabilization adopted, both in terms of the type of steel ribs, and in terms of amount fiber glass tubes present at the excavation face and at the contour.

To the application of each type advancement section correspond a well defined range of convergence and GSI, defined in this case by the direct observation of the excavation face (Figure 5).

Following, on the basis of the experiences deriving from the execution of the excavations until now completed, it is interesting to compare, for the investigated geological formation of Argille a Palombini, the expected behavior, modulated according to the guidelines indicated so far, and the one really found in the tunnels.

4 THE EXCAVATION OF THE ARGILLE A PALOMBINI: GEOMECHANICAL AND DEFORMATIVE FEATURES FOUND DURING CONSTRUCTION, COMPARISON WITH THE PROJECT.

Overall, in terms of the geomechanical and deformative behavior found in the rock mass and the consequent choice of the typical advancement sections to be adopted, the project forecasts have been widely respected, with a good correspondence between what was observed and what was expected. Instead, if the analysis of data collected is focused for very localized points, the evidence found during the excavation phase was not diagnostic with respect to the

VALICO SINGLE TRACK TUNNEL											
FORMATION	GROUP	GSI	ADVANCEMENT SECTION TYPE	CONVERGENCE ATTENTION THRESHOLD [cm]	CONVERGENCE ALERT THRESHOLD [cm]	EXCAVATION FACE CONSOLIDATION	CONTOUR CONSOLIDATION	BOTTOM SIDE CONSOLIDATION	MAXIMUM ADVANCEMENT STEP [m]	STEEL RIB	STEEL RIB STEP [m]
Ap – GR1	1	45-55	B0Lsb	4-6	6-8	NO	NO	NO	4.2	2IPN160	1,4
			B0Vsb			NO	n°25 steel tubes S355 Ø 88.9 Sp. 10 mm L ≥ 15.00 m, minimum overlay s ≥ 3.00 m	NO		HEB200	1,2
			B0/1sb	6-8	8-10	NO	NO	NO	3.6	HEB180	1,2
Ap – GR2a	2A	40-45	B1sb	6-8	8-10	NO	n 10-11 steel radial bolt Ø 24 mm B450C , L = 6.00 m, longitudinal step 1.20 m, transverse step 2.00 m	NO	1.2-1.0	2IPN220	1,2
			B2/1sb			n° 60 fibre glass tubes L≥ 24 m, overlay ≥ 9 m;	NO	NO		HEB200	1
			B2Vsb			n° 70 fibre glass tubes L ≥ 24 m, overlay ≥ 12 m	n° 25 steel tubes S355 injected Ø 88.9 Sp. 10 mm L≥ 15.00 m, minimum overlay s ≥ 3.00 m	NO		HEB240	1
Ap – GR2b	2B	30-40	B4/1sb	8-9	9-10	n° 60 fibre glass tubes, lunghezza ≥ 24 m, overlay ≥ 9 m	n 10-11 steel radial bolt Ø 28 mm B450C , L = 6.0 m, longitudinal step 1.00 m, transverse step 2.00 m	NO	1.0	2IPN240	1
Ap – GR3a	3A	30-35	C4sb	10-12	<15	n° 70 fibre glass tubes L≥ 24.00 m, overlay ≥ 9 m,	n° 70 fibre glass tubes, L ≥ 24.00 m, minimum overlay ≥ 9 m	6 + 6 fibre glass tubes injected L ≥ 24 m, overlay ≥ 9.0 m	1.0	HEB240	1
Ap – GR3b	3B	25-30	C2sb	10-12	<15	n° 55 fibre glass tubes, L ≥ 24 m, overlay ≥ 9 m	55 fibre glass tubes injected, L ≥ 24 m, minimum overlay ≥ 9.0 m	6 + 6 fibre glass tubes injected L ≥ 24 m, overlay ≥ 9.0 m	1.0	HEB240	1

Figure 5. Summary of the standard sections envisaged by the project for the area investigated, with details of the main construction features.

subsequent deformation behavior of the rock mass, especially in term of convergence values recorded. This aspect has sometimes led to a re-adaptation and a different management of the available design resources, without having to modify the project itself.

For this reason, following, two distinct examples of excavation in Argille a Palombini will be illustrated, in tunnel characterized by different overburden and tectonic stress size.

In detail, at the date of this paper, in the single-track section of the main tunnel (Figure 1) about 3 km of excavation was carried out within the formation of the Argille a Palombini by traditional method. In particular, along the route, the excavation was divided into two distinct sectors, consisting of the departure from the southern entrance of the tunnel (Figure 6), hereinafter called sector 1, and the Polcevera axis (Figure 7), located about 4 km northward, hereinafter called sector 2.

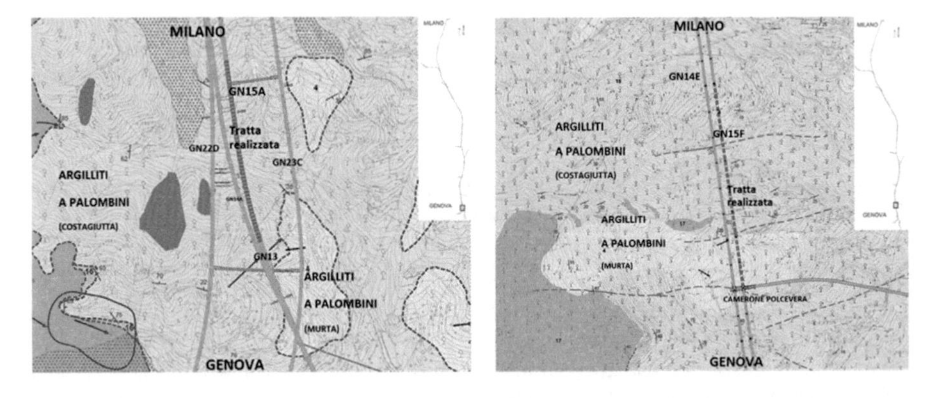

Figure 6-7. Sector 1 and Sector 2 geological sketch showing the section excavated in the Argille a Palombini.

Starting from the observations made during construction of main tunnel (Valico), in two analyzed setctors it is possible to notice some differences during the progress of work. The analysis will be inspired by two elements of judgment, strictly operational, useful to summarize the lithomechanical characteristics of the rock mass crossed:

- Vision of the excavation face with definition of a GSI value (geological strengh index).
- Analysis of tunnel deformations through the values recorded in the convergence measure points, located along the path.

4.1 *Sector 1*

This sector includes the starting part of the single-track tunnel and involves about 650 m of progress (Figure 6), with overburden between 20 m (Rio Ciliegia incision) and 100 m.

The progress in the section considered has underlined the presence of a rock mass always characterized by fair geomechanical conditions, to which was associated a deformation trend well conformed with what was detected at the excavation face, in terms of GSI values obtained and relative application of advancement type sections.

Figure 7 shows that the values of GSI for the section analyzed are always between 40 and 50, compared to a convergence maximum values between 5 cm and the complete stability (0.1 to 0.3 cm).

As can be seen from the same figure, there is a good concordance, consisting of an inverse proportionality, between the GSI values and the recorded average; as the GSI grows, the convergences decrease, as expected and, for GSI values > 45, it is possible to affirm, for the analyzed section, that the tunnel has a stable behavior in the short and medium term, in agreement with the project, which provide for the application of sections type B0 (B0L and B0/1).

In essence, therefore, the GSI values calculated for the rock mass in the considered tunnel segment, which presuppose the adoption of unconsolidated type sections, are correlated with comparable convergence values, lower than the design thresholds established for these sections.

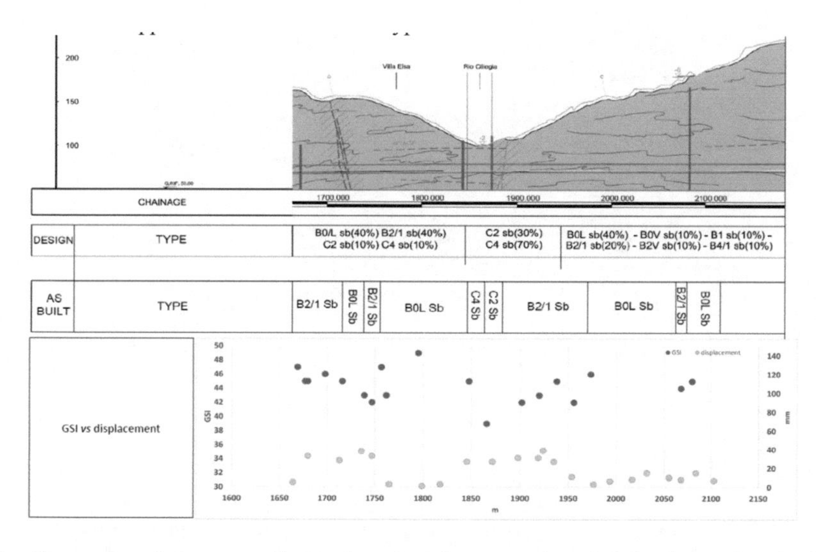

Figure 8. Comparison between applied and projected type sections for the Sector 1 and related GSI values and measured average convergences.

4.2 *Sector 2*

This sector includes the excavation of the Valico base tunnel, starting from the gallery of the Polcevera axis, located about 4 km northward as compared to the previous sector; the route of particular interest, due to the behavior demonstrated by the excavation in the rock mass, concerns about 800 m of advancement, with overburden between 330 m and 230 m (Figure 7).

Below, in the graph in Figure 7, similarly to what has been described for the Sector 1, the comparison between the GSI and convergences values, standard sections adopted for Sector 2 and those envisaged by the project, is reported.

The analysis of the graph shows some aspects that discriminate the sector under examination from the one described above (Sector 1):

- Convergences present, albeit for short localized distances, quite high deformation peaks (greater than 15 cm as average values), such as to exceed the design alarm threshold values for the most high-performing sections such as type C; these values, which envisage the application of consolidations on the excavation face and the contour, are normally associated with a range of GSI values of 30-35 (figure 4) and are not present in Sector 1.
- In correspondence with the convergence peaks described in point 1, the GSI values are always > 40, typical of the application of type B advance sections (short-term stable front) and to which the design links a range of GSI values of 40-45 (Figure 4).

On the basis of the foregoing, it should be noted that, in some cases located in the sector examined, there is no complete correspondence between the values of GSI detected in the rock mass and the convergence values registered. Although only on a local scale, there is a range of GSI values around 40-45 to which convergence associated values are locally > 15 cm, not compatible with the application of B-type sections as envisaged by the project. In this geomechanical context, generally coinciding in correspondence of fault lines, the excavation face view does not represent an exhaustive parameter in characterizing the typical section to be

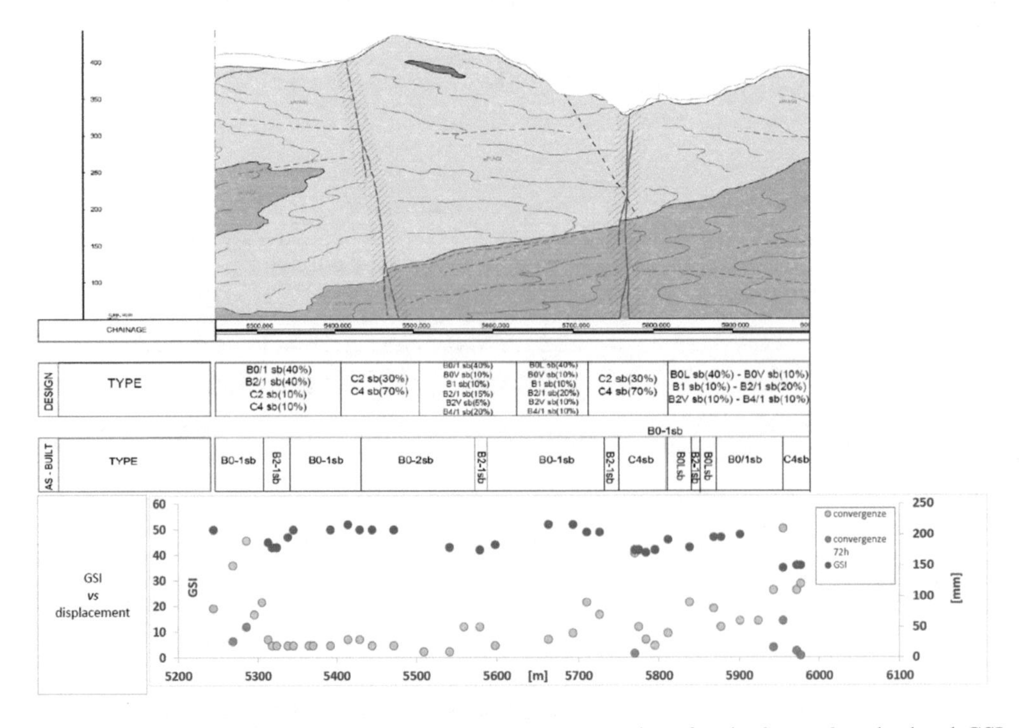

Figure 9. Comparison between applied and projected type sections for the Sector 2 and related GSI values and measured average convergences.

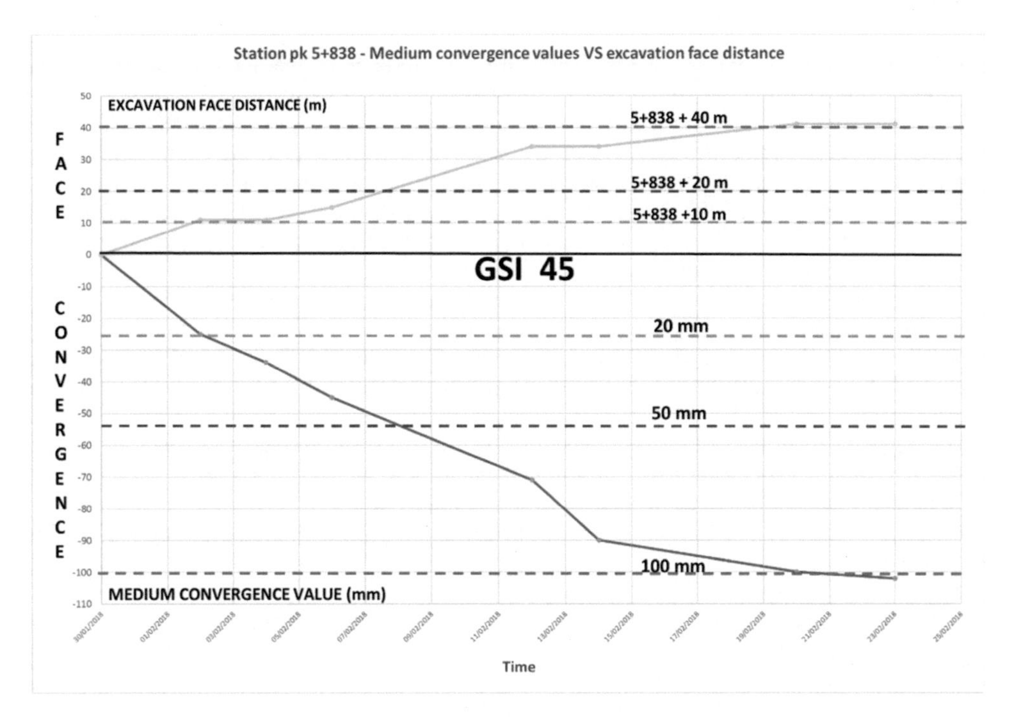

Figure 10. Comparison between medium convergences values vs excavation face distance at the station located pk 5+838.

adopted, because in the short term the rock mass does not show instabilities, whereas accentuated deformation trend in the medium term are gradually detected.

In this regard, to better understand the meaning of the previous statements, it is possible to report a relevant example, analyzing the data (figure 10) from the monitoring convergence station located at pk 5 + 838 ca (highlighted within the graph in figure 9 in its location along the path).

In figure 10 is reported the comparison between the evolution of the deformation trend of the mean convergences of the tunnel and the distance from the excavation face.

As can be seen from the data entered, after the installation of the monitoring station, the progress of the face continued for another 10 m, for about 4 days, according to the application of a type sectione B0L sb, with associated values of average convergences of 25 mm, well within the designed thresholds.

Following, through various work steps, the deformation trend has not gradually cancelled during the progress of the face but, conversely, continued its increase even with a distance of excavation face of 40 m, recording a medium convergence values of 106 mm.

Basically, in this example, initial deformation behavior of the tunnel was in line with what was found during the excavation phase (with a GSI > 40). Later, a very prolonged deformation "tail", takes place, still slightly active 25 days after the installation of the measuring station, with the excavation face placed almost 45 m away. In this way, the final values of the mean convergences found at the end of the described process are also slightly higher than the alarm threshold of the C-sections (figure 5).

5 INTERPRETATION OF THE DATA COLLECTED

Once the data of the two different sections, described in chapter 4, in different geomechanical contexts within the rock mass of Argille a Palombini have been analyzed, it is now necessary to define the reasons of difference recorded in geomechanical and deformative behavior.

A first criterion of analysis is represented by the presence of tectonics stress of the rock mass inside the fault alignments.

As evident and intuitive, the tectonic stress improve the weaking of geomechanical feature into the rock mass after the passage of the excavation face, triggering deformations, not detected in the remaining portions. In Sector 2, this deformation trend does not correspond to a marked deterioration of the geomechanical conditions of the material crossed, which often requires numerous hours of excavation to make the single excavation step, even in strong deformation contexts (convergence values > 10 cm).

Instead, the behavior of rock mass in Sector 1 appears to be much more linear. For this section, even inside faults alignments, assumed by the profile (Figure 8), the convergence values are always lower than those present in sector 2 (<5 cm in this case) and, therefore, are compatible with the application of sections B and with values of GSI > 40, recorded during the excavation.

With regard to the reasons for such behavior, it should be emphasized that the rock mass in Sector 2, located mostly within the hilly relief, appears to be more tectonized than sector 1, and above all, the size of the overburden present in different.

In fact, in Sector 2, overburden never fall below 200 m, while in sector 1 they reach 20 m, below the Rio Ciliegia.

In this regard, in chapter 2 we have highlighted how the presence of a low metamorphic degree and consequent maintenance of the original mineralogical composition of the clays in the metamorphosed rock, together with the presence of strong tectonic oriented stresses in the rock mass, can condition it geomechanical and deformative behavior.

In fact, these aspects (strong oriented stress and high overburden) contribute to making the excavation material particularly sensitive to the the presence of the tunnel, with considerable growth of deformation, even in presence of a well-manageable rock mass during excavation, with GSI values > 40, as highlighted in chapter 4.

Basically, therefore, based on the experience gained during the excavation of the analyzed area, for very local feature it seems that the shale material tends to resume a squeezing behavior, typical of a sedimentary rock mass with predominantly clayey component.

This behavior is only detected in sector 2, while is not detected in sector 1.

In this context, the next step of the research is now represented by the optimization of this analysis in relation to the design choices to be adopted in the continuation of the progress.

On the basis of the foregoing, for example, to obtain a better performance during the progress, it will be necessary to apply advancement type sections with a greater performance primary lining (steel ribs) and, at the same time, it will be possible to limit the pre-consolidation measures, always within the design guideline variability.

The phenomenologies described, indeed, are valid only on a local scale, in particular geomechanical contexts; in the majority of cases an excellent adherence between the planned design choices and what is actually applied during construction is present. However, the excavation of many kilometers of large underground sections in the same formation, has provided useful and original informations about geomechanical and deformative behavior of the Argille a Palombini formation, useful to define the best methods of advancement in difficult geomechanical condition in this kind of rock mass.

6 CONCLUSION

In a part of the underground excavation of the Milan-Genoa AC/AV line, the progress made in the formation of the Argille a Palombini highlighted a peculiar geomechanical and deformative behavior that, in some well-localized cases, introduced some technical difficulties, producing the optimization of the advance methods of excavation envisaged by the project.

In this context, the solutions proposed in the course of work, although always adhering to the design guidelines within the allowed variability, have undergone some modifications, essentially deriving from the presence of a deformation behavior not always aligned with the GSI values detected and, in essence, with the excavation behavior of the investigated tunnels.

In particular, although in a very local feature, corresponding to areas with high overburden and crossing tectonic alignments, the deformation behavior of the rock mass in medium term, before the positioning of definitive lining, showed conditions compatible with the application of C-type sections (unstable front), characterized by pre-consolidation on the excavation face front and contour. Otherwise, in the same area and same deformation contexts, the excavation face has shown a substantial stability e good performance during excavation, with GSI values often > 40, compatible with B-type advancement section.

REFERENCES

Lunardi, G. Cassani, G. Bellocchio, A. 2016. *Linea ferroviaria ad alta velocità Milano – Genova: analisi parametrica della risposta tenso deformativa delle "Argille a Palombini" durante lo scavo di gallerie.* Bologna: Convegno SIG "Le sfide per la realizzazione di grandi opere in sotterraneo: progettazione, costruzione e gestione di opere complesse e sfidanti" Expo tunnel

Lunardi, P. Cassani, G. Gatti, M. Zenti, C.L. 2016. The ADECO-RS approach and the recent European application experiences. Prague: Proceedings 13th International Conference Underground Construction Prague, 3rd Eastern European Tunnelling Conference

Lunardi, G. Barla G. 2014. *Full face excavation in difficult ground.* Salisburgo: Proceedings of the 63° Geomechanics Colloquy

Lunardi, P. Bindi, R. Cassani, G. 2014. The reinforcement of the core-face: history and state of the art of the Italian technology that has revolutionized the world of tunnelling. Some reflections. Iguassu Falls: Proceedings of the ITA/AITES World Tunnel Congress on "Tunnels for a better life

Tunnels and Underground Cities: Engineering and Innovation meet Archaeology, Architecture and Art, Volume 3: Geological and geotechnical knowledge and requirements for project implementation – Peila, Viggiani & Celestino (Eds)

Excavation of swelling rock by measuring the displacement under roadbed

K. Miyazawa & M. Fukushi
East Nippon Expressway Company Limited, Tokyo, Japan

T. Akiyama & H. Kinashi
Obayashi Corporation, Tokyo, Japan

ABSTRACT: Recently, in some in-service tunnels, roadbed heaving caused by swelling and squeezing has damaged the invert concrete. In many of these cases, it is presumed that roadbed displacement was undergoing during construction and the displacement should have been measured and managed. However, it is difficult to accurately grasp the roadbed displacement. Therefore, in the tunnel under construction, the displacement convergence of the whole tunnel is judged from the inner section displacement of the arch part and the like. Then, we newly developed a system which can accurately measure the roadbed displacement during construction, and applied it to the Sankichiyama Tunnel. It was consequently discovered that only the roadbed was significantly raised after inner section displacement of the arch part converged. In this paper, we report the case of roadbed displacement measurement using this new system, and also quantitatively evaluate the measures against heaving by analytical method using measurement data.

1 INTRODUCTION

The Tohoku-Chuo Expressway is a 268-km national automobile highway running from Soma City in Fukushima Prefecture through Fukushima City, Yonezawa City, and Yamagata City in Yamagata Prefecture, before arriving at its northern terminus in Yokote City, Akita Prefecture.

The Sankichiyama Tunnel (hereafter “this tunnel”) is a 2980m long provisional two-lane highway tunnel part of Tohoku-Chuo Expressway located 7–13 km to the south of Yamagata Station on the JR Ou Main Line. It is the longest tunnel in the section from Nanyo Takahata to-Yamagata Kaminoyama of the Expressway.

In this tunnel, localized pockets of springs and tuff breccia containing large quantities of expansive clay minerals were found during excavation from the tunnel face, and inner displacement of more than 300 mm occurred. Therefore, remedial construction work was conducted to converge the arch inner displacement and breakthrough such sections.

However, in some cases with other tunnels already in use, long-term roadbed bulging occurred as an effect of ground swelling, and the invert had to be reconstructed. To avoid such repair works after commencement of service, a newly developed invert displacement meter was additionally applied to this tunnel for the measurement of roadbed displacement during construction. This measurement revealed that the roadbed was rapidly bulging even when the inner displacement was moving toward convergence.

This paper reports the results of measurement of roadbed displacement during construction and the application of these results in remedial construction to rectify the roadbed bulges.

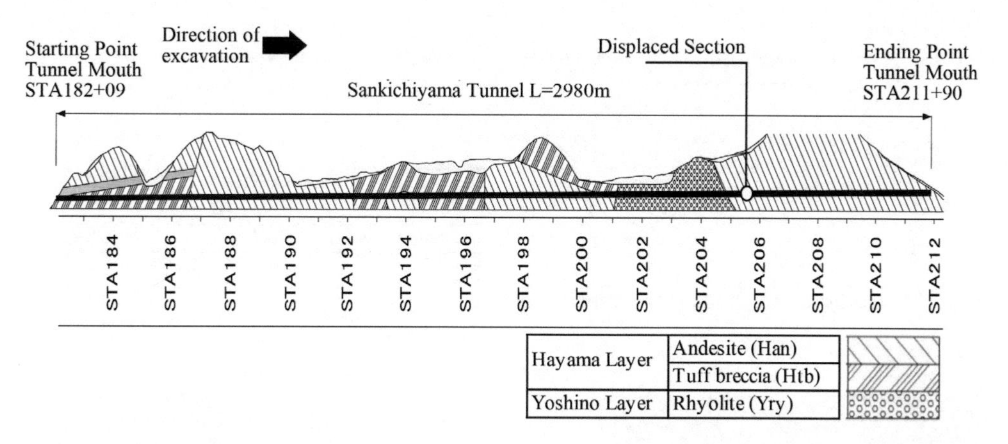

Figure 1. Geological profile.

2 SUMMARY OF GEOGRAPHICAL AND GEOLOGICAL CONDITIONS

This tunnel is located in the west wing of Mount Zao, a member of the Ou mountain range, which bisects the Tohoku region from east to west. "Mount Zao" is the general name for a belt of volcanoes along the boundary between Yamagata and Miyagi prefecture. The tunnel passes through the foothills of a mountain spur that splits off the west wing of the ridgeline of this volcano belt and the alluvial fan formed in the adjoining valley with a shallow rock covering. The rock covering the tunnel is 156.3 m deep at the thickest point and 14.5 m deep at the thinnest point, excluding the tunnel mouths.

The bedrock in the area is composed of the Yoshino Layer, dating from the Miocene Epoch of the Neogene Period, and the Hayama Layer, dating from the Pliocene Epoch, covered by a quarternary stratum on the surface. The Hayama Layer consists of strata of tuff breccia layers (Htb) and andesite layers (Han). The tunnel cuts through these layers; near the middle, the tunnel encounters extrusions of rhyolite (Yry) from the Yoshino Layer (Figure 1).

The andesite is charcoal grey in color and is slightly porous but very hard; it is used to develop cooling joints such as columnar joints and block joints. The lighter grey tuff breccia is a non-stratified block composed of andesite fragments, ranging in diameter from a few centimeters to a few dozen centimeters, locked in a weakly compacted matrix, and while the rock fragments themselves were soft, the bedrock was deemed favorable for supporting the tunnel face.

3 STATUS OF DISPLACED SECTIONS

3.1 *Topography/geology*

The displacement in this tunnel occurred along a 65.4-m extension (STA.205+16.80-STA.205 +82.75) approximately 2300 m from the tunnel mouth. The rock covering this section is 66 m deep at its thickest point, and during the design phase, it was predicted that this section would be composed of hard andesite, with an elastic wave velocity of 4.0–4.2 km/s. However, during excavation, localized protrusions of tuff breccia (Htb), which were not predicted during the design phase, appeared crossing the tunnel face (Figure 2). This tuff breccia was greenish grey in color, and the matrix had been transformed into clay through hydrothermal alteration, rendering its characteristics evidently different from those of the tuff breccia (Htb), which had appeared previously. In addition, approximately 200 L of spring water/minute flowed out of the juncture with the andesite (Han) layers, causing frequent cave-ins at the tunnel face and a steep drop in the stability of the cutting surface.

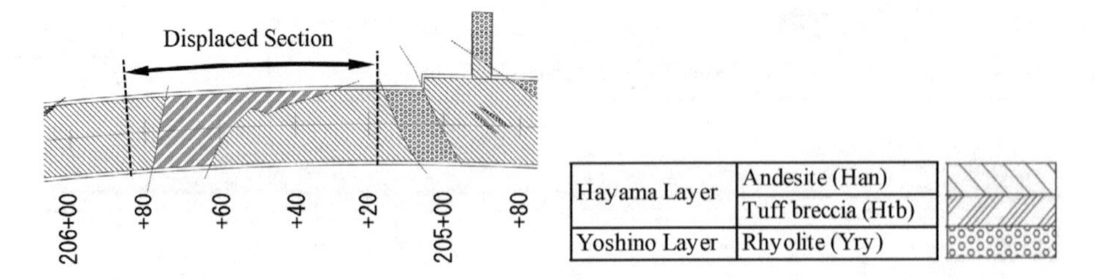

Figure 2. Geological condition in the displaced section.

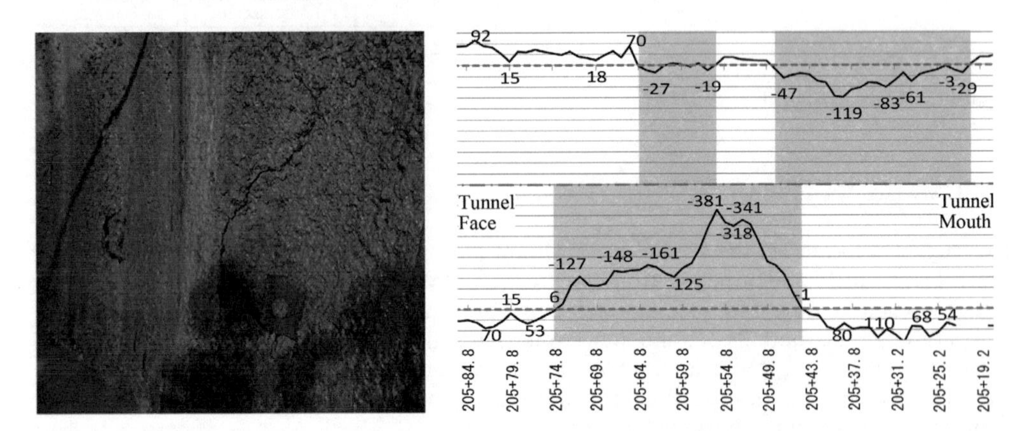

Figure 3. Deformations in the sprayed concrete.

Figure 4. Side wall displacement measurement results.

3.2 *Tunnel support distortion*

Although no significant distortion was detected in November 2016, soon after this section of the tunnel was excavated, after a while, various kinds of deformation were detected in the tunnel supports, including cracks in the sprayed concrete and distortion of the rock bolt plates (Figure 3). Most of these deformations were concentrated in the side walls, and when the location of the upper support feet of the steel arch was measured, a maximum deformation of 381 mm to the inner side was observed (Figure 4).

3.3 *Consideration of displacement status*

To investigate the causes of displacement in this section of the tunnel, the bedrock and support deformation status at the time of tunnel excavation were categorized as follows.

(1) The bedrock in this section of the tunnel was Neogene tuff breccia containing a large amount of spring water.
(2) There was a correlation between the distribution of the tuff breccia and the range of deformation in the steel arch supports.
(3) Displacement was concentrated in the side walls, and the inner space displacement along the horizontal traverse line exceeded the roof subsidence.

Accordingly, it is believed that the displacement in this section of the tunnel was caused by multiple factors, including stress release from excavation of the Neogene tuff breccia and swelling and slaking owing to the spring water.

3.4 *Displacement rectification work*

Rock bolts are effective against ground swelling, and can restrain the ground if poured as early as possible after excavation. However, it is considered that, in conjunction with the increased looseness of the ground, the pattern bolts (3.0 m) poured during excavation were not sufficiently effective, and longer rock bolts (6.0 m) were deemed necessary. As deformation was greater along the horizontal traverse line, three longer rock bolts were poured between each of the pattern bolts on the left and right side walls.

In addition, the sprayed concrete invert was installed at a thickness of 20 cm at the bot-tom edge of the invert concrete. While tunnel supports are not closed by invert concrete, structural resistance is not sufficient to retain the ground. Therefore, it was believed that the tunnel support needed to be closed as quickly as possible. Then, sprayed concrete in-vert, which is easily workable and can quickly develop support function, was used to close the cross-sectional support structure. The remedial construction is shown in Figure 5.

Figure 6 shows the results of the STA.205+65 measurements. Immediately after excavation, displacement from the inner displacement had already surpassed 50 mm (management level III). The additional longer rock bolts reduced the speed of the displacement slightly, but owing to an increase in the creep displacement regardless of the distance from the tunnel face, the sprayed concrete invert was introduced to stop the displacement.

Ultimately, the inner displacement showed a convergence trend, and the displacement did not increase even after tunnel excavation was resumed. Therefore, the ground was deemed stabilized, and back-stitching was conducted to secure the inner needed space.

3.5 *Swelling ground determination test*

Tunnels built in swelling ground indicate many specific behaviors different from regular tunnels, and hence, it is desirable that a detailed ground survey be conducted to gather da-ta for use during construction and after the highway is opened to the public.

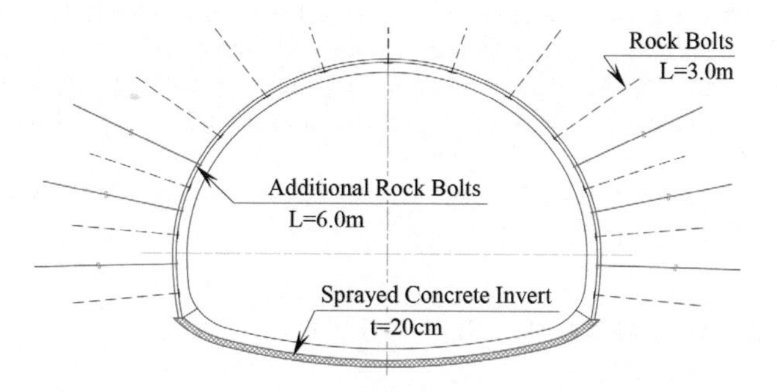

Figure 5. Diagram of remedial construction during tunnel excavation.

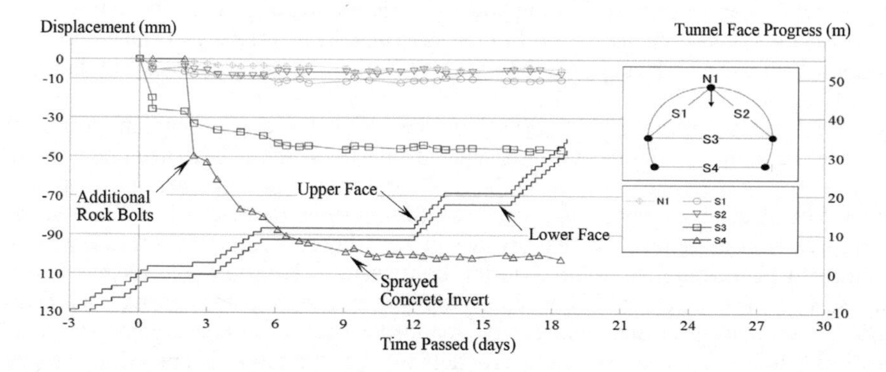

Figure 6. STA.205+65 measurement results.

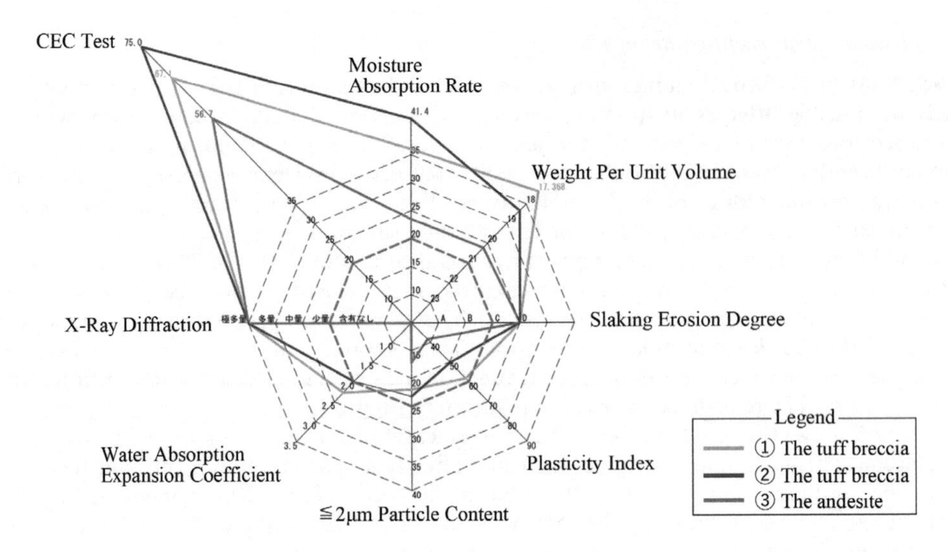

Figure 7. Octagonal diagram of laboratory rock test results.

Among methods for judging whether an invert should be installed in relatively hard CII bedrock, an index to determine swelling potential has been proposed (Ohshima et al. 2001). According to this proposal, the laboratory rock test was conducted for this tunnel to determine the swelling potential of the ground via the shape of an octagonal diagram showing the level of water impact and amount of clay minerals contained. Tests were conducted on two samples of tuff breccia (①, ②) and one sample of andesite.

The results of the tests are shown in Figure 7. The results of both X-ray diffraction and CE tests show very high levels of expansive clay minerals (montmorillonite: Ca-type) throughout the samples, and the slaking erosion degree also exceeds the standard values. In addition, the moisture absorption rate of the tuff breccia (①, ②) shows a value approximately equal to the standard, which is relatively higher than that of the andesite (③). Thus, it was judged that, for this section of the tunnel, the possibility that the tuff breccia would constitute swelling ground was high. While, it is believed that the andesite did not occur tunnel displacement owing to its high strength level of the rock fragments.

4 INVERT DISPLACEMENT MEASUREMENT

4.1 *Additional measurement process*

In some previous cases, on both highways and railways, roadbeds in public tunnels have bulged over long-term use owing to the effect of ground swelling, and remedial construction or invert repair has become necessary. In highway tunnels, invert repair or new installation is necessary, which has an adverse effect on road users owing to the need to restrict lanes or stop traffic (Miyazawa et al. 2016).

To prevent this problem, invert bulging must be assessed and remedial construction undertaken while the tunnel is still under construction. Conventionally, displacement convergence for the entire tunnel was estimated based on the roof subsidence and horizontal inner displacement measurements obtained as part of the usual A-measurements. Because it is difficult to measure invert displacement using methods that involve an everyday reference point, as heavy machinery and vehicles pass through the tunnel during construction. However, for tunnels in swelling ground, such as the one detailed in this report, even when displacement convergence is confirmed through A-measurements, displacement of the roadbed alone was observed to progress markedly. In most cases where roadbed displacement actualizes after the tunnel is opened to the public, it was estimated during construction that roadbed displacement might already be in progress.

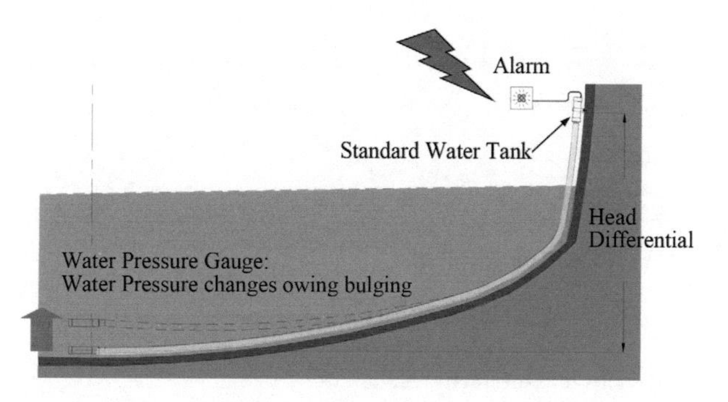

Figure 8. Outline diagram of invert displacement meter.

Owing to these factors, a system was developed to measure roadbed displacement during construction without posing a hindrance to the passage of heavy machinery and vehicles, and the new system was practically applied to this tunnel.

4.2 *Measuring system outline*

An outline of the newly developed invert displacement meter is shown in Figure 8. This meter is buried in the protection tubing around the water pipe connected to the water pressure gauge located under the roadbed during construction and measures the amount of bulging based on the head differential with the standard water tank installed above ground. This system allows for continuous, automatic measurement of vertical displacement of the ground under the invert. In addition, by measuring the height of standard water tank with a total station, it is possible to measure the level of absolute displacement.

4.3 *Meter installation process and measurement accuracy*

The invert area was excavated, and protective tubing was installed for later re-burying in earth or in sprayed concrete invert. After re-burying, the meter was inserted into the protective tubing, and electric lines were connected to the meter.

In advance, meter accuracy was verified through laboratory testing. The invert dis-placement meter was subject to an incrementally increasing level of displacement, and when the measurement values were compared with and verified against the measurements obtained from a high-precision CDP displacement meter, the margin of error of the invert displacement meter was less than 0.5 mm (0.5%) per 100 mm of deformation.

5 RESULTS OF MEASUREMENTS IN THIS TUNNEL

In this tunnel, invert displacement meters were installed at two locations (STA.205+55, STA.205+73) where large displacement occurred at the time of excavation (STA.205+55, STA.205+73). The results of the measurements are shown in Figure 9. Bulging exceeding 60 mm was recorded over a 50-day period, and cracks and other damage were observed when the status of the sprayed concrete invert was checked.

In addition, the system also included installation of a meter to measure the distribution of displacement, which attempted to measure vertical displacement every 0.5 m. The results, as shown in Figure 10, demonstrate that the greatest amount of displacement was observed in the center portion of the invert span. Thus, it was observed through this measurement that it would be appropriate to install a sensor in the center, where the displacement was greatest.

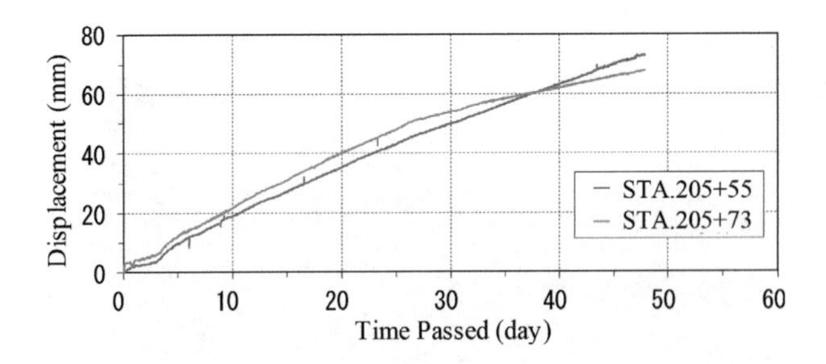

Figure 9. Invert displacement measurement results.

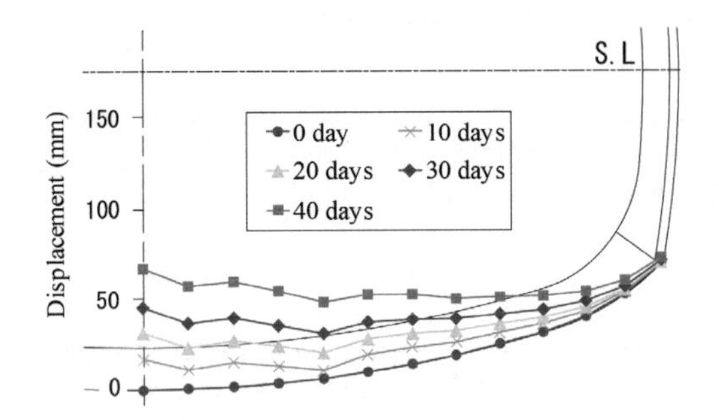

Figure 10. Invert area vertical displacement distribution.

6 REMEDIAL CONSTRUCTION FOR ROADBED BULGING

6.1 *Remedial construction selection*

Based on these measurement results, it was deemed that bulging was progressing in this section of the tunnel, and it was predicated that, if the tunnel was left as it was, remedial construction would be required after it was opened to the public. Therefore, remedial construction was undertaken.

To suppress the roadbed bulging, it was necessary to create a structure that could op-pose the load (swelling pressure) on the invert. First, past experience for correcting invert displacement were referred, and as shown in Figure 11, a design was chosen whereby the shape of the excavated cross-section would be made more circular, and after sealing off the area with high-spec steel supports (HH-100) and sprayed concrete invert (thickness 150 mm), the displacement would be suppressed by the weight of invert concrete with a maximum thickness of approximately 1.0 m.

6.2 *Remedial construction verification process flow*

Subsequently, an analysis of the appropriateness of the design was conducted through the process shown below.

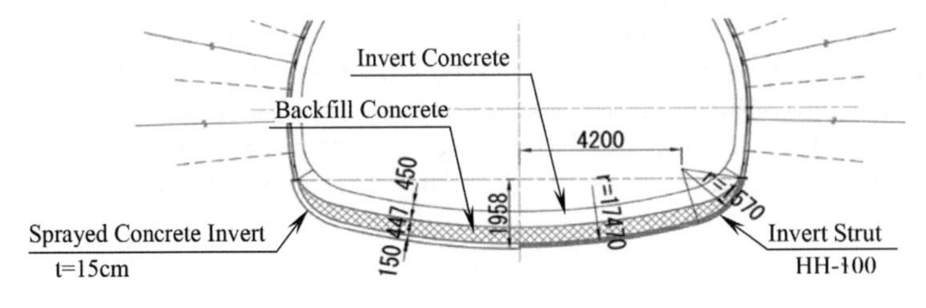

Figure 11. Diagram of remedial construction for roadbed puckering.

First, the residual displacement within the primary seal-off construction was calculated using the Voigt approximation formula. Second, the strength of future external force was predicted by frame structure analysis. Assuming that the external load would added uniformly on the arch area and invert, the load value was calculated inversely which occurred the residual displacement calculated above. Finally, the load was applied to the post-repair construction design to check the initiation stress of the components.

6.3 *Predicting the amount of residual displacement*

Based on the past measurement results, it was believed that the amount of invert deformation over time would converge to a certain value over an unlimited time period. These changes over time can be expressed through a Voigt model (Yamaguchi & Nishimatsu 1977), in which springs and dashpots are arranged in parallel. This model expresses time-dependent elastic deformation for a given amount of stress, as expressed by the following formula:

$$\sigma = E\varepsilon + \eta \frac{d\varepsilon}{dt} \tag{1}$$

where E represents the spring constant and η is the viscosity.

Time-dependent action under a given level of stress σ_0 is shown in the following formula:

$$\varepsilon = \frac{\sigma_0}{E}\left(1 - e^{-\frac{E}{\eta}t}\right) = \varepsilon_0(1 - e^{-at}) \tag{2}$$

where ε_0 shows strain at $t=\infty$. By using Equation 2, it is possible to predict the displacement of an invert in swelling ground over time. The forecast below was generated by substituting load–displacement in place of stress–strain.

The change in displacement speed over time based on the results of this round of measurements (Figure 9) is shown in Figure 12. The curves shown are matched as closely as possible to the actual data for the two locations.

Based on the measurements from cases where an invert was repaired after the tunnel was opened to the public, displacement speed ranges from a few millimeters per year to a few dozen millimeters per year, approximately 1/10 the speed of the displacement in this tunnel. Thus, it is believed that the amount and speed of displacement during construction were high and were therefore easier to measure.

When this displacement speed approximation formula is integrated to calculate a displacement–time relationship formula corresponding to Equation 2, a curve can be generated as shown by the dotted line in Figure 13. Based on these results, the highest value predicted for future residual invert displacement is shown to be 46 mm (Figure 14).

6.4 *Remedial construction verification results*

In the frame analysis, when multiple uniform loads were set to act on supports and an analysis was conducted, the relationship between acting load–displacement and acting load–strut

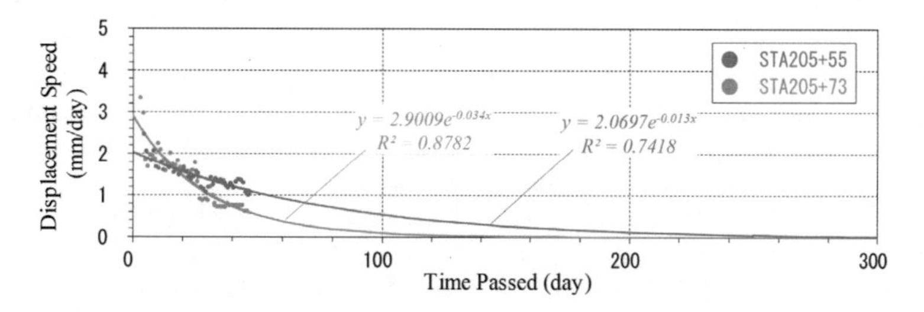

Figure 12. Change in displacement speed over time and best-fit curves.

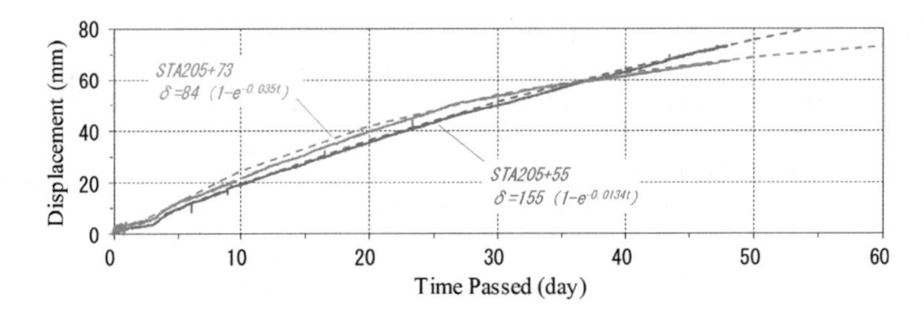

Figure 13. Change in displacement speed over time and best-fit curves.

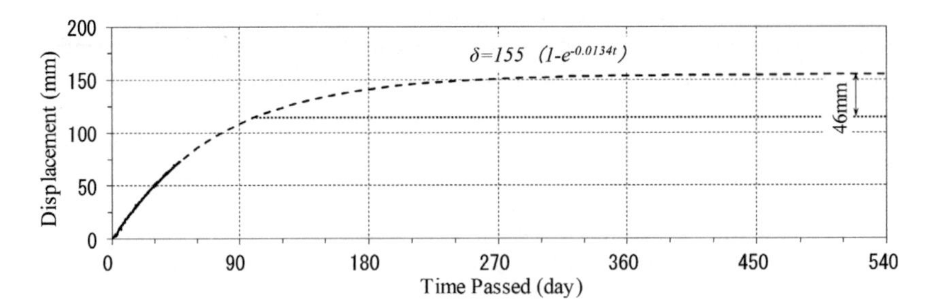

Figure 14. STA.205+55 residual displacement forecast results.

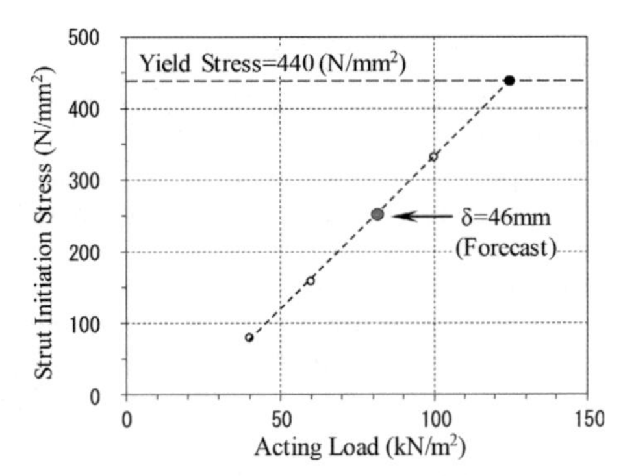

Figure 15. STA.205+55 strut initiation stress.

initiation stress was observed to be nearly linear (Figure 15). From these results, it was observed that the strut initiation stress in the event of the forecast maximum residual displacement of 46 mm was approximately 60% of the yield stress. In addition, the sprayed concrete initiation stress was approximately 20% of the design standard strength, thus verifying the appropriateness of this remedial construction.

7 MEASUREMENT RESULTS FOLLOWING ROADBED PUCKER REMEDIATION

Subsequently, remedial construction was undertaken according to the design in Figure 11. The spring water that appeared during the remedial construction work at the time of tunnel face excavation was drained, causing a drop in the underground water table, and hence, no spring water was detected in this round of construction.

After remedial construction on the roadbed bulging was completed, invert displacement was measured, along with invert strut stress. Figure 16 shows the measurement results for the significant bulging trend shown in Figure 9 alongside the measurement results following remedial construction on roadbed bulging. It can be observed that the remedial construction suppressed invert displacement to a significant degree. In addition, as shown in Figure 17,

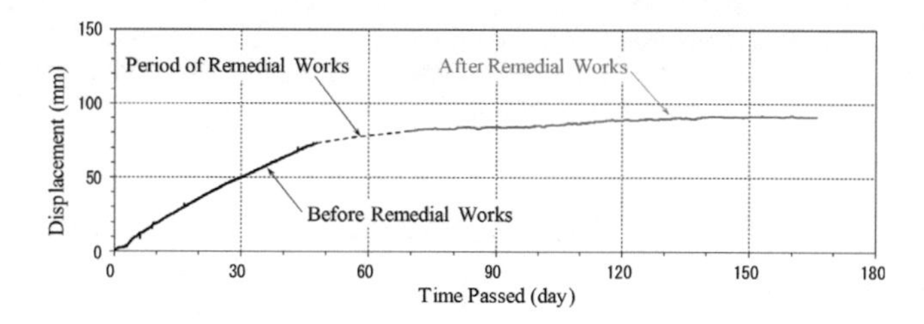

Figure 16. STA.205+55 invert displacement measurement results following roadbed pucker remediation.

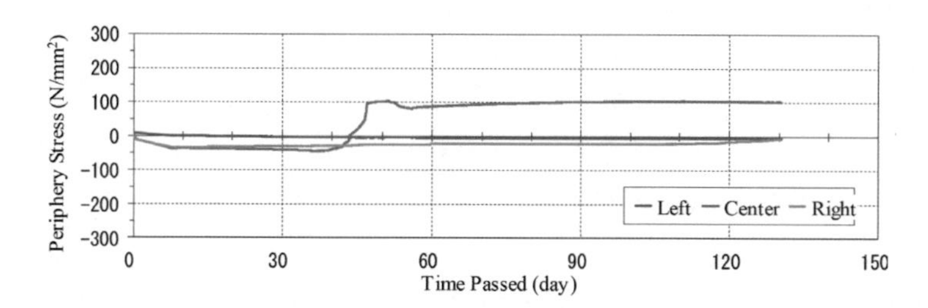

Figure 17. STA.205+55 strut initiation stress measurement results.

though a maximum bending tensile stress of approximately 100 N/mm2 is initiated on the center periphery invert struts, it is less than 25% of the yield stress, and this value is only approximately 38% of the predicted value shown in Figure 15, and hence, the remedial construction can be deemed sufficiently effective.

8 CONCLUSION

Significant roadbed bulging on swelling ground was measured during construction using the newly developed invert displacement meter. Although remedial steps were taken at this location at the time of tunnel excavation and inner displacement was brought to convergence, these measurements demonstrated that the roadbed would continue to bulge significantly over long-term use. Originally, ground displacement convergence was judged through A-measurements, but this case study fully demonstrates the importance of the in-vert displacement meter. In addition, it is believed to be very important for displacement to converge in order to prevent roadbed deformation after the public opening of the tunnel by measuring invert displacement at the points where large displacement occurs during tunnel excavation and undertaking remedial construction reflecting these results during construction.

Breakthrough on this tunnel was reached on April 26th, 2017. It is hoped that the knowledge gained from this study will help accelerate the opening of this tunnel to Tohoku Chuo Expressway traffic and be of use in the construction of future similar tunnels.

REFERENCES

Ohshima, K., Shiroma, H., Itoh, T., Muraji, E. & Kubota, T. 2001. Suggestions for Invert Installation Standards Based on a Cause Analysis of Deformation in Tunnels. *Proceedings of the Japan Symposium on Rock Mechanics* 11(1):329-334.

Miyazawa, K., Yasuda, K., Suyama, K. & Watanabe, A. 2016. Installation of New Invert Using the Whole-Surface Continuous Pressing Method under Day and Night Continuous Traffic Stoppage, *Tunnels and Underground* 47 (8):7-18.

Yamaguchi, U. & Nishimatsu, H. 1977. *An Introduction to Rock Mechanics*, Version 2, Tokyo: University of Tokyo Press.

Tunnels and Underground Cities: Engineering and Innovation meet Archaeology, Architecture and Art, Volume 3: Geological and geotechnical knowledge and requirements for project implementation – Peila, Viggiani & Celestino (Eds)
 ISBN 978-0-367-46583-4

Rockburst evaluation in complex geological environment in deep hydropower tunnels

A.M. Naji & H. Rehman
Department of Civil and Environmental Engineering, Hanyang University, South Korea

M.Z. Emad
University of Engineering and Technology, Lahore, Pakistan

S. Ahmed
Senior Engineer, National Development Consultants, Pakistan

J.-J. Kim
Korea Railroad Research Institute, 176 Cheoldobangmulgwan-ro, Uiwang-si, Gyeonggi-do, Republic of Korea

H. Yoo
Department of Civil and Environmental Engineering, Hanyang University, South Korea

ABSTRACT: Rockburst is the most dangerous phenomenon in deep underground excavation. It is inevitable when stresses are very high in hard rock mass. During deep hydropower tunnel excavation large faults are usually avoided in design but small scale joints, shear zones and structural planes are common which are very difficult to avoid. These structures sometimes become barriers to accumulate tangential stresses in underground which result in abnormal stress concentration which finally released in the form of huge rockburst energy. Additionally, the presence of geological structures like anticline and syncline can worsen the situation because the stresses are usually concentrated in synclines. Rockburst under these different conditions are really studied. In this paper two extreme rockburst events in recently completed Neelum-Jhelum and Jinping II hydropower projects, are discussed. The FLAC3D numerical simulation results show that stresses are concentrated near structure planes which result in displacement and itense rock-burst.

1 INTRODUCTION

Rockburst is most hazardous phenomenon resulting in severe form of disaster during deep excavation that not only causes damage to equipment and machinery but also cause damage to nearby structure due to its dynamic effect. In deep hard rock tunnels, the rockburst is influenced by geological structures. Many research work have done to investigate the influence geological structures on rockburst. Shepherd et al. (1981) have explained geological structures are the prime factors in rockburst occurrence and recumbent fold hinges are more rockburst prone regions where high stresses are concentrated. The recently completed Neelum-Jhelum hydropower project has witnessed this condition during construction. On the other hand, during deep excavations, geological structures are found nearby which causes unfavorable stress conditions. The stresses are usually concentrated near these structures. These structure planes might become barrier to adjust surrounding rock mass stress in deep tunnels lead to concentration of tangential stress between excavation boundary and structural plane along with huge amount of accumulated energy which finally result in rockburst. In Ontario hard rock mines rockburst conditions are prevalent when geological discontinuities are present near the opening (Hedley, 1992). Durrheim et al. (1998) investigated 21 rockburst cases in deep South African gold mines and found that the regional structures are the source mechanism for rockburst occurrence. In Creighton Mine, Canada, geological structures like shear zones experiences frequent microseismic and often macroseismic events (Snelling et al., 2013). More than 20 rockburst events associated with structural planes have been listed during the construction of Jinping-II headrace tunnels. Presence of faults and joints were the main reason behind the rockburst events of pilot tunnels of the Jinping II hydropower project in china (Jiang et al., 2010). Presence of shear zone was the main reason behind the most destructive rock burst event in Neelum-Jhelum hydropower project in Pakistan (Naji et al., 2018).

Rockburst occurrence is more destructive when there is presence of structural plane (Zhang et al., 2013). The small scale structure planes are very important in civil tunnels as compared to deep mines where structures (faults or discontinuities) are usually tens or hundred meters long. These small scale structures can be easily reactivated when they are present in the vicinity of tunnel and shear failure occurs. These planes have played a very important role during the construction of Jinping-II and Neelum-Jhelum hydropower projects by causing huge losses. The FLAC3D numerical simulation is carried out to investigate the impact of such structure planes during rock burst occurrence which is still unclear phenomenon during deep excavation. FLAC3D provides an interface element to model the shear zone type structures in a geologic medium. This element is a collection of triangular elements. It can be created at any location in the model. Interface element is characterized by Coulomb sliding. Different parameters like normal stiffness (k_n) and shear stiffness (k_s), friction angle (ϕ), cohesion (c), are assigned to interface element. FLAC (Group, 2012) explained the stiffness characteristics in an explicit way based on Barton and Choubey (Barton and Choubey, 1977) proposed relationships for k_n and k_s which have been used for analysis as discussed in FLAC manual:

$$k_n = \frac{EE_r}{s(E_r - E)} \tag{1}$$

where kn = joint normal stiffness; E = rock mass Young's modulus; Er = intact rock Young's modulus; s = joint spacing.

$$k_s = \frac{GG_r}{s(G_r - G)} \tag{2}$$

Where ks = joint shear stiffness; G = rock mass shear modulus; Gr = intact rock shear modulus

2 JINPING-II PROJECT DESCRIPTION AND OVERVIEW

The Jinping II Hydropower Station is located in Sichuan province, China, which has a power capacity of 4,800 MW. This project is built by using a 310-m natural drop along the 150-km-long river bend in Yalong River around Jinping Mountain, this station is designed to cut the river bend, as shown in Figure 1(a). The Jinping II Hydropower Station has the largest power capacity in the Yalong River Basin and the highest water head (Wu et al. 2010). The project has four headrace tunnels with 16.7-km length. Most of the tunnel section has overburden depth of 1700 m with the maximum overburden depth of 2,525 m. This project is considered to be one with the deepest hydraulic tunnels in the world. The headrace tunnels no.1 and no.3 are excavated using TBMs with the diameter of 12.4-m while the headrace tunnels no.2 and no.4 are excavated by using the

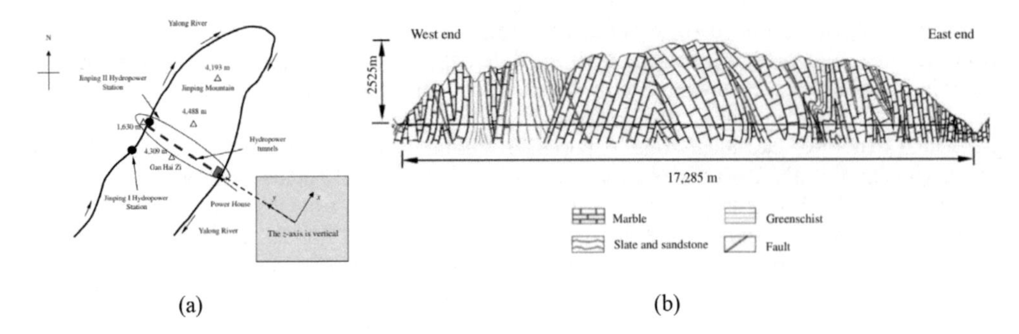

Figure 1. Jinping II hydropower project (a) project description (b) geological map.

Drill and blast method having a 13-m-diameter circular section. The center to center spacing among the four headrace tunnels is 60 m.

2.1 *Geological settings and in-situ stresses*

The project area is located in Qinghai-Tibet Plateau which is influenced by Eurasian and Indian plates collision. Figure 1(b) shows, the geological cross-section of the headrace tunnels. High strength and brittle marble is the main rock unit present in tunnels rocks while the west end of the tunnel crosses a small number of chlorite schist, green sandstone, and sandy slate. Due to high strength and brittle nature of marble, rockburst was inevitable in this area. Table 1 below shows the field measurements and back-analysis results of in situ stress conditions for the tunnel sections with an overburden depth of 1900 m and mechanical parameters of rock mass.

2.2 *Influence of structural planes on rockburst in Jinping II headrace tunnel*

During the construction of Jinping II headrace tunnels more than 400 rockburst events occurred. Rockburst influenced by geological structures are more dangerous and have intensity as compared to the one which are not. Here only rockburst caused by structural planes are discussed. These planes were exposed after the rockburst event in different directions in tunnel mostly in the walls, crown and on the working face. In case of rockburst, these planes also control the boundaries of damage pit. In this paper a structure plane approximately parallel to the tunnel axis is studied and discussed as shown in Figure 2. In this case deep damage pits are created with high intensity of rockburst (Feng, 2017). This event is numerically evaluated with help of FLAC3D simulation in subsequent section.

Table 1. The detailed information regarding in situ stresses and rock mass parameters of Jinping II.

1. In situ stress						
Depth	σ_x	σ_y	σ_z	σ_{xy}	σ_{yz}	σ_{xz}
1900	-48.54	-49.97	-51.46	-.35	-3.23	5.82
1. Mechanical parameters of rock mass						
Rock Type		Young's modulus (E)	Poisson's ratio (v)	Cohesive Strength(c)	Friction angle(ϕ)	Dilation angle(ψ)
Jinping II T2b marble		18.9	0.23	15.6	25.8	10

Figure 2. Structure planes parallel to tunnel axis.

2.3 *Numerical simulation of geological structure*

Rockburst is influenced by structural planes and its intensity of damage is more in their presence. A 3D numerical simulations via FLAC3D have been used to study the stability of long tunnels near geological structures. The interface elements have been used to model structure planes. In Jinping II hydropower tunnel, interface elements are parallel to tunnel axis same as in situ conditions. During numerical simulation, different parameters have been studied but here, the principal stress concentration and displacement are discussed. In Jinping-II tunnels, during initial conditions σ_1 and σ_3 have very high stress concentration near the structural plane area which is paralle to tunnel axis as shown in Figure 3(a) and (b). This explains that these planes act as barrier for stress concentartion and are the main cause of intense rockburst in deep Jinping II tunnels.on the other hand, during excavation stage there is also very high

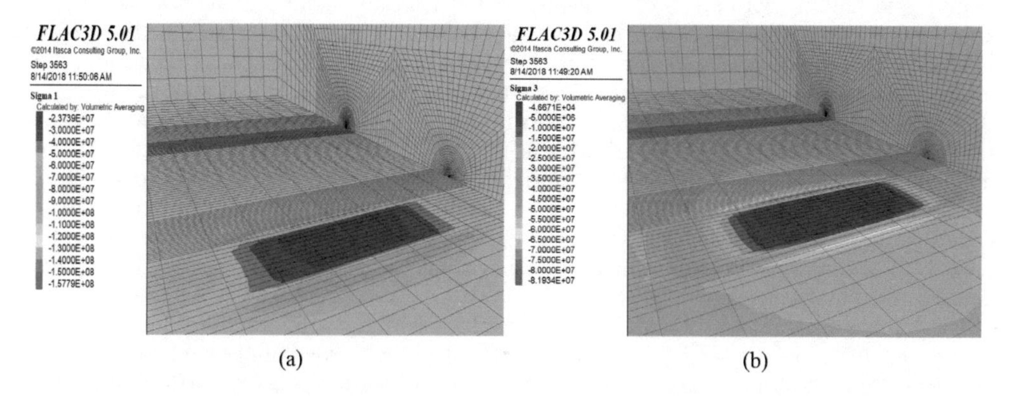

Figure 3. Principal stress (a) maximum (b) minimum.

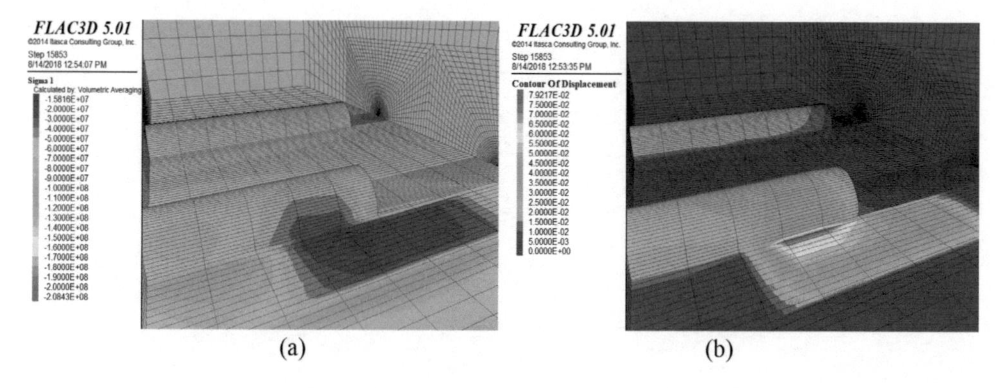

Figure 4. (a) maximum principal stress (b) displacement.

principal stress concenration near structure planes which result in displacement near this zone as shown in Figure 4 (a) and (b) respectively, while there is no such stress concentartion nearby tunnel.

3 NEELUM-JHELUM PROJECT DESCRIPTION AND OVERVIEW

Neelum-Jhelum Hydropower Project (NJHP) is located in Muzaffarabad district of Azad Jammu and Kashmir (AJK), in northeastern area of Pakistan. The project area is present in might mountains of Himalayas and its construction started in 2008. The 28.5 km long headrace tunnels are constructed to divert the water of Neelum River to Jhelum river for a total static gain of 420 m which finally drop on underground power house to produce the 969 MW energy. The headrace tunnels are excavated with drill & blast method and with TBMs (Tunnel Boring Machines) as a single and twin tunnels respectively. TBM excavated twin tunnels has initially center to center distance of 33 m before the major rockburst event in project history. This distance was increase to 66 m to avoid further inter-tunnel pillar bursting. The overburden was up to 2000 m which has potential of rock bursting in hard rock.

3.1 *Geological settings and in-situ stresses*

The area is highly folded and faulted due to tectonics. Main Boundary Thrust (MBT) which is major fault in the region is passing nearby the project area. Several thrust faults in the region are caused by northward movement of the Indian plate in Himalayan orogenic system and Muzaffarabad fault is one those faults which is passing through the headrace tunnel. Adverse folding and faulting with series of anticlines and synclines structures along with local faults and shear zones are present along the tunnel as shown in Figure 5. Normally bedding planes have favorable condition for construction, normally perpendicular to tunnel direction. High horizontal stresses is witnessed in previous studies in the area due active Himalaya (Wang and Bao, 2014). In-situ stress measurements performed by over-coring in sandstone beds have also reflected high horizontal stresses in TBM tunnels and the value of K_0 was up to 2.9, where the major principal stress was oriented sub-horizontally and nearly perpendicular to the tunnel azimuth. This high stress concentration and severe tectonics resulted in abnormal geological setting of the area as shown in Figure 6(a). Therefore, hard sedimentary rock under high stresses due to tectonics along stress concentrating synclinal structure, make the condition favorable for rockburst to happen.

The whole project is excavated in Murree formation with the stratigraphic sequence of alternative beds of sedimentary origin which comprise of different rock units; sandstone, siltstone and mudstone. Sandstone has uniaxial compressive strength (UCS) 86 MPa which is the

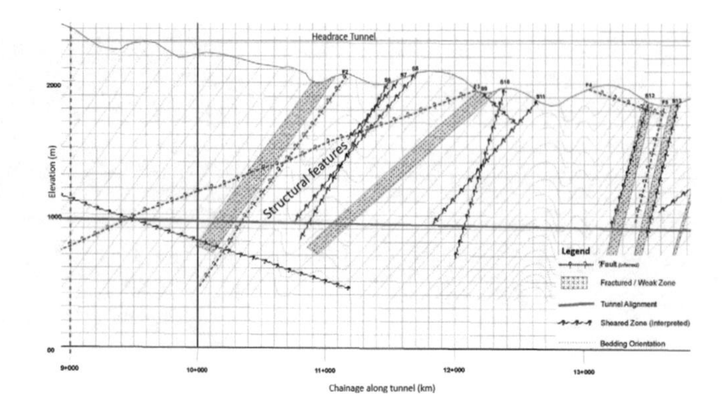

Figure 5. Geological Map of area.

Table 2. The detailed information regarding in situ stresses and rock mass parameters of NJHEP.

1. In situ stress						
Depth	σ_x	σ_y	σ_z	σ_{xy}	σ_{yz}	σ_{xz}
1900	-60.1	-35.8	-37.2	-1.3	-2.1	-5.7
1. Mechanical parameters of rock mass						
Rock Type		Young's modulus (E)	Poisson's ratio (ν)	Cohesive Strength(c)	Friction angle(ϕ)	Dilation angle(ψ)
Sandstone		20	0.25	4.2	42	0

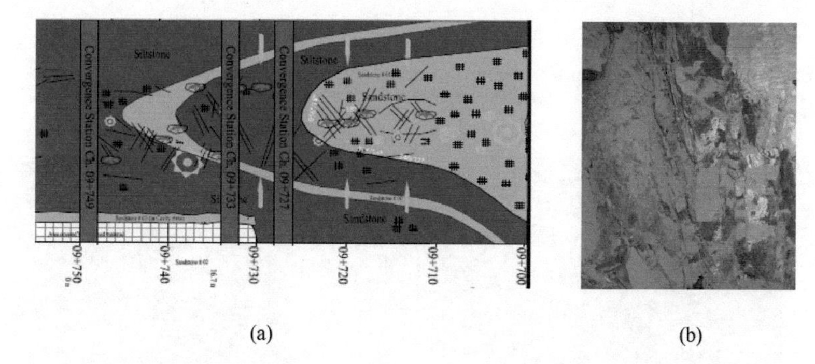

Figure 6. (a) geological section of abnormal folding in extreme rockburst area (b) shear zone in tunnel wall.

strongest rock unit and is thickly bedded and at places massive and blocky that why most of the rock burst events occurred in this rock. Siltstone is second rock unit having UCS of 66MPa which is medium strong unit of Murree formation. Mudstone has UCS of 42MPa and is the weakest rock unit which has very less potential for rockburst because of its very low strength properties. Engineering properties of these rocks show that the sandstone has the highest strength, as shown in Table 2.

3.2 *Influence of structural planes on rockburst in Neelum-Jhelum headrace tunnel*

During the construction of Neelum-Jhelum headrace, a total 879 rockburst events were recorded (Jack Mierzejewski, 2017). These rockburst were caused by both, the brittle failure of hard sedimentary sandstone under high horizontal stress and slip along structure plane. Here we only discussed the intense rockburst event of May 31, 2015. A shear plane was exposed in the wall of headrace tunnel after this intense rockburst, one of such structural plane was the main reason behind the accumulation abnormal stress nearby as shown in Figure 4(b). This stress concentration is numerically evaluated with FLAC3D numerical simulation in subsequent section.

3.3 *Numerical simulation of geological structure*

As said earlier rockburst is influenced by structural planes, a FLAC3D numerical simulations have been done for Neelum-Jhelum hydropower tunnels to study the stability near geological structures. In Neelum-Jhelum hydropower tunnels, interface elements are perpendicular to tunnel axis same as in situ conditions. During numerical simulation, different parameters have been studied but here, the principal stress concentration and displacement are discussed. In Neelum-Jhelum tunnels, during initial conditions σ_1 and σ_3 have very high stress concentration near the structural plane area which is perpendicular to tunnel axis as shown in Figure 7 (a) and (b). This explains that these planes act as barrier for stress concentartion and are the

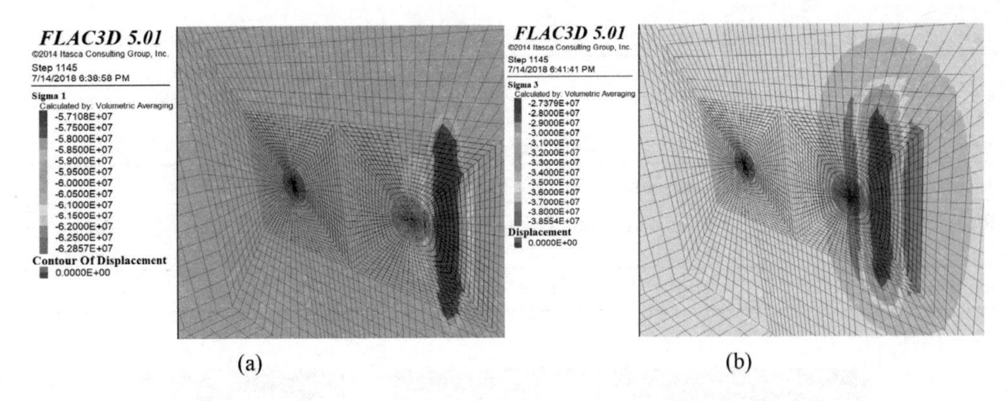

Figure 7. Principal stress (a) maximum (b) minimum.

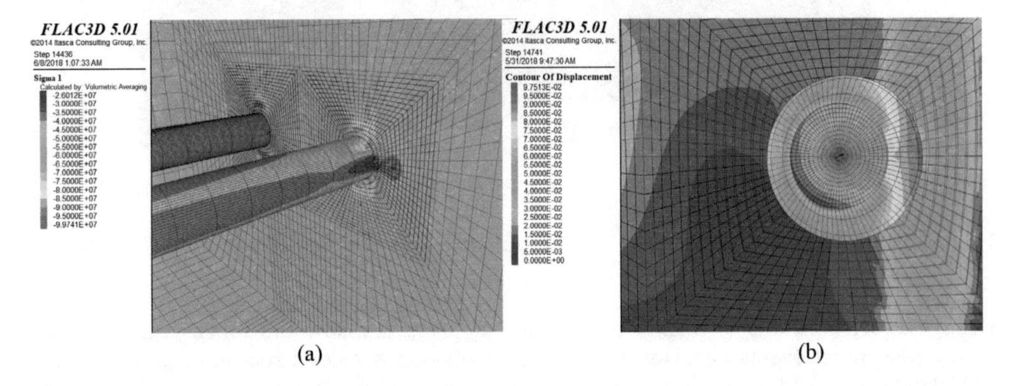

Figure 8. (a) maximum principal stress (b) displacement.

main cause of intense rockburst in deep same as Jinping II tunnels.On the other hand, during excavation stage there is very high stress concenration near structure planes which result in displacement near this zone as shown in Figure 8 (a) and (b) respectively, while there is no such stress concentartion nearby tunnel.

4 CONCLUSIONS

This paper has discussed unusual conditions under which most intense rockburst events occurred in Jinping II hydropower project, China and Neelum-Jhelum hydropower project, Pakistan. The damage cause by the presence of geological structural planes in the deepest hydroelectric tunnels in the world have been discussed here. These structures act as a barrier for normal distribution of in situ stresses which result in abnormal stress concentration nearby these structures. Additionally, synclinal, anticlinal and other fold structures also make condition worse in already stressed environment. The details of geological conditions for both projects have been discussed in this paper. The FLAC3D numerical simulations have been used to study the hazardous phenomenon of rockburst. These numerical simulations have shown that that stresses are concentrated nearby these geological structures which can be present parallel or perpendicular to axis of the tunnels. Such type of structures is generally over looked during tunnel construction which has proved to very dangerous. Therefore, it is recommended a focused geological structure mapping should be done during deep tunnel construction so that proper resistant support can be installed to minimize the risk of rockburst.

ACKNOWLEDGEMENT

This research was supported by 'Development of Design and Construction Technology for Double Deck Tunnel in Great Depth Underground Space(18SCIP-B089413-05)' from Construction Technology Research Program funded by Ministry of Land, Infrastructure and Transport of Korean government. The authors Abdul Muntaqim Naji and Hafeezur Rehman are extremely thankful to the Higher Education Commission (HEC) of Pakistan for HRDI-UESTPs scholarship.

REFERENCES

Barton, N. & Choubey, V. 1977. The shear strength of rock joints in theory and practice. *Rock mechanics*, 10, 1-54.

Durrheim, R., Roberts, M., Haile, A., Hagan, T., Jager, A., Handley, M., Spottiswoode, S. & ORTLEPP, W. 1998. Factors influencing the severity of rockburst damage in South African gold mines. *Journal-South African Institute of Mining And Metallurgy*, 98, 53-58.

Feng, X.-T. 2017. Rockburst: Mechanisms, Monitoring, Warning, and Mitigation, Butterworth-Heinemann.

Group, I. C. 2012. FLAC3D 5.0 manual. Itasca Consulting Group Minneapolis.

Hedley, D. G. 1992. Rockburst handbook for Ontario hardrock mines, Canmet.

Jack Mierzejewski, G. P., Bruce Ashcroft 2017. Short-Term Rockburst Prediction in TBM Tunnels. *Worl Tunnel Congress*. Bergen,Norway.

Jiang, Q., Feng, X.-T., Xiang, T.-B. & Su, G.-S. 2010. Rockburst characteristics and numerical simulation based on a new energy index: a case study of a tunnel at 2,500 m depth. *Bulletin of engineering geology and the environment*, 69, 381-388.

Naji, A., Rehman, H., Emad, M. & Yoo, H. 2018. Impact of Shear Zone on Rockburst in the Deep Neelum-Jehlum Hydropower Tunnel: A Numerical Modeling Approach. *Energies*, 11, 1935.

Shepherd, J., Rixon, L. & Griffiths, L. Outbursts and geological structures in coal mines: a review. International Journal of Rock Mechanics and Mining Sciences & Geomechanics Abstracts, 1981. Elsevier, 267-283.

Snelling, P. E., Godin, L. & Mckinnon, S. D. 2013. The role of geologic structure and stress in triggering remote seismicity in Creighton Mine, Sudbury, Canada. *International Journal of Rock Mechanics and Mining Sciences*, 58, 166-179.

Zhang, C., Feng, X.-T., Zhou, H., Qiu, S. & Wu, W. 2013. Rockmass damage development following two extremely intense rockbursts in deep tunnels at Jinping II hydropower station, southwestern China. *Bulletin of engineering geology and the environment*, 72, 237-247.

Tunnels and Underground Cities: Engineering and Innovation meet Archaeology, Architecture and Art, Volume 3: Geological and geotechnical knowledge and requirements for project implementation – Peila, Viggiani & Celestino (Eds)
© 2020 Taylor & Francis Group, London, ISBN 978-0-367-46583-4

Adapting tunnel construction to hydrogeological conditions in a karst region

M. Neukomm & M. Mercier
BG Consulting Engineers, Lausanne, Switzerland

P.-Y. Jeannin, A. Malard, D. Rickerl & E. Weber
Swiss Institute for Speleology and Karst Studies, La Chaux-de-Fonds, Switzerland

ABSTRACT: Hydrogeological risk assessment is an issue of significant importance for tunneling projects in karst regions such as Ligerz safety gallery located in the Bernese Jura, Switzerland. In such conditions, major changes in the project plan might be required either in advance or during the construction. Hydrogeological pre-studies revealed a significant probability of karst occurrences with discharge rates reaching up to 10 m^3/s. Risk mitigation measures were necessary to define security procedures to avoid large water inrush. The KarstALEA method was applied to identify sections with a high probability of karst occurrence and to characterize the associated problems. Finally, the combination of a hydraulic model of the karst flows with real-time measurements, make possible to estimate the expected inrush discharge and to define the alert threshold to temporary stop the excavation during the most critical situations.

1 INTRODUCTION

Communities of Twann and Ligerz (Bernese Jura, CH) are located along Lake Biel at the foot of the Jura Mountains (Figure 1), and are crossed by heavy traffic load. A 5 km long highway tunnel was planned in the 1980s. However, due to geological and hydrological difficulties – partly due to the karst hydraulics –, as well as to land-use conflicts, the tunnel had to be shortened to 2.5 km.

Owing to new safety standards, the Federal Roads Office (FEDRO) required to add a parallel safety gallery (SiSto, Figure 1) connected to the existing highway tunnel (Ligerztunnel) by 9 cross passages. In the future, the existing tunnel is assumed to be extended to get around Twann village (Tw. Tunnel on Figure 1), as it was initially planned.

During the drilling of the highway tunnel in the late 1990s, no major karst features were encountered, only inactive karst voids or filled with clayey materials. Moreover, water inrushes didn't exceed 200 l/s.

In addition, nearby communities are exclusively supplied by the Brunnmühle spring (Br. on Figure 1), which emerges from the karst aquifer, directly downstream of the existing tunnel and the planned galleries. Then, a more detailed assessment of the karst aquifer was mandatory prior to the construction of the safety gallery, especially about potential issues on the water quality. This assessment had therefore two main issues: (1) potential impact of the safety gallery on the tapped spring; (2) potential impact of the karst hydraulics on the safety gallery and tunnel construction. Mitigation measures had to be planned for both aspects, parallel to the TBM excavation of the safety gallery which started in mid-2014.

The KarstALEA method (Filipponi et al. 2012) has been applied on the site as it reveals the most appropriate method for assessing karst-related hazards (voids location, expected size and filling) along the tunnel trace and especially for locating zones where the conduits upstream of the permanent and overflow springs are supposed to develop. Besides

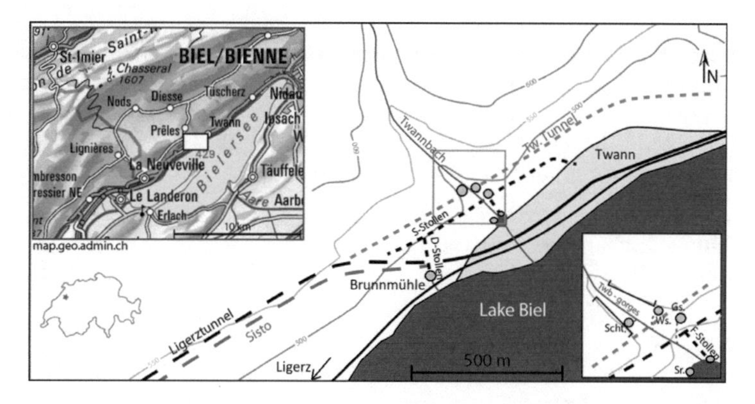

Figure 1. Overview of the site; *Br:* Brunnmühle spring, *Ws:* Wasserhooliloch cave, *Gs:* Gischeren cave, *Scht:* Schuttstein, *Sr:* Sauser spring.

KarstALEA, a hydraulic model was established, enclosing all natural and artificial aquifer outlets (permanent springs, drainage devices, overflow springs, wells, etc.). Geometry of the modelled conduit network was inferred from field and cave surveys, hydrological monitoring and local dye tracing tests, as well as from literature information. Based on this model, various scenarios of conduits intersection (diameter, hydrological stage, etc.) have been tested. These provide the range of the expected discharge rates in the gallery during the construction, which is crucial for dimensioning outflows devices and for initiating precaution measures.

2 CONTEXT AND PROBLEM

The drilling site is located along an anticline foothill in the Eastern part of the Jura Mountain (canton of Bern) along Lake Biel (see Figure 1 and Figure 2). Geologically, this area is composed by South-East dipping pile of Jurassic and Cretaceous limestone, underlain by Oxfordian Marls (aquiclude). The safety gallery develops approximately along the strike of bedding planes. The Brunnmühle karst spring emerges at the base of the massif, at the contact between the Portlandian limestone (top of the karst aquifer) and the Purbeckian marls, close to the elevation of Lake Biel. It is mainly fed by the Malm aquifer (upper Jurassic limestone, from

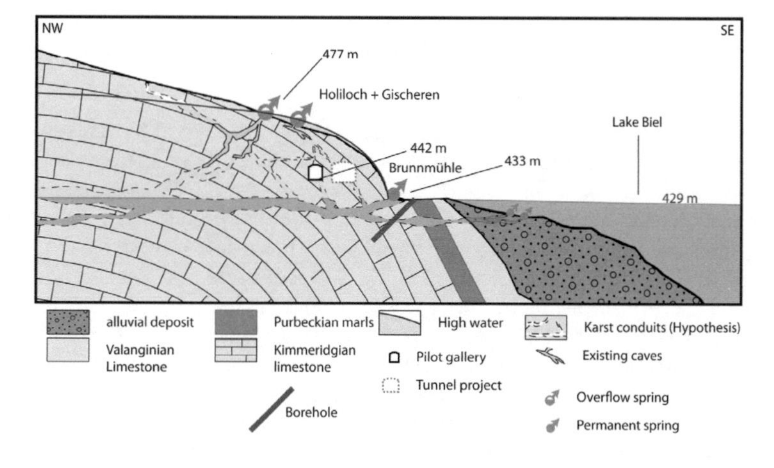

Figure 2. Overview of the geological organization (projected cross section perpendicular to safety gallery).

Sequanian to late Portlandian, ~600 m of thickness) which is dammed by the Purbeckian marls (early Cretaceous). The projected cross of Figure 2, which is perpendicular to the existing tunnel and safety gallery, gives a representative overview of the geological context. Several other springs emerge in the Twannbach gorges about 300 m to the north of the Brunnmühle. Their position is given on Figure 1, in the box down-right.

- The Sauser spring is considered as a permanent spring; it emerges at the foot of the gorges and its high-flow discharge rate is expected to reach 300-500 L/s
- Further upstream, the Wasserhooliloch, Gischeren and Schuttstein caves act as overflow springs.

While the Wasserhooliloch is active several times per year, Gischeren and Schuttstein do overflow less frequently. As their entrances are located 44 m above the elevation of the Brunnmühle spring, it suggests that the hydraulic head considerably rises for high-flow conditions. These caves are known along a few dozens of meters. Wasserhooliloch is the main cave and it is 33 m deep. All these outlets are expected to belong to a regional karst system, which extends up to the top of Chasseral mounts. The catchment area of the system extends over more than 30 square kilometers as previously assessed in the frame of the Swisskarst project (http://www.swisskarst.ch).

A part of the karst groundwater was drained by the exploration tunnel ("S-Stollen", Figure 1) drilled in the late 1980s, and was diverted from natural springs. In the eastern section of the exploration tunnel (S-Stollen), a part of the overflowing groundwater is diverted into the Twannbach gorges via F-Stollen, while in the western part groundwater is drained outside through D-Stollen. This multitude of outlets and drainage devices induces a high degree of complexity in the hydraulic scheme (Bollinger and Kellerhals 2007). Furthermore, no relevant measurements of discharge rates did exist at the beginning of the project for all of these outlets, excepting for those of the Brunnmühle spring and the D-Stollen since 2012. The monitoring was therefore expanded to all springs since 2015.

As indicated in Figure 1 and Figure 2, the trace of the Sisto develops between the Brunnmühle permanent spring and the group of springs emerging in the Twannbach gorges (Twannbach on Figure 1). It seems then obvious that active conduits are present in this domain and could be intersected by the gallery. The key-question is then related to the expected position of the conduits upstream of the Brunnmühle spring and the provenance of the groundwater from the upper part of the system.

Does this groundwater flow along the Twannbach gorges axis or may it come from another part of the aquifer? What are the consequences in terms of management of the water inrush risks – especially during drilling of the safety gallery and his 9 cross passages – and on what scale?

3 METHODS

3.1 *KarstALEA*

KarstALEA is a method developed for the prediction of karst-related hazards in underground works. It is published (Filipponi et al. 2012) as a practical guide of the Swiss Federal Roads Office (FEDRO).

It first bases on the observed fact that karst conduits preferentially develop along a restricted number of geological horizons, known as "inception horizons" (Filipponi 2009). Typically, 70% of the conduits may develop along 3 to 5 discrete horizons within the scale of a tunnel project. A second input factor for the assessment is that conduit density varies within the karst massif according to past and present hydrogeological conditions; there are commonly some concentrations of conduits at elevations corresponding to past base-levels (paleo-valley floors). Consequently, the identification of inception horizons and base-levels makes it possible to predict zones of highest probability of occurrences of karst conduit.

A 3D model of the geology, of the main aspects of the hydrogeology, as well as of the inception horizons and of past base-levels is required to elaborate the apply KarstALEA.

3.2 *Field Campaigns*

Field data regarding geology, hydrogeology, inception features and past base-levels were gathered. Geological data were mainly provided by existing maps and profiles, by field geological mapping and by observations from boreholes.

Hydrogeological data are mainly observed at springs (elevation, times series of discharge-rates, electrical conductivity and temperature), in boreholes (times series of water level), in caves (times series of water levels), and in the exploration gallery (S-Stollen). A series of tracer-tests were also carried out in order to validate some connections.

Inception features were looked for in outcrops (e.g. Twannbach gorges), in caves and in borehole (borescans, hydraulic tests). A fine analysis of digital elevation models of land-surface was also useful to assess fractures.

Past base-levels were assessed mainly using existing literature on quaternary deposits and landscape evolution, as well as on new observations in caves (passage morphology, cave deposits).

3.3 *Karst hydrology*

The KARSYS approach (Jeannin *et al.* 2013) was previously applied in this region within the framework of a project for the regional water authority (Malard *et al.* 2012). This approach is based on the 3D modelling of the geometry of the karst aquifers from which, by applying some general rules about the development of karst hydrogeological systems, main groundwater zones, flow-paths and catchment areas feeding a spring or a group of springs can be assessed.

The existing geological model was refined for the purpose of the tunnel investigation and the catchment area was delineated with a reasonable degree of reliability.

One complexity of this system is the presence of several springs. All springs were instrumented and the relationships between meteorological parameters and the total discharge rate of the system was analyzed. The detail of the behavior of each spring was investigated using a hydraulic model (see next paragraph).

3.4 *Hydraulic model*

Karst systems are characterized by caves, which are natural conduits draining water from the catchment area towards a spring (or a group of springs). Cave systems can be compared to pipe networks, with some of the pipes being saturated (pressurized flow), some being partially saturated (free-surface flow), some being flooded only during high-flow conditions,

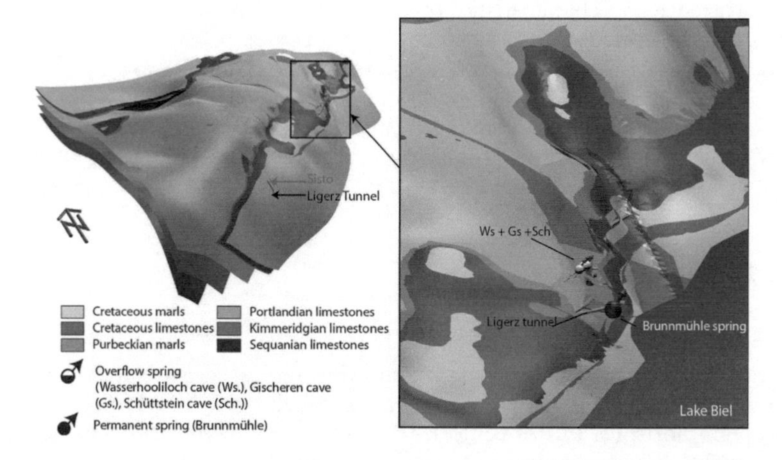

Figure 3. Perspective view of the 3D geological model; a large part of the Sisto develops within the Portlandian limestone, overlaid by the Purbeckian marls.

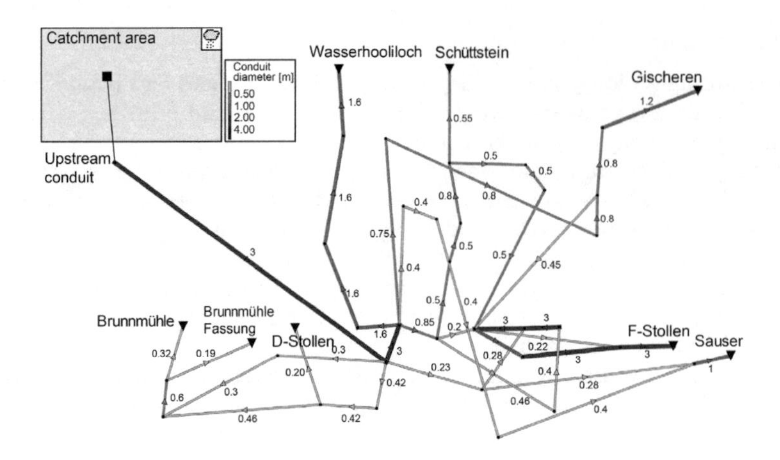

Figure 4. Schematic profile of the hydraulic model of the conduits feeding the respective springs of the Brunnmühle system. Colors and numbers indicate the diameter of the conduits (in m).

and some being never reached by water anymore (they were in the past, when the base-level was higher).

Pipe-flow modelling software can be applied to compute flow in cave networks (Jeannin 2001, Campbell and Sullivan 2002, Covington et al. 2009). In this type of approach, the Manning-Strickler formula is assumed for computing head-losses and 1D Saint-Venant equations are used for continuity and momentum. The SWMM package (Rossman 2004) was used for this purpose, and data of heads and discharge rates of all accessible conduits and springs were monitored in order to attempt the modelling of the conduit network.

A roughness of 0.05 $s \cdot m^{-1/3}$ was assumed for all conduits based on measurements in well-known caves (Jeannin 2001). The lengths of the respective pipes were assessed based on real distances. Heads and discharge rates were measured in all branches considered in the model domain.

The aim of the modelling was to understand the hydraulics of flow in the vicinity of all springs, through which the tunnel is expected to be dug. The whole range of conditions (from low-flow to very high-flow conditions) has to be covered, indicating heads and discharge rates in any conduit of the system.

The challenge is that most of the existing conduits are not known. The model is thus used to guess the existence and characteristics (mainly diameter) of all existing conduits, reproducing the data measured at all springs as well as heads in caves.

The model was thus built from the known points (springs and caves), by always assuming the simplest possible geometry. Building was started with low-flow conditions, when only the lowest conduits are active. The activation of the respective overflow springs and conduits for increasingly higher flow conditions produces very specific signature on the hydraulics, which makes possible to build up the model step-by-step.

4 RESULTS

4.1 *KarstALEA profile*

A profile was obtained from the 3D model, showing several karst-related characteristics all along the tunnel (Figure 5). The tunnel is almost parallel to the limestone beds, i.e. to stratigraphic inception horizons. In this context, the positioning is crucial because a slight change makes the tunnel to be along an inception horizon or besides.

Sixteen perpendicular profiles to the tunnel were also provided, and the 3D model could be used for producing profiles anywhere along the tunnel. According to the geological prognosis,

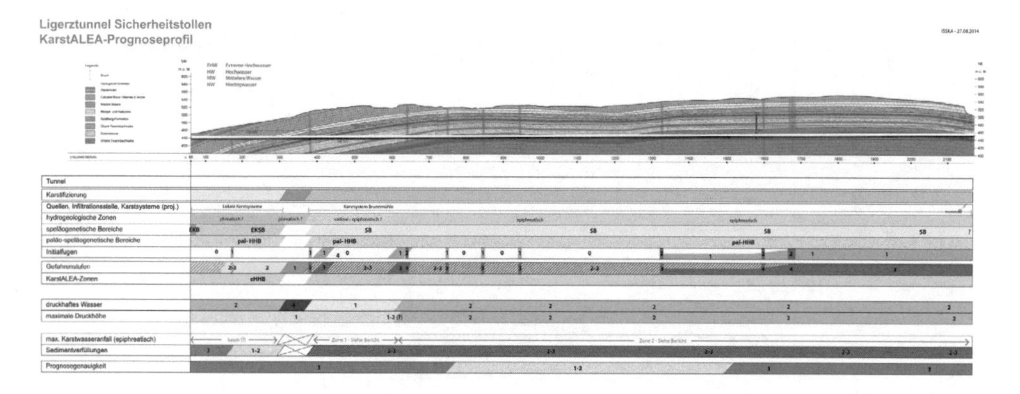

Figure 5. Profile along the tunnel showing the main characteristics related to karst (probability of occurrence, type of conduit, heads, potential rates, etc.).

the Eastern part (on the right) of the tunnel is expected to follow one inception horizon, i.e. has a high probability to cross natural conduits, potentially bringing a lot of water.

4.2 *Assessment of heads and discharge rates: observations and modelling*

The hydraulic model obtained after a careful calibration process is sketched in Figure 4. Three natural caves (Wasserhooliloch, Schüttstein and Gischeren) are overflow springs activated only during high-water conditions. F-Stollen and D-Stollen are artificial exploration tunnels, crossing natural caves and draining water. Sauser and Brunnmühle are the low water outlets of the system. Brunnmühle Fassung is a pumping well. Vertical scale is approximately 60 meters, and horizontal about 300 m. This network is the simplest network we could construct to explain head-discharge relationships we measured.

For any recharge feeding the "upstream conduit" the model provides heads and flow-rates in all the conduits and springs. The exact position of the inferred conduits is not known, meaning that the tunnel may cross any of them. The trickiest crossing is the upstream conduit, where the flow-rate and the head are the highest. Assuming that the tunnel may cross this conduit directly or through a side passage, several scenarios were calculated for assessing potential heads and flow-rates (Figure 6).

Flow-rates as high as 10 m^3/s could potentially flood the gallery (which is located at ~440 m a.s.l.). However, there are clearly conditions for which the tunnel can be dug safely, when head is lower than ~475 m a.s.l., and other conditions for which the crossing of a conduit larger than 0.5 m will produce considerable flow-rates (> 1 m^3/s).

Critical water level determination was defined by finding a compromise between (1) the potential number of days of excavation work interruption after exceeding the alert level and (2) the corresponding karst discharge rate.

Expected karst water inrushes in the gallery must be reliably evacuated to a receiving outlet - here Lake Biel - without causing damage on its way. In fact, between the portal of the tunnel and the lake are the national road No 5 and the Geneva-Basel railway line which is a major axis of the Swiss rail network. It was therefore inconceivable to put at risk these transport infrastructures. Furthermore, at a certain water level, karst flows at the surface through the overflow springs located in the Twannbach gorges (Figure 7) and the discharge then considerably increases, as shown in Figure 6 above. The alert level, leading to the interruption of all excavation works, was set at 476.50 m a.s.l., corresponding to a potential water inrush of max. 1500 l/s.

In addition to the problem of karst water inrush, it was also necessary to control the risk of pollution of the Brunnmühle spring by the exfiltration of dirty water to the karst conduit network during low water periods. When the water level in the karst aquifer is lower than the level of the different excavation faces, the construction works also had to be interrupted.

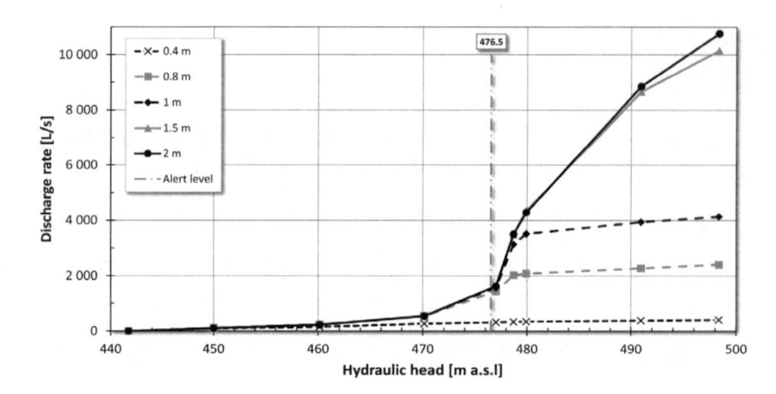

Figure 6. Expected discharge rate spilling in the safety gallery in case of crossing a karst conduit of a certain diameter (0.4 m to 2 m) within 100 m away from the main up-stream conduit of the system. The expected discharge rate is function of the measured hydraulic head in the aquifer.

Figure 7. Left side: Twannbach River and the Wasserhooliloch overflow spring/ Right side: view of the Wasserhooliloch overflowing.

4.3 *Prevention measures*

Considering that nearly 75% of the time conditions are acceptable for working safely, the tunnel could be constructed, as long as a careful management of the conditions was established. Therefore a hydrological forecast model of the flow system was designed in order predict conditions. The model is based on the real-time collection of meteorological data and forecasts and spring discharge measurements. From the real-time discharge rate, the model makes a prediction for the next five days using the hourly meteorological forecasts. Previsions of the hydrological model are updated every hour according to the update of the meteorological forecasts and the measurements on site. Alarms are sent by SMS to the head of the work-place as soon as critical conditions are identified. Conditions could thus be observed closer and evacuation of the tunnel could be timely decided and followed in real-time. The duration of the events were also assessed.

4.4 *Consequences for the safety gallery*

The continuation of the construction was then conditioned by the management of the human, material and financial risks. As the security gallery is linked by pedestrian connections every 400 m to the existing tunnel, its alignment had been planned parallel to the road tunnel. Its longitudinal profile thus had an ascending section over 1600 m, then descent over 400 m, with a slope of more than –3%. The interception of a karst conduit in the descending section would have resulted in flooding the tunnel boring machine within minutes. If the material risk could be acceptable, the human risk was not.

Different project scenarios were developed and compared according to a multi-criteria and risk analysis. Scenarios that do not fully control the human risk have been eliminated. The only

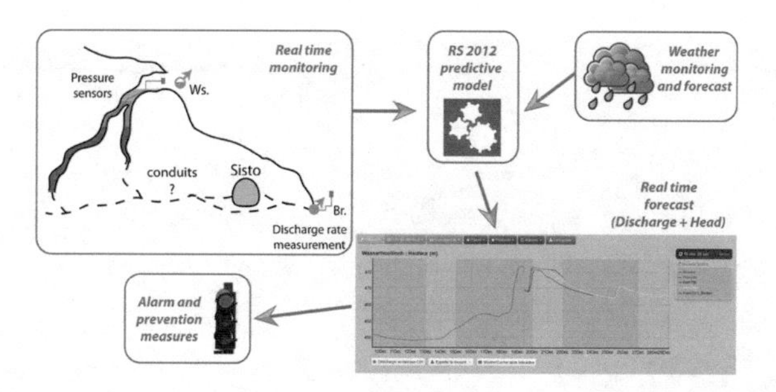

Figure 8. Sketch of the alarm device; using weather forecast and real-time measurements on site (discharge rates, hydraulic heads, etc.), a prediction of the system discharge and corresponding head for the next 5 days is updated hourly.

Figure 9. Works in progress at the top of shafts at bypass No. 8 and No. 9.

option in order to control the risks, while minimizing the impact on traffic in the road tunnel, was to excavate the entire gallery only on the way up. In the new project, the longitudinal profile of the security gallery is therefore different from that of the road tunnel. The last three pedestrian links present then a vertical drop of 5 to 14 m through 3 shafts to be excavated in the descent section from the bottom of the 2 km security gallery. This method of realization is technically not complicated, but poses many problems in terms of construction phasing logistics.

Safety procedures have been developed, including the interruption of all excavation works when the measured or predicted water level in the karst network exceeds 476.5 m.a.s.l. which corresponds to a potential water inrush of about 1500 l/s (see Figure 6 above).

In parallel with the adaptation of the project, several accompanying measures had to be put into place:

- Establishment of a water outlet between the gallery portal and the lake, while crossing a national road and a major national railway line. In order to evacuate a discharge of nearly 1500 l/s, it was necessary to excavate a gallery with micro-TBM Ø 0.8 m under the national road and the railway line that were to be kept into operation.
- Modification of the excavation method of the last pedestrian cross links, including the shafts. The choice was to use the traditional excavation method: drilling and blasting. It is the only method that keeps workers safe during excavation and thus potential water inrush. The proximity of the blasting to the existing tunnel required the establishment of a follow-up and monitoring of the shockwaves during blasting.

Figure 10. Safety gallery (SiSto) at the end of underground works.

- The flow-rate of 1500 l/s in the gallery under construction represents a water height of more than 60 cm. To allow the safe evacuation of personnel in the event of water intrusion, a raised escape walkway was built at the back of the tunnel boring machine.
- Some measures and procedures have also been put in place to prevent pollution of the lake water in the event of a massive water spill.

5 DISCUSSION

The characterization of karst-related hazards was based on a consistent approach including 3D modelling, hydrological and hydraulic simulations, which could be developed and applied on the site in order to offer pragmatic criteria for assessing potential voids and water inrushes. We mainly considered geological and hydrogeological data which were acquired before the beginning of the study, and we completed the dataset by observations we could carry out during the limited time duration of the pre-study.

One main uncertainty of the proposed approach is related to the precision of the geological information. The tunnel is dug along the strike of the bedding dip, meaning that it follows more or less the geological beds for long distances. In these conditions, a slight change in the vertical position of the inception horizons induces a significant shift in the intersection along the tunnel profile. A comparison between the predicted profile and the observed one was conducted after the tunnel had been drilled, and it clearly showed that the uncertainty of the prediction method is lower than that of the geological setting. This geological uncertainty is mainly related to the precision of the geological observations (e.g. in drillholes), and to the exact the positioning of inception horizons within the stratigraphic pile. This aspect should be better taken into account all along investigations in relation with a tunnel digging.

Another weak aspect in the present study was the lack of discharge rate measurements (short or incomplete hydrographs) of the different outlets at the beginning of the project, especially the main ones (Sauser springs, F-Stollen and Wasserhooliloch). Many springs were never properly gaged, even punctually. A series of discharge rate measurements were conducted in hurry, sometimes in bad conditions and the obtained results were not always really satisfying. However, some apparent inconsistencies observed in the measured rates could be better understood after the modelling of the pipe network. A monitoring system was installed afterwards, providing validation data. It turned out that discharge rates assessed for the first modelling were 20 to 30% lower than real ones. The model could thus be updated, but the overall picture of the system did not significantly change.

This example illustrates how karst hydraulics can be strongly non-linear, appearing sometime almost as chaotic. It can however be assessed and modelled as far as sufficient data are available. A real effort must be dedicated from the beginning on of a tunnel project in order to obtain reasonable measurements of discharge rates of all the system outlets. When it is possible, head measurements in caves (and drillholes) must be recorded, as they are very informative, providing direct information about heads in the whole karst system. Head and discharge values from the beginning of a project or even before the beginning are also very important for assessing the environmental impact of the tunnel on groundwater and nearby springs. This is therefore of highest significance for planning and designing a tunnel project in a karst region.

6 CONCLUSION

Excavation works of the safety gallery and cross-links to the road tunnel were completed in summer 2016 after 2 years of works. In total, the excavation works of the gallery and cross-links were to be interrupted for about 35 days and no active karst conduit was encountered.

Despite the difficult hydrogeological context and the many unknowns, the realization of the safety gallery could be carried out without any incident. This satisfactory result was made possible thanks to the close collaboration of all participants as well as the combination of different engineering branches such as geology, hydrogeology and civil engineering. The implementation of real-time monitoring and warning techniques, including the monitoring and prediction of the karst flows and the monitoring of the road tunnel shakings during blasting, made it possible to control the risks throughout the entire duration of the works.

REFERENCES

Bollinger, D. & Kellerhals, P. 2007. Umfahrungstunnel Twann (A5): Druckversuche in einem aktiven Karst (Twann tunnel A5: pressure tests in active karst system). *Bulletin für angewandte Geologie* 12(2): 49–61.

Campbell C.W., Sullivan S.M. 2002. Simulating time-varying cave flow and water levels using the Storm Water Management Model. *Engineering Geology*, 65: 133–139.

Covington M.D., Wicks C.M., Saar M.O. 2009. A dimensionless number describing the effects of recharge and geometry on discharge from simple karstic aquifers. *Water Resources Research*, 45: W11410, doi:10.1029/2009WR008004.

Filipponi M. 2009. Spatial Analysis of Karst Conduit Networks and Determination of Parameters Controlling the Speleogenesis along Preferential Lithostratigraphic Horizons. Ecole polytechnique fédérale de Lausanne (EPFL), Switzerland, PhD dissertation: 305.

Filipponi, M., Schmassmann, S., Jeannin, P.Y., Parriaux, A. 2012. KarstALEA: Wegleitung zur Prognose von karstspezifischen Gefahren im Untertagbau (KarstALEA, a practical guide for the prediction of karst-related hazards in underground works) Forschungsprojekt FGU 2009/003 des Bundesamt für Strassen ASTRA. *Schweizerischer Verband der Strassen- und Verkehrsfachleute VSS*. Zürich: 200.

Häfeli, C. 1964. Die Jura/Kreide-Grenzschichten im Bielerseegebiet (Kt. Bern) (The Jurassic-Cretaceous boundary formation in the Biel lake massif). Dissertation, University of Bern.

Jeannin, P.Y. 2001. Modeling flow in phreatic and epiphreatic karst conduits in the Hoelloch cave (Muotatal, Switzerland). *Water Resources Research* 37(2): 191–200.

Jeannin, P.Y., Eichenberger, U., Sinreich, M., Vouillamoz, J. & Malard, A. et al 2013. KARSYS: a pragmatic approach to karst hydrogeological system conceptualisation. Assessment of groundwater reserves and resources in Switzerland. *Environ Earth Sciences* 69(3): 999–1013.

Kellerhals, P. 1982. N5 Ligerztunnel. Geologie (N5 Ligerz tunnel, Geological study). Bern.

Weber, E., Jordan, F., Jeannin, P.Y., Vouillamoz, J. & Eichenberger, U. et al. 2011. Swisskarst project (NRP61): Towards a pragmatic simulation of karst spring discharge with conceptual semidistributed model. The Flims case study (Eastern Swiss Alps). *In: Proceedings 9th conference on limestone hydrogeology H2Karst*. Besançon: 483–486.

Tunnels and Underground Cities: Engineering and Innovation meet Archaeology, Architecture and Art, Volume 3: Geological and geotechnical knowledge and requirements for project implementation – Peila, Viggiani & Celestino (Eds)
 ISBN 978-0-367-46583-4

Excessive deformations induced by twin EPBM excavations in soft and shallow grounds: The case of Otogar-Kirazli metro, Istanbul

I. Ocak
Independent Consultant, Istanbul-Turkey

C. Rostami
Colorado School of Mines, CO, USA

ABSTRACT: In metro tunnel excavations, it is important to control surface settlements observed before and after excavation, causing damages to the surface structures. Otherwise, metro tunnel cannot perform the task expected and the advantages of metro tunnel are lost. For this purpose, in this study, short term surface settlements and its effect on buildings are examined in three zones for twin tunnels which are to be excavated between Otogar and Kirazlı stations of Istanbul Metro line, which is 5.8 km in length and 6.5 m in diameter. Geology in the study area is composed of fill, very stiff clay, dense sand, very dense sand, and hard clay, respectively starting from the surface.

During tunnel excavations, 28 buildings with 214 apartments have demolished by Istanbul Metropolitan Municipality because of the damages. New apartments are given to the aggrieved people in return for demolished ones. Also, in total for 364 apartments, together with the apartments that are demolished and evacuated in terms of precaution during tunnel excavation, rent and removal expenditures have been paid off. Demolished buildings, repaired buildings, rent payments and other expenses are in total 35.6 million dollars. Total project cost is 225 million dollars. Consequently, project cost increased 15.8% because of weak and shallow ground conditions. Also, duration time of the EPBMs occurred 27.7% for Lovat and 29.3 % for Herrenknecht.

1 INTRODUCTION

Increasing demand on infrastructures increases attention to shallow soft ground tunneling methods in urbanized areas. Especially in metro tunnel excavations, due to their large diameters, it is important to control the surface settlements observed before and after excavation, which may cause damage to surface structures (Ocak and Seker, 2013).

Basic parameters affecting the ground deformations are ground conditions, technical/ environmental parameters and tunneling or construction methods (O'Reilly and New, 1982; Arioglu, 1992; Karakus and Fowell, 2003; Tan and Ranjit, 2003; Minguez et al., 2005; Ellis, 2005; Suwansawat and Einstein, 2006). These basic parameters affecting the ground deformations can be summarized as Table 1 (Ocak, 2013).

The primary reason of ground movements above the tunnel, which are also known as surface settlements, is convergence of the ground into the tunnel after excavation, which changes in-situ stress state of the ground and results in a stress relief. Convergence of the ground is also known as the ground loss or volume loss. The volume of the settlement on the surface is usually assumed to be equal to the ground (volume) loss inside the tunnel (O'Reilly and New, 1982).

In this study, Otogar- Kirazlı twin metro line was studied in terms of deformation occurred in three regions. The line was constructed by Gulermak- Dogus joint venture.

Table 1. Main factors affecting the surface settlement (Ocak, 2013).

Category	Parameters
Excavation and support method	Excavation method (NATM, TBM, EPBM etc.)
	Excavation type (full face or sequential mining)
	Support methods (anchoring, shotcrete, steel sets, lining etc.)
	Shield operation factors for EPBM (face pressure, penetration rate, pitching angle, tail void grouting pressure, percent tail void grout filling) etc.
	Effect of worksite conditions
Tunnel geometric properties	Tunnel depth
	Tunnel diameter
	Number of tunnel (single or twin)
	Distance between two tunnels etc.
Ground properties	Modulus of elasticity
	Unit weight
	Cohesion
	Friction angle
	Poisson's ratio
	Lateral earth pressure coefficient
	Effect of groundwater
	Permeability etc.

1.1 *Excavation and Support*

The metro line has twin tunnels and between Otogar and Kirazlı district. It has totally 5.77 km. Metro line consist of 3.87 km tunnel, 0.62 km cut and cover station and 1.28 km at grade crossing. The excavation of the line was started in April 2006 and finished June 2008. This metro line is integrated Kirazlı- Başaksehir- Olimpiyat Koyu Metro Project which is 15.8 km. At the same time, Otogar and Kirazlı Metro Line is integrated Aksaray- Ataturk Airport light metro line (Fig. 1).

Totally 2 EPBMs are used for excavation of the tunnels (Figure 2). The metro lines in the study area are excavated by a Herrenknecht EPBM in the left tube and a Lovat EPBM in the right tube. One tube excavation has followed around 100 m behind the other tube. Lovat EPBM was started excavation at 1 April 2006 and Herrenknecht EPBM was started excavation at 13 May 2006. But during the excavation process first EPBM was changed several times (Figure 3). Which tunnel is the first and which tunnel is the second is important in terms of occurring of deformations. Because, in the excavation of twin tunnels, the first tunnel excavates undisturbed soil.

However, the excavation process for the first tunnel affects the soil. Therefore, the second tunnel is excavated in disturbed soil. In the project, it was reported that volume loss value of

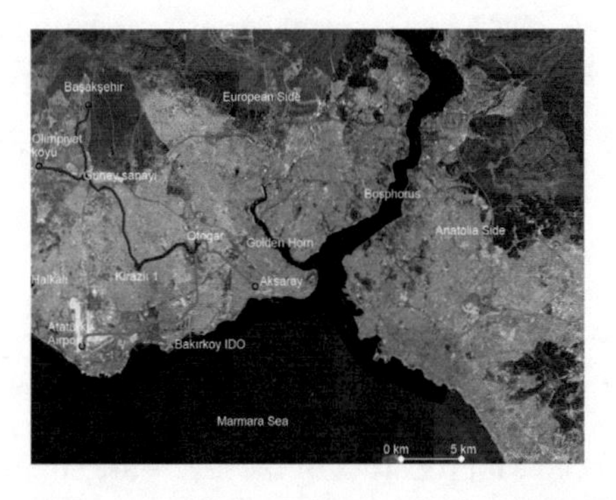

Figure 1. Main route of Otogar- Kirazlı metro line and location (Ocak, 2013).

Figure 2. Tunnel boring machines used in excavation; (a) Herrenknecht (b) Lovat (Ocak, 2013).

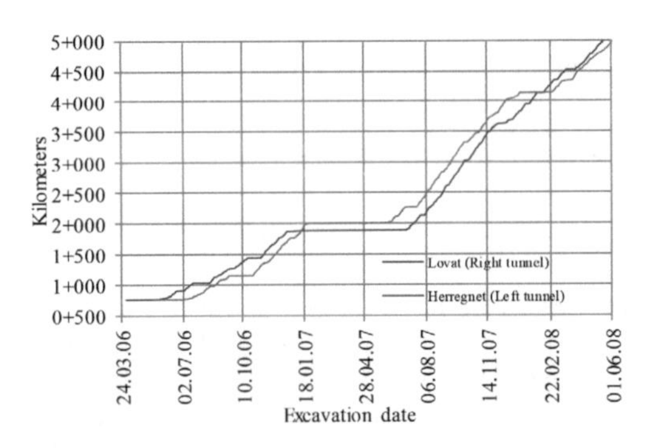

Figure 3. Change of excavation sequence of EPBMs.

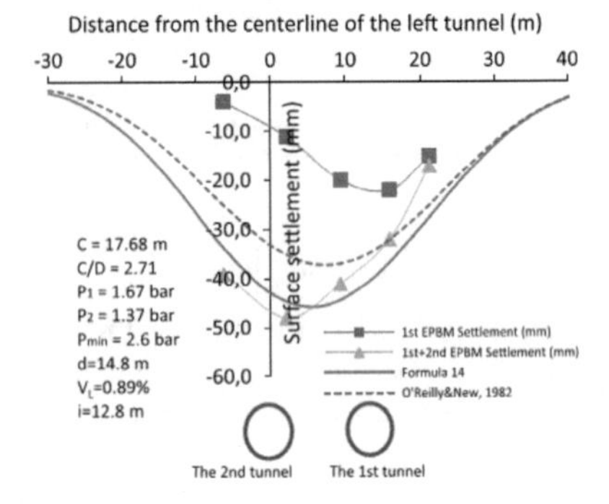

Figure 4. Interaction of surface settlements curve in twin tunnel excavation (Ocak, 2014).

the second tunnel was 48.6 % greater than that of the first tunnel (Ocak, 2014) (Figure 4). Some of the technical features of the machines are summarized in Table 2.

Excavated material is removed by auger (screw conveyor) through the machine to a belt conveyor and then loaded to rail cars for transporting to the portal. Since the excavated ground bears water and includes stability problems, the excavation chamber is pressurized by 300 kPa and conditioned by applying water, foam, bentonite and polymers through the injection ports. Chamber pressure is continuously monitored by pressure sensors inside the chamber and auger. Installation of a segment ring with 1.4 m length (inner diameter of 5.7 m and outer diameter of 6.3 m) and 30 cm thickness is realized by a wing type vacuum erector. Ring is configured as five segments plus a key segment. After installation of the ring, the excavation restarts and the void between the segment outer perimeter and excavated tunnel perimeter is grouted by 300 kPa pressure through the grout cannels in the trailing shield. This method of construction is proven to be minimizing the surface settlements (Ercelebi et al, 2011).

1.2 *The Geology of the Study Area*

The study area includes the twin tunnels of the change between km 0+860 - 1+000 (zone 1), km 1+850 - 2+250 (zone 2), km 4+200 - 4+450 (zone 3) between Otogar and Kirazlı stations. Gungoren Formation of the Miosen age is found in the study area. Laboratory and in-situ tests was applied to define the geotechnical features of the formations that tunnels pass through. Some of the geotechnical properties of the layers are summarized in Table 3 (Ayson, 2005). Fill layer consists of sand, clay, gravel and some pieces of masonry. Sand layer is

Table 2. Some technical features of the EPBMs (Ocak, 2013).

	Herrenknecht	Lovat
Excavation diameter (m)	6.500	6.564
External diameter (m)	6.30	6.30
Internal diameter (m)	5.70	5.70
Segment thickness (m)	0.30	0.30
Average segment length (m)	1.40	1.40
Shield outside diameter (m)	6.45	6.52
Maximum cutter head (rpm)	0 – 2.5	0 – 6.0
Total installed power (kW)	963	1622
Cutter head power (kW)	630	900
Maximum torque (tm)	435 (2.5 rpm)	445 (1.9 rpm)
Maximum thrust (kN)	35,200	54,000

Table 3. Some geotechnical properties of the study area (Ayson, 2005).

Strata	Unit weight (kN/m^3)	Modulus of Elasticity (kN/m^2)	Cohesion (kN/m^2)	Poisson ratio	Angle of Friction
Fill	18.0	5,000	1	0.30	10
Sand	18.3	25,000	1	0.25	35
Very dense sand	18.5	30,000	1	0.30	30
Clay (Güngören fr)	16.5	20,000	20	0.35	14
Hard clay (Güngö-ren fr.)	17.2	28,000	25	0.40	20

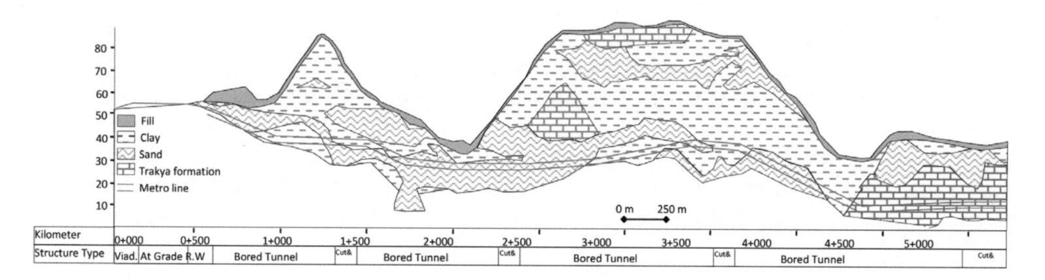

Figure 5. The geology of the study area (Ocak, 2013).

brown in upper levels and greenish yellow in lower levels, consisting clay, silt and mica. Very dense sand is greenish yellow, consisting mica. Clay layer is grayish green in color, consisting gravel and sand. The base layer of the tunnel is hard clay which is dark green, consisting shell. Underground water table starts about at 4-5 m below the surface at three study zone (Ocak, 2013). Tunnels are 14.2-23.5 m, 15.2-24.0 m, 14.8-22.6 m below the surface in zone 1, zone 2, zone 3 respectively. Geology of the line is given in Figure 5.

2 DEFORMATION OF BUILDINGS AT CRITICAL AREAS

2.1 *Surface Settlement and Building Deformation Measurements*

Contractors performed measurement points with intervals approximately each 3- 10 meters at building measurement point (BMP) and surface settlement measurement point (SMP) in each 3-5 meters depending on building situations on the surface. Both measurement readings

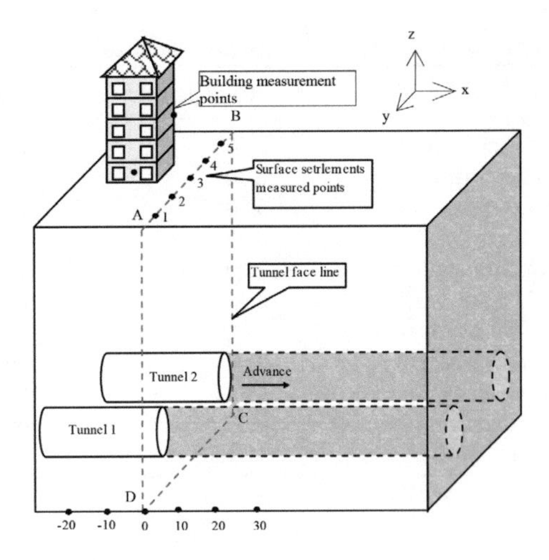

Figure 6. Surface settlement and building deformation measurement points (Ocak, 2009).

started before excavation at that section and continued up to allowed limits. Surface settlement measurements have been performed at 5 points on the surface including both tunnel sections. Building deformation measurements were performed with BMP bolts installed at different points installed on building. BMP and SMP geometry are given in Figure 6.

2.2 *The changes of the deformations by EPBM progress*

In this study, three critical regions, where metro tunnel passes through, was studied. These three regions have hard conditions in terms of tunneling. These conditions are poor ground structure, shallow depth, interference of twin tunnels and having construction in the city. At each three regions, tunnels were bored in clay or sand-clay.

When 1. tunnel comes near to SMP and BMP point, both deformations start. When Tunnel excavation comes to the measurement point, deformations are increasing. When Tunnel excavation passes the measurement point, deformations continues for a while. The actual deformations, that occur both at surface and in the buildings, are formed when 2. EPBM comes to the area which was passed through and disturbed by 1. EPBM. Both deformations stopped

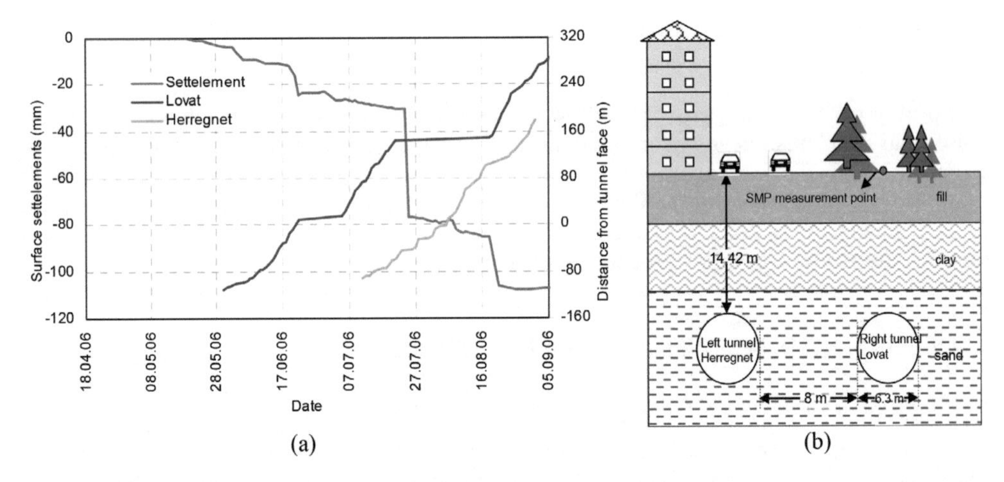

Figure 7. (a) The relationship between surface settlement and EPBMs position at km 0+879 (b) Position of SMP30 at km 0+879 (Ocak, 2009).

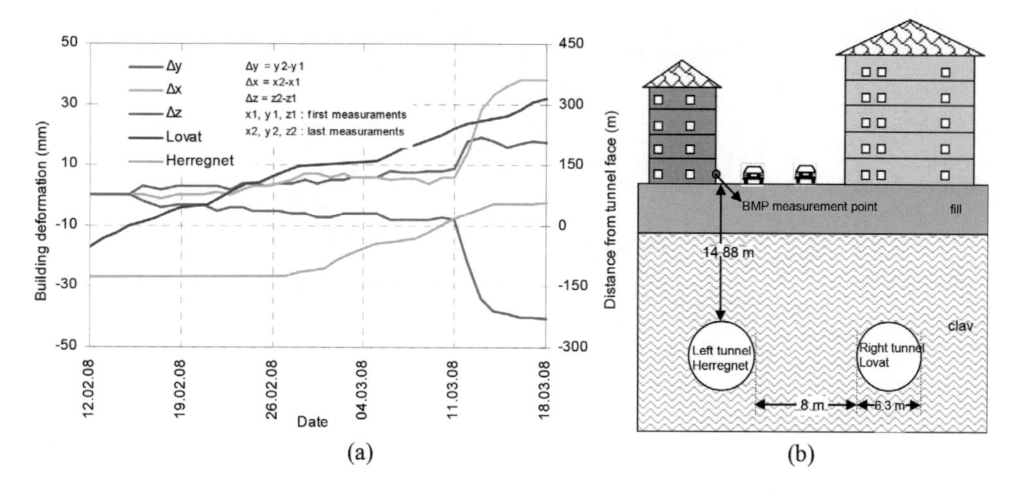

Figure 8. (a) The relationship between building deformation and EPBMs position at km 4+278 (b) Position of BMP at km 4+278 (Ocak, 2009).

after continuing for a while after 2. EPBM passed. Figure 7 shows the change of the surface settlements of a sample point according to the EPBM positions and the location of this SMP point. Figure 8 shows the change of the building deformations of a sample point according to the EPBM positions and the location of this BMP point (Ocak, 2009).

2.3 *Deformations Occurred at the Study Area*

Approximately 3.8 km of the line is bored by EPBMs. Only at the three regions serious deformations were occurred on the buildings and surface. These regions are zone 1, zone 2 and zone 3. Zone 1 is between km 0+875 and 0+963, zone 2 is between km 1+839 and 2+172 and zone 3 is between km 4+226 and 4+480. These three region are composed of mostly sand and clay. Depth of the tunnel is between 10.7 m and 15.3 m in zone 1, between 12.7 m and 15.5 m in zone 2 and between 16.0 and 18.0 in zone 3.

In Figure 9, total building deformation (Δδ) are given at these three region. Total deformations are computed below formula;

$$\Delta\delta = \sqrt{x^2+y^2+z^2}$$

Where;

$x=x_0-x_n$, x is total deformations of direction x, x_0 is initial deformation measurement and x_n is the last deformation measurement.

$y=y_0-y_n$, y is total deformations of direction y, y_0 is initial deformation measurement and y_n is the last deformation measurement.

$z=z_0-z_n$, z is total deformations of direction z, z_0 is initial deformation measurement and z_n is the last deformation measurement.

As can be seen in Figure 9, total building deformations in zone 1 changed from 11.1 mm to 29.7 mm and average deformation was 21.6 mm. Total building deformations in zone 2 changed from 2.3 mm to 42.2 mm and average deformation was 20.8 mm. The biggest deformations were occurred in zone 3. Total building deformations in zone 3 changed from 14.6 mm to 97.9 mm and average deformation was 46.7 mm. Al building deformations using here is selected only above of the tunnels.

According to literature (Rankin, 1988), these deformation values are to bigger in terms of building deformations (Table 4). And therefore extremely big building damages were occurred in these three regions (Figure 10).

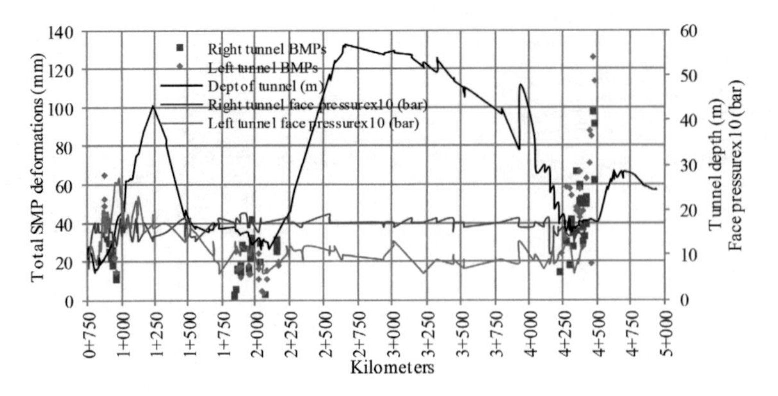

Figure 9. Surface settlements measured at tunnel line.

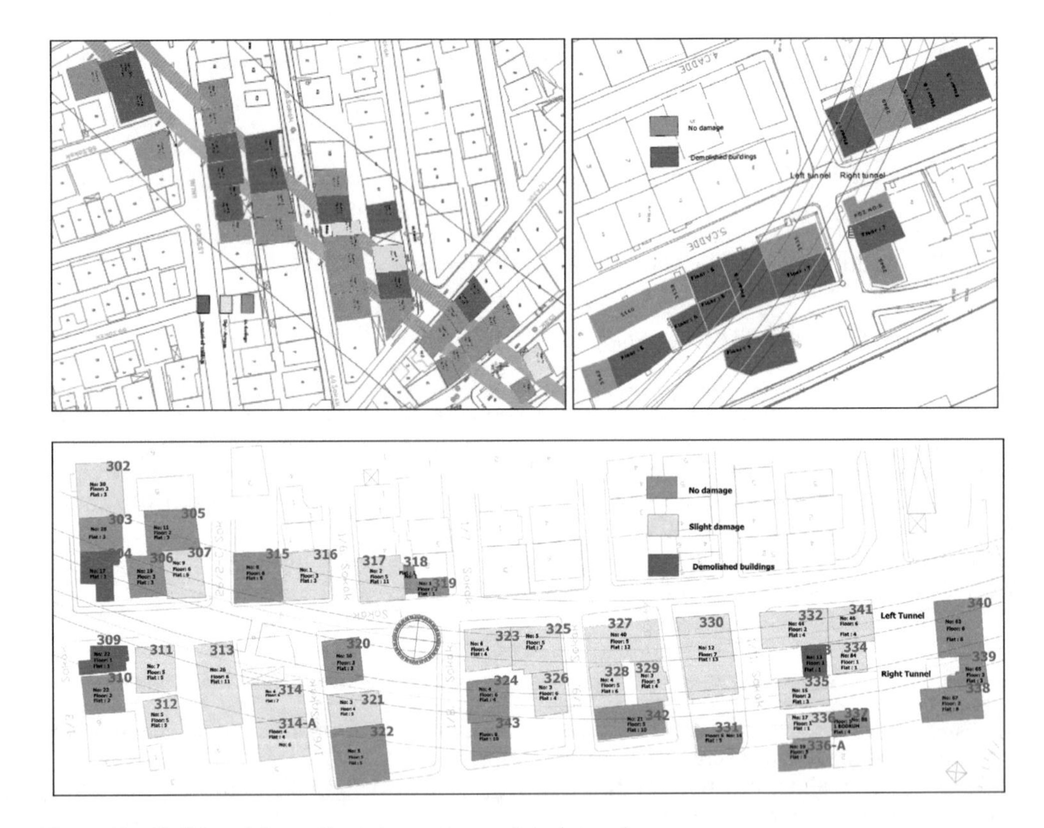

Figure 10. Building deformations at zone 1, zone 2 and zone 3.

2.4 *Financial Costs of the Deformations*

During tunnel excavations, 28 buildings have demolished by Istanbul Metropolitan Municipality because of the damages. At these buildings in total 214 apartments exist. New apartments are given to the aggrieved people in return for demolished ones. Also, in total for 364 apartments, together with the apartments that are demolished and evacuated in terms of precaution during tunnel excavation, rent and removal expenditures have been paid off. Demolished buildings, repaired buildings, rent payments and other expenses are in total 35.6 million dollars. Total project cost is 225 million dollars. Consequently, project cost increased 15.8% because of hard ground conditions (Ocak, 2009).

Table 4. Description of risk according to settlements and building slope (Rankin, 1988).

Risk category	Maximum slope of building	Maximum settlement of building (mm)	Description of risk
1	Less than1/500	Less than10	Negligible: superficial damage unlikely.
2	1/500 - 1/200	10-50	Slight: possible superficial damage which is unlikely to have structural significance.
3	1/200 – 1/50	50-75	Moderate: expected superficial damage and possible structural damage to buildings, possible damage to relatively rigid pipelines.
4	Greater than 1/50	Greater than 75	High: expected structural damage to buildings. Expected damage to rigid pipelines, possible damage to other pipelines.

Table 5. Money paid for demolished and damaged buildings (Ocak, 2009).

Type of payment	Apartment number	Money paid (USD)
Demolished buildings	214	27,820
Repaired buildings	150	1,395
Renting payments	364	4,828
Carrying payments	364	182
Good payments	8	240
Other payments		1,200
Total		35,665

Table 6. Overall performance of the EPBMs (Ocak, 2009).

	Herrenknecht (Left tunnel)	Lovat (Right tunnel)
Excavation started time (day)	13.05.2006	01.04.2006
Excavation finished time (day)	04.06.2008	20.05.2008
Total study time (day)	753	780
EPBM study time (day)	532	564
EPBM waiting time (day)	221	216
Percentage of waiting (%)	29.3	27.7

Another damage that occurs because of hard ground conditions is the standby of EPBM without excavating. In every project, there are assured costs like staff, energy consumption, taxes, site expenses. These expenses are always constant whether EPBMs work or not. So, each day EPBMs don't work, increase the cost. Duration time is by order for Lovat 27.7 % and for Herrenknecht 29.3 % (Table 6) (Ocak, 2009).

3 CONCLUSIONS

Control of surface settlement by tunneling in urbanized areas is very important in terms of preventing any damage to existing buildings and infrastructures. In this study, surface settlements are examined for twin tunnels which was excavated between Otogar and Kirazlı in Istanbul. Geologic structure of area can be classified as soft ground. Tunnel depth at critical areas is 14-23 m (Ocak, 2009).

The optimal excavation method for soft and shallow grounds is EPBM. NATM method is too difficult at such a ground and also settlement values occur too much. With a diameter of 6.5 m and a center-to-center spacing of approximately 14 m, the following conclusions were obtained:

a) The volume loss value of the second tunnel was approximate 50 % greater than that of the first tunnel (Ocak, 2014). So, building deformations above second tunnel is bigger than above first tunnel.
b) With EPBM excavation method especially at the shallow depths the damages that are given to the environment increased the project cost 15.8 % more. Also, soft and shallow tunnel conditions increased the project time 29.3 % (Ocak, 2009).
c) It was seen that, especially less than 2 tunnel diameter in soft soils it may be very big surface settlement and building deformations.

REFERENCES

O'Reilly, M.P., New, B.M., 1982. Settlement above tunnels in the United Kingdom - their magnitude and prediction. Proceedings of the Tunneling 82 Conference, Brighton, pp. 173-181.

Arioglu, E., 1992. Surface movements due to tunneling activities in urban areas and minimization of building damages. Short Course, Istanbul Technical University, Mining Eng. Dept. (in Turkish)

Karakus, M., Fowell, R.J., 2003. Effects of different tunnel face advance excavation on the settlement by FEM. Tunneling and Underground Space Technology. 18 (2003), 513–523.

Minguez, F., Gregory, A., Guglielmetti, V., 2005. Best practice in EPB management. Tunnels and Tunnelling International, Nov., pp. 21-25

Ellis, D., 2005. High standards in Heatrow's art. Tunnels and Tunnelling International, Semp., pp. 29-34

Ercelebi, S., Copur, H., Ocak., I., 2011. Surface settlement predictions for Istanbul Metro Tunnels excavated by EPB-TBM, Environmental Earth Sciences, 62(2): 357-365.

Suwansawat, S., Einstein, H.H., 2006. Artificial neural networks for predicting the maximum surface settlement caused by EPB shield tunnelling. Tunnelling and Underground Space Technology. 21 (2006), 133–150.

Tan, W.L., Ranjit, P.G., 2003. Parameters and considerations in soft ground tunnelling. The Electronic Journal of Geotechnical Engineering., pp. 344

Ocak, I., Seker SE, 2013. Calculation of surface settlements caused by EPBM tunneling using artificial neural network, SVM, and Gaussian processes, Environmental Earth Sciences, 70(3):1263-1276

Ocak, I., 2013. Interaction of Longitudinal Surface Settlements for Twin Tunnels in Shallow and Soft Soils: The Case of Istanbul Metro, Environmental Earth Sciences, 69(5): 1673-1683.

Ocak, I., 2009. Environmental Effects of Tunnel Excavation in Soft and Shallow Ground with EPBM: The Case of Istanbul, Environmental Earth Science, 59(2): 347-352.

Ocak, I., 2014. A new approach for estimating the transverse surface settlement curve for twin tunnels in shallow and soft soils, Environmental Earth Sciences, 72(7):2357-2367.

Ayson Drill Research and Build A.S., 2005. Otogar-Bagcılar Station Geological Geotechnical Report. July (in Turkish).

Rankin, W. J., 1988. "Ground movements resulting from urban tunneling: Predictions and effects." Engineering geology of underground movements, F. G. Bell, M. G. Colshaw, J. C. Cripps, and M. A. Lovell, eds., Geological Society, London, 79–92.

Tunnels and Underground Cities: Engineering and Innovation meet Archaeology, Architecture and Art, Volume 3: Geological and geotechnical knowledge and requirements for project implementation – Peila, Viggiani & Celestino (Eds)

The geological portal of the Turin-Lyon project

M. E. Parisi, L. Brino & D. Paletto
TELT sas, Turin, Italy

M. Cussino
gd test srl, Turin, Italy

ABSTRACT: The Cross-Border Section, the first functional phase of the New Turin-Lyon Railway Line, concerns the mountainous territory between Saint-Jean-de-Maurienne (France) and Susa-Bussoleno (Italy). From the early project stages up to the start of the 2000s, the data produced have been collected in a specialized structure for the purpose of rapid and organized consultation. After implementing the portal, data were received from the excavation of the survey adit tunnels. A large volume of data is expected to arrive once the future construction sites become active. These will be sites where sections excavated in the traditional way will be followed by many mechanized sections. In order to achieve near real-time use, data organization for the purpose of filing and accessibility must be done in a structured and simple manner. With a single touchscreen on a personal smartphone it is now possible to follow the progress of the Federica TBM, which is excavating 9 km of gallery in French territory, and it will shortly be possible to follow the progress of a further 7 TBMs.

1 INTRODUCTION

1.1 *The geographic and planning context*

The Cross-Border Section, starting from the west, consists of the crossing of the open air sector of Saint-Jean-de-Maurienne, the 57.5 km Montcenis Base Tunnel which crosses the Western Alps, the open air crossing of Susa, the Interconnecting Tunnel and the entrance at Bussoleno station. To investigate the rock mass along the projected route of the future railway tunnel, in addition to 65 km of boreholes and 260 km of geophysical explorations, preliminary survey works were carried out, consisting of three intermediate adit tunnels in France over a total length of 9 km and an exploratory tunnel in Italy of around 7 km; from west to east they are:

- the “Saint Martin la Porte” adit of 2,329 m;
- the “La Praz” adit of 2,480 m;
- the “Villarodin Bourget/Modane” adit of 4,036 m;
- the “La Maddalena” exploratory tunnel of 7,020 m.

In 2015, construction of the first underground works began, along the axis of the Base Tunnel, on the Saint Martin La Porte site, the main works of which include establishing a complementary adit tunnel of 1,800 m (completed) and the survey gallery between Saint Martin La Porte and La Praz of 8.8 km (Brino & al. 2013). The excavation of this latter gallery, placed in axis and with the same section of the southern tube of the future base tunnel, started with a TBM in summer 2016 and is currently in progress, having advanced 4,800 m to date (Figure 1).

Figure 1. The Cross-Border Section of the New Turin-Lyon Line.

1.2 *The reasons for the geological portal: in the study phase and in the works phase*

Since the start of the study phase in the early 2000s, there has been a need to develop and make available to the company, and beyond, a system that would receive the data collected in the field in a single centralized system which could then be used in a simple and efficient manner.

First Alpetunnel and LTF, then TELT, availing themselves of external consultants and gathering suggestions from the technicians in the field, undertook to develop an information system able to collect the huge volume of data produced in such a project. Today, in the works phase, the need to structure, collect and manage the data as simply and rapidly as possible has become essential (Parisi & al. 2007).

Therefore, with the first soils investigations, the geological, hydrogeological, geomechanical and cartographic data produced have been collected in a specialized structure for the purpose of rapid and organized consultation. The data emerging from the excavation of the adits, for example the rock face reliefs, geotechnical and hydrogeological monitoring data, and progress reports, arrived and were implemented in TELT's geological portal. With the onset of mechanized excavations, the excavation parameters also flow into the specific portal. Now that the main excavation phase has started, every future site is expected to produce a large volume of data, which will be collected in TELT's Geological Portal.

2 THE GEOLOGICAL PORTAL

2.1 *IT structure of the geological portal*

The hardware structure provided has always been first-rate, using the best and most advanced technologies available on the market, such as a high-performance relational database (Oracle© DB) and a front end for graphic display and the processing of data collected based on Microsoft servers, using dedicated on-site systems with ultra-fast SSD-type disks (solid-state disks which guarantee high reading and writing performances) and high-speed data lines secured by VPN tunnels which connect all the sites to the data center where they are processed and inserted into the system (Oracle 2010; Microsoft 2017).

The upgrading of the IT systems has ensured that the platform keeps abreast of new technologies, gradually providing the necessary upgrades. This has included transferring the entire structure from dedicated systems to geographically redundant virtual platforms, guaranteeing greater data security and scalability in terms of performance in accordance with increased computational requirements, to the point of being able to merge two different platforms while maintaining a common database, in the order of 5 Terabytes (Figure 2).

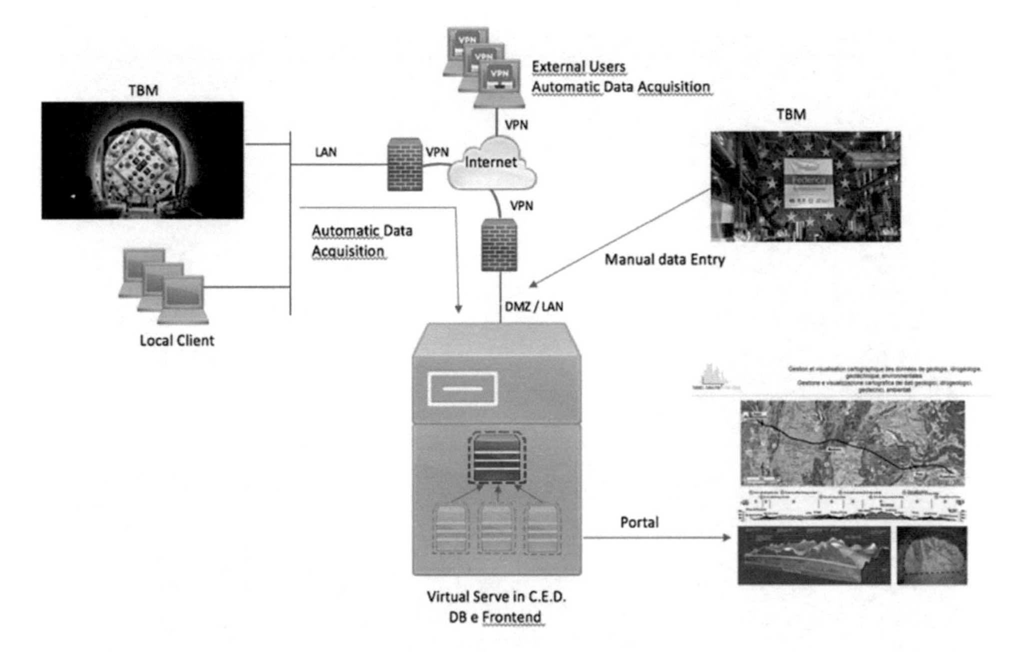

Figure 2. Diagram of IT structure of TELT's geological portal.

A key factor were the communication lines, which also constantly improved their performance and that are widely distributed throughout the territory, which enables data to be collected and received in real time via a VPN tunnel that guarantees their security.

In areas considered difficult to reach and link up, all the resources at the disposal of the various local providers were put to work, including the projected installation of radio links for transmitting the data to the servers or the use of systems equipped with a 4G modem onboard the automatic survey control units or portable survey devices.

Internal and external access to the portal utilizes fiber optic connections, in order to guarantee fast access to the data in all contexts.

2.2 *Geological portal content*

Year 2006 saw the start of development of a restricted-access portal (Figure 3) which would enable data access, insertion and searches, complementing it with cartographic plans showing the location of the measurements and works. The portal is protected by an access system using TELT's private VPN and it is hosted on servers within the TELT infrastructure that are separate from the rest of the office servers.

The database uses a relational geodatabase, initially built with Oracle © technology and accompanied for geographic data by the functionalities provided by ESRI ® ArcSDe software, whereas now, following developments in information technology, the database management has been entirely migrated to a platform that uses Microsoft ® SQL Server products and the latest version of ESRI products, configuring the cartographic plans to be made available through the ArcGis ® Server (Figure 4) (ESRI 1999–2014).

The geodatabase collected the geological investigation data and the designs delivered during the various study phases, integrated with the data collected in monitoring points together with those coming from the excavation works.

The portal enables the management of all data types available during the excavation of an underground work, both with mechanized and traditional excavation. The following data are thus filed in the geodatabase:

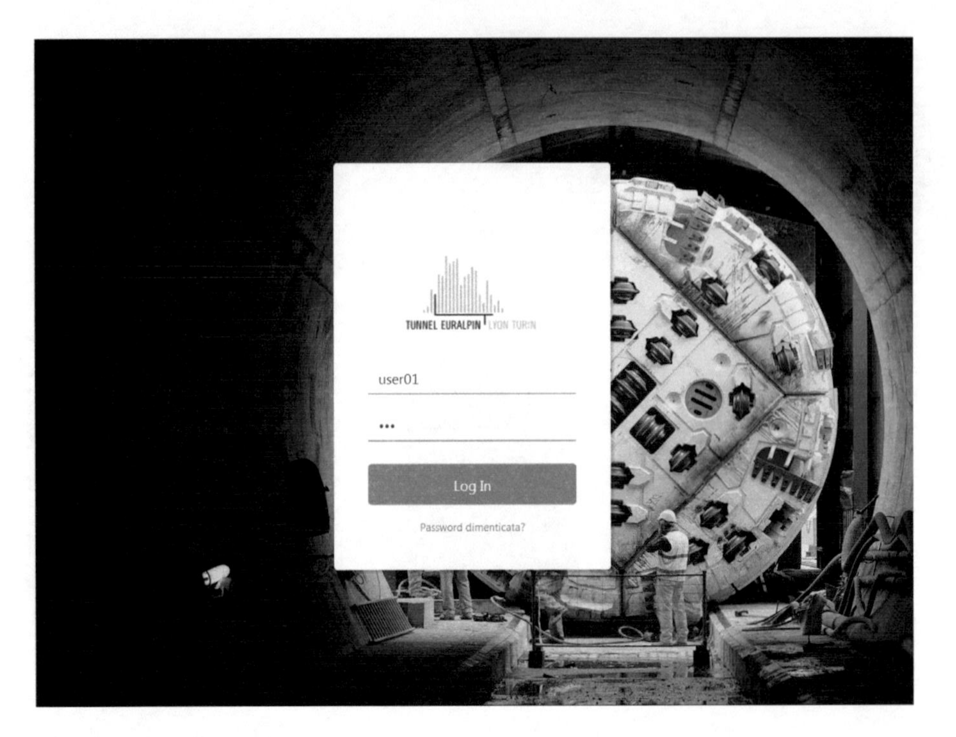

Figure 3. Access to the portal.

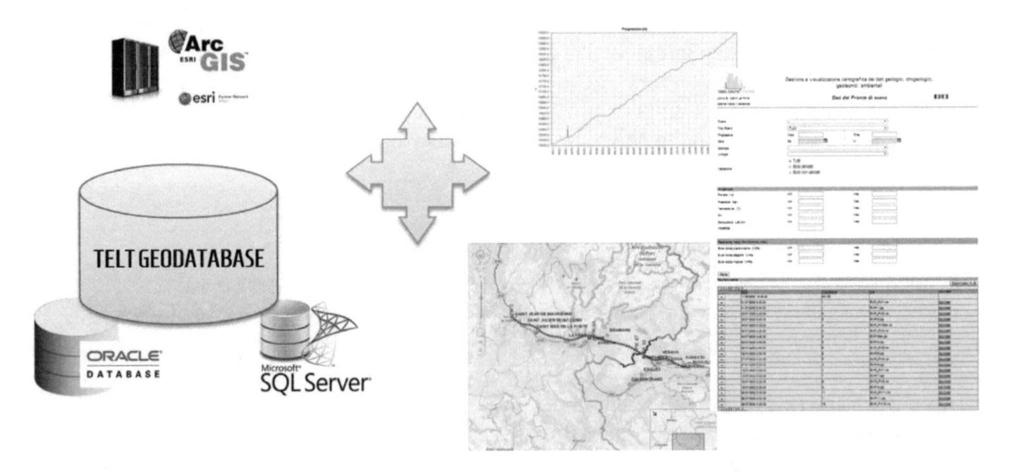

Figure 4. Software used.

- Geological and geotechnical surveys (stratigraphy, laboratory tests, on-site tests, diagraphs, etc.)
- Geophysics (geophysical stringing, cartography, etc.)
- Springs and water courses (on-site measurements, water analyses, monographic datasheets, etc.)
- Thematic cartography (geology, hydrogeology, geomechanics, etc.)
- Underground works (excavation progress, rock face reliefs, monitoring, progress surveys, instrumented sections, documentary reports, etc.)

2.2.1 *Management methods*
Access to the portal requires users to be identified with a dedicated login depending on their profile, which may be of three kinds:

- Insertion
- Validation
- Visualization

The data upload process allows users with a validation profile to verify and approve all the measurements and documents inserted by the users responsible for data uploading. Data can be loaded either manually, by compiling a guided input mask, or automatically.

Some standard procedures, such as loading data relating to rock face samples, have been automated with the development of *ad hoc* procedures which enable the user to insert a new analysis without having to type in manually all the data present in the completed sample form. The same form used for loading the data is thus also optimized for printing and delivery, and it is filed in the TELT portal as an attachment (Figure 5).

The procedure interfaces the loading model, an excel file, with the geodatabase, imports from the model the data contained in a few specific fields by means of a few customized "macro" procedures and allows the user to attach the complete analysis as an additional document.

The data available in the geo-database can be consulted with *ad hoc* search masks from the menu bar, and can be exported in formats compatible with the Excel/CSV format.

For some types of data the values are also available in the form of graphs, for example those relating to hydrogeological monitoring (Figure 6).

The geodatabase was also fed during the different planning phases by loading all the cartographic data delivered by the planners, so that the database could also be used as a catalogue of all the data available. On the portal the data present are collected in a section dedicated to designs; all available designs can be consulted with a webgis visualizer configured for each design (Figure 7).

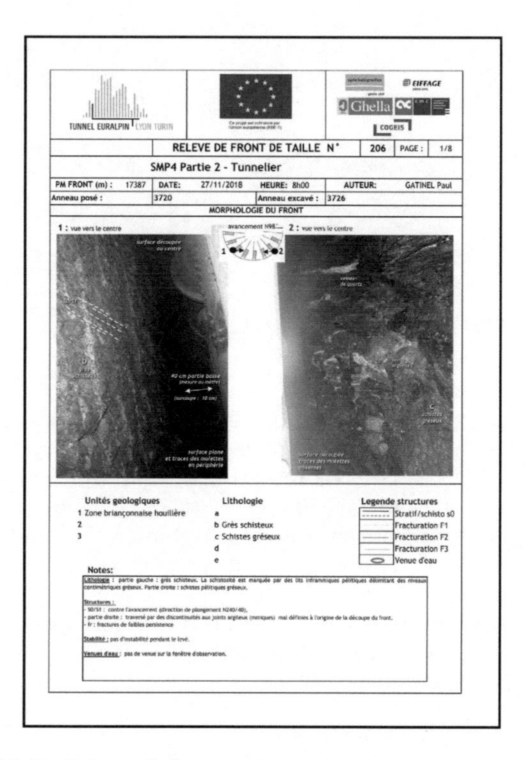

TUNNEL EURALPIN LYON TURIN

	RELEVE DE FRONT DE TAILLE N°		206	PAGE :	1/8
SMP4 Partie 2 - Tunnelier					
PM FRONT (m) : 17387	DATE: 27/11/2018	HEURE: 8h00	AUTEUR:	GATINEL Paul	
Anneau posé :	3720	Anneau excavé :	3726		
MORPHOLOGIE DU FRONT					

Unités geologiques	Lithologie	Legende structures
1 Zone briançonnaise houillère	a	Stratif/schisto s0
2	b Grès schisteux	Fracturation F1
3	c Schistes gréseux	Fracturation F2
	d	Fracturation F3
	e	Venue d'eau

Notes:

Lithologie : partie gauche : grès schisteux. La schistosité est marquée par des lits inframmiques pélitiques délimitant des niveaux centimétriques gréseux. Partie droite : schistes pélitiques gréseux.

Structures :
- S0/S1 : contre l'avancement (direction de plongement N240/40),
- partie droite : traversé par des discontinuités aux joints argileux (mmiques) mal définies à l'origine de la découpe du front.
- fr : fractures de faibles persistence

Stabilité : pas d'instabilité pendant le levé.

Venues d'eau : pas de venue sur la fenêtre d'observation.

Figure 5. Loading model, Rock face relief.

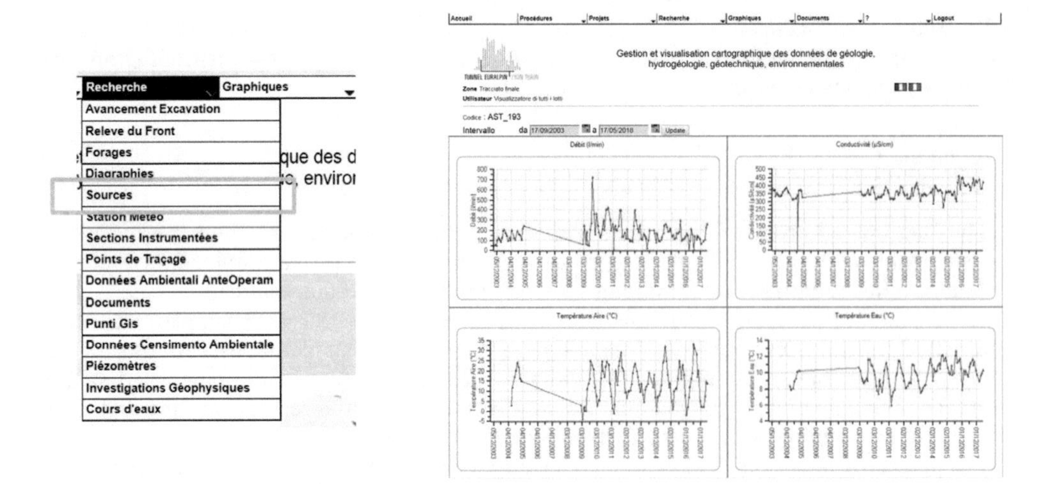

Figure 6. Graphic searches and visualization.

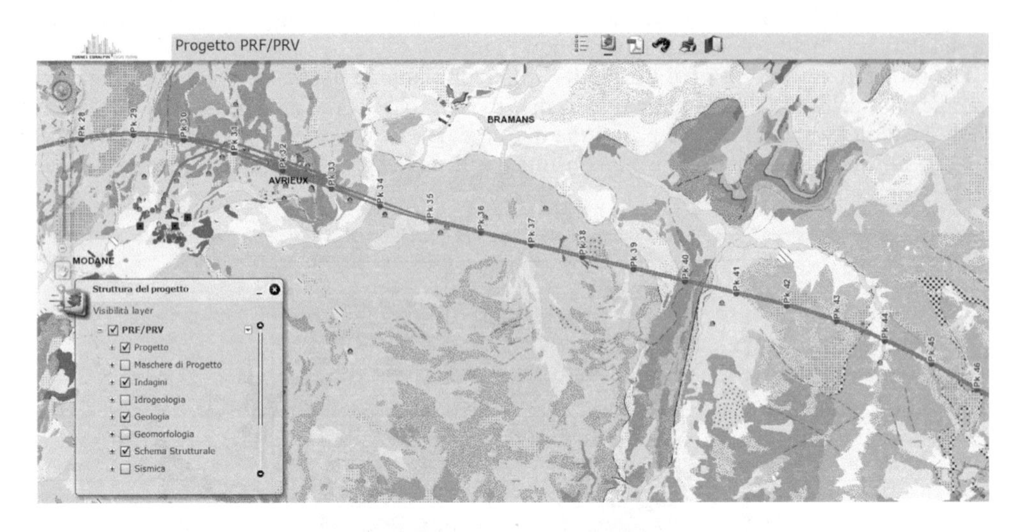

Figure 7. Cartographic design, geology, Final Baseline Design phase.

All the points present in the data bank are georeferenced. The coordinates are relative to the International system UTM WGS 84 and to the local system LTF2004c, set up specifically by LTF/TELT. The latter is the geodetic and altimetry georeferencing system common to the two countries for the design works. The requirement of developing a "local" georeferencing system arose in order to overcome the fundamental differences in the respective national geodetic referencing systems and to minimize deformations of the cartographic display model.

Each measurement point can thus be located on the cartography and searched for in the geo-database by code. The system updates dynamically as and when new points or new measurements linked to the points are added.

2.2.2 *Management methods of TBM parameters*

The data relating to mechanized excavation produced by the TBMs are filed in the geodatabase in order to provide real-time consultation and visualization on graphs. For the excavation of the La Maddalena Exploratory Tunnel (Chiomonte site) and for the Saint-Martin-La-Porte site, two *ad hoc* uploading procedures were developed on the base of the raw data available.

Each procedure was optimized for the management of the data volume produced by a mechanized excavation that is able to record 300 different parameters with frequencies of 1 data set every 5–10 seconds.

This also called for an appropriate dimensioning of the databank to enable it to contain data amounting to an average of 1 GB/month.

For the Chiomonte site, a specific procedure was installed on a computer present on site and connected directly to the origin folder of TBM data. This allowed to set up an upload process that updated the TBM geodatabase directly with the machine data at regular 10-second intervals. The computer was connected to the TELT network via a dedicated VPN and remained active for 24h/24h. A copy of the machine data was also kept locally in case they could not be filed immediately in the geodatabase.

For Saint-Martin-La-Porte, the data are recorded by the excavation system every 5 seconds, but are made available at hourly intervals, because they are not acquired directly from the control system of the machine, but rather from an interchange area where they are exported periodically. The frequency of updating of these data provides for 1 update every hour, with the production of an export file with all the data recorded during the last hour. It is shared in a specific area dedicated only to TBM data. A procedure was therefore set up in the TELT servers that recovers and imports data from this dedicated area with the same frequency as that in which the data are updated. The same procedure was also adopted for acquiring the progress reports on the loops and the loop construction reports from the site.

From the TELT portal, for the purpose of consulting the machine data, the user has access to the consultation in the form of a graph, where he can select the parameters of interest to be shown on the graph and indicate a specific time or progressive interval (Figure 8).

The procedures for loading mechanized excavation data are configured to update the cartography of the project automatically and to show the state of progress of the excavation in real-time.

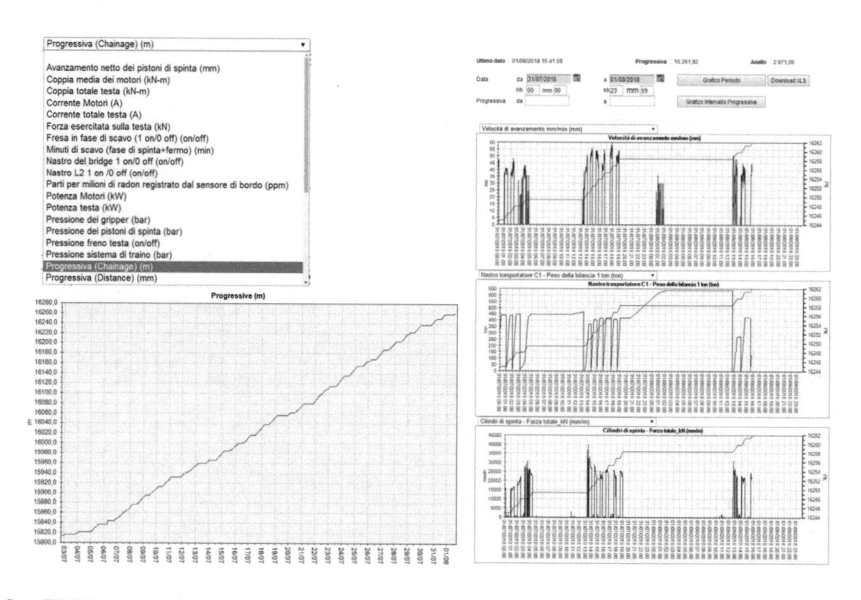

Figure 8. TBM parameters.

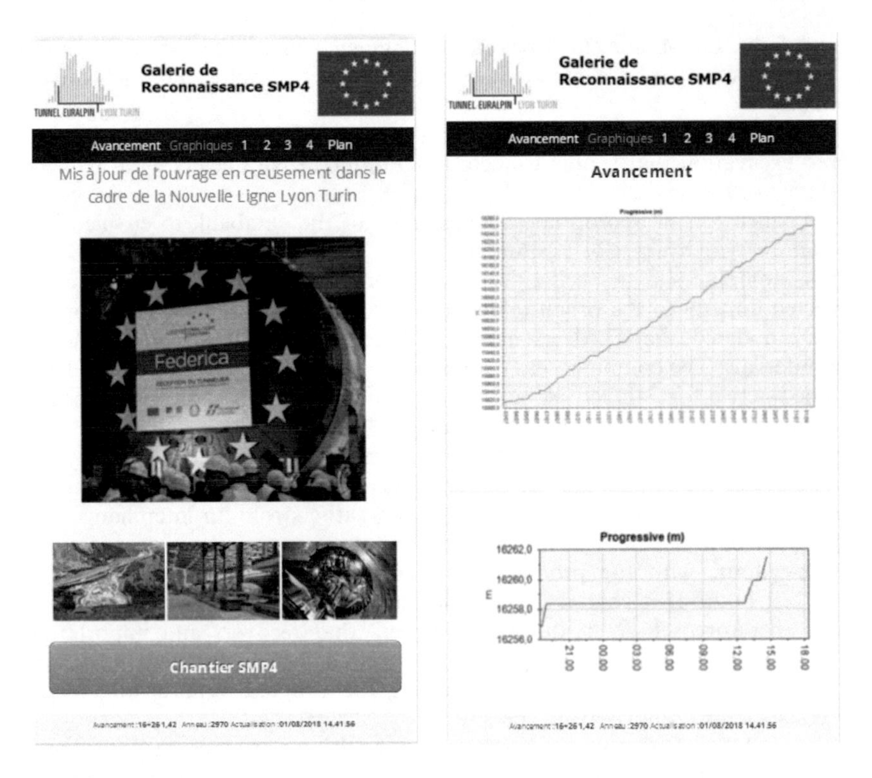

Figure 9. TBM mobile app.

Furthermore, in order to follow the progress of the site in real-time, a smartphone "app" was created, characterized by preconfigured and automatically updated instrument panels which show some significant parameters and the progress of the excavation on a planimetry. The procedure is available on the mobile devices (smartphones, tablets, etc.) of the technicians involved, subject to access via VPN to the TELT protected network (Figure 9).

For the excavation of the survey tunnel of the SMP4 site, the screen shows the last available update on graphs which display:

- The progress of the excavation over one month;
- The performance over the last 24h of parameters such as excavation progress (m), excavation rate (mm/min), conveyor belt balance weight (ton/cycle), excavation head torque (kN).

2.2.3 *Statistical data*

In order to give an idea of the data volume currently present in the portal, below are some statistical figures:

- 4,000 measurement points (boreholes, geophysics surveys, sources etc);
- 926 surveys;
- 1,300 water points;
- 240,000 manual mesures;
- 4,000 rock face samples;
- 14,000 attached documents;
- 20 million lines of TBM data.

And more than 400 logins were created to allow access to users at their request.

3 CONCLUSIONS

The TELT Geological Portal will be implemented with data from the future sites which will be started up in the next few years. The preparatory work sites will be followed by those of the main sites. The work models, the existing guided masks for insertion of data in the portal, will be made available to the new sites. Where appropriate, these tools will be adapted in order to better manage all the data, including those specific to each site; but the loading will always use procedures that verify the data structure is correct. The data of all the future TBMs which will mechanically excavate some parts of the tunnels will be integrated in the portal and the monitoring of these excavations will take place in real-time by means of instrument panels dedicated to each excavation, both on desktops and on mobile apps, as is already the case with the excavation of the Saint-Martin-la-Porte survey gallery.

In order to be able to manage as efficiently as possible not only the geological and construction data deriving from the construction sites but also environmental data, TELT is providing a unique platform which will contain both the geological and environmental portals. The advantages which will derive from this, in addition to that of optimizing the IT machines by refining the hardware resources, are also linked to facilitating the interface between the two portals, eliminating the risk of data redundancy and improving the usability of data managed by the different services for the benefit of more efficient control of site activities by the Client and by external control bodies. Finally, in order to exploit the large quantity of cartographic data present a catalogue of the data and metadata will be produced so that they can be clearly identified and easily reused during the remainder of the works.

REFERENCES

Brino, Monin, Fournier, Bufalini, 2013. "Nuova Linea Torino-Lione - Ritorni d'esperienza", Congresso Internazionale Società Italiana Gallerie - Gallerie e spazio sotterraneo nello sviluppo dell'Europa, Bologna, 17–19 ottobre 2013, pp. 595–604, Pàtron Editore

Carrieri, Cussino 2007 *"Il Sistema Informativo Territoriale del Portale di Commessa"* MONDOGIS n. 59

ESRI 1999. *Understanding ArcSDE®*. Redlands, CA: ESRI, http://downloads.esri.com/support/documentation/sde_/706understanding_arcsde.pdf

ESRI 2009. *ArcGis Server installation guides*. Redlands, CA:,ESRI https://enterprise.arcgis.com/en/documentation/install/

ESRI 2014. *ArcGIS Viewer for Flex*. Redlands, CA: ESRI. http://resources.arcgis.com/en/communities/flex-viewer/

Microsoft 2017. *Microsoft SQL Server*. Redmond, WA: Microsoft. https://www.microsoft.com/en-gb/sql-server/default.aspx

Monin, Parisi, Brino, Cussino, Mocco, 2008 *"Suivi géologique de l'avancement des ouvrages du Lyon-Turin Ferroviaire par la mise en place d'une géodatabase"*, Congrès International AFTES 2008, Monaco, 6–8 ottobre 2008, pp. 351–356, Edition Spécifique

Monin, Brino, Darmendrail, 2011 « Le projet ferroviaire Lyon-Turin – 20 ans d'etudes et de reconnaissances techniques pour la conception du tunnel de base », Géologie de l'ingénieur, Hommage à la mémoire de Marcel Arnould, Journée scientifique internationale, CFGI (Comité Français de Géologie de l'Ingénieur et de l'Environnement) et IAEG (International Association for Engineering Geology and the environment), Paris, 12 ottobre 2011, pp. 25–40, Presse des Mines

Monin, Brino, Chabert, « *Le tunnel de base de la nouvelle liaison ferroviaire Lyon-Turin : retour d'expérience des ouvrages de reconnaissance* », Congrès International AFTES 2014, Lyon, 13–15 ottobre 2014

Oracle, *database requirements for ArcGIS, 2010 e aggiornamenti successivi.*

Parisi, Monin, Darmendrail, Brino, Mocco, Cussino, Parboni, 2007 *"Gestione e visualizzazione cartografica del nuovo collegamento ferroviario Torino-Lione"*, MondoGIS, maggio-giugno 2007, n. 60, pp. 21–25, Editore MondoGIS

Parisi, Monin, Brino, Bufalini, Fournier, 2010 *"Approccio metodologico per determinare le previsioni idrogeologiche e le venute d'acqua nell'ambito della progettazione della Nuova Linea Torino Lione – Ritorno di esperienza della Discenderia di La Praz"*, Le acque di superficie e sotterranee e le infrastrutture di trasporto dalla pianificazione all'esercizio, Convegno Nazionale Associazione Idrotecnica Italiana,

Istituto di Ricerca per la Protezione Idrogeologica IRPI-CNR, Società Italiana di Geologia Ambientale SIGEA, Roma, 6–7 maggio 2010, Collana La Sintesi - Scienze Ambientali, pp. 455–466
Parisi, Farinetti, Brino, 2018 *"Measurement of Massif temperatures through the Maddalena Exploration Tunnel"*, AITES-ITA 2018 World Tunnel Congress, Dubai, 21–26 aprile 2018
Parisi, Brino, Chabert, 2018 *"Il Tunnel di Base del Moncenisio: bilancio del test di scavo con fresa nel Cunicolo Esplorativo de la Maddalena di Chiomonte – L'opera preliminare"*, Swiss Tunnel Congress 2018, Lucerna, 13–15 giugno 2018

Tunnels and Underground Cities: Engineering and Innovation meet Archaeology, Architecture and Art, Volume 3: Geological and geotechnical knowledge and requirements for project implementation – Peila, Viggiani & Celestino (Eds)

Design and construction of hydraulic tunnels at Gibe III dam, Ethiopia

G. Pietrangeli, A. Pietrangeli, A. Cagiano de Azevedo, G. Pittalis & C. Rossini
Studio Ing. G. Pietrangeli s.r.l., Rome, Italy

ABSTRACT: The paper describes the main aspects of the design of the Power-waterways of Gibe III Hpp (Ethiopia). The plant includes a 250 m high RCC gravity dam, two 11 m diameter and 1,2 km long power waterways excavated in the left bank of the river. The project commenced in 2006, the dam was impounded in 2015. At present the plant is completed and in operation. The maximum reservoir level reached at date is 215 m over the foundation (90% of maximum head). The design was challenging in reason of the huge size of the works, to their complexity and considering that these power waterways operate with heads up to 210 m of water column, with a total discharge up to 950 m^3/s. The tunnel lining was one of the main topics of design, especially in relation to the control of leakage and potential risk of rock mass hydro-fracturing and hydro-jacking.

1 POWERWATERWAYS LAYOUT

1.1 *General*

The general layout of the waterways is illustrated in the Figure 1. The waterway system includes two separate lines (named PTR and PTL) on the left abutment, each one feeding five turbines, comprising the following structures:

- intake with trashracks ($A_{net} = 330\ m^2$)
- wet gate shaft (D=14 m, H=120 m) controlled by No. 2 bulkhead and No. 2 wheel gates (4.5 x 11 m)
- a large headrace tunnel (L = ~1 km, D = 11 m, flow section A=95 m^2, slope s=0.5%)
- a surge shaft (H = 147 m, D = 18 m)
- penstocks shaft (D = 7.7 m, L= 250 m)
- manifolds with 5 branches (D= variable from 7.7 to 4.2 m. L ~ 65 m)

The power waterways are designed to safely conduct water from reservoir to Power House units, and to cope with internal water pressure (depending from water reservoir levels) and possible external ground/water pressure, in both full and empty conditions.

1.2 *Intake towers*

The intake towers are designed in order to:

- guarantee a minimum submergence of the Power Tunnel in any operating condition;
- minimize head losses at the tunnel entrance, providing a smooth transition for the water velocity at the inlet and a water velocity through the trash racks of no more than 3 m/s;
- provide a sustain to a trash rack to prevent entrance inside the tunnel of large elements transported by the waters;
- protect and sustain the portal of the tunnels;
- provide possibility, in emergency conditions and with reduced water reservoir levels, to close by means of stoplogs the first stretch of tunnel.

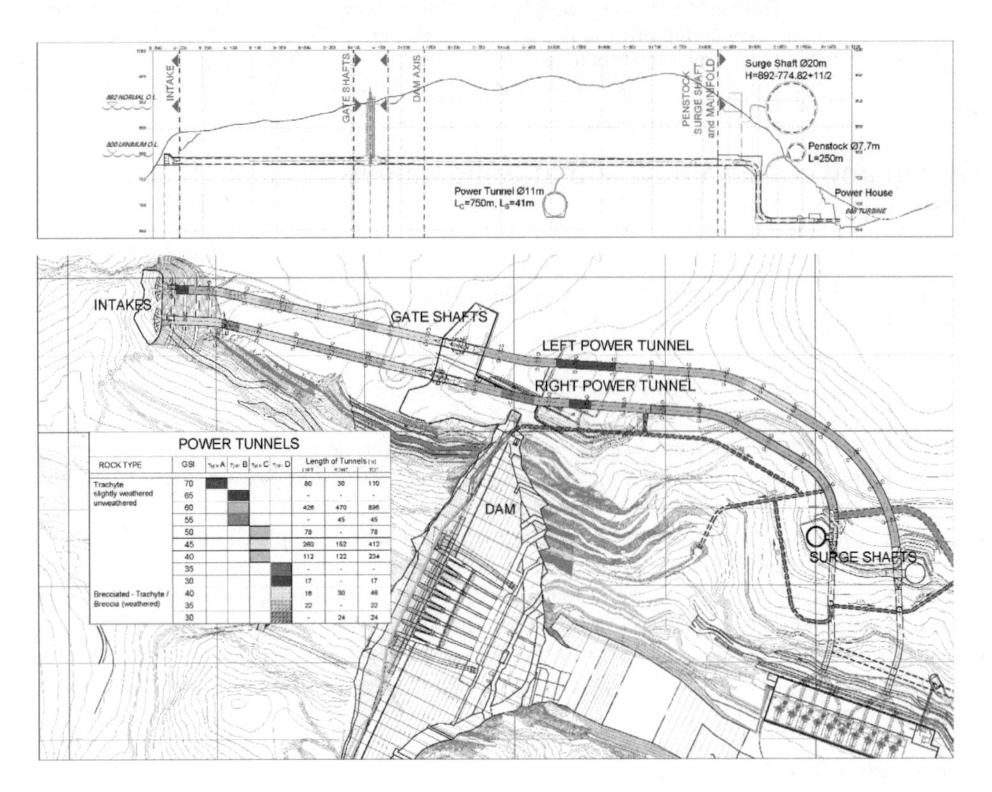

POWER TUNNELS

ROCK TYPE	GSI	A	B	C	D	Length of Tunnels (m) Left	Right	Tot
Trachyte slightly weathered unweathered	70	■				80	30	110
	65		■			-	-	-
	60		■			426	470	896
	55		■			-	45	45
	50			■		78	-	78
	45			■		260	152	412
	40			■		112	122	234
	35				■	-	-	-
	30				■	17	-	17
Brecciated - Trachyte / Breccia (weathered)	40				■	18	30	48
	35				■	22	-	22
	30				■	-	24	24

Figure 1. Power Tunnels general layout.

Each tower has a maximum height of 18 m, a width of about 30 m and a depth of about 25 m. The shape of the transition downstream of the bell mouth is designed in order to have a gradual variation of the water velocity. Trashracks at the entrance prevent accidental entrance of large submerged debris that could damage turbine parts, which is however unlikely because of the submergence always higher than 50 m during the operational life of the plant. There is also a predisposition for insertion of stoplogs downstream of the trash racks slots, to close exceptionally for inspection also the first stretch of tunnel upstream of the gates shafts.

1.3 *Gate shafts*

The gate shafts are designed in order to allow closure of each Power Tunnel by operating gates in wet conditions, and gates maintenance in dry area above el. 892 m a.s.l., where a control building is located. The closure of the power tunnel, for extraordinary maintenance, is possible by means of two sets of two gates (bulkhead and wheel, both 4.5 x 11 m) operated through the shafts. Shafts have height of about 118 m and internal diameter of 17 m, include the air vent pipes for aeration during tunnels emptying or opening of the wheel gates. The excavation of the upper portion of the shaft has been conceived with a telescopic enlargement of the excavation to allow to build a concentric stepped concrete wall acting as temporary support of excavation in a zone interested by alluvium and ignimbrite.

1.4 *Power tunnels*

The key features of the two power tunnels are:

- total length of about 1000 m, each;
- a longitudinal slope of approximately 0.5%;
- circular section with 11 m internal diameter;

- normal flow velocities up to about 4 m/s, exceptionally up to 5.4 m/s.
- internal water pressure up to about 1.1 MPa, with an exceptional maximum of 1.4 MPa nearby surge shaft

The main design topics, including geotechnical assessment, static design and construction issues of power tunnels are described in detail in the next paragraph of the paper.

1.5 *Surge shafts*

Cylindrical Surge shafts, one for each line, are envisaged to control the hydraulic transient dampening the water oscillations and reducing maximum pressures on power tunnels and penstocks caused by turbines manoeuvres, in order to prevent a water overflow in the shaft and air entrance in the penstocks. The dimensions and shape of the 5.8 m throttle orifice connecting to the Power Tunnels improve the dampening effect of the shaft while limiting the overpressure in the tunnel, at the base of the orifice, and controlling the potential cavitation risk. The shaft diameter was reduced from the initial value of 24 m to 18m, to minimize geo-mechanical and structural problems of the associated excavation, compatibly with the hydraulic requirements. The main dimensions of the left surge shaft are:

- 18 m Shaft internal diameter
- 1–1.5 m Reinforced concrete lining thickness
- 892.0 m a.s.l. Normal = Maximum Operating level
- 781.17 m a.s.l. Throttle axis elevation
- 5.80 m Throttle diameter
- 150 m Total height of Surge Shaft (out of which about 40m above ground)

1.6 *Penstocks*

The steel lined penstocks, with an internal diameter variable from 11 to 7.7 m together with relevant manifolds complete the power waterways system feeding the ten turbines in the powerhouse. The penstocks diameter has been selected to maintain maximum flow velocity in the range of about 10 m/s. The steel lining thickness ranges between 34 and 40 mm.

At the lower end of each penstocks a manifold is envisaged with five branches of diameter variable from 7.7 to 4.2 m, intercepted by butterfly valves, feeding the generating units.

Diameter gradually decreases to ensure maximum design velocities always in the range of 10 m/s, thus minimizing head losses in the bifurcations while maintaining a smooth hydraulic flow and controlling potential vibration and cavitation phenomena.

Penstocks are also equipped with a Ø1600 mm ecological discharge pipe.

2 GEOLOGICAL AND GEOTECHNICAL SETTINGS

2.1 *Geological settings*

The power waterways were excavated through the rock mass which outcrops on the left bank of the Omo River. The stratigraphy of the area is characterized by the presence of a sequence of volcanic and volcano-sedimentary rocks, which comprises, from top to bottom: Columnar Basalts, layer of pyroclastic rocks and Trachytic body, formed of:

- Slightly weathered Trachyte: porphyric, grey to yellow coloured, fine grained, often fractured and weathered, locally characterized by solution cavities;
- Unweathered Trachyte: light grey and dark grey trachyte flows).

The excavation of the power tunnels crosses for almost its entire length the unweathered and slightly weathered trachyte, classified as a strong or very strong rock. Locally worst rock conditions (about 100 m on a total length of about 2 km), presence of brecciated trachyte, have been found along the left and right power tunnels.

2.2 *Geotechnical settings*

The mechanical behaviour of the trachytic rock mass is modelled by the generalized Hoek–Brown criterion and Hoek-Diederichs equation, based on of the mechanical parameters of intact rock, the quality of the rock mass and disturbance induced by excavation. The geotechnical parameters of the intact rock, summarized in Table 1, were assessed based on extensive laboratory tests, including uniaxial and triaxial tests.

The mechanical quality of rock mass was evaluated using the GSI, estimated during the power tunnels excavation (Figure 1).

As described in the next paragraph, the tunnels excavation was executed using the standard heading-and-bench partial-face drill & blast method. A thickness of about 3 m of disturbed zone (D=0.5) is considered in the geotechnical model and in the stability calculations in order to take into account the disturbance due to the blasting operation.

Moreover, extensive campaign of in situ tests, including dilatometer, plate load and flat jack tests, were performed to assess the deformability of the rock mass and the in-situ stress state and calibrate the deformability parameters estimated by the empirical method suggested in bibliography.

The identification of joint sets in the rock mass surrounding the power tunnels was performed from the geological mapping made during the tunnel excavation. The trachytic rock mass along the power tunnels, but in general at the Gibe III site, is characterized by the presence of three main joint sets:

- two joint sets (named K1 and K2) with medium to high dip (from 60 to 90°) and high persistence, oriented respectively parallel and orthogonal to the Omo river;
- one joint set (named J1) with medium to low dip (< 45°) and low persistence.

Joint sets were generally moderately/widely spaced with high persistence, with joint surface generally slightly rough, slightly weathered, tight to partly open (0.1÷1 mm) and filled with soft gouge.

2.3 *Hydro-fracturing assessment*

The rock mass geological conditions and particularly the orientation of the main joints sets, combined with the high water pressure in the tunnel and its parietal position (i.e. close to the deep valley slopes) caused a significant hydro-fracturing risk. A test program was carried out included No. 47 hydro-fracturing tests (Figure 2). Results showed in situ stresses in the range of 1–1.2 MPa, slightly lower than the water pressure inside the tunnels normally up to 1.1 MPa and exceptionally up to 1.4 MPa. Consequently, in case of water leakages from the tunnel lining, the hydro-fracturing risk along the left bank is significant.

The Don Deere geomorphological criterion was applied to individuate the tunnel stretches with higher risk of hydro-fracturing. As a safety factor, upon Client's request, we have used the empirical geometrical criterion known as "Norwegian Rule", as reported in USACE manual "Tunnel and Shafts in Rock". The trend of relevant computed safety factors for the two tunnels, with water level at maximum operating conditions, is shown in Figure 3. The safety factor is

Table 1. Intact rock geotechnical parameters.

Rock type	Unit weight kN/m^3	Uniaxial compressive strength MPa	Young's modulus GPa	Hoek-Brown parameter m_i -
Unweathered Trachyte	24.5	98	25	14
Slightly Weathered Trachyte	23.9	65	22	20
Moderately weathered trachyte				
Trachytic breccia	24.5	50	20	19

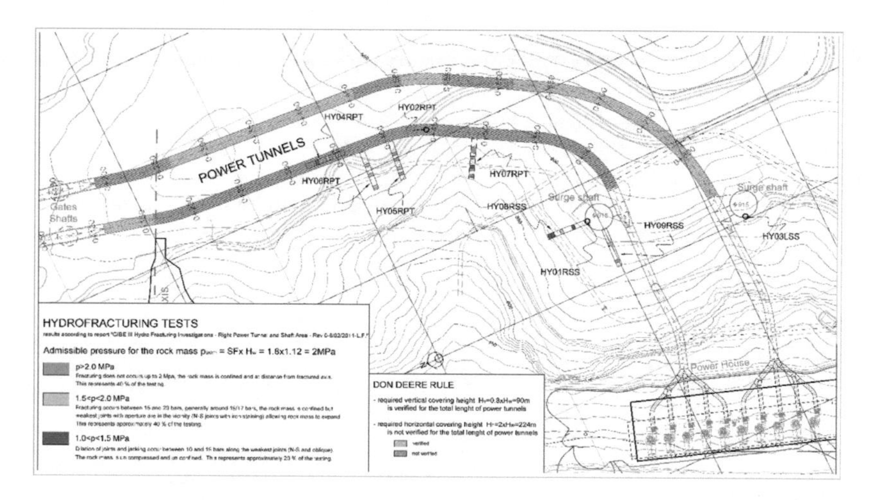

Figure 2. Tunnel stretches with potential risk of hydro-fracturing (in red).

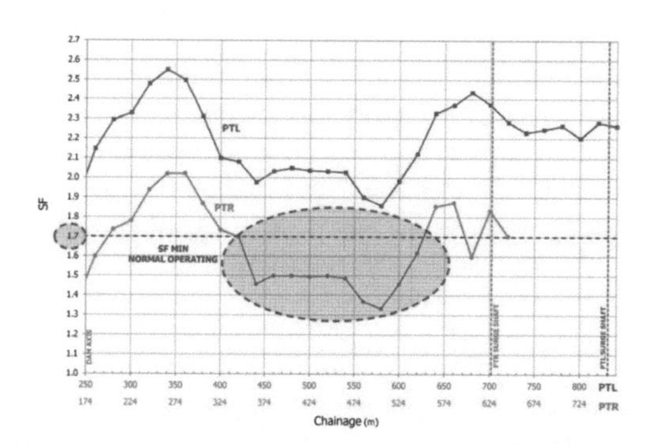

Figure 3. Hydro-fracturing safety factors along the power tunnels.

obviously lower for the right tunnel (PTR) closer to the valley slope surface with a minimum in correspondence of the middle small valley, which provides less rock cover to the tunnel.

This analysis is of course indicative and may be applied for a preliminary study only, since it considers just the geometric aspects without taking into account neither the rock mass nor the joints sets characteristics. The strictly application of such "rule" would result in an excessively conservatively approach, not justifying an extension of the tunnel steel lining.

3 POWER TUNNELS EXCAVATION AND TEMPORARY LINING

3.1 *Excavation method*

The Power tunnels excavation has been executed using the standard heading-and-bench partial-face drill & blast method. The cross-section has a modified horseshoe shape with a variable height ranging from 12.3 and 13.4 m and a maximum width ranging from 12.2 to 13.2 m, depending on the tunnel stretch and rock-mass quality. The excavation cycle included:

- drilling of 51 mm blasting holes using a fully computerized Jumbo (Tamrock Axera T12 Data) that guaranteed the precise execution of the memorized hole's pattern by means of a Laser Beam control system.
- loading of the blast holes with explosive cartridges and electric detonators;
- connecting all the detonators with the firing cable;

- checking of all the wirings and connecting the fire cable with exploder;
- detonating the blast;
- ventilating to remove blast fumes by means of a fan line plastic duct, blowing air inside tunnel at a position approximately 50 to 75 m to the front;
- scaling crown and walls to remove loosened pieces of rock;
- mucking and loading on dump truck using also lateral niches excavated approximately every 150 m along the tunnel for reverse and manoeuvre of vehicles.

Soon after mucking the following activities were carried out:

- topographical survey of the excavated section;
- geological mapping of the excavation front area;
- protection and support of the exposed rock, according to the results of the geological mapping and as per the design drawings;

3.2 *Temporary lining*

Design of temporary lining was based on the geotechnical data from the available boreholes, geo-structural stations, laboratory tests and field inspections carried out in the inlet/outlet areas and in the adjoining temporary access tunnels.

Since the conditions encountered during the excavation could locally differ from the ones assumed in the design, it was decided to apply the so-called "observational method", performing systematically a detailed geological and geotechnical mapping of the excavation front after each blasting activity. The geotechnical team, devoted to this mapping activity, was charged to evaluate the geo-mechanical characteristics of the rock-mass, identify the supports to apply and, if necessary, adapt and integrate the systematic support system as needed.

The stability analysis of the Power Tunnels, for the definition of the temporary support and lining, comprised general stability of the excavated section, analysing the possible development of plastic zones around the opening and structurally controlled stability of the crown and sidewalls, resulting from the development of wedges.

The general stability and temporary support system has been preliminary assessed with the "Q" method and the "characteristic lines" method. A more detailed analysis have been performed through a bi-dimensional finite element model which allow to take into account also

Table 2. Power Tunnels, Temporary supports system.

Section type	Lithology	GSI range	Shotcrete	Bolts	Steel Sets
A	Grey-trachyte	> 70	-	-	-
B	Grey-trachyte	56...70	Min. 10 cm thickness, reinforced with welded mesh ϕ 5 mm @ 10x10 cm	No. 14 ϕ25, L = 6 m, @ 2.5 m	-
C	Grey-trachyte	36...55	Min.15 cm thickness, reinforced with welded mesh ϕ 5 mm @ 10x10 cm	No. 23 ϕ25, L = 6 m, @ 1.8 m	-
D	Grey-trachyte Trachytic breccia	23...35 33...46	Min. 25 cm thickness, reinforced with welded mesh ϕ 5 mm @ 10x10 cm	No. 23 ϕ25, L = 6 m, @ 1.5 m	-
E	Trachytic breccia	23...32	two layers with minimum thickness of 5 and 25 cm, each one reinforced with welded mesh ϕ 5 mm @ 10x10 cm	-	Steel sets type HEB 180 @ 1.0 m (installed after the first shotcrete layer)

the shape of the tunnel, the excavation sequence and the in situ field stress (and in particular the possible interaction between the two parallel tunnels).

The wedge analysis has been assessed by modelling the shear strength of rock discontinuities using the classical Barton-Bandis or Mohr failure empirical criterion.

Depending on the GSI class the temporary lining and support works illustrated in Table 2 were prescribed. In all the section types spot bolting were applied, where required, in order to control the specific local structural instabilities identified during the mapping. Generally, a thin layer of shotcrete was always prescribed as protection against rock weathering and/or local ravelling phenomena. Where required by site conditions (wet or dripping conditions), drain holes (Ø 60 mm, 2 to 5 m long; spaced at 2/3 m) where drilled after the shotcrete application.

4 POWER TUNNELS FINAL LINING

4.1 *General*

Based on the results of geological assessment and the hydro-fracturing risk analysis, described in the previous paragraph, the tunnel lining was carefully designed in order to provide three lines of defence to control the risk of leakages: 1) reinforced concrete or steel lining, 2) systematic consolidation grouting and 3) drainage tunnels.

4.2 *Tunnels lining*

The final lining of the power tunnels is constituted by a reinforced concrete ring for the entire length of the left tunnel and for the first stretch (600 m long) of the right tunnel. The steel lining was adopted only for a limited stretch of right power tunnel (about 300 m upstream of surge shaft) where higher risk of hydro-fracturing is present.

The reinforced concrete, with thickness ranging from 50 to 70 cm, was designed to withstand internal and external water pressures (Figure 4), tailored stretch by stretch based on rock characteristics, expected water loads, and local risk of hydro-fracturing. This steel lining, 36 mm thick and backfilled with concrete, was designed to be self-standing against internal and external loads.

The design criteria required a crack width below of 0.1 mm, under normal operating conditions, and 0.2 mm under exceptional conditions. This required a heavy reinforcement, which caused construction challenges during bars installation and subsequent concrete pouring. Applying standard procedures for the crack analysis, in order to maintain the crack width below the limits, the steel bars would had worked well below the allowable tensile stresses and the calculated lining deformation was not compatible with the expected deformation of rock mass-lining system. Moreover, the required reinforcement could have caused locally bars congestion and

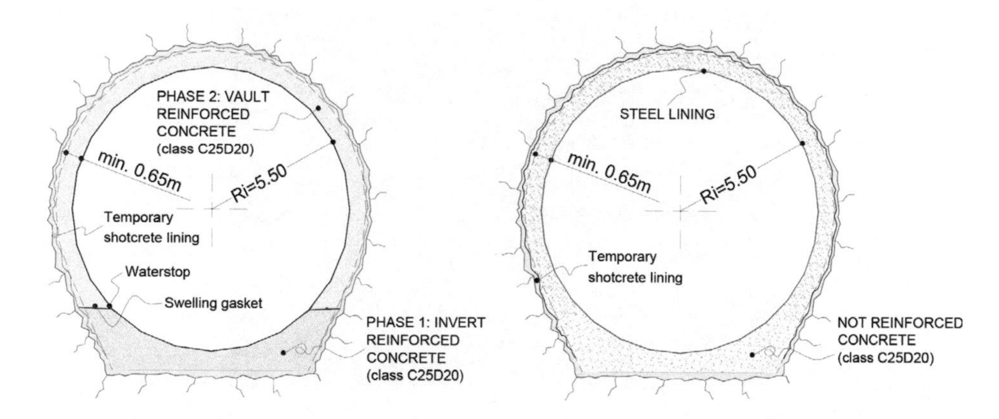

Figure 4. Typical lining sections. Left: Reinforced concrete section, Right: Steel lining section.

consequent difficulties in concrete pouring and vibration, highly increasing the risk of local lining defects. Therefore, it was applied the design approach proposed by Hendron et al., controlling bars design by crack analysis, reducing the total amount of steel bars of about 20%.

The placement of reinforcing steel in the tunnel was carried out by hand and using a special scaffold on wheels, based on that used for lining erection support. The tunnel was lined in two stages, starting with the invert first, that had a flat floor to facilitate the movement of formwork and scaffolding on wheels. Despite the measures adopted by the designer, the problem of steel congestion could not be completely avoided. This issue was faced by taking special care during concrete placing and vibration. Additional external vibrators were applied to the metal formwork in the sidewalls and crown of the tunnel, and even though some repairs of the lining surface had to be carried out after removing the formwork, the final result was consistently satisfying throughout the tunnel.

All longitudinal and transversal construction joints were provided with external PVC waterstops (200 mm wide, 2 mm thick) housed in preformed shallow trenches, bolted and sealed with epoxy resins (Figure 7)

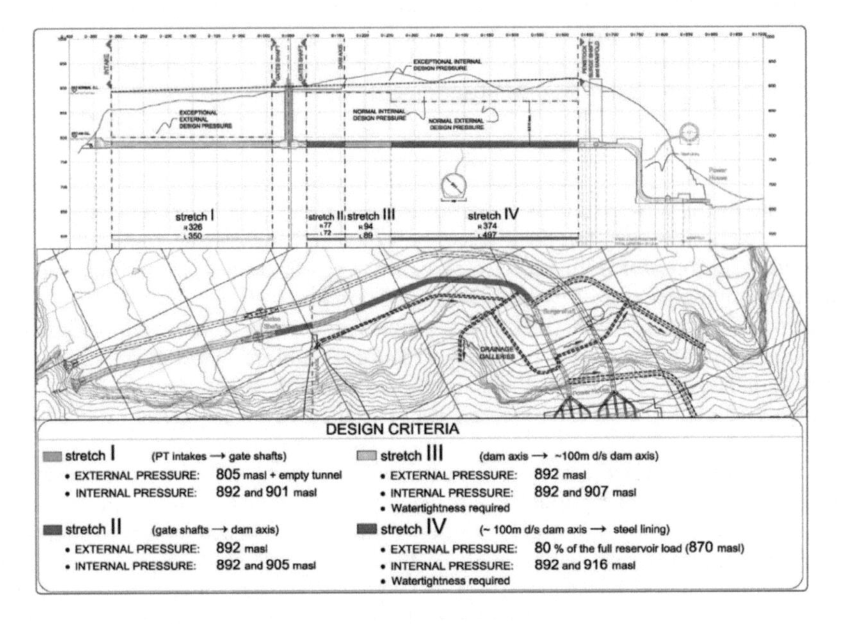

Figure 5. Typical lining layout (right tunnel) and design criteria.

Figure 6. Reinforcement and concrete placement.

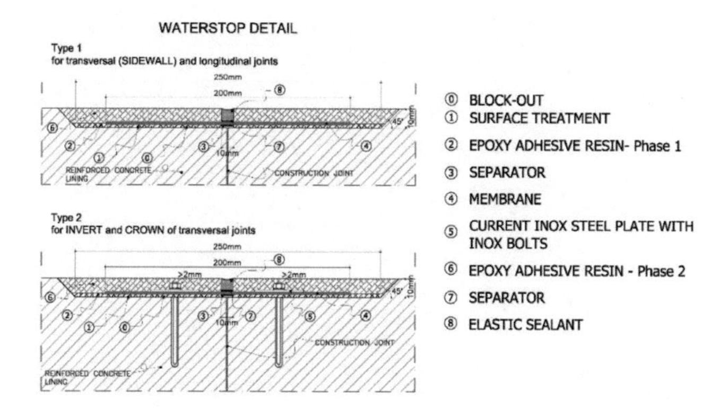

Figure 7. Power tunnels, details of longitudinal and transversal joints.

4.3 *Consolidation grouting*

Consolidation grouting was done systematically all along the tunnel, after placing the final lining. The main purposes of consolidation grouting work were:

- improve the quality and resistance of rock disturbed by blasting operations;
- filling and classing possible open joints around the tunnel, with benefit impact in contrasting the possible passage of water and the possible instable wedges formation.

The radial injections were made in two stages: 1) contact: low pressure (0.5 bars) grouting to fill the possible gap between the final lining and the rock mass 2) consolidation: high pressure (10 Bars) and deep grounding to improve the rock mass characteristics around the tunnel.

The holes spacing has been chosen in according with the spacing of main persistent sub-vertical joint sets. Sleeve pipes were inserted inside the lining to avoid perforation of the reinforced concrete at the moment of the grout works. Where the steel lining was adopted phase 2 was omitted with the exception of the first 25 m after concrete lining section

4.4 *Drainage galleries*

A special drainage gallery system (Figure 8) was added parallel to the right tunnel, closer to the slope, in order to protect the valley slope, in case of leakages, against the risk of hydro-fracturing and slope instabilities caused by seepages (especially on the cliff above the Power House and Plunge Pool). Along these tunnels there are deep drainage holes (n.4–5 drainages per section, L=12m, ϕ = 51 mm, s = 2 m spacing along tunnel direction). The Figure 9 shows the leakages measured along the drains and in the V-notches installed along the drainages galleries during impounding (almost completed). We can observe that the total leakage flow is very low, about 2.0 l/s. Moreover, leakages from the galleries appear to be influenced by the tunnels opening and/or by the raising of the reservoir water level above the elevation of the drain gallery invert.

5 CONCLUSIONS

The design of the Power waterways of Gibe III Hydroelectric Plant was challenging in reason of the huge size of the works, the construction/operation constraints and the site characteristics (the geomorphology of the valley and rock-mass characteristics).

Such difficulties required to adapt the project during the construction by implementing specific countermeasures (including portals relocation, large drainage and lining works, etc.) for the successful completion of the works. The monitoring of project performance during its first years of operation indicates the soundness of the design solutions.

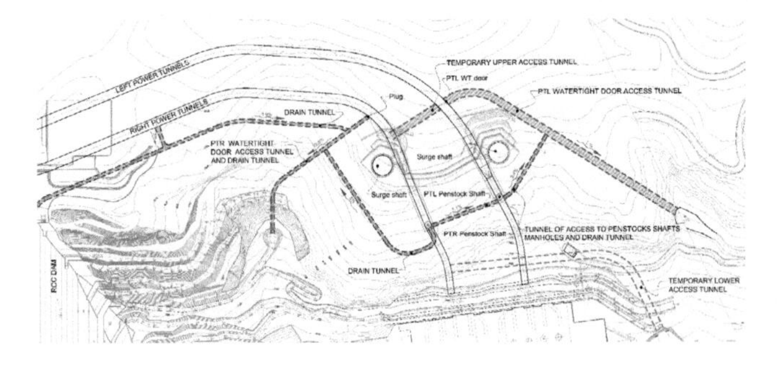

Figure 8. Drainage tunnels layout.

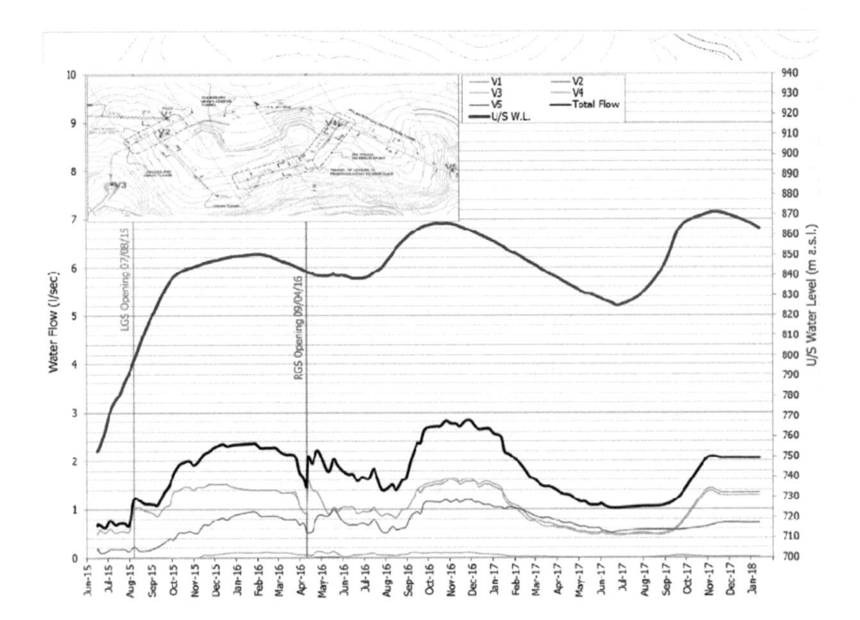

Figure 9. Power tunnel drain gallery, leakages.

REFERENCES

Barton, N. (2002). Some new Q-value correlations to assist in site characterisation and tunnel design. Int. Journal of Rock Mech & Min Sci, 39.

Fernandez, G. (1194). Behaviour of Pressure Tunnels and Guidelines for Liner Design. ASCE Journal of Geotechnical Engineering, Vol. 120, No. 10.

Hendron, A. J., Fernández, G., Lenzini, P. & Hendron, M.A. (1987). Design of pressure tunnels. The art and science of geotechnical engineering at the dawn of the twenty first century. Prentice-Hall, Englewood Cliffs, N.J., 161–192.

Hoek, E., Carranza-Torres, C. & Corkum, B. (2002). Hoek-Brown criterion – 2002 edition. Proc. NARMS-TAC Conference, Toronto.

Hoek, E. & Diederichs, M.S. (2006). Empirical estimation of rock mass modulus. *International Journal of Rock Mechanics and Mining Sciences.*

USACE (2005). Tunnels and shafts in rock, University Press of the Pacific.

Tunnels and Underground Cities: Engineering and Innovation meet Archaeology, Architecture and Art, Volume 3: Geological and geotechnical knowledge and requirements for project implementation – Peila, Viggiani & Celestino (Eds)
© 2020 Taylor & Francis Group, London, ISBN 978-0-367-46583-4

Salamonde II undergound hydroelectric complex in the North of Portugal. Design and construction

R.S. Pistone
COBA, Lisboa, Portugal

N. Plasencia
EDP Produção, Porto, Portugal

L. Gonçalves
EPOS, Lisboa, Portugal

ABSTRACT: In the North of Portugal, a new hydroelectric scheme was carried out. The new hydropower plant is a repowering of the Salamonde facility functioning since 1953. This repower consists of an energy improvement up to 386 GWh annual production, by the installation of additional 224 MW of production capacity. Part of this new scheme included the construction of a new powerhouse cavern, 165 m deep, on the Gerês granitic rock mass, a large surge chamber, a mainly vertical shaft headrace circuit and a 1930 m long, unlined, tailrace tunnel. In this paper the geological and geotechnical characteristics are summarized, the underground work design is presented as well as the most relevant aspects of the powerhouse cavern and tailrace construction phase, including the results of the monitoring system installed. EDP Gestão da Produção de Energia (EDPP) is the owner of this repowering project, the designer was COBA Group and the company "Construsalamonde, ACE" (a joint venture composed by Teixeira Duarte-Engenharia e Construções, S.A., EPOS-Empresa Portuguesa de Obras Subterrâneas, S.A. and SETH-Sociedade de Empreitadas e Trabalhos Hidráulicos, S.A.) was the contractor.

1 INTRODUCTION

The Salamonde II hydropower project, consists on the repowering of the existing Salamonde hydropower plant through the installation of a reversible Francis unit with an installed capacity of 224 MW, under a nominal head of 118 m and a design flow of 200 m^3/s (Figure 1).

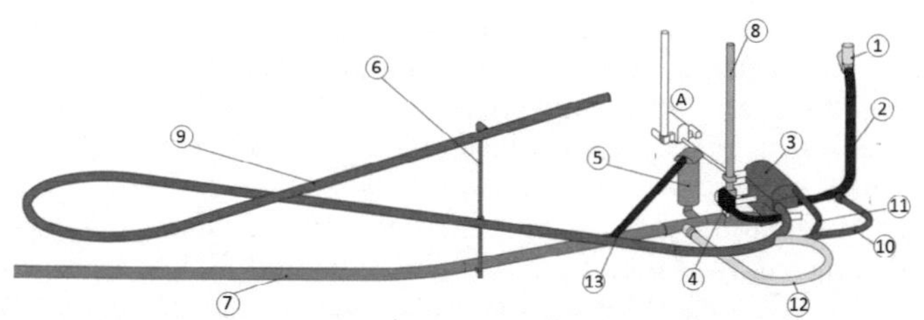

Legend: 1 - Water intake; 2 – headrace shaft; 3 – Powerhouse; 4 - Gate chamber; 5 - Surge tank; 6 - Pumping shaft of the emptying system; 7 - Tailrace tunnel; 8 - Busbar shaft; 9 - Access tunnel to the power plant; 10 - Access tunnel to the headrace tunnels; 11 - Auxiliary gallery the top of the powerhouse; 12 - Access tunnel to the tailrace tunnel; 13 - Auxiliary gallery to the top of the surge tank; A – Salamonde I underground powerhouse

Figure 1. Layout of the Salamonde repowering project (Salamonde II).

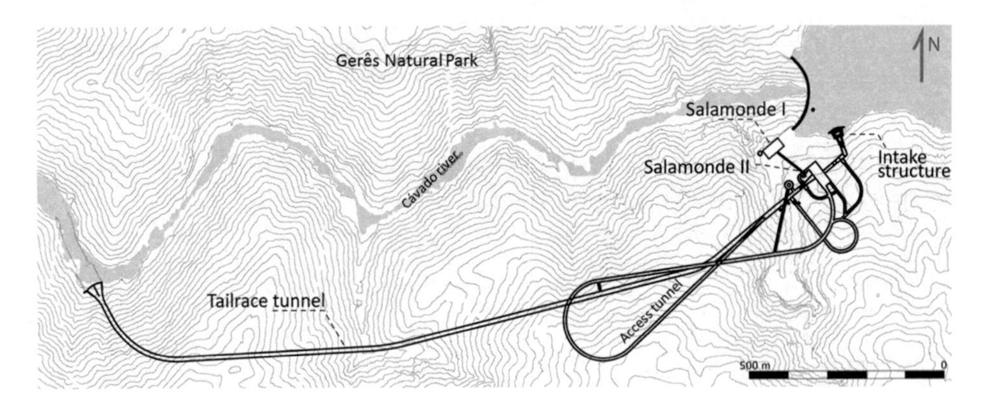

Figure 2. Salamonde II general layout.

The project is located in the surroundings of the Gerês Natural Park, nearby the village of Salamonde, at the North of Portugal. The layout of the 2.2 km-long hydraulic system is entirely underground and surrounds the existing powerhouse, on the left bank of the Cávado river, including a new powerhouse in cavern, under an overburden of approximately 165 m, on the Gerês granitic rock mass.

The water intake is located on the left bank of Salamonde reservoir – FSL at elevation (270.36) around 120 m upstream the left abutment of the existing dam. The high pressure circuit has a total length of about 200 m comprising, from upstream to downstream, an initial short sub-horizontal stretch, a 90° vertical curve, a 90 m high vertical shaft with 8.3 m inside diameter, a second vertical curve, a transition piece from 8.3 m to 5.8 m and a final horizontal stretch of tunnel that leads to the powerhouse (Santos et al. 2017). The high pressure circuit is concrete lined, with a maximum flow velocity of 3.7 m/s.

The underground powerhouse is in the upstream zone of the hydraulic tunnels system. The cavern is 66 m long and 27 m wide and has a height ranging between 27.5 m in the South (assembly hall) and 44.7 m in the North area where the generation unit is installed (Figure 3).

The low pressure circuit, or tailrace tunnel, initiates in the turbine draft-tube, extends for about 2030 m and ends in the tailrace structure at Caniçada reservoir. The first stretch of this circuit, between the draft-tube and the access tunnel, with a maximum flow velocity of 4 m/s, has around 100 m long and is concrete lined. The tailrace tunnel is shotcrete lined.

2 GEOLOGY AND GEOTECHNICS

The hydropower project is within the NW sector of the large tectono-stratigraphic unit called "Zona Centro Ibérica" – Central Iberian Zone – (Galicia-Trás-os-Montes Subzone), and the project is constructed on the Gerês granitic rock mass. This granitic formation presents, in general, coarse facies in the borders and medium to coarse porphyroid towards the inside of rock mass. The region is densely fractured, the fractures being often filled by quartz veins. Regionally, the predominant fracturing systems are: N-S to NNE-SSW and NNW-SSE. Figure 5 shows the isodensity diagram of the joints measured during design phase, compared with joint sets surveyed during the construction.

The geological investigation program consisted of a surface survey and specific geological mapping, including the study of discontinuities (fractures and faults), their geotechnical characterization as well as of a detailed geo-structural survey of the rock mass, taking advantage of the existing Salamonde powerhouse and the surrounding unlined tunnels.

The geotechnical investigation program (Sarra Pistone et al., 2010), allowed to establish a basic geological model (Figure 4), from which it was expected that the conditions of the rock mass improve with depth, prevailing the slightly weathered to unweathered rock mass (W2 to W1–2) and widely to extremely wide spaced fractures (F2 to F1–2).

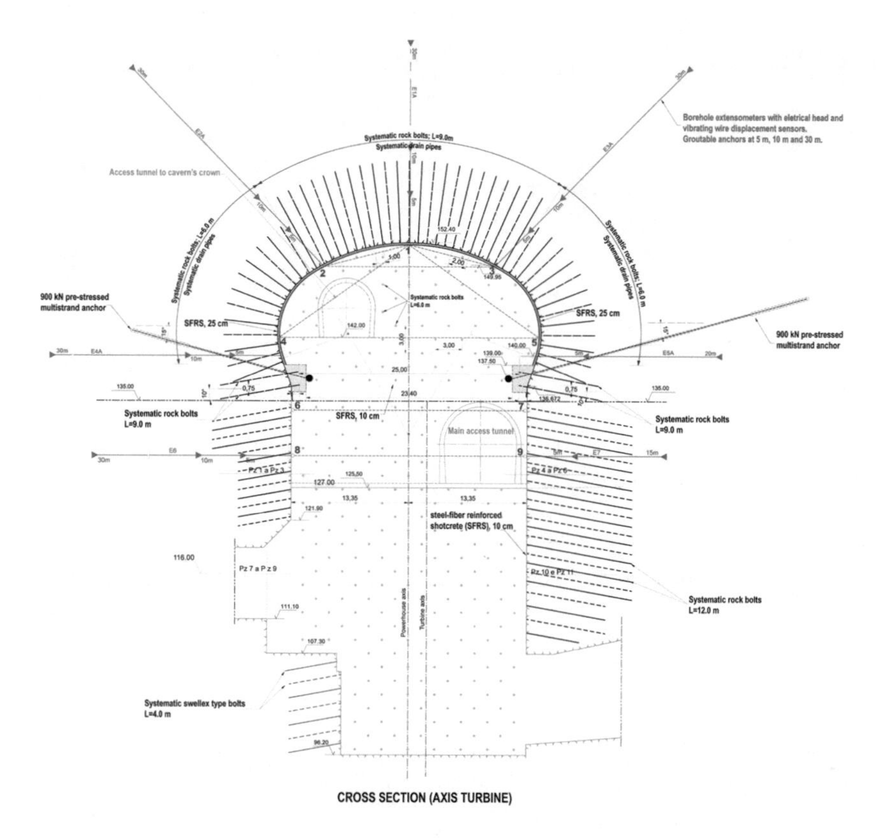

Figure 3. Cross section of the underground cavern by the axis of the unit, including support and monitoring devices.

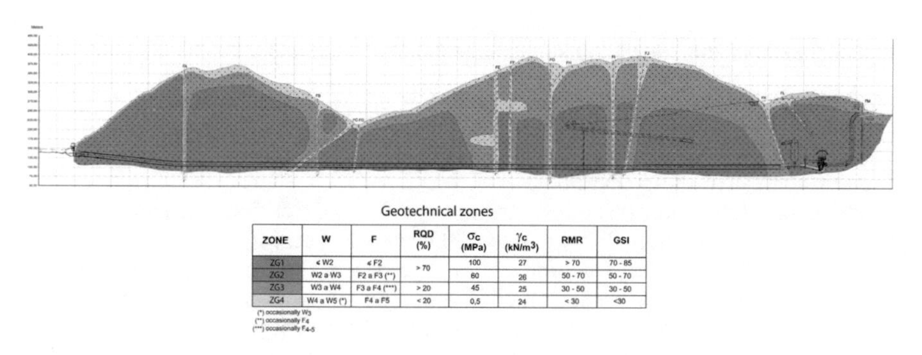

Geotechnical zones

ZONE	W	F	RQD (%)	σ_c (MPa)	γ_c (kN/m^3)	RMR	GSI
ZG1	≤ W2	≤ F2	> 70	100	27	> 70	70 - 85
ZG2	W2 a W3	F2 a F3 (**)		60	26	50 - 70	50 - 70
ZG3	W3 a W4	F3 a F4 (***)	> 20	45	25	30 - 50	30 - 50
ZG4	W4 a W5 (*)	F4 a F5	< 20	0,5	24	< 30	<30

(*) occasionally W_3
(**) occasionally F_4
(***) occasionally $F_{4\text{-}5}$

Figure 4. Geological and geotechnical longitudinal profile of the hydraulic circuit.

The geomechanical parameters, estimated from the results of the tests carried out, showed for the cavern area a slightly weathered to unweathered rock mass of approximately elastic behavior: E = 50 GPa, ν = 0,21. The shear strength parameters of the intact rock were estimated at c = 16 MPa, φ = 58°. The shear strength parameters of the discontinuities were estimated at: c = 0 - 0.35 MPa, φ = 24° - 33° for an applied normal stress of σ_N = 0.1 to 6.3 MPa.

As for the water circulation in the rock mass, it was considered that the rock mass was very impervious and that the water could circulate essentially through the discontinuities (Figure 5). It should be noted that, on plan, the Salamonde reservoir is located 60 m away from the cavern vault (North top). The normal water level is (273) and the cavern roof arch top is at level (152.40). So, in spite of the joint conditions, some water could be expected during excavation and operation.

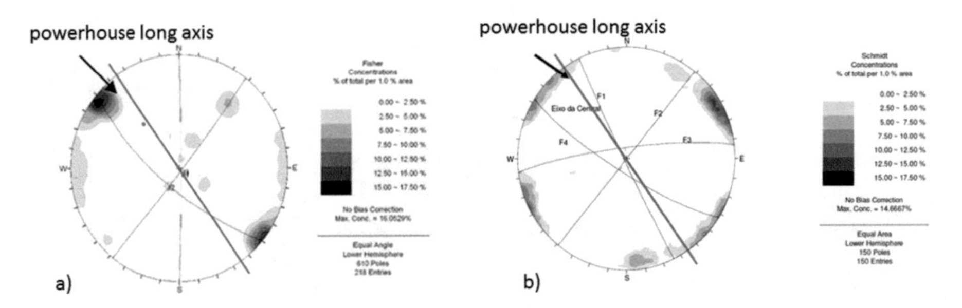

Figure 5. a) Isodensity diagram of main discontinuities during site investigation; b) joint sets measured during construction.

3 GEOTECHINAL DESIGN

3.1 *General considerations and geotechnical parameters*

The typical geomechanical parameters of the excavated rock mass were assigned according to the estimation of the GSI parameter combined with the values of the uniaxial compressive strength of the rock, *σci*, using the elastoplastic model based on the Hoek-Brown failure criterion for the material. Equivalent parameters for the Mohr-Coulomb failure criterion were obtained from the previous ones for depths from 100 m (ZG2) to 150 m (ZG1).

In the specific zone of the powerhouse, 6 STT (Stress Tensor Tube, three-dimensional deformeter) tests were carried out by LNEC (National Civil Engineering Laboratory) in the S8 and S13 boreholes (Figure 6). From the results of the STT tests, the value of k=σh/σv equal to 0.7 used in the geotechnical design was considered appropriate. The behavior of the excavation was also checked for a value of k equal to 1.5.

3.2 *Numerical models for the analysis of excavation and support*

In the stability analysis of the excavated areas and design of supports, continuous and discontinuous numerical models were used. In the discontinuous model, the size and kinematic feasibility of probable blocks were analyzed using the program "Unwedge" (Figure 7), from the geo-structural surveys of the excavations.

The analysis of relevant plastic zones was made on 2D and 3D finite element models, using the calculation programs "Phase2" and "Plaxis 3D Tunnel", respectively.

For the 2D model, a 60% decompression factor of the rock mass has been considered, allowing important displacements and the consequent redistribution of the stresses installed in the rock mass before the execution of the support. In the powerhouse 3D model, it was assumed that the support system would be operating immediately after the start of the excavation in order to absorb the totality of displacements

Table 1. Summary of calculation parameters of the rock mass and discontinuities.

Geotechnical zone	σci (MPa)	Em (GPa)	Geo Strength Index GSI	Mohr-Coulomb		Hoek-Brown		
				c (kPa)	φ(°)	mi	mb	s
ZG1	100	30	70–85	1500	55	34	12	0,05
ZG2	60	10	50–70	800	50	32	6,5	0,01

Angle of friction of discontinuities: ϕ (°)	25
Cohesion: c_a (kPa)	10

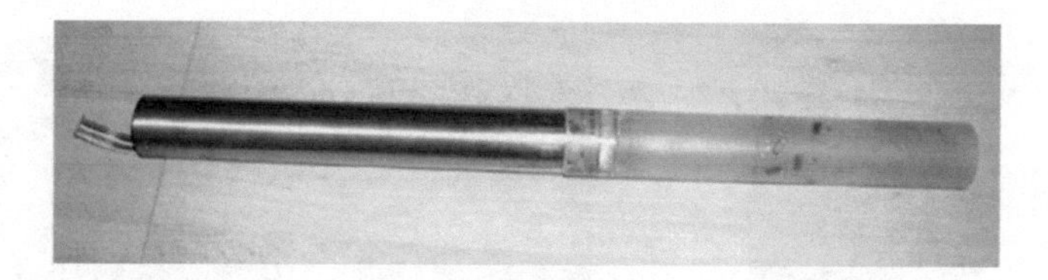

Figure 6. Three-dimensional deformeter tests for determining the initial stress condition (STT).

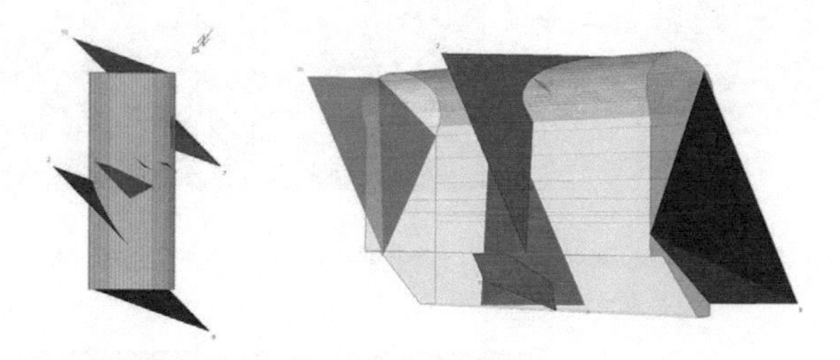

Figure 7. Analysis of probable blocks in the powerhouse cavern.

resulting from the stresses release on the excavated areas. The definition of the support was made taking into account the most critical situations, according to the geotechnical zoning of the rock mass.

As definitive support, the powerhouse roof arch was supported exclusively by fully grouted rockbolts (A500NR, galvanized steel, φ25 mm and φ32 mm) and fiber reinforced shotcrete, whilst on sidewalls the shotcrete lining was not considered, taking advantage of the good geotechnical characteristics of the rock mass and into account that half of the excavated walls height would be concreted (Figure 2). The powerhouse shows the rock mass walls as an architecture value.

According to the construction schedule, the activities associated with the excavation, concreting and equipment assembly presented such a phasing and incompatibility over the time, that they would require, at a provisional stage, the support of both crane beams when loaded (carrying the electromechanical equipment), using prestressed anchors instead of reinforced concrete pillars.

4 POWERHOUSE CAVERN CONSTRUCTION ASPECTS

The cavern excavation began in February 2012 and ended in March 2013, about 1 year later. Excavations were carried out by the drilling and blast method. The primary support applied in the various top heading fronts of the access galleries or in the access to the cavern and in the various hydraulic tunnels consisted, in general, of "Swellex" type provisional bolts and steel fiber reinforced shotcrete (density and length of the rockbolts depending on the geological and geotechnical characteristics of the actual rock mass, such as the shotcrete thickness to be applied. Definitive fully grouted rockbolts and shotcrete lining were performed in a sequential stage.

The excavation of the powerhouse cavern was phased and, in general, complied with the planned construction phasing (Figure 8). Firstly, the construction of two longitudinal galleries on each side of the arch (A), in order to install the first extensometers.

After completing the excavation of the entire roof arch and respective support, the crane beams and the corresponding supports were executed (Figure 9).

The tunnels and galleries crossing the cavern were constructed in full-face excavation.

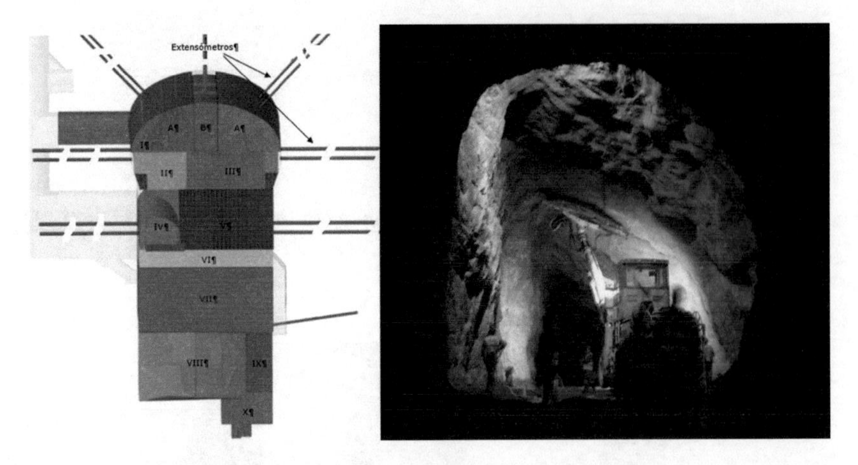

Figure 8. Excavation phases of the main cavern. Photograph shows the stage A.

Figure 9. Excavation of the roof arch and execution of the crane beams. Ramp and excavation of the lower level and installation of lateral extensometers.

5 TAILRACE TUNNEL CONSTRUCTION ASPECTS

The shafts of the hydraulic tunnels system, busbars and the surge shaft were carried out using the "raise-boring" technology and were later widened to the final diameter, through vertical drill and blast excavation. The final lining of the hydraulic system consisted of cast in place concrete in tunnel and pressure shaft, and of flexible lining (steel fiber reinforced shotcrete and fully grouted rockbolts) in tailrace tunnel, except in specific situations with weak geological and geotechnical characteristics, where concrete was also used.

The lining of the 1930 m long tailrace tunnel (Figure 10) was defined with grouted rockbolts and reinforced shotcrete in three different classes and a reinforced concrete lining solution, applied accordingly with the geological and geotechnical conditions.

The excavation method was drilling and blast in the full 12 meters diameter section (Figure 11). As defined in the contractual conditions, a team of site geologists of the contactor was responsible for all the geological and geotechnical surveying and monitoring of Salamonde II excavations, with the main goal of acquiring more information as the excavation goes along, as well to suggest and adapt support measures if different conditions were encountered. The observational method was applied in a wide way and the contractor was responsible for the progress of the excavations in safety.

Figure 10. Final lining of the tailrace tunnel. Left: rockbolts and shotcrete. Right: reinforced cast in place concrete.

Figure 11. Drilling at the tailrace tunnel face.

The geological and geotechnical follow-up work of underground excavations was one of the most important activities developed by the experts in the field of geology. From this activity resulted a set of important information. The geological and geotechnical mapping of all tunnels, shafts and caverns was performed, as well the geotechnical and hydrogeological monitoring in the underground excavations. From this activity resulted a set of important information useful to adapt and refine the design.

Before the final definition about the final lining to be considered for each zone of the tailrace tunnel, the geological information collected from the excavations mapping was analyzed together with the observed behavior registered by the monitored convergences. Taking into account the hydraulic purpose of the tunnel, based on the above analyses, and considering the presence of geological fractures parallel to the tunnel walls, geo-structures with potential erosion and sub-horizontal joints, some additional support elements were applied accordingly with Figure 12 types of additional support. Despite the low water velocity and the moderate hydraulic head, the sudden water level variation inside the surge chamber had to be considered on the final lining of the tailrace tunnel (Plasencia et al. 2016).

6 MONITORING

The monitoring program of the tailrace tunnel, was constituted, in exclusivity, by a systematic convergence measurement performed by a half second monitoring total stations (Figure 13).

The monitoring was most useful for evaluating the reaction of the support applied. As a rule, the numerical modelling was confirmed, but in some cases, although rare, it was

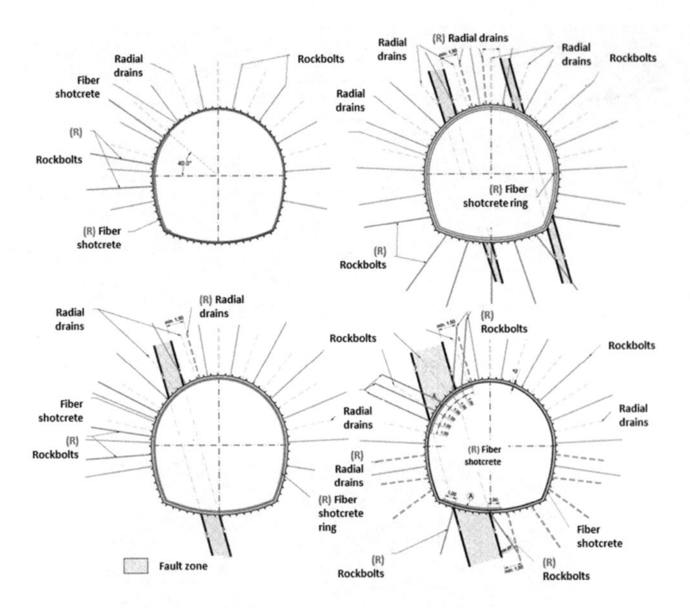

Figure 12. Types of additional support measures (R) applied accordingly geological conditions and behavior observed by monitoring convergence to complete the final lining.

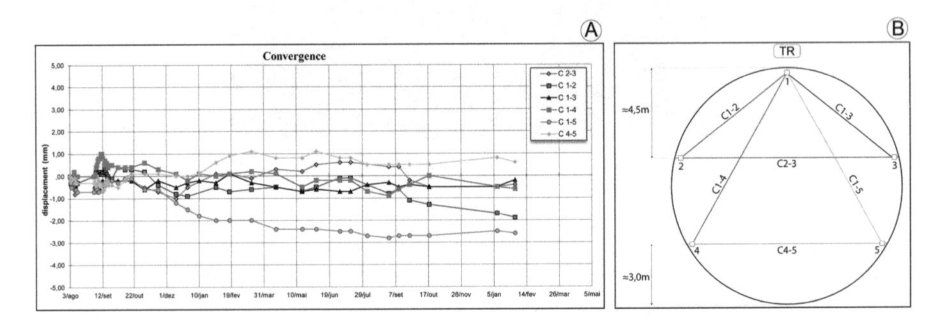

Figure 13. A – Convergence monitoring of 0+730 section at tailrace tunnel. B – Scheme of the position of topographic targets in the tailrace tunnel.

advisable the reinforcement of the support applied accordingly with the geomechanical classification and mapping of the excavations and, consequently, it was necessary the reformulation of the final lining.

The control of deformations inside the excavation of the powerhouse cavern is monitored through 12 rod extensometers and 6 convergence sections with 9 points (Figure 14). The measurement of convergences in the rock mass was carried out by means of optical methods that allow the visualization of the targets, despite the movements of the crane.

Furthermore, 11 piezometers and 7 load cells were installed in order to know the hydrostatic pressures in the surrounding rock mass as well as the variation of loads in the prestressed anchors during the excavation and use of the crane, respectively. A relevant aspect related to this project and to the construction methods adopted for the excavations is the constraints imposed as regards the limitation of vibrations caused by blasting.

Given the moderate state of compression of the roof arch, the installed extensometers practically did not register deformations. On the contrary, the horizontal extensometers placed in the boot of the walls, as expected, showed some deformations with the advance of the excavation.

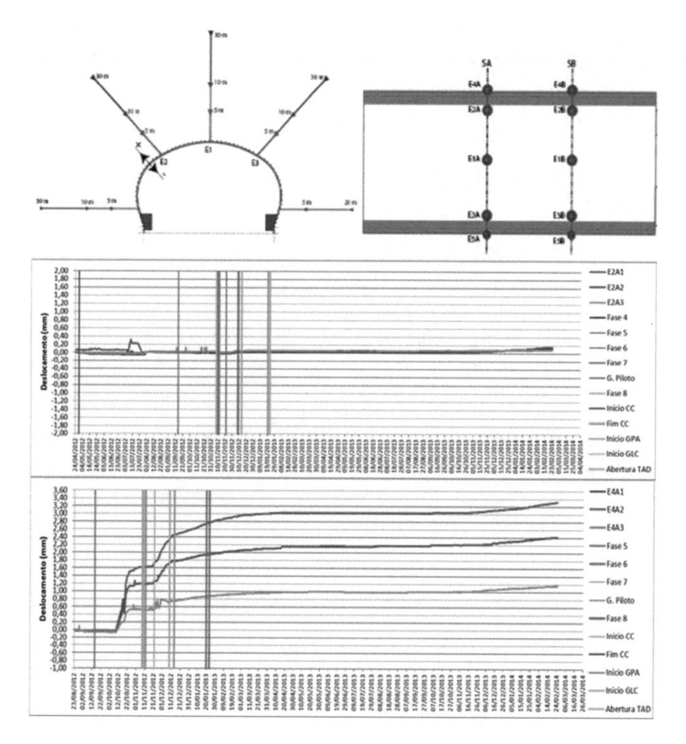

Figure 14. Displacements recorded in the extensometers E2A (top arch) and E4A (horizontal) installed in the cavern vault during the construction phase.

7 CONCLUSIONS

This project presents some relevant characteristics, of which mention is made to the relative proximity between the new hydraulic tunnels and the existing underground facilities, dam and reservoir, and the large volumes of excavation that, together with the vibration limits imposed by the current normative, constrained the construction sequence and the methodologies. During the excavations, the blasting plans were permanently adapted in order to minimize eventual damages to the existing structures. Regarding to the constructive method it was a challenge to keep in operation the existing powerhouse and protect the generators from any vibrations induced by blasting

The actual response of underground excavations was monitored and compared with the numerical predictions and careful observation made, demonstrating, as usual, the key importance of the monitoring in the safety control.

The geological surveying (mapping and geomechanical classification) during the excavations led to an increasing knowledge of the geological conditions.

Considering the use of the tailrace tunnel and the actions in the exploitation phase, the final executed lining took into consideration the characteristics of the geological structures and the behaviour of the underground space. Depending on the results of the analysis of the information obtained during the excavations, some linings had to be reviewed and the final lining had to be adapted to the actual conditions.

Due to the proximity to the Gerês Natural Park, the scheme went underground as far as possible. For example, the main access tunnel, had its length doubled during the construction, in order to reduce as much as possible the environmental impact.

For the connection of the new facility with the old powerhouse, a tunnel with a very low inclination was also constructed, using the raise-boring technique performed between the two caverns.

The rock material resulting from the underground excavation was used in the manufacture of concrete and the remaining volume was placed in a topographic depression resulted of an existing quarry from the previous hydropower construction, properly integrated to the landscape.

Finally we must praise the spirit of cooperation between all parties involved, which led to the success of the project.

REFERENCES

Plasencia, N. & Lima, C. 2016. Geoengenharia, Estruturas Geológicas e Túneis Hidroelétricos. *15CNG Congresso Nacional de Geotecnia.* Porto

Santos, P.G., Gusmão, L., Fangueiro, H. & Costa, J.S. 2017. The Salamonde II hydroelectric project. *International Hydropower association. Hydro 2017.* Seville.

Sarra Pistone, R., Silva, M. M., Lima, C., Bento, J., Plasencia, N. & Oliveira, S. 2010. Estudo Geotécnico da Central em Caverna para o Projeto do Reforço de Potência do Aproveitamento Hidroelétrico de Salamonde. *Encontro Nacional sobre o Espaço Subterrâneo e sua utilização. CPT. LNEC.* Lisboa

Tunnels and Underground Cities: Engineering and Innovation meet Archaeology, Architecture and Art, Volume 3: Geological and geotechnical knowledge and requirements for project implementation – Peila, Viggiani & Celestino (Eds)
 ISBN 978-0-367-46583-4

Turning water into energy: A geological and geotechnical overview of the Venda Nova III hydropower project construction phase

N. Plasencia
EDP - Gestão da Produção de Energia SA, Porto, Portugal

P. Matos
Spie batignolles Génie Civil, Nanterre, France

ABSTRACT: The Venda Nova III hydropower project is a key component of a hydroelectrical scheme, in the North of Portugal. This underground hydraulic circuit consists of headrace and tailrace unlined tunnels with 12m diameter, amounting to 5km in length, several construction galleries and a new powerhouse, reaching over 1000000m^3 of underground excavations. Those characteristics, associated to the high-water head, that can reach 480m, imposes many challenges of great complexity, during the geological survey and namely during the construction phase. A relevant fault zone intersected the main circuit as well some of the construction galleries and brought several challenges during the construction phase. Was the geological investigation during the studies phases enough? An overview of the Venda Nova III Hydropower Project experience is performed; a review of the state of research to overcome these persistent geological issues is presented, as well as proposals for future studies on these subjects.

1 INTRODUCTION

The goal of this document is to revise the geological and geotechnical studies and works developed in the scope of Venda Nova III (VN III) project and bring to the discussion some reflections that the authors have made throughout their joint work. EDP Gestão da Produção de Energia (EDPP) is the owner and designer of this project, and the joint venture composed by the companies MSF, Mota-Engil, Somague and Spie batignolles (named "Reforço de Potência da Barragem de Venda Nova III, ACE") was the contractor.

It should be noted that subjects such as construction methods and strategies will not be discussed in this paper.

2 DESCRIPTION OF THE PROJECT

Venda Nova III hydroelectric power plant is located in the north-western region of Portugal, within the Cávado river mountainous basin, the Portuguese region with the highest rainfall values (Figure 1, upper left corner). It is the second repowering project of the original scheme that is operating since 1951 (Oliveira et al. 2011). Like the first repowering scheme, Venda Nova II/Frades (Lima et al. 2002), VN III is mostly built underground in the granite rock mass of the Rabagão river left bank. Both hydraulic circuits are represented in Figure 1, 2 and 6. Venda Nova III consists essentially of a large underground cavern (103 m x 25 m x 53 m) designed for two reversible power units (756 MW) and transformers, an access tunnel (1.5 km) and underground hydraulic unlined tunnels (5.6 km) with an excavation diameter of 12 m, located near and almost parallel to the existing repowering scheme of Venda Nova II (VN II). The distance between the two powerhouses is around 300 m. The foreseen average yearly power generation is over 1000 GWh. The volume of excavations was around 1000000m^3.

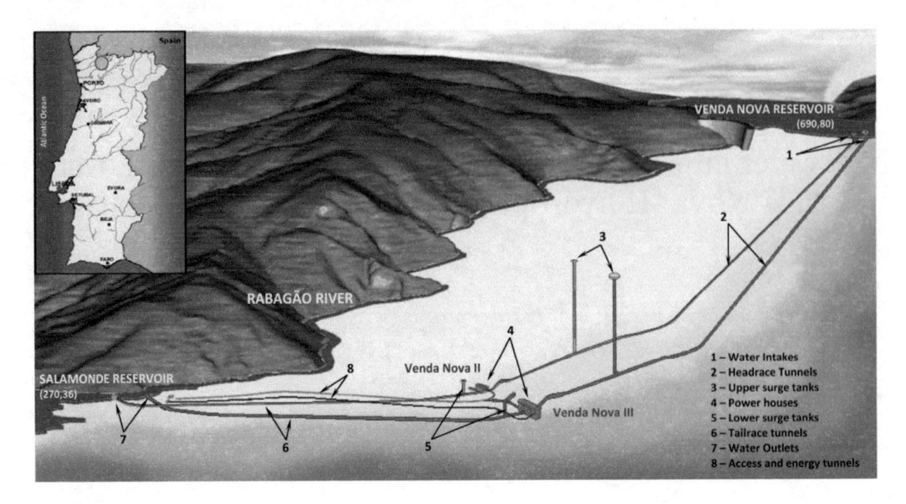

Figure 1. VN II and VN III. Hydropower schemes. Location and 3D representation (Marques et al. 2017).

3 THE VENDA NOVA III CONTRATUAL CONDITIONS

Regardless the planning of the geological surveying works, geological and geotechnical information management is approached differently by each of the actors involved in the construction. Underground conditions are often the argument used to justify delays, costs overruns that usually ends in claims and contractual disputes between the owner and the contractor with no real advantage for the project development or for the quality of the works

In this kind of projects, the current trend in the management of "Know-Unknown" risks (or uncertainties arising from eventual and possible insufficient characterization in the study phase – Palmieri 2012, so, unexpected geological conditions) recommends flexible contractual mechanisms, which allow the adaptation of the project or the construction methods to the faced reality, thus minimizing the risk of complaints by the contractor.

Following this principle, in last decade EDPP's projects, the type of contract was based in a *bill of quantities*, with EDPP defining the expected quantities and types of work, based on the information collected during the studies phases (Plasencia et al. 2017). The contractor's remuneration is then determined on the basis of the actual quantities actually executed, allowing the execution project to be adapted to the actual conditions encountered and to minimize the likelihood of a claim by the executing entity. The contract also foresees a mechanism to adjust the contractual duration with the volume of work performed visa foreseen work and the rules of payment the contractual extra permanence of the contractor on site.

However, to improve geological risk management, international trends are to the include a GBR (Geological Baseline Report) (ITA 2009) on the contract terms, which will establish the contractual limits of the conditions for construction, clarifying the expected geological conditions and setting the baselines for the contractual management of the geological risk. They also point to the inclusion of contractual clauses to manage geological uncertainties, defining the specific compensations associated with the occurrence of those not clarified situations.

With the goal of minimize the "Unknown-Unknown" risks (Palmieri 2012), also known as unpredicted geological conditions, the VN III contract indicated a continuous geological interpretation resulting from the analysis of the geological mapping and survey of excavated areas, to anticipate the geological conditions, but also geological investigation during the construction phase, made by the contractor and payed by the owner (Plasencia et al. 2017).

In view of geological risk management and aware that the unforeseen exists, EDPP assumed a contractual architecture of its own, imputing the responsibility for execution, making available the geological-geotechnical information to the contractor, on the assumption that this entity, being responsible for the progress of the excavations in safety, should also be responsible for obtaining the information of the surrounding geological environment. It was also contractually envisaged that, to obtain additional geological-geotechnical information, the contractor should have a team responsible for geological mapping, geotechnical monitoring and interpretation of information, but also ensure the immediate mobilization of drilling equipment whenever necessary. These contractual obligations are reinforced by attributing the geological risk of the execution phase to the contractor, who is primarily responsible for "any delays or suspension of work, partial or total, as well as additional costs, due to unfavorable geological or geotechnical conditions".

However, despite all these measures, the works were impacted by unforeseen situations directly or indirectly related with geological conditions.

4 GEOLOGICAL AND GEOTECHNICAL DATA DURING DESIGN PHASE

4.1 *Geological and geotechnical surveying*

The topographic model and the geological characteristics of the rock mass were decisive factors for selecting the location of the VN III powerhouse cavern, looking for adequate confinement conditions as well as compatibility with the most known important geologic structures, in accordance to the required submergence of the reversible units. Once the approximate powerhouse location was selected, the other hydraulic structures configuration and placement were also defined.

The surface geological mapping (Figure 2) and a site investigation consisting on a field program and in-situ tests were of major importance to the geological and geotechnical studies. The topography and the geomorphology, but also nowadays the environmental legislation, impacts on the site investigation program. In VN III project some places at the terrain were inaccessible.

Table 1 presents a summary of the investigation works that were made during the design phase of VN II and VN III that represent 0,7% of investment when compared with the civil construction investment of the project.

The VN II site investigation and the geological mapping of VN II excavations (Figure 3) included the group of information considered to geological and geotechnical studies of VN III. The excavation mostly cut middle to coarse grained porphyritic granite and highly fractured and weathered Gerês facies granite, as well as migmatites, hornfels and metagraywackes (Cotelo Neiva et al. 2000, Lima et al. 2002). Orientation of regional major faults was

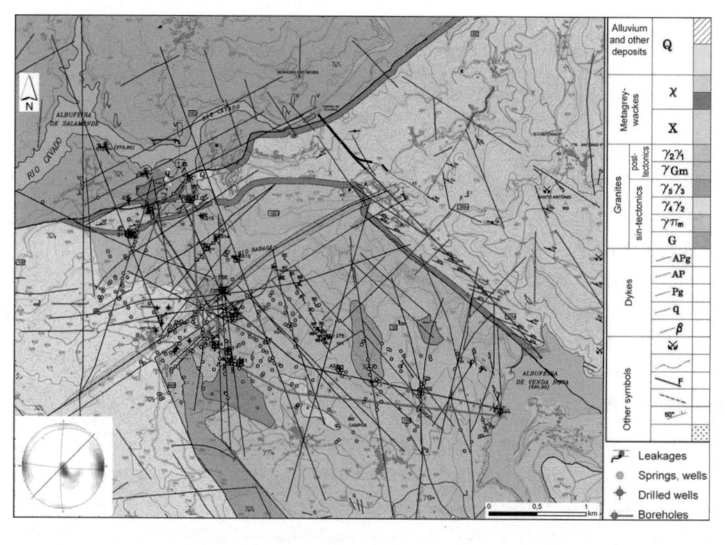

Figure 2. Geostructural and lithological mapping (developed for the first repowering studies and updated for more recent projects, by EDP), monitoring surface water points (springs, wells and boreholes (blue spots)) and the site investigation. VN II (black) and VN III (red) layouts at the left bank. Isodensity diagram of powerhouse cavern joint set (lower hemisphere projection), where red line represents the powerhouse caverns axis (Plasencia et al. 2015).

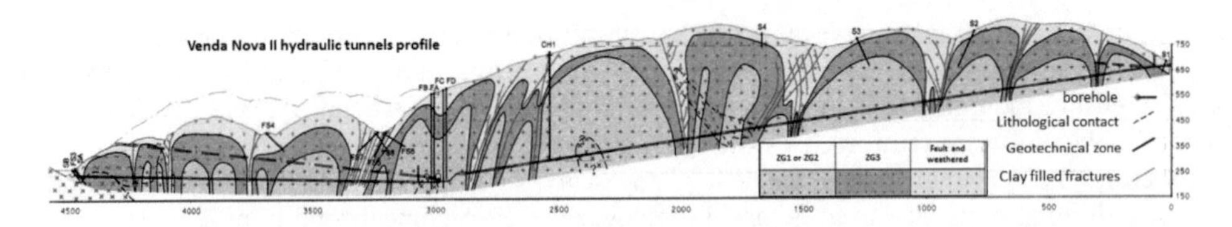

Figure 3. Postconstruction geological model of VN II, considering the information obtained during excavation (Plasencia 2003).

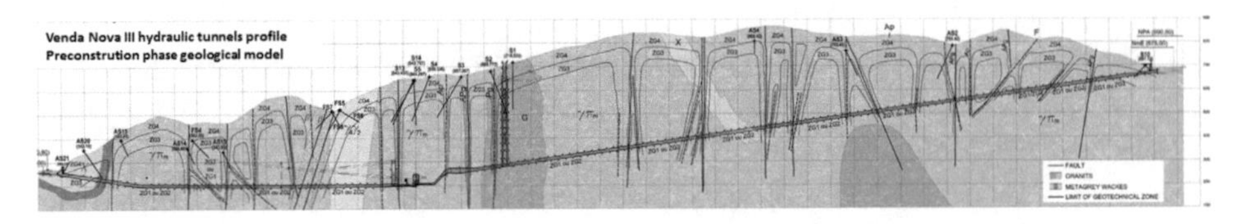

Figure 4. Geological model of VN III before excavation (EDPP 2009b).

Table 1. VN II and VN III site investigation for preconstruction phase (Plasencia 2003, EDPP 2009b)

	VN II - Feasability, design	VN III – design phase
Geophysical survey	4 profiles; tomography between 4 long and vertical boreholes	6 profiles
Boreholes (nb.)	36	17
Borehole samples	4980 m	1990 m
Lugeon tests (nb)	923	364
Powerhouse cavern survey gallery	Yes	Yes
BHD*, LFJ**, SFJ-**and STT***	Yes	Yes
Lab tests	Yes	Yes

* - with the boreholes campaign; ** - after the excavation of the powerhouse survey gallery; *** - with the boreholes campaign and after the excavation of the powerhouse survey gallery.

confirmed as NE-SW and NW-SE and also trending NNE-SSW to N-S. A very important NE-SW fault corridor (the so-called Botica fault zone) was crossed by the tailrace tunnel.

Based on site investigation and on the known information of VN II, the VN III hydraulic layout was classified in three geotechnical zones, whose main geotechnical rock mass description are presented in Table 2 (following the ISRM criteria, 1981) and Figure 4.

4.2 *Hydrogeological surveying*

As usually expected for crystalline rocks, the permeability is quite low (Table 2). Nevertheless, relatively high hydraulic conductivity was expected to be found in fault zones and near the surface. One important characteristic of this region is that infiltration and recharge occur with some regularity due to the distribution of rainfall along the year (average precipitation is 2116 mm/year). The observed behavior of groundwater during VN II excavations allowed predicting no major impacts on water levels during VN III excavations (Plasencia et al. 2015). One year before the construction, during the excavations, and two years after the hydraulic tunnels filling, 270 water points were monitored (Figure 2, chapter 4.1).

Table 2. Geotechnical zoning of the rock mass along the hydraulic layout of VN III (EDPP 2009b)

Geotechnical Zone	W, F	RQD (%)	P-w (m/s)	E (GPa)	UCS (MPa)	Permeability (UL)			
						> 10	10–2	2–1	<1
ZG3	W5, F5 W3, F4	0–70	800–2800	< 20	< 30	36%	29%	9%	26%
ZG2	W3, F3 W2, F1-2	70–95	3000–600	20–40	70–100	1%	27%	9%	63%
ZG1	W1-2, F3 W1, F1	95–100	4800	40–70	110	2%	2%	4%	92%

Legend: W, F - Degree of weathering and of fracturing; RQD – Rock Quality Designation; P-w – P-waves velocity; E – Young's modulus; UCS – Uniaxial compressive strength; UL – Lugeon unit.

5 GEOLOGICAL AND GEOTECHNICAL SURVEY DURING CONSTRUCTION PHASE

The geological and geotechnical follow-up work of underground excavations is one of the most important activities developed by experts in the field of geology, if permanent staff is provided on site. From this activity results a set of important information, often depending on the contractual requirements of the owner of the project. For the VN III project, a team of engineering geologists was present on site 24/7 for the geological survey, so that the progression of all excavations could be monitored and reported. The geological and geotechnical mapping of all tunnels, shafts and caverns was performed, as well the geotechnical and hydrogeological monitoring in the underground excavations.

5.1 *Geological and geotechnical surveying*

As defined in the contractual conditions (EDPP 2009a), a team of site geologists of the contactor team was responsible for all the geological and geotechnical surveying and monitoring of VN III excavations, with the main goal of acquiring more information as the excavation goes along, as well to suggest and adapt support measures if different conditions were encountered. This surveying started in 2010 with the access and construction tunnels, executed before the hydraulic tunnels. It included the works that are described in Table 3.

This means that the access and construction galleries worked as big boreholes, giving more detailed information at deeper depths and near the hydraulic layout (tunnels and powerhouse), just like the operating circuit of VN II gave for the studies phase of VN III. The observational method (Peck 1969, EC7 NP EN 1997-1 2010) was applied in a wider way, always connecting the numerical information of the excavation behavior, collected from the monitoring methods, with the geological and geotechnical characterization of the surrounding rock mass.

The geological and geotechnical surveying during the construction phase of VN III confirmed the good rock mass quality, where approximately 75% of the hydraulic tunnels were excavated in the higher quality geotechnical zone (ZG1 and ZG2, see Table 2 and Figure 5). In structural and statistically terms, two more joint sets were highlighted due to the increased number of measured joints, but with no negative impacts in the rock quality, and so, no negative impacts for the construction phase. The lower quality geotechnical zone (ZG3) included some fault zones detected during the studies phase, as zones of water inflow and outtake zones, given their lower depth. The well-known major fault zone, called Botica fault, was intersected, as predicted and causing no delays, first at the access and energy tunnel, and later in the tailrace tunnel of VN III.

However, a few meters after intersecting the Botica fault during the excavations of the access and energy tunnel, a fault zone (named fault **F** in this document) with a trend related with some of the most regional geostructures but not undoubtedly identified at the design phase, was present as a joint set not very representative or persistent, and not characterized

Table 3. VN II and VN III surveying works during construction phase (Plasencia 2003, Matos 2014).

	VN II (from 1998 to 2002)	VN III (from 2010 to 2015)
Geological mapping	Projection plane, hand-made	Surface development, in CAD basis
Rock mass classification	RMR_{89}, Q	RMR_{89}, Q
Geo-structural analysis	Stereograms, characterization and statistical analysis of discontinuities	Stereograms, characterization and statistical analysis of discontinuities
Rock mass description	Lithology, W, F, RQD	Lithology, W, F, RQD
Rock mass testing	Point load tests (every 10 m)	Point load tests (every 10 m)
Convergence monitoring	Yes (every 20 m, in average)	Yes (every 20 m, in average)
Boreholes with sampling	Yes (access tunnel, tailrace tunnel)	Yes (water intake, surge shafts, access tunnel, headrace tunnel)
In situ tests	Yes *	No
Exploratory boreholes	No	Yes (every 20 m, in average)
Synthesis rapports	Yes	Yes

Legend: RMR_{89} – rock mass rating; Q – Q system; CAD – Computer aided design; W, F - Degree of weathering and of fracturing; RQD– Rock Quality Designation; *- related to the design phase, see Table 1.

with such an important aperture and fillings with mylonitic clay and breccia, came to show that it could intersect the powerhouse access tunnel, the tailrace tunnel, the headrace tunnel and both surge shafts. Later the geological mapping, developed during the excavation of the access tunnel near the powerhouse and of the powerhouse cavern roof, provided geological information that came to confirm that this fault zone was indeed persistent and composed by a set of surfaces/branches, undulated, with the average direction NNW-SSE (see the example of a conceptual model of a major fault zone that represents this description on Figure 6).

Far ahead, the geological mapping excavation confirmed the intersection of this major fault zone, fault **F**, in the headrace tunnel lower adit and in the tailrace tunnel and allowed to conclude that some surfaces/branches of the fault zone would, with no doubt, intersect the shafts. At the Figures 5 and 7 the major fault **F** is indicated.

This geological fault had to be overpassed and alternative strategies were set to excavate, treat and support the affected part of the shaft walls inside the fault zone, using specific methodologies, which brought delays and extra costs to the endeavor (Marques et al. 2017).

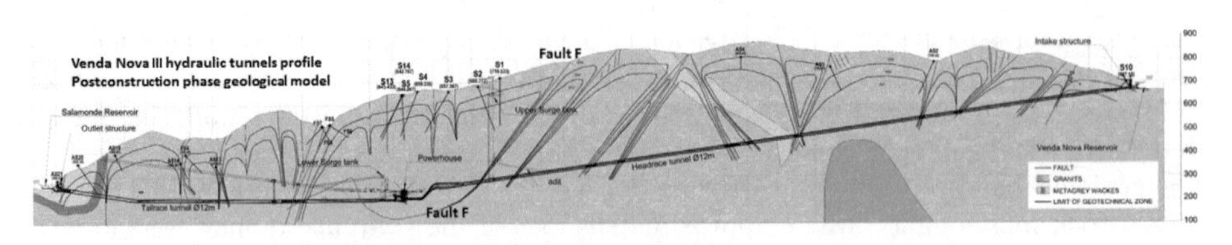

Figure 5. Geological model of VN III, considering the information obtained during excavation (post-construction). The major fault zone F is indicated (EDPP, 2015).

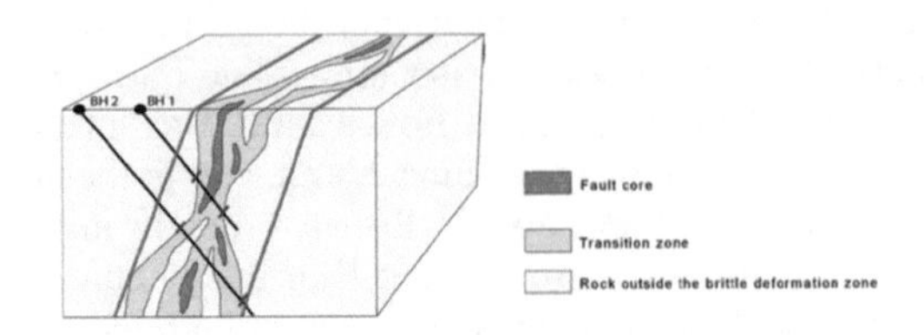

Figure 6. Conceptual model of a major fault zone (adapted from Caine et al. 1996.)

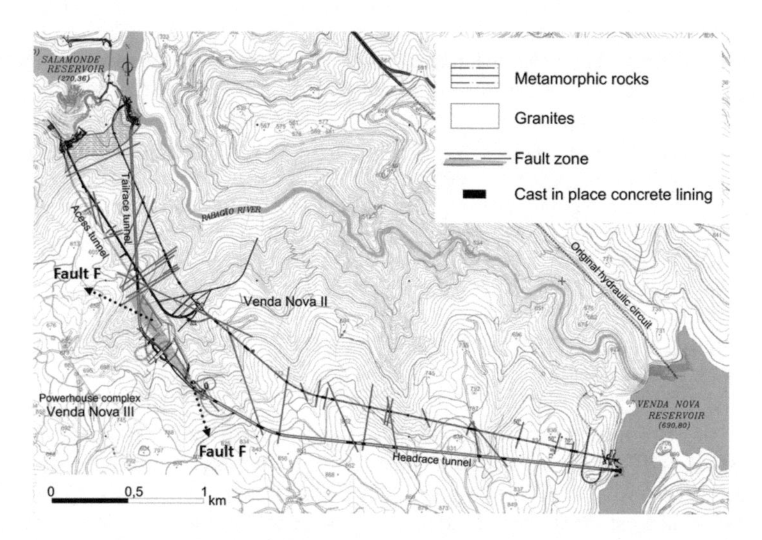

Figure 7. Venda Nova II (in blue) and Venda Nova III (in red). Cast in place concrete lining applied on the headrace and tailrace tunnels. The fault zones are schematically indicated (adapted from Esteves et al. 2017).

This geologic structure, as well all fault zones intersected, and also the results of convergence measurements in the tunnels, but also water inflows, were considered on the final lining solutions of the hydraulic circuit (Figure 7) and the filling procedures (Esteves et al. 2017).

Based on site investigation and considering the new information obtained during the excavations of VN III, a new geological model is presented in Figure 5. It is possible to compare this model with the one for the design phase, showed in Figure 4.

5.2 *Hydrogeological surveying*

The excavation of VN III was carried out nearby the hydraulic tunnels of VN II in operation. At the design stage, the 300 m of distance between the two powerhouses was considered enough to minimize inducing potential effect on the flow rates from VN II to VN III tunnels during its excavations. Nevertheless, during VN III construction, in addition to the monitoring of surface water points that monitored the impacts at the surface, all leakage to the excavations were controlled by measuring water inflows and monitoring water quality (chemical and physical analyses) to find out the nature of the water inflows, if it was from the VN II hydraulic tunnels (from the reservoir) or if it was natural groundwater, stored in the rock mass (Esteves et al. 2017). For that purpose, a contractual methodology of systematic exploratory boreholes was planned every 20 meters during the excavation of the hydraulic circuit (Figure 8).

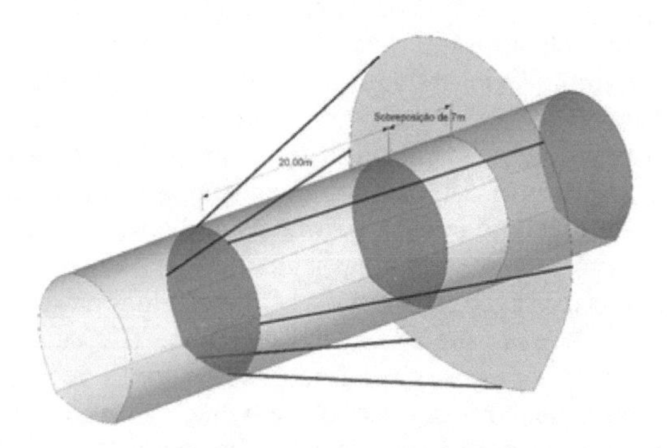

Figure 8. Scheme of the exploratory boreholes for hydrogeological monitoring in the hydraulic tunnels (EDPP 2009a).

No significant inflows have been detected, even on the most permeable (fractured) zones, and the analyses revealed that all occurred inflows correspond to existing groundwater.

The intermediate-lower quality and more permeable geotechnical zones were concentrated at the superficial regions, close to the water intakes and the upper part of the upstream surge chamber, as well in the intersection of as major fault zones.

Excluding the surface areas where the groundwater level had been identified with high probability of been affected, observed behavior suggests that VN III excavations had negligible impact on the water table.

6 GEOLOGICAL SURVEY PROGRAM: HOW TO EVALUATE, WHEN IS IT ENOUGH?

"You pay for the site investigation whether you have one or not" (ICE 1991). Thirty years have passed since ICE published this report, and still nowadays the final costs of endeavors and the accomplishment of deadlines continue to be influenced by the quality of the site investigation studies. These studies are intended to provide information to designers so that they are able to interpret ground conditions that will lead to the underground project. The "value" of a geological-geotechnical study, that includes the factual data and the interpretation analysis, will have a direct influence on the degree of knowledge within the design phase as well for the construction phase. For this, it is important to control the quality or strengths of geological and geotechnical information in order to implement a set of techniques and measures to ensure that such information is reliable and appropriate.

Controlling the geological information does not replace risk management systems to be implemented during construction, nor eliminates the need of monitoring, and mitigation measures in case of unexpected situations. The goal of this "quality control" is to minimize such occurrences by introducing requirement criteria and classification of the studies. The control of the information quality does not intend to do the impossible, that is remove the inherent unreliability of the geological environment, neither eliminate the scale effect that this kind of studies are always subjected to, nor alleviates the need for criticism on the way that these issues are seen and dealt with, but rather to evaluate the strength of the geological information that is used as a model for the engineering design development. Some quantitative guidance exists in the literature on assessing the confidence level of geological and geotechnical studies, and some successful attempts to describe what level of geotechnical data is required have been made by several authors (Jaska et al. 2003, Dunn et al. 2011, Miranda 2007, Read 2013, Thomas 2013).

6.1 *A database production*

In 1984, a subcommittee of the U.S. National Committee on Tunneling Technology (USNC/TT) made a comprehensive study of exploration practices in the United States to determine if a greater level of geotechnical investigation effort could reduce the final constructed cost of tunnel projects. The approach adopted was to examine completed projects for which the results of the preconstruction site investigation could be related to the construction history. The procedure was designed to permit in-depth study of a large number of these projects, their respective site investigation programs, and the construction problems and unanticipated costs, or lack thereof, as a means of determining the nature and significance of the relationship between investigation programs and project problems and costs. It was found that claims for unexpected subsurface conditions were a significant part of the total cost of a tunnel. Final payments in result of claims averaged nearly 12% of the original basic construction cost. Some as-completed costs were 50% over the engineer's estimate. The subcommittee's conclusions and recommendations convey a philosophy of genuine desire to get a job done right the first time while being fair to both the contractor and the owner. As well noted by Parker (1996), it is often difficult to convince owners to implement the recommended amount (and cost) of studies. It is well documented that special investigation programs developed specifically to reduce such contingencies have been instrumental in reducing the bid costs by more than ten

(10) times the cost of geological and geotechnical studies. Even greater savings are accrued to the owner because such investigations can lead to minimization of construction delays and of potential conflicts and claims from contractors. Savings in the bid price have been achieved on the order of 5 to 15 times the cost of investigation programs.

However, the cost of a site investigation is only part of the story. It is the quality of the investigation that is vital to understand and overcome (Littlejohn et al. 1994). One promising opportunity is examination of the geotechnical site investigation process for proposed construction sites. For that purpose, a creation of a worldwide underground works database could be organized by listing the types and extent of geological and geotechnical studies at the site, carried out in the design and construction phases, as well any changes during the construction phase, cost overruns and delays which are usually associated to geotechnical and geological issues.

6.2 *A database analysis*

This valuable database could then be analyzed to identify common and important parameters (either quantitative or qualitative variables) that determine the nature, significance and uncertainty of the relationship between investigation programs and project problems and costs, by applying data mining techniques and other robust statistical tools. From this point, by determining the main variables that had influenced the quality of the geological studies, it will be also possible to apply decision-analysis tools for measuring the attractiveness of a geological and geotechnical survey study by categorical based evaluation techniques that should include engineering judgment by a board of experts.

7 CONCLUSIONS

The geological studies for the Venda Nova III project were developed in four phases: the geological and geotechnical studies for Venda Nova II preconstruction phase, the geological survey of Venda Nova II during construction phase, the geological and geotechnical studies of Venda Nova III preconstruction phase, and the excavation surveying during the construction phase of Venda Nova III. This has led to an increasing geological knowledge from one to other phase, but also to an increased investment that overcame largely the values for geological investigation proposed by the geotechnical community. For these facts, VN III was a particular case in terms of geological knowledge, but even so it wasn't possible to eliminate completely the geological risk. Is it possible that more investment in geological investigation would reduce the so called "unknown" situations in underground projects within areas difficult to access? If so, how much more?

Each tunnel is a different tunnel, and each geological and geomorphological scenario dictates different approaches and a specific investigation program. But it could be of major interest if more information, with the geological investigation conducted in all civil underground projects developed around the world as an output to decision analyses for future studies, would be shared, or a database created with such information.

In such underground projects, as the VN III, the geological survey during construction, the acquisition of geological information as the excavation goes along, is fundamental because it complements the geological survey for the preconstruction phase, reducing the "Known-Unknown" geological risk, and it can identify geological aspects, not found at the preconstruction phase, also mitigating the "Unknown-Unknown" geological risk. Considering the importance of the information obtained during construction - geological mapping, geotechnical monitoring; etc. - it should be mandatory to create a database containing such information. This database assumes a relevant paper for new underground urban excavations and major publics works.

At VN III, associated to the proximity of VN II in operation, systematic exploratory boreholes for hydrogeological monitoring in the hydraulic tunnels were executed. For the management of the geological risk, the contractor was responsible for the safety progress of the execution in terms of construction methodologies and identification of geological conditions during excavation. EDPP defined the expected quantities and types of work, based on the information collected in the studies of the design phases and payed the executed quantities, allowing the design to be adapted to the encountered conditions. It can be said that the geological risk was shared.

REFERENCES

Caine, J.S., Evan, J.P. & Forster, C.B. 1996. Fault zone architecture and permeability structure. *Geology*, 24, 11: 1025–1028.

Cotelo Neiva, J.M., Plasencia N.S. & Lima, C. 2000. Geological and geotechnical characteristics of the rock mass of the Venda Nova II hydraulic circuit (in Portuguese). In Proceedings *VII Portuguese Geotechnical Congress, Porto*, pp. 113–122.

Dunn, M.J., Basson, F.R. & Parrot, T.T. 2011. Geotechnical data – a strategic or tactical issue? in Proceedings *Fourth International Seminar on Strategic versus Tactical Approaches in Mining*, Australian Centre of Geomechanics, Perth.

EDP Gestão da Produção de Energia, 2009a. Condições Técnicas. *Volume III – Caderno de Encargos.* Reforço de Potência de VN III. Not published document (in Portuguese), Porto, Portugal.

EDP Gestão da Produção de Energia, 2009b. *Estudos de Caracterização Geológica. Volume I, Parte 1 – Geologia e Geotecnia. Aditamento ao Projeto – Estudos de especialidades e elementos complementares.* Not published document (in Portuguese), Porto, Portugal.

EDP Gestão da Produção de Energia, 2015. *Venda Nova III – Primeiro Enchimento do Circuito Hidráulico – Especificação - PRT-2015-01378 (Rev. C).* Not published document (in Portuguese), Porto, Portugal.

Esteves, C., Plasencia, N., Pinto, P. & Marques, T. 2017. First filling of hydraulic tunnels of VN III hydropower scheme. WTC2017. Bergen, Noruege.

Institution of Civil Engineers 1991. *Inadequate Site Investigation.* London: Thomas Telford.

International Society of Rock Mechanics 1981. Basic geotechnical description of rock masses. *Int. Journal of Rock Mechanics, Mining Sciences and Geomechanics Abstracts*, Vol. 18, No. 1, 85—110.

International Tunneling Association 2009. General Report on Conventional Tunneling Method. *ITA Working Group Conventional Tunneling.* ITA report nº002.

Jaska, M.B., Kaggwa, W.S., Fenton, G.A. & Poulos, H.G. 2003. A framework for quantifying the reliability of geotechnical investigations. *Applications of Statistics and Probability in Civil Engineering.* Rotterdam: Der Kiureghian, Madanat & Pestana, Millpress.

Littlejohn, G.S., Cole, K.W. & Mellors, T.W. 1994. Without Site Investigation Ground is a Hazard. *Proceedings of the Institution of Civil Engineers, Civil Engrg*., Vol. 102, May: 72–78.

Lima, C., Resende, M., Plasencia, N. & Esteves, C. 2002. VN II hydroelectric scheme - powerhouse geothecnics and design. In *ISRM News Journal*, pp37–41.

Marques, T., Plasencia, N., Esteves, C., Pinto, P., Oliveira, M.A. & Lima, C. 2017. The surge shafts of Venda Nova III hydropower plant's surge tank–geological conditions and constructive solutions. In WTC 2017, Bergen.

Matos, P. 2014. Hydrogeochemical evolution in the granitic rock mass of hydraulic circuit of Venda Nova III repowering boost (Vieira do Minho, Nord Portugal). MSc thesis (in Portuguese), University of Minho, Braga, Portugal.

Miranda, T., 2007. Geomechanical parameters evaluation in underground structures. Artificial intelligence, Bayesian probabilities and inverse methods. PhD Thesis, University of Minho, Dept. of Civil Engineering, Guimarães.

NP EN 1997-1 2010. Eurocódigo 7 Projeto geotécnico, Parte 1: Regras gerais.

Oliveira, M.A., Esteves, C. & Duarte, F. 2011. Venda Nova III repowering project. In Proceedings *HYDRO 2011.* Prague.

Palmieri, A. 2012. *Managing financial risks for uncertainty.* In Proceedings *HYDRO 2012.* Bilbao.

Parker, H.W. 1996. *Geotechnical investigation. In Tunnel Engineering Handbook.* Parsons Brinckerhoff, NY, USA.

Peck, R. B. 1969. Advantages and limitations of the observational method in applied soil mechanics. *Géotechnique*, 19(2), 171–187. http://doi.org/10.1680/geot.1969.19.2.171

Plasencia, N. 2003. *Contributions of engineering geology for conception and design of underground structures.* MSc thesis (in Portuguese), Technical University of Lisbon.

Plasencia, N., Carvalho, J.M. & Cavaco, T. 2015. Groundwater monitoring impacts of deep excavations: hydrogeology in the VN repowering schemes. *Environmental Earth Sciences.* pp 2081–2095.

Plasencia, N., Alves, S., & Sá, P. 2017. Imprevistos Geológicos: Risco Sistémico nos Aproveitamentos Hidroelétricos? In *EDPP - Estudos de Caso nº4* (in Portuguese). Not published document, Porto.

Read, J., 2013. Data gathering, interpretation, reliability and geotechnical models. In Proceedings *Slope Stability 2013*, pp81–89, Australian Centre of Geomechanics, Perth.

Thomas, R.D.H., 2013. A statistical approach to account for level of uncertainty during geotechnical design, in *Proceedings Slope Stability 2013*, pp 325–335, Australian Centre of Geomechanics, Perth.

U.S. National Committee on Tunneling Technology 1984. Geotechnical Site Investigations for underground projects - Volume 1, overview of practice and legal issues, evaluation of cases, conclusions and recommendations. Washington D.C.: National Academy Press.

Tunnels and Underground Cities: Engineering and Innovation meet Archaeology, Architecture and Art, Volume 3: Geological and geotechnical knowledge and requirements for project implementation – Peila, Viggiani & Celestino (Eds)
 ISBN 978-0-367-46583-4

A Rosette stone between geotechnical engineer and project manager for TBM tunneling

G. Potgieter & O. Yeni
Glencore Copper, Mount Isa, Australia

ABSTRACT: During planning and design the geotechnical engineer must anticipate the behavior of the tunnel under several scenarios. His ability to predict this behavior is critical to cost estimation as well as on time delivery. A knowledge gap is almost always present between what is obtained by the principal during the design phase and what is required by the contractor for equipment purchase and construction. Many principals and contractors underestimate the im-portance of the geotechnical input this does not become apparent until the project starts to run into cost and time overruns. This paper sets out to provide both a reference to project managers as to when and what input they should seek from geotechnical engineers and give guidance to geotechnical engineers on how to present information to project managers so that they can make smart decisions around project management, reducing disputes and ultimately providing better outcomes.

1 INTRODUCTION

Tunnelling is a serious engineering project. Geology and geotechnical properties play a very important role. Any adverse and unforeseen geotechnical conditions may influence the safety of tunnels, construction efficiency, time and costs.

An experienced geotechnical engineer can clarify the unforeseen geotechnical conditions as the project matures. The collected data by an experienced geotechnical engineer must be translated to the real-time practical knowledge that can be useful by the project manager for the possible contract negotiations while avoiding conflicts before the project run into cost and time overruns.

2 TBM SELECTION STAGE

TBM selection is critical to the success of the project. Once made this decision cannot easily be changed. Tunnel purpose, type and design are not the only criteria to select TBM.

TBM Quick selection chart Figure 1 which is used by a lot of contractors is not the recommended method to select the TBM.

There are a few variants to these main types, e.g. "Mechanical Excavation Type", "Beam Type", and "Mixed Shield Slurry Type" however Table 1 is enough to demonstrate the idea.

As you can see from Table 1, prior knowledge of expected ground conditions is critical. Hard rock may be indicated during feasibility studies based on limited core drilling; this would result in the choice of a Shield-type TBM. If a lot of faults are present in the rock, this would mean much of the ground will be weak or variable, in which case it would have been better to choose an EPB TBM. This type of mistake is costly. The choice of TBM should be a collaborative exercise between the Geotechnical Engineer and the procurement/tender team.

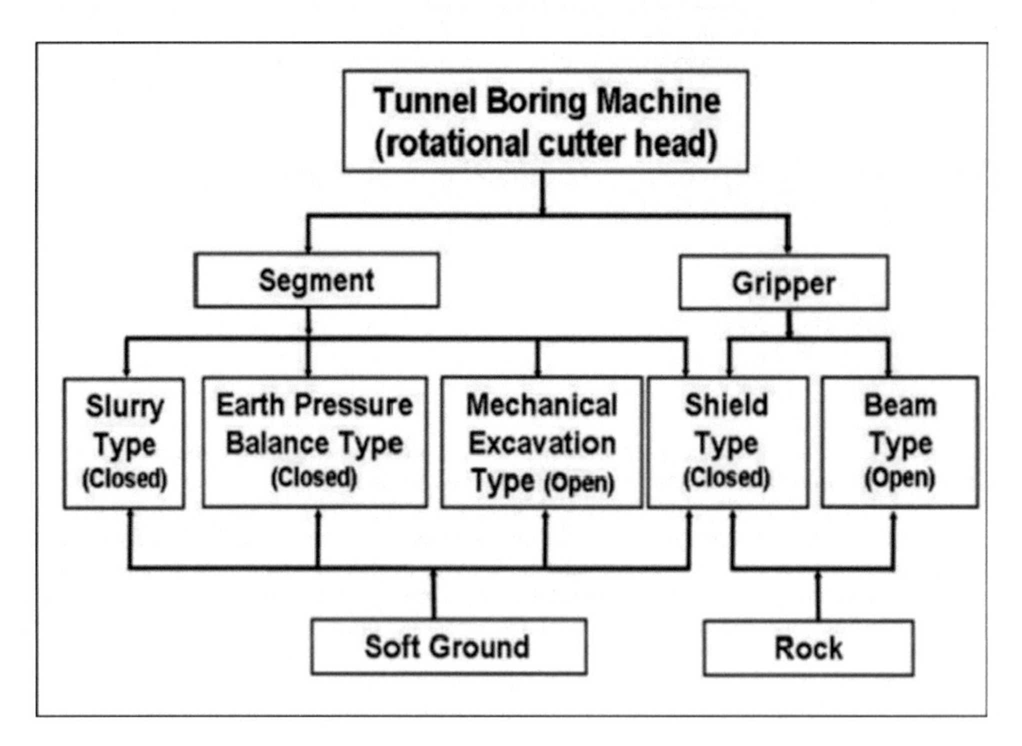

Figure 1. TBM quick selection chart (General).

Table 1. Basic comparison of TBM types.

Type	Cons	Best For
Slurry Type	• Require significant backup systems and power • Large stones and boulders can hinder drive	• Allows soft, wet, or unstable ground to be tunnelled • Suitable for ground with high water pressures • Limits ground settlement and produce a smooth tunnel wall
Earth Pressure Balance	• Not suitable for ground with high water pressures • Not suitable for hard rock conditions	• Allows soft, wet, or unstable ground to be tunnelled • Limits ground settlement and produce a smooth tunnel wall
Shield type	• High capital cost • Difficult to transport • Very sensitive for weak ground conditions • Limited options to deal with groundwater	• They offer a continuous and controlled means of tunnelling capable of high rates of advance under favourable conditions

As the project matures, the amount of information will increase, however as decisions are made, and capital has spent the ability to change will decrease Figure 2. Both Geotechnical Engineers and Project Managers need to understand this dynamic and the pressures it creates.

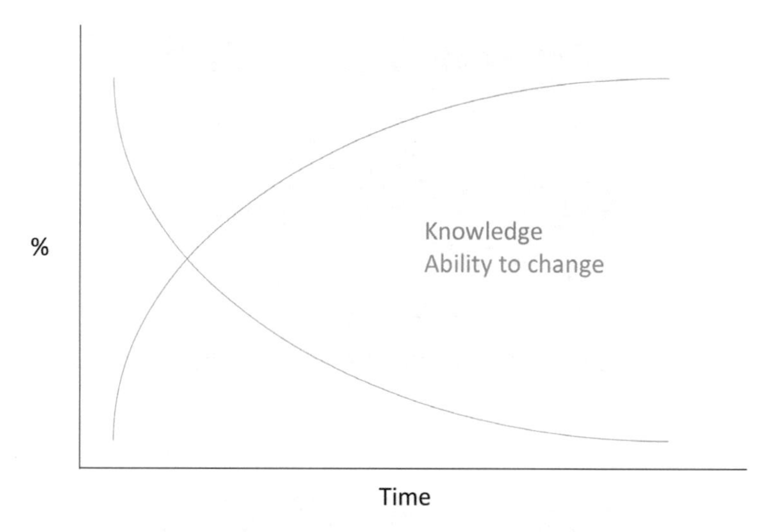

Figure 2. Knowledge vs. Ability to change over time.

3 TBM DESIGN STAGE

After selecting the TBM type then the geotechnical engineer should still be actively engaged in the design and manufacture as many variables will affect the performance of the TBM. Examples include the design of cutter head, maximum trust and torque, over-boring, probing and ground improvement capabilities.

Of secondary importance is the fact that during this stage any further geotechnical assessment will allow for optimisation of the TBM which will increase performance.

There will always be pressure to reduce costs however understanding consequences of each compromise must be appreciated in each instance. Saving money on a probe drill may make it impossible to conduct adequate probe drilling ahead of the cutter head during maintenance periods creating the situation where you need to either mine without enough cover, delay mining to achieve the required cover or purchase and fit better drills during excavation. The third choice may not always be possible.

The question should not be, “Are all requested designs in place?” but, “Are all requested designs in place and optimised to the predicted ground conditions?”

4 TBM OPERATIONAL STAGE

Nobody likes surprises. TBMs do not like surprises. Project Managers do not like surprises. Any Geotechnical Engineer that wants a long career in tunnel boring would do well to remember this. Conditions in the tunnel can change quickly transitioning from minute to minute as shown in Figure 3 very wet and Figure 4 very dry or from week to week as shown in Figure 5 hard rock.

To prevent surprises the Geotechnical Engineer needs to be present at the job site. Working remotely is not an option. The advantages of having a Geotechnical Engineer on site always are;

- Prediction of ground conditions ahead of the cutter head both good and bad along with expected impacts on advance.
- Providing solutions to allow the advance rate to be maintained.
- Advise as to when mining can or cannot be stopped and advise if this will require a contract variation from the client.

Figure 3. Tunnel face inspection – Fault (Yeni, O., KARGI HEPP Project Notes, and 2011–2013).

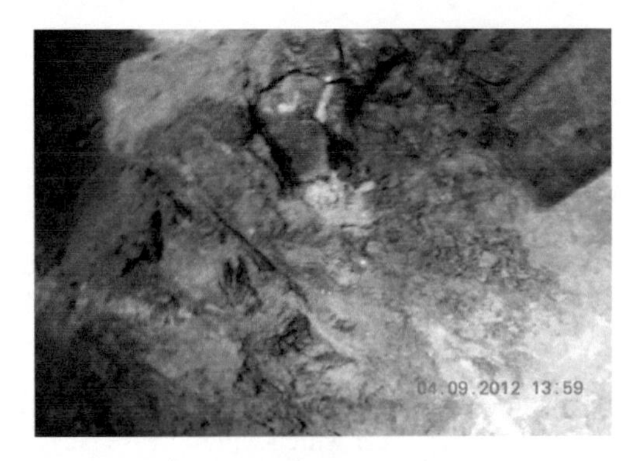

Figure 4. Tunnel face inspection – Fault (Yeni, O., KARGI HEPP Project Notes, 2011–2013).

Figure 5. Tunnel face inspection – Basalt (Yeni, O., KARGI HEPP Project Notes, 2011–2013).

- Analyse the performance of the TBM and provide understanding to the Project Manager as to why variation is occurring.
- Monitor ground conditions and responding to mining and updating the geotechnical model to enable better predictions.

Project Managers are specialists in their field they are not Geotechnical Engineers. The jargon used by Geotechnical Engineers is often confusing and does not provide the knowledge required to make decisions. Project Managers constantly must weigh up conflicting priorities and advice as per Figure 6. Geotechnical Engineers need to understand this and provide information in a way that makes sense.

Recommendations must be free of jargon and must be given in context. For example, looking at Figure 7, the Engineer detects increased cutter head torque at chainage 11+100 due to

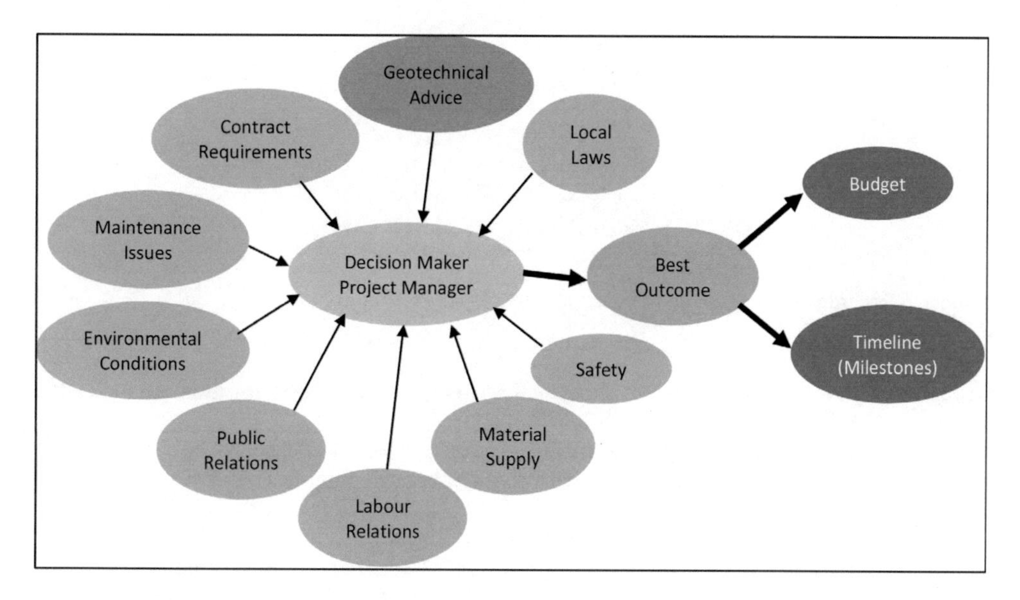

Figure 6. Considerations.

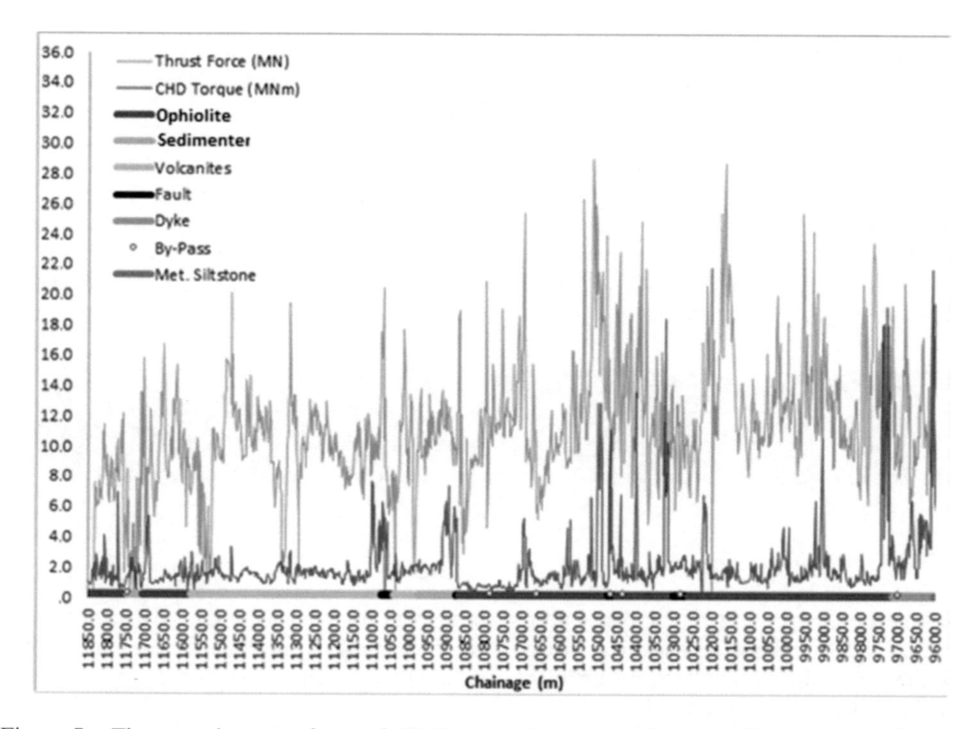

Figure 7. Thrust and torque force of TBM vs. geological formations while advancing (Yeni, O., KARGI HEPP Project Notes, 2011–2013).

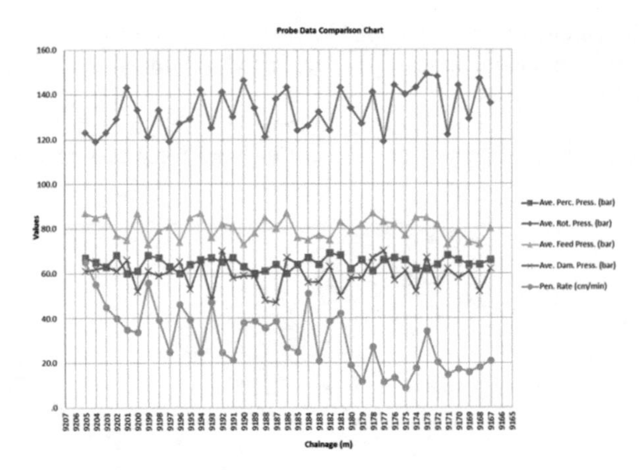

Figure 8. Recorded probe drill data (Yeni, O., KARGI HEPP Project Notes, 2011–2013).

mining through a fault and is aware that the TBM is required to stop for maintenance within the next few hours.

Do not tell the Project Manager the torque is high, so he cannot stop; this does not provide context and will create a direct conflict with the Mechanical Engineer, and Contract requirements so will probably be ignored.

If the Geotechnical Engineer explained that stopping now would result in the cutter head being stuck and that if the stop was delayed for 50m or so when better ground conditions are expected, the TBM could safely be stopped for maintenance. A variation would need to be obtained from the client, and the maintenance schedule would need to be adjusted. The geotechnical recommendation would be far more likely to be followed, and this would produce a better result for the project.

In addition to making the recommendations clearer, i.e. in terms, the Project Manager is familiar with, the Geotechnical Engineer should collect supporting data from additional sources such as sampling and analyse rocks from the stockpile and review probe drill data as per Figure 8 and use these to build a convincing argument.

In the modern highly mechanised world of TBM tunnelling, the Geotechnical Engineer has access to many methods of collecting raw data from dozens of sources both on the TBM and around the site. The Geotechnical Engineer needs to translate the data into knowledge and wisdom and share that with the Project Manager. Data and information while critical to the Geotechnical Engineer are useless to the Project Manager and will only confuse the issue if not presented in a clear and concise manner.

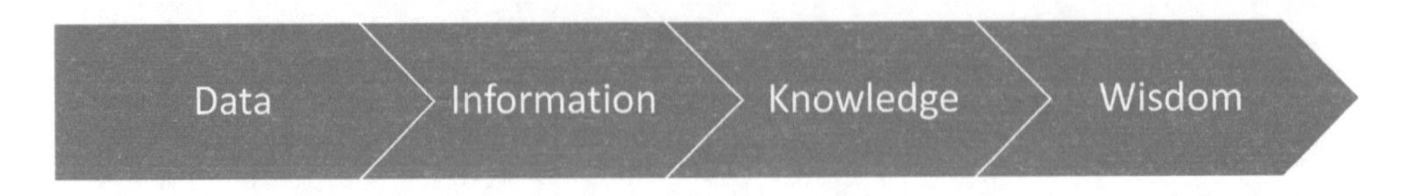

5 CONCLUSION

No project no matter how much time and money is spent up front will ever have all the information. By getting good geotechnical advice early in the project and then continuing to work with the Geotechnical Engineer during TBM selection and design, the Project Manager will be more likely to be able to succeed.

An experienced Project Manager understands that the competence and experience of the Geotechnical Engineer can make the difference between meeting the project goals or incurring delays and additional costs. The ability of these two people to understand each other and work together during the construction of the tunnel is critical.

The Geotechnical Engineer must try to understand the pressures that the Project Manager is under the conflicting priorities that he must resolve daily. Deliver knowledge and share wisdom and becoming an ally rather than just another source of conflict that must be resolved.

REFERENCE

Yeni, Omer, 2013. KARGI Hydroelectric Power Plant Project. Personal Notes. Turkey.

Tunnels and Underground Cities: Engineering and Innovation meet Archaeology, Architecture and Art, Volume 3: Geological and geotechnical knowledge and requirements for project implementation – Peila, Viggiani & Celestino (Eds)
 ISBN 978-0-367-46583-4

Application of deflection curves, trend lines and displacement vector orientation in the evaluation of surface displacements in tunneling

A.L.C. Rissoli & A.P. de Assis
University of Brasilia, Brasilia, Brazil

ABSTRACT: In order to increase the applicability of displacement curves and trend lines, the surface displacements of shallow tunnels excavated in rock were evaluated. For this, several geologies were simulated by the authors using the finite element software PLAXIS 3D Tunnel. Nineteen simulations include homogeneous and nonhomogeneous geologies with zones of different materials, with better or worse properties, with different lengths inside the rock mass. The behavior of the surface displacement components curves was similar to the walls displacements curves. However, displacement vector orientations curves presented some considerable differences. A dependence of the curves to the distance from face that the trend is been evaluated is evident. Furthermore, the shape and peaks are different from crown displacement vector trend lines. Notably, it occurs in the near-face portion of excavation by the stress distribution that occurs from depth of the excavation to surface.

1 INTRODUCTION

Wall displacement evaluation in underground excavations, especially in tunneling, was extensively investigated in Austria in the researches developed at Graz University Institute of Rock Mechanics and Tunneling. These studies consolidated new forms of data evaluation with data that were already available by conventional 3D displacement instrumentation. These forms of evaluation include the tracking of deflection curves and trend lines of the displacement components and the displacements vector orientation trend lines (Schubert et al., 2005).

Because of the geological features where these indicators were initially studied, they were essentially evaluated in deep rock tunnels. Therefore, they were only applied to tunnel walls. Consequently, in order to increase the applicability of the deflection curves and trend lines, especially the displacement vector orientation trend lines, surface displacements of shallow tunnels with two diameters overburden excavated in rock were evaluated. For this, several geologies were simulated by mean of the finite element software PLAXIS 3D Tunnel.

The application of these new forms of displacement interpretation was applied not only for the forecast of the material ahead of the excavation, but also for predicting a possible collapse.

2 EVALUATION AND INTERPRETATION OF MONITORING DATA

The monitoring data evaluation and interpretation are essential during the construction. The data become information for the understanding of the mechanical process, identification of "normal" behavior, detection of deviations in the expected behavior and future behavior prediction (OGG, 2014). Is it thus a tool for stability evaluation and construction optimization (Moritz et al., 2011).

The following are some of the most recent forms of field instrumentation assessment.

2.1 *Deflection curves and trend lines*

Deflection curves are constructed by connecting the displacements measured along the tunnel at a certain time (OGG, 2014). Plotting these lines for several time intervals facilitates the spatial visualization of the development of each component of the displacements and the influence of the excavation advance in the sections behind the face (Schubert & Grossauer, 2004). For vertical displacements, curves with onion-shell shape are considered "normal". Deviations from this geometry or increase of area between successive curves indicate different ground characteristics ahead (Figure 1).

In turn, trend lines are constructed by connecting deflection curves values that are at a constant distance from the face (OGG, 2014). Horizontal trend lines indicate "normal" behavior, and any deviation may suggest different conditions in the mass ahead (Figure 1).

In trend lines evaluation, it is important to note that each point corresponds to the displacement at some distance from the face. So, if the point is (100 m; 5 mm), it means that the monitoring point is at chainage 100 m and the face is at 100 + a, where a corresponds to the distance of the face in which the trend line is being evaluated, in which can assume a positive or negative value, referring to points ahead or behind the excavation face, respectively.

Deflection curves are already adopted in the evaluation of the displacements occurring on the surface. It is though less frequently used than the time-displacement and distance-displacement diagrams, which are more traditional forms of monitoring data evaluation. Trend lines are still rarely used for surface displacements evaluations.

2.2 *Displacement vector orientation*

Displacement vectors can be presented in the tunnel cross section, plotting the radial displacements, or in the longitudinal, plotting vertical and longitudinal displacements (OGG, 2014), as can be observed in Figure 2. The vectors allow evaluating the influence of the mass structure, system response and anisotropic displacements in the presented section (OGG, 2014). They are evaluated by two of its components, as shown in Figure 3, with L/S form being the most usual form.

Not only the vectors variation in the section are evaluated, but also the vectors orientation. Deviations in the "normal" orientation are related to changes in the characteristics of the mass ahead of the excavation.

Vector displacements (L/S) oriented against the excavation direction have positive orientation angles. On the other side, vectors displacements oriented in the direction of the excavation have negative angles. Jeon et al. (2005) consider the "normal" displacement vector

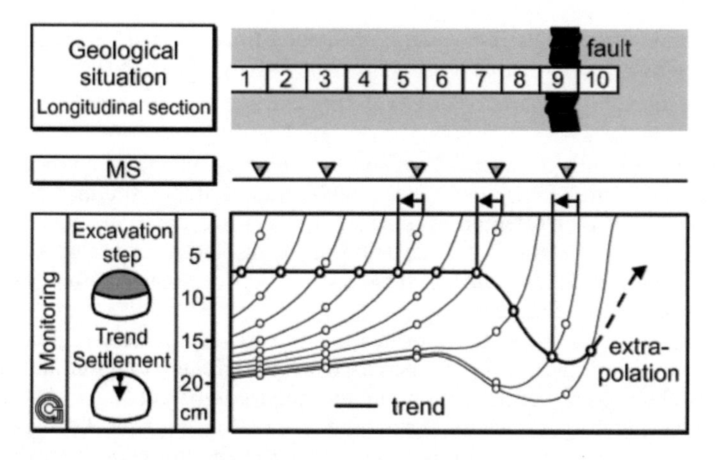

Figure 1. Deflection lines and trend line approaching a fault zone (Steindorfer, 1998).

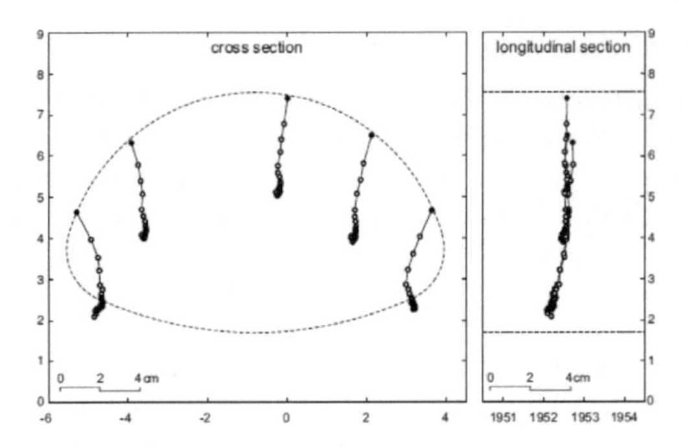

Figure 2. Displacement vectors in cross section and longitudinal section (Grossauer, 2009).

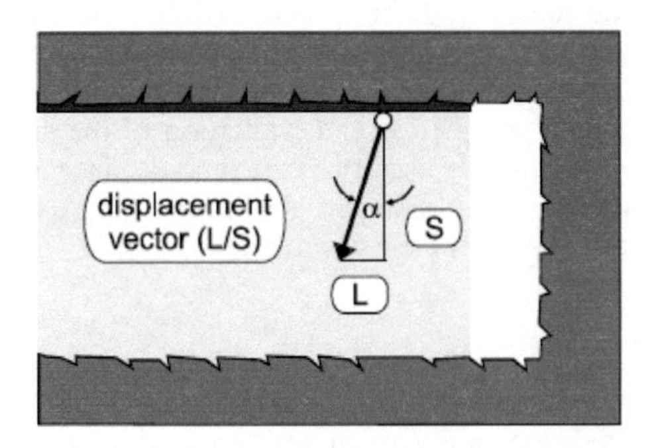

Figure 3. Displacement vectors components (Steindorfer, 1998).

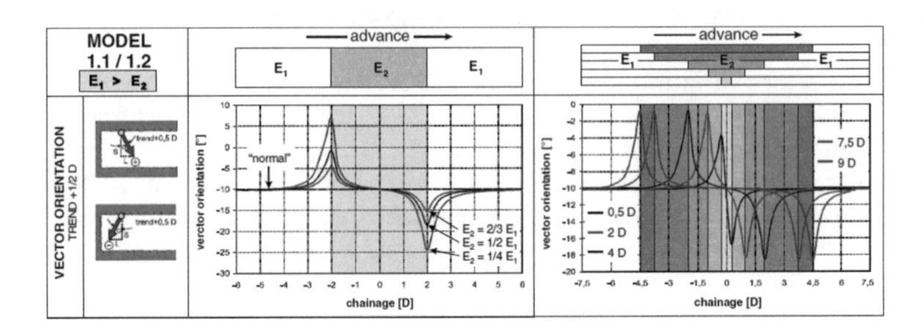

Figure 4. Deviation of the displacement vector orientation from "normal" in response to crossing a weakness zone of different stiffness contrasts and zone lengths (Grossauer, 2001).

orientation to be 15 ° with the vertical against the direction of the face advance. Studies by Schubert et al. (2005) and Grossauer et al. (2008) show that, for the materials studied, the normal displacement vector orientation is about 10 °. However, what indicates variation in the mass behavior is not the absolute orientation of the displacement vector itself, but its variation. This variation is related to the stiffness contrast between materials and extension of the different materials regions (Figure 4).

As is evident in Figure 4, when approaching the lower stiffness region, the displacement vector orientation angle (L/S) is increased by the greater longitudinal displacement. It occurs due to the stress concentrations in the most competent mass. The opposite tendency, of decreasing angle, occurs when the excavation approaches greater stiffness region. In sum, displacement vector is oriented in the direction of the excavation when the mass ahead is more competent, and against the excavation when the mass ahead is less competent.

3 NUMERICAL SIMULATIONS

This study covered nineteen different types of geologies, with four types of materials. Unlined rock tunnels with 10 m diameter and 20 m overburden were simulated. The geologies are homogeneous and nonhomogeneous with different material zones with 1, 3 and 5 diameters of length (zone 2) with vertical transition planes, as shown in Figure 5.

In order to reduce model dimensions, the symmetry of the tunnel was explored, simulating only one of its halves. The dimensions of the computational model are 250 m on *z*-axis and 50 m on both *x* and *y* axis. Longitudinally, the first 50 m were excavated in larger steps of 10 m, and subsequent steps of 2.5 m in the internal part of the model, as can be observed in Figure 6. All output data were extracted in the internal 150 meters of the model, discarding the 50 meters closest to the boundaries.

3.1 *Materials*

Several authors have adopted the linear-elastic constitutive model to simulate materials behavior to evaluate the behavior of transitions between materials, for example Tonon & Amadei (2000) and Yong et al. (2013). Some of them, such as Jeon et al. (2005) and Grossauer (2001), adopted the Mohr-Coulomb model only for the layer of different material within the mass. However, these studies focused only on the prediction of the change of behavior of the mass, without studying the collapse, which is the objective of this work. Thereat, it was necessary to adopt a constitutive model with an adequate criterion of failure for rocks, the material to be simulated. The Hoek-Brown criterion was adopted, and it was necessary to convert these parameters to the Mohr-Coulomb criterion, model available in the PLAXIS 3D Tunnel software.

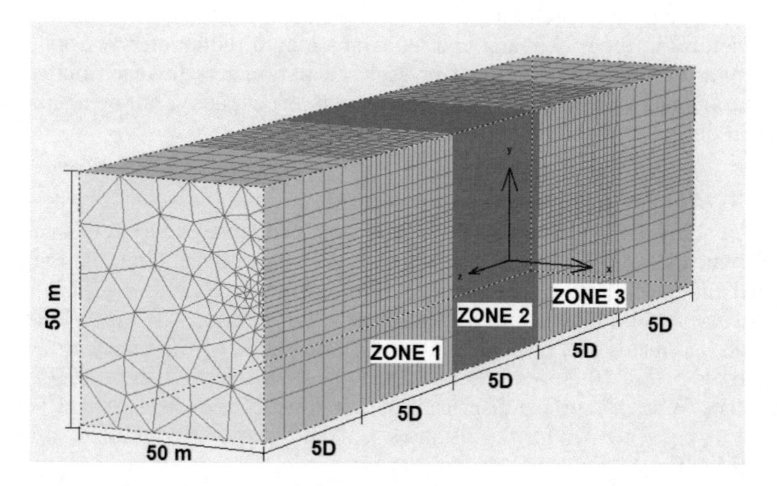

Figure 5. Computational model indicating zones 1, 2 and 3.

Table 1. Materials parameters.

Material	K_0	γ (kN/m³)	mi	D	GSI	σ_{ci} (MPa)	E (GPa)	c (MPa)	Φ(o)
Good	1	27	10	0	80	5.0	12.574	0.2802	49.6
Average	1	27	10	0	40	5.0	1.2574	0.0761	40.5
Bad	1	27	10	0	20	5.0	0.3976	0.0465	33.1
Poor	1	27	10	0	20	2.5	0.2812	0.0363	28.2

Table 2. Cases of study.

Case	Material 1	Material 2	Zone 2 length	Chainage of collapse (m)
TRB	Good	-	-	
TRBM1D	Good	Average	1D	
TRBM3D	Good	Average	3D	
TRBM5D	Good	Average	5D	
TRBR1D	Good	Bad	1D	
TRBR3D	Good	Bad	3D	
TRBR5D	Good	Bad	5D	
TRBP1D	Good	Poor	1D	
TRBP3D	Good	Poor	3D	115.0
TRBP5D	Good	Poor	5D	115.0
TRMB1D	Average	Good	1D	
TRMB3D	Average	Good	3D	
TRMB5D	Average	Good	5D	
TRMR1D	Average	Bad	1D	
TRMR3D	Average	Bad	3D	
TRMR5D	Average	Bad	5D	
TRMP1D	Average	Poor	1D	
TRMP3D	Average	Poor	3D	117.5
TRMP5D	Average	Poor	5D	117.5

The following parameters were estimated for the application of the Hoek-Brown criterion: specific weight (γ); modulus of elasticity of the rock mass (Em); Poisson's ratio (v); uniaxial compressive resistance of intact rock (σc); parameter of intact rock (mi); Geological Strength Index (GSI); Disturbance parameter (D). For the in situ stress state, the horizontal stress was assumed to be equal to the vertical stress.

The simulated materials are named as Good, Average, Bad and Poor Rock, nomenclatures related to the competence of each one. Geological Strength Index (GSI) reduction was applied to account fractures in the rock mass, and uniaxial strength reduction was applied to account for rock strength deterioration. Table 2 presents all parameters for each material. Bad and Poor Rock parameters were selected to have poor support capacity, not being able to be excavated for great distances without support.

3.2 *Cases evaluated and output*

The 19 cases of study are presented in Table 2, in which collapse occurred in 4 cases.

The displacements at the crown of the tunnel and at the surface along the longitudinal axis were extracted and the deflection curves were generated for each component. Also, the trend lines for the displacement components and displacement vector orientations were plotted. The trend lines for the displacements and displacement vector orientation of the tunnel crown were generated for distances 5, 7.5, 10, 15 and 20 m behind the face, referring to 0.5, 0.75, 1, 1.5 and 2 tunnel diameters. Whereas surface displacement components and displacement vector orientations trend lines were generated for the distances 5, 10 and 15 m behind the excavation face and 5 and 10 m ahead of face.

4 RESULTS

4.1 *Deflection curves and trend lines*

As was expected, the crown deflection curves changed its shape after the interface between materials, as is evident in Figure 6. However, these lines were not adequate to predict the change in the material in front of the face, as substantial changes only occurred after the entrance of the excavation in zone 2.

The association of collapse to displacement values is not an innovative way of evaluating instrumentation. In addition, in the case of these numerical simulations, larger displacements were expected in situations of collapse (Figure 7) due to the smaller elastic modulus. The surface deflection curves presented similar behavior to the crown curves, although smoother curves were observed in the transition between materials, as in Figure 8. In Figure 9, it can be observed increasing distance between curves before collapse.

As the deflection curves, the trend lines were not able to predict material change (Figure 10 and Figure 11). However, more abrupt and steep changes in the trend lines of the vertical crown displacements may suggest states close to collapse, as shown in Figure 12 and Figure 13.

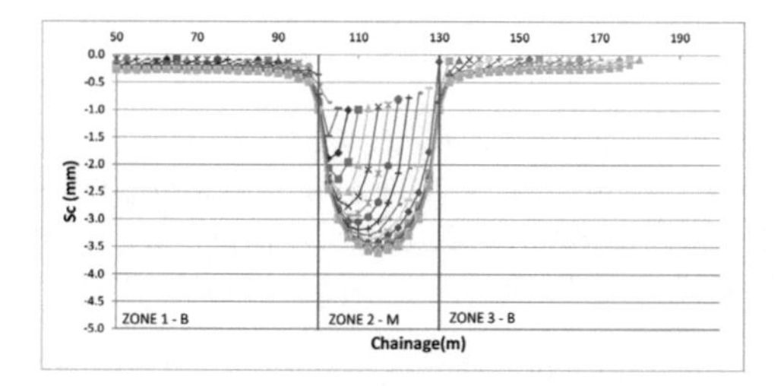

Figure 6. TRBM3D crown vertical displacements (Sc) deflection curves.

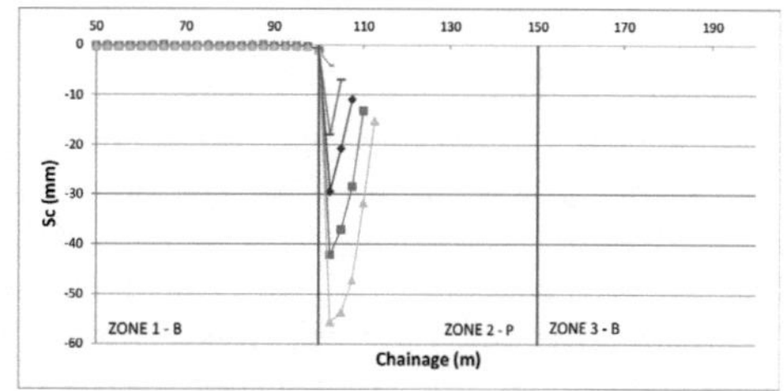

Figure 7. TRBP5D crown vertical displacements (Sc) deflection curves.

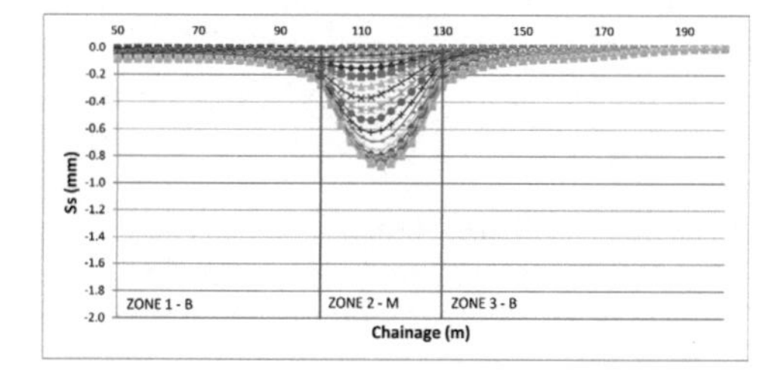

Figure 8. TRBM3D surface vertical displacements (Ss) deflection curves.

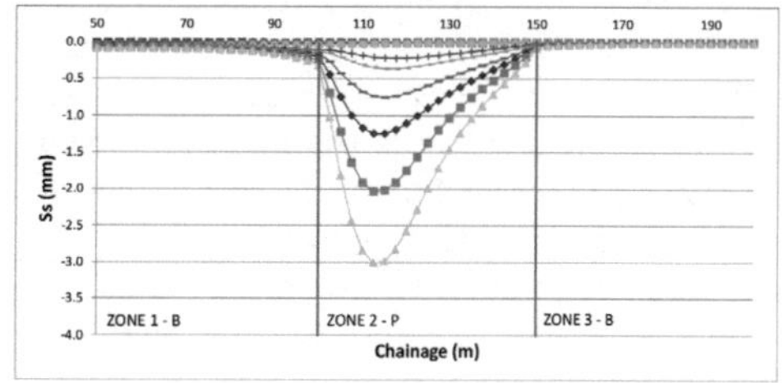

Figure 9. TRBP5D surface vertical displacements (Ss) deflection curves.

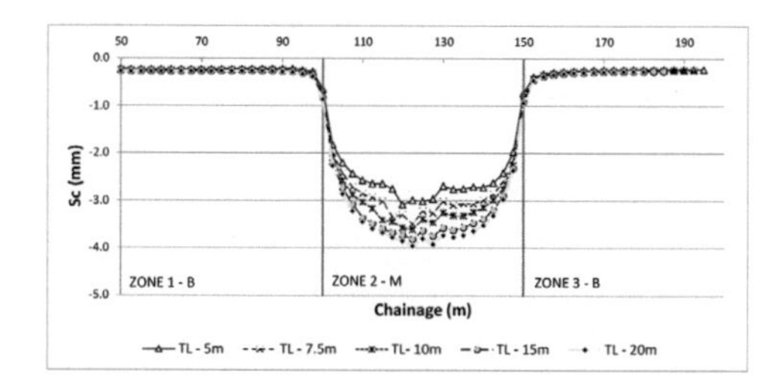

Figure 10. TRBM5D crown vertical displacements (Sc) trend lines.

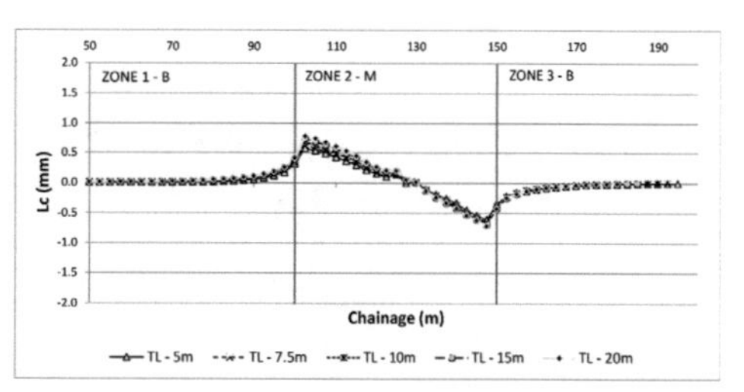

Figure 11. TRBM5D crown longitudinal displacements (Lc) trend lines.

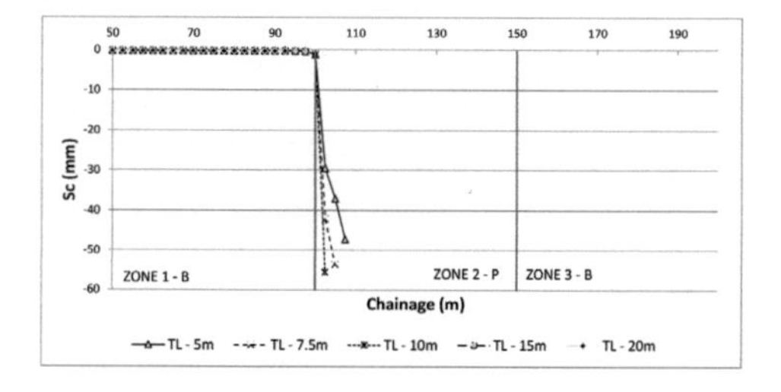

Figure 12. TRBP5D crown vertical displacements (Sc) trend lines.

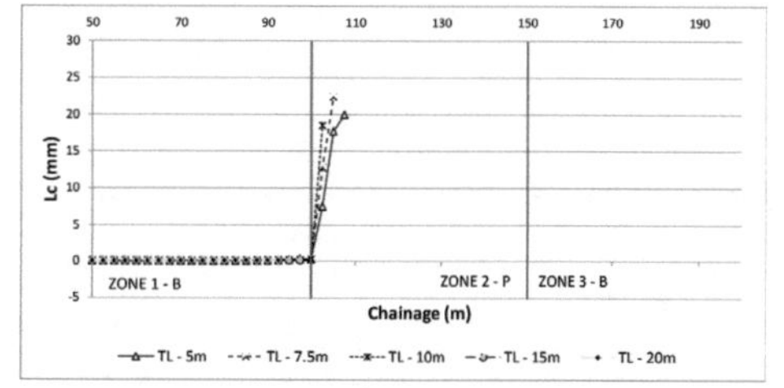

Figure 13. TRBP5D crown longitudinal displacements (Lc) trend lines.

Like the deflection curves, the trend lines of displacements occurring on the surface are very similar to those of the tunnel walls, as observed in Figures 14 and 15.What can be noticed differently is that, in cases in which there is no collapse, as in Figure 15, the trend lines behind the face are more representative of the material change. They exhibit greater distancing from the "normal" behavior, referring to the behavior of the excavation in zone 1. However, in the occurrence of collapse, the trend lines with a smaller face distance, 5 m behind the face, and 5 and 10 m ahead of the face, present a larger slope (Figure 15). This behavior is smoothed in crown trend lines.

4.2 *Displacement vector orientation*

The "normal" crown displacement vector orientation was 0° for the Good rock (Figure 16) and 15° for the Average rock, values aligned with those found by Sellner & Steindorfer (2000), Schubert et al. (2002) and Jeon et al. (2005). However, for the surface displacement vector orientation, it was observed a very different behavior in relation to crown, presenting orientations with larger and different angles between trend lines.

For trend lines evaluated ahead the face, the greater is the distance, the greater are the angles, reaching a value of about 70° for the trend line 10 m ahead of the face (Figure 17). This behavior can be related to the occurrence of almost only vertical displacements (y-axis) in the tunnel crown excavated in homogeneous material. Also, due to the 3D distribution of the stresses, there is the occurrence of higher longitudinal displacements (z-axis) against the direction of the excavation in relation to vertical displacements (y-axis). In this way, the increase of the displacement vector orientation ahead of excavation face can be related to the mass movement towards the face. This behavior can be verified in Figure 18, which presents the displacement vectors in the simulated rock mass.

In accordance with Jeon et al. (2005) and Orsini (2017), the displacement vector orientation was able to predict different mass behavior one diameter before the material transition, as is evident in Figures 19 and 20. The displacement vector orientation in collapsing tunnels has shown an increase of degrees even after the entrance of the excavation in zone 2 (Figure 20), which is an opposite behavior to that observed by Schubert & Budil (1995) in the collapse of the tunnel Galgenberg Tunnel.

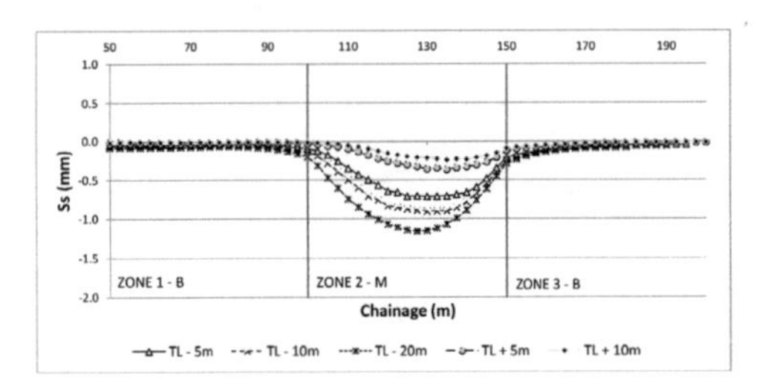

Figure 14. TRBM5D surface vertical displacements (Ss) trend lines.

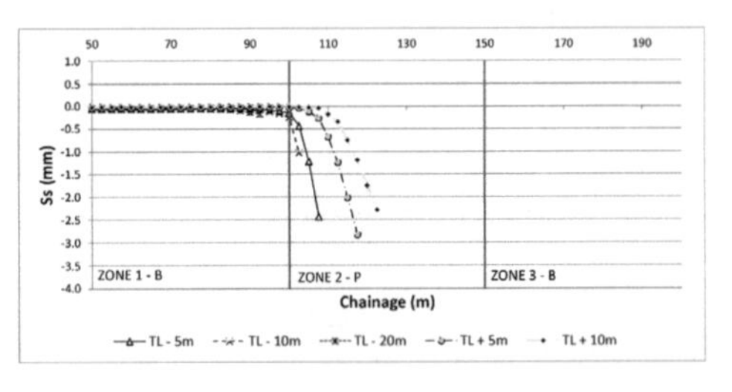

Figure 15. TRBP5D surface vertical displacements (Ss) trend lines.

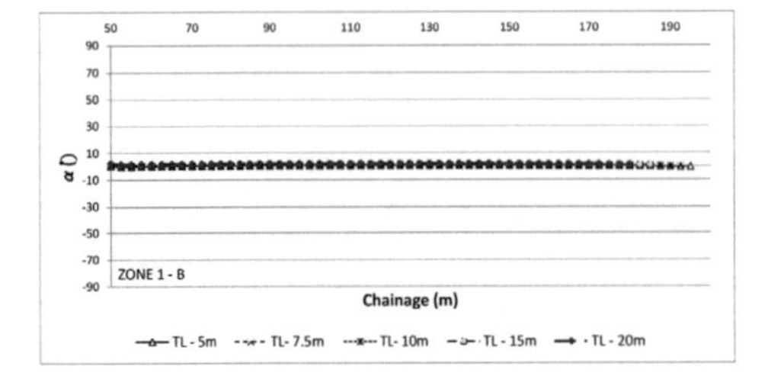

Figure 16. TRB crown displacement vector orientation (α) trend lines.

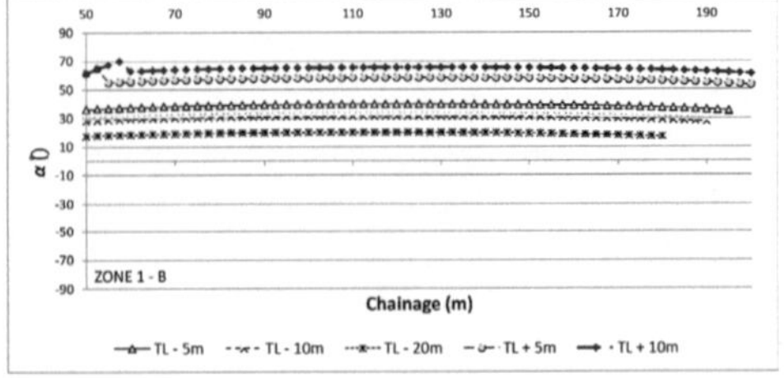

Figure 17. TRB surface displacement vector orientation (α) trend lines.

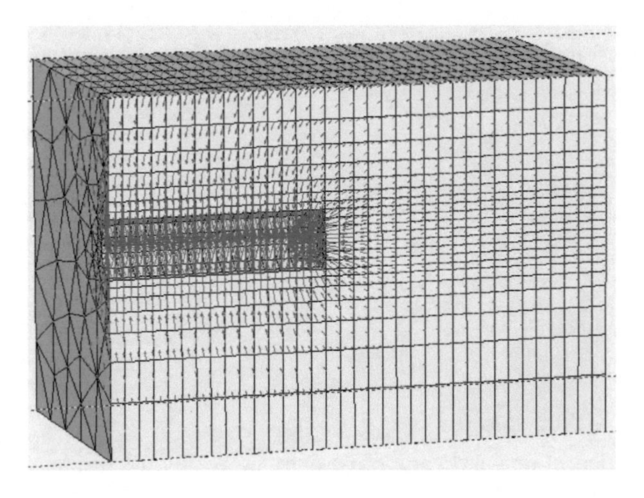

Figure 18. TRB displacement vectors.

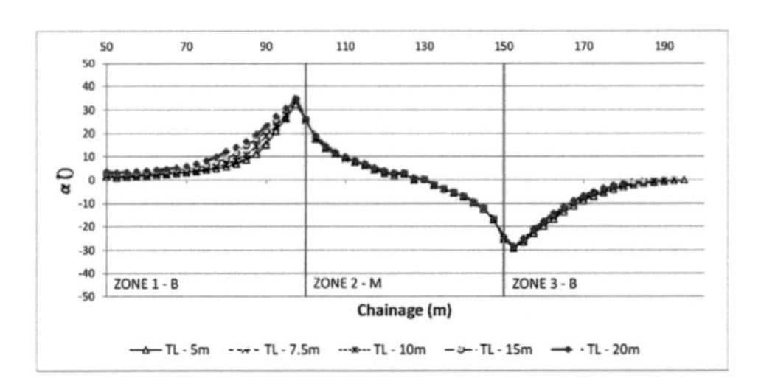

Figure 19. TRBM5D crown displacement vector orientation (α) trend lines.

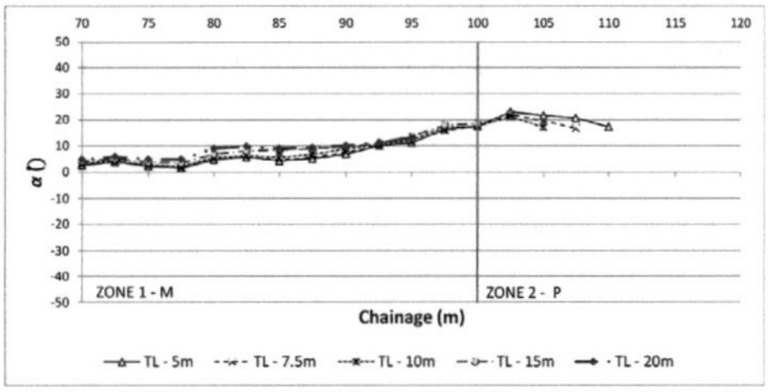

Figure 20. TRMP3D crown displacement vector orientation (α) trend lines.

This behavior can be associated with a non-stabilization of the mass. The behavior in materials where the excavation is stable throughout zone 2, shortly after entering the material of worse quality, or even of better quality, shows that the displacement vector orientation tends to return to its "normal" value. Then, the orientation changes again with the approach of the new interface, now between zone 2 and zone 3 (equal to zone 1). However, in situations where there is collapse, the orientation of the displacement vector continues to shift against the direction of the excavation because the excavated material is not strong enough to withstand excavation. This behavior is a promising feature to be studied, but caution should be taken in interpreting these increases as it may be associated with collapses or irregularities of trend lines.

The surface displacement vector orientation trend lines were different from those observed in the tunnel crown. Firstly, orientation angle values higher than those observed in the displacement crown vector orientation lines are observed, as can be seen by comparing Figure 21 to Figure 19. In addition, the surface displacement vector orientation trend lines have different shapes compared to crown trend lines. Peaks before entry of the different material, in the lines behind the face, and negative values at the entrance in zone 2 are evident in Figure 21. As for the trend lines before collapse (Figure 22), there is no change in behavior close to destabilization, compared to that in stable tunnels (Figure 21). Neither a distancing behavior between trend lines was observed, comparing Figures 21 and 22. Thus, surface displacement vector orientations were not able to predict collapse. That may be related to small displacements occurring on surface, about 1 mm, compared to wall displacements, indicating a wall collapse that does not reach the surface.

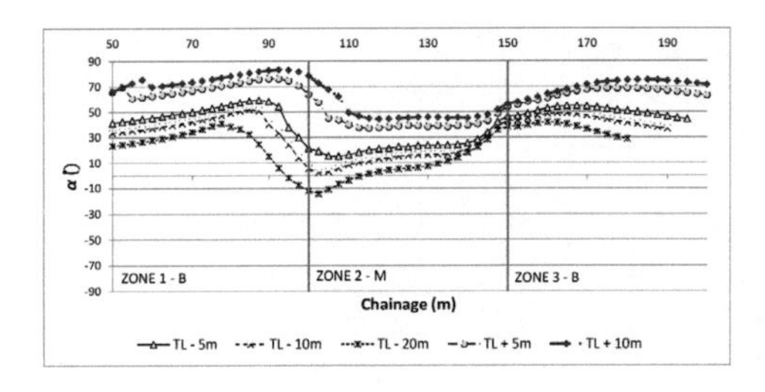

Figure 21. TRBM5D surface displacement vector orientation (α) trend lines.

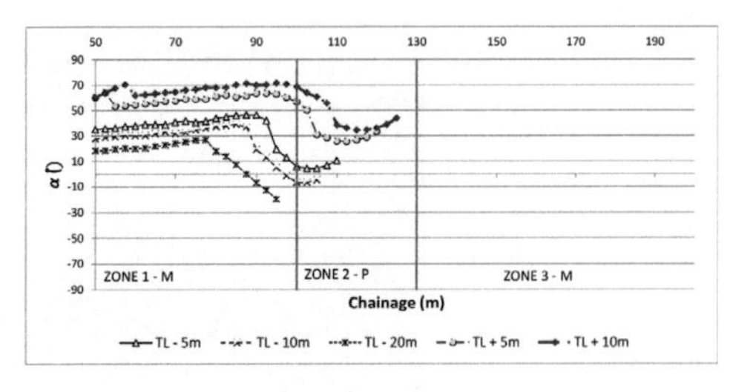

Figure 22. TRMP3D surface displacement vector orientation (α) trend lines.

5 CONCLUSIONS

The application of deflection curves and trend lines, especially the surface displacement vector orientation, still needs to be investigated. However, the knowledge on displacement curves and trend lines of the displacement components in the excavation walls can be used for the evaluation of the displacements on the surface. It is due to the occurrence of very similar behaviors. However, the behavior of the surface displacement vector orientation still needs to be further explored, especially by the stress distribution that occurs from depth of the excavation to surface. The stress distribution ahead of face makes the displacement vector orientation trend lines very dependent to the distance of face that they are evaluated.

REFERENCES

Grossauer, K. 2001. Tunnelling in Heterogeneous Ground - Numerical Investigation of Stresses and Displacements. Diploma Thesis, Graz University of Technology, Graz, Austria, 56 p.

Grossauer, K. (2009). Expert System Development for the Evaluation and Interpretation of Displacement Monitoring Data in Tunnelling. Doctoral Thesis, Department of Civil Engineering, Technical University Graz, Austria, 124 p.

Grossauer, K. & Schubert, W. 2007. Methods for the evaluation and interpretation of displacement monitoring data in tunnelling. *1st Canada-US Rock Mechanics Symposium*, Vancouver, Canada. CRC Press, Vancouver, Canada, 1149–1156.

Grossauer, K., Schubert, W. & Lenz, G. 2008. Automatic displacement monitoring data interpretation – the next step towards an expert system for tunneling. *42nd US Rock Mechanics Symposium and 2nd U.S.-Canada Rock Mechanics Symposium*, San Francisco, CA, USA.

Hoek, E., Carranza, C. & Corkum, B. 2002. Hoek-brown failure criterion – 2002 edition. *Narms-Tac*, 267–273.

Jeon, J.S., Martin, C.D., Chan, D.H. & Kim, J.S. 2005. Predicting ground conditions ahead of the tunnel face by vector orientation analysis. *Tunn. Undergr. Sp. Technol.* 20, 344–355.

Moritz, B., Koinig, J. & Vavrovsky, G.M. 2011. Geotechnical safety management in tunnelling – an efficient way to prevent failure. *Geomech. und Tunnelbau* 4, 472–488.

OGG. 2014. Geotechnical Monitoring in Conventional Tunneling Handbook. Austrian Society for Geomechanics, 90 p.

Orsini, B. de O. 2017. Inovação em Analise de Risco e Tomada de Decisão em Escavações em Túneis. Dissertação de Mestrado, Departamento de Engenharia Civil e Ambiental, Universidade de Brasília, Brasília, DF, 96 p. .

Schubert, W. & Grossauer, K. 2004. Evaluation and interpretation of displacements in tunnels. *14th International Conference on Engineering Surveying, Zurich*, 1–12

Schubert, W., Grossauer, K. & Sellner, P. 2005. Advances in the observational approach in tunnelling by new techniques of monitoring data evaluation. *10th ACUUS International Conference. Moscow State University of Civil Engineering, Moscow*.

Schubert, W. & Steindorfer, A. 2004. Selective displacement monitoring during tunnel excavation. Felsbau, Vol. 14, p. 93–97.

Schubert, W., Steindorfer, A. & Button, E. 2002. Displacement monitoring in tunnels. *Felsbau* 20, 7–15.

Sellner, P. J., & Steindorfer, A. F. 2000. Prediction of displacements in tunnelling. *Felsbau* 2.

Steindorfer, A. (1998). Short Term Prediction of Rock Mass Behaviour in Tunnelling by Advanced Analysis of Displacement Monitoring Data. Doctoral Thesis, Department of Civil Engineering, Technical University Graz, Austria, 127p.

Tonon, F. & Amadei, B. 2000. Detection of rock mass weakness ahead of a tunnel - a numerical study. *Pacific Rocks 2000* 1, 105–111.

Yong, S., Kaiser, P.K. & Loew, S. 2013. Rock mass response ahead of an advancing face in faulted shale. *Int. J. Rock Mech. Min. Sci.* 60, 301–311.

Tunnels and Underground Cities: Engineering and Innovation meet Archaeology, Architecture and Art, Volume 3: Geological and geotechnical knowledge and requirements for project implementation – Peila, Viggiani & Celestino (Eds)
© 2020 Taylor & Francis Group, London, ISBN 978-0-367-46583-4

Geo-mechanical behaviour and monitoring system in the Ceneri Base Tunnel

M. Ruggiero
AF TOSCANO SA, Lugano, Switzerland

A. Malaguti
LOMBARDI SA, Minusio, Switzerland

F. De Martino
AF-CONSULT ITALY S.r.l., Milan, Italy

ABSTRACT: The Ceneri Base Tunnel is part of the new Trans-Alpine Railway Line (NRLA). The NRLA axis of the Gotthard also includes the Gotthard Base Tunnel and connections to existing lines. The excavation of the 40 km of tunnels, caverns and cross-link, with maximum rock cover of 800m, involved the crossing of various geological formations consisting of Ortogneiss and Paragneiss rock, and disturbed areas, consisting of cataclasite and kakirite rock, in different percentages. The geo-mechanic behaviour of the different areas, investigated with geological surveys, and monitoring tools, was decisive for the choice of excavation and rock support systems. This paper will examine the types of monitoring applied, and the analysis of the data performed, making correlations between the rock mass, the excavation technique and the geo-mechanical behaviour.

1 INTRODUCTION

The Ceneri Base Tunnel (CBT) is part of the new Trans-Alpine railway line (NRLA) which is the most important construction project in Switzerland. The NRLA axis of the Gotthard includes, in addition to the Ceneri Base Tunnel, the Gotthard Base Tunnel and connections to existing lines.

The Ceneri Base Tunnel, 15.4 km long, is the third longest railway tunnel in Switzerland, after the Gotthard Base Tunnel and Lötschberg Base Tunnel. For the construction of it, a total of 40 km of tunnels and cross-link between the two tubes were excavated. The need to

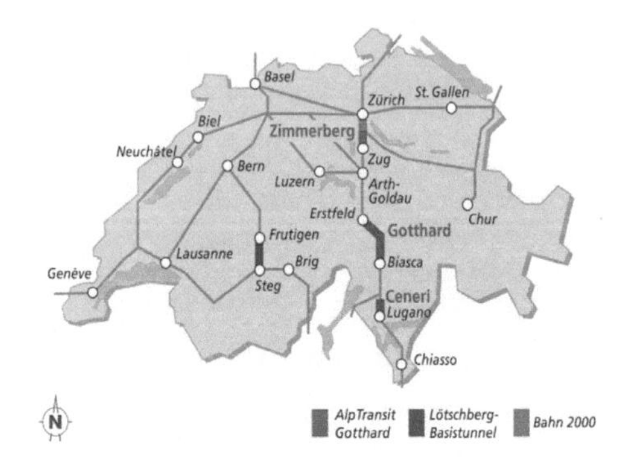

Figure 1. NRLA Works framework (Source: ATG).

make large interconnecting caverns made it necessary to carry out specific constructive interventions, both during the excavation phase and during the lining phase

The maximum rock cover is 800 m, and progress was made by drill & blast from the North and South portals and the intermediate section of Sigirino.

From a geological point of view, the tunnel alignment passes through areas consisting of Orthogneiss and Paragneiss rock. In the prognosis, 46 disturbed areas have been catalogued, consisting of different percentages of Cataclasite and Kakirite rock. The most important disturbed area crossed is the Val Colla Line (LVC), an intensely tectonized zone with a length equal to 658m in the East and 529 m in the West tube, the effects of which occurred more than 200m before the interception in the tunnel.

2 GEOLOGICAL CHARACTERISATION

During the project phase the different geological formations were divided into tectonic units as described below [1]

2.1 *North Ceneri Zone*

The North Ceneri Zone extends itself from the North of the CBT for a length of about 4.2km. It is affected by the general East - West trend of the main structures in the rock mass, dominated by the PGC (Ceneri Paragneiss). The rock cover reaches 820 m, the highest value of the whole CBT.

The rock mass of the zone is dominated by fragile cataclastic faults transversally to the north-west and south-east direction, that are different from the regional schistosity, with sub-vertical inclination (greater than 70 °) and real spacing between 5 and 25m.

However, since it is a system of considerable tectonic imprint (total lateral extension of at least 10 km to the south and north of the Insubric Line) and taking into account the sub-vertical layout of the fault planes, these disturbed zones are deep enough to affect the CBT axis.

2.2 *Intermediate Ceneri Zone*

The intermediate Ceneri Zone starts from the end of the North Ceneri Zone described above and extends itself for a length of about 3.8 km at the axis of the CBT. The zone is characterised by the regular and constant progress of the main structures (schistosity, lithological contacts) direct South-West and North-East, and immersed in the South-East. Towards the southern end of the domain, all the main structures rotate progressively until they assume a North - South direction, in correspondence with a very open fold, and set along an axis oriented North-West and South-East. The presence of large-scale isoclinal folds has also been highlighted, as well as significant disturbed zones coherent with the main structures and therefore very unfavourably disposed, as indeed are all the main structures, with respect to the CBT axis.

In this sector, the rock covers above the CBT are about 500 – 700 m with a maximum value of 780.

The lithologies of this zone's domain are mainly constituted of OGC (Ceneri Ortogneiss), PGC, B-UBC (Ceneri Basit and Ultrabasit)

2.3 *South Ceneri Zone*

This area is litologically dominated by the GGium (Giumello Gneiss). In this zone the covers are of the order of 650 – 700 m above the GbC with a progressive decrease towards the South of up to 400m. This area is parallel to the schistosity and lithological contacts with the CBT axis, and entirely located in the GGiums.

The trend of schistosity turns almost North - South and diverges mainly towards the East, with angles of 60–80°. The presence of folds from medium to markedly narrow (isoclinal), and large-scale on the surface locally, disturbs the immersion of schistosity.

Similar to the intermediate CZ, the rock mass in the South CZ is dominated by a system of cataclastic faults with a north-east and south-west course, immersing the North-West (averaging 320/70). These faults are unfavourably disposed with respect to the CBT axis and have real spacing between 5 and 15 m, corresponding to effective lengths on the CBT axis of 35 to 50 m.

2.4 *Val Colla Line*

The VCL (Val Colla Line) represents a section of high constructive relevance, in that it is made up of rocks originally deformed in the ductile field (pre-Alpine age), and is subsequently superimposed as a destructive tectonic, composed of mainly kakiritic and cataclasitic faults. The geotechnical properties are strongly micaceous and the tectonized milonites are poor. These faults constitute a system oriented around North-East and South-West and are immersed at high angle towards North-West. The length of this section on the CBT axis is in the order of 600 m and the cover over the CBT varies from 200 m to the South, to 350 m to the North. To the North, the border is with GGiums and to the South with the GStab.

From a structural point of view, the orientation of the planes of milonitic schistosity varies considerably within the LVC, but on average shows a north-west dive, with angles of 40 - 70°. From the lithological point of view three main types can be distinguished: type 1 characterised by rocks that are more or less intact and not overlapped by cataclastic processes; type 2 characterised by the alternation of cataclasitic and kakiritic rocks; type 3 characterised entirely by kakiritic rocks.

2.5 *North Val Colla Zone*

The North Val Colla tectonic unit extends from the southern limit of the LVC. The dominant geological formation of this tectonic domain is represented by the GStab. To a lesser extent there are some lentiform bodies from metric to decametric B-SBVC (Basist and semibasites of the Val Colla) and OGBern (Ortogneiss Bernardo).

Near the LVC, the main structures, such as schistosity and lithological contacts, first plunge in the direction from the North-West to the North-East, and then rotate and immerse in the South-West direction.

The contact between VCZ Nord and VCZ South defines an important structural change, characterised by the transition from an inclined schistosity to a generalised sub-horizontal trend.

2.6 *South Val Colla Zone*

Also, in this tectonic domain, the dominant geological formation is represented by the GStab, with some bodies of OGBern and minor passages of B-SBVC. In the South VCZ the roofs are between 35 – 140 m, progressively decreasing towards the South until reaching minimum values (13m). The orientation of the South ZVC's planes of schistosity is generally sub-horizontal and variable with a more likely inclination from 0 to 40°.

The Sarè interconnection caverns were planned in the GStab and OGBern (see chapter 5.2 for constructive method detail).

3 TECHNICAL EXCAVATION SYSTEM

For the excavation of the 40 km of tunnels and caverns a drill & blast system was applied by the implementation of 10 class excavation types.

Each class was different from anchor number, steel profile, shotcrete thickness and deep forward. The excavation in good rock was conducted with a class from 1 to 6, the excavation in fault or disturbed rock mass with a class from 7 to 10.

All the rock supports applied for each excavation class are shown in Tables 1 and 2.

The classes were adapted to the different rock masses, but mainly to the rock mass behaviour.

Figure 2 is an example of a class number 7 applied for the excavation of a medium fault zone, with displacements of about 10 cm, and the potential risk of face instability.

Table 1. Rock Support for normal forward.

Excavation class	Anchors	Shotcrete thickness	Fibre-reinforced shotcrete	Electro welded mesh	Steel Profile
1	Friction anchors (Ftk >120 kN, L = 3 m)	5 cm with fibre and 10 cm with welded mesh	30 kg/m^3	K196	None
2	Friction anchors (Ftk >240 kN, L = 4 m)	5 cm with fibre and 10 cm with welded mesh	30 kg/m^3	K196	None
3	Friction anchors (Ftk >240 kN, L = 4 m)	5 cm with fibre and 10 cm with welded mesh	30 kg/m^3	2 x K196	None
4	Friction anchors (Ftk >240 kN, L = 4 m) Bar anchors (Ftk > 340 kN, L = 6 m).	5 cm with fibre and 10 cm with welded mesh	30 kg/m^3	2 x K196	None
5	Bar anchors (Ftk >250 kN L = 4 m) Bar anchors on the face (Ftk > 340 kN L = 6 m)	5 cm with fibre and 12 cm with welded mesh	30 kg/m^3	K196 + K283	TH 25, pitch 1.5 m
6	Bar anchors on the lining and on the face (Ftk > 340 kN L = 6 m)	5 cm with fibre and 12 cm with welded mesh	30 kg/m^3	K196 + K283	TH 25, pitch 1.2 m

Table 2. Rock Support for fault rock forward.

Excavation class	Anchors	Shotcrete thickness	Fibre-reinforced shotcrete	Electro welded mesh	Steel Profile
7	Bar anchors (Ftk >340 kN L = 5 m) Injectable anchors on the face (Ftk > 320 kN L = 8 m)	5 cm with fibre and 15 cm with welded mesh	30 kg/m^3	2xK283	TH 29, pitch 1.20 m full round
8	Bar anchors (Ftk >340 kN L = 6 m) Injectable anchors (Ftk > 460 kN L = 8 m) Injectable anchors on the face (Ftk > 460 kN L = 8 m)	5 cm with fibre and 15 cm with welded mesh	30 kg/m^3	2xK283	None
9	Bar anchors (Ftk >340 kN L = 6 m) Injectable anchors (Ftk > 460 kN L = 8 m) Injectable anchors on the face (Ftk > 460 kN L = 8 m)	5 cm with fibre and 15 cm with welded mesh	30 kg/m^3	2xK283	TH29, pitch 1.00 m full round with displacement slot
10	Injectable anchors (Ftk > 460 kN L = 8/10 m) Injectable anchors on the face (Ftk > 460 kN L = 10 m)	5 cm with fibre and 15 cm with welded mesh	30 kg/m^3	K283	TH 29, pitch 1 m full round with displacement slot

The rock mass behaviour was analysed by geological and scanner surveys and displacement measurements. For example, anisotropic behaviour required special applications to study it [2] and to control it during the excavation.

The rock support system applied was a mix of the complete knowledge at the time and the surveys, in order to grant a displacement of the system, which is different for each class (Table 3).

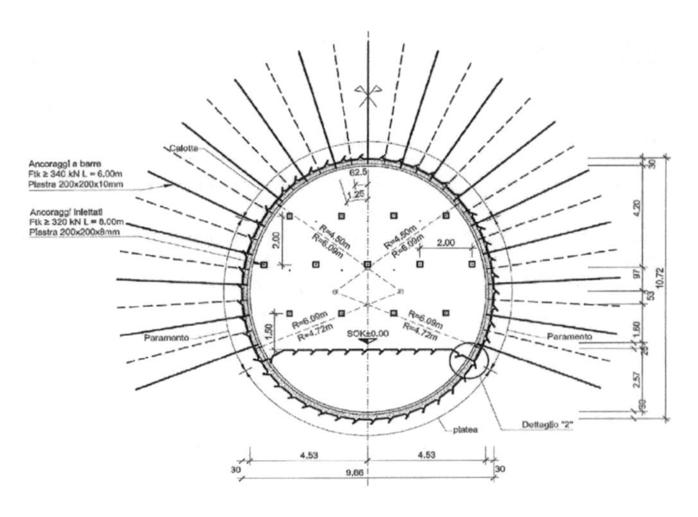

Figure 2. Excavation class number 7 (Source: ITC).

Figure 3. Slot displacement system (Source: ITC).

The displacement of the system, and its perfect control, was necessary in order to apply a final ring with as low thickness as possible, as required by the project, and to keep the necessary safety factor.

For the highest excavation classes, the displacement was permitted by the addition of special slots realised in the shotcrete profile. They are in a precise place where the steel rib can also deform.

The example in Figure 3 allows for a 30 cm displacement of the ring.

4 GEOMECHANICAL BEHAVIOUR MONITORING TOOLS

4.1 *Scanner Survey*

After each excavation a survey of the profile was made with the laser scan technique.

An example of output is shown in Figure 4. The green colour means that the excavation was made correctly without under-profile and it's possible to add sprayed concrete and, if required, to install the steel profile in the correct position. A successful laser scan survey, considering the

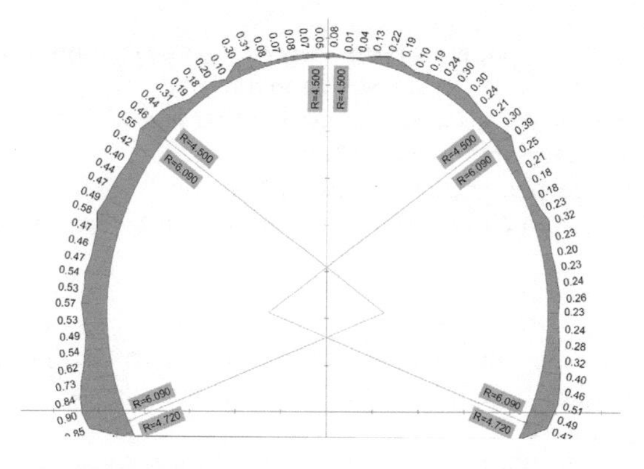

Figure 4. Laser scan graphic output (Source: ITC).

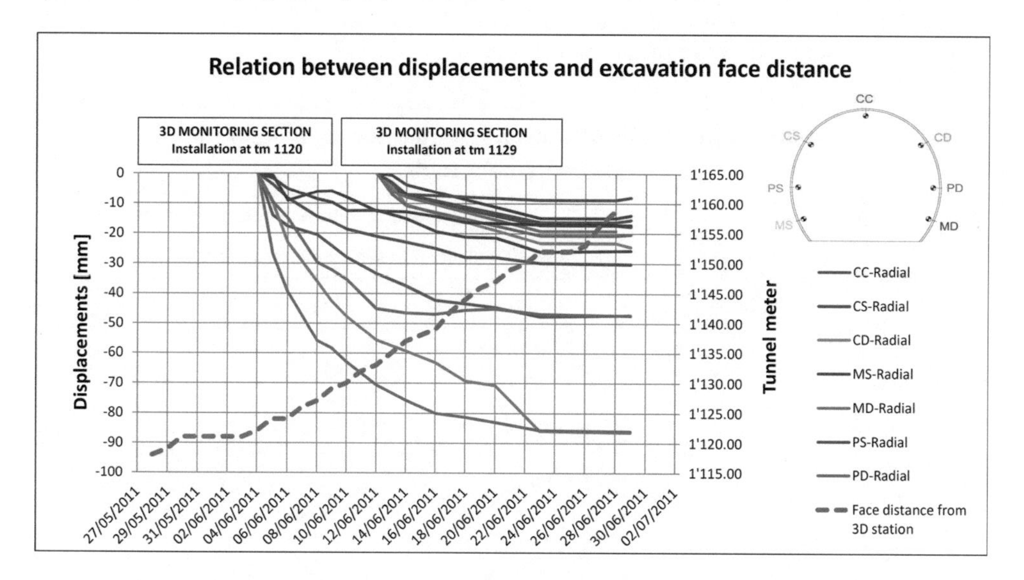

Figure 5. Relation between displacement and excavation face distance (Source: ITC).

sprayed concrete's final thickness, allows one to determine the radial displacements of the excavation class sustain system. The laser scan survey was made for each metre of tube excavation.

4.2 *3D Analysis points*

The System exists in several points, usually from 3 to 7, around the tunnel's internal ring, and is installed after excavation, as close as possible to the face.

The radial displacements, obtained by a total topographic station for each point, was considered for the application of the different forward classes.

Analysis of the data is shown in Figure 5. The graph represents the radial displacement recorded in relation with the excavation face's distance. In the example the results of the two near sections are very different, demonstrating the great heterogeneity of the rock mass.

With this kind of data many other graphs were made in order to better understand the rock mass' behaviour in relation with the class system applied. In chapter 5.1 we present an example.

4.3 *Multi-base extensometer*

The multi-base extensometer indicates how the rock mass deforms deeply around the tunnel.

It consists of a tool with a number of fibre optic cables at each point of measurement and at different distances from the profile. It is installed in a radial perforation.

The head of the extensometer has a control system that registers the movement of the different bases, at different distances and in 3 directions.

This system shows the deep impact of the excavation.

The result is important for a correct evaluation of the charge over the support system.

4.4 *Sliding micrometre*

A sliding micrometre provides similar information to the multi-base extensometer. The difference is that sliding micrometre is a manual tool and provides displacement data for each metre of the total length.

This tool was only tested once in the Ceneri Base Tunnel for a comparison with a multi-base extensometer. The two instruments were installed in the same place at a close enough distance that allowed for a mutual non-influence.

The results of it are shown in the graph below:

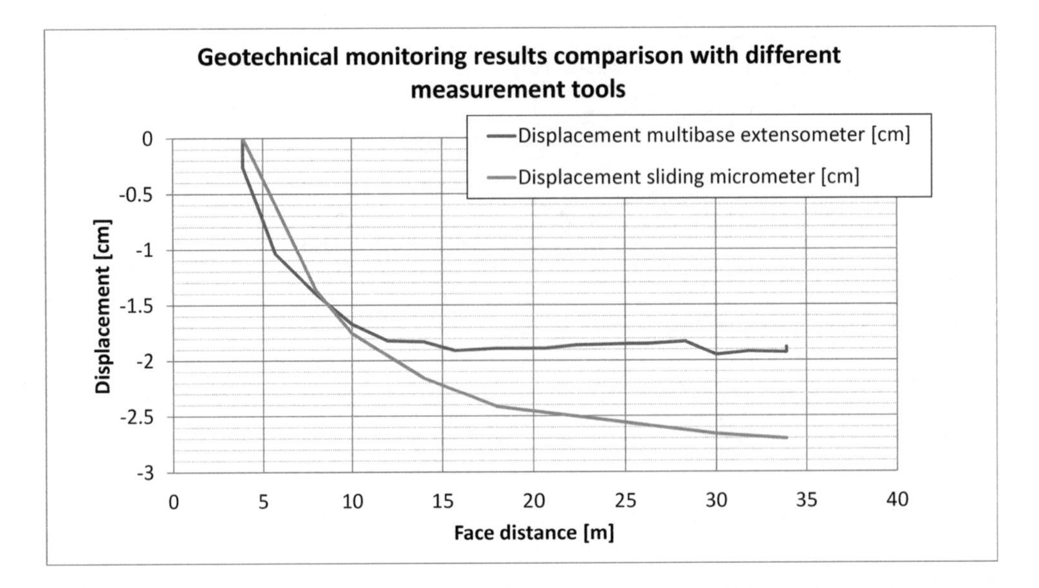

Figure 6. Multi-base extensometer and sliding micrometer results comparison (Source: ITC).

The data analysed in the graph shows that the results of the tools are similar, especially in the first 10 metres from the face. After this distance the slide micrometer records a major displacement. All the data is referred to the lining of the tunnel, and the comparison is made by the total displacement at this point. The other displacements were compared with the displacement of the sliding micrometer, at the same place as the base of extensometer, and the results are similar. For greater clarity, it was decided to represent only the lining displacement that is the highest and the most significant.

The decision to apply only the multi-base extensometer was based on the ability to read the results automatically, and therefore faster, than the sliding micrometer.

4.5 *Geological Survey*

Each excavation face was reported in a geological survey by a specialist geologist at the place after drill & blast. The information collected by the geologist is combined with other information to rebuild the rock mass' behaviour. Figure 7 is an example of a face excavation that is ready to be surveyed, that has a cataclastic fault in the rock mass.

Figure 7. Excavation face ready for survey (Source: ITC).

The information of the geological survey is extremely important and every day in the Ceneri Base Tunnel a face pro forward was surveyed. This mean about 14,000 geological surveys that allowed the team to rebuild the entire tunnel's geology.

5 DATA ANALYSIS AND GEOMECHANICAL BEHAVIOUR

5.1 *Application of 3D analysis points*

The different systems shown in the previous points are combined to find the most realistic behaviour of the rock mass, in order to apply the correct system of excavation and support for the excavated tunnel. Each excavation class represented in Tables 1 and 2 have a range of deformability as a result of this calculation. This deformability allows a stress decrease in the lining and permits a correct dimensioning of the whole system. The range of displacements of each class is represented in the following table:

Table 3. Excavation class and displacements permitted.

Excavation class	displacements permitted (cm)
1	0 – 5
2	0 – 5
3	5 – 10
4	5 – 10
5	5 – 10
6	5 – 10
7	10 – 15
8	15 – 25
9	25 – 30
10	25 – 30

The graph in Figure 8 is an example of the control of displacements. A similar graph was made for the whole tunnel for all 40 km.

In the abscissa line each 3D station is represented with 5 or 7 points, depending on the number of monitoring points for each station. The plotting isn't dependent on the installation place in the tunnel and every point represents one station. This graph allows the team to control the project in respect of the displacements and to make the required corrections to the excavation

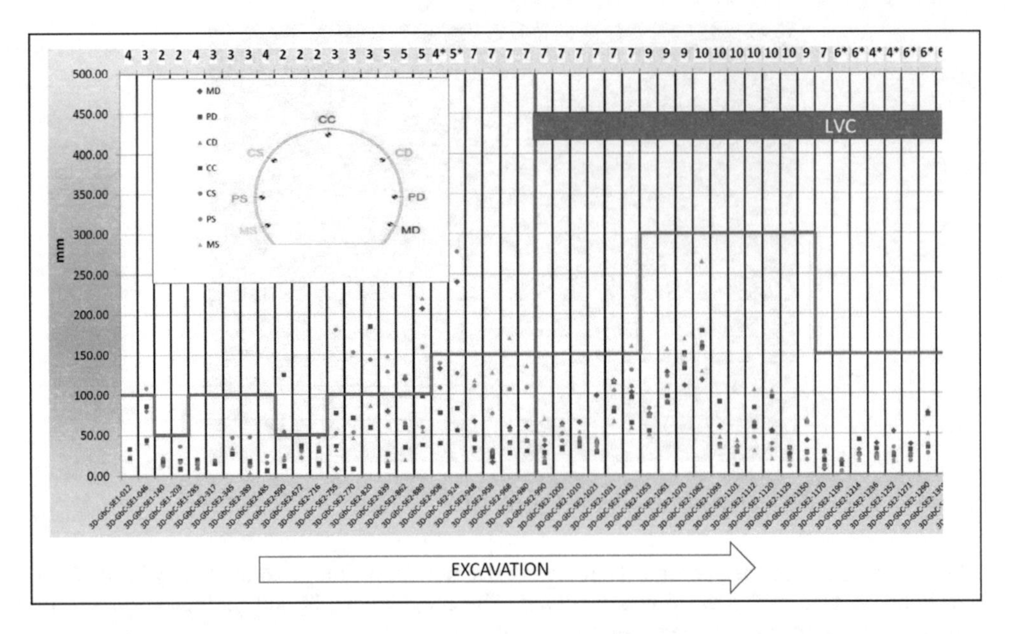

Figure 8. Graph for displacements control and excavation class applied; the red line represents the maximum displacement allowed for the excavation class and the blue line is the border of the Val Colla tectonic unit described in 2.4 (Source: ITC).

classes. The base line represents the displacements in mm and at the top the excavation class is applied. The red line shows the maximum displacement allowed by project. In the first part of graph it is clear that the excavation class applied isn't correct because there are several points over the red line. With successful corrections it was possible to obtain a good application of the excavation classes according the allowed displacements. The blue label means that the excavation is in the Val Colla Line, the tectonical unit described in chapter 2.4.

5.2 *Excavation of large cavern*

Many kinds of large caverns were excavated in the Ceneri Base Tunnel to grant interconnections between the different railway lines. Two of those were named the Sarè Interconnection Caverns and were two twin caverns in the southern part of tunnel. The Sarè Caverns will be very important for future links to the new Alp Transit Line to the South. These caverns allow a direct link to the actual railway line. In the future it will be the place where the South tunnel links to the Ceneri Base Tunnel. The Sarè Interconnection Caverns were planned in the GStab and OGBern (Chapter 2.6). The excavation of a large tunnel with about 240m^2 of face excavation represented a difficult challenge for every engineer. All the monitoring systems described were vital to the understanding of the geo-mechanical behaviours. The excavation system and the monitoring systems are represented in Figure 9. The black label represents a 3D displacement station and the blue label is one or more extensometer multi-base. Full face, half face or pilot tunnel, with different combinations, is represented with different colours in the picture.

The excavation system was adapted by the monitored results. The east tunnel, at the top of the picture, has an anisotropic behaviour that required the larger part to be near the interconnection with the west tunnel, which is a pilot tunnel, coupled with the west side excavation. The west tunnel had isotropic behaviour, but the larger displacements recorded required a pilot tunnel and an east and west side excavation.

All the systems were correctly applied thanks to the monitoring system and allowed the teams to complete the two caverns without problem, and in the required time.

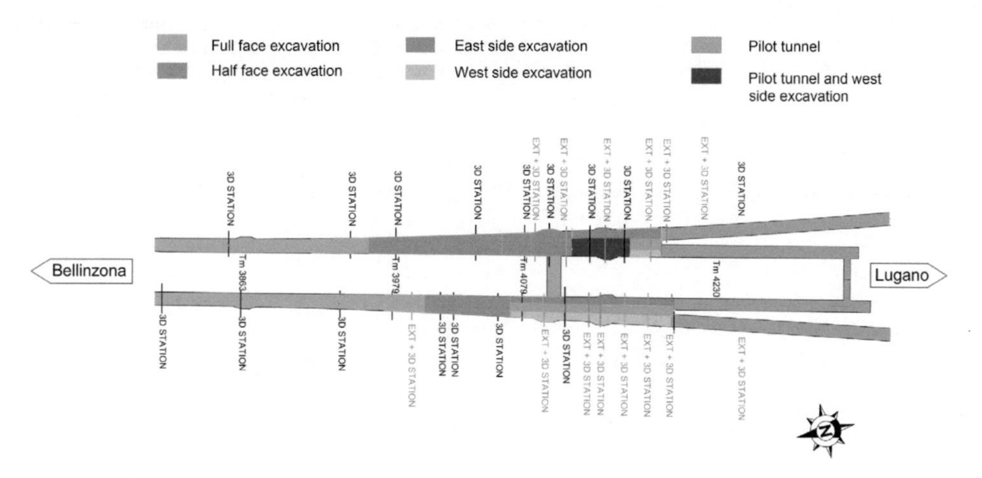

Figure 9. Sarè Caverns excavation system (Source: ITC).

6 CONCLUSIONS

The project, and excavation systems based on displacements controls, needs an accurate and flexible displacement system, as described. For example, the completion of a definitive lining could take place only at the end of the displacement phenomena. Some 3D stations or multi-base extensometer stations were in place and ready for a year before the complete exhaustion of the phenomena. In some cases, it was also necessary to integrate anchors or sprayed concrete in order to stabilise the displacements. Finally, we can conclude that to build a tunnel, a geotechnical project does not finish on arrival in the field, but needs to be continuously updated, according to the results of the monitoring.

ACKNOWLEDGEMENTS

Special thanks to Alp Transit Gotthard AG for providing the data analysed and colleagues from the Ceneri Base Tunnel Site Supervision for their technical support.

Special thanks for their contributions to Engineers Alberto Del Col and Alessandro Bagattini (Alp Transit Gotthard AG) and Adrian Fontana, Matteo Genini, Stefano Morandi, Jean Pierre Hürzeler (ITC consortium).

REFERENCES

[1] ITC Team 2018 *Geological, geotechnical and idrogeological report*. Lugano

[2] Malaguti A., Morandi S., Stocker D. 2013. *Anisotropic geomechanical behaviour of tunnelling*. Val Colla Line, Ceneri base Tunnel, Switzerland. WTC, Geneva

Tunnels and Underground Cities: Engineering and Innovation meet Archaeology, Architecture and Art, Volume 3: Geological and geotechnical knowledge and requirements for project implementation – Peila, Viggiani & Celestino (Eds)
© 2020 Taylor & Francis Group, London, ISBN 978-0-367-46583-4

Drained and undrained seismic response of underground structures

E.A. Sandoval
Purdue University, USA
Universidad del Valle, Colombia

A. Bobet
Purdue University, USA

ABSTRACT: Two-dimensional dynamic numerical analyses are conducted to assess the seismic response of deep tunnels located far from the seismic source, under drained or undrained loading conditions. It is assumed that the liner remains elastic and that plane strain conditions apply. An elastoplastic constitutive model is proposed to simulate the nonlinear behavior and excess pore pressures accumulation in the ground with cycles of loading. The results show negligible effect of input frequency on tunnel distortions for frequencies smaller than 5 Hz; that is, for ratios between wavelength and tunnel opening larger than about eight to ten. The results also show that undrained conditions, compared with drained conditions, tend to reduce deformations for flexible liners and increase them for stiffer tunnels, when no accumulation of pore pressures with cycles of loading is assumed. However, when pore pressures increase with number of cycles, the differences between drained and undrained loading are reduced.

1 INTRODUCTION

Underground structures must be able to support static overburden loads as well as to accommodate additional deformations imposed by seismic motions. Soil-structure interaction and stress and displacement transfer mechanisms from the ground to the structure, during a seismic event, have been explored extensively in the last few years. For most tunnels, perhaps excluding submerged tunnels, it seems that the most critical demand on the structure is produced by shear waves traveling perpendicular to the tunnel axis (Bobet, 2003; Wang, 1993; Merritt et al., 1985; Hendron and Fernández, 1983), which distort the tunnel's cross section. Such distortions are called ovaling for a circular tunnel and racking for a rectangular tunnel, as shown in Figure 1, and produce axial forces and bending moments to the liner, in addition to those under normal working conditions.

In the past, the seismic response of underground structures has been assessed under the assumption that the structure follows the free field deformations of the ground (i.e. without the presence of the tunnel), and therefore the structure must accommodate such deformations without loss of integrity. In such case, the tunnel distortions are obtained as those of the perforated ground (Hendron and Fernández, 1983; Kuesel, 1969; Merritt et al., 1985; Newmark, 1967). This assumption however may result in extremely conservative designs, especially for stiff structures in a soft medium. A more realistic approach, which is used in this paper, is that the tunnel modifies the deformation of the surrounding ground and thus demand and response depend on the relative stiffness between the ground and the tunnel support.

There is consistent evidence in the technical literature that shows that the dynamic amplification of stress waves impinging on a cavity is negligible when the wavelength (λ) of the seismic peak velocities is at least eight times larger than the width of the opening (e.g., Paul, 1963; Mow and Pao, 1971; Yoshihara, 1963; Hendron & Fernández, 1983; Merritt et al., 1985). This

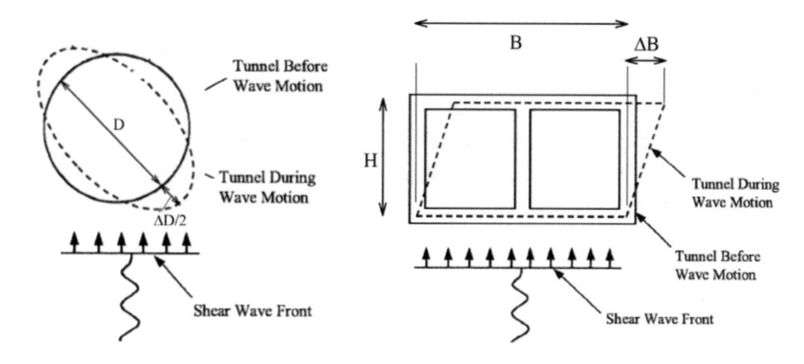

Figure 1. Ovaling and racking deformation of the tunnel cross section (after Owen and Scholl, 1981).

is usually the case for most structures located far from the seismic source. According to Dowding (1985), this happens when the seismic source is located between 10 and 100 km from the tunnel, where the predominant frequencies are generally between 0.1 and 10 Hz. In these cases, pseudo-static numerical analyses and closed-form solutions seem appropriate (Bobet, 2003, 2010; Huo et al., 2006; Penzien, 2000; Wang, 1993). The closed-form solutions are suitable for deep tunnels (i.e., those where the geostatic stress gradient with depth has negligible effect) placed in an infinite, linear-elastic, homogeneous and isotropic medium. A plane strain condition, which is reasonable for tunnel sections located far from the face of the tunnel, is also assumed.

Previous work has found that the most important parameter that determines the distortions of the tunnel cross section is the relative stiffness between the ground and the liner, represented by the flexibility ratio, F. The ratio F is a dimensionless parameter that quantifies the resistance of tunnel and ground against distortion, under a state of pure shear, and so it is often used as a measure of the relative stiffness between the underground structure and the surrounding ground. Other parameters such as depth and shape of the structure have second-order effects (Sandoval & Bobet, 2017; Bobet, 2010, Wang, 1993). While all this has been well-studied for structures under drained conditions, there is little information regarding the behavior of buried structures under undrained conditions, i.e., when excess pore pressures generated are not dissipated.

This paper provides results of 2D plane strain dynamic numerical analyses conducted using FLAC 7.0 (Itasca, 2011a) for deep circular tunnels subjected to cyclic shear. The liner is assumed to remain in its elastic regime without relative displacement with the ground, i.e., with a tied interface. Nonlinear ground behavior under drained and undrained loading is considered. The effects of the input frequency and the flexibility ratio on the distortions of the cross section of the tunnel are investigated for both drained and undrained loading. For comparison purposes, results of linear-elastic analysis for drained loading using one of the closed-form solutions (e.g. Bobet, 2010) are also included.

2 CYCLIC NONLINEAR ELASTOPLASTIC MODEL

The constitutive model used in the paper is built based on the work by Jung (2009) and later by Khasawneh et al., (2017). The model incorporates the well-known Masing's rules, where the response of soil under cyclic loading follows a hysteretic behavior (Hardin & Drnevich, 1972; Pyke, 1979; Ohsaki, 1980). The model is implemented in FLAC 7.0 (Itasca, 2011a) and is verified by comparing its predictions with results from different laboratory tests under drained and undrained loading. The model is rate-independent, as it is generally assumed for the seismic response of most geomaterials, and is defined within a small-deformation framework for incremental plasticity theory. The model includes four main characteristics: i) nonlinear stress-strain and hysteretic behavior during cyclic loading; ii) dependence of the very small strain shear modulus on confinement; iii) coupled shear-volumetric strains with excess pore pressures accumulation; and iv) a yield criterion and non-associated plastic flow rule.

The nonlinear behavior is simulated through a hyperbolic formulation. Equation 1 provides the stiffness reduction of the ground as a function of the modified hyperbolic strain (γ_h), as suggested by Hardin & Drnevich (1972), and shown in Equation 2.

$$G = \frac{d\tau}{d\gamma} = G_0 \frac{1}{(1+\gamma_h)^2} \tag{1}$$

$$\gamma_h = \frac{\gamma}{\gamma_r}\left[1 + a exp^{-b\left(\frac{\gamma}{\gamma_r}\right)}\right] \tag{2}$$

where, $\gamma_r = \tau_f/G_o$ is the reference shear strain, τ_f is the shear stress at failure, G_o is the shear modulus at very small strains, γ is the current shear strain, and a and b are constants that determine how the stress-strain relation deviates from the hyperbolic function.

Equation 3 gives the degradation of the shear stiffness in the three-dimensional space (after Jung, 2009 and Pyke, 1979). In the equation, a factor $n = 1$ is used for the initial loading curve, a factor $n = 2$ reproduces the Masing's formulation, while $n > 2$ and $n < 2$ account for cyclic hardening or softening, respectively, during unloading and reloading. The variables needed in Equation 3 are given in Equations 4 to 9.

$$G = G_o \frac{1}{\left[1 + \frac{1}{n}|\Gamma - \Gamma_{rev}|\right]^2} \tag{3}$$

$$\Gamma = \frac{\gamma_{oct}}{\gamma_{oct,r}}\left[1 + a exp^{-b\left(\frac{\gamma_{oct}}{\gamma_{oct,r}}\right)}\right] \tag{4}$$

$$\Gamma_{rev} = \frac{\gamma_{oct,rev}}{\gamma_{oct,r}}\left[1 + a exp^{-b\left(\frac{\gamma_{oct,rev}}{\gamma_{oct,r}}\right)}\right] \tag{5}$$

$$\gamma_{oct} = 2\sqrt{\frac{2}{3}}\sqrt{\frac{1}{2}e_{ij}e_{ij}} \tag{6}$$

$$\gamma_{oct,r} = \frac{\tau_{oct,r}}{G_o} = \frac{\sqrt{2}}{3G_o}\left(\sigma'_m tan\alpha - \kappa\right) \tag{7}$$

$$e_{ij} = \varepsilon_{ij} - \frac{1}{3}\varepsilon_{kk}\delta_{ij} \tag{8}$$

$$\varepsilon_{ij} = \frac{1}{2}\left(\frac{\partial u_i}{\partial x_j} + \frac{\partial u_j}{\partial x_i}\right) \tag{9}$$

where n is a scaling factor for the hysteresis loop, σ'_m is the effective mean stress, e_{ij} is the deviatoric strain tensor, ε_{ij} is the Lagrangian strain tensor, and u_i, u_j are the displacements along the x_i, x_j axes. The terms α and κ represent strength parameters and are discussed later.

The dependence of the ground stiffness on the effective mean stress has been extensively discussed in the literature (e.g., Hardin & Drnevich, 1972; Hardin, 1978, Porovic & Jardine, 1994). Such behavior is considered in Equation 10, which is used to obtain the incremental small-strain shear modulus (dG_o) as a function of the incremental effective mean stress ($d\sigma'_m$).

$$dG_o = \frac{1}{2}\frac{G^2_{o(ref)}}{\sigma'_{m(ref)}}\frac{1}{G_o}d\sigma'_m \tag{10}$$

The model provides a formulation that couples the shear and volumetric plastic strains, needed to account for excess pore pressures accumulation during undrained cyclic loading. The formulation is based on Byrne (1991), after the work by Martin et al. (1975). Equation 11 is used to obtain the plastic volumetric strains produced by shear strains and is formulated in 3D. The tendency to generate compressional volumetric strains during drained cyclic loading leads to positive excess pore pressures accumulation during undrained loading, due to the constraint for the material to not change volume.

$$\Delta\varepsilon^p_{v(1/2cycle)} = \frac{1}{2}(\gamma_{oct} - \gamma_{oct-th})c_1 exp\left(-c_2\frac{\varepsilon_v}{\gamma_{oct} - \gamma_{oct-th}}\right) \tag{11}$$

where $\Delta\varepsilon^p{}_{v(1/2cycle)}$ is the incremental increase in plastic volumetric strain per ½ cycle of loading, γ_{oct} is the octahedral shear strain, $\gamma_{oct\text{-}th}$ is a threshold octahedral shear strain, ε_v is the accumulated volumetric strain from previous cycles, and c_1, c_2 are constants for the coupling between shear and volumetric strains that depend on the relative density or consistency of the ground.

The model also includes an elastoplastic formulation with a yield criterion and non-associated flow rule to evaluate plastic strains. Although the model is intended to be used for small to moderate strains, local yielding may occur due to the decrease in confinement caused by the excess pore pressures accumulation. The total strain increment is decomposed into its elastic and plastic components ($d\varepsilon_{kl} = d\varepsilon_{kl}^e + d\varepsilon_{kl}^p$). The stress increment ($d\sigma_{ij}$) is computed from the generalized Hooke's law using Equation 12. Yielding is defined with the Drucker-Prager (D-P) criterion, and the plastic strains are determined from a strain hardening law with a non-associated flow rule. Equations 13 and 14 show the yield function and the plastic potential, respectively. Equations 15 and 16 are used to determine the plastic strains, once yielding has occurred, invoking the consistency condition.

$$d\sigma_{ij} = C^e_{ijkl}d\varepsilon^e_{kl} \tag{12}$$

$$f = \sqrt{3J_2} + \frac{1}{3}I_1\tan(\alpha) - \kappa \tag{13}$$

$$g = \sqrt{J_2} + \frac{1}{3}I_1\tan(\psi) \tag{14}$$

$$d\varepsilon_{ij} = d\lambda\left(\frac{\partial g}{\partial\sigma_{ij}}\right) \tag{15}$$

$$d\lambda = \frac{\frac{\sqrt{3}}{\sqrt{J_2}}GS_{ij}d\varepsilon_{ij} + Kd\varepsilon_{kk}(tan\alpha)}{\sqrt{3}G - K(tan\alpha)(tan\psi)} \tag{16}$$

where C_{ijkl} is the elastic modulus tensor, $d\varepsilon_{kl}^e$ is the incremental elastic strain tensor, $J_2 = 1/2\,(S_{ij}S_{ij})$ is the second invariant of the deviatoric stress tensor, $I_1 = \sigma_{kk}$ is the first invariant of the stress tensor, α is the D-P friction angle, κ is the D-P cohesion, and ψ is the interlocking component of strength, i.e., interference and dilation.

The capabilities of the model are verified by comparing results of simulations with different cyclic laboratory tests under drained or undrained loading. As an example, Figure 2 shows experimental results and the simulations for an undrained cyclic simple shear test conducted on Ottawa sand by Ishibashi et al. (1985). The hysteresis loop and the normalized excess pore

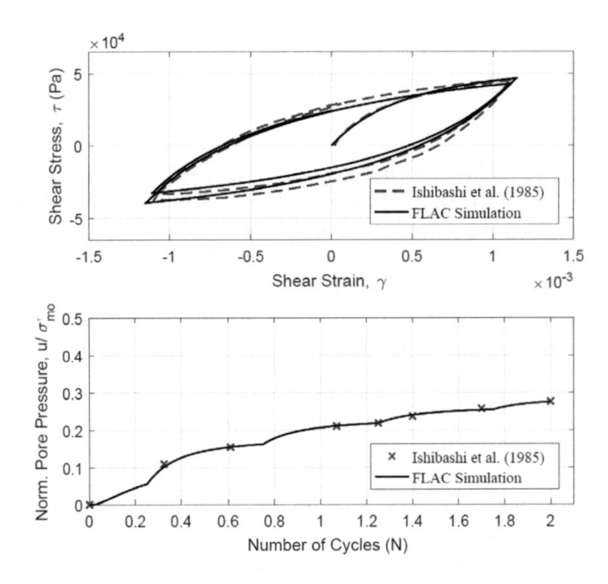

Figure 2. Comparison between simulations and results of laboratory undrained cyclic simple shear tests on Ottawa sand, from Ishibashi et al. (1985).

Table 1. Model parameters for undrained cyclic simple shear test on Ottawa sand, from Ishibashi et al. (1985)

$G_{o(ref)}$ (MPa)	$\sigma'_{m(ref)}$ (kPa)	ν	α (°)	κ (kPa)	Ψ (°)	a	b	n_1	n_2	γ_{th}	c_1	c_2
127	138	0.20	41	0.10	5	-0.02	0.80	2.05	2.15	0.00003	1.58	2.52

pressures are compared in Figure 2. As seen in the figures, a good approximation is obtained with the constitutive model. Table 1 shows the input parameters used in the simulations. As seen in the table, the model includes 13 input parameters, of which 6 represent any physical property and the others are fitting parameters that are also related to the physical properties of the soil.

3 EFFECT OF INPUT FREQUENCY ON TUNNEL DISTORTIONS

Previous research has found that the dynamic amplification of stress waves impinging on a tunnel is negligible when the rise time of the pulse is larger than about two times the transit time of the pulse across the opening; that is, when the wavelength (λ) of peak velocities is at least eight times larger than the width (B) or diameter (D) of the opening. Such statement was proposed and verified in the past for linear-elastic media (e.g., Paul, 1963; Yoshihara, 1963; Mow, 1971; Hendron Jr. & Fernández, 1983; Merritt, 1985; Bobet, 2010), and has been recently verified for nonlinear ground (Sandoval & Bobet, 2017). However, such observation has been reached for dry ground or for drained loading, and it is not clear if it can be applied for undrained loading conditions.

Distortions for circular tunnels with 4.0 m in diameter and 0.40 m liner thickness are obtained for nonlinear ground under drained and undrained loading. The mesh has dimensions 300 m x 100 m and the tunnel is placed at 75 m depth. The depth of the model is selected following the recommendations from Itasca (Itasca, 2011b) to avoid the effect of waves reflection from the free surface, especially for input frequencies equal or lower than 1 Hz. Reflections from the bottom and sides are minimized by placing quiet and free-field boundaries.

These are absorbing boundaries that use independent dashpots in the normal and shear directions to avoid or decrease energy radiation (Itasca, 2011b; Lysmer & Kuhlemeyer, 1969). A dynamic input velocity with amplitude equal to 0.1 m/s and frequencies ranging from 0.25 to 15 Hz is imposed at the bottom of the model. These frequencies correspond to λ/D ratios ranging from 193 to 2.9, given the dimensions of the tunnel. The wavelength (λ) is obtained as the ratio between the shear wave velocity in the ground (C_s) and the frequency of the dynamic input (f).

For the ground, a shear modulus (G) and a Poisson's ratio (v) equal to 80 MPa and 0.25 are assumed, respectively. For the liner, a value of 0.15 is used for the Poisson's ratio (v_s). A (relative) stiff tunnel is obtained with a Young's modulus (E_s) equal to 1.56×10^5 MPa, and a flexible tunnel with E_s equal to 2.61×10^3 MPa. With these values, the initial flexibility ratios (F) are 0.25 and 15 for the stiff and flexible tunnels, respectively; see Equation 17 after Peck et al. (1972). Due to the stiffness degradation of the ground during dynamic loading, the flexibility ratios decrease with strain, and at peak tunnel distortion the values are F= 0.18 to 0.21, and F= 11.2 to 13.9 for the rigid and flexible tunnels, respectively. The values are variable because are related to the ground stiffness, which changes between drained and undrained loading, and between the different cycles in undrained loading.

$$F = \frac{E_m/(1+v)}{6E_sI_s/\left(R^3\left(1-v_s^2\right)\right)} \tag{17}$$

where E_m is the Young's modulus of the medium, I_s is the moment of inertia of the liner per unit length, and R is the radius of the tunnel.

Figure 3 shows results of the analyses in terms of normalized distortions of the tunnel versus input frequency for the stiff and flexible tunnels, for both drained and undrained loading. The normalization is done with respect to the distortions that would occur in the free field without the cavity. Note that three sets of values are reported for undrained loading: one at peak distortion during the first cycle, another at peak distortion during the 5th cycle and the third one, also at peak distortion, during the 10th cycle. The normalized excess pore pressures are $\Delta u/p'_o$ = 0.02, 0.21 and 0.32, respectively. Δu is the excess pore pressures in the free field and p'_o is the initial effective confinement stress. Figure 4 shows the tunnel normalized distortions as a function of the λ/D ratios. Note that, in Figure 3, frequency increases along the horizontal axis while in Figure 4 the normalized wavelength decreases. In Figure 3, small changes

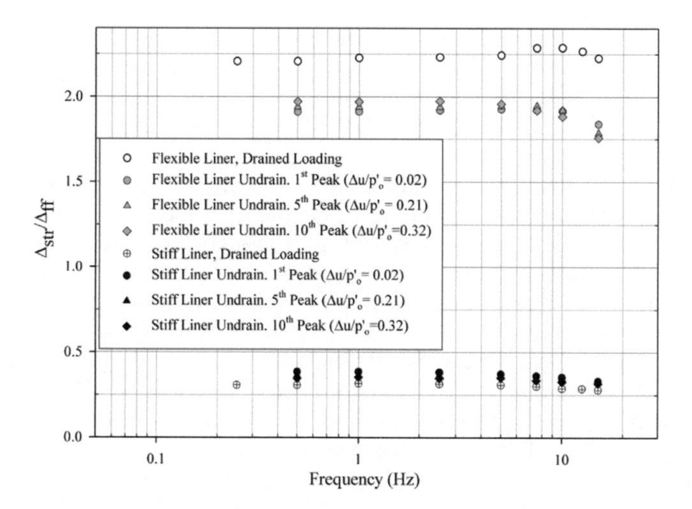

Figure 3. Effect of input frequency on distortions of circular tunnels under drained or undrained loading.

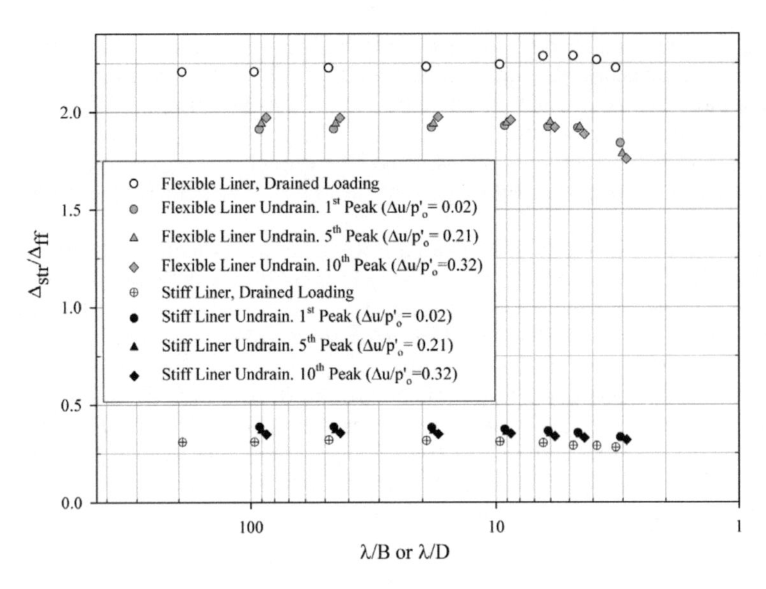

Figure 4. Effect of λ/D ratios on distortions of circular tunnels under drained or undrained loading.

of tunnel distortions are observed for frequencies smaller than 5 Hz for both flexible and stiff structures and for drained and undrained loading. Figure 4, which is a plot of the same results but using normalized wavelength, shows that negligible changes are observed for λ/D ratios larger than 9.5 for drained loading and larger than 8.5 to 9.0 for undrained loading. Note that for the same input frequency in Figure 3, different λ/D ratios are seen in Figure 4. This is due to the reduction in wavelength as the excess pore pressures increase, i.e., as the ground degrades more.

Thus, for both drained and undrained loading, there is no effect of the input frequency on the seismic response of tunnels for ratios between the wavelength of the seismic ground motions and the size of the tunnel larger than about 8.5 to 9.5. Such relation is usually satisfied for tunnels located far from the seismic source (distances larger than about 10 km, according to Dowding, 1985). In other words, for tunnels located far from the epicenter of the earthquake, a static analysis seems to be sufficient for both drained and undrained loading conditions. The differences in the magnitude of the distortions between drained and undrained loading depend on the tunnel's stiffness, and are discussed in Section 4.

4 EFFECT OF RELATIVE STIFFNESS ON TUNNEL DISTORTIONS

Full dynamic numerical analyses, for both drained and undrained loading, are performed to evaluate the effect of relative stiffness between the liner and the ground on the tunnel distortions. For the undrained loading, two scenarios are considered: one where no accumulation of excess pore pressures occurs, and the other with pore pressures accumulation. For comparison purposes, results from the closed-form solution for drained linear-elastic ground provided by Bobet (2010) are included. The size of the mesh, amplitude of the dynamic input, tunnel geometry and ground properties are the same as those used to evaluate the effect of input frequency in the preceding section. An input frequency f = 2.5 Hz is selected, such that there is no wave amplification due to the cavity, as shown in Figure 3. The initial flexibility ratios range from 0.25 to 15, which are reduced during the dynamic input to values from 0.18 to 0.21, and from 8.9 to 13.9, respectively. The results are shown in Figure 5. The inset in the figure contains results for stiff tunnels, i.e. for $F \leq 1$. As seen in the figure, for the nonlinear ground under drained or undrained loading, the normalized distortions increase with the increase in the flexibility ratio, as has been previously reported from linear-elastic analysis (e.g., Bobet, 2010;

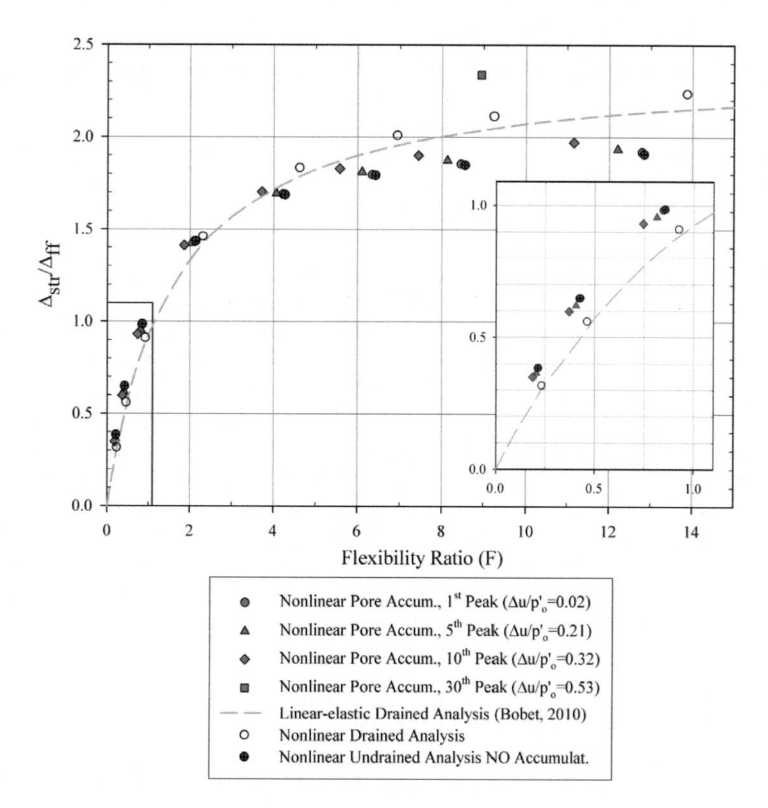

Figure 5. Effect of flexibility ratio on distortions of circular tunnels under drained or undrained loading.

Wang, 1993). It can also be seen in Figure 5 that larger normalized distortions are observed for drained loading using the nonlinear ground model than for the linear-elastic model. This result is expected, as the nonlinear ground deforms more and so there is a larger deformation demand on the tunnel.

It is also seen in Figure 5 that undrained conditions, with no pore pressures accumulation, tend to reduce deformations for flexible structures and increase them for stiffer structures, compared to drained loading. However, when pore pressures in the ground increase with cycles of loading, the tunnel distortions increase for flexible liners and decrease for stiff liners. Depending on the magnitude of the excess pore pressures accumulation, i.e. depending on the number of cycles, normalized distortions for flexible/stiff tunnels can be even larger/smaller than those from the drained loading.

The results highlight the importance of considering in the analyses the excess pore pressures accumulation in the ground. If excess pore pressures are not considered, undrained loading increases distortions for stiff tunnels and decreases them for flexible tunnels; however, as seen in Figure 5, the behavior is more complex, since the outcome depends on the combined effects of the flexibility ratio and excess pore pressures accumulation.

5 CONCLUSIONS

The paper evaluates the drained and undrained seismic response of deep circular tunnels. Numerical simulations are done using the commercial code FLAC 7.0, where it is assumed that plane strain conditions apply to any cross section perpendicular to the tunnel axis. A new elastoplastic hyperbolic constitutive model is developed, implemented and verified in FLAC.

The model considers the stiffness degradation of the ground with deformation, the dependency of the small-strain shear modulus on confinement, the coupling between shear and volumetric strains, and yield with associated plastic deformations.

The effects of the input frequency and the flexibility ratio on the distortions of the tunnel cross section are investigated for drained and undrained loading conditions. The results show that there is no effect of the input frequency on the seismic response of a tunnel when the input frequency is smaller than 5 Hz. That is, for λ/D ratios larger than 8.5 to 9.5, depending on the magnitude of excess pore pressures. This is usually the case for most structures located far from the seismic source, at distances larger than about 10 km from the epicenter. The normalized tunnel distortions increase when the flexibility ratio (F) between the tunnel and the ground increases. Larger distortions are obtained for nonlinear than for linear-elastic ground, with differences increasing as the tunnel becomes more flexible, i.e., when F increases. Undrained loading conditions, with no pore pressure accumulation, tend to reduce deformations of flexible structures and increase them for stiffer structures, compared to drained loading. However, when pore pressures in the ground accumulate with cycles of loading, distortions in flexible liners increase and in stiff liners decrease, so the differences with those from drained loading are reduced. Depending on the magnitude of the excess pore pressures, normalized distortions for flexible/stiff tunnels can be larger/smaller than those obtained from drained loading. These results highlight the importance of considering excess pore pressures accumulation in the ground when undrained conditions apply.

REFERENCES

Bobet, A. 2003. Effect of pore water pressure on tunnel support during static and seismic loading. *Tunnelling and Underground Space Technology* 18(4): 377–393. http://dx.doi.org/10.1016/S0886-7798(03)00008-7.

Bobet, A. 2010. Drained and undrained response of deep tunnels subjected to far-field shear loading. *Tunnelling and Underground Space Technology* 25 (1): 21–31. http://dx.doi:10.1016/j.tust.2009.08.001.

Byrne, P. M. 1991. A cyclic shear-volume coupling and pore pressure model for sand, *Proc. Second International Conference on Recent Advances in Geotechnical Earthquake Engineering and Soil Dynamics, 47–55, St. Louis, Missouri, USA, 11–15 March 1991.*

Dowding, C. 1985. Earthquake response of caverns: empirical correlations and numerical modeling, *Proc. 1985 Rapid Excavation and Tunneling Conference, 1: 71–83, New York, New York, USA, 16–20 June 1985.*

Hardin, B. O. 1978. The nature of stress-strain behavior for soils, *In: Specialty Conference on Earthquake Engineering and Soil Dynamics ASCE, 3–90 Pasadena, California, USA, 19–21 June 1978.*

Hardin, B. O. & Drnevich, V. P. 1972. Shear modulus and damping in soils: design equations and curves. *Journal of the Soil Mechanics and Foundations Division* 98 (7): 667–692.

Hendron Jr., A. J. & Fernández, G. 1983. Dynamic and static design considerations for underground chambers, *In: Seismic Design of Embankments and Caverns, ASCE Symposium, 157–197, Philadelphia, Pennsylvania, USA, 16–20 May 1983.*

Huo, H., Bobet, A., Fernández, G., & Ramírez, J. 2006. Analytical solution for deep rectangular structures subjected to far-field shear stresses. *Tunnelling and Underground Space Technology* 21 (6): 613–625. http://dx.doi:10.1016/j.tust.2005.12.135.

Ishibashi, I., Kawamura, M., & Bhatia, S. 1985. *Effect of initial shearing on cyclic drained and undrained characteristics of sand.* Technical report, Cornell Geotechnical Engineering Report 85-2.

Itasca. 2011a. *FLAC Fast Lagrangian Analysis of Continua, Ver. 8.0.* Minneapolis, MN, USA: Itasca Consulting Group.

Itasca. 2011b. *FLAC* Fast Lagrangian *Analysis of Continua*, Ver. *8.0, Dynamic* Analysis *(Fifth ed.)*. Minneapolis, MN, USA: Itasca Consulting Group.

Jung, C. M. 2009. *Seismic loading on earth retaining structures.* PhD thesis, Purdue University, West Lafayette, Indiana, USA.

Khasawneh, Y., Bobet, A., & Frosch, R. 2017. A simple soil model for low frequency cyclic loading. *Computers and Geotechnics* 84: 225–237. http://dx.doi:10.1016/j.compgeo.2016.12.003.

Kuesel, T. R. 1969. Earthquake design criteria for subways. *Journal of the Structural Division* 6: 1213–1231.

Lysmer, J. & Kuhlemeyer, R. L. 1969. Finite dynamic model for infinite media. *Journal of the Engineering Mechanics Division* 95 (4): 859–878.
Martin, G. R., Finn, W. L., & Seed, H. B. 1974. *Fundamentals of liquefaction under cyclic loading.* Technical report, The University of British Columbia, Soil Mechanics Series No. 23.
Masing, G. 1926. Eigenspannungen und verfestigung beim messing, *In: Proceedings, Second International Congress of Applied Mechanics, 332–335, Zürich, Switzerland, 12–17 September 1926.*
Merritt, J. L., Monsees, J. E., & Hendron Jr, A. J. 1985. Seismic design of underground structures. *In Proc. 1985 Rapid Excavation and Tunneling Conference, 1: 104–131, New York, New York, USA, 16–20 June 1985.*
Mow, C. C. & Pao, Y. H. 1971. *The diffraction of elastic waves and dynamic stress concentrations.* Technical report, R-482-PR, United States Air Force Project Rand.
Newmark, N. M. 1967. Problems in wave propagation in soil and rock. *In: Proc. International Symposium on Wave Propagation and Dynamic Properties of Earth Materials, 7–26, Albuquerque, NM, USA, 23–25 August 1967.*
Ohsaki, Y. 1980. Some notes on Masing's law and non-linear response of soil deposits. *Journal of Faculty of Eng., The University of Tokyo Ser B* 105 (4): 513–536.
Owen, G. N. & Scholl, R. E. 1981. *Earthquake engineering of large underground structures.* Technical report, NASA STI/Recon Technical Report N.
Paul, S. L. 1963. *Interaction of plane elastic waves with a cylindrical cavity.* PhD thesis. University of Illinois, Urbana Champaign, Illinois.
Peck, R. B., Hendron, A. J., & Mohraz, B. 1972. State of the art of soft-ground tunneling, *Proc. 1972 Rapid Excavation and Tunneling Conference, 1: 259–286, Chicago, Illinois, USA, 5–7 June 1972.*
Penzien, J. 2000. Seismically induced racking of tunnel linings. *Earthquake Engineering and Structural Dynamics* 29(5): 683–691.
Porovic, E. & Jardine, R. 1994. Some observations on the static and dynamic shear stiffness of Ham River sand. *In: Proceedings of the International Symposium on Pre-failure Deformation Characteristics of Geomaterials, 25–30, Sapporo, Japan, 12–14 September 1994.*
Pyke, R. 1979. Nonlinear soil models for irregular cyclic loadings. *Journal of the Geotechnical Engineering Division* 105(GT6): 715–726.
Sandoval, E. & Bobet, A. 2017. Effect of frequency and flexibility ratio on the seismic response of deep tunnels. *Underground Space* 2 (2): 125–133.
Wang, J. N. 1993. Seismic design of tunnels. *A simple state-of-the-art approach.* Technical report, Parsons Brinckerhoff Monograph 7.
Yoshihara, T. 1963. *Interaction of plane elastic waves with an elastic cylindrical shell.* PhD thesis. University of Illinois, Urbana Champaign, Illinois, USA.

Tunnels and Underground Cities: Engineering and Innovation meet Archaeology, Architecture and Art, Volume 3: Geological and geotechnical knowledge and requirements for project implementation – Peila, Viggiani & Celestino (Eds)

How to strengthen a geotechnical campaign for tunnel's secondary lining. Implementation and results

J.M. Santa Cruz, L. Roldan, G. Dankert & L. Sambataro
National Department of Transportation, Buenos Aires, Argentina

A.O. Sfriso & R.D. Bertero
Universidad de Buenos Aires, Buenos Aires, Argentina

ABSTRACT: As a result of a reliability analysis done for the "Red de Expresos Regionales – RER" project in Buenos Aires, permanent or secondary lining of tunnel models built under the New Austrian Tunneling Method have shown to be very sensitive to Young's modulus for small deformations and lateral earth pressure at rest coefficient. Results from this analysis are shown, as well as an experimental program implemented in some of the project's most critical sections

1 INTRODUCTION

The "Red de Expresos Regionales – RER" is a new 33 km tunnel project in downtown Buenos Aires, aiming to connect Argentina's over 800 km railroad network. The construction methodology expected are the New Austrian Tunneling Methodology – NATM, which has been used in the last 25 years in the expansion of the Subway network, and Earth Pressure Balance Tunnel Boring Machines – EPB-TBM, being used in the past 5 years in important sanitation projects such as the Maldonado and Vega streams piping projects and Riachuelo river sewers systems.

NATM tunnels are designed with a temporary or primary lining made of shotcrete and a permanent or secondary lining of cast-in-place reinforced concrete.

As a result of a reliability analysis, secondary lining of NATM tunnel models have shown to be very sensitive to Young's modulus for small deformations, E_o and lateral earth pressure at rest coefficient, K_0 (Roldan, et al., Unpublished).

Some of the project's most critical sections include the main station located beneath the Obelisco monument, a double cavern 400 m long and 20 m wide.

A typical soil profile can be observed in Figure 1, with a superficial and anthropic layer in the first 6 m, followed by the Pampeano Formation up to 33 m deep and ended with the Puelchense Formation.

2 STATE OF PRACTICE

Local geotechnical practice bases most of its calculations (even state of the art finite element analysis) to an extensive campaign based mostly on SPT boreholes, routine laboratory test, back-analysis methods and experience.

This is probably due to the strong confidence in the local geotechnical conditions. The city of Buenos Aires is located on top of the Pampeano formation, a 30 – 50 m thick overconsolidated soil by desiccation and cemented modified loess that resembles more to a weak rock than a cohesive soil (Núñez, 1986). The risk of failure in these tunneling projects usually reside in two main aspects (not evaluated in this paper):

– Face stability due to sand or silty sand lenses undetected in SPT boreholes.
– Heave uplift due to the confined aquifer of the Puelchense formation, a dense and clean sand underneath the less permeable Pampeano formation.

This situation has render the design of the secondary lining to a less important matter, even though its importance in the project's final cost is very relevant.

The state of practice in geotechnical investigation for tunneling projects usually include SPT (with 90% energy rate) approximately every 100 m, some prebored PMT and sometimes horizontal PLT in excavation pits later used for ventilation or evacuation of the project. In spite of the availability of alternative geophysical test (e.g. Down-Hole, Cross-Hole, SASW and MASW), their use is not common practice in tunnel projects.

Regarding laboratory tests, the routine includes Atterberg limits, #200 sieve and moisture content tests in every sample obtained during the SPT and unit weights in representative samples from the same source. Unconsolidated–undrained triaxial test from this samples were usually done, but they are slowly disappearing due to the large degree of alteration caused

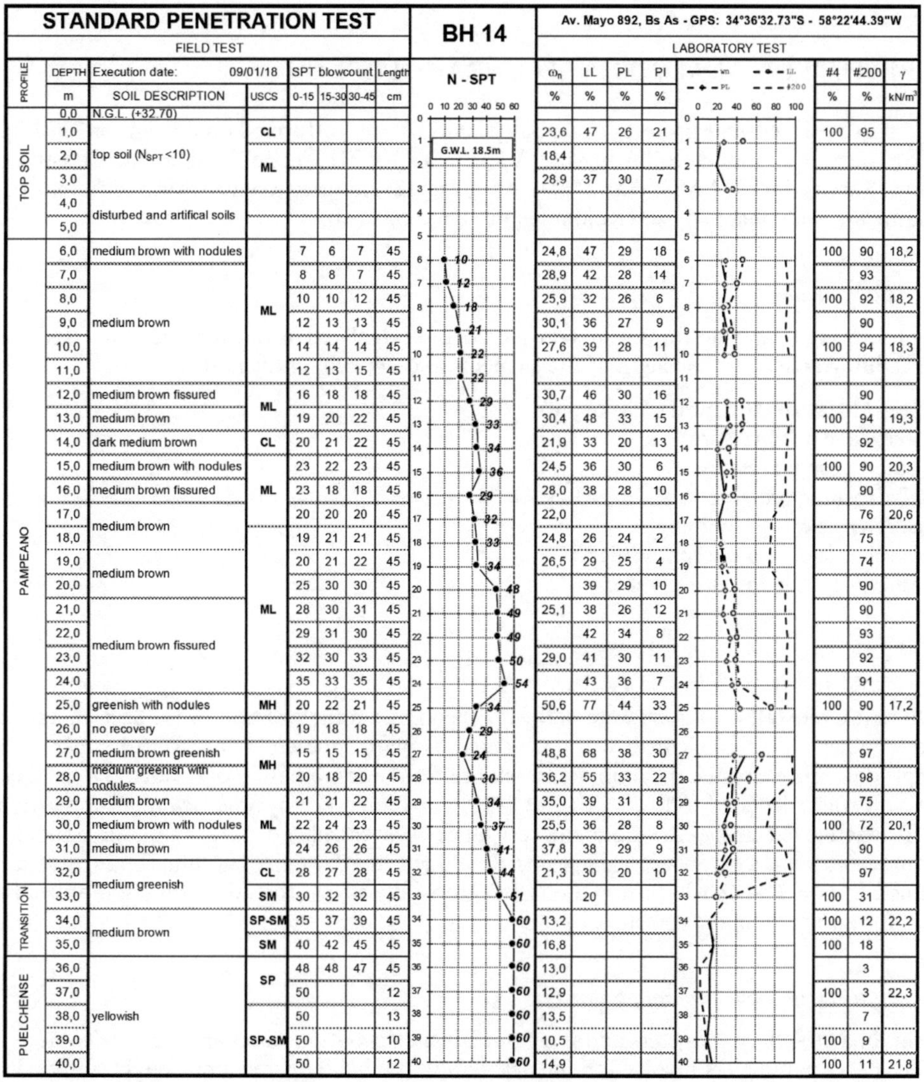

STANDARD PENETRATION TEST — **BH 14** — Av. Mayo 892, Bs As - GPS: 34°36'32.73"S - 58°22'44.39"W

FIELD TEST / LABORATORY TEST

Execution date: 09/01/18 — G.W.L. 18.5m

PROFILE	DEPTH (m)	SOIL DESCRIPTION	USCS	SPT blowcount 0-15	SPT blowcount 15-30	SPT blowcount 30-45	Length (cm)	N - SPT	ω_n (%)	LL (%)	PL (%)	PI (%)	#4 (%)	#200 (%)	γ (kN/m³)
	0,0	N.G.L. (+32.70)													
TOP SOIL	1,0	top soil (N_{SPT} <10)	CL						23,6	47	26	21	100	95	
	2,0		ML						18,4						
	3,0								28,9	37	30	7			
	4,0	disturbed and artifical soils													
	5,0														
PAMPEANO	6,0	medium brown with nodules	ML	7	6	7	45	10	24,8	47	29	18	100	90	18,2
	7,0	medium brown		8	8	7	45	12	28,9	42	28	14		93	
	8,0			10	10	12	45	18	25,9	32	26	6	100	92	18,2
	9,0			12	13	13	45	21	30,1	36	27	9		90	
	10,0			14	14	14	45	22	27,6	39	28	11	100	94	18,3
	11,0			12	13	15	45	22							
	12,0	medium brown fissured	ML	16	18	18	45	29	30,7	46	30	16		90	
	13,0	medium brown		19	20	22	45	33	30,4	48	33	15	100	94	19,3
	14,0	dark medium brown	CL	20	21	22	45	34	21,9	33	20	13		92	
	15,0	medium brown with nodules	ML	23	22	23	45	36	24,5	36	30	6	100	90	20,3
	16,0	medium brown fissured		23	18	18	45	29	28,0	38	28	10		90	
	17,0	medium brown		20	20	20	45	32	22,0					76	20,6
	18,0		ML	19	21	21	45	33	24,8	26	24	2		75	
	19,0	medium brown		20	21	22	45	34	26,5	29	25	4		74	
	20,0			25	30	30	45	48		39	29	10		90	
	21,0	medium brown fissured		28	30	31	45	49	25,1	38	26	12		90	
	22,0			29	31	30	45	49		42	34	8		93	
	23,0			32	30	33	45	50	29,0	41	30	11		92	
	24,0			35	33	35	45	54		43	36	7		91	
	25,0	greenish with nodules	MH	20	22	21	45	34	50,6	77	44	33	100	90	17,2
	26,0	no recovery		19	18	18	45	29							
	27,0	medium brown greenish	MH	15	15	15	45	24	48,8	68	38	30		97	
	28,0	medium greenish with nodules		20	18	20	45	30	36,2	55	33	22		98	
	29,0	medium brown	ML	21	21	22	45	34	35,0	39	31	8		75	
	30,0	medium brown with nodules		22	24	23	45	37	25,5	36	28	8	100	72	20,1
	31,0	medium brown		24	26	26	45	41	37,8	38	29	9		90	
	32,0	medium greenish	CL	28	27	28	45	44	21,3	30	20	10		97	
TRANSITION	33,0		SM	30	32	32	45	51		20			100	31	
	34,0	medium brown	SP-SM	35	37	39	45	60	13,2				100	12	22,2
	35,0		SM	40	42	45	45	60	16,8				100	18	
PUELCHENSE	36,0	yellowish	SP	48	48	47	45	60	13,0					3	
	37,0			50			12	60	12,9				100	3	22,3
	38,0		SP-SM	50			13	60	13,5					7	
	39,0			50			10	60	10,5				100	9	
	40,0			50			12	60	14,9				100	11	21,8

Figure 1. Typical SPT profile and classification tests of the project site.

during sampling. Consolidated – undrained triaxial test (with pore pressure measurement) from undisturbed samples taken by Denison sampler are not common but commercially available, mainly due to the difficult and more expensive sampling techniques required. When carried on, the axial strain measurement in these tests is external (no local measurement available to the author's knowledge). No direct measurement of K_0 is known to have been carried on.

The primary lining and the safety of the construction stages are analyzed using 2D and 3D complex finite element models, whereas the secondary lining is designed using simpler analysis of beam or slab on elastic and linear foundation, subjected to distributed "loads" due to soil weight, groundwater level and superficial surcharges.

Since most of the soil information comes from the geotechnical campaigns described previously, most of the parameters are estimated via correlations, experience and back-analysis. Therefore, it's hard to stablish a correlation between measured parameters and observed behavior of geotechnical structures.

There are several publications about the Pampeano formation's mechanical behavior (see (Bolognesi, 1975), (Núñez & Micucci, 1986), (Sfriso & Codevilla, 2011)), but few have been found that seem to have studied E_0 or K_0.

3 RELIABILITY ANALYSIS IN SECONDARY LINING

As a part of the current project, a reliability analysis was performed on a beam on elastic and linear foundation model, following the AASHTO's LRFD Tunnel Design and Construction Guide Specification (AASHTO, 2017). As a result, importance factors for different input parameters were obtained, see (Roldan, et al., Unpublished).

3.1 *Load model*

The following load conditions were considered and calculated as recommended in (AASHTO, 2012), integrated over the length of the elements and applied on the nodes.

- Dead loads of structural components
- Vertical earth pressure
- Lateral earth pressure
- Hydrostatic pressure
- Surcharge due to buildings
- Surcharge due to internal vehicles
- Surcharge due to surface vehicles

3.2 *Limit state function*

Different limit state functions were defined depending on the position of the section analyzed and whether the lining element was considered as plain concrete or reinforced concrete. For simple concrete elements, the nominal resistance for the combination of flexure and axial compression was calculated following the ACI-318 equations. For symmetrical and non-symmetrical concrete elements, the nominal resistance was calculated as a simplification of the flexural and axial compression diagrams developed in ACI-318. The limit state function is defined as $G = R - Q$, where R is the resistance of the structure and Q are the moments and axial forces acting on the structure.

3.3 *Statistical distribution of parameters*

In this analysis, the parameters and their statistical distributions used are those in Table 1

The other parameters involved in the calculation (such as concrete unit weight, internal surcharges, etc.) were considered deterministic ones, since a preliminary analysis showed their importance was not relevant, using their mean values.

Table 1. Statistical distribution of parameters used (Roldan, et al., Unpublished).

Parameter	Type	μ	σ	L1	L2	Description
K_0 [-]	Beta	0.65	0.09	0.40	1.20	Lateral earth pressure coefficient
Es [MPa]	Log-normal	500	100	-	-	Modulus of elasticity of the soil
γs [kN/m3]	Beta	18.5	0.27	17.0	20.0	Unit weight of the soil
zw [m]	Normal	16.5	1.00	-	-	Depth of the groundwater level
fc [MPa]	Log-Normal	43	6.45	-	-	Concrete compression resistance
t [cm]	Normal	40	0.40	-	-	Concrete thickness

Where μ= mean, σ=standard deviation, L1=inferior limit, L2=superior limit.

3.4 *Analysis results and calibrated coefficients*

The case analyzed consisted of a 21m cover deep tunnel. Three sections were studied, considering simple concrete for the top heading and reinforced concrete for the bench and bottom. Two failure modes were identified: flexural and axial yielding, resulting in different failure probabilities depending on the concrete thickness.

In Table 2 the sensitivity analysis and resulting importance factors can be observed. The two most important parameters (higher α, in absolute values), which have the greatest impact on the calculations, are E_s and K_0.

Table 2. Reliability analysis results (Roldan, et al., Unpublished).

Section	Yield mode	Parameter	α	δ
Heading	Comp.	K_0	0.28	0.15
"	"	Es	0.15	0.20
"	"	γs	-0.12	0.02
"	"	zw	-0.03	0.06
"	"	fc	0.94	0.15
Heading	Flex.	K_0	0.39	0.15
"	"	Es	0.89	0.20
"	"	γs	-0.12	0.02
"	"	zw	-0.17	0.06
"	"	fc	-0.12	0.15
Bench	Comp.	K_0	-0.32	0.15
"	"	Es	0.15	0.20
"	"	γs	-0.11	0.02
"	"	zw	-0.07	0.06
"	"	fc	0.92	0.15
Bench	Flex.	K_0	0.25	0.15
"	"	Es	0.90	0.20
"	"	γs	-0.03	0.02
"	"	zw	0.05	0.06
"	"	fc	-0.33	0.15
Bottom	Comp.	K_0	-0.38	0.15
"	"	Es	0.31	0.20
"	"	γs	-0.10	0.02
"	"	zw	0.00	0.06
"	"	fc	0.86	0.15
Bottom	Flex.	K_0	-0.46	0.15
"	"	Es	0.84	0.20
"	"	γs	-0.02	0.02
"	"	zw	0.05	0.06
"	"	fc	-0.26	0.15

Where α = importance factor, δ =variation coefficient. Parameter "t" is not included in the table as its importance is low enough to yield in amplification factors of 1.00 in every case.

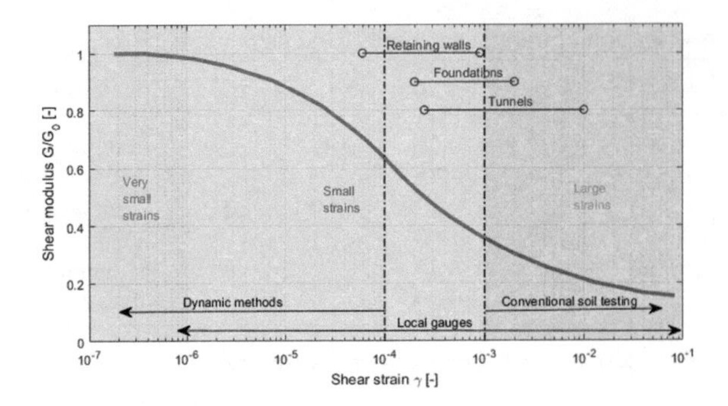

Figure 2. Characteristic stiffness-strain behavior of soil with typical strain ranges for laboratory tests and structures, after (Atkinson & Sallfors, 1991) and (Mair, 1992).

4 SMALL-STRAIN STIFFNESS IN SOILS

It is known since the 1970's that the stress-strain relationship in soils is highly non-linear and so the small-strain stiffness, which has a great impact in soil-structure interaction problems such as tunnel's lining (Atkinson, 2000).

Based on Figure 2 and the previously described state of practice, most stiffness measurements correspond to shear strains larger than 10^{-3} %, whereas observed shear strains in local tunnels are well below that value (around 10^{-4} %).

To take this situation into consideration, the authors propose the measurement of very small-strain stiffness through dynamic methods, more specifically with MASW tests, and adjust the measured stiffness to the corresponding one according to the expected final shear strain of the tunnel.

This adjustment can be done using the curve in Figure 2, approximating it with the Hardin and Drnevich's expression modified by (Santos & Correia, 2001):

$$\frac{G}{G_0} = \frac{1}{1 + a\left(\frac{\gamma}{\gamma_a}\right)} \tag{1}$$

where G = shear stiffness; G_0 = initial or small-strain shear stiffness; a = 0.385; γ = shear strain; and γ_a = the threshold shear strain at which G decays to $0.7G_0$.

γ_a (also known as $\gamma_{0.7}$) can be approximated by (Stokoe & Darendeli, 2001) expression as:

$$\gamma_a = \gamma_{0.7\ ref} + 5\ 10^{-6}\ PI\ OCR^{0.3} \tag{2}$$

where $\gamma_{0.7\ ref} = 10^{-4}$; PI = plasticity index; and OCR = overconsolidation ratio.

5 DYNAMIC PROSPECTION OF SOILS

5.1 *Seismic waves*

If an excitement is generated in the surface of the earth, two families of seismic waves are generated: body waves and surface waves (Aki & Richards, 1980).

Body waves travel through the interior of the material and can be decomposed in compressional or P waves and shear or S waves, as shown in Figure 3. P waves motion is parallel to the wave motion itself, whereas S waves motion is perpendicular to it, in both vertical and horizontal directions (Aki & Richards, 1980).

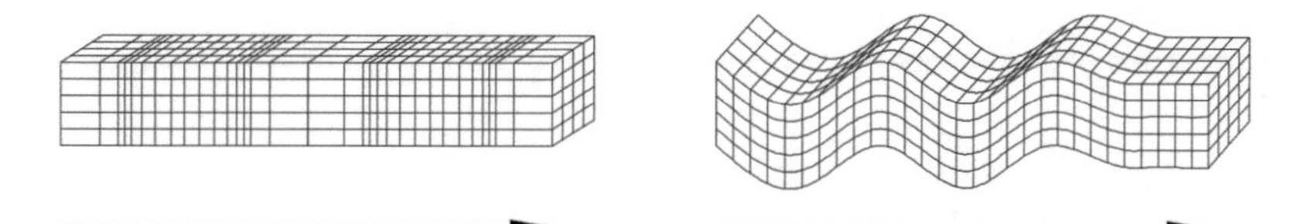

Figure 3. a) Particle motion associated with P waves; b) Particle motion associated with S waves (Bolt, 1976).

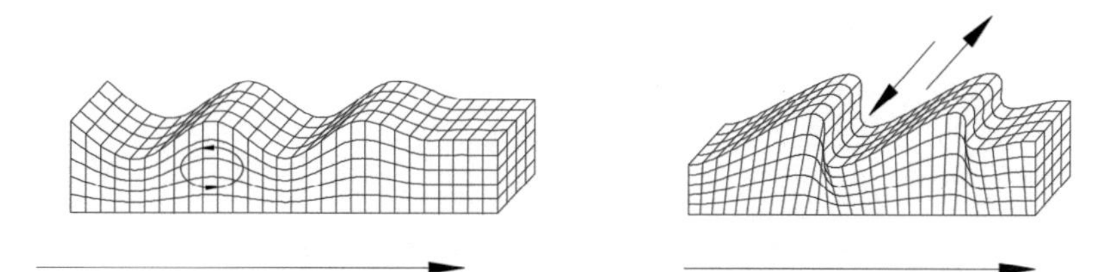

Figure 4. a) Particle motion associated with Rayleigh waves; b) Particle motion associated with Love waves (Bolt, 1976).

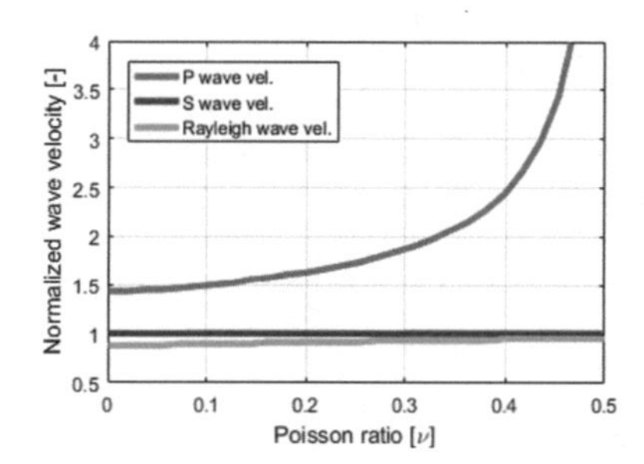

Figure 5. Variation of compressional, shear and Rayleigh wave propagation velocities in a homogeneous medium with Poisson's ratio (Ólafsdóttir, 2016).

Surface waves travel through the interfaces between different materials and can be decomposed in Rayleigh or R waves and Love or L waves, as shown in Figure 4. Rayleigh waves are a combination of P and vertical S waves, whereas Love waves are a combination of P and horizontal S waves (Kramer, 1996).

67% of the seismic energy from a vertical seismic source is imparted into Rayleigh waves, 26% to S waves and 7% into P waves (Woods, 1968). Even further, the variation of the relation between Rayleigh's and S wave's phase velocity on a wide range of Poisson ratio is approximately 0.93 (see Figure 5) (Ólafsdóttir, 2016).

It is then possible to correlate the S wave velocity to the Young's and shear modulus with the following expressions (ASTM D2845, 2008):

$$G_0 = \rho \ V_s^2 = \frac{\gamma_s}{g} \ V_s^2 \tag{3}$$

where ρ = soil density; V_s = S wave velocity; γ_s = soil's total unit weight; and g = gravity acceleration.

5.2 *Surface wave analysis methods*

Surface wave analysis methods are based on the dispersive properties of surface waves in a vertically heterogeneous medium (Ólafsdóttir, 2016). Hence, different frequencies reflect material properties of soil layers at diverse depths. The main steps of this methods are divided into (Socco, et al., 2010):

- Field measurements of surface waves
- Data processing to extract dispersion curves
- Estimation of soil properties by inversion analysis of the dispersion curves

The current state of practice in Argentina offers both the Spectral Analysis of Surface Waves (SASW) and the Multichannel Analysis of Surface Waves (MASW) as the main sources for field measurement and data processing, with the use of a sledgehammer and geophones with low frequency (<10Hz).

5.2.1 *Spectral Analysis of Surface Waves*
In the SASW method the surface waves are generated with an impulsive source and detected by geophones (Ólafsdóttir, 2016). A reliable estimate of the shear wave velocity profile down to around 20 m depth can be obtained (Kramer, 1996).

Typically, a sledgehammer and between two and twelve geophones are used for field measurement. The geophones are either lined up with equal spacing on the surface of the test site or with varying spacing in a symmetrical line-up (Ólafsdóttir, 2016).

For a given site, multiple measurements are carried out by varying the distance between the impact load point and the first receiver in the geophone line-up, in order to excite waves with different frequency contents (Ólafsdóttir, 2016). A second receiver is chosen so that the pair is to an equal or similar distance to the first receiver than the impact source.

The acquired surface wave data are analyzed in the frequency domain to determine a dispersion curve. Dispersion curves for multiple pairs of receivers are determined and combined, and their average used as an estimate of the phase velocity of Rayleigh wave components belonging to the given frequency range (Ólafsdóttir, 2016). The average dispersion curve is then used as a basis for the computation of a shear wave velocity profile as a function of depth for the given site (Ólafsdóttir, 2016).

Due to the necessity of repeated computations, the data processing involved in the SASW method is time intensive. The analysis must be carried out for each receiver pair separately and the results for each pair examined manually in order to evaluate their quality. Moreover, as time series from only two receivers are used at a time, difficulties can arise in distinguishing reliable surface wave signal from noise, such as inclusion of body waves and/or higher modes (Park, et al., 1999). This may cause errors both in the dispersion curve and ultimately in the shear wave velocity profile.

5.2.2 *Multichannel Analysis of Surface Waves*
MASW surveys can be divided into active and passive surveys based on how the surface waves required for analysis are acquired (Park, et al., 1999).

In active MASW surveys, geophones are lined up in an equally spaced line on the surface of the test site. Surface waves are generated actively by impulsive or vibrating seismic sources that are applied at one end of the receiver line-up. A single multichannel surface wave record is sufficient for analysis (Park, et al., 1999), although present experience with applying the MASW method indicates that it is beneficial to combine results from several different records prior to the inversion analysis (Ólafsdóttir, 2016). In passive MASW surveys, the source is of natural or cultural origin, for example traffic (Park, et al., 1999).

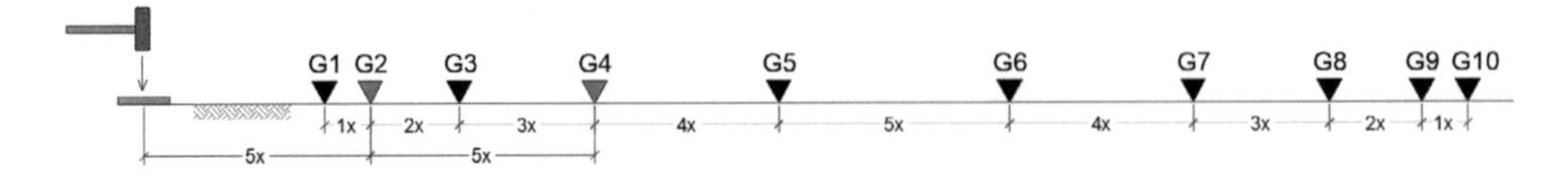

Figure 6. SASW data processing. Choice of receiver pairs for analysis (Ólafsdóttir, 2016).

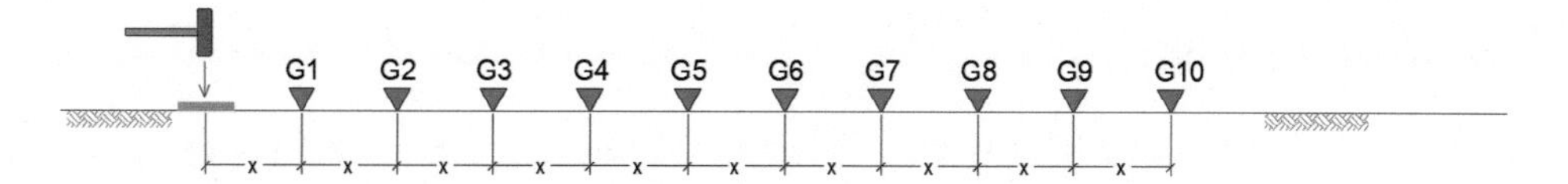

Figure 7. MASW field array and choice of receiver for analysis (Ólafsdóttir, 2016).

The field measurements are followed by the dispersion analysis, in which dispersion curves are extracted from the acquired data. At least three methods are known and feasible when active testing (Socco, et al., 2010):

- Slowness-frequency ($p - \omega$) transform (McMechan & Yedlin, 1981).
- Frequency-wave number ($f - k$) transform (Yilmaz, 1987).
- Phase shift method (Park, et al., 1999).

The three methods have been compared and the conclusion was that the phase shift method is a robust and computationally effective method that provides accurate fundamental mode phase velocities, even when data from as little as four geophones are available (Dal Moro, et al., 2003). Some improvements to this method have been accomplished since (Park, 2011).

To obtain the shear wave velocity profile, an inversion back analysis of the experimental dispersion curve obtained is performed. A theoretical dispersion curve is obtained based on an assumed number and thickness of soil layers and assumed material properties for each layer. Different sets of parameters are inserted into the theoretical model in an iterative way in search of the theoretical dispersion curve that is the most consistent with the observed dispersion curve (Ólafsdóttir, 2016).

5.2.3 *Advantages of the MASW method*

The MASW method was developed in order to overcome some of the weaknesses of the SASW method. The main advantages of the MASW method over the SASW method are the following (Ólafsdóttir, 2016):

- Data acquisition in the field is more effective as compared to the SASW method. For application of the SASW method, multiple measurements are carried out using different seismic sources and by varying the source offset. This is required in order to excite waves with different frequency contents.
- The dispersion analysis involved in MASW is much faster and easier to automate. Data from all receivers is processed at once, instead of repeated computations for multiple pairs of receivers as in the SASW method.
- Noise sources, such as inclusion of body waves and reflected/scattered waves, can more easily be identified and noise eliminated as compared to the SASW method (Park, et al., 1999). Reduction of noise leads to increased accuracy in the dispersion analysis and ultimately a more precise shear wave velocity profile.
- The MASW technique can provide more investigation depth than the SASW method, given the same impact load. The investigation depth that can be achieved by the (active) MASW method is generally around 30m, whereas the SASW method can in general provide an estimation of the shear wave velocity profile down to around 20m depth.
- The MASW method can be used to analyze passively generated surface waves. Surface waves that are generated by passive sources have lower frequencies (longer wavelengths) than waves generated by impact (active) loads. Hence, passive MASW surveys can provide more investigation depth than active surveys.

5.2.4 *Shear wave velocity of the Pampeano Formation*

Based on (Rinaldi & Clariá, 2006), the Pampeano Formation's V_s measured from Cross-Hole tests ranges between 800 and 1000 m/s. Considering Eq. (3) and $\gamma_s = 20\ kN/m^3$, the resulting G_0 ranges from 1300 and 2000 MPa, considerably larger than the usual data found on local geotechnical reports.

6 DESIGN PROCESS

Once G_0 is measured with a MASW test assume a shear strain γ seed value from Figure 2 and use Eq. 1. An expected value of G can be obtained and elasticity theory can then be applied to obtain the soil's Young modulus:

$$E_s = 2G\,(1+\nu) \tag{4}$$

where E_s = soil's Young's modulus; and ν = drained Poisson's ratio.

Once the soil modulus is known, Eq. 7.7.5.4-1 and 2 from (AASHTO, 2017) can be used to obtain both radial and tangential modulus of subgrade reaction:

$$k_r = \frac{E_s}{R_{eq}(1+\nu)} \tag{5}$$

$$k_t = \frac{k_r}{2(1+\nu)} \tag{6}$$

where R_{eq} = tunnel's equivalent radius.

Once a model is ran, a new value for γ can be obtained and a new iteration starts, similar to the simple design process proposed in (Atkinson, 2000).

7 RESULTS

Using the design methodology proposed and the soil parameters in Table 3, the results shown in Table 4 and Table 5 were obtained.

Table 3. Soil parameters used.

Parameter	γ_s	V_s	PI	OCR	ν	R_{eq}
Value	20 kN/m^3	800 m/s	10	3.5	0.3	5.6m

Table 4. Results obtained with $\gamma = 10^{-4}$.

Parameter	G_0	$\gamma_{0.7}$	G	E	k_r	k_t
Value	1305 MPa	1.7 10^{-4}	1067 MPa	2775 MPa	381 MN/m^3	146 MN/m^3

Table 5. Results obtained with $\gamma = 10^{-3}$.

Parameter	G_0	$\gamma_{0.7}$	G	E	k_r	k_t
Value	1305 MPa	1.7 10^{-4}	404 MPa	1051 MPa	144 MN/m^3	55 MN/m^3

8 CONCLUSIONS

It is possible to approximate the S wave's velocity of homogenous soils by measuring Rayleigh waves, easier to generate and measure. Therefore, using an SPT borehole for profiling and the MASW technique described previously, already available in the local market, a reliable and cost-time efficient measurement of soil's shear wave velocity (even a 2D profile) is possible and, consequently, the stiffness modulus for small deformations.

Using the small deformation modulus and their decay curve, a simple iterative process can be performed to analyze most geotechnical problems solved using beam on lineal elastic foundation, including secondary lining in tunnel's final stage design.

9 FURTHER WORKS

There is currently an ongoing geotechnical campaign at the project's locations, including two different MASW surveys, one with a sledgehammer and geophones and the other one with a vibrations inducing device and accelerometers. The objective is to compare the quality of the results obtained with both types of equipment.

ACKNOWLEDGEMENTS

To all of the Department of Transportation staff members, especially to G. Bussi, E. Zielonka, H. Tymkiw, D. Correa, S. Wolkomirski, C. Cogliati and I. López Godoy, for their constant support and encouragement.

To the RER – Design Criteria team, F. Bissio from Universidad Nacional de La Plata and P. Fernández from SRK Consulting Argentina S.A.

To M. Codevilla from AOSA S.A. for his expertise in geotechnical investigation.

REFERENCES

AASHTO, 2017. *LRFD Road Tunnel Design and Construction Guide Specifications.* 1st ed. Washington, D.C.: s.n.

Aki, K. & Richards, P. G., 1980. *Quantitative seismology: Theory and methods (Vol 1)*, San Francisco: W. H. Freeman and Co.

ASTM D2845, 2008. *Standard Test Method for Laboratory Determination of Pulse Velocities and Ultrasonic Elastic Constants of Rock*, West Conshohocken: ASTM International.

Atkinson, J. H., 2000. Non-linear soil stiffness in routine design. *Geotechnique*, 50(5), pp. 487–508.

Atkinson, J. H. & Sallfors, G., 1991. *Experimental determination of soil properties*, Florence: Proc. 10th ECSMFE, Vol 3.

Bolognesi, A., 1975. *Compresibilidad de los suelos de la Formación Pampeano*, Buenos Aires: V PCSMFE, V: 255–302.

Bolt, B. A., 1976. *Nuclear Explosions and Earthquakes: The Parted Vail*, San Francisco: W. H. Freeman and Co.

Dal Moro, G., Pipan, M., Forte, E. & Finetti, I., 2003. *Determination of Rayleigh wave dispersion curves for near surface applications in unconsolidated sediments*, Dallas: SEG International Exposition and Seventy-Third Annual Meeting.

Kramer, S. L., 1996. *Geotechnical Earthquake Engineering*, Upper Saddle River: Prentice-Hall.

Mair, R. J., 1992. *Developments in geotechnical engineering research: application to tunnels and deep excavations*, s.l.: Proceedings of Institution of Civil Engineers, Civil Engineering.

McMechan, G. A. & Yedlin, M. J., 1981. *Analysis of dispersive waves by wave field transformation*, s.l.: Geophysics.

Núñez, E., 1986. *Geotechnical conditions in Buenos Aires City*, s.l.: V ICIAEG.

Núñez, E. & Micucci, C., 1986. *Cemented preconsolidated soils as very weak rocks*, s.l.: V ICIAEG.

Ólafsdóttir, E. A., 2016. *Multichannel analysis of surface waves for assessing soil stiffness*, University of Iceland: School of Engineering and Natural Sciences.

Park, C. B., 2011. *Imaging Dispersion of MASW Data—Full vs. Selective Offset Scheme*, s.l.: Journal of Environmental and Engineering Geophysics.
Park, C. B., Miller, R. D. & Xia, J., 1999. Multichannel analysis of surface wave. *Geophysics*, 64(3), pp. 800–808.
Rinaldi, V. & Clariá, J. J., 2006. *Aspectos geotécnicos fundamentales de las formaciones del delta del Río Paraná y del estuario del Río de la Plata*, s.l.: Rev. Int. de Desastres Naturales, Accidentes e Infraestructura Civil, Vol 6 (2).
Roldan, L. A. y otros, Unpublished. *LRFD Coefficient calibration for tunnel design in Buenos Aires*, Buenos Aires: Department of Transportation.
Santos, J. A. & Correia, A. G., 2001. Reference threshold shear strain of soil. Its application to obtain an unique strain-dependent shear modulus curve for soil. *15th International Conference on Soil Mechanics and Geotechnical Engineering*, 1(1), pp. 267–270.
Sfriso, A. O. & Codevilla, M., 2011. *Actualización de la información geotécnica de los suelos de la Ciudad de Buenos Aires*, s.l.: Pan-Am CGS Geotechnical Conference.
Socco, L. V., Foti, S. & Boiero, D., 2010. *Surface-wave analysis for building near-surface velocity models - Established approaches and new perspectives.*, s.l.: Geophysics.
Stokoe, K. H. & Darendeli, M. B., 2001. *Development of a new family of normalized modulus reduction and material damping curves*, Austin: University of Texas.
Woods, R. D., 1968. *Screening of surface waves in soils*, s.l.: Journal of the Soil Mechanics and Foundation Division.
Yilmaz, Ö., 1987. *Seismic Data Processing*, Tulsa: Society of Exploration Geophysicists.

Tunnels and Underground Cities: Engineering and Innovation meet Archaeology, Architecture and Art, Volume 3: Geological and geotechnical knowledge and requirements for project implementation – Peila, Viggiani & Celestino (Eds)
© 2020 Taylor & Francis Group, London, ISBN 978-0-367-46583-4

Seismic prediction of the rock quality at the Brenner Base Tunnel

C. Schwarz
Brenner Basis Tunnel BBT SE, Innsbruck, Austria

ABSTRACT: Seismic prediction is performed in a 15 km long stretch of the exploratory tunnel of the Brenner Base Tunnel (BBT). The great overburden makes it evident, that the initial geological-geotechnical model has to be updated alongside while heading. Several techniques are performed to do so. The seismic prediction method is sensitive to the elastic material properties of the material the seismic wavelet travels in. The results obtained are 3D seismic models based on compression (P) and shear (S) wave velocity which are jointly interpreted to grade the rock mass quality quality in the light of the ongoing tunnel heading. The seismic measurements are performed with the Tunnel Seismic Prediction (TSP) 303 system from Amberg Technologies (AT). Discussing one geophysical seismic measurement campaign as a case study it will be shown that structural as well as lithological changes affect the seismic image and how these are interpreted.

1 INTRODUCTION

1.1 *The Brenner Base Tunnel*

The Brenner Base Tunnel (BBT) is one joint tunneling project of Austria and Italy connecting the two countries via the new base tunnel under the Brenner Pass crossing the Alps. The entire project comprises 230 km of various tunnel sections, including the two 55 km long main rail tubes and the continuous exploratory tunnel (Bergmeister, 2011).

In the lot H33 Tulfes-Pfons i.a. the 15 km long tunnel boring machine (TBM) heading of the exploratory tunnel started in October 2015 (Lussu, et al., 2019) and the two main tubes are heading southwards for 1.6 km and 1.2 km respectively since December 2017 as drill and blast headings (Figure 1). The TBM of choice is an open Gripper Herrenknecht TBM. For geological exploration and in order to perform several tests and measurements on the rock mass the direct accessibility to the tunnel wall is essential.

Even though the purpose of the exploratory tunnel is the exploration of the rock mass for the upcoming main tubes the tunnel is a performance heading. Its future function will be a logistic tunnel for a following construction lot and later in the operating phase it will be used as a drainage and service tunnel. To avoid downtime of the TBM the measurements have to be smoothly integrated in the entire workflow on the TBM heading.

1.2 *Predictions at BBT*

In the entire project area different investigation campaigns were performed in all stages of the project.

Before the start of the excavation work several mapping and drilling campaigns were performed in order to obtain structural and lithological information to derive a geological model of the project area of the BBT (Brandner, et al., 2008). Nevertheless the model has its uncertainties. The mapping strongly depends on the present of outcrops and their accessibility which is partly limited in the alpine environment. Furthermore the geological findings of the mappings on the surface have to be extrapolated onto the tunnel axes.

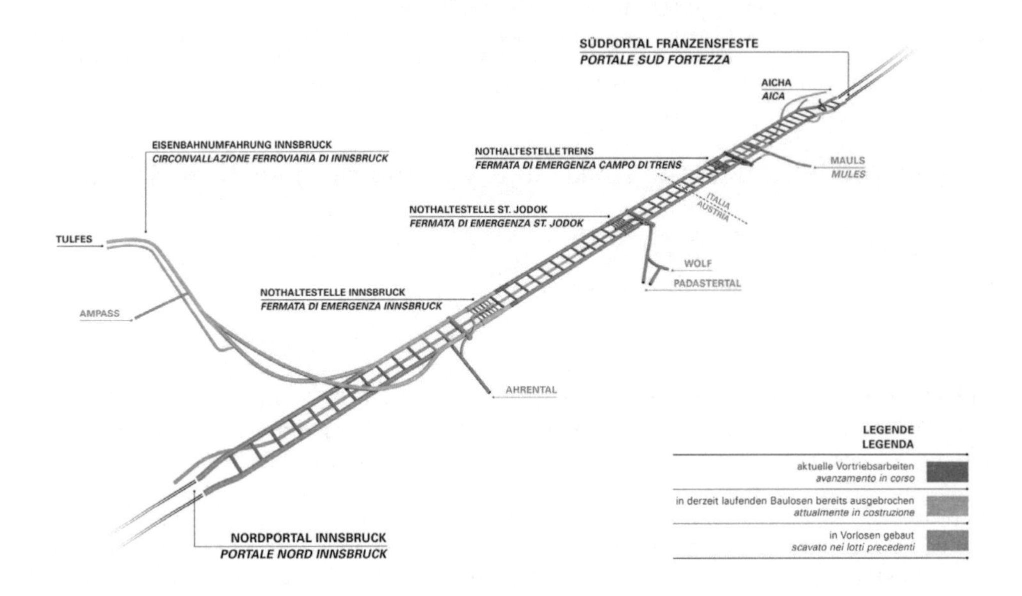

Figure 1. Location and relative position of the headings in the Brenner Base Tunnel system.

Knowing that the overburden reaches about 1800 m in the project area and 1300 m in the area of the lot H33 and that those geological features can vary, vanish or appear makes the limitations evident. This can be partly compensated by deep drillings fixing a certain geological feature in space, especially depth, orientation and appearance. This method can only be performed on selected points and delivers a one-dimensional result often not down to the tunnel level.

This makes a short-term prognosis while excavation necessary. The BBT SE has decided to follow a holistic approach comprising geological, geotechnical and geophysical methods to benefit from the advantages by combining them.

In the exploratory tunnel an elaborated set of different prediction methods from various fields are performed (quasi)continuously. The geological prediction is done with a percussion drilling and its geological analysis, the geotechnical prognosis is made with the data of the TBM drive obtained from sensors in the cutter head. The geophysical prediction is obtained by performing reflection seismic measurement campaigns and tunnel specific processing (Reinhold, et al., 2017).

2 GEOPHYSICS

2.1 *Geoseismic fundamentals*

The reflection seismic method is an active geophysical method investigating the rock mass with elastic waves. Three main parts are essential in any geoseismic acquisition to rule the outcome: a) the seismic source emitting a wavelet, b) the propagation of the wave in the investigated medium and c) the receiver recording the waveform (Kearey, et al., 2002).

a) A seismic source generates a seismic wavelet, ideally an impulse at a given time. The source wavelet must be coupled into the medium of investigation.
b) The seismic wave propagation is controlled by the elastic properties of the medium passed through. The geometrical spreading of the released energy over the surface of an increasing sphere with distance leads to a decreasing amplitude. The intrinsic attenuation is strongly material dependent and leads to a loss of amplitude due to

not fully reversible energy exchange as the wave propagates in the medium. Furthermore the energy splits on material boundaries into a transmitted and a reflected part. The threshold of transmission and reflection is controlled by the seismic impedance contrast between the two media.

c) The seismic receiver is the tool which reads the waveform travelling in the medium. Here the waveform has to be decoupled from the medium but its further propagation may not to be affected.

Seismic method is a data intense method since many source and receiver locations are necessary to illuminate the rock mass properly. Each wave from a source-receiver-pair travels a unique way and illuminates therefor a different part. Consequently all travelpaths from each source-receiver-pair are reflected on the same material boundary, but on different reflection points (depending on the boundary in relation to the source and receiver location). By connecting all reflection points the boundary is sketched.

Critically to know is the 3D-location of source and receiver and the traveltime. Therefor the source time has to be triggered. Knowing these parameters the seismic velocity and seismic boundaries can be calculated.

2.2 *Geoseismic at BBT*

At the Brenner Base Tunnel the seismic measurements are performed with the Tunnel Seismic Prediction (TSP) 303 system from Amberg Technologies (AT). The acquisition is done in a quasi-continuous manner. This means, that seismic campaigns are performed and by the end of the prediction area of one campaign the subsequent campaign is performed.

The TSP 303 of AT uses a non-destructive blast as a seismic source because a blast is very close to an impulse. The electric ignition impulse for the blasting is used as a trigger for the seismic recording.

The receivers used in the seismic measurement are 3-component receivers. The recording of the incoming wave in three orthogonal components allows a full detection no matter which direction of propagation and displacement is describing the wave (Schwarz & Schierl, 2017).

3 CASE STUDY

3.1 *Layout and acquisition*

The further discussed seismic campaign was performed on May 18th 2017 in the exploratory tunnel. The face at this date was stationed at tunnel meter (TM) 6010.78. This is located below the south slope of the Arz valley. The overburden at the face location is about 700m with increasing trend towards the south (Figure 2).

The seismic layout has to be fitted into the front part of the TBM in order to fulfill the geophysical as well as the construction logistic needs. At the BBT two shotlines are used on both sides of the accessible invert in the L1 area of the TBM (Atzl, et al., 2013). Each shotline should consist of 18 individual shots which are placed about 1.4 m deep into the tunnel wall. Due to geological constrains it can be essential to reduce the number of shots. This leads to a shotline on the left side of 15 shots and on the right side of 18 shots. The shot spacing is about 1.3 m. The four receivers are placed in the L2 area of the TBM in two sections in both sides. The sections are about 10 m apart and the receivers are located about 1.5 m in the tunnel wall. Thus the layout is about 69 m long (measured from the face). The geological situation in this area is *Innsbrucker Quartzphyllite* with some divisional surfaces (Figure 3).

The acquisition is done in the multi-shot-recording (MSR) mode. In this mode a series of 1000 ms delayed shots are fired in order to reduce the acquisition time dramatically. By the time the second shot is fired the data from the previous one has already decayed below noise level. In this campaign four shot groups of seven shots, one shot group of four shots and one

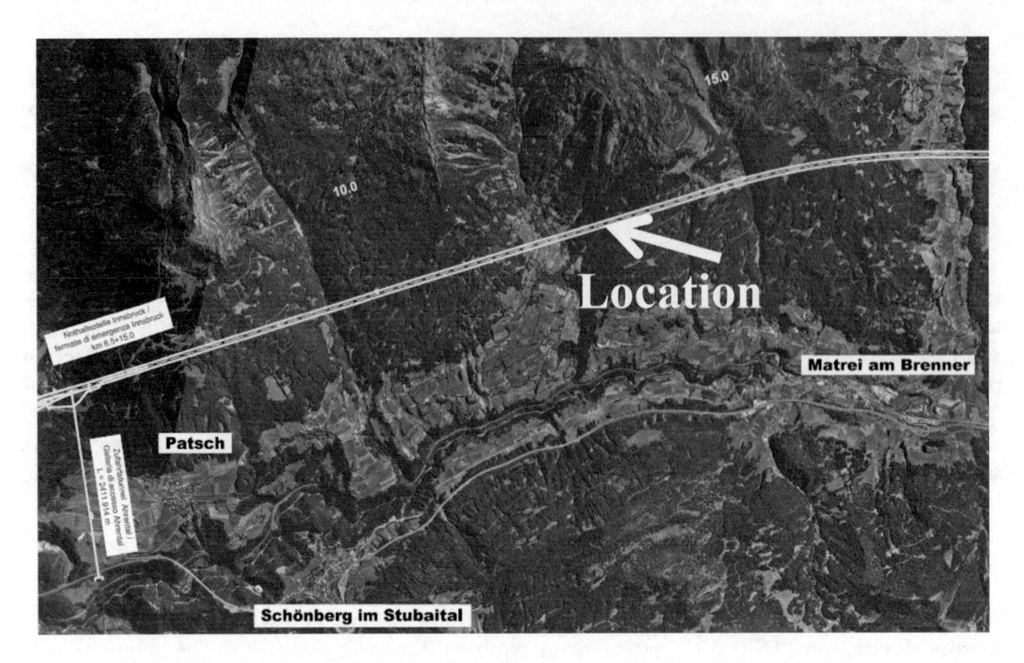

Figure 2. Location and relative position of the headings in the Brenner Base Tunnel system.

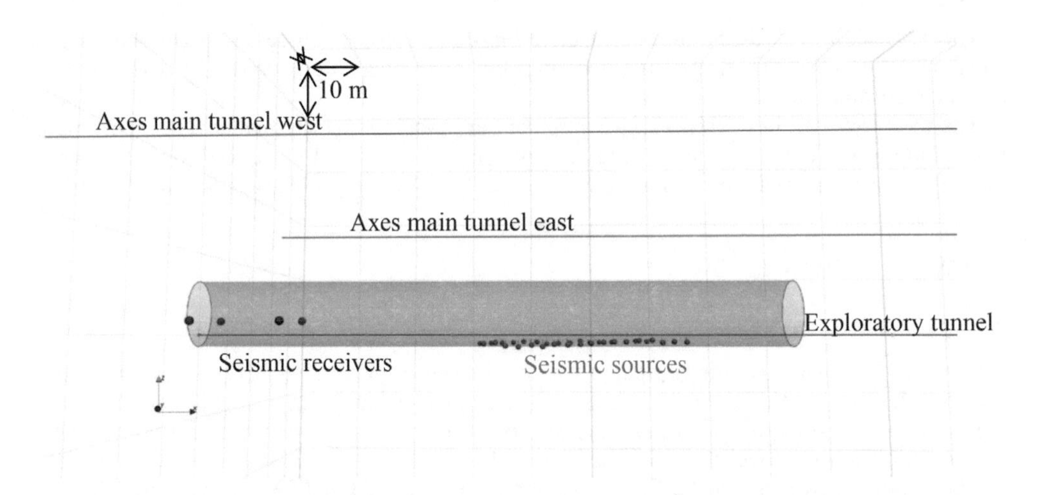

Figure 3. Seismic layout of the campaign.

single shot were fired. In each shot group the zero-delay detonator is triggered, the others have to be recalculated.

3.2 *Data processing*

The data acquired for the processing consists of 132 source-receiver-pairs (33 shots recorded at 4 receivers) and all are recorded in 3 orthogonal components which leads to 396 seismic traces.

Before the actual processing a data quality check is performed in order to exclude potentially harmful traces from the processing. Basically two criteria are analyzed to determine the quality of a seismic trace:

a) The first parameter is the strength of the signal which is directly measured. It indicates the quality of the coupling of the source into the rock and the attenuation of the direct wave.
b) The second parameter is the noise level which is the unwanted recorded data. Noise is immanent but its level should be suppressed to a minimum. Immanent noise is created as tunnel wave from the blast and avoidable noise is generated by other works in the tunnel or due to bad coupling of the receiver to the rock. Bad coupled receivers can be vibratingly excited by the passing wave.

Some seismic traces fail to fulfill the quality criteria for example shot 2 on the right side recorded on the receiver at the rear left and front right as well as many shots from the left shot line recorded at the front right receiver. Therefore 17 traces are excluded. All other seismic traces have good signal strength and a sufficient low noise level and are appropriate for the following processing.

Since the receivers are individually mounted and therefore vaguely aligned the recorded traces have to be reoriented in one common coordinate system.

Defining the first break is crucial for the seismic processing since data is recorded in time and the first breaks describe the fix points to transfer from time into space in a later stage. The first break has to be defined for all source-receiver pairs individually and the delayed shots are fixed in time with the help of the triggered ones.

The following steps line by line in Figure 4 are especially designed to decompose the wavefield in order to perform the final calculations on its smaller well-tailored parts of the data: a) noise suppression, b) separate the waves according their direction of propagation i.e. arriving from ahead and aside, c) separate both according to their direction of polarization i.e. into the compressional (P) wave, horizontal shear (SH) wave and vertical shear (SV) wave. Thus the number of traces is now multiplied by factor six (132 source-receiver pairs; 17 source receiver pairs of insufficient quality; 2 directions of propagation; 3 directions of polarization; (132-17) x 2 x 3=690). The noise stays unchanged and is deleted.

As final step the processed seismic traces are migrated from time into 3D space. Each boundary is represented as a deflection in each seismic trace. In the migration process the reflections from all seismic traces are placed in space and at the true boundary location they interfere constructively and describe the boundary. This leads to a 3D-velocity model and a 3D-reflectivity model for each wave type.

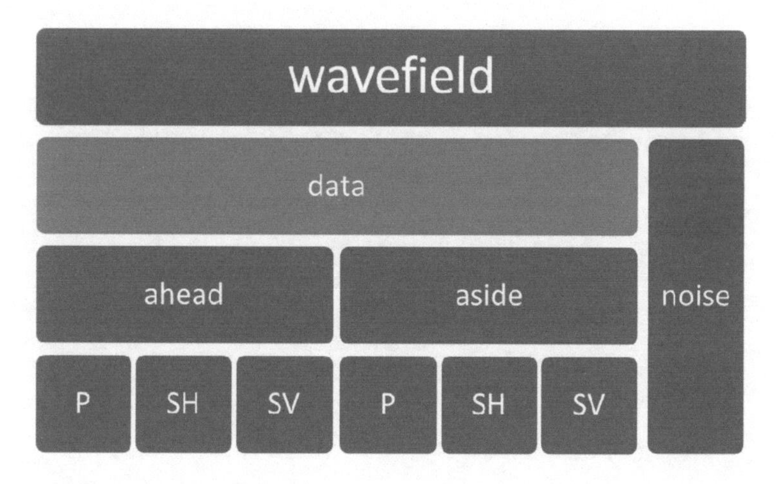

Figure 4. Geoseismic processing flowchart.

3.3 *Results*

The base for the interpretation is the 3D-velocity and 3D-reflectivity models of all wave types together with the 3D-reflector model. In this discussion only the results from ahead of the face will be considered since no relevant reflections from the sides are observed.

The 3D-velocity model displays the seismic velocity in the calculated cuboid. It is obtained by locally adapting the seismic velocity to optimize the constructive interference of the deflections in the seismic traces. The 3D-reflectivity model displays the strength of the reflectivity in the calculated cuboid. It is obtained by performing the migration of the seismic traces in the updated seismic velocity model. For the 3D reflector model all the models are merged by extracting dominant reflectors and checking the coherence across the wave types. The reflectors segment the model in blocks which are assigned to seismic as well as elastic rock mass properties.

The seismic velocity measured from the direct wave (the wave directly traveling from the source to the receiver without reflection) is used as initial velocity model for the prediction area. The initial P-wave velocity is $v_P = 5030\,\mathrm{m/s}$und the initial S-wave velocity is $v_S = 2908$ m/s. Both are intermediate values and in good agreement with the observed geological conditions in the layout. The calculation cuboid extends 300 m x 100 m x 100 m thus the prediction length extends 230 m.

The P-wave velocity model (Figure 5) shows a strong velocity decrease (1) from about TM 6029 to TM 6036. The velocity drops by more than 10 % below 4500 m/s. The low velocity area dips steeply in heading direction and is nearly perpendicular to the tunnel axes.

Another low velocity area (2) in the P-velocity model starts at TM 6180. The drop is not as strong as the previous one. It reaches 5 % and its values are below 4750 m/s.

The part from TM 6040 to TM 6165 (3) is characterized by high P-wave velocity. The increase is up to 15 % to 5750 m/s. Nevertheless this area cannot be characterized as homogeneous, locally it even shows a velocity decrease of 10 %.

The S-wave velocity model shows a rather homogeneous high velocity with an increase up to 10 %.

The P-reflectivity shows a rather dense pattern until TM 6040 and two distinct reflections at TM 6067 and TM 6192. The SH- and SV-reflectivity show similar pattern.

The reflector model (Figure 6) supports the results shown. It shows a dense reflector pattern from TM 6024 to TM 6040, from TM 6178 to TM 6180, from TM 6190 to TM 6194, from TM 6208 to TM 6210 and from TM 6221 to TM 6223. A sparse reflector pattern is shown from TM 6052 to TM 6161. It additionally clearly shows that the reflection points are located above right of the tunnel axes until TM 6092 and mainly related to the P-wave. The reflectors in the end of the prediction area are perpendicular and additionally related to the S-waves.

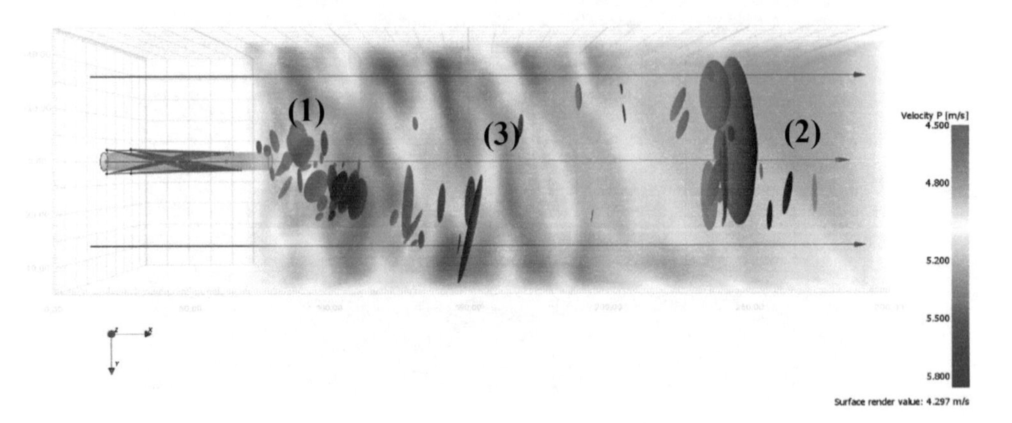

Figure 5. Geoseismic 3D P-wave velocity model and extracted reflectors.

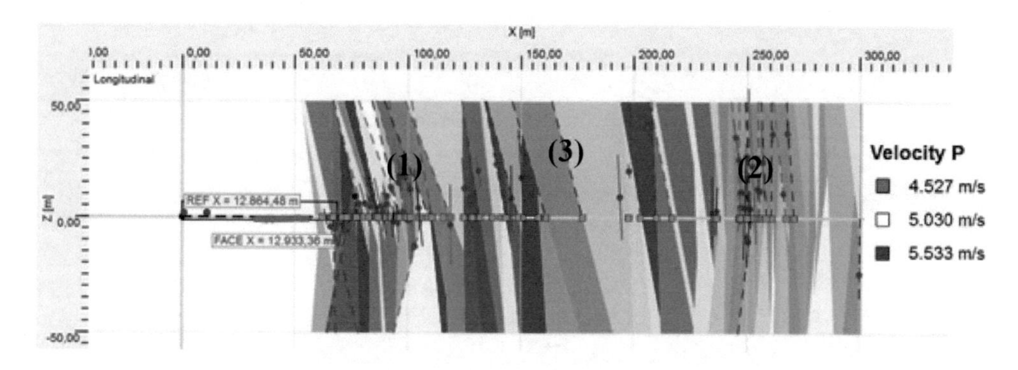

Figure 6. Seismic vertical 2D cross section of the reflector model with assigned P-wave velocity.

Table 1. Rock mass quality index, top assigning the interpreted parts, middle the indication (green=good, red=poor), below the TM.

	(1)			(3)				(2)							
6011-6024	-6029	-6036	-6040	-6067	6067	-6165	-61 78	-6180	-6190	-6194	-6208	-6210	-6221	-6223	-6242

The reflectors are steeply dipping in heading direction and are sub-perpendicular to the tunnel axes. Thus the main dip direction is SSW.

Referring low velocity to soft and weak rock, a strong reflectivity to a strong material contrast and a dense reflector pattern to an accumulation of dividing surfaces and vice versa a rock mass quality estimation is established (Table 1).

3.4 *Interpretation*

The interpretation of the findings in terms of geological equivalent and especially geotechnical impact on the heading is very challenging and requires experience in geophysics and training in the actual tunnel on site. As important as the detection of hazardous conditions ahead is the exclusion of these or the information that a weakened rock mass is thin and is going to improve quickly.

Area (1) is interpreted as a fault zone of moderate thickness and of minor influence on the heading.

Area (2) is interpreted as zone of reduced rock quality without clear evidence of a fault zone. Since this area is on the end of the prediction length the quality and the resolution is reduced and a detailed statement is not possible

Area (3) is interpreted as a very compact zone with very good rock quality, which is pervaded by discontinuous. Such high values have not been seismically observed in the *Innsbrucker Quartzphyllite* over such a long section.

In the following the geophysical interpretation will be compared with the geological documentation. The geological documentation provides vertical and horizontal cross-sections along the tunnel axis based on continuous peripheral mapping (Schwarz & Schierl, 2017). The results for the tunnel section of this seismic campaign are shown below (Figure 7)

In area (1) several faults with some decimeters of fault zone material appear from TM 6029 to TM 6035. As predicted in the geophysical interpretation the orientation is steeply dipping in heading direction.

Area (2) is characterized by several lithological changes from quartzphyllite and to imbricated limestone phyllite which is slightly sheared. The lithologies are separated by faults and

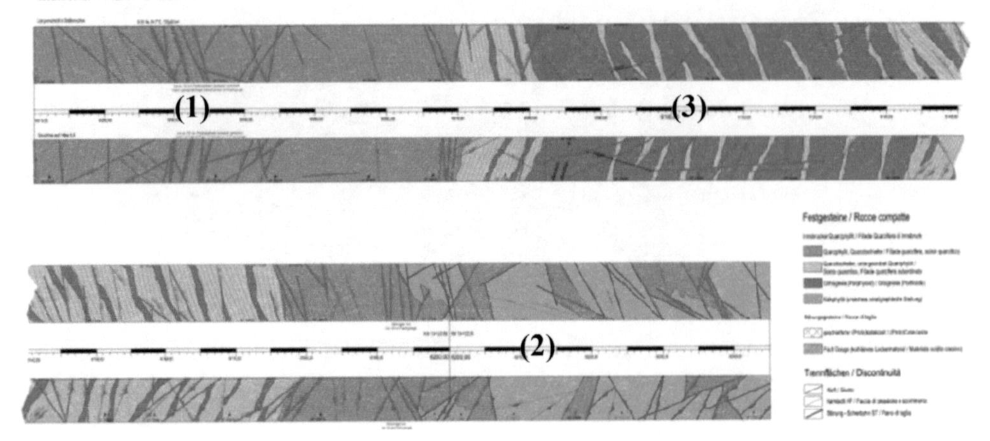

Figure 7. Geological horizontal and vertical cross-sections from TM 6010 to TM 6245 based on geological mapping.

from TM 6190 to TM 6195 some faults with fault zone material are documented. The lithological difference is not detectable geophysically, nevertheless the structural information is present in the geophysical data.

Area (3) comprises an intact part of quartzphyllite from TM 6040 to TM 6067. From TM 6070 onwards the lithology changes to quartzite schist and orthogneiss. These rocks are extremely strong and have an estimated (EN ISO 14689-1) uniform compression strength (UCS) of 50-250 MPa.

4 CONCLUSION

The BBT SE has decided to follow for the construction of the exploratory tunnel a holistic prediction approach comprising geological, geotechnical and geophysical methods to benefit from the advantages by combining them. The geological prediction is done with a percussion drilling and a geological analysis, the geotechnical prognosis is made with the data of the TBM drive obtained from sensors in the cutter head. The geophysical prediction is obtained by performing reflection seismic measurement campaigns and tunnel specific processing.

For the latter it was shown in this article that the geophysical seismic method applied at the BBT, Tunnel Seismic Prediction (TSP) 303 system from Amberg Technologies, is capable of detecting structural as well as lithological variations in the rock mass ahead of the face. The seismic prediction describes well the variations of the rock mass ahead and was able to exclude the presence of hindering rock conditions, which is equally important as their detection. So an on-site observed decrease in rock quality can be diminished with the help of the seismic information. The described section of about 230 m was excavated without difficulty in 8 days (29 m/24 h).

These findings are further used to update and refine the geological and geotechnical model for the future main tubes headings (Bergmeister & Reinhold, 2017).

REFERENCES

Atzl, G. et al., 2013. Richtlinie für die geotechnische Planung von Untertagebauten mit kontinuierlichem Vortrieb, Salzburg: Österreichische Gesellschaft für Geomechanik.

Bergmeister, K., 2011. Brenner Basistunnel - Brenner Base Tunnel - Galleria di Base del Brennero. Innsbruck/Bozen: Tappeiner Verlag.

Bergmeister, K. & Reinhold, C., 2017. Learning and optimization from the exploratory tunnel - Brenner Base Tunnel. Geomechanics and Tunnelling, October, pp. 467–476.
Brandner, R., Reiter, F. & Töchterle, A., 2008. Überblick zu den Ergebnissen der geologischen Vorerkundung für den Brenner Basistunnel. Geo. Alp, pp. 165–174.
Lussu, A., Kaiser, C., Grüllich, S. & Fontana, A., 2019. Innovative TBM transport logistics in the constructive lot H33 - the Brenner Base Tunnel. ITA-AITES World Tunnel Congress 2019.
Reinhold, C., Schwarz, C. & Bergmeister, K., 2017. Development of holistic prognosis models using exploration techniques and seismic prediction. Geomechanics and Tunnelling, December, pp. 767–778.
Schwarz, C. & Schierl, H., 2017. Integration of reflection seismic data into the documentation during construction of the Brenner Base Tunnel. Geomechanics and Tunnelling, October, pp. 552–560.

Tunnels and Underground Cities: Engineering and Innovation meet Archaeology, Architecture and Art, Volume 3: Geological and geotechnical knowledge and requirements for project implementation – Peila, Viggiani & Celestino (Eds)
© 2020 Taylor & Francis Group, London, ISBN 978-0-367-46583-4

Geological investigation for urban tunnel portal in made ground

G. Senol, G. Alan Jatta, S. Aydogan & E.A. Tekyildiz
EMAY International Engineering and Consultancy Inc., Istanbul, Turkey

ABSTRACT: Istanbul is one of the most crowded and traffic-intensive cities in the world. It is planned to construct 2-3 lane highway tunnels in order to reduce the traffic load, time, fuel loss and damage to the environment. Levazim Tunnel, one of the most important of these projects, is located on the Dolmabahçe-Kilyos route. In this study, the entrance portal of the Levazim Tunnel (T2) was investigated. Intensive structures were found in the upper sections of the portal section, and made ground were identified in the field surveys. There are intensive structures in the upper sections of the entrance portal and made ground were determined in the site investigations. Geological and geotechnical surveys are very important in order to eliminate potential problems. In this paper, planning and implementation of the site investigations required for design, preparing of geological model and geological sections with the proposal of solution are included.

1 INTRODUCTION

The Dolmabahçe-Kilyos Highway Tunnels tendered by Istanbul Metropolitan Municipality and projected by Emay International Engineering and Consultancy Inc. is planned with 2x3 lanes in order to reduce the traffic load as an alternative to existing roads in Istanbul (Figure 1, Figure 2). The total length of the 6 tunnels is 25000 m and Levazim Tunnel with the length of 3550 m is second of these aforementioned tunnels. Majority of the tunnels pass below intense settlement areas. Within the extent of the study, the Levazim Tunnel Entrance Portal was investigated. Intensive structures (buildings) were present at the upper sections of the entrance portal section, and made ground were identified in the conducted field surveys. The interaction between the made ground with the observed thickness of approximately 24 meters and the buildings located in the upper sections of the tunnel has been examined and explained in detail.

Staged site investigations commenced after office studies. Site investigations are planned to characterize the made ground, soil and rock formations located at the portal section and to support the classification of geological units. Interpretation of these units is critical for the development of geotechnical risks that can be realized during construction of tunnel portal. Within the scope of site investigations, drilling, geophysical techniques and on-site tests (SPT, pressuremeter etc.) were performed. SPT samples obtained from soil units and samples obtained from rock units were collected and classification and strength experiments were carried out in the laboratory.

Through evaluation of data obtained from drillings and geophysical studies, cross and longitudinal geological-geotechnical sections of the portal section were produced. In particular, made ground and dissociated rock levels were determined, and these levels were associated with the tunnel. Through the evaluation of data obtained from field and laboratory studies, engineering parameters related to geological units were obtained and stability analyzes were carried out according to these data.

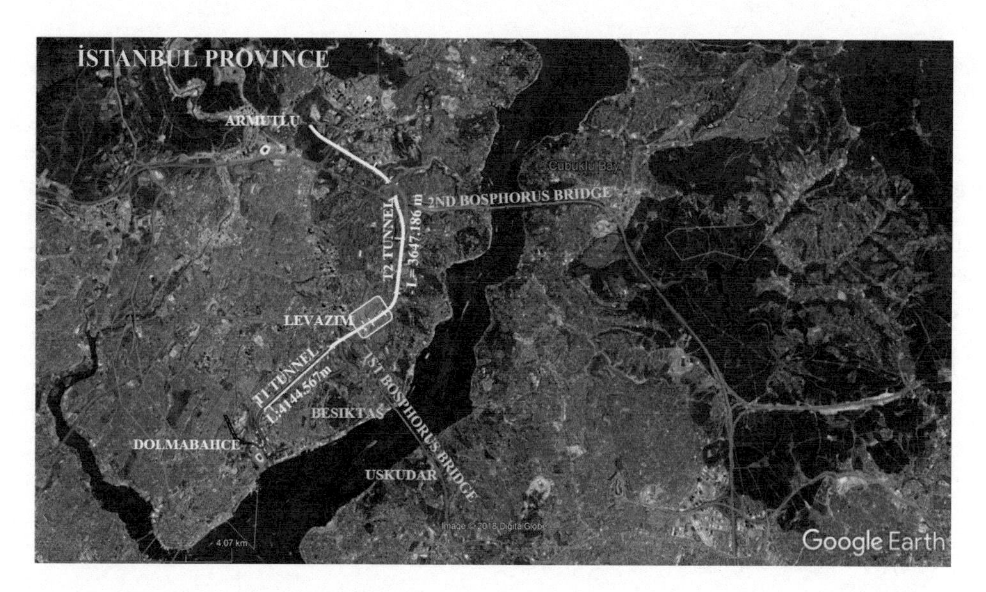

Figure 1. Plan view of the Levazim Tunnel.

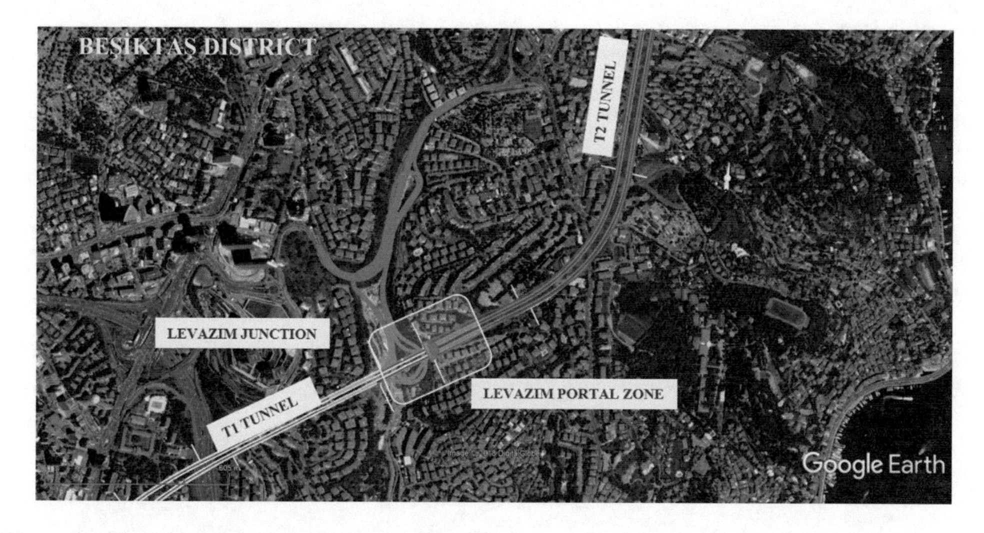

Figure 2. Plan view of the Levazim portal Tunnel.

2 METHOD OF OPERATION

The Made Ground is located at and around of the Tunnel Portal which has been determined as a critical section in the western portal of the Levazim tunnel where the project studies and site investigations are carried out simultaneously. Since this made ground is contemporary, this unit is not included in the literature studies conducted earlier. Residual Material from the main excavation studies of a large shopping mall was filled into Levazim valley between the years of 1990–1993. In the following years, construction has become prevalent in this area and a large number of buildings have been constructed on the made ground which has a maximum depth of 45 meters.

During the field trips and surface geology examinations, the thickness of the made ground, interaction between the tunnel portal and surrounding buildings and the risk of that interaction was determined. For this reason, drilling locations which will be conducted for determination of the thickness of the made ground layer around the tunnel portal and its surroundings was determined and studies have been started with on-site experiments.

The made ground obtained from the drilling studies is identified to be the same as the main rock unit, and it is also determined that the made ground also exhibits similar characteristics with the weathered rock. In addition, it was determined that the thickness of the made ground changes with very small intervals. For this reason drilling studies around the location of the tunnel besides drilling studies conducted on upper sections of the tunnel were planned and thickness of made ground were examined in detail. In addition to these studies, MASW studies, which are a geophysical method, along 4 lines, were carried out and data obtained from these studies were evaluated together with the data obtained from the drillings.

Because of the changes in thickness of made ground level even at drilling locations very close to each other, the number of drillings to be conducted had been increased for determination of thickness of the artificial filler unit where the buildings located just over the tunnel excavation limit settled on and in addition geophysical studies had been conducted and the geology of the area to be worked on has been tried to be determined in detail.

3 REGIONAL GEOLOGY

The project area lies within the geological region known as the İstanbul Zone. There are two large rock-stratigraphic units within the Istanbul zone. One of these two communities separated by a large tectonic line is the Istıranca Massif and the other is the Istanbul Zone, which also includes the project area (Figure 3).

The Paleozoic rocks of the Istanbul Zone consist of a transgressive sequence extending from the Ordovician to the Carboniferous (488–299 million years ago). Paleozoic sediments are represented by different geological formations ranging from the Ordovician aged Kurtköy Formation to the Carboniferous Trakya Formation and representing various sedimentary environments.

In the project area, there are Carboniferous rock units which are called Trakya Formation. Carboniferous aged Trakya Formation, which is commonly found in Istanbul, is a flysch formation and consists of sandstone, siltstone and shale units.

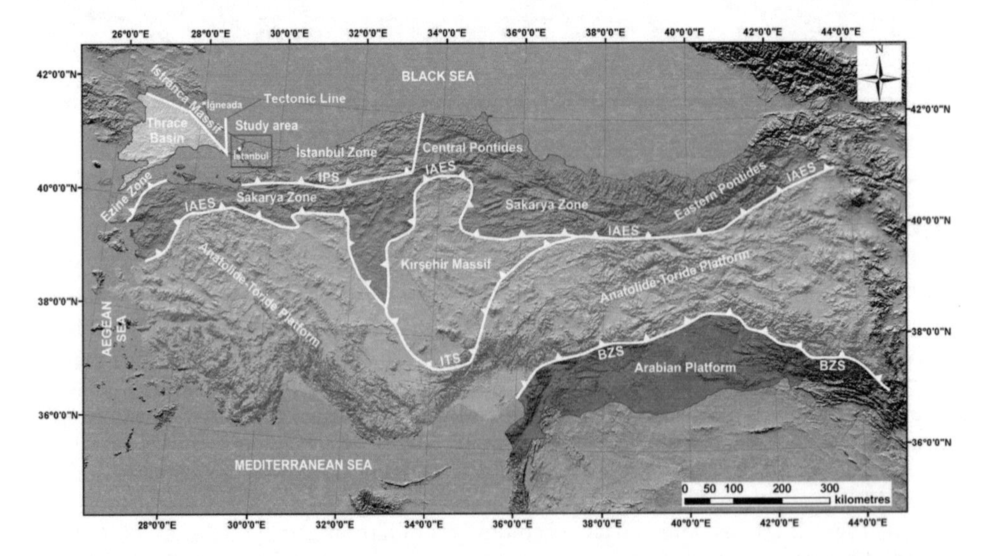

Figure 3. Tectonic map of Turkey and surrounding area with major suture zones (modified from Okay & Tüysüz 1999).

Figure 4. View of Levazim portal zone.

3.1 *Local geology conditions*

Levazim tunnel portal is located on the right side of the Levazım valley in the district of Beşiktaş in Istanbul Province (Figure 4). There are sandstone-siltstone-shale units of the Trakya Formation and Made Ground around the project area (Figure 5). Two different levels of Trakya Formation units were observed during field trips, drilling and geophysical studies. These two units, which are divided according to the characteristics of engineering geology, are named as moderately weathered sandstone-shale and slightly weathered siltstone-shale.

Moderately weathered sandstone-shale rock unit; weak to very weak strength, yellowish brown, very closely to closely spaced (Figure 6). The slightly weathered siltstone-shale rock unit is of medium strong-strong, bluish gray color and moderately widely spaced. These two levels are shown in the geological section studies and the other engineering geology features are shown in Table 1.

Figure 5. View of T2 Tunnel route.

Figure 6. The sandstone-shale unit in the study area.

Table 1. Engineering Geology Properties of Rock Masses

Rock Mass Identification Criteria	Moderately weathered Sandstone-Shale	Slightly weathered Siltstone-Shale
Strength	Weak-Very weak	Medium strong-strong
Structure	Thinly (6-20 cm)	Medium (20–60 cm)
Colour	Yellowish brown	Bluish Grey
Texture	Regular Bedding	Regular Bedding
Grain size	Fine to Medium Grained	0,001–0,0063 mm
Weathering	Moderately weathered	Slightly weathered
Discontinuities		
Orientation	135/25	135/25
Spacing	Very closely to closely spaced	Moderately widely spaced
Persistence	Low (1-3 m)	Medium (3-10 m)
Aperture	Moderately wide (2,5-10 mm)	Open (0,5-2,5 mm)
Roughness	Planar-rough	Undulating-rough
Filling	Clay and silt	Carbonate and Silica
Seepage	Medium flow (0.5 l/s-5.0 l/s)	Small flow (0.05 l/s-0.5 l/s)

4 SITE INVESTIGATION

Within the scope of field studies 14 boreholes which reaches the total depth of 316.40 meters were conducted at T2 tunnel. 8 of these boreholes were conducted for determination of the made ground/rock boundary. Besides, 31 pressuremeter tests are conducted at 6 wells. In addition, MASW studies were carried out in 4 different sections in order to determine the spread and depth of the made ground thickness.

4.1 *Borehole studies*

A total of 14 boreholes were conducted with the total depth of 316.40 meters at T2 Tunnel. 6 of these boreholes were made for geotechnical purposes and a geological model is constructed through the determination of natural rock boundary with 8 of these boreholes (Table 2).

Table 2. List of boreholes information.

Borehole	Borehole Km	Borehole Purpose	Depth (m)	Elevation	Ground Water
DLB-33	T2 left tube 4+150	Tunnel Portal	30.00	69.00	21.00
DLB-33A	T2 left tube 4+200	Tunnel Portal	31.00	76.30	13.50
DLB-33B	T2 left tube 4+250	Tunnel Portal	16.50	78.60	21.00
DLB-33C	T2 left tube 4+268	Tunnel Portal	18.00	-	24.30
DLB-33D	T2 left tube 4+170	Tunnel Portal	15.40	-	-
DLB-34	T2 left tube 4+660	Tunnel Portal	30.00	69.00	19.95
DLB-34A	T2 right tube 4+730	Tunnel Portal	30.00	75.40	7.00
DLB-34 B	T2 right tube 4+810	Tunnel	27.00	83.80	-
DLB-34C	T2 right tube 4+703	Tunnel	18.00	72.00	9.20
DLB-34 D	T2 right tube 4+740	Tunnel	15.00	78.00	6,90
DLB-34E	T2 right tube 4+760	Tunnel	22.00	84.00	9.70
DLB-34F	T2 right tube 4+787	Tunnel	25.50	89.00	12.40
DLB-34G	T2 right tube 4+767	Tunnel	18.00	88.00	9.50
DLB-34H	T2 right tube 4+705	Tunnel	20.00	76.00	10.70

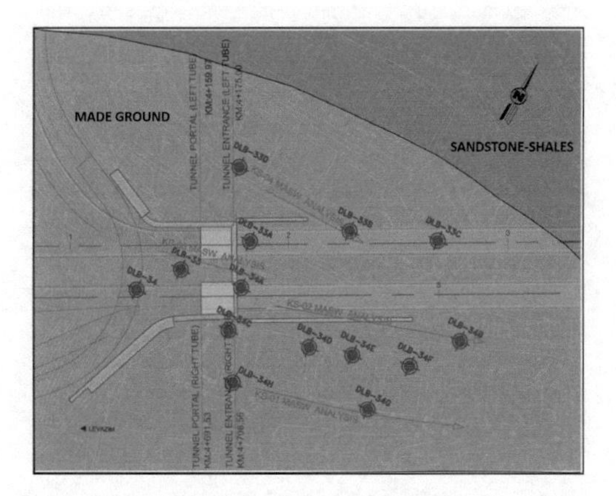

Figure 7. Boreholes layout plan of Levazim Tunnel portal.

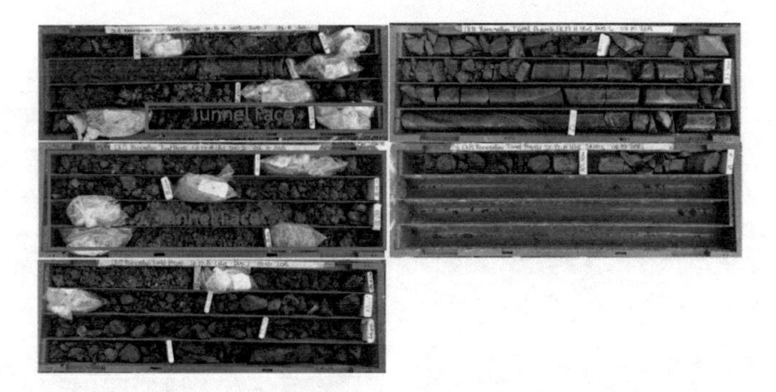

Figure 8. Core tray photo of DLB-33A borehole.

TCR, SCR, RQD values, made ground thicknesses, dissociation grade and weak zones were determined from the data obtained from the boreholes. The geomechanical properties of the rock were determined by performing point load test in the laboratory, single axis pressure experiments, elasticity and poisson ratio tests (Figure 7, Figure 8).

4.2 *Geophysical Studies*

It has been determined at the drilling studies that a considerable portion of the tunnel entrance portal passed into made ground (Table 3). However, geophysical investigations have been conducted for determination of thickness and extent of the made ground level. MASW studies were carried out in 4 different sections in order to determine the boundaries of the made ground/rock units observed from cross-sectioned obtained from the field (Figure 9, Figure 10).

Table 3. Geophysical studies.

Line	Line length	Operation Mode	Purpose
KS-01	92	MASW	Tunnel Portal
KS-02	92	MASW	Tunnel Portal
KS-03	46	MASW	Tunnel Portal
KS-04	46	MASW	Tunnel Portal

Figure 9. Seismic boreholes points on the route.

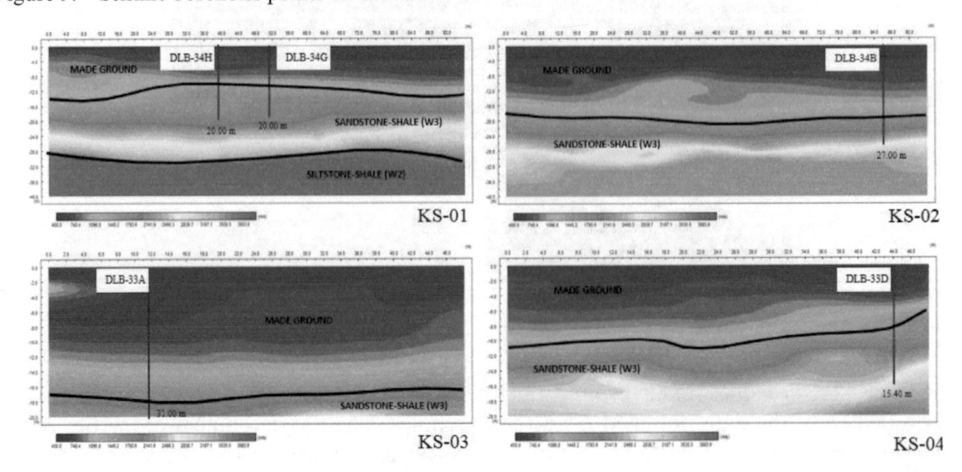

Figure 10. KS-01, KS-02, KS-03 and KS-04 lines MASW analysis results.

Table 4. Pressuremeter test data.

Borehole	PMT pcs.	PMT (Depth)	Borehole Depth (m)
DLB-33	1	18.00	30
DLB-33A	6	3.00–18,00	31
DLB-34	11	2.00–24.00	30
DLB-34A	5	3.00–16.00	30
DLB-34C	4	3.00–12.00	18
DLB-34E	4	3.00–12.00	22

4.3 *Pressuremeter test*

In order to obtain the load-deformation values of the loose units observed in the cross-sections obtained from drilling studies at the entrance portal of the tunnel, pressuremeter experiments were carried out (Table 4).

5 TUNNEL PORTAL GEOLOGICAL MODEL

T2 Levazim Tunnel entrance portal will be opened in the place where intense urbanization is present. For this reason, the supervision of the ground/structure interaction has become one of the most important parameters that should be controlled during the opening of the entrance portal (Figure 11, Figure 12 and Figure 13).

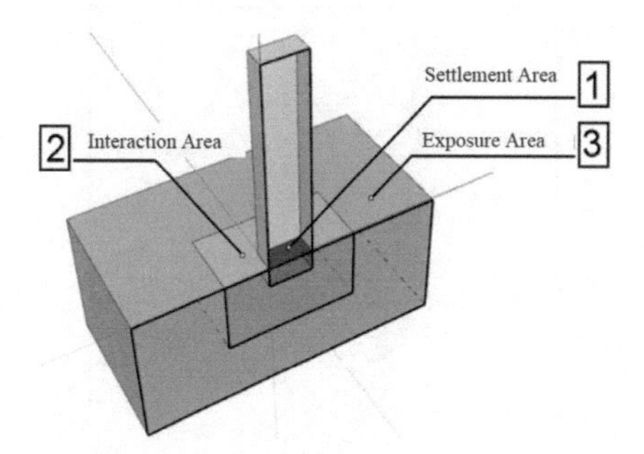

Figure 11. Model structure according to the 3-zone evaluation principle in the field of project design (Vardar, M., 2015).

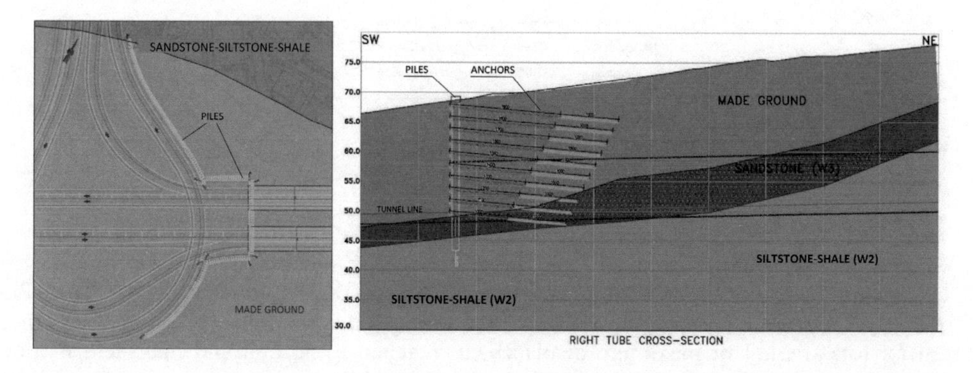

Figure 12. Excavation plan of the T2 Tunnel entrance portal (left), T2 Tunnel right tube portal cross-section (right).

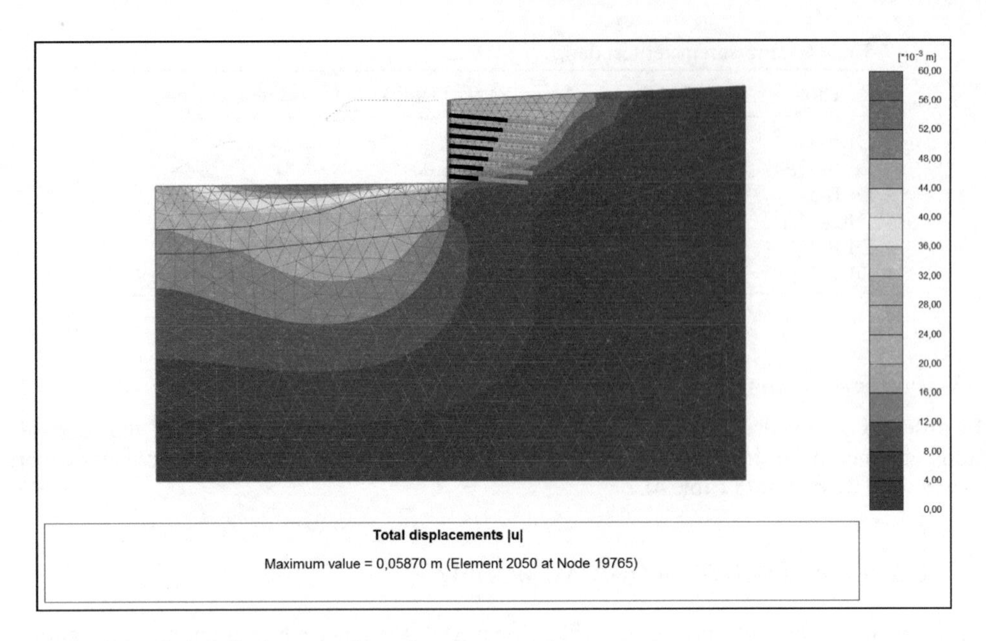

Figure 13. Maximum displacement formed on the ground.

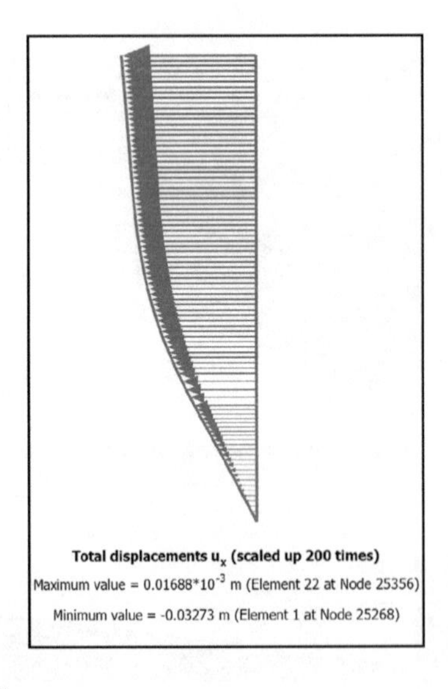

Figure 14. Displacement per pile.

When all available data were evaluated, the most negative slides for portal excavation were determined and portal stability analyzes were performed on these sections using the Plaxis 2D 2016 computer program. At the most critical section which is the right tube tunnel entrance portal mirror, amount of maximum displacement reached to 50 mm and max deformation amount per pile reaches up to 0.016 mm (Figure13, Figure14).

6 CONCLUSIONS

Soil improvement is required for the portal structure to be constructed at the entrance portal area of the T2 Tunnel. For this reason, it is deemed appropriate to start excavations after soil improvement is conducted with piles whose details were provided at relevant map sections. For this reason, geotechnical assessments of the units located at the portal areas have been carried out during the field studies.

As a result of the studies carried out in the field and the site investigations, the dominant unit observed in the entrance portal area is Sandstone-Shale (W3), Siltstone-Shale (W2) and Made Ground. The analyses conducted for these units are based on laboratory test results and borehole logs. The geomechanical parameters are calculated according to the results of the unconfined compressive strength test.

A deep excavation - pile wall system was chosen for the soil improvement work to be performed. The piles to be manufactured are planned to be supported with anchors in the rock unit so that they are not affected by the rock layer on the back. The piles in the rock unit will have a diameter of 120 cm and they will be applied with 1.40 m horizontal distance. Also, in some places where anchorage can not be done, double-row console piles are considered depending on elevation.

With regard to conducted analyzes, entrance portal is assessed under the sections of right walls, right tube mirror, left tube mirror and left walls. Analyzes were conducted by using Plaxis 2D.

REFERENCES

EMAY International Engineering and Consultancy Inc. 2017. *Geological and geotechnical investigation report of Levazim-Armutlu (T2) Tunnel.*

EMAY International Engineering and Consultancy Inc. 2017. *Entrance portal design report of Levazim-Armutlu (T2) Tunnel.*

Ozgul, N. et al. 2011. Geology of Istanbul province area. *Istanbul Metropolitan Municipality Department of Earthquake and Soil Research.* Istanbul.

Ozgul. 2012. *Stratigraphy and Some Structural Features of the Istanbul Paleozoic.*

Ulusay R., Sonmez H., 2006. Engineering Properties of Rock Masses. *TMMOB Geology Chamber of Engineers Publications.*

Vardar, M. 2010–2011. Rock Mechanics Master Course Notes. *Istanbul Technical University Faculty of Mines.*

Yuzer E., Vardar M., 1986. Rock Mechanics. *Istanbul Technical University Foundation.*

Okay & Tüysüz, 1999. Tectonic map of Turkey and surrounding area with major suture zones

Tunnels and Underground Cities: Engineering and Innovation meet Archaeology, Architecture and Art, Volume 3: Geological and geotechnical knowledge and requirements for project implementation – Peila, Viggiani & Celestino (Eds)
 ISBN 978-0-367-46583-4

Importance of physical-mechanical properties of rocks for application of a raise boring machine

A. Shaterpour-Mamaghani & H. Copur
Mining Engineering Department, Istanbul Technical University, Istanbul, Turkey

E. Dogan
Eczacibasi Esan, Balikesir, Turkey

T. Erdogan
Sargin Construction and Machinery Industry Trade Inc., Ankara, Turkey

ABSTRACT: Raise boring machine is one of the preferred mechanical excavation machines that used for drilling/excavation of different shafts in mining and tunneling industries. The geological and geotechnical knowledge is one of the important factors for implementation of the shaft projects. This study aims to analyze the effects of uniaxial compressive strength, Brazilian tensile strength, and elasticity modulus on the performance of a raise boring machine (daily advance rate, consumed gross reamerhead torque and power, net pulling force, instantaneous penetration rate, unit penetration, and field specific energy) during the reaming operation of two shafts in the Balya lead-zinc underground mine. The field investigation indicated that besides the physical-mechanical properties of the excavated rocks, the other factors such as mechanical (machine related) and operational parameters should be considered for performance evaluation of a raise boring machine.

1 INTRODUCTION

The duration of the individual construction activities such as shaft excavation should be taking into account in the feasibility and design stages of the mining and tunneling projects. The detailed and optimized implementation schedule could be increased the efficiency of the mining and tunneling operations. It is well known that excavation of shafts and other vertical structures is a difficult job taking quite long time to realize. In this case, the selection of the best method for shaft excavation plays an important role in the projects implementation and completion. In recent years, the continuing trends has been towards to mechanization of shaft excavation. Mechanization offers safer, faster, more efficient, and more environment friendly operations. Raise Boring Machine (RBM) is one of the mechanical miners that is used to drill/excavate a circular hole between two levels of underground structures without the use of explosives. RBMs are used for different purposes such as excavation of ventilation shafts, ore transport shafts and manways in the mines, the surge in hydropower plants, inclined penstock lines (generally vertical), switching lines between underground subway tunnels (horizontal), stairs tunnels (inclined), and ventilation shafts in the road and railway tunnels.

In the conventional raise boring method, the machine is set up on the upper level of the two levels. Then a small diameter hole (pilot hole) is drilled by a tricone bit to the lower level. Once the drill string has broken into the opening of the target level, the pilot drill bit is removed and a reamerhead with roller cutters, of required diameter of the excavation, is connected to the drill string and raised back up towards to upper level. In the pilot drilling, the cuttings are removed from the hole with the aid of classical borehole flushing. However, in the reaming stage, excavated material fall by gravity to the bottom of the hole and from where the muck is continually are removed. The safety, faster operation, and product quality are the main benefits of using RBMs in the shaft drilling/excavation. The excavated hole is more stable than a hole opened by drilling and blasting method and has better airflow, making it ideal for ventilation shafts. However, unfavorable ground conditions can cause the loss of drill strings (rod) and reaming tools, and deviation from the hole alignment; in addition, the initial/capital cost of RBMs is higher.

Like the other mechanical excavation machines, such as Tunnel Boring Machine (TBM), roadheaders etc., geological and geotechnical characteristics of the excavated formation are the important factors for the implementation of shaft excavation projects with RBMs. In addition, as stated by Shaterpour-Mamaghani and Copur (2017), these two characteristics are one of the factors that affects the design, selection, and performance estimation of the RBMs. They stated that an appropriate level of geological/geotechnical investigation had to be performed prior to the consideration of the use of RBMs. This investigation should cover the information about rock mass properties and physical-mechanical properties of the intact rocks as Rock Quality Designation (RQD), identification of main faults zones, broken ground, and layered ground, hydrogeology (ground water, water content, water ingress), strength (uniaxial compressive strength, Brazilian tensile strength, elasticity modulus, etc.), abrasivity of drilled/excavated ground, mineralogical and petrographical properties, and cuttability properties (Everell 1972, Bilgin et al. 2014). It should be kept in mind that additional rock testing such as indentation test should be carried out to predict the accurate advance rates of the RBMs. In the literature, there are many paper related to indentation tests and its application for performance estimation of the RBMs (Bilgin 1989, Dollinger et al. 1998, Copur et al. 2003, Shaterpour-Mamaghani et al. 2016).

This paper aims at investigating the effects of some of the physical-mechanical properties of the rocks excavated (uniaxial compressive strength, Brazilian tensile strength, and elasticity modulus) on the performance of a RBM (daily advance rate, consumed gross reamerhead torque and power, net pulling force, instantaneous penetration rate, unit penetration, and field specific energy) during the reaming operation of two shafts in the Balya lead-zinc underground mine.

2 OVERVIEW OF THE MINE AND SHAFT PROJECTS

Balya lead-zinc underground mine is located in the 50 km northwest of Balikesir city in western Turkey. Eczacibasi Esan industrial group has been operating the Balya mine since 2007. The sublevel stoping method is used for extraction and total depth of mine still keeps increasing. An incline ramp with a slope of 14.5% is used as the mine main gallery to minimize the transportation distance from the ore production area to the main gallery.

Since 2013, the shaft excavation has been done with RBM in the Balya lead-zinc underground min. In 2013, the first shaft with 197.10 m length was excavated with a RBM in 38 days. The second shaft with 198.44 m length was excavated in 37 days in the same year. After these successful operations, the management decided to continue using RBM for excavation of another ventilation shafts. In the next year, the third shaft with 198.88 m length was excavated in 35 days. In 2015, the excavation of the forth shaft (195.80 m) was completed in the 84 days. Finally, in 2017, the excavation of the fifth shaft (335.32 m) was successfully completed in 101 days. The pilot and final diameters of the all shafts were 0.31 and 2.44 m, respectively. In this study, the excavation of the fourth and fifth shafts are investigated. The fourth shaft was excavated between the levels of 1005 and 805 m; and the fifth shaft was excavated between the levels of 805 and 505 m. The general information about these two shafts are summarized in Table 1.

Table 1. General information about the raise bored shafts.

Project Name	Year	Length (m)	Diameter (m)		Inclination (°)	Construction time (working day)	
			Pilot Drilling	Reaming		Pilot Drilling	Reaming
RBM-4	2015	195.80	0.31	2.44	90.0	31	53
RBM-5.2	2017	331.36	0.31	2.44	67.9	38	63

3 THE GEOLOGY OF THE MINE AND PHYSICAL-MECHANICAL PROPERTIES OF THE ROCKS BORED

Permian limestone, Triassic shale and sandstone, and unconformably overlying Tertiary dacite and andesitic volcanics are the main rock formation in Balya deposit (Ciftci et al. 2013). Paleozoic and Triassic rocks are characterized by recrystallized limestone, metasandstone, meta clayey sandstone and chloritic clayey siltstone. Two different fault zones exist in this area. The first fault is related to mineralizations and the second is essentially devoid of mineralization. Due to skarn types of mineralization, the ore deposit goes to deeper levels in uniform shape and this allows for high amount of concentrate production. Fig. 1 shows geological map of the Balya mine and surroundings.

A series of laboratory tests were carried out to characterize the intact rock samples obtained during site investigation. The rock samples were selected from one of the closest exploration boreholes to the shafts mentioned in this study. Four different rocks including limestone, andesite, dacite, and metasedimentary were encountered during the excavation of the fourth shaft. However, the eight different rocks and geological zones including dacite, hornfels, fault zone, pyrite ore, limestone-1, metasedimentary, limestone-2, and skarn were encountered during the excavation of the fifth shaft.

Uniaxial compressive strength, Brazilian tensile strength, and static elasticity modulus tests were performed according to the suggested methods by International Society of Rock Mechanics (ISRM 2007). Uniaxial compressive strength tests were performed on the grinded core samples with a length to diameter ratio of around 2.5 and the loading rate was 0.5 kN/s. Brazilian tensile strength tests were performed on core samples at 0.25 kN/s loading rate and the ratio of length to diameter of around 0.5. The laboratory tests were repeated at least 3 times for each rock sample. The results of physical-mechanical property tests along with RQD, which is a rock mass characteristic, are summarized in Table 2. It should be noted that it was not possible to obtain limestone samples from the closet borehole to the shaft 4.

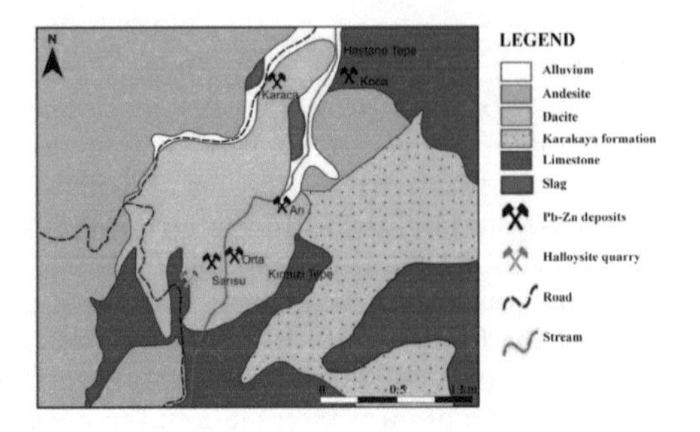

Figure 1. Geological map of the Balya Mine and surroundings (Ciftci et al. 2013).

Table 2. Summary of physical-mechanical properties of the rock samples along with RQD.

	Rocks	Geological Origin	RQD (%)	ρ (g/cm^3)	σ_c (MPa)	σ_t (MPa)	E_{sta} (GPa)	v_{sta} (-)
Shaft 4	Andesite	Igneous	-	2.65	116.63	8.44	30.40	0.23
	Dacite	Igneous	-	2.47	94.64	8.50	11.90	0.11
	Metasedimentary	Sedimentary	-	2.57	79.24	7.00	21.58	0.10
Shaft 5	Limestone	Sedimentary	-	-	-	-	-	-
	Dacite	Igneous	93	2.58	74.34	8.00	32.12	0.23
	Hornfels	Metamorphic	52	2.65	85.74	2.40	16.35	0.14
	Fault Zone	-	51	2.62	211.40	13.48	34.10	0.18
	Pyrite Ore	-	92	3.44	126.69	5.41	23.62	0.27
	Limestone-1	Sedimentary	53	2.67	74.52	6.91	21.42	0.15
Table 2. Continued.								
Shaft 5	Metasedimentary	Sedimentary	91	2.61	133.17	10.57	19.37	0.13
	Limestone-2	Sedimentary	89	2.73	115.76	5.41	28.27	0.19
	Skarn	Metamorphic	49	3.14	53.32	-	-	-

RQD: rock quality designation, ρ: density, σ_c: uniaxial compressive strength, σ_t: Brazilian tensile strength, E_{sta}: static elasticity modulus, v_{sta}: static Poisson's ratio.

4 RBM USED IN THE PROJECT AND GENERAL PERFORMANCE

Sandvik Rhino 1088 DC raise boring machine was used to excavate the shafts. This machine is a hydraulic driven rig with maximum 4000 kN operating thrust. The machine is capable of operating at a torque of 300 kNm in pilot drilling and 160 kNm in reaming operation. In addition, it is capable of operating at a rotational speed of 60 and 21 rpm in pilot drilling and reaming operations, respectively. Figure 2 shows the pilot bit, reamerhead, and RBM used in excavation of the shafts. Pilot hole drilling was based on rotational methods using a tricone roller bit of 311 mm diameter. The reamerhead is equipped with 14 button cutters, of which seven cutters consist of five rows of tungsten carbide inserts and other seven consist of four rows inserts.

The fourth shaft pilot drilling started on 11 February 2015 and completed on 30 March. The reaming operation started on 03 April and the 195.80 m length shaft enlargement finished on 1 June 2015. In the fifth shaft excavation, the RBM commenced drilling pilot hole on 4 February 2017 and completed on 23 March. Due to some difficulties in the attachment stage of reaming head, the 3.96 m of the bottom of hole was excavated with drill and blast method. After this preparation, the reaming operation started on 7 April and the 331.36 m length shaft enlargement finished on 28 August 2017. However, on 16 June, it was detected that some of the reamerhead cutters had to be replaced due to observing some metal tools in the excavated material. The 37 drill strings were added to the RBM and the reamerhead was lowered down to the level +722. Then, a new gallery (24 m length) was excavated to reach the shaft level and from the bottom of the shaft in where the reamerhead was dismounted. The four cutters of reamerhead were replaced with the new cutters and the reaming operation was continued on 18 August.

Figure 2. Photographs of (a) Raise boring machine, (b) tricone bit and (c) reamerhead.

Table 3. General performances of the RBM during pilot drilling and reaming operations.

Daily advance rate (m)	Shaft 4		Shaft 5	
	Pilot Drilling	Reaming	Pilot Drilling	Reaming
Minimum	1.52	1.24	0.92	1.52
Mean	6.32	3.69	8.86	5.26
Maximum	12.16	9.12	19.76	9.12

General performances of the RBM during pilot drilling and reaming operations of the fourth and fifth shafts are given in Table 3. The values cover the minimum, mean, and maximum daily advance rates included with stoppages. The working pattern was 12 h/shift, 1 shifts/day, and 6 days/week. One operator, one chief, and two workers were used for excavation of these two shafts..

5 EFFECTS OF PHYSICAL-MECHNICAL PROPERTIES AND FAULT ZONE ON THE PERFORMANCE OF THE RBM

RBM performance have been obtained during excavation of the shafts that were excavated in igneous, sedimentary, and metamorphic rocks. The recorded (operational) parameters from the machine's data acquisition system included excavation time, excavation length, rotational speed, consumed gross reamerhead torque, and net pulling force. The calculated performance parameters including instantaneous penetration rate, unit penetration, consumed gross reamerhead power, and field specific energy. The mean measured values of RBM operational parameters and calculated performance parameters in the reaming operation of these two shafts are summarized in Table 4.

Excavation of the fourth shaft (only reaming operation) was completed in 53 working days with average daily advance of 3.69 m/day. Excavation of the fifth shaft was completed in 63 days with an average daily advance rate of 5.26 m/day. The net pulling force was 713 kN in the fifth shaft and 940 kN in the fourth shaft. The comparison of these values indicated that at almost the same mean rotational speed (3.8 rev/min) excavation of the fourth shaft was more difficult than of the fifth shaft. The most RBM performance parameters during reaming showed superiority in the fifth shaft.

As stated in Section 2, the length of the fifth shaft was about twice of the fourth shaft. In addition, the inclination of the fifth shaft was 67.9° (difficulties might be expected in excavation), while it was 90° for the fourth shaft (easier to excavate compared to inclined shaft). In addition, four types of lithological units were excavated in the fourth shaft, while eight types of lithological units were excavated in the fifth shaft (more rock types with higher strength range including a very weak fault zone of 36.5 m).

Table 4. Measured/calculated operational and performance parameters in reaming operation.

Parameters	Shaft 4	Shaft 5
Excavation time (min.)	246	244
Rotational speed (rev/min)	3.8	3.8
Consumed reamerhead torque (kNm)	45.0	48.5
Consumed reamerhead power (kW)	18.17	19.58
Field specific energy (kWh/m^3)	9.27	9.62
Instantaneous penetration rate (m/h)	0.60	0.76
Unit penetration (mm/rev)	2.66	3.58
Net pulling force (kN)	940	713

The RBM excavated the fourth shaft in limestone, dacite, andesite, and metasedimentary rocks. During the excavation of limestone, the instantaneous penetration rate was 0.89 m/h, the net pulling force was 880 kN, and the field specific energy was 5.62 kWh/m^3. However, the mentioned values were 0.37 m/h, 909 kN, and 11.83 kWh/m^3 respectively in the excavation of andesite rocks. The detailed investigation on the reaming operation of the fourth shaft indicated that the RBM was faced with unusual condition between 131 and 128 m depths. The 87th drill string was back reamed after 880 minutes. In this drill string, the rotational speed was decreased to 1.5 rpm; and at the same time, the pulling force was decreased from 140 kN to 40 kN. In addition, the 86th drill string was back reamed after 1815 minutes. In this drill string, the rotational speed was decreased to 1.3 rpm; and at the same time, the pulling force was decreased to 36 kN. This unusual condition might be related to the transition zone that RBM excavated more than one lithological units during reaming operation. In such a case, when RBM commenced excavation of high strength rocks after the excavation of low or medium strength rocks, the operational-performance values suddenly decreased. When performance evaluation in terms of uniaxial compressive strength, the andesite rocks had the maximum value with 116.63 MPa, and the metasedimentary rocks had the minimum value with 79.24 MPa. In andesite rocks, instantaneous penetration rate was 0.37 m/h; however, this rate was 0.66 m/h in sedimentary rocks. Apparently, when the strength of rocks decreased the instantaneous penetration rate increased. In addition, when performance evaluation in terms of static elasticity modulus, the andesite rocks had the maximum value with 30.40 GPa, and the dacite rocks had the minimum value with 11.90 GPa. In andesite rocks, instantaneous penetration rate was 0.37 m/h; however, this rate was 0.48 m/h in dacite rocks. Seemingly, when the s static elasticity modulus decreased the instantaneous penetration rate increased (Figures 3 and 5). Moreover, the Brazilian tensile strength values are close to each other; therefore, it is not logical to evaluate the performance of RBM in terms of this property. Lastly, it should be kept in mind that additional data about the limestone samples are needed to evaluate the overall performance of RBM during the fourth shaft excavation.

The same RBM excavated the fifth shaft in dacite, hornfels, fault zone, pyrite ore, limestone-1, metasedimentary, limestone-2, and skarn zones. The field investigations indicated that the fault zone had different characteristics among the other geological zones. The rock samples of this zone were weak/highly fractured/disturbed. Roughly, 36.50 m of the shaft excavation was completed through the fault zone. Although the compressive strength of the samples from the fault zone had the highest value (211 MPa), the field studies indicated that the instantaneous penetration rate and unit penetration rate in this zone were comparatively the highest being 1.54 m/h, and 9.11 mm/rev, respectively. On the other hand, the rotational speed and net pulling force were the lowest being 3.1 rev/min and 326 kN, respectively. The upper level of the fault zone included the dacite rocks and the lower level included

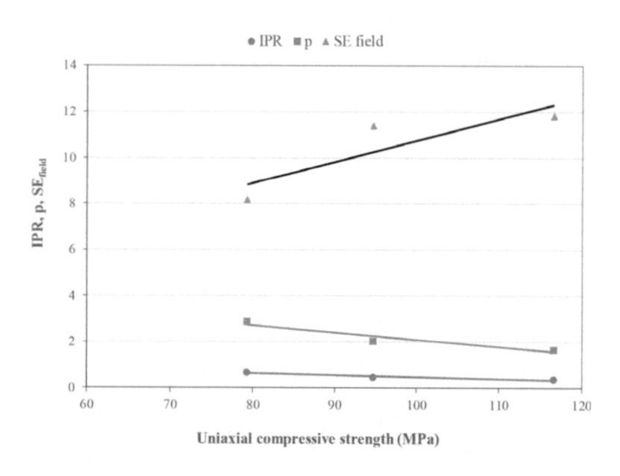

Figure 3. Relationships between uniaxial compressive strength and instantaneous penetration rate, unit penetration, and field specific energy in shaft 4.

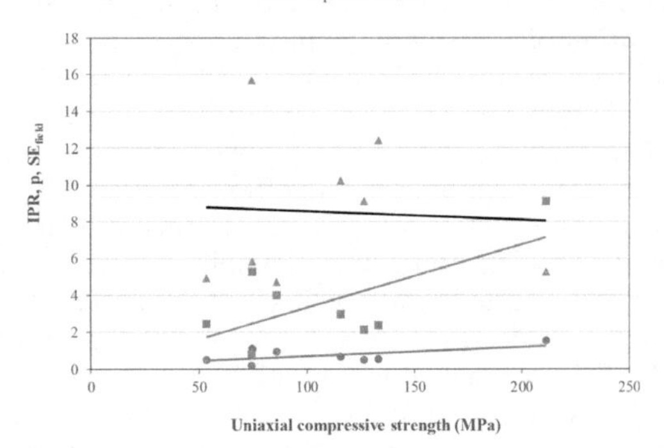

Figure 4. Relationships between uniaxial compressive strength and instantaneous penetration rate, unit penetration, and field specific energy in shaft 5.

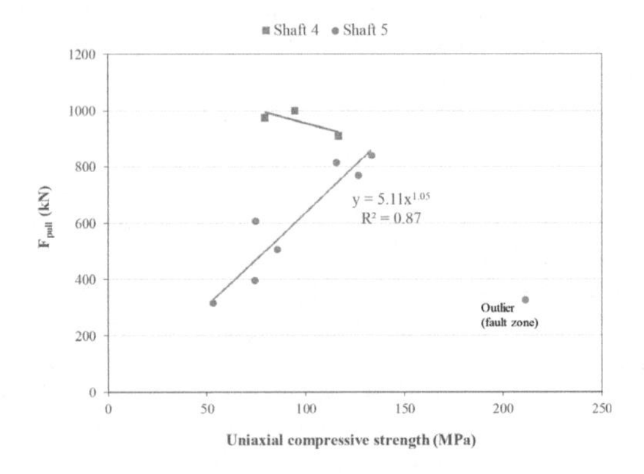

Figure 5. Relationship between uniaxial compressive strength and net pulling force in shaft 4 and 5.

metasedimentary rocks. These two rock types show different characteristics; the dacite samples were moderately hard and metasedimentary samples were hard (Figures 4 and 5).

The detailed investigation on the reaming operation of the fifth shaft showed that limestone-1 zone had the maximum instantaneous penetration rate as 1.10 m/h. The hornfels zone had the 0.95 m/h instantaneous penetration rate. However, as considering the net pulling force values, the skarn zone had the minimum value as 317 kN; and the dacite zone had the second minimum value as 395 kN. In addition, as considering the field specific energy values, the hornfels zone had the minimum value as 4.73 kWh/m^3; and the skarn zone had the second minimum value as 4.90 kWh/m^3 (in these comparisons, the fault zone values were not considered).

6 CONCLUSIONS

The effects of physical-mechanical properties of the rocks on the performance of a RBM during the reaming operation of two shafts in the Balya lead-zinc underground mine is investigated in this study. Some of the results of the field studies are summarized as follow:

- In the fourth shaft, the instantaneous penetration rate was 0.60 m/h. However, this parameter in the fifth shaft was 0.76 m/h.
- The net pulling force values were recorded as 940 and 713 kN in the fourth and fifth shafts, respectively.

- In the fourth shaft, the consumed gross reamerhead torque was 45.0 kNm. This parameter in the fifth shaft was 48.5 kNm.
- The field specific energy values were calculated as 9.27 and 9.62 kWh/m^3 in the fourth and fifth shafts, respectively.

This investigation showed that various machine performance parameters such as instantaneous penetration rate, net pulling force, as well as field specific energy are directly influenced by physical and mechanical properties of the rocks (uniaxial compressive strength and static elasticity modulus) and geological parameters (such as fault zone). During the reaming operation of the fourth shaft, RBM was faced with unusual condition between 131 and 128 m depths. The rotational speed was decreased to 1.5 rpm; and at same time, the pulling force was decreased from 140 kN to 40 kN. This case might be related to the transition zone that RBM excavated more than one lithological units during reaming operation. In addition, during the fifth shaft reaming operation, the fault zone showed the maximum instantaneous penetration rate value as 1.54 m/h. Additional work is required to improve the database of this study in both field and laboratory. As a conclusive remark, the other important parameters such as mechanical (machine related) and operational parameters (such as shaft length and inclination) should be considered with geological and physical-mechanical parameters in the investigation of RBM performance.

ACKNOWLEDGEMENT

This study summarizes some of the results of Ph.D. research work carried out by Aydin Shaterpour-Mamaghani. The authors are grateful to the support of Eczacibasi Esan lead-zinc underground mine and Sargin Construction and Machinery Industry Trade Inc.; this work would be impossible without their support.

REFERENCES

Bilgin, N. 1989. *Applied Rock Cutting Mechanics for Civil and Mining Engineers*, 1st ed. Istanbul: Birsen (In Turkish).

Bilgin, N. Copur, H. & Balci, C. 2014. *Mechanical Excavation in Mining and Civil Industries*. New York: CRC Press.

Ciftci, Y. Revan, M.K. Sen, P. & Zimitoglu, O. 2013. The basic geological aspects of Eskisehir, Balikesir, Tepeoba, Izmir, and Usak ore deposits. International Workshop on Base and Precious Metals, *Field Trip Guide Book*, May 20-27, MTA, Ankara, Turkey. pp. 22-27.

Copur, H. Bilgin, N. Tuncdemir, H. & Balci, C. 2003. A set of indices based on indentation tests for assessment of rock cutting performance and rock properties. *Journal of the South African Institute of Mining and Metallurgy*, 103: 589-599.

Dollinger, G.L. Handewith, H.J. & Breeds, C.D. 1998. Use of the Punch Test for Estimating TBM Performance. *Tunnelling and Underground Space Technology*, 13 (4): 403-408.

Everell, M.D. 1972. Performance of raise borers as a function of geology and rock properties, *In: Proceeding of the 8th Canadian rock mechanics symposium*, Toronto, November-December, pp. 83-100.

ISRM 2007. The complete ISRM suggested methods for rock characterization, testing and monitoring: 1974-2006. *In*: Ulusay R, Hudson JA *(eds.) Suggested methods prepared by the ISRM commission on testing methods, compilation arranged by the ISRM Turkish National Group*. Kozan Ofset, Ankara, p 628.

Shaterpour-Mamaghani, A. & Copur, H. 2017. Factors Affecting the Selection and Performance of Raise Boring Machines (RBMs) and Case Studies from Turkey. *In: 26th International Symposium on Mine Planning & Equipment Selection Conference*, Luleå, Sweden, 153-161 (ISBN: 978-91-7583-935-6).

Shaterpour-Mamaghani, A. Bilgin, N. Balci, C. Avunduk, E. & Polat, C. 2016. Predicting Performance of Raise Boring Machines Using Empirical Models. *Rock Mechanics and Rock Engineering*, 49 (8): 3377-3385.

Visser, D. 2009. Shaft Sinking Methods Based on The Townlands Ore Replacement Project-Rise Boring. *In: The Southern African Institute of Mining and Metallurgy Shaft Sinking and Mining Contractors Conference*, 13p.

Tunnels and Underground Cities: Engineering and Innovation meet Archaeology, Architecture and Art, Volume 3: Geological and geotechnical knowledge and requirements for project implementation – Peila, Viggiani & Celestino (Eds)

Practical solutions to geotechnical problems related to Ituango hydropower tunnels, Colombia

M.C. Sierra, J. Ramos, D. Jurado & J.D. Herrera
INTEGRAL S.A Consultant Engineers, Medellín, Colombia

ABSTRACT: Ituango Project is the largest hydroelectric development in Colombia, with an installed generation capacity of 2400 MW. The rock mass of interest is located on a lithology corresponding to schistose gneisses. In the main works around thirty kilometers of underground excavation have been constructed. Regarding the infrastructure of the project, twelve road tunnels have been excavated, totalizing around four kilometers of underground excavations. The behavior of the rock mass under construction was predicted by empirical and analytical analysis based on the 8000 m of drills and 2000 m of adits. Despite following the rigorous procedures set by the designers, some problems of falling wedges and convergences occurred, related to the heterogeneity of the rock mass and the normal problems of construction. The purpose of this paper is to report the main problems under construction and describe the causes to the biggest problems occurred during the underground excavation.

1 INTRODUCTION

The stability of deep underground excavations is a common issue in a variety of rock engineering fields, including hydroelectric power projects. The powerhouse rock mass behavior is strongly affected by complex interaction mechanisms between multiple adjacent caverns and the final support of rock pillar between caverns.

The excavation of underground tunnels can alter the natural stress field of the surrounding rock depending of its geometry not only causes decompression of the floor and increases sidewall stress, but also leads to stress concentration at the corners of the excavation, which influences the stability of the rock in a very complex manner that can trigger some fall-outs close to the crown corners.

From a rock mechanics perspective, this issue implies that excavation and support of underground works (tunnels, powerhouse complex) become a challenging task. Among other key factors features as geometry, geological conditions, portal access.

There are many steps involved in taking a hydroelectric project from concept to construction. The purpose of this paper is to clearly present the geotechnical cases during its excavation.

The predicted behavior in the design were based on the results of the intense field exploration and laboratory and these was like the real measures on the construction, despite that there were issues to falling wedges and deformations. The most important causes were the variation of the mass rock across the site and the impossibility to do better predictions, in the same way about the deformation base on the measurements reported by the instrumentation. Some of the challenges of this project was to predict the effect of the Tocayo and Mellizos faults which cross the caverns and road tunnel, and the high horizontal stress relation (k=2) and its variation along the cave.

The behavior of the rock mass under construction was the order of magnitude expected in caverns although the variation along the space was heterogeneous. The irregular behavior of the rock mass was governed by the parameters of the rock mass and the issues by construction effects, and expected wedges.

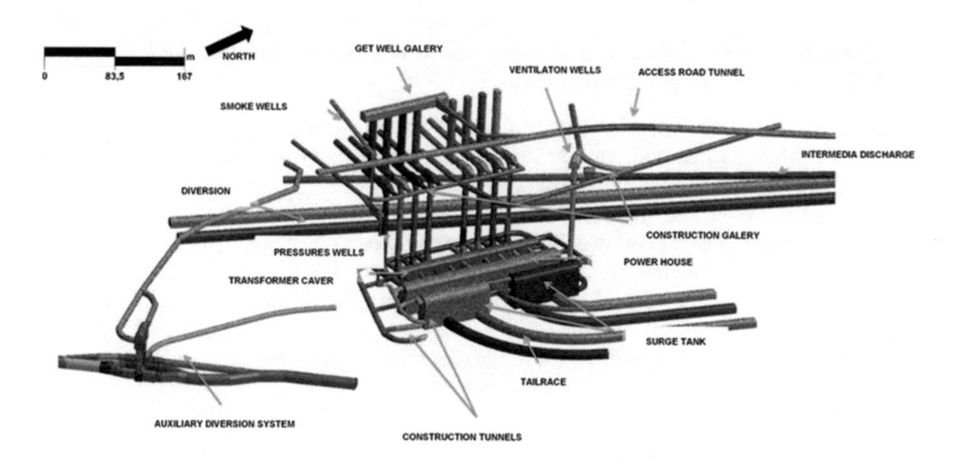

Figure 1. Layout Ituango Hydropower Project.

2 ITUANGO HYDROELECTRIC PROJECT LAYOUT

The project is in the northwestern region of Antioquia. The project is a part of the initiative undertaken to exploit the hydroelectric potential of the Cauca River in its middle stretch. The project was designed by INTEGRAL S.A and it is being developed by Colombian multi-utility group EPM (Empresas Públicas de Medellin).

The Ituango preliminary construction works comprehends river diversion and access road construction; the construction of this works started since September 2011. Two parallel diversion tunnels, both one kilometer long and 14 m in diameter were excavated to divert water from the Cauca River. Those tunnels did not have floodgates because the schedule and it was built another diversion tunnel call auxiliary gallery to diversion.

The underground works comprise:

The cave complex consists of the Transformer cavern (21, m high, 16 m wide and 218 m long), the Powerhouse cavern (49 m high, 23 m wide and 240 m long) and two Surge chambers (48 m high, 18 m wide and 100 m long one for every four generating units). The tunnels drift, discharges and auxiliary drift are 14 m high and 14 wide and 1000 m long approximately. There are eight surges tunnels and eight conduction wells. The access tunnel has 1,3 km and many kilometers of constructions tunnels.

3 GENERAL GEOLOGY

The works are being constructed entirely in quartz feldspar paragneisses, locally with schistose texture, covered occasionally by colluviums on the slopes and a torrential-alluvial deposit on the base of the right slope in front of the Ituango river mouth.

The gneisses, mainly because their sedimentary protolith has a high textural variability over very short distances. They range from coarse granular rocks where are thickly interspersed with zones of quartz and feldspar with dark layers of biotite and amphibole, up to fine-grained textures with thin interlaminations of similar composition. Set of quartz-feldspar (Pznf) and alumina (Pznl) gneisses, with a folded variable structure between schistose, gneissic and migmatitic. They have a remarkable mineralogical and textural differentiation due to the variability of metamorphism and the heterogeneity of the sedimentary origin.

The main structural feature of the rock are the planes of weakness created by the foliation, of general trend N10°-30°E/15°-30°SE, although occasional and specific processes such as folding varies the dip. This structure has allowed the releases of efforts resulting in shear zones parallel to them, which are located so random with very different persistence throughout the excavations and are characterized by a thickness that varies from a few centimeters to 1,5 m thick and sudden disappears. The principal structures are:

- The main structural system of discontinuities is the metamorphic foliation. This system is the texture, when committed fine-grained material is smooth, while on coarse-granular areas is rough, both from planar to wavy.
- A system with subvertical dips and NS direction, always converging to the Cauca River. This system defines the canyon walls and the river course control.
- A system of fewer occurrences has EW trend and vertical to subvertical dips and serves to control the course of some local streams like Tenche, Burundáa and other smaller creeks.
- A system NS and dip 60°W control the falling blocks on the right bank of the river.

4 GEOTECHNICAL PROPERTIES

For stress strain analysis the rock mass is considered which the following parameters defined from the tests carried out during the exploration and research phase of the Project:

Table 1. Properties rock.

Property	Value
– Unit weight of the rock γ (kN/m^3):	26–27
– Resistance to unconfined compression σci (MPa):	90 ± 16
– Modulus of elasticity of the intact rock Ei (GPa):	15 – 20
– Poisson module μ (adim):	0,3
– GSI quality index (adim):	55–70
– Petrographic parameter mi (adim):	28
– Dilatancy Ψ (°)	14–30

An affectation value of 0,3–0,7 is considered only in the first two meters of the mass adjacent to the excavation, for the rest it is taken as 0.

At the bottom of the slope the quality of the rockmass reaches high values more rapidly than the upside. Thus, at the bottom of the canyon GSI values are reached above 60 at depths of up to 20 m and on the top GSI over 60 appears at most at 60 m depth. Weathering horizons found are II (Horizon IIB Patton Deer) with high recoveries (95% <Rec <90%), joints with oxide, average resistance of unconfined compression of 93 MPa with a standard deviation of 16 MPa. And the horizon III (IIA Deer and Patton) with recoveries between 10 and 90 %. The resistance for the horizon IIA is taken as the mean minus standard deviation (68–15 = 53) taking a value of 50 MPa, given the high variability in the condition of the rock in this stratum.

5 GEOTECHNICAL CONDITION ON EXCAVATIONS

The net effect of an excavation is the weakening of rock is a movement toward the tunnel and forms wedges of rock with the structures. [Terzaghi, 1946]

An exhaustive study was carried out for the underground works in the project based on a detailed geological-geotechnical study. It covers research on initial stresses in situ, mechanical properties of rock, tunneling conditions, groundwater conditions, eventual erroneous practices and others. All the conditions mentioned above were applied to the problems identified during the construction.

It is necessary to implement is divided into four types according to the quality of the rock mass and the dimensions of the excavation. These types have the following characteristics:

Type I: the unconfined compression strength of the rock is medium or higher (greater than 50 MPa), the rock mass is massive or slightly fractured mass, which can advance without the need of placing support at the front of the excavation.

Type II: the unconfined compression strength of the rock is medium or higher (greater than 50 MPa), the mass rock is broken in the roof and walls slightly fractured, you can move without putting support in the face of the excavation. This required systematic treatment with rock bolts and shotcrete reinforced roof and reinforced shotcrete walls and eventually some bolts.

Type III: The unconfined compression strength of the rock is medium or higher (greater than 25 MPa), and the solid is fractured and altered the roof and walls can be put forward without support in the face of the excavation. In this type of soil infiltration stability problems can occur due to the degree of fracturing of the rock. In this type of rock is required systematic treatment with rock bolts and shotcrete reinforced in the ceiling and walls.

Type IV: The material is composed of highly fractured soil or rock. The unconfined compressive strength of the rock is low. The mass rock is highly fractured and eventually soil is found into fractures, ground sections with lengths greater than 5 m in soil or sites located in areas where the soil or rock roof is thin and is loaded with soil, the tunnel progress may require placing support at front of the excavation. In this type of soil infiltration can lead problems of stability. The treatment is systematic excavation with steel, rock bolts and shotcrete in the ceiling and walls. The approximate percentage of each type of terrain are:

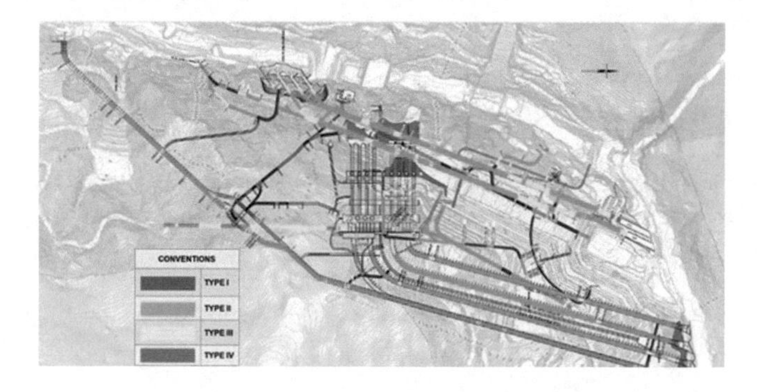

Figure 2. Type of support; Type I: 22%; Type II: 62%; Type III: 15%; Type IV:2%.

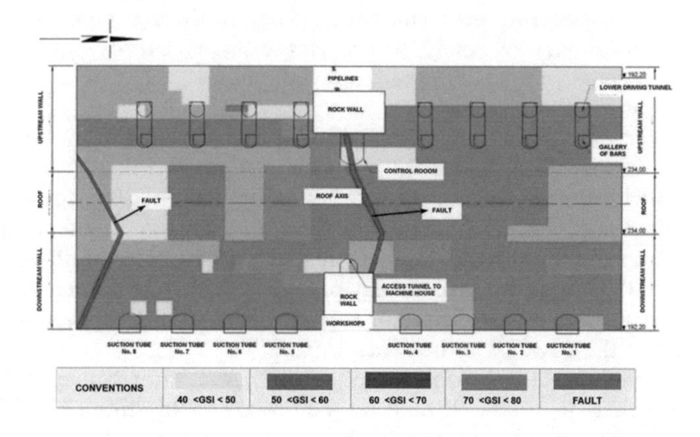

Figure 3. Quality of rock in power house.

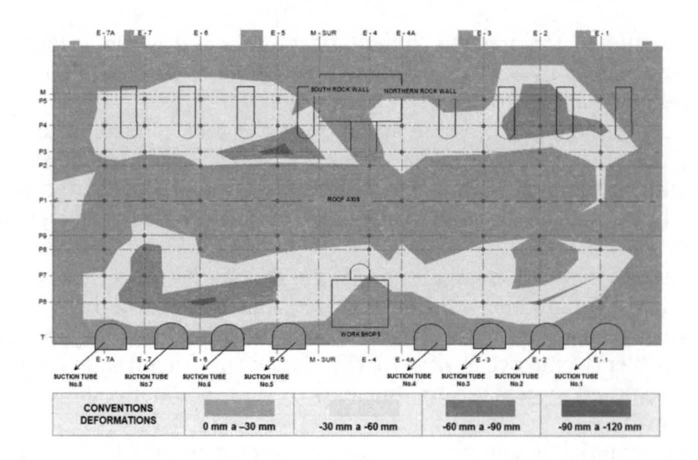

Figure 4. Deformation measure.

The main strain in E-W direction is about 2 in the north and half in the south, the minor is about 1 in north and half in the south. k the strain relation is about 2.

K south=[1,000,551,00] main direction 90° with north

K north=[1,901,301,00] main direction 105° with north

In general, the magnitude of the measured deformations is like those estimated by numerical methods. The distribution of the deformations is not homogeneous, mainly due to wedges weakened by the effects of blasting and to over-excavations.

6 FAILURES OF ROCK MASS ON CONSTRUCTION

Despite the knowledge, analysis and monitoring of the excavations there were problems in the construction, these were, two fallen wedges in the caverns and a high convergence in the access tunnel.

In addition, in road tunnels there were failures due to low coverage and low resistance terrain. Below are some of these cases with the causes of the failures so that they serve as lessons learnt for other projects.

6.1 *Powerhouse Cavern*

The only severe rock-engineering problem, which occurred during the construction, was the rock fall presented on January 22th of 2015 between the chainage 0+017 and 0+031. However, the problem was restricted to a 10–20 m section of the tunnel, the general view of the north-end crown fall is shown in figure 5.

As part of the instrumentation there was an extensometer located in the roof in the fault, it was installed in September 2013. Its record is shown in figure 6. Initially, a high deformation speed was detected with the advance of the excavation until about 20 m forward and then the deformation rate decreased to 0,02 mm/day until July 2014; for that moment, the accumulated deformation was 17,5 mm; as of this moment it had a slight increase, but it was still low. Increased the rate to 0,03 mm/day until the day of failure. In January a deformation of 23 mm was carried. In conclusion, the deformation rate was low, as the deformation. It was considered normal due to the advance in the excavations.

The excavation generated a redistribution of forces k = 2, an irregular surface with over-excavation in the corner, the mass affected by the process of blasting (initial blasting of the cavern), all of these allowed the wedge to be deconfined and slip over the foliation plane. In this way the wedge did not fail by gravity but by cutting; the bolts worked as shear dowels and not to tension as they were designed. This was verified with the analyzes performed on the steel of the bolts which failed by shear.

Figure 5. Fall-out occurred on north-end powerhouse complex. Wedge of 96 m^3 in 132 m^2.

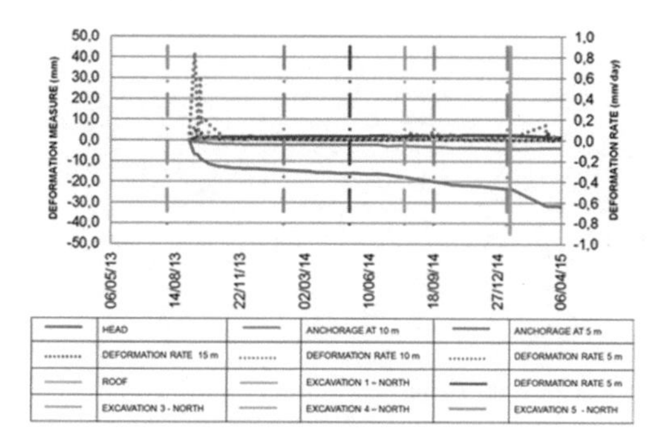

Figure 6. Deformation and deformation rate extensometer roof.

The original support consisted of shotcrete thickness 0,1 m combined with reinforced welded-mesh with nominal area 3,35 cm^2/m, rockbolts type SRL with a length of 12,0 m@ 2,0 m. The reinforcement involved:

Area of fallen: Installation of rock bolts N 8. of 12 m in length, spaced at 1 m. Additionally, three (3) layers of grout concrete of 0,05 m of thickness each layer with double steel mesh D158, installed between the first and second layers and between the second and third layers, the latter being reinforced with steel fibers.

- Afferent area to the fallen: Rock bolts N.8 of length 12 m in length spaced 2 m, the final mesh of bolts installed every 1 m. Additionally, two grout concrete layer of 0,05 m thick each one; reinforced with steel mesh D158 for a final conformation with the existing support of four grout concrete layers of 0,05 m thick.

The remedial measures required filling the cavern again to do the work in the roof. All those took about six months.

6.2 *Transformer Cavern*

The fall in the crown occurred ahead of the advancing main excavation. Attributing factors considered are:

- Increased stress concentration ahead of the advancing face accentuated by delayed installation of the rock support,
- Slippage of the mechanical anchorage due an erroneous installation of steel rock bolts with resin (SRL),

Figure 7. Fall presented on Transformer chamber.

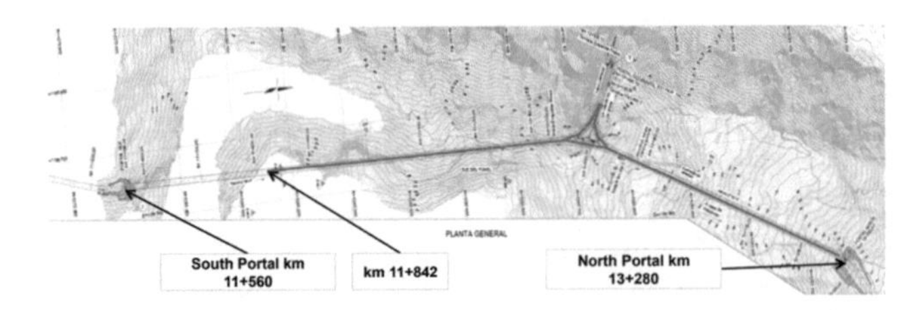

Figure 8. Plan View of the CH 12+000 Tunnel

The remedial measures used to stabilize the section was successful and consisted on the application of an additional layer of shotcrete (0,05 m) with a reinforced welded mesh and as a complement, the installation of rock bolts type SRL of 12,0 m length intercalated with the ones placed with the primary support with an average spacing of 1,0 m.

6.3 *Road way tunnel in chainage 12+000*

The Tunnel located is a road tunnel of access to the hydroelectric project with a length of 1,3 km and a maximum overburden of 200 m with a horseshoe geometry consisting in 10 m wide and 7,3 m high excavated in Schist gneiss with the drill and blast technique.

The quality of the rock mass found at the tunnel face was not as good as expected, compared to the geological conditions found during the excavation of the north side and considering the previous investigations and the experience from other tunnels of the main works excavated in the same type of rock. The rock mass was characterized by presenting intrafolial shears and the maximum width detected was 1,5 m on the left margin at the foundation of the dam. But, in this case the shear was presented with a width of approximately 10 m, something completely unpredictable according to the exploration and excavations carried out. This band started to appear from the lower part of the tunnel wall and its thickness increased with the advances, up to a point where it covered the entire tunnel section, and suddenly convergence started at the lower part of the walls, along a sector of the tunnel influenced by a shear zone concordant to the foliation planes. The convergence increased rapidly, reaching values up to 1 m at the most critical section in only three days. To control the deformation a concrete slab was built on the floor; this was successful and allowed to make the recovery of the tunnel.

To continue the excavation, it was necessary to used frames with sliding joints to dissipate stress through deformation until the gaps were completely closed. The load assumed by the support needed be much lower than the initially transferred by the rock mass to not damage the steel sets or the shotcrete.

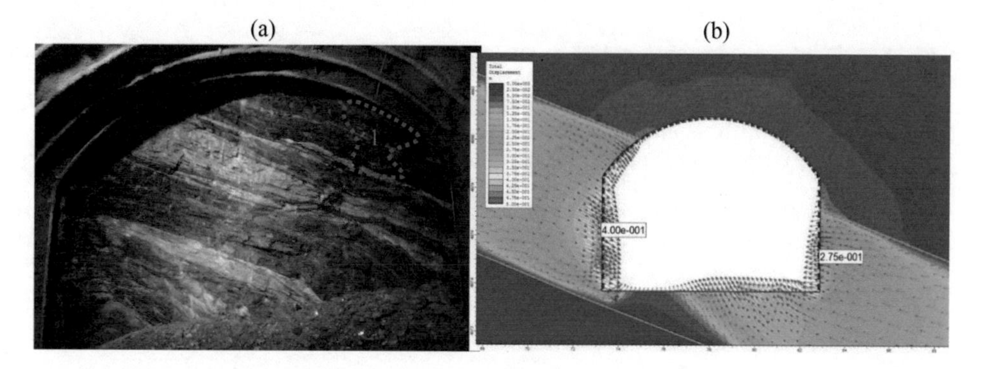

Figure 9. (a) Foliation structures with shear zone, (b) Deformation analysis stress relation k=1,5.

A definite support, consisting in steel frame type TH36 type with invert and a 0,15 m thick layer of shotcrete applied in the whole cross section, except for the sliding joint, waiting for them to have the expected displacement. This solution successfully controlled the displacements caused by the high horizontal stresses

7 CONCLUSIONS

At the Ituango hydroelectric project an extensive exploration and characterization campaign of the mass rock was carried out to support the design. The excavation showed good rock quality, about the 80% was type I and II (GSI >60). The behavior during construction and the deformations were in accordance with the provisions especially in the caverns.

Despite all this and the installed instrumentation, there were two wedges that failed in the caverns and other instabilities in the road tunnels.

There were two types of problems, one related to the change in the quality of the rock mass during construction and defects in excavate surface and others due to unforeseen changes in the geological conditions.

It is essential to note that all the geological features of importance for the construction of tunnels and caverns were identified and the measured deformations were of a magnitude like that of the design. Despite this, in the road tunnel an intrafoliar shear was found with an enormous width (10 m wide) completely different to the maximum detected of 1,5 m wide.

In the power house, the fallen wedge was created by several causes such as the mass rock affected with the blasting that was starting in the cavern, un over-excavation in the site where a wedge was defined, redistribution of adverse strain because the geometry of surface after excavation. All this caused the wedge slipped and the bolts to failed by cutting. The strength of bolt is high to tension and very low to cutting.

The transformer cavern also combined several causes such as a very fractured rock, bolts not well tightened, changes in the installation process and there was not tested their effectiveness before continuing.

The main conclusion is that in spite just there were only three failures in about 30 km of tunnel excavations these involved delays with high cost.

REFERENCES

Bickel, J.O., Kuesel, T.R. & King, E.H. 1996. Tunnel Engineering Handbook (Second Edition), Chapman & Hall: 544 pages.

Clayton, C.R.I., Matthews, M.C. & Simons, N.E. (1995). Site Investigation (2nd ed.), http://www.geotechnique.info.

IHA. 2017, International Hydropower Association, Hydropower Status Report, http://www.hydropower.org.

Jurado, D et al. 2017. Sudden convergence at the lower part of the walls of the Ituango Hydroelectric access road tunnel, 10 pages.
Liu, B.G. & Sun, J. 1998. Identification of rheological constitutive model of rock mass and its application. Journal of Northern Jiaotong, vol. 22, no. 4. pp. 10–14. [In Chinese].
Murcia et al. 2017. Deformation analysis for large underground caverns in a moderately jointed rock mass, 10 pages.
Terzaghi, K. 1946. Rock defects and loads on tunnel supports. In Rock tunneling with steel supports, (eds. R. V. Proctor and T. L. White) 1, 17–99. Youngstown, OH: Commercial Shearing and Stamping Company.
Yao, X.M. 2005. Studying the rock destroy mechanism induced by surrounding rock unloading. Northeastern University, Shenyang. [In Chinese].

Tunnels and Underground Cities: Engineering and Innovation meet Archaeology, Architecture and Art, Volume 3: Geological and geotechnical knowledge and requirements for project implementation – Peila, Viggiani & Celestino (Eds)
 ISBN 978-0-367-46583-4

Geotechnical investigations governing shaft excavations for metro projects – an approach for secant pile termination depths

A. Sindhwani, M.K. Chaudhary, A.K. Dey & A.H. Khan
L&T STEC JV, Mumbai, India

V.M.S.R. Murthy
Department of Mining Engineering IIT(ISM), Dhanbad, India

ABSTRACT: Urban mass transportation system in India requires huge investment and a massive upgradation. With the exponential rise in population and metro cities being center of attraction for working population of the far flung rural areas in the vicinity, a cost effective, scheduled and comfort mode of public transport was a need to meet the requirement of masses. Metro Rail Transportation System proves to be a boon to the society and cater the mobility of urban India. Underground metro projects aligned beneath the ground, require deep excavations in the order of 25–40 m in depth in urban populated areas which impose many challenges, especially, predicting the ground behavior.

This paper showcases a case study of an ongoing underground metro project construction covering the detailed site investigations proposed depending upon the outcome of preliminary investigations. It also presents how geological variations can occur within shaft excavation and further provides guidelines on fixing a geotechnical testing program for ground predictions and to decide upon secant pile termination depths in hard rock. Estimated ground conditions during investigations and actual conditions during piling were in close agreement thus allowing timely and cost-effective completion of secant piling job.

1 INTRODUCTION

Mumbai, the financial capital of India is one of the fast-growing metropolitan regions in India, experiencing a rapid growth of population and employment opportunity. In order to improve the overall traffic, transportation scenario and also to cater the future travel needs, Mumbai Metro Line-3 is one of the key project being implemented by Mumbai Metro Rail Corporation (MMRC).

Mumbai Metro Line-3 is a 33.5 Km long underground corridor running along Colaba-Bandra-Seepz. It is a fully underground corridor having 27 key stations. The execution of Line 3 is divided into 7 contract packages (UGC-01 to UGC-07) and both the extreme packages viz. UGC-01&UGC-07 are being awarded to L&T-STEC JV. UGC-01 project is located in south Mumbai with the scope of Design and Construction of underground stations at Cuffe Parade, Vidhan Bhavan, Churchgate and Hutatma Chowk and associated tunnels together with two tunnel sidings to Cuffe Parade.

Tunnel Boring Machines (TBM) is planned to be lowered from Launch Shaft located at north end of Cuffe Parade Station and this paper covers launch shaft geotechnical investigations, geological variations, secant pile terminations and guidelines for geotechnical testing program.

The Shaft (Length = 60 m, Breadth = 30 m and Depth=25 m) is a part of Cuffe Parade station (400 m long) and to be constructed by Cut and Cover method with Bottom-up approach in rocky strata. Retaining wall with Secant Piles shall be bored up to hard rock level which acts as a cut off wall and excavation can be carried out after installing strutting/anchor

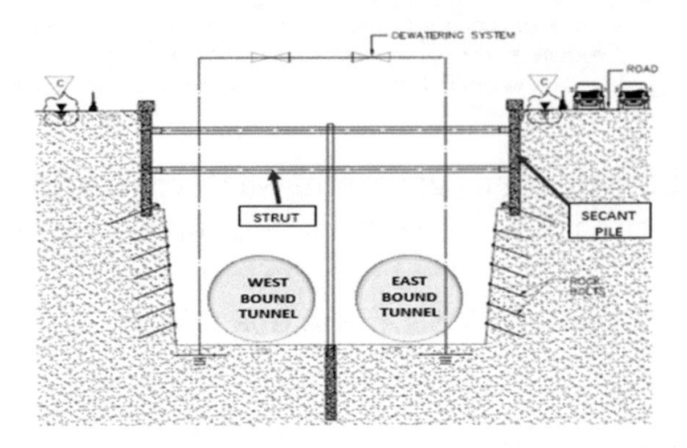

Figure 1. Typical Sectional Layout of Launch Shaft for Tunnel Boring machine (TBM).

systems to ensure sufficient resistance against the soil pressure outside the shaft excavation. Beyond hard rock level, an open cut excavation can be done as shown in (Figure 1).

Presence of ground water at shallow depths along with backfilled material underlain by hard rock poses many challenges in design and construction of deep excavation system however secant piles seems to be best fit in these conditions as the overlapping of primary pile (without reinforcement) and secondary pile (with reinforcement) acts as a water tight temporary structure.

2 GEOLOGICAL SETTING OF PROJECT AREA AND PLANNING OF GEOTECHNICAL INVESTIGATIONS

Geologically, Mumbai presents complex lithological combination showing large heterogeneity. Inter-trappean beds represents geological breaks in the tectonic volcanic activities. Basaltic flows and inter-trappean beds shows westerly dip of 5–10 degrees.

Mumbai region is located in the great volcanic formation building up the Deccan plateau. A wide variety of Basalts and associated rocks such as volcanic breccia, shale etc. occur in the area covered by Deccan trap basalts. Mostly Basalts are either compact i.e. with no gas cavities, or amygdaloidal with gas cavities filled with secondary minerals and vesicular Basalts with empty gas cavities, Zeolites are the commonest secondary minerals filling gas cavities, through silica, calcite and chlorophacite also occur as infillings.

Literature study coupled with surface and sub surface investigations were formed the backbone in understanding and developing the geological model of the area in regards to lithology, structure, stratigraphy and geotechniques as well as arriving at sustained design inputs. The ground parameters form the inputs for all the softwares' like RocLab, Wallap, Plaxis, Phase2 to analyses retaining wall design, tunnel design, expected ground settlements, water seepage analysis, excavation support systems (Struts, Anchors and excavation support system in rock).

Geotechnical investigations are planned in such a way that number and location of boreholes shall be optimized without compromising with the extent of information required to carry out detailed design analysis. Investigations to be done in two phases: Preliminary investigations to be done in phase 1 focusing on getting all the field and lab testing for carrying out the design and phase 2 includes detailed investigations based upon the results of preliminary investigations and to finalize the depth of rock socketed piles which acts as a retaining wall of the shaft.

3 INVESTIGATIONS IN PHASES

3.1 *Phase 1: Preliminary Investigations*

Two boreholes (BH-01 and BH-02) at 40–50 m spacing have been planned as a part of preliminary investigation drilled 6 m below the desired excavation depth as the objective is mainly to understand the ground (type of rock/soil) that needs to be excavated, strength parameters of ground to design the secant pile, retaining wall support (Strut, Waller and Anchor) and excavation support system. Grade of Rock mass during investigations is being classified based upon the IS 4464 code.

Boreholes BH-01 and BH-02 revealed that the ground condition is almost similar consisting of filled up material at the top underlain by completely to moderately weathered Basalt. A black Shale layer with Tuff bands (5–6 m thick) is encountered which is sandwiched between moderately weathered basalt and slightly weathered, strong to very strong, greyish colored Basalt (Figure 3-Left).

Since the piles to be socketed in G-III/better weathered grade of Basalt, there is a difficulty in deciding whether to terminate the pile in Basalt layer which is encountered above Shale layer or the Basalt layer which is encountered below Shale Layer.

From the results of Phase (1), it was understood that the rock layers are slightly dipping towards the head wall of shaft (since the recovered rock types are at slightly different levels in individual boreholes though the ground level is same) and also a shale layer is recovered which is sandwiched between G-III/better Basalt (Upper Layer) and G-II/better Basalt (Lower Layer).

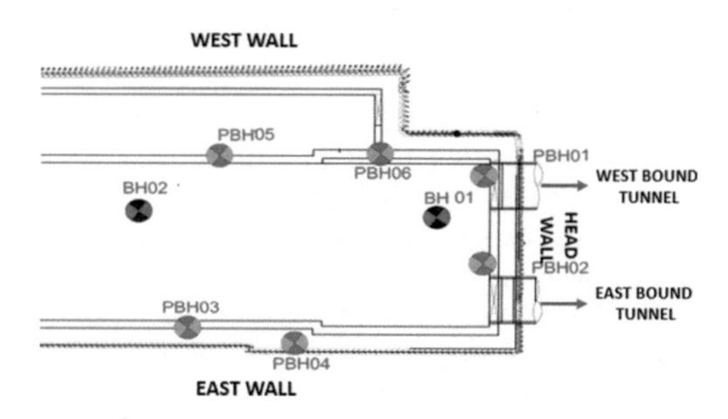

Figure 2. Borehole Layout of Launch Shaft (Head Wall, West & East Wall).

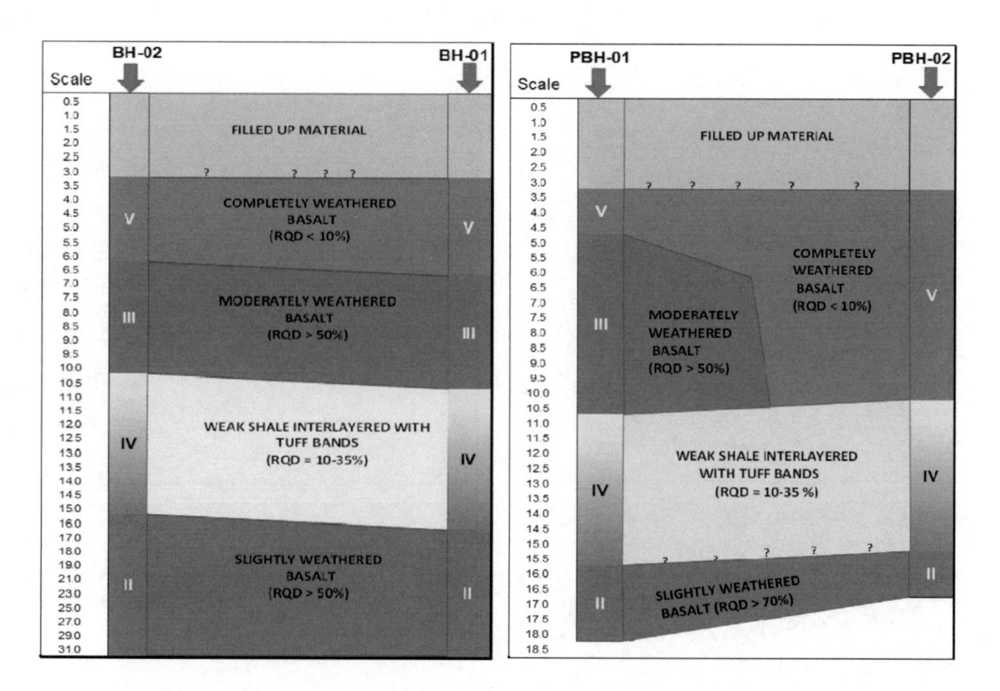

Figure 3. Geological profile of BH-01 &02 (Left) and PBH-01 & 02: Head Wall Transition Zone (Right).

Confirmatory boreholes as a part of Detailed Investigations is being considered as a critical path before start of piling works since there is an uncertainty of whether the piles should be socketed in lower layer of Basalt (below shale layer) or in Upper Layer of Basalt (above shale layer). If in case the piles have to be terminated in latter case, the extent of G-III Basalt should be consistent along all the side walls of shaft which is still an uncertainty.

3.2 *Phase 2: Detailed Investigations*

These are confirmatory Boreholes done on basis of the results from Phase-1 investigations. Shale layer with tuff bands recovered in BH-01 & 02 indicates that there is a geological break between two different volcanic flows during Deccan volcanism. Since the strata is likely to be dipping westwards as per regional geology, boreholes for phase-2 have been planned on both side walls as part of detailed investigations to understand the extent of ground variation i.e Upper G-III/better basalt extent to finalize the pile depths; rock profile longitudinally and laterally along the width of shaft to decide upon pile termination toe levels as piles to be socketed in G-III/better Basalt in addition to record total and solid Core recoveries, rock quality designation (RQD), uniaxial compressive strength and other physio-mechanical properties.

Six boreholes have been planned namely PBH-01, 02, 03, 04, 05, 06 of which PBH-01&02 are on Head Wall; PBH-03& 04 on East Wall and PBH-05 & 06 are on West Wall of the shaft. (Figure 2).

PBH-01, PBH-05, PBH-06 (Phase 2) are in line with BH-01 & BH-02 (Phase 1). Along entire West wall, 2–3 m thick layer of G-III/Better Basalt has been recovered above Shale layer with tuff bands (Figure 4) where it was decided to terminate the piles in Upper Layer of Basalt (0.5–0.8 m socketing) which is categorized as Zone – 1 shallower depth piles of approx. 6–7 m deep (Shown as red in Figure 6).

PBH-02, PBH-03, PBH-04 (Phase 2) are though similar wrt each other but totally different from PBH-01, 05, 06 boreholes. Along entire East wall, Upper Basalt layer was recovered as completely/highly weathered (G-V/IV) and then Shale layer underlain by G-III/Better Basalt layer (Figure 5). As per design requirement, piles cannot be socketed in completely/highly weathered Basalt. Thus, it was decided to terminate piles only in bottom layer of Basalt (0.5–0.8 m socketing) which is categorized as Zone-2 deeper piles of approx. 15–16.5 m deep (Shown as Blue in figure 6).

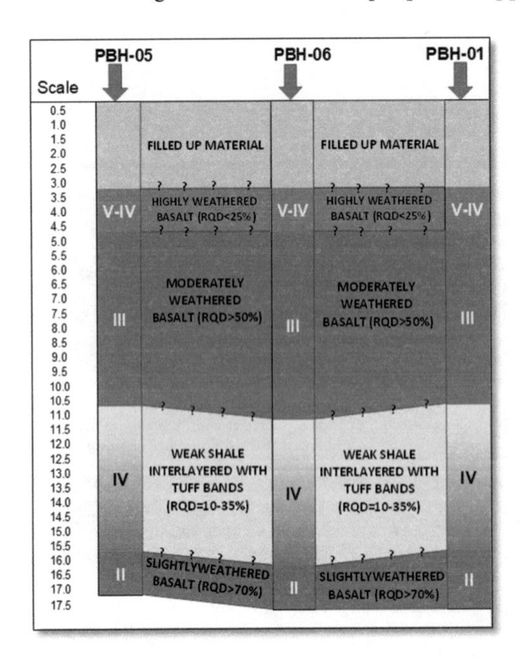

Figure 4. West side Boreholes.

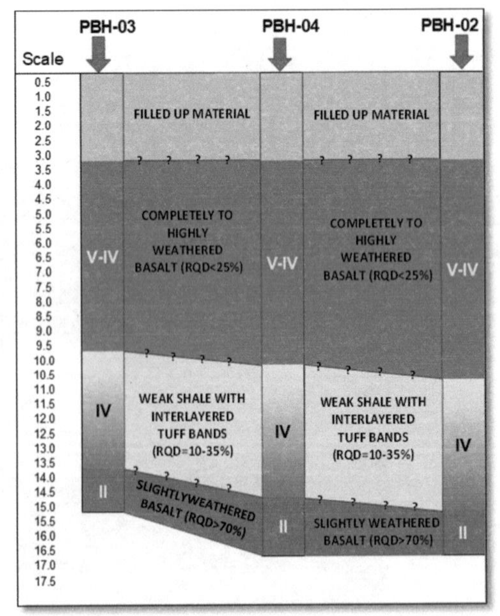

Figure 5. East Side Boreholes.

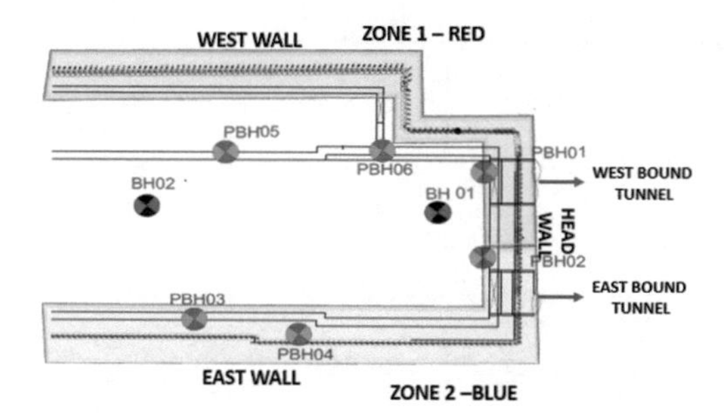

Figure 6. Pile Type Zonation.

Detailed investigations revealed an important information on ground variation and its extent, identified the transition zone on head wall side (Figure 3) where due to extensive weathering effect, rock mass changed from moderately weathered basalt to completely/highly weathered basalt. Phase 2 investigations also helped in deciding upon the pile termination levels on site which further helped in achieving desired progress by estimating the number of rigs, power of rigs needed to complete the required number of piles. It reduces uncertainties in selection and sequencing of piling rigs and also to avoid any ground failure during excavation.

4 SECANT PILE METHODOLOGY AND TERMINATION APPROACH

Secant pile wall (0.8 m diameter) involves a combination of reinforced and unreinforced piles which are interlocked with each other. Firstly, two soft piles are bored and casted using concrete mix and then hard pile are bored and casted using reinforcement and higher grade of concrete (Figure 7). These hard piles provide structural strength to the wall and since these piles are overlapped, it acts as a water tight structure. It also acts as a temporary cut off wall for soft ground, water and also to support the excavation in soil and weak rock around the periphery of the shaft which is a must requirement considering the shaft in vicinity of high rise buildings.

While designing the secant pile wall, many parameters has been considered like the surcharge load due to construction activities, nearby buildings load, TBM lowering cranes load, and presence of ground water table near ground. These secant piles are cast in situ and constructed by using rotary type piling rigs and core barrels are being used while cutting good grade of Basalt rock to recover the rock core and also to reduce the vibrations since working near the buildings.

Pile lengths are recommended based on the geotechnical investigations parameters mainly Total Core recovery (TCR), Solid Core Recovery (SCR), Rock Quality Designation (RQD) and UCS (Uniaxial Compressive Strength) and pile to be terminated in moderately weathered Basalt having min threshold limits as shown in Figure 8. However, the rock samples recovered during piling is being verified by the geologist and pile termination notes were followed.

Termination notes includes the piles to be terminated in G-III/better Basalt having core recovery of minimum 0.5 m, Uniaxial compressive strength > 50 MPa and rate of penetration being decided for individual rock layers based on the calibration done by each type of piling

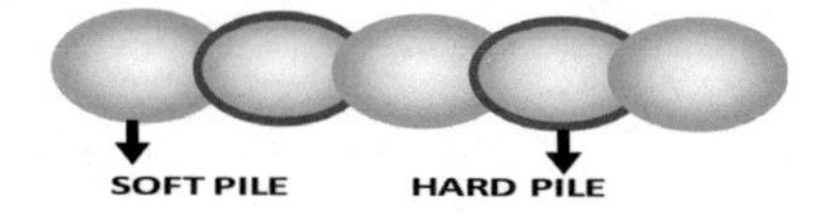

Figure 7. Secant Pile General Layout.

ROCK TYPE	WEATHERING GRADE	TCR (%)	SCR (%)	RQD (%)	UCS (Mpa)
BASALT	I/II (Slightly to Fresh)	80-100	80-100	60-100	60-100
	III (Moderately Weathered)	60-100	40-80	40-60	40-60
	IV (Highly Weathered)	40-60	20-40	20-40	20-40
	V (Completely Weathered)	0-40	0-20	0-20	0-20

Figure 8. Parameters for deciding pile termination and their threshold limit.

Figure 9. Core Recovery during Pile Termination in G-III/better Basalt.

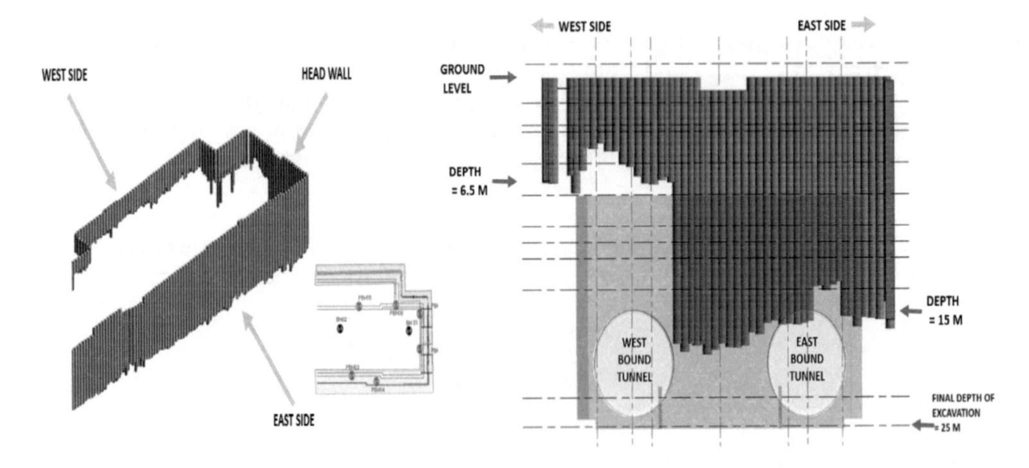

Figure 10. As Built Pile Termination Levels of Launch Shaft 3-D View (Left) and Head Wall (Right).

rig. (Example: ROP of 0.05 mm/min using Bauer BG-25 rig in G-III Basalt). This termination criteria is successfully implemented on site as seen good core recoveries (Figure 9) and this approach helped in successful completion of secant piling in and may also be helpful in creating standards while deciding the termination criteria of rock socketed piles.

Detailed investigations not only helped in planning number and sequencing of piling rigs, concrete estimation, reinforcement quantities based on pile depths but also provided zonation based on pile depths and successful execution of piling since actual termination levels on site are in line as per the recommended zonation based on investigation. As built pile termination 3-D view and as built levels are shown in Figure 10–11.

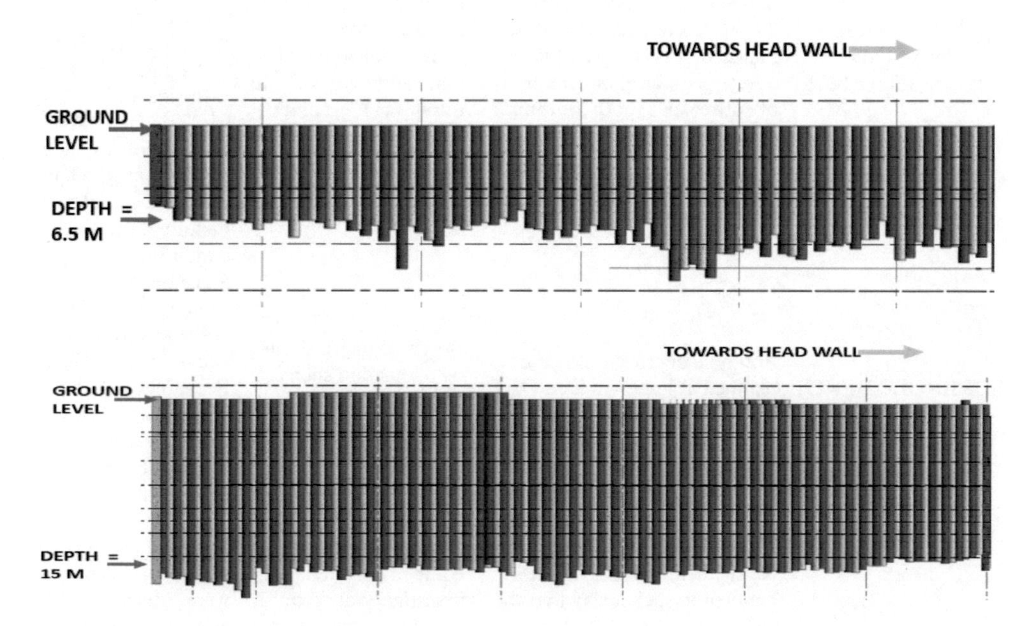

Figure 11. As Built Secant Pile Termination Levels of West Wall (Top) and East Wall (Bottom).

Table 1. Comparison of Estimated V/s Actual Ground Conditions.

Estimated Ground Conditions based on Investigations							Actual Ground Conditions based on Pile Termination	
Borehole No	Location	Total Depth	Depth of G-III Basalt	Ground Water Table	Estimated Pile Depth (Meter below GL)		(Average)	
BH-01	Head Wall	30	6	3	West and Head Wall	6–7	West and Head Wall	6.5
BH-02	South End	30	6.5	2.5				
PBH-01	Head Wall	17.5	5.5	2.8	West wall and partially Head Wall	5–6	West wall and partially Head Wall	6.5
PBH-05	West Wall	17.5	5	3.1				
PBH-06	West Wall	17.5	5	2.6				
PBH-02	Head Wall	16.5	15.5	3	East Wall and partially Head Wall	14–15.5	East Wall and partially Head Wall	15
PBH-03	East Wall	16.5	14	3				
PBH-04	East Wall	16.5	14.5	3.2				

Table 2. Planning of investigations for any urban infrastructure underground project- An approach.

PARAMETERS	PRELIMINARY INVESTIGATIONS	DETAILED INVESTIGATIONS	CASE STUDY SPECIFIC EXAMPLE
BOREHOLE SPACING (LONGITIDINAL)	Minimum distance to be maintained between two boreholes is 40–50 m. Location should be finalized in such a way that it covers geological information of entire shaft (both Start and end point).	Spacing to be decided based upon outcome of preliminary investigation and as per the regional geology. Boreholes should be planned to covers both the sides (West and East) as lateral variations can also be encountered in ground.	Initially, BH-01 and BH-02 were planned at a distance of 40–50 m as part of preliminary investigations. Further PBH-05,06 planned on west side and PBH-03,04 planned on East side and PBH-01,02 on Head Wall at an average spacing of 12–15 m which has given the complete picture of ground conditions.
BOREHOLE DEPTH	Borehole depth should be planned at least 6 m below the desired excavation depth.	Borehole depth should be planned in such a way that it should suit the objective.	Since the objective is to understand the lateral and vertical extent of shale layer, Boreholes were terminated 1 m below the weak layer (Shale) and in bed rock Basalt.
FIELD TESTING (PERMEABILITY AND PRESSURMETER TESTING)	All the field tests should be planned in such a way that the tests should be covered in all rock types and in different weathering grades which helps in deciding the sustained design parameters.	Testing shall be planned based on outcome of preliminary investigations. If in case preliminary test results were not sufficient enough to decide the design parameters, extensive testing shall be performed in detailed investigations.	Since the preliminary investigations results were sufficient enough to decide the design parameters, there was no field testing proposed in detailed investigations.
LAB TESTING (DENSITY, POROSITY & WATER ABSORPTION, UCS, PLI, TRIAXIAL STRENGTH, BTS, CERCHAR ABRASIVITY AND HARDNESS TEST, PETROGRAPHIC ANALYSIS)	Predetermined frequency of individual tests shall be decided. Main objective of lab testing is to obtain the rock properties at different locations in few selected boreholes as mentioned below: (i) In Overburden (ii) Near Tunnel Level (Between Crown and Invert)	All the lab tests should be planned in such a way that overburden ground properties shall be covered and also the ground characteristics at tunnel location in shaft and also below tunnel invert shall be tested in all of the planned individual boreholes.	Lab testing has been performed only in weak layers of Rock in phase 2 boreholes (Completely to highly weathered Basalt, Shale Layer) so as to understand the properties and behavior of weaker ground.

UCS: Uniaxial Compressive Strength; PLI: Plate Load Index Test; BTS: Brazilian Tensile Strength

5 GUIDELINES FOR PLANNING GEOTECHNICAL INVESTIGATIONS

In urban cities, only way of understanding the local geology is by conducting the geotechnical investigations as the area in general is devoid of natural rock outcrops. These should be extensive enough to understand and should not leave out any subsurface information near to sensitive/heritage/high rise structures if tunnel alignment is planned below these buildings.

There are many parameters in geotechnical investigations which should always be optimized/adjusted like Spacing of Boreholes, Location of Boreholes with respect to Shaft Location/Tunnel alignment, Depth of Boreholes, Number of Boreholes, Insitu testing and Lab Testing types and frequencies to best predict the ground. Design of retaining walls, temporary support (Struts/Wallers) and excavation support systems can also be optimized in case detailed investigation reports are made available.

Based on the extensive geotechnical investigations carried out, estimated ground conditions and actual conditions encountered during piling are in close agreement as shown in table 1 and accordingly an approach has been suggested in table 2 that can be followed during planning of investigations for any urban infrastructure underground project and can also be helpful for timely and cost-effective completion of investigations. Investigations should not be limited to only drilling of boreholes and one should also plan for geophysical investigations at least in critical locations of tunnel stretch and further the results can be correlated with geotechnical investigations so as to minimize the risk.

6 CONCLUSION

Predicting ground conditions in urban tunneling is one of the most complex but important activity as it is useful in planning, design and construction of any infrastructure project. Though costs allocated for Geotechnical Investigations in general are very less but if investigated properly, one can avoid cost over runs, time loss and contractual disputes. Majority of the project failures have been recorded all over the world due to inadequate investigations, lack of sound geological model and remedial measures.

Proper planning with due consideration to invest more in subsurface investigations should be worked out to reduce/minimize the ground uncertainties. In case of excavation planned below the buildings, one must provide boreholes by finding a suitable location or to mobilize small portable rigs inside the buildings. If drilling is not possible, geophysical investigations should be planned so as to avoid mis-interpretation of ground. Before to start tunneling, probe holes shall always be drilled on head wall (launching face) to cross check and validate the expected ground conditions.

Involvement of experienced geologist/geotechnical engineer must be ensured in mobilization, execution and managing Geotechnical/Geophysical investigations so as to provide inputs related to construction risks involved. Investigations should always be planned in two stages named as Preliminary and Detailed and one should always take help from the preliminary investigations and regional geology to decide the scope of detailed investigations which will further help to interpret the ground parameters, ground movement due to excavation and a decision to arrive at secant pile termination depths.

Suggested guidelines for geotechnical testing program and approach followed for secant pile termination has been successfully implemented and helped in timely decision with regards to the extent of detailed investigations to be carried out and thus rationalize the cost & time involved.

ACKNOWLEDGEMENTS

Authors are thankful to all the team members of L&T STEC JV UGC-01 for their support and this work forms as a part of main author's research work being pursued at IIT (ISM), Dhanbad.

REFERENCES

Austrian Society of Geomechanics (2010). Guideline for the Geotechnical Design of Underground Structures with Conventional Excavation. Austrian Society for Geomechanics: Salzburg

Codes of Indian Standard (1985). IS 4464: Code of practice for presentation of drilling information and core description in foundation investigation. Indian Standard Institution. Manak Bhavan, New Delhi.

Goldsworthy, J. S. (2006). "Quantifying the Risk of Geotechnical Site Investigations," PhD, University of Adelaide: Adelaide, Australia

Indian Geotechnical Society (2016). General guidance for Geotechnical Investigations. Chennai: IGC

Jaksa, M. B. (2000). "Geotechnical Risk and Inadequate Site Investigations: A Case Study." Australian Geomechanics, 35(2),39–46.

Jaksa, M. B., Goldsworthy, J. S., Fenton, G. A., Kaggwa, W. S., Griffiths, D. V., Kuo, Y. L., and Poulos, H. G. (2005). "Towards Reliable and Effective Site Investigations." Geotechnique, 55 (2),109–121.

Jaksa, M. B., Kaggwa, W. S., Fenton, G. A., and Poulos, H. G. (2003). "A framework for quantifying the reliability of geotechnical investigations." 9th International Conference on the Application of Statistics and Probability in Civil Engineering, San Francisco, USA, 1285–1291.

Parker, Harvey W. (1996), Geotechnical Investigations, Chapter 4 of Tunnel Engineering Hand book, 2nd Edition, edited by Kuesel & King, Chapman & Hall, New York, 1996.

U.S. National Committee on Tunneling Technology (1984), Geotechnical Site Investigations for Underground Projects. National Research Council: Washington, DC.

Whyte, I. L. (1995). The Financial Benefit from Site Investigation Strategy. Ground Engineering, Oct., pp. 33–36.

Tunnels and Underground Cities: Engineering and Innovation meet Archaeology, Architecture and Art, Volume 3: Geological and geotechnical knowledge and requirements for project implementation – Peila, Viggiani & Celestino (Eds)
© 2020 Taylor & Francis Group, London, ISBN 978-0-367-46583-4

Angat water transmission and improvement project for Metro Manila – Philippines: Hydraulic tunnel design under severe seismic conditions

A. Socci & P. Galvanin
Alpina S.p.A. – Milan - Italy

G.W. Bianchi
EG Team -Turin - Italy

G. Bay
CMC di Ravenna - Italy

ABSTRACT: The Angat Water Transmission Improvement Project (AWTIP) aims to improve the reliability of the raw water supply for Metro Manila through rehabilitation of the transmission system from Ipo to La Mesa and the introduction of water safety, risk and asset management plans. The Italian contractor CMC di Ravenna has been awarded for the Design and Construction of the Project with Alpina and EG Team as Design Consultants. The Project involves the construction of a new hydraulic tunnel including a new intake structures at Ipo reservoir and channel connecting the tunnel outlet to existing aqueducts. The tunnel has a length of 6.45 km with an internal diameter of 4.2 m and will be excavated with a Double Shield TBM. The challenge of this project lies in the fact that the structures are located in a highly seismic zone near an active fault system capable of generating earthquakes of magnitude up to 7.7.

1 INTRODUCTION

According to the last review of the Global World Urbanization Prospects, issued by United Nations: "55 % of the world's population lives in urban areas in 2018; by 2050, 68 % of the world's population is projected to be urban". Urban growth is closely related to three dimensions of sustainable development: economic, social and environmental. Undoubtedly water is the most immediate and critical limiting factor to both human and environmental well-being.

For the above reasons, the municipality of Metro Manila started a remarkable investment plan to improve the reliability of the raw water transmission system towards the city, including seismic hazard mitigation. One of the most important projects is the so called AWTIP (Angat Water Transmission Improvement Project) aims to improve the reliability of the raw water supply for Metro Manila through partial rehabilitation of the transmission system from Ipo dam to La Mesa and the introduction of water safety, risk and asset management plans. The Italian construction company CMC di Ravenna has been awarded for the Design and Construction of the AWTIP Project on may 2016 with Italian Alpina and EG Team as Design Consultants.

The present paper, after a brief description of the main project features, focuses on some seismic aspects with reference to tunnel structures, since they are located in a highly seismic zone near an active fault system capable of generating earthquakes of magnitude up to 7.7.

In the past, seismic design of tunnel structures received considerably less attention than that of surface structures, perhaps because generally the tunnels had a good performance during the earthquakes compared to above ground structures behaviour. Data collected worldwide demonstrate that tunnels can suffer damages under certain conditions, especially in zones with high seismicity. Awtip Tunnel is not shallow and it is not excavated in soft soils: notwithstanding the favorable geotechnical conditions described in this paper, a correct evaluation of stresses inside the final revetment under seismic actions was considered necessary to optimize the tunnel seismic design and comply the safety requirements.

2 MAIN PROJECT FEATURES

2.1 *The new Tunnel n° 4*

The Project involves the construction of a new hydraulic tunnel (named Tunnel No. 4) having a length of approximately 6.45 km, with an internal diameter of 4.2 m, crossing volcanic and sedimentary formations consisting of basalts, andesites and tuffs, with a maximum coverage of 150 m.

The tunnel, under construction, is excavated for its entire length with a double shield Tunnel Boring Machine (TBM). The final lining consists of precast segments with a thickness of 250mm based on a trapezoidal left and right ring pattern. The tunnel section is sized in order to convey a flow rate of 19 m3/s, ensuring a free surface flow.

2.2 *The Intake structures*

The intake structure is located upstream the Ipo dam and provides a conduit between the Ipo dam reservoir and the tunnel inlet portal. Such structure comprises a rectangular box with nominal dimensions of 10 m wide and 18 m long with a transition structure, allows for transition from the rectangular section to the circular profile at the tunnel. The intake structure invert is set nearly 5 m below the normal reservoir operating water level. The intake structure includes a trash rack to control debris entering the tunnel, stoplogs for closure, and control or sluice gates to regulate the flow to the tunnel. As shown in Figure 3, the existing slope, where the construction of the intake structures should be placed, extends upstream the basin level; the total height of the slope is about 35-40 m, and

Figure 1. AWTIP tunnel alignment.

Figure 2. Precast segmental lining for Tunnel No. 4 – a) 3D view of the design ring b) Test ring executed at the segment mould factory.

Figure 3. Ipo Dam reservoir at the Inlet structure and existing slope where the Intake structure is located

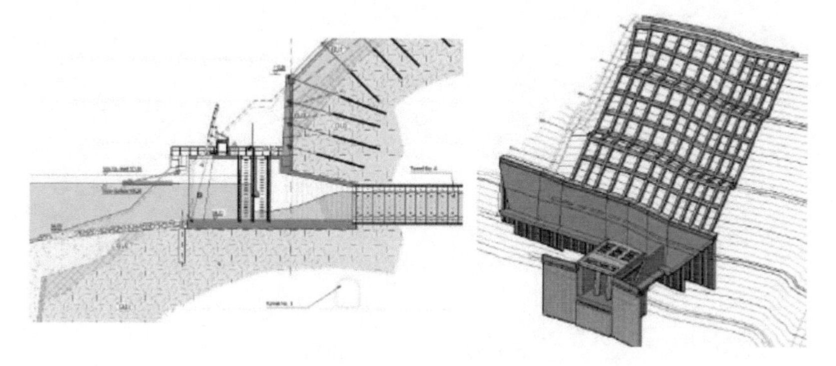

Figure 4. Design of the intake structure and slope stabilization.

it is inclined roughly at 45°. Detailed geotechnical investigations, executed on site during different campaigns, highlighted the occurrence of critical conditions along the slope surrounding the tunnel inlet area. The design solutions proposed to overcome the geotechnical problems were based on the following principles:

- the Intake structure is placed on the deep intact rocky strata inside the slope;
- retaining walls and slope stabilization are provided for deep excavation and long term seismic stabilization.

The slope in long term and seismic conditions is retained by anchors and by the final concrete revetment together with a grid of reinforcing beams and the anchors. The installed anchors are considered permanent ground anchors: to assure their durability over time, the anchors have a double corrosion protection (DCP-anchors) provided by means of a factory pre-grouted encapsulation of the bars within a corrugated plastic sheath, which ensures a comprehensive protection to all parts of the anchor.

2.3 *The Outlet structure*

The outlet structure is located at the end of Tunnel No.4 and provides a transition between the tunnel cross-section and the channel to existing aqueducts. The outlet structure comprises a rectangular channel with a width of about 4 m and a height of 4.5 m and an uncovered rectangular basin 4 m wide and 7 m long.

3 GEOLOGIC AND SEISMOTECTONIC SETTING OF THE PROJECT AREA

3.1 *Geological setting*

The project area is characterized by occurrence of a sequence of volcanic and sedimentary rocks ranging in age from Mesozoic (Cretaceous) to Cenozoic (Plio-Pleistocene). Two main

Figure 5. New tunnel no. 4 – Construction progresses at Outlet tunnel portal.

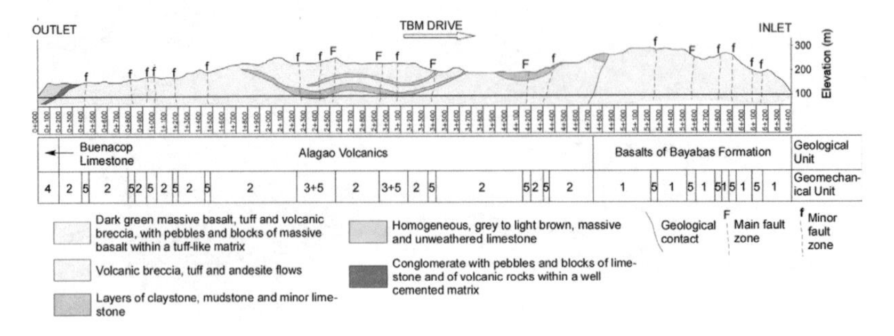

Figure 6. Longitudinal geotechnical profile along the axis of Tunnel 4.

formations are identified along the tunnel alignment: the Madlum Formation and the Bayabas Formation. The Madlum Formation of Middle Miocene age consists of two members:

- the Alagao Volcanics which corresponds to a sequence of volcanic breccia, tuff and andesite flows with local layers of claystone, mudstone and minor limestone;
- the Buenacop Limestone, a sequence of rather homogeneous, grey to light brown, massive and unweathered rocks. The basal sections of this member is a conglomerate made of pebbles and blocks of limestone and of volcanic rocks (i.e. tuff, basalt and diorite) within a well cemented matrix of tuff and sand.

The Bayabas Formation is dated to be Late Eocene to Early Oligocene. This formation is composed of dark green massive basalt, of tuff and of volcanic breccia, with pebbles and blocks of massive basalt within a tuff-like matrix.

From the structural point of view, three main sets of faults have been identified, based on analysis of aerial images and of field survey. The fault zones have been classified as main of minor faults according to the thickness of the core zone as well as to the total thickness: main faults are characterized by a thickness of the core zone greater than 5 m and a total thickness greater than 10 m; minor faults show a core zone of about 1 to 3 m and a total thickness of about 2 to 5 m.

Foreseen geological and geotechnical conditions along the axis of Tunnel 4 are illustrated in Figure 6.

3.2 *Seismotectonic setting*

The seismotectonic setting of the region is dominated by two main tectonic structures, the Philippine Fault and the Valley Fault System:

- The Philippine Fault represents the most relevant structural feature on Luzon Island consisting of a NW-SE trending sinistral strike-slip fault zone, extended through the entire Philippine archipelago. The Philippine Fault has been associated with at least two and probably three large historical earthquakes with M>7 (Koo et al., 2009), including the M7.6 Luzon earthquake of 1990 (Yoshida & Abe, 1992).

- The tectonic structure closer to the project site is the Valley Fault System (PHILVOLCS, 1999). It consists of two sub-parallel, northeast-trending faults: the West and the East Marikina Valley Faults (WMVF and EMVF, respectively). The project area is located in the close proximity of the northern termination of the West Marikina Valley Fault, a dominant dextral strike-slip fault with vertical slip component, consisting of several fault segments (Rimando & Knuepfer, 2006): considering the maximum displacement observed in the field and deduced from the geomorphic analysis and the length of the fault segments, these authors estimated the maximum magnitudes for each segment, assigning values between M 7.0 and M 7.7 to Segment I, the closest to the project area. These values are consistent with the M 7.4 estimated by PHILVOLCS (1999). The West Valley Fault represents the nearest active fault to the Tunnel 4, located at a distance between 1 km and about 6 km, and must be considered as a potential seismic source that is capable of generating earthquake with magnitude between 7.4 and 7.7.

4 SEISMIC DESIGN APPROACH

The Project comprehends underground structures (i.e. The Tunnel 4) as well as surface structures (intake and outlet structures) and the design criteria has to be defined considering the specificity of these two kinds of structures. Accordingly, two different design approaches were adopted:

- Force Method for Surface Structures. For aboveground structures, the seismic loads are largely expressed in terms of inertial forces. The traditional methods generally involve the application of equivalent or pseudostatic forces in the analysis or better with acceleration response spectra, which represents the response of a damped single degree of freedom system to ground;
- Deformation Method for Underground Structures. The design and analysis for underground structures should be based, however, on an approach that focuses on the displacement/deformation aspects of the ground and the structures, because the seismic response of underground structures is more sensitive to such earthquake induced deformations.

The intensity of earthquake ground motion is described by several important parameters, including peak acceleration, peak velocity, peak displacement, response spectra, duration and others. For aboveground structures, the most widely used measure is the peak ground acceleration and the design response spectra, as the inertial forces of the structures caused by ground shaking provide a good representation of earthquake loads.

Since the underground structures are supposed to be rigidly connected with the surrounding ground, the PGA at time $t = 0$ is the basic parameter to be defined for structural verifications. Peak ground acceleration is not necessarily a good parameter, however, for earthquake design of underground structures such as tunnels, because tunnel structures are more sensitive to the distortions of the surrounding ground than to the inertial effects. Such ground distortions are the ground deformations/strains caused by the traveling seismic waves. Applying the deformation method to tunnel design, the response of tunnel to seismic shaking motions may be demonstrated in terms of three principal types of deformations (Owen & Scholl, 1981): axial, curvature and ovaling (for circular tunnels) or racking (for rectangular tunnels such as cut-and-cover tunnels).

The definition of these deformations needs the definition at least of these parameters:

- the effective wave propagation velocity;
- the peak ground particle velocity;
- the peak ground particle acceleration;

The peak velocity and acceleration can be established through empirical methods, field measurements, or site-specific seismic exposure studies. The effective wave propagation velocity in rock was determined by mean of geophysical investigations (i.e. refraction and reflection seismic profiles along the tunnel alignment and cross-hole investigations in the intake area) performed in the project area.

5 SEISMIC HAZARD ASSESSMENT

5.1 *Definition of design seismic levels*

After the definition of the seismic design approach for main structures of AWTIP Project, the seismic hazard assessment for the site was carried out in accordance with the design requirements and with the ICOLD Design taking into account the specificity of the design structures.

Three different Design Seismic Levels were defined:

- Level 0 (Construction Earthquake, CE).
- Level I (Design Basis Earthquake, DBE).
- Level II (Defined as Maximum Credible Earthquake (MCE)

5.2 *Type of ground response*

The evaluation of ground response to shaking can be divided into two groups:

- ground failure;
- ground shaking and deformation.

The first assumption for seismic hazard assessment report focuses on ground shaking and deformation: it was assumed, for the AWTIP tunnel - on the basis of literature data, distance from the fault and surveys - that the ground does not undergo large permanent displacements.

5.3 *Horizontal PGA_H definition*

5.3.1 *Seismic Level 0 - Construction Earthquake*

The CE PGA at T= 0 was determined from the acceleration coefficient contour maps for 50-years return period, derived from the probabilistic seismic hazard analysis (PSHA) conducted by Japan International Cooperation agency (JICA). The maximum values shown in these maps is equal to 0.08 g and it was used for design analysis.

5.3.2 *Seismic Level 1 - Design Basis Earthquake*

For the Level I design earthquake the ground motion corresponds to a minimum 475-year return period ground motion. The peak ground acceleration at T=0 for Level I earthquake is determined from the acceleration coefficient contour maps for 500-years return period, derived from the probabilistic seismic hazard analysis (PSHA) developed by Japan International Cooperation agency (JICA).

5.3.3 *Seismic Level 2 - Maximum Credible Earthquake*

The MCE corresponds to the event which produces the largest ground motion expected at the project site on the basis of the seismic history and the seismo-tectonic setting of the region. The peak ground acceleration for the MCE has been estimated on deterministic earthquake scenarios.

The main seismic source consists in the West Valley Fault characterized by several segments the nearest two located 1 km towards southeast and east of the inlet area, capable of generating a maximum earthquake with magnitude between 7.4 and 7.7. The horizontal peak ground acceleration at the project site was determined on the basis of the attenuation model of Fukushima & Tanaka (1990). A design earthquake is assumed to occur at a point along the causative fault that is nearest to the site, represented, in this case, by the segment of the West Valley Fault located at a distance of 1 km from the inlet area. The attenuation model of Fukushima and Tanaka is written as:

$$Log_{10}A = 0.41M - \log_{10}\left(R + 0.032 \times 10^{0.4M}\right) - 0.0034R + 1.30 \tag{1}$$

where: A = mean peak acceleration (cm/sec^2); R = shortest distance between the site and the fault rupture (km); M = surface-wave magnitude (also referred to as Ms).

Correction factors are applied to the to the mean peak acceleration depending on the type of foundation material: rock, 0.6; hard soil, 0.87; medium soil, 1.07; and soft soil, 1.39.

The below table summarizes the MCE PGA at the minimum and maximum distances from the West Valley Fault.

5.4 *Vertical PGAv definition*

The near-fault ground motions are generally characterized by long-period horizontal pulses and high values of the ratio between the peak value of the vertical acceleration, PGA_V, and the analogous value of the horizontal acceleration PGA_H. Due to lack of information and recordings for the West Valley Fault, in order to assess the ratio between vertical and horizontal acceleration in near fault regions, it is necessary to use relationship proposed by international researchers. Some authors presented a relationship between horizontal and vertical acceleration based on 110 near field recordings (d< 15 km) relatively large (surface wave magnitude M>6), shallow (h< 20 km). They suggest that V/H slightly exceeds 1 for source to site distances less than about 5 km for large thrust fault ruptures and for moderate to strong strike-slip events. According to the seismic assessments developed in the above paragraphs, the PGA values at T=0 were defined for main underground and buried structures as shown in Table 2.

6 SEGMENTAL LINING DIMENSIONING UNDER SEISMIC CONDITIONS

6.1 *Analytical approach for ovaling effects estimate*

Three types of deformations express the response of underground structures to seismic motions: axial compression and extension, longitudinal bending, ovaling effects. Axial deformations in tunnels are generated by the components of seismic waves that produce motions parallel to the axis of the tunnel and cause alternating compression and tension. Bending deformations are caused by the components of seismic waves producing particle motions perpendicular to the longitudinal axis. Design considerations for axial and bending deformations are generally in the direction along the tunnel axis (Wang, 1993).

Table 1. Design seismic MCE PGA at T=0.

Distance R (km)	Magnitude (-)	Log10A (-)	A (cm/s2)	a_{gh}/g
1	7.7	2.86	719.9	0.73
4	7.7	2.81	653.56	0.67
6	7.7	2.79	614.5	0.63

Table 2. Design seismic PGA at T=0.

Seismic Level	Retrurn Period	PGA_h (T=0)	PGA_v (T=0)	Reference site category	Approach	Reduction/amplification site factors
CE Construction Earthquake DBE	50 years	0.08g	Negligible	Soft Rock-AASHTO	Probabilistic	Site factors for T= 0 (F_{PGA}) - AASHTO - DPWH-BSDS
Design Basis Earthquake	500 years	0.45g	$=PGA_h$	Soft Rock-AASHTO	Probabilistic	
MCE Maximum Credible Earthquake	Not defined	0.63-0.73g	$= PGA_h$	General	Deterministic Fukushima e Tanaka	Site factors for T= 0 (F_{PGA}) According to Fukushima and Tanaka

Hereinafter, a special focus is made on ovaling or racking deformations in a tunnel structures developed when shear waves propagate normally or nearly normallly to the tunnel axis, resulting in a distortion of the cross-sectional shape of the tunnel lining. Design considerations for this type of deformation are in the transverse direction. The general behavior of the lining may be simulated as a buried structure subject to ground deformations under a two-dimensional plane strain condition. Closed form solutions for estimating ground-structure interaction for circular tunnels have been proposed by many investigators and they are extremely useful – especially in the first design stages – to carry out an estimate of the effects induced by seismic actions on the underground structures. These solutions are commonly used for static design of tunnel lining. They are generally based on the assumptions that:

- the ground is an infinite, elastic, homogeneous, isotropic medium.
- the circular lining is generally an elastic, thin walled tube under plane strain conditions.

The main assumptions used in these kind of approach generally are:

- full-slip or no-slip conditions exist along the interface between the ground and the lining depending on the tunnel construction method;
- loading conditions are to be simulated as external loading (overpressure loading) or excavation loading.

The resulting expressions for maximum thrust, bending moment, and diametric strain, were presented by several Authors for full-slip and not slip conditions.

Nevertheless, when the foreseen seismic actions are quite severe, as in the case of AWTIP Tunnel, it becomes essential to investigate the role of the interface conditions between the lining and the rock: as a matter of fact, under a severe seismic shock the no slip conditions could provide an excessively conservative estimate of the internal actions (above all positive or negative axial forces); conversely, the full slip conditions could provide a not safe prediction. This is more evident for tunnels excavated with TBM where the annular gap between the surrounding rock mass and the segmental lining is filled with pea-gravel, a material which usually exhibits low shear stiffness. To overcome the aforesaid limits, the equations normally used for full-slip or no-slip conditions were modified as proposed by Kyung-Ho Park et al. (2009), by means of the introduction of a shear parameter "D" (PTTO method hereinafter). Using these modified equations the ovaling effects in different tunnel cross sections were estimated on the basis of the following main assumptions:

$$T_{PTTO} = \frac{(1-v_s)E_s\gamma_c R}{(1+v_s)\Delta''}\left\{2F + (1-2v_s)C + 4 + D\frac{4E_s}{R(1+v_s)}\right\}\cos 2\left(\theta + \frac{\pi}{4}\right) \tag{2}$$

$$M_{PTTO} = \frac{(1-V_s)E_s\gamma_c R^2}{(1+v_s)\Delta''}\left\{(1-2v_s)C + 2 + D\frac{4E_s}{R(1+v_s)}\right\}\cos 2\left(\theta + \frac{\pi}{4}\right) \tag{3}$$

$$\Delta'' = CF(1-2v_s) + F(3-2v_s) + C\left(2.5 - 8v_s + 6v_s^2\right) + 6 - 8v_s + 2D\frac{(2F+5-6v_s)E_s}{R(1+v_s)} \tag{4}$$

Where: C: compressibility ratio; F: flexibility ratio; D: the spring-type shear flexibility coefficient; γ_c: average free-field strain of the soil; E_s, v_s = modulus of elasticity and Poisson's Ratio of medium. As extreme cases, D=0, represents a no-slip interface condition, while D→∞, corresponds to full-slip interface condition, which means the lining is subjected to sliding along the interface. The 1/D values correspond to shear stiffness of materials: setting an appropriate value for D it is possible to predict the stress resultants in the tunnel lining taking into account the relative movements between the lining and the rock mass.

Table 3 shows a rational approach to estimate the D values on the basis of mechanical properties of materials, taking into account the variability of shear modulus from concrete (upper value) to weak materials (lower value). The pea gravel, used to fill the gap between the lining and the rock mass in the AWTIP tunnel after segments laying, cannot guarantee a

Table 3. Variability of shear modulus G and Flexibility Coefficient D

Type of shear connection between lining and surrounding rock	Shear modulus G (kPa)	flexibility coefficient D (1/G)
NO slip	∞	0.00E+00
Concrete	1.31E+07	7.62E-08
Jointed rock	3.50E+06	2.86E-07
Weak rock	4.00E+05	2.50E-06
Pea Gravel	9.60E+04	1.04E-05
FULL slip	0	∞

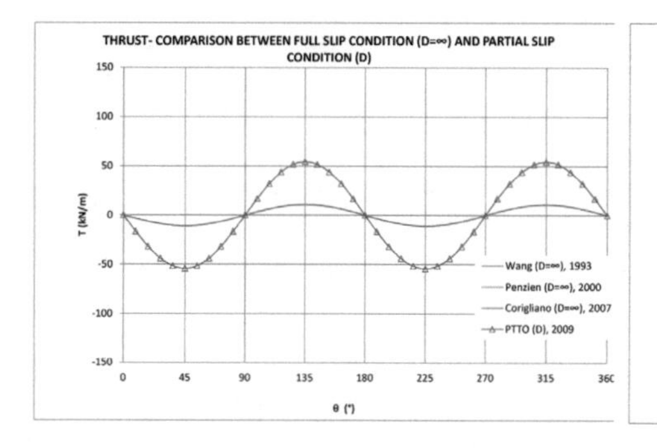

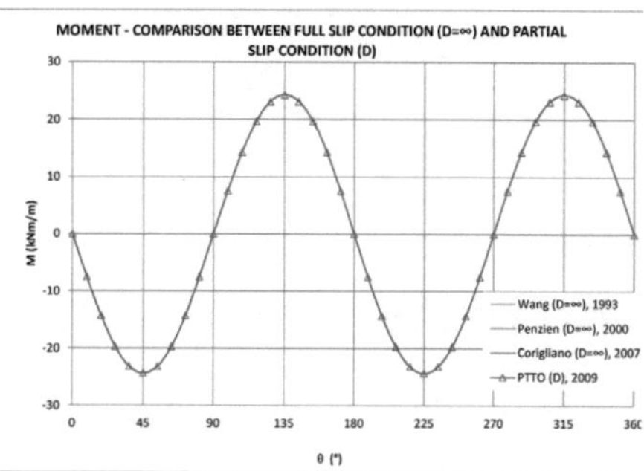

Figure 7. Typical analysis results for ovaling effects estimate, using the PPTO method.

strong adhesion between the lining and the surrounding rock (no slip condition) and the shear strength is mainly limited by friction forces along the lining.

Thus, the shear strength will be similar to that of loose rock/slightly cemented soil as shown in the table. Once estimated the D values, an extensive parametric analysis was carried out for different sections and materials to analyze the bending effects and thrust forces on the internal revetment; the following figures show typical results of the parametric analysis and the comparison between the proposed approach and the standard full-slip/no slip conditions

6.2 *Numerical approach for ovaling effects estimate*

The PPTO procedures was then validated through some 2D FDM numerical analyses, assuming plane strain conditions, following an excavation procedure and a gradual staged construction, The calculation code used was FLAC 2D (Fast Lagrangian Analysis of Continuous), developed by ITASCA Consulting Group. The material behaviour in the models was described by means of linear elastic constitutive laws in the seismic stage, and the seismic loads are applied using a quasi-static approach. The annular gap was simulated through the grid elements of the mesh: to no-slip condition were simulated assigning the parameters of the intact rock to the gap material; conversely, full slip conditions were simulated assigning a very soft material. Following the approach suggested by several authors ovaling deformations due to seismic loading were assigned as inverted triangular displacements along the lateral boundaries of the model and uniform lateral displacements along the top boundary (see Figure 8). The value of the assigned displacement at the top of the model is related to the maximum shear strain γ_{max} and to the height of the model. The horizontal base is restraint in all directions. As explained above, the proposed solution ignores seismic inertial interaction effect. The model was used also for static loads evaluation applied before the stage in which ovaling deformation, due to seismic loading, is assigned: this is done by means of a load staged conventional approach.

The results were compared with the PPTO and Wang (1993) solutions, assuming different stiffness values of the material inside the annular gap: the final results of the numerical models were in good accordance with the analytical predictions.

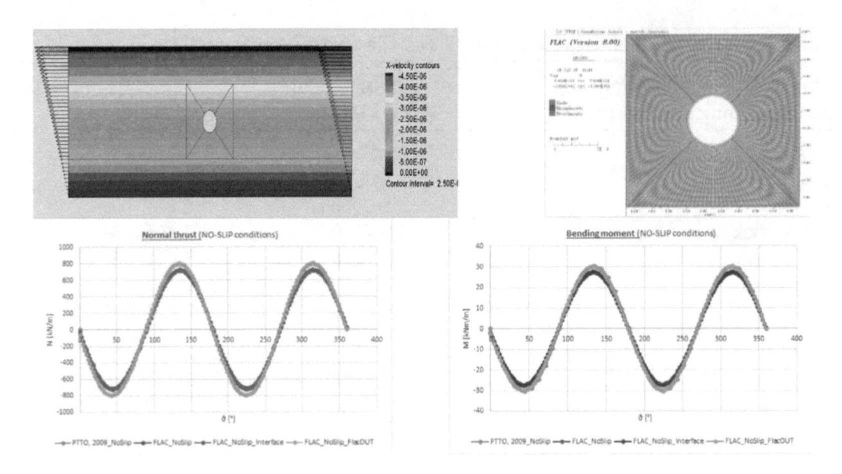

Figure 8. Quasi static approach using numerical analysis: comparison between analytical and numerical solutions

7 CONCLUSIONS

The AWTIP project in the Philippines, aimed at improving the reliability of the raw water transmission system towards the city, was described in this paper: after a brief presentation of the main project features, the paper focused on some seismic aspects with reference to the new water Tunnel n° 4 and its final internal revetment, since the tunnel is located in a highly seismic zone, near an active fault system able to generate earthquakes of magnitude up to 7.7.

After a geological description of the whole area interested by the project and the seismic hazard assessment, the strategy used for the design of the tunnel internal lining was described: analytical and numerical models were developed, paying special attention to the role played by the annular gap filling material on the ovaling response of the revetment. Analytical models, able to take into account the shear stiffness of the filling material, were used to determine internal stresses for the structural design of the revetment; FDM models allowed to check the results obtained with the analytical models, achieving a final good agreement for the different conditions analyzed in the design stage.

REFERENCES

Einstein, H.H. and C.W. Schwartz, 1979. Simplified analysis for tunnel supports. J. Geotechn. Eng. Division, 105: 499–518.

Fukushima Y. and Tanaka, T. 1990. "A new attenuation relation for peak horizontal acceleration of strong earthquake ground motion in Japan" Bull. Seism. Soc. Am. 80, 4, 757–783.

Koo, R., Mote, T., Manlapig, R. V., & Zamora, C. 2009. Probabilistic Seismic Hazard Assessment for Central Manila in Philippines. *In Australian Earthquake Engineering Society Conference.*

Kyung-Ho Park, Kullachai Tantayopin, Bituporn Tontavanich, Adisorn Owatsiriwong, 2009 Analytical solution for seismic-induced ovaling of circular tunnel lining under no-slip interface conditions: A revisit, Tunnelling and Underground Space Technology.

Kyung-Ho Park, Kullachai Tantayopin, Bituporn Tontavanich, Adisorn Owatsiriwong 2008 - School of Engineering and Technology, Asian Institute of Technology, P.O. Box 4, Klong Luang, Pathumthani.

Owen, G.N. and Scholl, R.E. 1981. Earthquake engineering of large underground structures. *Report no. FHWA_RD-80_195. Federal Highway Administration and National Science Foundation.*

PHILVOLCS Active Fault Map at scale 1:50,000 in the Bulacan and Rizal areas. Sheets 4-5-6-7 3230 IV.

Rimando, R. E., & Knuepfer P. L. 2006. Neotectonics of the Marikina Valley fault system (MVFS) and tectonic framework of structures in northern and central Luzon, Philippines. *Tectonophysics (1)*: 17–38.

Wang J. 1993 "Seismic Design of Tunnels: A Simple State-of-the-art Design Approach", Monograph 7, Parsons, Brinckerhoff, Quade and Douglas Inc, New York.

Yoshida, Y., & Abe, K. 1992. Source mechanism of the Luzon, Philippines earthquake of July 16, 1990. *Geophysical Research Letters, (6)*: 545–548.

Tunnels and Underground Cities: Engineering and Innovation meet Archaeology, Architecture and Art, Volume 3: Geological and geotechnical knowledge and requirements for project implementation – Peila, Viggiani & Celestino (Eds)
 ISBN 978-0-367-46583-4

Correlation between TBM performance and RMR

S. Son, T. Ko, Y. Pak & T. Kim
SK engineering & construction, Seoul, South Korea

ABSTRACT: The productivity of TBM tunneling can be determined by its advance rate, proper maintenance, material and equipment logistics, workmanship of operating manpower, etc. Among mentioned factors, the Advance rate is the dominant factor of the TBM performance. This study tried to correlate the Specific Penetration and Field Penetration Index with RMR to figure out the regression between TBM performance and ground condition with TBM operating parameter.

1 INTRODUCTION

Even with high thrust force acting at the tunnel excavation face, penetration rate can be low value depending on the ground condition. Since the penetration rate, which is dominant parameter in excavation performance, is a function of thrust force and ground condition, the relationship between the parameters and geological conditions need to be analyzed for determining the optimal range of operation parameters.

Some studies correlate between the rock mass classification and TBM parameters to predict TBM performance. Cassinelli et al. (1982) and Innauato et al. correlated penetration rate and Rock structure rating(RSR). Barton (2000) introduced QTBM and correlated with penetration rate. Ribacchi and Lembo-Fezio (2005) correlated Specific Penetration and RMR. Hassanpour et al.(2009) correlated Field Penetration Index and several rock mass classification system which are BRMR, GSI and Q-system.

This study focused on TBM performance which can be evaluated with Specific Penetration(SP) and Field Penetration Index(FPI) depending on the ground condition represented by rock mass rating(RMR). The TBM excavation data such as penetration, cutter rotation speed and thrust force was derived from 11.5km length tunneling with double shield TBM. The ground condition is evaluated with RMR value by geologist during TBM stoppage for daily maintenance.

2 TBM TUNNELING OF THE PROJECT

2.1 *Tunneling overview*

Both NATM and TBM method are applied for 13.6km of headrace tunnel excavation. With considering the total length of the headrace tunnel with limited period of construction, TBM is planned to get the high performance in tunneling. Especially, double shield TBM is considered for excavating 11.5km with taking into account the geological formation. To optimize the construction period for tunneling, partial section at both end of the headrace tunnel is planned to be excavated with NATM which is 2.1km length.

2.2 *TBM specification*

Mobilization, uncertainties on the geological formation along the alignment and tightness of the project schedule are taken into account for planning the TBM. The specification of the

Table 1. TBM specification and segment detail.

Parameter	Values
Cutterhead	
Type	Dome Type
Support Type	Intermediate Support Type
Excavation Diameter	5,740mm
Power	2000kW(8Nos. 250kW)
Speed	0.3~7.0rpm
Rotation Direction	Clockwise
Disc Cutter	33Nos. Single Disc of 17" Back Loading Type 4Nos. Twin Disc of 17" Back Loading Type
Trust Cylinder	
Max total thrust of main Cylinder	30,000kN at 350bar
Number of main Cylinder	18
Number of Auxiliary Cylinder	18
Gripper	
Strocke of the Cylinder	300mm each
Gripper force	15000kN at 300bar each
Number of Gripper shoe	2
Segment erector	
Angle	220° each direction
Speed	0.71 to 1.43rpm
Segment	
Type	Hexagonal
Outer diameter	5,500mm
Inner Diameter	5,000mm
Thickness	250mm
Width	1,500mm
Number of Segment per Ring	4

Double Shield hard rock TBM which is used for excavation of the headrace tunnel is as followed.

The TBM for the tunnel is equipped 42nos of 17inches disc cutters and the maximum thrust acting on each cutter is 250kN. The cutterhead rotation speed varies between a minimum of 0 and a maximum of 7 RPM. The total power which is installed on the TBM is 2000kW, given by 8 drives and being 250kW power of single drive. The maximum torque assured by the drives is 6,800kNm, while the breakout torque has a value of 8,200kNm

3 GEOLOGICAL CONDITION

3.1 *Geomechanical property*

Stratigraphically, the headrace tunnel alignment is located in two geological formations which are "Champa Formation" and "Tholam Formation".

Table 2. Geomechanical properties.

	Mudstone	Siltstone	Sandstone
UCS intact rock (MPa)	10–30	40–70	70–150
Quartz contents (%)	15–25	40–60	60–85
RQD	20–100	30–100	60–100
Cerchar Index	0.5–1.0	0.8–2.0	3.8–4.6
Range RMR	30–49	46–64	74–89

Sandstone

Siltstone

Mudstone

Figure 1. Rock specimen from TBM tunneling.

The "Champa Formation" is formed between Upper Jurassic and Cretaceous. The compositions of the formation are siltstones, sandstones and conglomerates, whereas The "Tholam Formation" which is formed between Lower and Middle Jurassic, principally contains mudstones and silt stones. Erosion and flood channels of several meters which are filled with clay were identified. "Tholam Formation" is mainly encountered on the southern side of the headrace tunnel (around 1/5 of the length) and the "Champa Formation" is present in the remaining portion of the tunnel.

During TBM tunneling, three kinds of sedimentary rocks are discovered as governing materials which are sand stone, Siltstone and Mudstone as shown in below.

3.2 *RMR distribution*

Bieniawski (1976) developed the rock mass classification with obtained data from civil engineering excavation in sedimentary rocks in South Africa called the Geomechanics Classification or the Rock Mass Rating (RMR) system. The analyzed TBM tunnel project has daily maintenance to check the TBM especially cutterhead and cutting tool condition and geological condition of tunneling face. The RMR value which is used as representative value of ground condition for this study has evaluated by the geologists during the daily maintenance.

Total 10.95km length of tunneling face condition has evaluated by RMR value and distributed as shown below.

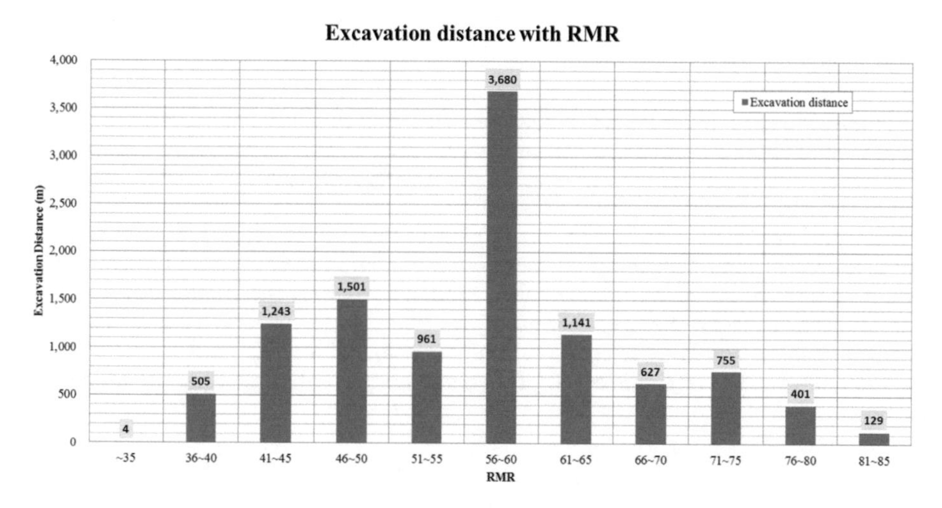

Figure 2. RMR distribution.

Among the evaluated RMR values which are varied from 35 to 83, More than 90% of the ground condition is evaluated 41~75 of RMR value and especially 3.68km (34%) of tunneling faces are evaluated within 56~60.

4 CORRELATION BETWEEN TBM PERFORMANCE AND RMR

4.1 *Specific Penetration, Field Penetration Index*

The penetration rate varies with the normal force acting on disc cutter and the ground condition. To correlate these three factors which are penetration rate, normal force acting on disc cutter and ground condition, two composite indices are proposed in this study which are Specific Penetration, "SP" (Alber 2001) and Field Penetration Index "FPI" (Klein et al. 1995)

Since the TBM performance can be also influenced by the TBM diameter and number of cutters, to focus on the TBM performance depending on the ground condition, in this study, the TBM performance is measured by the indices of "SP", and "FPI". "SP" is calculated penetration rate per normal force acting on disc cutter whereas, "FPI" is normal force acting on disc cutter per penetration rate.

$$SP = \frac{P}{R_c \times F_n} = \frac{PR}{F_n} \tag{1}$$

Where, *SP*: Specific Penetration (cm/ rev/MN); *P*: Penetration (cm/min.); R_c: Cutter Rotation Speed (rev/min.); F_n: Normal force acting on Cutter (MN); *PR*: Penetration ratio (cm/rev)

$$FPI = \frac{F_n}{PR} \tag{2}$$

Where, *FPI*: Field Penetration Index (kN/mm/ rev); F_n: Normal force acting on Cutter (kN); *PR*: Penetration ratio (mm/rev)

4.2 *Normal force acting on each disc cutter*

Thrust force consists with 6 factors which are ①Friction between shield and ground, ② Chamber pressure acting on bulkhead, ③Driving force caused by direction changed in curved alignment, ④Friction between segments and tail seals, ⑤Hauling force of back-up facility, ⑥Penetration force of cutting tools into the ground.

Since the analyzed TBM tunnel is excavated by open-face double shield TBM in straight alignment, four factors of thrust force which are bulkhead pressure, curved alignment, friction of tail seals and hauling can be disregarded among the six factors. Therefore, the total thrust force consists with only two factors which are the penetration force and friction force between shield and ground.

Normally double shield TBM can be operated with two modes which are single shield mode and double shield mode depending on the ground condition. The tunneling data utilized for this study is only operated by double shield mode. Therefore, the friction force created by only front parts of the TBM shield in front of main thrust cylinder which consists with cutterhead, forward shield and inner telescopic shield is considered for this study

The friction force between shield and ground can be calculated by using the following simplified equation.

$$F_r = \mu_f \times (G_f + W_s) \tag{4}$$

Table 3. Total weight of shield for calculating friction force.

Items	Weight(kN)	Remark
Cutterhead	588.0	Including Disc Cutters
Main Drive with Bearing	490.0	
Cutter Driving Motors(8EA)	525.3	Including Support Frame
Forward Shield	666.4	
Inner Telescopic Shield	245.0	Including Articulation Cylinders(4EA)
Total weight	2514.7	

Table 4. Skin friction coefficient for sliding and static friction(i.e., during TBM advance and for restart respectively) with and without lubrication by bentonite (Gehring, 1996).

Status	Sliding friction (continuous excavation)		Static friction (restart after a standstill)	
Lubrication	Not lubricated	Lubricated	Not lubricated	Lubricated
Rock	0.25 ~ 0.30	0.10 ~ 0.15	0.40 ~ 0.45	0.15 ~ 0.25
Gravel	0.25 ~ 0.30	0.15	0.40 ~ 0.45	0.20 ~ 0.30
Sand	0.35 ~ 0.40	0.15	0.45 ~ 0.55	0.20 ~ 0.30
Silt	0.35 ~ 0.40	0.10	0.30 ~ 0.50	0.15 ~ 0.20
Clay	0.30 ~ 0.35	0.10	0.20 ~ 0.55	0.15 ~ 0.20

Where, F_r : Friction force between shield and ground; μ_f : Coefficient of friction; G_f : Ground force around the shield; W_s : Shield weight

From the value of total thrust force which is stored by the TBM data logging system, the bearing thrust force which is actual thrust force acting on the main bearing can be calculated by subtracting the friction force from the total thrust force. The range of the coefficient of friction can be obtained by the table which is 0.25~0.30. That is due to the ground excavated is rock condition and there was no lubrication such as bentonite injection during excavation. Moreover, since the value of the total thrust force which is obtained from the data logging system is average value during the excavation, the friction is considered in sliding condition.

The actual normal force acting on each disc cutter can be simply calculated by dividing the bearing thrust force into number of disc cutters.

$$F_n = \frac{TH_b}{N_c} \tag{3}$$

Where, F_n: Normal force acting on each disc cutter; TH_b: Bearing thrust force; N_c: Number of disc cutter

4.3 *Regression Analysis*

Khademi Hamidi et al. (2010) showed that TBM performance is maximum in fair rock mass condition (RMR: 40–70) while slower penetration is experienced in both too bad and too good rock masses. In this study, since majority of the tunnel sections are consisted with RMR value from 41 to 75 (more than 90%), the regression of SP and FPI with RMR is analyzed in the section of RMR value from 41 to 75.

The friction force which affects the normal force acting on disc cutter can be varied by the friction coefficient. In this study, two friction coefficients 0.25(μ_1) and 0.3(μ_2) are applied for correlation.

4.3.1 *Specific Penetration(SP) and RMR*

The result of regression analysis between SP and RMR is shown in Table 5. The SP has tendency to get decreasing with increasing value of RMR as shown in Figure 3 and 4.

Table 5. Correlation between SP and RMR.

Correlations (SP and RMR)	Shield Friction coefficient (μ)	Regression coeff. (R^2)
$SP = 17.805e^{-0.016RMR}$	0.25	0.65
$SP = 18.436e^{-0.016RMR}$	0.30	0.65

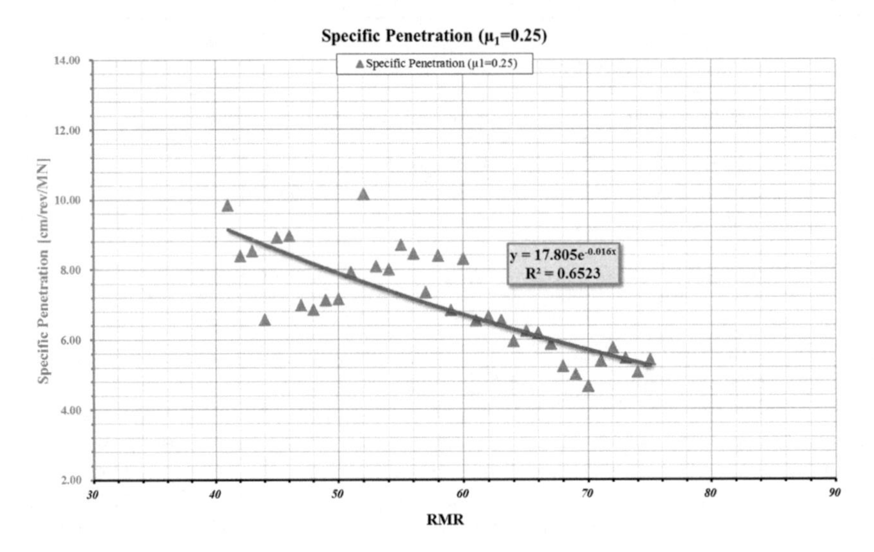

Figure 3. Specific Penetration and RMR (μ_1=0.25).

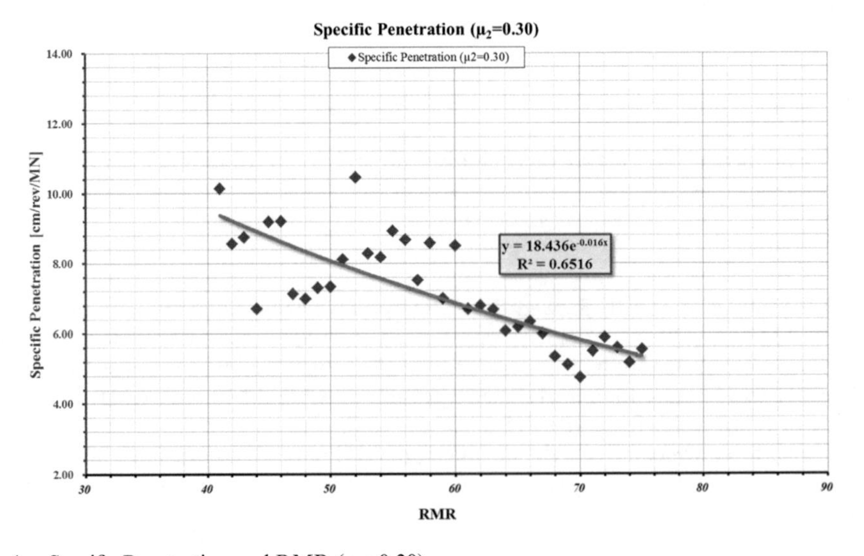

Figure 4. Specific Penetration and RMR (μ_2=0.30).

4.3.2 *Field Penetration Index(FPI) and RMR*

The result of regression analysis FPI and RMR is shown in Table 6. The FPI has tendency to get increasing with increasing value of RMR as shown in Figure 5 and 6.

Table 6. Correlation between FPT and RMR.

Correlations (FPI and RMR)	Shield Friction coefficient (μ)	Regression coeff. (R^2)
$FPI = 0.011RMR^2 - 1.0218RMR + 37.025$	0.25	0.76
$FPI = 0.0108RMR^2 - 1.0049RMR + 36.34$	0.30	0.76

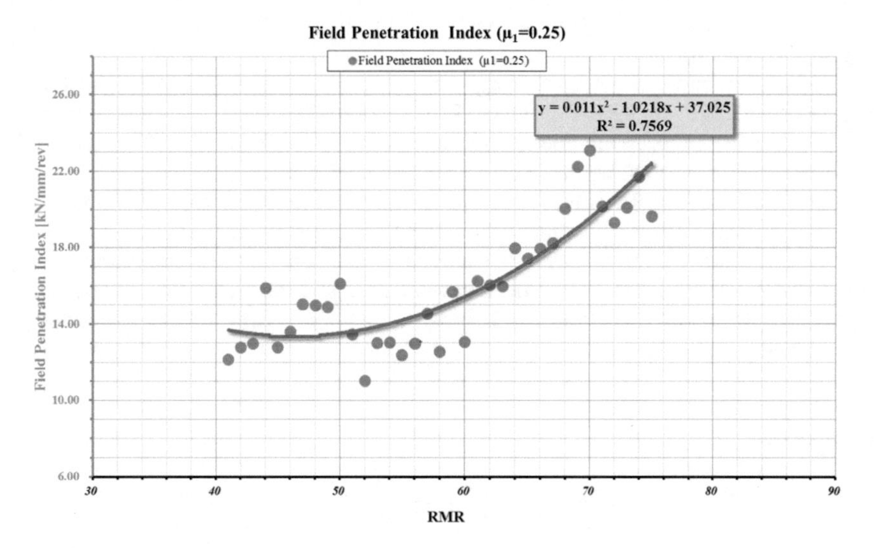

Figure 5. Field Penetration Index and RMR (μ_1=0.25).

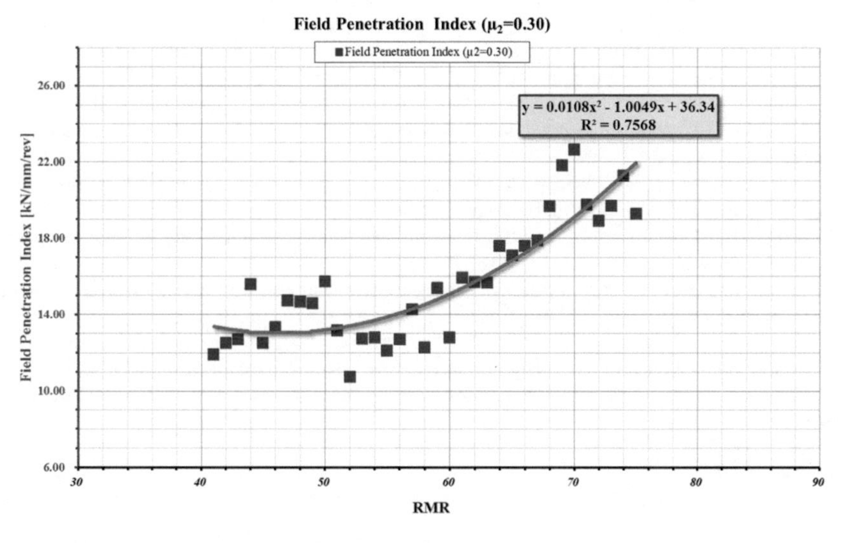

Figure 6. Field Penetration Index and RMR (μ_2=0.30).

5 CONCLUSION

This paper analyzed TBM excavation data from the hydro power project headrace tunnel which excavated with double shield TBM especially focused on the TBM performance and its correlation with ground condition

The analyzed ground consists with sedimentary rocks mainly sandstone, siltstone, mudstone. The RMR value of the ground distributed from 35 to 83, and in the majority of sections (more than 90%), RMR value from 41 to 75 is identified along the tunnel alignment. This RMR value range(RMR 41 to 75) is considered for this study

The Specific Penetration(SP) and Filed Penetration Index(FPI) are adopted for correlation analysis with RMR which is representative of ground condition for this study

The SP has tendency to get decreasing with increasing value of RMR. The correlations between SP and RMR are "$SP=17.805e^{-0.016RMR}$" in shield friction coefficient 0.25 with 0.65 of regression coefficient, and "$SP=18.436e^{-0.016RMR}$" in shield friction coefficient 0.30 with 0.65 of regression coefficient

The FPI has tendency to get increasing with increasing value of RMR. The correlations between FPI and RMR are "$FPI=0.011RMR^2–1.0218RMR+37.025$" in shield friction coefficient 0.25 with 0.76 of regression coefficient, and "$FPI=0.0108RMR^2–1.0049RMR+36.34$" in shield friction coefficient 0.30 with 0.76 of regression coefficient

Estimating the TBM performance is one of the key factors for success in TBM tunnel project. This study starts during the tunnel construction to estimate tunneling productivity for reducing construction duration. Since the ground condition was not changing rapidly, with evaluated RMR value, TBM excavation parameter could be determined with considering target productivity. Hope this study can help to estimate TBM productivity for excavating in typical sedimentation rock.

REFERENCES

Bieniawski, Z.T. 1976. Rock mass classification in rock engineering, In Exploration for rock engineering. proc. Of the symp. (ed. Bieniawski, Z.T.) Vol. 1. pp. 97–106. Cape Town: Balkema

Cassinelli F., Cina S., Innaurato N., Mancini R., Sampaolo A. 1982. Power consumption and metal wear in tunnel boring machines: analysis of tunnel-boring operation in hard rock. Tunnelling '82. London. Inst. Min. Metall. 73–81

Klein S., Schmoll M. & Avery T. 1995. TBM performance at four hard rock tunnels in California. Rapid Excavation & Tunnelling Conference. pp. 61–75. Chapter 64

Gehring, K.H. 1996. Design criteria for TBMs with respect to real rock pressure. In: *Tunnel Boring Machines - Trends in Design & Construction of Mechanized Tunnelling*, International Lecture Series TBM Tunnelling Trends. Hagenberg. A.A.Balkema Rotterdam Brookfield. pp.43–53

Barton N. 2000. Rock mass classification for choosing between TBM and drill-and-blast or a hybrid solution. In: *Proceedings of the International Conference on Tunnels and Underground Structures*. Singapore

Alber M. 2001. Advance rates of hard rock TBM drives and their effects on project economics, In: Proceedings of Tunneling & Underground Space Technology, Elsevier, pp.55–64

Sapigni M. 2002. TBM performance estimation using rock mass classification, In: *International Journal of Rock mechanics and mining Sciences*. 39(6):771–788

Ribacchi R., Lembo-Fezio A. 2005. Influence of rock mass parameter on the performance of a TBM in a gneissic formation (Varzo Tunnel). Rock Mech. Rock Eng. 38 (2) pp.105–127

Hassanpour J., Rostami J., Khamechiyan M., Bruland A. 2009. Developing new equations for TBM performance prediction in carbonate-argillaceus rock: a case history of nowsood water conveyance tunnel. *Geomechanics and Geoengineering: An International Journal* Vol. 4 No. 4 December 2009. 287–297

Khademi Hamidi J., Shahriar K., Rezai B., Rostami J. 2010. Performance prediction of hard rock TBM using Rock Mass Rating(RMR) system. Tunnelling and Underground Space Technology. Vol. 25 No. 4 pp.333–345
Bilgin, N., Copur, H., Balci, C. 2013. Mechanical Excavation in Mining and Civil Industries. CRC Press. pp.224~225

Tunnels and Underground Cities: Engineering and Innovation meet Archaeology, Architecture and Art, Volume 3: Geological and geotechnical knowledge and requirements for project implementation – Peila, Viggiani & Celestino (Eds)
 ISBN 978-0-367-46583-4

Efficient communication of geologically related data and 3D models in tunneling projects

M. Svensson & O. Friberg
Tyréns AB, Sweden

B. Brodic & A. Malehmir
Uppsala University, Sweden

ABSTRACT: During the design and the construction of a tunnel there is a constant need for updated geological/rock mechanical models, often visualized in specific software within the geological discipline. Tunnel design engineers handle the design in CAD software. Often there is a communication issue between those two disciplines. The GeoBIM concept is developed for efficient communication of any geo-related underground space data. The aim is to constantly supply all team members with quick access to the same data and models at the same time. The core of the GeoBIM concept is an efficient and flexible database that can import any data in any format and make it easily accessible. From the database data can be exported for further use. Within the GeoBIM concept it is possible to publish full 3D models, which is all navigable by using only the web browser. The GeoBIM concept is exemplified by two large Swedish tunnels.

1 INTRODUCTION

The underground space is increasingly used for a wide range of purposes, with urbanization being one of the main driving forces (Lindblom et al., 2018). Continuously increased urbanization rates inevitably impose the need for underground infrastructures and hence an improved city planning mindset, including both the surface and the underground space (Broere, 2015; Liu and Chan, 2007). A fundamental need regardless of the design of the various facilities is that they all need to handle the geological, hydrogeological and geotechnical conditions below the ground surface. The processes for realization of those facilities also engage many disciplines and organizations – engineers, designers, architects, clients, contractors, stakeholders, authorities. All these different parties need to communicate, amongst others, on the underground data and the interpreted underground space conditions. In many aspects this is not always an obvious and quality assured process. Some of the reasons for this, include, for example:

- lack of data formats and their inconsistency for geotechnically related data (including rock mechanics), making efficient quality assured joint interpretation complicated and cumbersome,
- advanced software for interpretation of geotechnical data and modelling, resulting in long processing time (partly due to the lack of data formats) and higher cost,
- expensive licenses for visualizing data and models, resulting in low accessibility to data and models for many stakeholders,
- lack of transfer formats for data and models between different disciplines and stages along the process, making the workflow inefficient and quality assurance sometimes low.

Since the society across the globe is heading towards digitalization (e.g., the concept of "smart cities") there is a lot to gain if different geotechnically/rock related data formats and geotechnically related data and object transferring could be standardized. These types of hinders could be classified as communication obstacles, which when they are overcome, will make the whole process of investigating, interpreting and communicating the underground space conditions both more efficient and quality assured than today and ease communication with other engineering disciplines related to urban underground infrastructure.

This paper's aim is to clarify the need for better communication tools in the tunneling, geotechnical and the infrastructure design discipline and to suggest several solutions and ways forward on the aforementioned issues. We present a few examples where the GeoBIM concept has facilitated and provided better environment for which this communication has successfully been done.

2 GEOTECHNICAL MODELLING HINDERS AND REQUIREMENTS

The everyday mission for the geologist and the geotechnical engineer is to find out and in the best possible way define and describe what the underground space looks like, i.e. mechanical properties and geometry, and is sometimes called geotechnical modelling. In this work, a lot of data is handled, and several different software used. It is not uncommon that the number of methods used exceeds one hundred, including core drilling and sampling, hydraulic tests, geophysical surveys such as ERT, seismics and GPR, environmental sampling and testing, hydrogeology and geotechnical laboratory testing, and others. One of the biggest challenges is to make use of all the data available during the interpretation and the modelling stages. In Scandinavia, since the early 1990's, there is a standardized data format (SGF, 2012) used for geotechnical sampling and sounding data. The usage of this format makes it almost straightforward and convenient to handle and jointly interpret these data types, regardless of which drill rig or data collection unit is used. However, with most of the methods used, data and models are far from standardized and bringing different data types together is a challenging and often time-consuming task. Therefore, it is common that the full potential of the combined data set is not reached, or simply formulated: there is a communication issue.

The traditional way of joint interpretation includes numerous drawings, plots, diagrams and tables, spread on desktops, walls, floors, screens, etc. In the digital world that has evolved during the last 10 years, many new opportunities for joint interpretation have opened. With the BIM requirements in infrastructure design, including also underground data and models, flying around the globe, the industry driven initiatives showing the potential of communicating for example geophysical data and interpretations in 3D models together with designed facilities are being brought to spotlight. The GeoBIM concept is one of those tools capable of integration, joint interpretation and visualization of all aforementioned. In project meetings, application of this concept has clearly shown to help clients and other stakeholders to understand the great potential of geophysics for reaching a proper 3D geo-model. An example of the tool used to acquire geophysical data (seismic) and final results obtained integrated in GeoBIM with different geo-data and models from Varberg railway tunnel site in south Sweden is shown in Figure 1.

2.1 *Existing data formats*

The most widely spread standards for geotechnically and rock investigation related data are the AGS format for geotechnical and geoenvironmental data (www.ags.org.uk) and the LAS data format for geophysical borehole logging data (CWLS). In Scandinavia and the Baltic countries, the data format for geotechnical sounding and sampling defined by the Swedish Geotechnical Society is widely used (SGF, 2012). For seismic data, official standard data exchange format is the SEGY format (Norris and Faichney, 2001). However, seismic acquisition equipment commonly uses the SEG2 format, or in the case of modern seismic equipment, improved data acquisition formats (offering more header information and capability to record important data acquisition information) such as SEGD (see examples in Malehmir et al. (2015) or Brodic et al. (2015)).

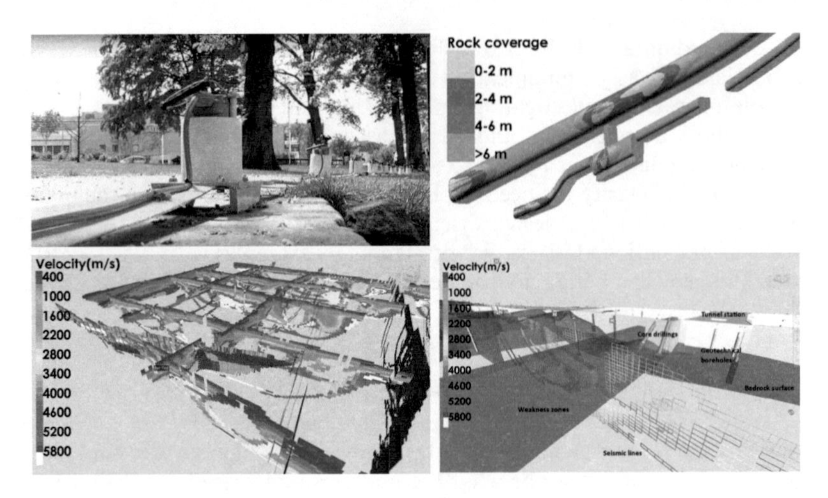

Figure 1. Integration of different data using GeoBIM for theVarberg 6 km railway tunnel project, Sweden. Photo showing a state-of-the-art digital-based seismic landstreamer (Brodic et al., 2018; Malehmir et al., 2018) used for data acquisition (UL). Rock coverage for current tunnel design (UR), geophysical profiles (seismics and DCIP) (LL) and fracture zones, core drillings, seismic profile and top of bedrock model (LR).

Once the data format hinders mentioned above are overcome within the tunneling discipline itself, a more efficiently closed digital process would be reached. With this, the possibilities for better joint interpretation and generation of the most probable geo-model, including both geometry and properties of the underground space in 3D, would drastically increase.

3 USE AND ADMINISTRATION OF UNDERGROUND GEO RELATED DATA IN A LIFETIME MANAGEMENT PERSPECTIVE

Geophysical and other geotechnically related data are produced at many stages in a project along its road to the archive for long time management (typically over 100 years or through the expected life of the infrastructure). The different stages can be separated as:

- Storing data (during project)
- Modelling
- Designing
- Visualizing
- Data management (long term/archiving).

3.1 *Infrastructure design phase*

The final aim of investigating and interpreting the underground space modelling is to deliver a geo-model to the design team for the actual facility. This is a continuous process commonly requiring daily updates of geotechnical conditions available for design of foundations, handling of settlements etc. Those design disciplines most often use design software of a CAD-type. For an efficient workflow, this requires that the geotechnical modelling tools can communicate with the design software. For quality assurance purposes, it is quite beneficial if all sounding, monitoring and sampling data can be visualized in 3D in the software itself. Integration of geophysical and geotechnical data from two tunnel projects in Sweden are shown in Figure 2. Generation of this type of models requires data formats capable of communicating with each other, which is nowadays not always the case.

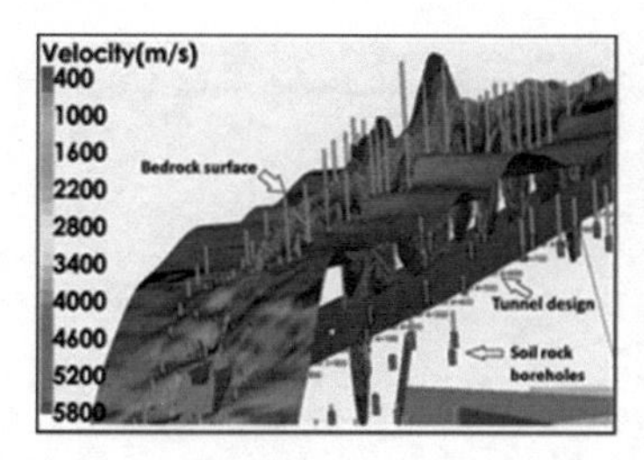

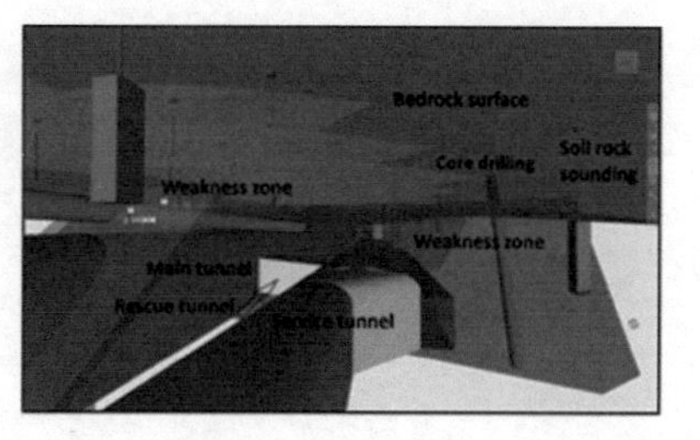

Figure 2. Various types of geo-related data and models are combined for joint visualization together with the current tunnel design in the CAD design tools for the Norrbotniabanan railway tunnel, Sweden (left). A close up of a complicated area in the Varberg railway tunnel project, where interpreted weakness zones, core drillings, geotechnical soundings and the tunnel design are visualized together for further decisions on the design (right). (Autodesk Navis works).

To work according to a BIM vision within an unbroken digital information supply chain, the geo-model and its objects need to be delivered to the design departments in a format which they can instantly use to steer their design. The information process therefore must be streamlined for many different tasks, which requires information interpreted and evaluated by all geotechnically related engineering expertise.

Each project has its own geotechnical challenges since the geological conditions and the requirements of the construction vary. Therefore, the choices made during the modelling are also project-specific. As a result, the model is constantly updated, or several different models are often produced as the knowledge of the geotechnical conditions increase. To handle the model updates or changes, a rational information management and modelling methodology is required, as the time schedule of projects is often tight.

Transferring the geo-model to the designer/structural engineer (importing geotechnical data and geo-models into the design software) is today often done via the internal CAD software formats (e.g., dwg, dxf). However, this procedure does not fulfil the proper BIM requirements. To fully comply with the BIM requirements, other transferring formats like IFC, InfraGML and CityGML have been suggested for above the ground data and objects concerning structural engineering. So far, there is little evidence they will work satisfactory for geophysical or other geo-related underground data and models.

3.2 *Visualizing data and 3D models*

Traditionally, geotechnical and rock mechanical engineers, geologists and hydrogeologists are those working with the geotechnically related data, reporting the final models in 2D drawings or 3D models and/or in written reports for further use in the design process. These models seldom reach the full potential in terms of how much value there is in the geo-model. With modern tools capable of integration of all of them, much more can be gained from the geo-models at all stages of a project, supporting the entire design process in a better way than presently used, aiming at a more optimized project. By using modern 3D visualization tools, better quality assurance of all data is reached, and larger group of stakeholders/actors can be engaged or involved during different project stages. The geo-model can be communicated in a pedagogic way and better and more optimized understanding and design can be gained, see Figure 3. In large scale projects, communication is the key factor for success, and the GeoBIM concept has proved to be successful in communicating all geo-related data and models in infrastructure planning projects (Svensson and Friberg, 2018).).

3.3 *Data management (long term/archiving)*

Data and models from infrastructure projects are managed differently in different projects and different countries. On a national level in Scandinavia, governmental databases are missing. All data that is collected most often stays with the contractor. Among the reasons for not

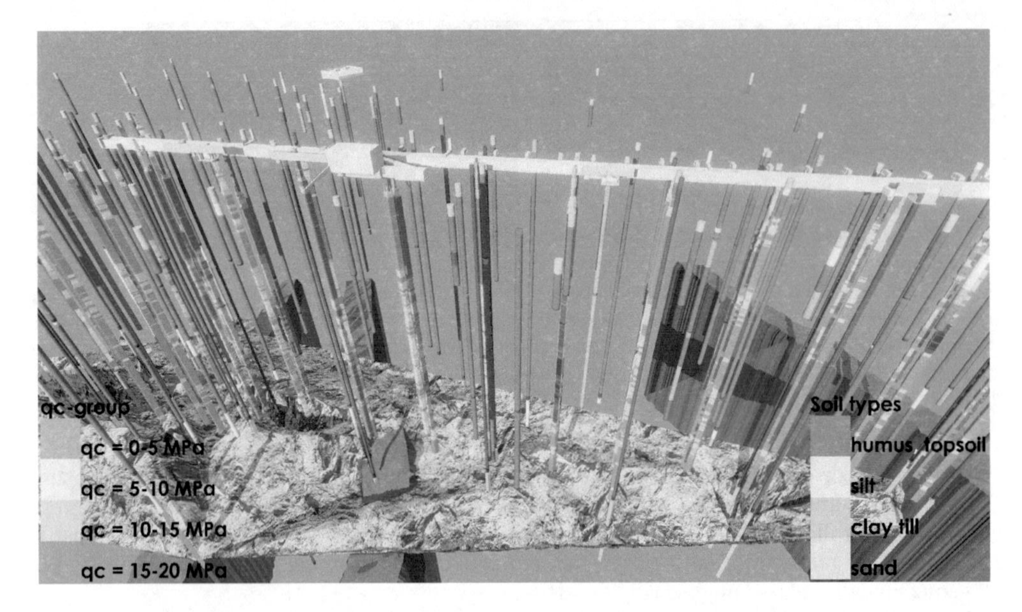

Figure 3. A geotechnical 3D model visualized together with core drilling data and parts of the current design of the facility in a software using computer game technique to get photo realistic images. A powerful tool to communicate a geotechnical model to the non-geo public. Example from the European Spallation Source (ESS), Lund (www.ess.se; Svensson, 2015).

keeping any national databases are the lack of standardized data formats and transformation formats, and tools for accessing the data and the models, see below. Another reason is the legal aspect, it is not always clear in what way data could be used, and by whom. However, it is obvious that national databases of geophysical and geotechnical data and models accessible for the whole industry would gain a lot to the industry on an everyday basis and save a lot of money for the society in general.

4 THE GEOBIM CONCEPT

The core of the GeoBIM concept is a database capable of handling - importing, storing, exporting – all geotechnically related data that are used in an infrastructure project, hence including data from geophysics, geotechnical sounding and sampling, rock cores, borehole geophysics, laboratory tests, groundwater and contaminated soil investigations. Figure 4 shows a block diagram of the GeoBIM concept. The GeoBIM concept relies on the database configuration and information accessibility.

4.1 *The database configuration*

Essentially, the database stores point information with location and eventual relation to other points. The point carries a value and this value is related to a measurement (information of measurement method etc.) as well as an unlimited number of dimensions (what has been measured i.e. resistivity, velocity, time etc). This approach ensures that information from new methods may easily be added, and large amounts of data can be handled rationally since data are handled as a point cloud. The database model is implemented in a PostgreSQL database located in the Cloud.

4.2 *Database access*

To access the information, a database viewer and a web map has been connected, which are accessible through a GeoBIM Portal (www.geobim.se) where project members can login to

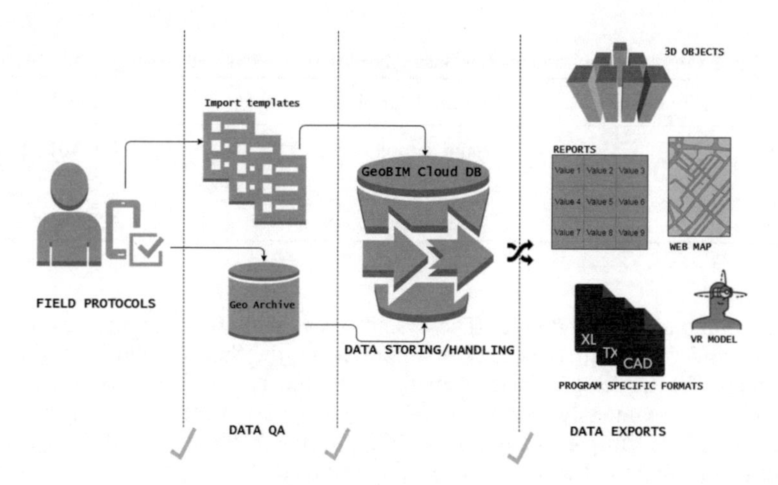

Figure 4. The GeoBIM concept process, enabling a completely closed digital chain from field data collection to 3D geo-model.

their projects. This enables the project member to get an overview, as well as a detailed view, of the data and related documents (e.g. seismic profiles, borehole information, etc.). From the GeoBIM Portal, project members are also able to import data from investigations and export data in various data formats depending on the task being performed and software they will be using. As an example, a project member responsible for the geotechnical modelling might be interested in all geo-data that can improve the interpretation of a rock surface model (e.g., Figure 1), while a geo-chemical engineer might want data to analyze a pollution propagation. The output from both these tasks is then used combined with raw field data in a BIM model, to review and share the results, see Figure 1 and Figure 2.

When geo-models are produced by skilled staff and according to design standards (e.g. Eurocode, 1997), making use of all available data and reviewed by cross discipline experts ensuring their interrelationships seem correct, the best possible interpretation can be performed. These models may thereafter be used for more accurate design and, for example, optimize mass balance calculation or reinforcement design, among others.

5 ONGOING AND FUTURE WORK

The next generation of the infrastructure design process focuses on full BIM and LCC (Life Cycle Costs) as the governing decision tool. This means that the maintenance phase is gaining a lot more interest than before, resulting in a need for systems where the facility and the accompanying data and information need to be accessible for 120 years or so (design criteria). This is part of the global digitalization and there is a great opportunity and potential for a quicker and more precise process for producing 3D models of the underground space and aid in infrastructure planning process at all stages. As the development of both hardware and software in the geophysical community is continuously improving, old data can be revisited and reprocessed or modelled. However, if this is to become reality, more standardized data formats and exchange formats need to be developed and widely accepted. The tunneling and the geotechnical community needs to come up with a sustainable solution similar to those used by hydrocarbon industry but involving a variety of applications and targets in civil engineering projects.

In a recently started project the CoClass system is under development for underground objects, aiming at giving all geo-related underground objects – points, lines, areas, volumes – a certain code, which will give all systems in the infrastructure process a possibility to identify that certain object (Table 1). The CoClass system is a Swedish initiative built on the

Table 1. Example of preliminary CoClass classification system for underground objects.

Type of principal object	Example of object	Example of code
Point	Sand sample	A.BA01.ZGA.10
Line	Seismic line	A.BA01.ZGB.03
Area	Interpreted bedrock surface	A.BA01.CBC.01
Volume	Contamination plume	A.BA01.CBE.01

international standard ISO 12006-2. The system (version 1.0) is already set for structures above ground and is now introduced and supported by the Smart Building organization [ref smart building]. When CoClass is implemented for underground objects, for example geophysical models can seamlessly start to be used by all parties in the design process. A contractor on site could, for example, easily find a DCIP profile in the project database, bring it up on the screen and compare it with on line drilling results from a tunnel, in real time, regardless of what software or tool she/he is using.

6 CONCLUSIONS

Presently, numerous different methods (over 100s) for delivering data in various formats are used in tunneling and the underground infrastructure industry. Among others, these include core drilling, geological mapping and geophysical data providing models and results often not supported by commonly software in the design process. This means the full potential of a joint interpretation of all available data is set aside at different stages. To make an optimal use of all available data, a proper database and an efficient workflow is needed to communicate both data and models for the end user – the design team. The GeoBIM concept suggests such a tool and process, capable of efficient digital handling of all geotechnically related data from field to long time management. The end user will be able to access data via the web. The core of the concept is the GeoBIM database, from which data can easily be exported for 3D modelling purposes, visualized in 2D/3D, or the metadata analyzed, among others.

The GeoBIM concept has been implemented in four large infrastructure projects in Sweden and more than ten other smaller projects. The experience obtained suggests that the GeoBIM methodology should be considered as a state-of-the-art tool and solution for communication of underground geo-data, models and interdisciplinary between parties involved in the design, management or the overview process in any underground infrastructure project.

In the near future, there is a lot of efficiency and preciseness to gain, if common data formats and exchange formats of all geo-related data properly compatible with CAD software are developed, and national databases of geotechnically related data are put up and made easily accessible. In a global digitalization perspective, the communication formats/methods/skills for a larger group of parties involved (stakeholders) than today must be developed, to make use of the full potential of geo-related data and models in a facility's lifetime perspective. The development of the CoClass system for underground geo-related objects is such an initiative. The tunnelling community should act better and proactively adopt and support such a development. Acquiring and reviving all geo-related data, if not expensive, it would sometimes be impractical or impossible to do at a later stage, and hence the great value they offer needs to be secured at least for the design life time of any facility, for example a tunnel.

ACKNOWLEDGEMENTS

The GeoBIM project was part of the TRUST umbrella as TRUST 4.1 (http://trust-geoinfra.se/) financed by Sven Tyréns Stiftelse and Formas (project 2012-1919). Uppsala University contributed to this work through collaborations within the Varberg double-train underground planning project and Trust 2.2 (http://trust-geoinfra.se/) financed by Formas (project 2012-

1907), BeFo (project 340), SUB-Skanska, SGU and Sven Tyréns Stiftelse. The BIM/GIS team at Tyréns AB is greatly acknowledged, as well as Henrik Möller Geokonsult.

REFERENCES

Broere W. (2016). Urban underground space: Solving the problems of today's cities, *Tunnelling and Underground Space Technology* 55 (2016) 245–248

Brodic, B., Malehmir, A., Svensson, M., and Jonsson, J., (2018), Feasibility of 3D random seismic arrays for subsurface characterizations in urban environments. *Near Surface Geoscience, Porto-Portugal*, September 2018

Brodic, B., Malehmir, A., Juhlin, C., Dynesius, L., Bastani, M., and Palm, H., (2015), Multicomponent broadband digital-based seismic landstreamer for near surface applications. *Journal of Applied Geophysics*, 123, 227–241

Eurocode 7, (2007), EN 1997-2:2007

Lindblom, U., Ericsson, L. O., Winqvist, T., Tengborg, P., Håkansson, U. (2018). *Sweden Underground*, Stockholm, BeFo Rock Engineering Research Foundation and Swedish Rock Engineering Association

Liu, L., & Chan, L. S. (2007). Sustainable urban development and geophysics. *Journal of Geophysics and Engineering*, 4(3), 243

Malehmir, A., Lindén, M., Friberg O., Brodic, B., Möller, H., and Svensson, M., (2018). Unraveling contaminant pathways through a detailed seismic investigation, Varberg southwest Sweden. *EAGE Near Surface Geoscience, Porto-Portugal*, September 2018

Malehmir, A., Zhang, F., Dehgahnnejad, M., Lundberg, E., Döse, C., Friberg, O., Brodic, B., Place, J., Svensson, M., and Möller, H., (2015), Planning of urban underground infrastructure using a broadband seismic landstreamer—Tomography results and uncertainty quantifications from a case study in southwest of Sweden. *Geophysics*, 80, B177–B192

Norris M. W. & Faichney A. K. (2002). *SEG Y rev 1 Data Exchange format*, SEG Technical Standards Committee. Society of Exploration Geophysicists

SGF (2012), SGF:s dataformat, *SGF Rapport 3:2012* (in Swedish), Swedish Geotechnical Society, Linköping

Svensson M. and Friberg O., (2018), Communication of geophysics in underground infrastructure projects, *Proceedings of the 31st Symposium on the Application of Geophysics to Engineering and Environmental Problems*, Nashville, TN, March 25–29

Svensson M. and Friberg O. (2017). GeoBIM for infrastructure planning, *EAGE Proceedings of the 23rd Near Surface Geoscience conference, Malmö* 3–7 September

Svensson M., (2016), GeoBIM for optimal use of geotechnical data, *Proceedings of the 17th Nordic Geotechnical Meeting, Reykjavik* 25th – 28th of May

Tunnels and Underground Cities: Engineering and Innovation meet Archaeology, Architecture and Art, Volume 3: Geological and geotechnical knowledge and requirements for project implementation – Peila, Viggiani & Celestino (Eds)
 ISBN 978-0-367-46583-4

A finite element analysis of tunnel response to permafrost thaw

D. Vo, L. Chung, J.S. Cho & Y. Salem
California State Polytechnic University, Pomona, California, USA

ABSTRACT: Rising global temperatures are accelerating the rate of thaw in permafrost soils, resulting in losses of soil volume and shear strength. For tunneling in freezing climates, examination of tunnel structure interactions with permafrost is pertinent. This research analyzes a tunneling scenario in Anchorage, Alaska. A tunnel with precast concrete liner is created to serve as a mode of transportation underneath Elmendorf Air Force Base. Alaska is chosen due to the freezing temperatures that facilitate permafrost soil and its proximity to the Ring of Fire where seismic waves originate. Methodology entails gathering relevant soil properties of the Anchorage region to examine changes in displacement and stress of the tunnel. Static analysis using Finite Element Method (FEM) is performed using RS2 software by Rocscience for frozen and thawed conditions in pursuance of identifying the impacts of melting permafrost on tunnels.

1 INTRODUCTION

Permafrost or related ground ice is defined as permafrost-soil or rock-in ground that remains at or below 0°C for at least two years (Murton, 2009). However, permafrost can also exist in ground that contains little to no moisture, which is referred to as dry permafrost. The more prevalent cases of permafrost found in nature consist of pores that are either completely filled with ice or partially filled with a mixture of ice and unfrozen water. This research deals with the latter of the two cases. Another important distinction to establish is the difference between permafrost and seasonally frozen soil. Seasonally frozen soil is known as the active layer which lies at the surface of the ground. Depending on geographic location the frost line can penetrate to a certain depth and define the thickness of this active zone. A general depiction of the active zone and permafrost are depicted in Figure 1.

As the mean temperature increases, the active layer becomes thicker, intruding the permafrost. This issue is especially significant in Anchorage, as the mean annual temperature has climbed 3.6 degrees Celsius since 1950 (Stuefer, 2016). While the rising temperature is a large factor of thermokarst, ablation cannot be solely attributed to global warming, as the removal of flora for development is also a component (Cooper, 2014). Regardless of the causes, it is certain that the climate will continue to warm during the next century, provoking thaw, which reduces the soil strength and consolidates the soil volume from loss of excess ice (Murton, 2009). This can result in subsidence, surface water percolation, and thermal or mechanical erosion from permafrost infused groundwater. From there, geotechnical problems, landscape evolution, and sedimentary disturbance may arise, affecting the existing subsurface infrastructure. All signs point to the unbreakable sequence of events, as the existence of thermokarst is a sensitive indicator of environmental deviants, while warming temperatures is a key element of the thermokarst phenomenon (Murton, 2009).

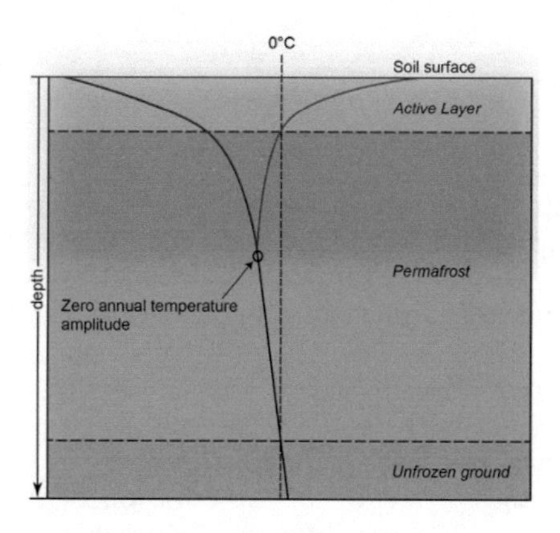

Figure 1. Typical active and permafrost layers (CEN: Center for northern study).

Looking deeper into the characteristics of permafrost, when the pores of soil are saturated by unfrozen water the soil has a set volume. If the soil freezes the water expands and an extra volume of excess ice is present in the frozen soil. The National Snow & Ice Data Center (NSIDC) defines excess ice as "the volume of ice in the ground which exceeds the total pore volume that the ground would have under natural unfrozen conditions." The process of excess ice thaw is called thermokarst. Thermokarst is problematic in that soil strength is reduced due to thawing of cementitious ice, and reductions in soil volume due to loss of excess ice. (Murton, 2009). As mentioned before, the process of thermokarst is expedited as arctic annual mean temperatures rise at a rate twice as fast as the rest of the planet.

The region of interest for the tunnel excavation is Anchorage, Alaska, under the Elmendorf Air Force Base. Eighty percent of Alaska is underlain by permafrost (Muller, 2008). The permafrost in Anchorage has a discontinuous span from 0-15 m below surface, and an active layer of 2 to 3 meters, as seen in Figure 2.

Finite element modeling within the scope of frozen soils versus thawing soils is not a thoroughly researched area. Thus, this research examines the differences in tunnel excavation response due to permafrost thaw. The proposed tunnel will require a 7.3-meter diameter excavation, and the center of the tunnel is 21.9 meter below surface, as displayed in Figure 3. Parameters of interest will be points of high displacement and stress between the frozen and thawed model.

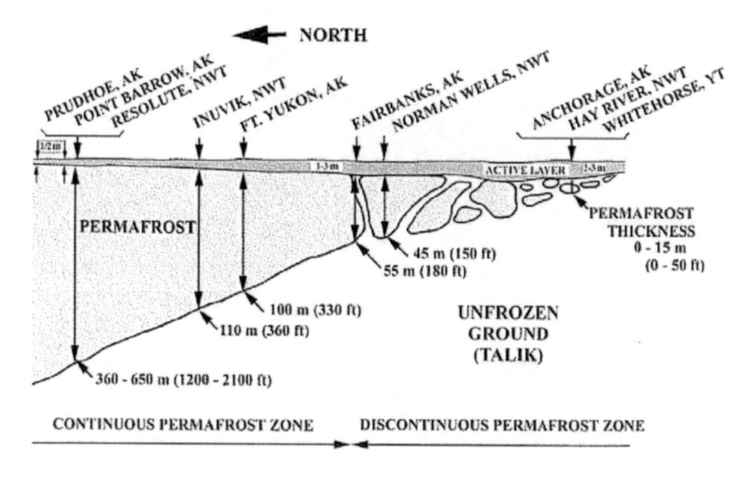

Figure 2. Active layer and Permafrost thickness in Alaska (Davis, 2001).

2 METHODOLOGY

2.1 *Geotechnical Site Conditions*

A generalized geologic map of Anchorage details the surficial deposits over the project site as Alluvium of the Anchorage Plain-gravel and sand-that is well bedded and evenly sorted (Miller & Dobrovolny, 1959).

Bootlegger Cove Clay underlies much of the Elmendorf Air Force Base, areas around the international airport, and the outlying suburbs of Anchorage (Miller & Dobrovolny, 1959). The report referenced relevant soil properties of the clay as shown in Table 1. While the soil layer data dates back to 1959, its accuracy is adequate for the purpose of this research, due to lengthy soil cycle. It takes hundreds to thousands of years for soil to form, so 60 years is simply a fraction of the duration of the formation process. Further, by using soil data from 1959, the difference between frozen and thawed conditions is more practically represented, as the thawing process requires an extended amount of time.

Well Logs completed by U.S. Corps of Engineers in the Miller report under Elmendorf Air

Force Base allowed the stratigraphy of the soil profile underlying the base to be envisioned as shown in Figure 3. Soil layers include gravelly sand, sandy gravel, medium sand, gray clay, and a waterline at 35 feet, per USGS report.

Without readily available laboratory testing information on local site soil, this research presumes critical failure in the soil, and idealizes some portions of soil properties to be chosen to minimize thawed soil strength to represent the worst-case scenario. Table 2 presents a summary of frozen to thawed soil parameters for each layer. Previous testing and research on relationships of unfrozen and frozen sands and clays was used to convert typical unfrozen soil parameters from Das (Das, 2009) to their frozen counterparts. Moreover, Youssef and Hanna found that frozen sands tend to increase shear strength

Table 1. Soil properties of bootlegger cove clay underlying Elmendorf Air Force base.

Liquid Limit	Plastic Limit	Plastic Index	Field Moisture (Percent)	Friction Angle (°)	Cohesion C (kPa)
39	22	17	35	22	55.06

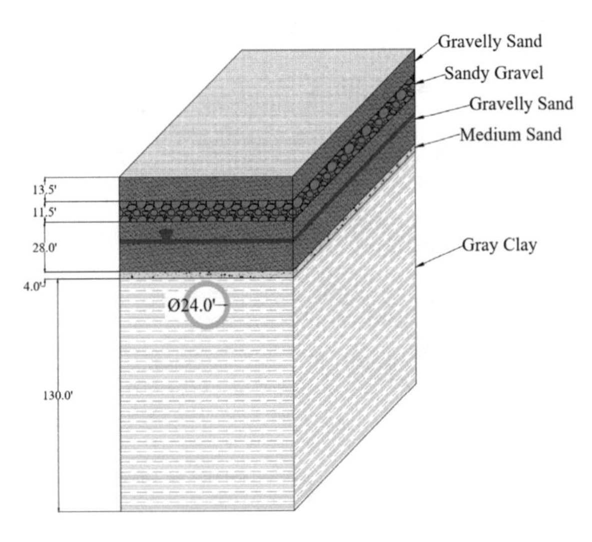

Figure 3. Soil Profile under Elmendorf Air Force Base (units in feet).

Table 2. Idealized Soil Properties.

Idealized Soil Properties		Frozen		Unfrozen	
General	Soil Type	Gravelly Sand/ Sandy Gravel	Gray Clay	Gravelly Sand/ Sandy Gravel	Gray Clay
	Number of Layers	4	1	4	1
	Unit Weight (kg/m^3)	1601	1922	1601	1922
	Failure Criterion	Mohr Coulomb	Mohr Coulomb	Mohr Coulomb	Mohr Coulomb
Strength	Material Type	Elastic	Elastic	Elastic	Elastic
	Tensile Strength (kPa)	1	1500	21	30
	Peak Friction Angle (deg)	42	22	42	22
	Peak Cohesion (kPa)	7000	330	0	55
	Type	Isotropic	Isotropic	Isotropic	Isotropic
Stiffness	Poisson's Ratio	0.3	0.3	0.3	0.3
	Young's Modulus (kPa)	400,000	80,000	100,000	20,000

and modulus of elasticity by factors of 2.5 and 4, respectively (Youssef & Hanna, 1988). In another study by Czurda and Homann, it is found that on average, cohesion of clays increased by a factor of 6 when testing from an unfrozen to a closed frozen system (Czurda & Hohmann, 1997).

2.2 *Finite Element Analysis*

A two-dimensional plane stress analysis was performed using RS2 by Rocscience. A frozen 2D model was constructed using previously gathered geotechnical engineering properties to be analyzed for displacement of the tunnel face and points of high stress in the excavation of the tunnel. Soil material type was chosen to be elastic and the Mohr-Coulomb failure criterion was selected to reflect a conservative, linear analysis of soil strength (Griffiths, 1990).

Creation of the model began with accurately creating a soil profile to reflect the soil stratigraphy underneath the project site as shown in Figure 4. The model width and depth are chosen to be several magnitudes larger than the 7.3 m. diameter tunnel excavation to avoid having boundary constraints which could affect any stress bulbs or displacements local to the excavation. A fine graded mesh with 11,468 six-noded triangle elements was used to approximate the soil domain, shown in Figure 4.

Properties for the precast concrete tunnel lining are displayed in Table 3.

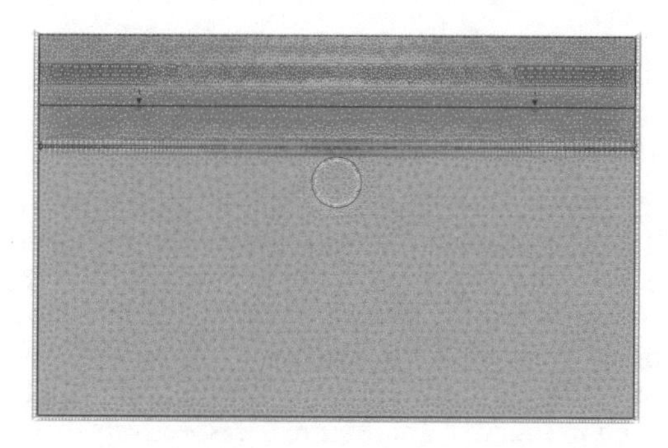

Figure 4. 2D Mesh of Frozen Soil in RS2 (right).

Table 3. Liner Properties to support model excavation.

Concrete Liner Properties	
Liner Type	Standard Beam
Young's Modulus (kPa)	20,684,272
Poisson's Ratio	0.2
Thickness (m)	0.3
Material Type	Elastic
Unit Weight (kg/m^3)	2403

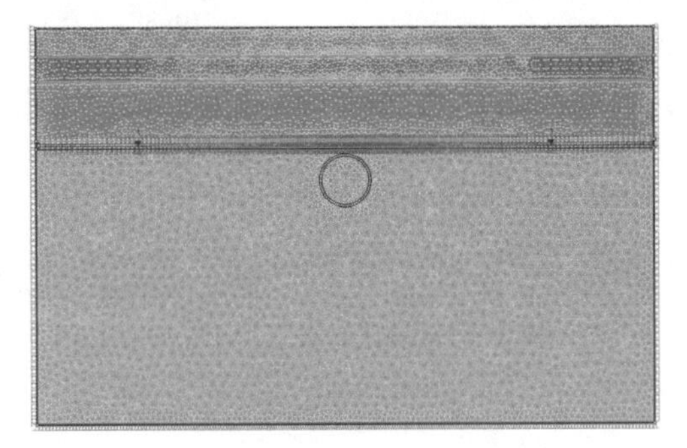

Figure 5. Thawed RS2 model with drained waterline.

In order to gradually display the displacements of the excavation and high stress areas a 10-stage analysis was performed. Stage 1 of the analysis is the unexcavated soil with body forces included. Stage 2 presents the tunnel excavation of 7.3 m. diameter and center at 21.9 m. below surface. In stage 2, the tunnel internal pressure has a factor of 1 so as to be equivalent to induced stresses on the tunnel surface. As the model progresses from stage 3-9 the internal pressure relaxes with factors of 0.8, 0.6, 0.4, 0.2, 0.1, 0.05, and 0 respectively. Once the internal pressure factor is 0 in stage 9 the tunnel excavation is complete. One-foot thickness pre-cast concrete tunnel liner is then installed in stage 10 and displacements and stresses can be considered and graphically displayed. A similar process is performed for the unfrozen model with the exception of varying geotechnical properties and the shifting of the waterline to the beginning of the clay layer. Waterline is placed at the impermeable clay layer to reflect the drained condition of the porous nature of sand and gravelly sand in the 4 layers above, shown in Figure 5.

3 RESULTS

3.1 *Comparison of Frozen and Thawed Model Displacements*

Upon a query of the excavation boundary in RS2, it is discovered that the local area of highest displacement is at the roof of the excavation. On the displacement contour map in Figure 6. the region is bounded by red. This will be true for the frozen and unfrozen analysis.

When plotting the point of highest displacement (top of the tunnel face) versus stage number, the progression of displacement as the excavation is completed for the frozen and unfrozen model can be seen in Figure 7 and Figure 8, respectively.

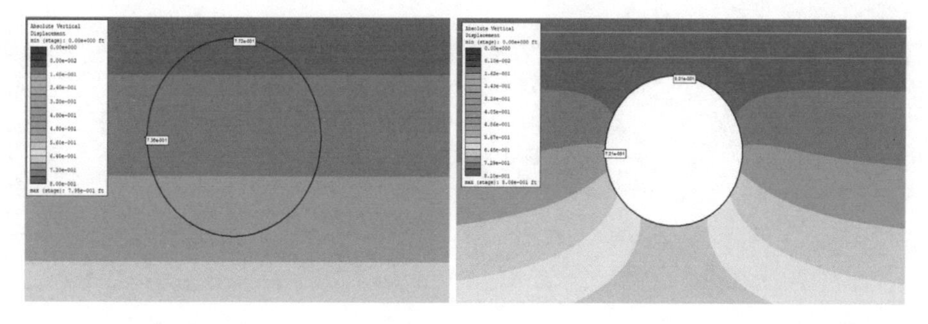

Figure 6. Absolute vertical displacement contours for pre-excavation (left) and post-excavation (right).

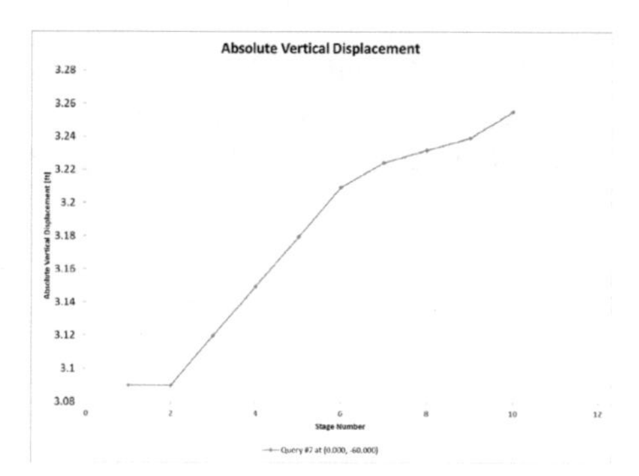

Figure 7. Stage number vs absolute vertical displacement for frozen model.

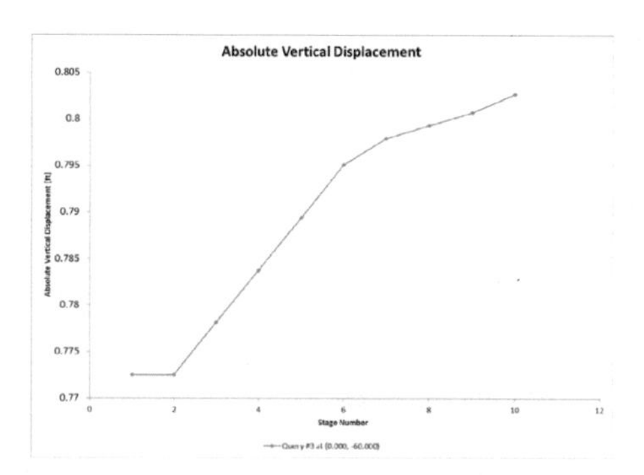

Figure 8. Stage number vs absolute vertical displacement for unfrozen model.

From stage 1 to stage 2, the soil profile settles to its natural state due to the inclusion of body forces. This research defines the difference in absolute vertical displacement from stage 2-10, as effective displacement due to this value of interest is the displacement that occurs when excavation and liner installation is performed. Table 4. displays values for the effective displacement of the tunnel roof.

Table 4. Effective Displacement for Frozen and Unfrozen Model.

Frozen Effective Displacement (mm)	Thawed Effective Displacement (mm)
10.16	45.97

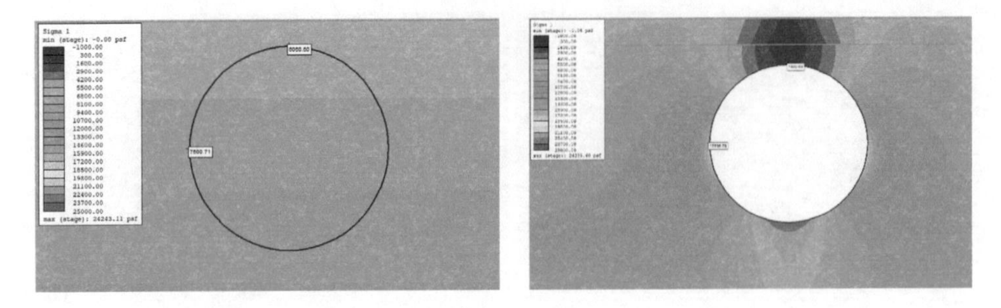

Figure 9. Principal stress contours for pre-excavation (left) and post-excavation (right).

3.2 *Comparison of Frozen and Unfrozen Principal Stress*

The principal stress contour map in Figure 9. displays this region in green for pre-excavation, and yellow-green for post-excavation. Another query for principal stress on the excavation boundary. RS2 shows that the local areas of highest stress occur at the sides of the excavation for both frozen and unfrozen. The highest points of stress are typically 45 degrees below the horizontal in unfrozen stiff clays (Galli & Grimaldi, 2004).

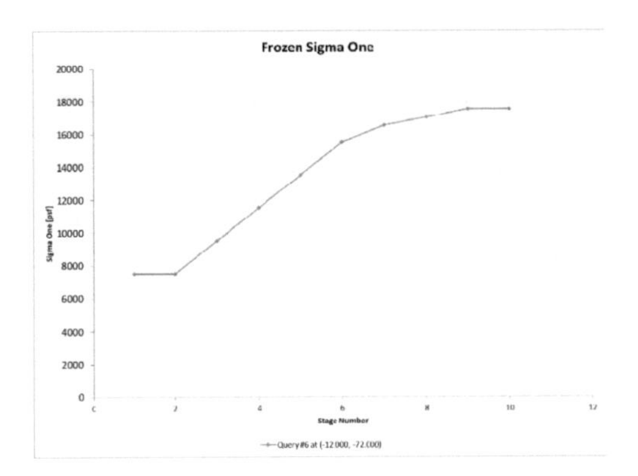

Figure 10. Stage number vs frozen principal stress.

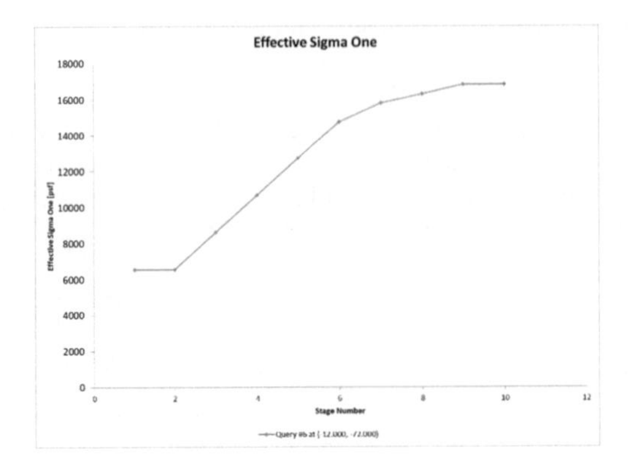

Figure 11. Stage number vs unfrozen effective principal stress.

Table 5. Stress Change for Frozen and Unfrozen Model.

Frozen Stress Change (psf)	Unfrozen Effective Stress Change(psf)
480	490

When plotting for a point in the region of highest principal stress (side of tunnel face) versus stage number, the progression of principle stress as the excavation is completed for the frozen and unfrozen model is shown in Figure 10 and Figure 11.

From stage 1 to stage 2 the soil profile experiences no stress change because no excavation has been performed and internal pressure factor has not been reduced. Stress change occurs from stage 2–10, the excavation and liner installation occurs. Table 5. displays values for stress change from stage 2–10. The unfrozen model has a waterline at the clay layer to represent the drained condition, and experiences a slightly higher effective stress due to pore water pressure.

4 CONCLUSION

Tunnel responses between frozen and thawed soil layer conditions were explored with the aid of RS^2 by Rocscience. Preliminary research was completed on the what-if location of Anchorage, Alaska to approximate site conditions in regards to soil layer thickness, groundwater level, strength parameters, etc. Conclusions to be made include:

1) The maximum effective displacement on the tunnel face was found to be at the top of the tunnel face and the difference in displacements of the frozen model to thawed model are drastic. The tunnel face in thawed soil experiences a much higher magnitude of displacement and this falls in-line with the loss of shear strength and thawing consolidation that is a function of losing the cementitious nature of excess ice in its pores.
2) Regions of high stress on the tunnel face appear on the sides. Tunnel stress appears to be relieved at the top and bottom of the tunnel face according to the principal stress distribution contours. Magnitude of effective principal stress by the tunnel in the frozen and thawed model are virtually identical except for stress changes from pore water pressure differences apparent in the drained condition of sand and gravel.
3) The tunnel lining installed in stage 10 of analysis did not relieve stress to a significant magnitude.

Future work on the model can include a dynamic model to reflect the high seismic activity apparent in the Alaskan region, as well as an analysis of heaving or subsidence. An examination of the brittle nature of ice can be modeled in response to seismic loading.

REFERENCES

CEN: Center for northern study, *Laval University*: http://www.cen.ulaval.ca/adapt/communications/permafrost101.php

Cooper, E.J. 2014. Warmer Shorter Winters Disrupt Arctic Terrestrial Ecosystems. *Annual Review of Ecology, Evolution, and Systematics* 45 (1): 271–295.

Czurda, K.A. & Hohmann, M. 1997. Freezing effect on shear strength of clayey soils. *Applied Clay Science* 12: 165–187.

Das, B. 2009. *Principles of Geotechnical Engineering 7th Edition*. Boston: Cengage Learning.

Davis, N. 2001. *Permafrost: A guide to Frozen Ground in Transition*. Fairbanks: University of Alaska press.

Galli, G. & Grimaldi, A. 2004. Three-Dimensional Modeling of Tunnel Excavation and Lining. *Computer and Geotechnical*. 31(3): 171–183.

Griffiths, D. 1990. Failure Criteria Interpretation Based on Mohr-Coulomb Friction. *Journal of Geotechnical Engineering* 116 (1): 986–999.
Miller, R. & Dobrovolny, E. 1959. *Surficial Geology of Anchorage and Vicinity Alaska*. Washington: United States Government Printing Office.
Muller, S.W. 2008. *Frozen in Time. Permafrost and Engineering Problems*. Reston: American Society of Civil Engineers.
Murton, J.B. 2009. Global Warming and Thermokarst. In Margesin R. (eds), *Permafrost Soils. Soil Biology* (16): 185–203.
Stuefer, M. 2016. Temperature Changes in Alaska. *The Alaska Climate Research Center*.
Youssef, H. & Hanna, A. 1988. Behavior of Frozen and Unfrozen Sands Triaxial Testing. *Transportation Research Record* 1190: 57–64.

Tunnels and Underground Cities: Engineering and Innovation meet Archaeology, Architecture and Art, Volume 3: Geological and geotechnical knowledge and requirements for project implementation – Peila, Viggiani & Celestino (Eds)

Acoustic emissions from flat-jack test for rock-burst prediction

A. Voza, L. Valguarnera & S. Fuoco
Brenner Basis Tunnel BBT-SE, Bolzano, Italy

G. Ascari
Akron S.r.l., Bovisio-Masciago, Italy

D. Boldini
DICAM, University of Bologna, Bologna, Italy

D. Buttafoco
B.T.C. Brenner Tunnel Construction, Rome, Italy

ABSTRACT: Rock-burst is a local instability problem affecting compact rock-masses in deep tunnels: it can generate important over-breaks, but even more, safety issues for workers in the area where it occurs. The experimental study, aimed at predicting rock-burst occurrence during the excavation of the Brenner Base Tunnel, was carried out in massive granite below an overburden of 1000–1200 m. The experiment consisted in compressing a portion of the rock-mass at the tunnel side-wall by means of a couple of flat-jacks, measuring simultaneously the generated acoustic emissions with velocimeters and accelerometers. Recorded data clearly show a peak in the acoustic emission energy few instants before the failure of the rock-mass. This value, after careful site validation based on the continuous monitoring of acoustic emissions during the excavation, has been implemented as an alarm threshold for the interpretation of measurements and for the adoption of appropriate countermeasures.

1 INTRODUCTION

Rock-burst is defined as the damage to an excavation that occurs in a sudden or violent manner and is associated with a seismic event (Hedley 1992, Kaiser et al. 1996).

Rock-burst phenomena occur frequently in deep tunnels excavated in hard rock-masses and can cause safety problems, construction delays and economic losses. With increasingly mining and tunnelling depth, the intensity and frequency of rock-burst occurrence are becoming higher, imposing serious threats to the safety of underground constructions (Kaiser & Cai 2012).

Two main damage mechanisms, characterised by different level of severity, are recognisable (i.e. Diederichs 2007): the spalling, consisting in the development of fractures parallel to the excavation boundary, and the strain-burst, associated to a violent rupture and expulsion of rock wedges.

Risk associated to rock-burst occurrence is nowadays managed by a continuous monitoring of the micro-seismicity detectable during the excavation (i.e. Feng et al. 2015, Feng et al. 2017, Yu et al. 2017), whose characteristics depend on the failure processes ongoing in the rock-mass and whose interpretation can be used to anticipate large-scale and violent events.

This paper describes an experimental activity carried out in the Brenner Base Tunnel for rock-burst detection. In fact, few mild rock-burst phenomena affected the site in the past and there is a pressing need of preventing similar events by early-stage assessment and implementation of appropriate countermeasures.

The experiment consisted in a specifically designed double flat-jack test conducted in a massive granite under an overburden of about 1000 m. During the test, acoustic emissions were measured showing a peak in intensity and a well-defined frequency content before failure was reached in the rock-mass.

These findings were employed to set a new, physically-based, alarm threshold for the interpretation of the continuous micro-seismic measurements, higher than that previously adopted. As far as the new limit was fixed, the recorded acoustic emissions have been maintained below the threshold and the tunnel excavation has progressed without any rock-burst episode.

2 ROCK-BURST IN THE BRENNER BASE TUNNEL

The Brenner Base Tunnel (hereafter called also as BBT) is an important project between Austria and Italy supposed to shift a considerable part of European freight traffic from road to rail. It runs from Innsbruck to Fortezza (55 km). Considering also the Innsbruck railway bypass, which is already completed, the entire tunnel system through the Alps will be 64 km long.

The BBT is a complex tunnel system (Fig. 1); besides the two main tubes, each 8.1 m in diameter and running 40-70 m apart from one another, and the exploratory tunnel, there are bypasses between the two main tubes every 333 m, side tunnels and emergency stops plus four lateral access tunnels located every 20 km. The exploratory tunnel lies between the two main tunnels and about 12 m below them and with a diameter of about 6 m is noticeably smaller than the main tubes. This tunnel is meant to provide information about the rock-mass quality, thus reducing construction costs and times of the main tubes. It will be essential for drainage when the BBT becomes operational.

Rock-burst phenomena have occurred in the past during the construction of the Trens access tunnel (GA) and of the main tunnels southwards (GLS). In both cases the excavation was carried out in the Brixner granite (uniaxial compressive strength of over 100 MPa and a RMR value of over 60), typically at overburdens larger than 800 m. In fact, high overburden, high rock-mass strength and other excavation works ongoing nearby are the main factors leading to rock-burst.

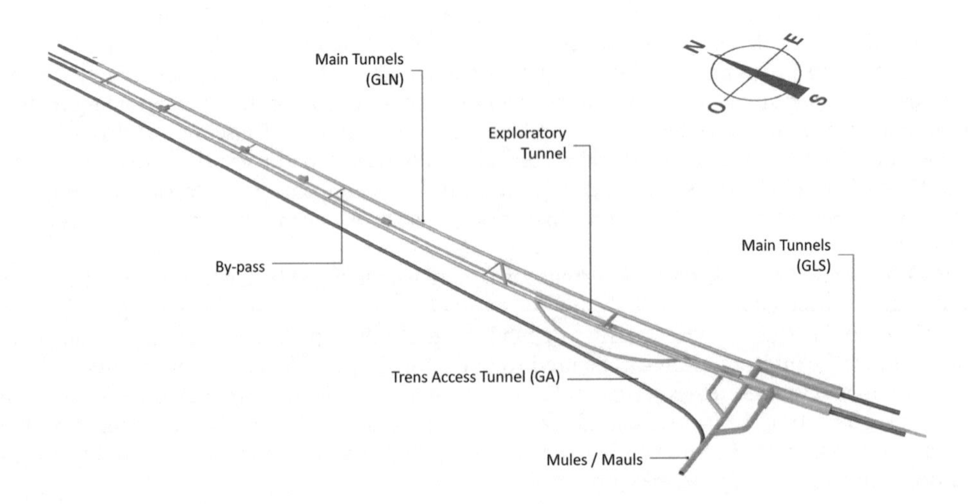

Figure 1. The complex BBT tunnel system.

3 MONITORING OF ACOUSTIC EMISSIONS

Monitoring of acoustic emissions (AE) is a passive technique that shows several similarities to seismic monitoring. The first studies on acoustic emissions, carried out on metals, date back to Kaiser (hence the name "Kaiser effect"). He observed that an increase in the state of stress is associated to the emission of high frequency vibrations.

A similar phenomenon occurs in rocks: the sudden increase in circumferential stress at the tunnel walls due to the excavation, particularly severe at high overburdens, generates such acoustic emissions.

Acoustic emissions are a vibratory phenomenon characterized by frequency values significantly higher than those related to anthropic phenomena commonly detected in tunnels and that obviously have nothing to do with rock-burst occurrence.

Compared to metals, rock-burst in tunnels is affected by a number of complex factors, related to the anisotropy and inhomogeneity of the rock-mass as well as to the space and time variability of the state of stress.

Generally, acoustic emissions generated by rock-burst phenomena occur at 4 different stages of the rock-mass deformative response to excavation. They are:

1. closure of the pre-existing fractures resulting from the excavation of the cavity;
2. triggering of new fractures due to stress accumulation;
3. coalescence of the new fractures;
4. triggering of rock detachment phenomena.

These different stages are characterised by acoustic emissions of different amplitudes and energy content. In all cases, the frequency content has a spectrum range above 500 Hz.

Each excavation displays a peculiar AE response in relation to the local conditions of the rock-mass, the lithostatic stress and the tunnel geometry. As such, the monitoring system as well as the approach for data interpretation should be specifically designed for each case-history.

The BBT system for AE monitoring consists of 6 sections, each equipped with 3 accelerometric sensors located at progressive intervals of 25-50 m from the tunnel face (Fig. 2). The recording is continuous and characterised by a sampling rate of 2000 Hz.

A first type of approach for data interpretation consists in the counting of acoustic emission number during the time unit; this methodology, that is simple if applied to metal studies, while it is associated to several uncertainties in the determination of the AE frequency content and wavelength cycles when adopted for the rock-burst analysis.

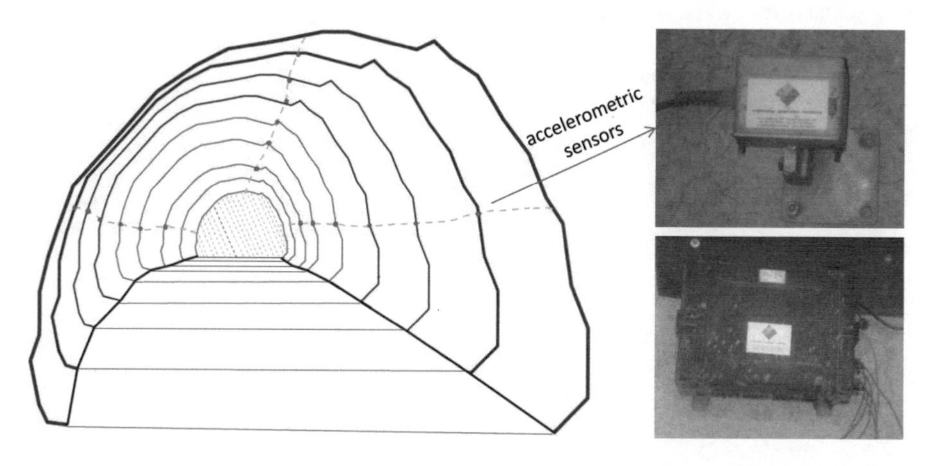

Figure 2. Position of the monitoring sections along the tunnel axis and on-site equipment.

Therefore, a different procedure, based on the power spectrum density (PSD) in the range of 500-1000 Hz, was adopted at the BBT site. It is based on the registration of the maximum power spectrum density (hereafter indicated as PSD_{max}) every acquisition interval of 10 minutes. A related quantity over the 12 hours, indicated as power spectrum density summation (PSDS), is then calculated as follows:

$$PSDS = \sum_{i=1}^{n} PSD_{max.i} \tag{1}$$

where n is the total number of PSD_{max} collected data, equal to 72 in this specific case. The PSDS is not influenced by other vibration phenomena related to the tunnel activities since the frequency band taken into account is higher than the frequencies associated to the machineries and to the excavation works (including blasting).

This quantity (PSDS) represents the amount of energy released by the rock-mass as acoustic emissions. The energy emitted as acoustic emissions is significant when the fractures propagate and interact up to a limit value that corresponds de facto to the failure of the material. It can therefore be considered as a rock-burst activation index. The main objective of the test is the definition of the acoustic emission energy limits by numerical correlation of the reference parameter used to measure the EA (PSDS) and of the rock-mass breaking phenomena due to the increasing stress. The PSDS threshold values need to be defined in order to implement alarm systems and burst countermeasures at the tunnel site.

4 FIELD TEST

The PSDS threshold value for rock-burst occurrence at BBT was recently updated based on a specifically designed in situ test carried out in the Brixner granite formation. More specifically, the test consisted in the compression, up to failure, of a portion of rock-mass at the tunnel side-wall and in the simultaneous measurements of acoustic emissions. The compression of the rock-mass was performed with two flat-jacks, while the acoustic emissions were measured with velocimeters and accelerometers. The use of two different types of instruments for AE monitoring was deliberately pursued in order to assess which sensor, i.e. active or passive, performs better in consideration of the electric noise in the tunnel.

The test was performed in a by-pass tunnel (2.1 m in radius) so as to not interfere with ordinary construction activities (Fig. 3). In order to minimise the influence of the local geological, hydrogeological and geomechanical characteristics of the rock-mass, the area selected for the test the test site was identified to fulfil the following criteria: i) high RMR values, ii) no significant fractures or discontinuities and iii) no water inflows.

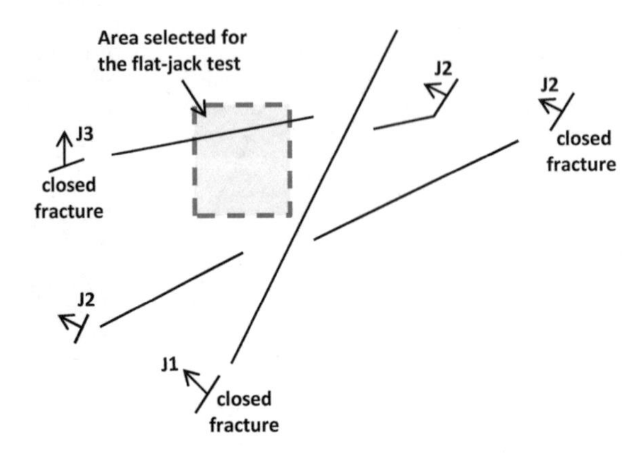

Figure 3. Sketch of the tunnel side-wall area selected for the double flat-jack tests.

4.1 *Geomechanical characterization*

The geomechanical characterization of the Brixner granite formation was based on laboratory tests on samples of intact rock, field tests during the pre-excavation prospection campaigns and tunnel face surveys in the exploratory tunnel in the immediate vicinity of the investigated area. The average values and the standard deviations of the main geomechanical parameters in the area of interest (from chainage location 49+000 to 50+300) are summarised in Tables 1-2. They were used to compute the analytical distributions of the state of stress around the tunnel assuming an ideal-fragile elasto-plastic behaviour for the rock-mass (the calculation was performed considering an unsupported circular tunnel, a uniform lithostatic stress field and plane-strain conditions). The circumferential stress (σ_9) at the tunnel wall analytically estimated is equal to 6 MPa.

The same quantity was also estimated trough a preliminary single flat-jack test. A recovery pressure of 4.25 MPa was measured, consistently with the analytically estimated one (Fig. 4).

Table 1. Parameters of the rock material.

	Average value	Standard deviation	Number of tests
γ*(MN/m^3)	0.0267	± 0,00015	28
σ_{ci**}(MPa)	115	± 19	31
*mi****(-)	24	± 4	16

* Unit weight of the intact rock.
** Uniaxial compressive strength of the intact rock.
*** Constant of the Hoek-Brown criterion for the intact rock.

Table 2. Parameters of the rock-mass.

	Average value	Standard deviation	Number of tests
*RQD**(%)	86	±14	210
*RMR***(-)	73	±11	210
*GSI****(-)	77	±13	208

* The rock quality designation.
** The rock-mass rating.
*** The geological strength index.

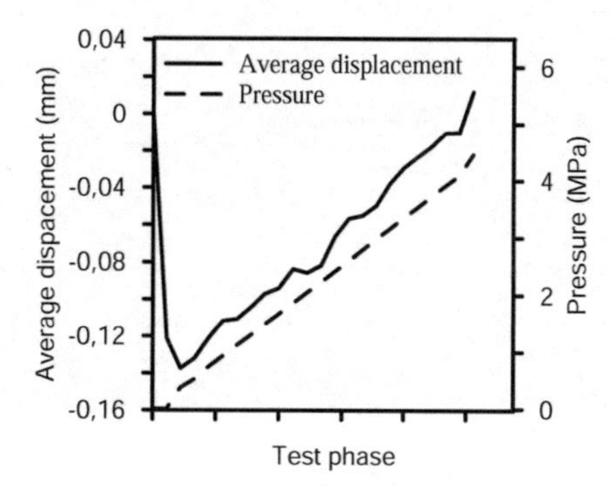

Figure 4. Results of the preliminary single flat-jack test.

4.2 *Test execution*

The in situ test was carried out in the following steps:

- single jack-test for the estimation of the state of the stress at the tunnel side-wall;
- installation of acoustic emission measurement arrays (accelerometers and velocimeters according the scheme shown in Fig. 5);
- execution of the double flat-jack test up to rock-mass failure (i.e. until cracks occurrence, Fig. 6) and simultaneous acquisition of accelerometric and velocimetric data at a high sampling rate (2000 Hz).

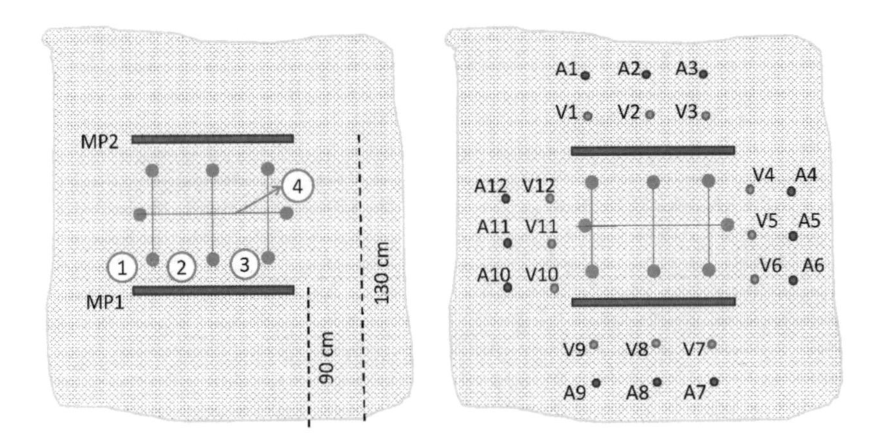

Figure 5. Arrangement of AE measuring arrays and location of the flat-jacks.

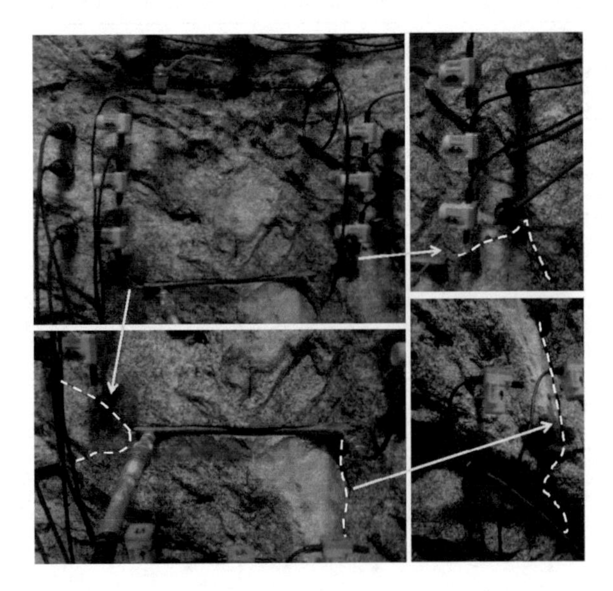

Figure 6. Occurrence of cracks in the rock-mass.

5 RESULTS

The recorded data were processed as follows:

- determination of the frequency content of the signals;
- notch filtering in the 50 Hz band to eliminate the not desired frequency intervals (and multiple tonal components);
- PSD_{max} determination over 30 second intervals;
- correlation of the energy content of the measured acoustic emissions with the displacements imposed by the flat-jacks.

The analysis of measurements clearly shows a peak in the PSD_{max} values shortly before the rock-mass failure (Fig. 7), followed by their decrement in the post-failure stage. In terms of frequency content, the AE registrations indicate a predominant frequency around 700 Hz before failure (Fig. 8) and around 200 Hz immediately after crack occurrence (Fig. 9).

In detail, the observation of PSD_{max} evolution in time allows to distinguish 4 different phases:

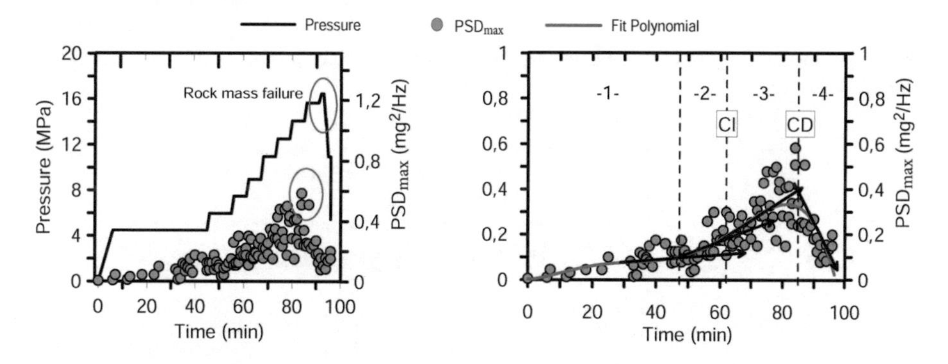

Figure 7. Variation of the PSD_{max} during the double flat-jack test.

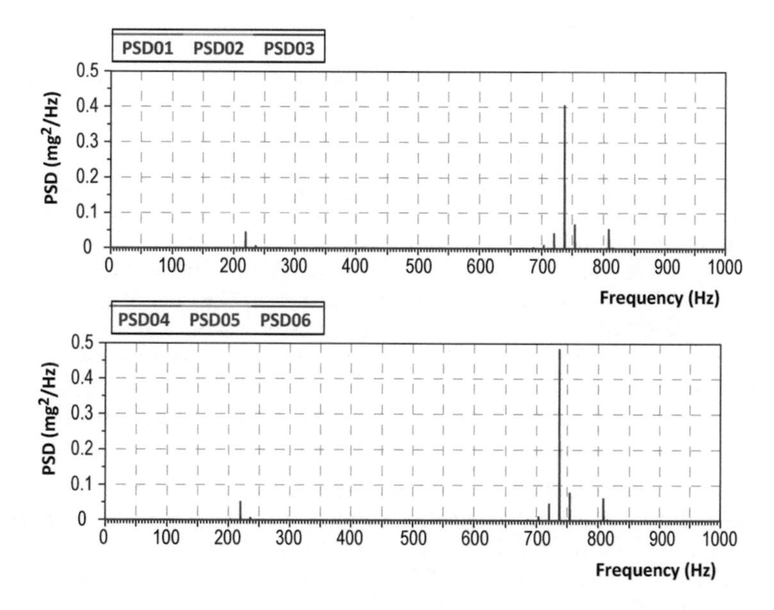

Figure 8. Frequency content of the AE registrations before rock-mass failure, (i.e. at a value of the applied pressure of 12.53 MPa).

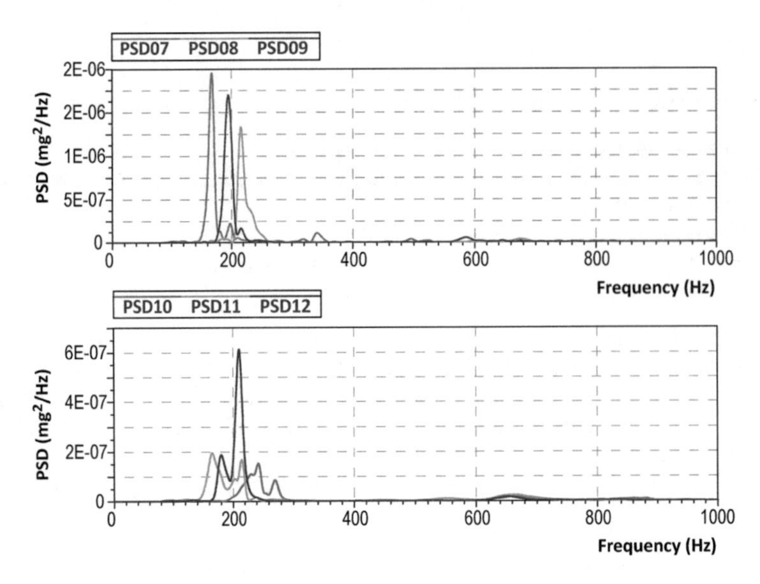

Figure 9. Frequency content of the AE registrations after rock-mass failure (i.e.at a value of the applied pressure of 16.44 MPa).

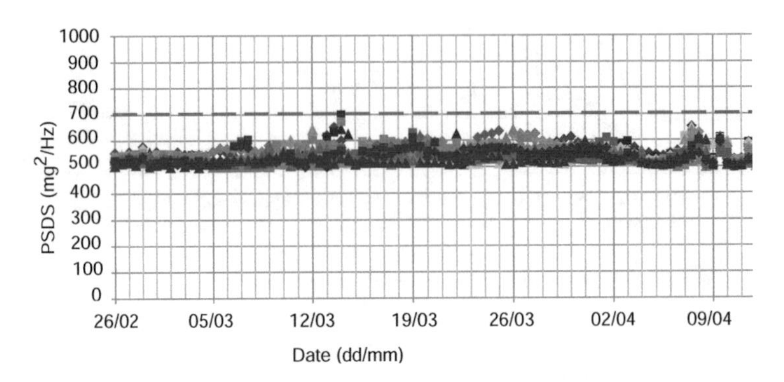

Figure 10. Comparison between the PSDS values of the acoustic emissions recorded in the Brenner Base Tunnel and the new alarm threshold (values calculated over 12-hour time intervals).

- the early stage of the test is characterised by a slight increase in the PSD_{max}: the small amount of acoustic emission is due to the closure of pre-existing cracks in the rock-mass;
- the following stage, dominated by the Kaiser effect, presents higher values of PSD_{max};
- the sudden increase in PSD_{max}, observed in the third stage, is due to the stable propagation of new fractures. The start of this phase is thus defined as the "crack initiation threshold". Peak values for the PSD_{max} are reached at the pressure value known as "crack damage threshold", at which the fractures formed in the rock-mass begin to interact and spread in an unstable manner;
- finally, a significant drop in the PSD_{max} values is observed in the fourth stage of the test, when the pressure is realised.

The analysis of the acoustic emission data acquired during the test indicates a peak PSD_{max} value, calculated using a 30-second acquisition window, between 0.5 and 0.6 mg^2/Hz. The corresponding value of the accumulated energy value (PSDS) over a 12 hour period is then equal to 700 mg^2/Hz. This value was adopted at BBT as the new alarm threshold for rock-burst triggering for similar tunnel conditions (Fig 10).

5 CONCLUSIONS

Several studies indicate that the risk of rock-burst in tunnels with high overburdens can be managed by monitoring the acoustic emissions during the tunnel excavation.

The paper describes a preliminary study conducted in the Brenner Base tunnel in a compact granite with a 1.000-m overburden. An in situ test was specifically designed with the aim of identifying the alarm threshold for rock-burst occurrence. The test consisted in the compression, up to failure, of the rock-mass at the tunnel side-wall trough two flat-jacks and in the simultaneous measurements of acoustic emissions. Test results indicate that a correlation exists between the energy and frequency content of the acoustic emissions and the rock-mass failure. More in detail, acoustic emissions were characterised by a predominant frequency around 700 Hz immediately before rock-mass failure and around 200 Hz in the post-peak regime. The maximum power spectrum density over a 30-second time interval reached values higher than 0.5 mg^2/Hz immediately before failure.

These observations proved the presence of high-frequency precursor emissions before rock-mass failure, in the frequency band 500-1000 Hz, consistently with the frequency interval already considered in the Brenner Base Tunnel for the interpretation of acoustic emission monitoring. Following the experimental investigation a new alarm threshold, equal to a value of PSDS of 700 mg^2/Hz over a 12 hour period, was implemented at the construction site.

Even though the proposed in situ test is representative only of a specific area of the tunnel, it is believed that its significance is much higher from a methodological point of view. In fact, the experimental strategy seems appropriate to define threshold values for rock-burst risk, thereby increasing the safety level inside the tunnel. As such, further tests need to be carried out to consolidate the experimental procedure and to provide a larger experimental database. If standardised, the proposed in situ test could be implemented since the preliminary stages of tunnel construction in order to characterise the response of homogeneous portions of the rock-mass.

It is worth saying that any risk threshold needs to be supported by the study of the geological and geomechanical properties of the rock-mass under investigation as well as of the characteristics of the in situ stress field.

REFERENCES

Diederichs, M.S. 2007. The 2003 Canadian Geotechnical Colloquium: *Mechanistic interpretation and practical application of damage and spalling prediction criteria for deep tunnelling. Canadian Geotechnical Journal* 44: 1082–1116.

Feng, G.L., Feng X.T., Chen, B.R. & Xiao, Y.X. 2015. *Microseismic sequences associated with rockbursts in the tunnels of the Jinping II hydropower station. International Journal of Rock Mechanics & Mining Sciences* 80: 89–100.

Feng, G.L., Feng X.T., Chen, B.R. & Xiao, Y.X. 2017. *Performance and feasibility analysis of two microseismic location methods used in tunnel* engineering. *Tunnelling and Underground Space Technology* 63: 183–193.

Hedley, D.G.F. 1992. *Rockburst handbook for Ontario hardrock mines*. CANMET SP92-1E.

Kaiser, P.K. & Cai, M. 2012. Design of rock support system under rockburst condition. *Journal of Rock Mechanics and Geotechnical Engineering* 4 (3): 215–227.

Kaiser, P. K., Tannant, D.D. & McCreath, D.R. 1996. *Canadian rockburst support handbook*. Sudbury, Ontario: Geomechanics Research Centre, Laurentian University.

Yu, Y., Chen, B.R., Xu, C.J., Diao, X.H., Tong, L.H. & Shi, Y.F. 2017. Analysis for microseismic energy of immediate rockbursts in deep tunnels with different excavation methods. *International Journal of Geo*mechanics 17(5): 10 pages.

Valguarnera, L. 2018, Metodo sperimentale per la determinazione della soglia di rischio rockburst nella Galleria di Base del Brennero. Master thesis in Ingegneria Civile, University of Bologna, Bologna, Italy.

Tunnels and Underground Cities: Engineering and Innovation meet Archaeology, Architecture and Art, Volume 3: Geological and geotechnical knowledge and requirements for project implementation – Peila, Viggiani & Celestino (Eds)

An attempt to derive a method to estimate velocity during exploration ahead of the tunnel face

H. Yamamoto & M. Nakaya
Hazama Ando Corporation, Minato-ku, Tokyo, Japan

S. Imamura & K. Ohta
4D Geotek LLC, Saitama-city, Saitama, Japan

ABSTRACT: A tunnel face tester (TFT exploration) is developed a forward facing exploration system based on reflection method elastic wave exploration using excavation blasting as a vibration source in mountain tunnel construction. In this study, an algorithm for estimation of velocity behind tunnel face is developed and implemented in this system. The algorithm is based on the ratio between the amplitude of forward going wave and reflecting wave. In addition, it was applied to the actual tunnel site, compared with the actual excavation results, and good results were obtained.

1 INTRODUCTION

Geological surveys are conducted prior to the construction of a mountain tunnel including geophysical explorations such as boring surveys and elastic wave exploration and the results are used as the basis for designing support patterns for the tunnels. However, geological surveys like these are usually conducted from the ground level and the number of exploration sites is limited. In addition, with elastic wave exploration, the exploration accuracy is known to decrease if the overburden is large and the geological structure is complicated.

When prior surveys are considered insufficient or the exploration accuracy is thought to be low, exploration ahead of the tunnel face in construction is important. Of the methods for exploration ahead of the face currently available, survey boring is known to be the most reliable and accurate. In particular, long distance boring surveys require a long interruption during boring. Even in the case of the TSP method which has been frequently used at various sites, it is necessary to provide a large facility for exploration blasting as a vibration source, which has been used in many cases of exploration (Sattel et al.1992).

Accordingly, we are working on the development of a "tunnel face tester (TFT exploration)," a system of exploration ahead of the tunnel face using the seismic reflection method, which allows safe and quick exploration in tunnels without affecting the construction cycle by using excavation blasting as a vibration-triggering point (see Figure 1) (Nakaya et al. 2016).

However, it has been shown that although the method is able to identify the reflective position as a qualitative evaluation, it is not capable of determining the quantitative indices of faults, geological boundaries and strengths which are deemed to be those of reflection surfaces. To deal with this issue, we developed a trial algorithm to calculate the elastic wave velocity ahead and behind the reflection surface for studying a quantitative evaluation method of the ground ahead of the face. (Imamura et al. 2016).

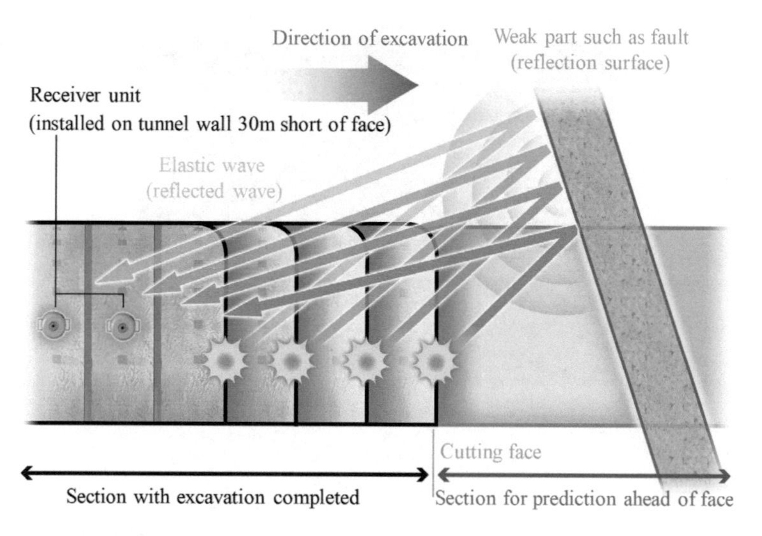

Figure 1. Caption of a typical figure.

This paper presents the system configuration and quantitative evaluation method and reports on the results of verification by application to an actual tunnel site and comparison with the excavation results.

2 SYSTEM CONFIGURATION

Figure 2 shows the system configuration of the TFT exploration, Figure 3 installation of a receiver and Table 1 the specifications of the equipment.

Each of the receivers shown in Figure 2 for recording elastic wave data integrates a seismometer, logger and battery and measurement at two points is the standard method. Receivers are installed using existing support rock bolts as waveguides and special fixing jigs are used to ensure installation in close contact with the tunnel wall.

For measurement, the equipment is started by remote operation (transmission frequency 426 MHz, communication range 150 m max.) of the start switch when taking refuge from blasting, which stops automatically after recording. This allows reduction of power consumption and continuous exploration for about one month with reduced power supply. The quantity of data

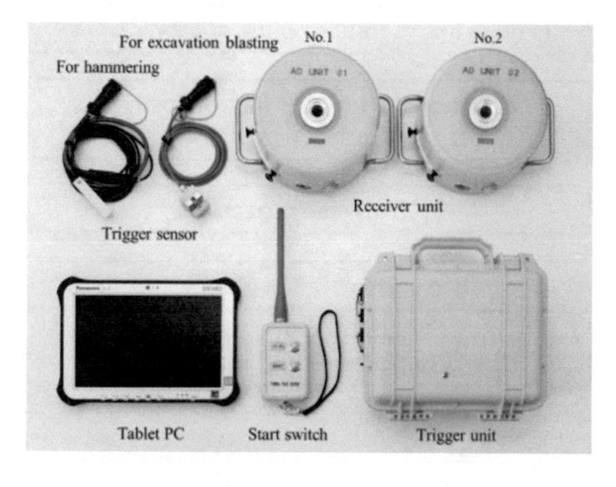

Figure 2. TFT exploration system configuration.

Figure 3. Installation of receiver unit.

Table 1. Specifications of receiver unit.

Component	1 component (along tunnel axis)
Sensor	GS-20D (28 Hz)
Sampling rate	20 kHz
A/D resolution	24 bits
External dimensions	Base diameter: 150 mm Max. diameter (handle): 244 mm) Thickness: 117 mm

required for analysis in principle is 20 measurements (data for about three days), during which equipment is not moved.

For recording the data, when the trigger sensor mounted on the blasting leading wire detects a vibration signal, the trigger unit sends a radio signal (transmission frequency in the 800 MHz band, communication range 110 m max.) to the receiver, and initiates recording. Data are collected and analyzed by using a tablet PC with the dedicated software installed and, after data are collected wirelessly by Bluetooth, analysis for about one hour makes the results of prediction available for output.

The next step is analysis, in which based upon the analysis flow in Figure 4, the said tablet PC calculates elastic waves in the vicinity of the tunnel face from directly arriving waves, and

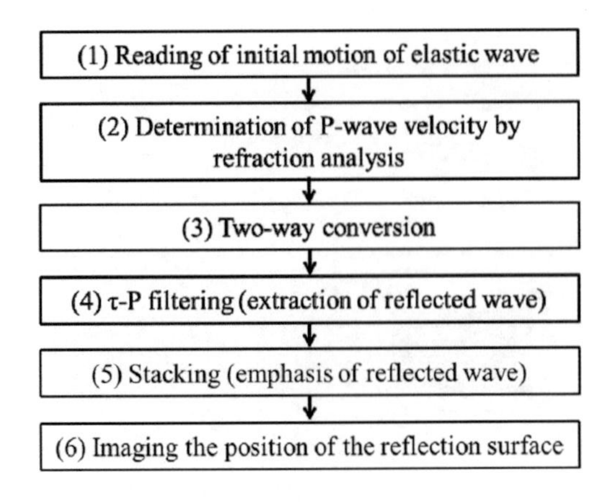

Figure 4. Flow of analysis by exploration ahead of face.

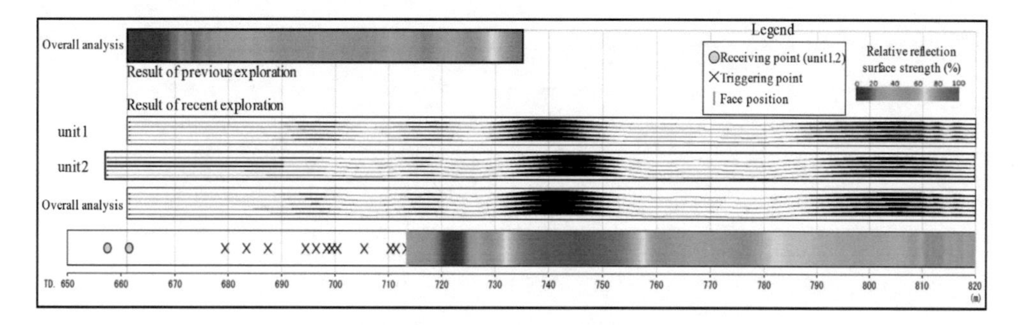

Figure 5. Example of results of exploration ahead of tunnel face.

extracts reflective waves which are contained in the blasting data to predict the location of reflective waves ahead of the tunnel face. Figure 5 shows an example of exploration ahead of the tunnel face obtained in this way.

Regarding the results of the exploration shown in Figure 5, the reflection surface strength is drawn by color contouring, in which sections in warm colors indicates reflection surfaces, that is the predicted positions of poor ground such as faults and fracture zones.

3 ESTIMATION OF ELASTIC WAVE VELOCITY AHEAD OF THE FACE

3.1 *Method of estimation of elastic wave velocity*

In TFT exploration, two receivers are installed at locations 20 to 30 m short of the face to record elastic waves from excavation blasting. As the face moves forward, the vibration-triggering point (excavation blasting point) moves forward, which makes it possible to obtain waveforms of vibration generated at many points. If waveforms of vibration from many points are obtained, the wave field can be divided into direct and reflected waves and the point of generation of the reflected wave is regarded as the position of the surface of velocity discontinuity (see Figure 1).

The velocity from the face to the first surface of velocity discontinuity can be regarded as the same as that obtained from the gradient of the initial travel time but the velocity further ahead is the question.

Accordingly, we attempted to estimate the velocity ahead of the face from the reflected wave amplitude information as follows. Generally, with vertical incidence to a plane boundary, the reflected wave amplitude A_R is represented as shown in Equation (1), where Z_1 is the impedance on the incidence side and Z_2 the impedance on the transmission side. The right side of Equation (1) is the reflection coefficient.

$$A_R = A \cdot \frac{Z_2 - Z_1}{Z_2 + Z_1} \tag{1}$$

Based on Equation (1), when $Z_2 < Z_1$ (impedance is lower ahead of the surface of velocity discontinuity), the reflected waveform is reversed in polarity. Figure 6 shows the relationship between the velocity structure, reflection coefficient and assumed waveform. Let the incident wave amplitude and the reflected wave amplitude in Layer i be $A_{T(i)}$ and $A_{R(i)}$ respectively and $Z_{(i+1)}$ can be calculated from Equation (2).

$$A_{(i+1)} = Z_{(i)} \cdot \frac{A_{T(i)} + A_{R(i)}}{A_{T(i)} - A_{R(i)}} \tag{2}$$

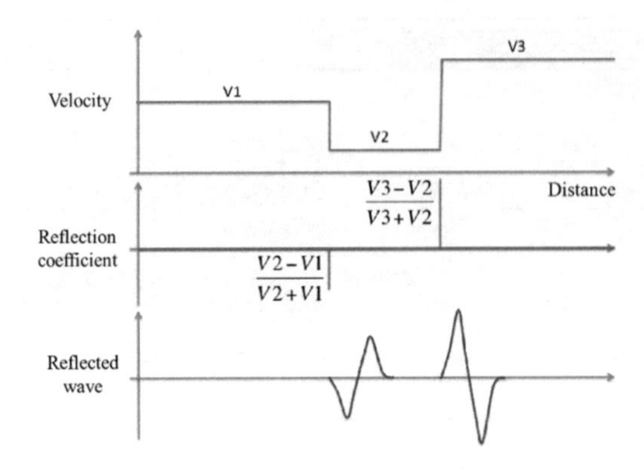

Figure 6. Conceptual diagram of velocity distribution and reflected wave (with constant density).

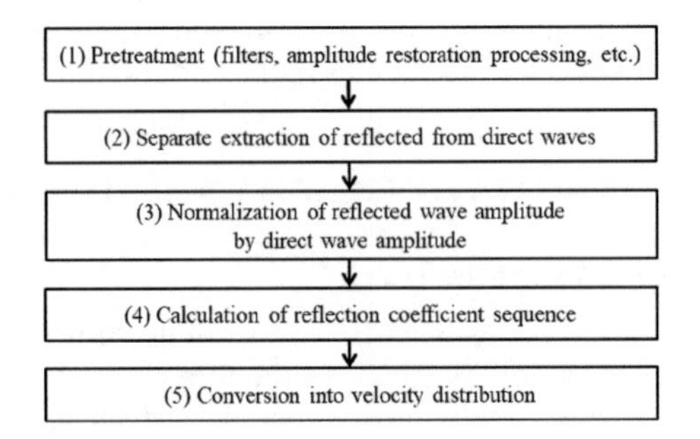

Figure 7. Basic flow of elastic wave velocity estimation.

That is, with the density assumed to be constant, *V1* is given by the gradient of the initial travel time and *V2* can be calculated if the amplitude ratio between the reflected and direct waves can be obtained. Figure 7 shows the basic flow of elastic wave velocity estimation.

The Pretreatment (1) shown in Figure 7 includes velocity and band-pass filters and geometric attenuation correction and (2) includes FK and τ-P filters and various stacking processes. Carrying out (3) is expected to correct variation of the vibratory force. For (4) and (5) can be obtained through, the elastic wave velocity inversion by setting a forward model such as a convolution model.

3.2 *Simplified method of elastic wave velocity estimation*

From the basic idea of velocity estimation method presented in (1), we simplified the method for TFT exploration and tried a method of elastic wave velocity estimation based on the flow shown in Figure 8 that does not involve trial-and-error analysis by changing the analysis parameter. The algorithm shown in Figure 8 features the following strategies

- Use of an envelope for extracting major events
- Use of displacement waveforms for facilitating evaluation of the reflection phase
- Calculation of the reflection coefficient using the envelope ratio
- Use of the correlation coefficient of waveforms for evaluation of the reflection phase

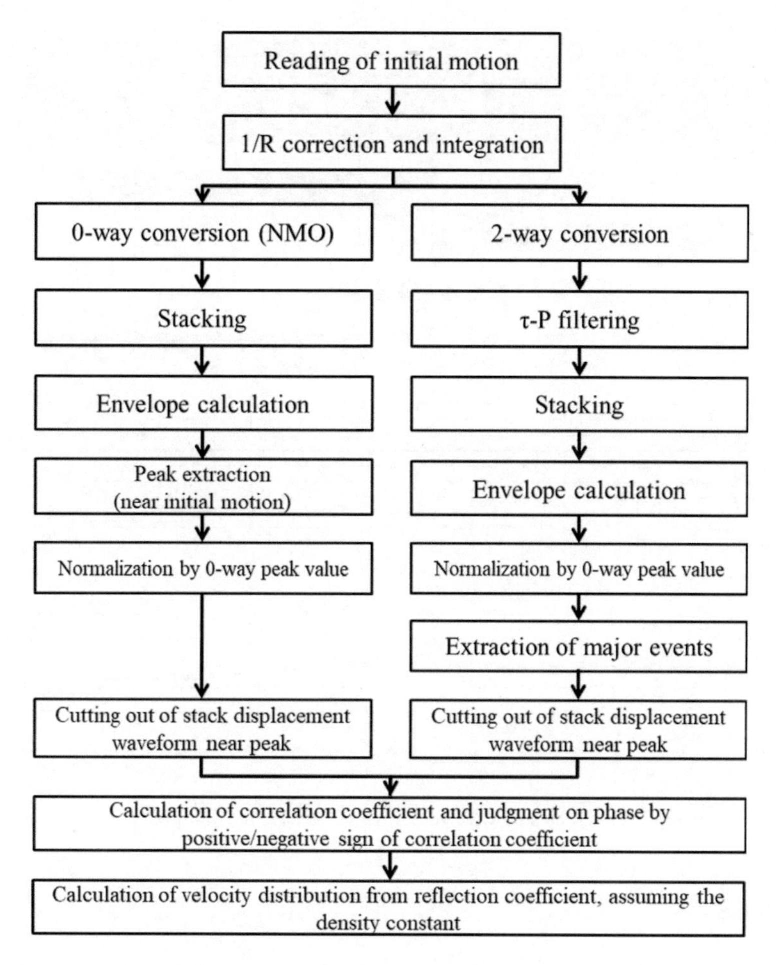

Figure 8. Flow of elastic wave velocity estimation in TFT exploration.

4 EXAMPLE OF APPLICATION TO ACTUAL TUNNEL SITE

The relationship between the elastic wave velocity acquired by using the trial algorithm of this study on a tunnel site under construction and the face evaluation points (actual result) is shown in Figure 9. For analysis, excavation blasting data at 18 locations in sections between TD.496.6 and 535.6 m was used to calculate the elastic wave velocity in sections ahead of the face sections between TD.535.6 and 639.0 m.

The geological structure of the tunnel includes Quaternary andesite (massive and autobrecciated), tuff breccia and lapilli tuff and Figure 9 shows photos of the actual faces together with the results of TFT exploration conducted three times successively (reflection surfaces shown in red).

Based on Figure 9, the support pattern adopted offers an elastic wave velocity of about 3.0 km/s as compared with that of the DI pattern (design elastic wave velocity: 2.5 to 3.2 km/s), which is roughly equivalent. A change in the elastic wave velocity value is detected at a position where the face evaluation point changes, which is assumed to be effective as an indicator of quantitative evaluation.

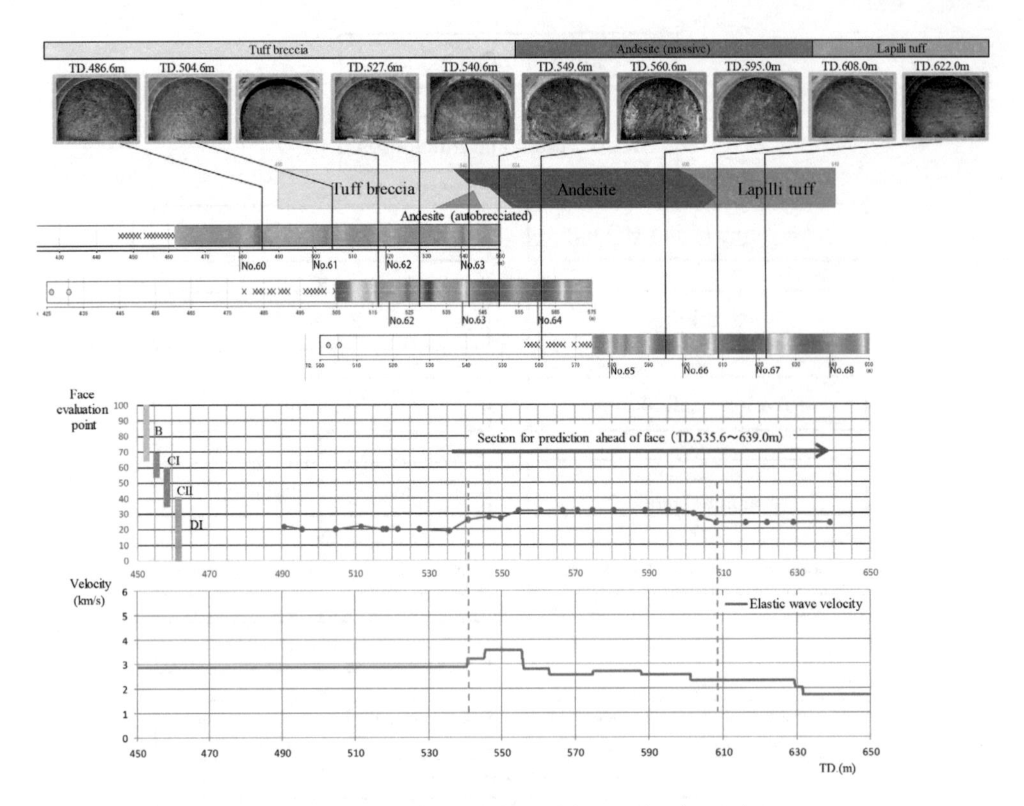

Figure 9. Relationship between elastic wave velocity and face evaluation point.

5 CONCLUSION

For the "tunnel face tester (TFT exploration)," a system for exploration ahead of the tunnel face by means of seismic reflection method using excavation blasting as a vibration-triggering point, we developed a trial algorithm to calculate the elastic wave velocity. And it was applied the algorithm to an actual tunnel site for comparison with the excavation results.

As a result, generally satisfactory results have been obtained as compared with excavation results. In the future, further data verification is required for improving the trial algorithm to achieve higher accuracy of prediction ahead of a tunnel face.

REFERENCES

Sattel, G., Frey P. & Amberg, R. 1992. Prediction ahead of the tunnel face by seismic methods, First Break, Vol.10, No.1

Nakaya, M., Onuma, K., Yamamoto, H., Nishikawa, A. & Niitsuma, H. 2016. Study on the evaluation of geological condition and the prediction ahead of tunnel face by seismic survey using blasting for excavation, Journal of JSCE F1 (Tunnel Engineering), Vol. 72, No. 2, pp. 53-66. (in Japanese)

Nakaya, M., Yamamoto, H., Kirihara, A., Tendo, R. & Suzuki, M. 2016. Development and application of cableless seismic survey system in mountain tunneling, Proceedings of Tunnel Engineering, JSCE, Vol. 26, I-38. (in Japanese)

Imamura, S. Yamamoto, H. Nakaya, M. & Ohta, K. 2018. Simplification of a method of estimating the velocity ahead of tunnel face in exploration ahead of tunnel using blasting at the tunnel face as a vibration source, Proceedings of the 138th SEGJ Conference, No. 25, pp. 95-98. (in Japanese)

Tunnels and Underground Cities: Engineering and Innovation meet Archaeology, Architecture and Art, Volume 3: Geological and geotechnical knowledge and requirements for project implementation – Peila, Viggiani & Celestino (Eds)

A probabilistic approach to assess the risk of liner instability when tunnelling through karst geology using geotechnical baseline reports

K. Yau & C. Paraskevopoulou
University of Leeds, Leeds, United Kingdom

S. Konstantis
Ruler Consult Ltd, London, United Kingdom

ABSTRACT: Probabilistic methods can provide more insight on risk allocation and can be used in Geotechnical Baseline Reports (GBR). A novel, probabilistic approach of the spatial distribution of karstic cavities is presented to assess the risk of liner instabilities when tunnelling through karstic limestone. The presented research work is based on the following geotechnical baseline statement: "There is a 10% chance of encountering a sediment-filled cavity, with a maximum size of $2m^3$", which was assumed to be correct for the analysis. A Matlab script generates a random distribution of karst cavities, modelled as a circular shape, which is transferred into FEM 2D and 3D numerical analysis. The measured variable is the index for capacity utilization of linings in tunnels (CULT-I), which act as proxy for liner stability assessment. The methodology presented provides a framework for further investigations to assess the risk from karst cavities to tunnelling works.

1 INTRODUCTION

Tunnel construction faces many technical challenges, most of which stem from the heterogeneity in the ground surrounding the tunnel. The risk of uncertainty in material properties and spatial variability need to be assessed and estimated prior to project construction, as encountering unforeseen conditions can lead to delays, cost overrun (Paraskevopoulou & Benardos, 2013) and can potentially be catastrophic. An example of such uncertainty comes from the spatial variability of solution cavities, when tunnelling through karstic limestone, which comprises of approximately 10–15% of Earth's ice-free continental surface (Ford & Williams, 2013).

Methods for karst risk evaluation have largely been qualitative or semi-quantitative, and are limited to risk classification. Quantitative approaches may assess cavity parameters but not in a probabilistic fashion. Probabilistic approaches could be implemented for karst tunnels, as these can provide insight on risk allocation for geotechnical baseline reports (GBRs). GBRs are becoming more widespread for tunnelling projects as they act as a risk allocation and transfer mechanism between the contractor and owner and act as a basis for bid proposals and project disputes. Therefore, having a more elaborated risk evaluation is vital for such projects.

The primary goal of this paper is to assess and evaluate the applicability of probabilistic methods to quantify the risk of encountering karstic cavities in a tunnel alignment and the potential effect on liner instability. More specifically, this research work focuses on producing a generalized model to quantify the probability of encountering solution cavities during tunnelling through karst prone areas, using simplified assumptions and a baseline statement: "There is a 10% chance of encountering a sediment-filled cavity, with a maximum size of $2m^3$".

2 BACKGROUND

2.1 *Karst*

Limestone is a good medium for tunnelling but when karstified it can deteriorate the construction process. Karst can be described as a type of landscape that occurs from the dissolution of highly soluble rocks, such as limestone and gypsum by water movement. This process forms extensive water networks made up of cavities and discontinuities that induce secondary porosity (fractures) and tertiary porosity (conduits). The latter contributes to spatial uncertainty as the formation of pathways is both a function of rock dissolution rate and the geological structure. In addition, karst systems continually evolve over time because of the feedback loop between the dissolution and material property. Continual flow can erode pathways and wash out fines, leading to material degradation (Marinos, 2001).

Most dissolution happens at or near the bedrock surface, resulting in surface karst features. However, underground karst can be present despite the absence of any surface indications and karst should always be assumed to be present in a carbonate terrain unless proved otherwise (Ford & Williams, 2013).

Ground investigation in karst limestone present some of the largest challenges in foreseeing the complex characteristics of the system. Geophysical surveys can be used to detect voids and caverns for further investigation but can be limited by survey depth, resolution and extent. Hydraulic and tracer testing can lack proper assessment of the groundwater network. Also, probes usually need to be closely spaced for a sufficient model of the subsurface cavities. For a 90% probability of encountering one cavity of 2.5m diameter, there would need to be 2500 probes per 10,000 m^3 (Waltham & Fookes, 2003). Karst cavities can be infilled by air, water, sediment (Figure 1) or a mixture, which can lead to potential issues, such as water inflow, mud flow and tunnelling through weak fill material or void.

The main problems when tunnelling through karst arise from groundwater and any voids that intersect the tunnel alignment. Groundwater may seep into underground works, posing threats of contamination and disturbance of surrounding structures (Song et al. 2012). Even minor leakage into the tunnel can be proven to be problematic, especially if groundwater is contaminated. Tunnel induced settlements are not usually a major concern from karst tunnels unless tunnelling is taking place in an urban environment, where structures may be sensitive to ground deformations and/or changes to the ground water level (e.g. timber piles)

2.2 *Uncertainty, hazard and risk*

Uncertainty can be defined as either aleatoric or epistemic. Aleatoric uncertainty arises from temporal or spatial, natural variation, such as discontinuity structure, material properties and voids in karst terrain. Epistemic uncertainty stems from a lack of fundamental understanding

Figure 1. Example of karstic void partially filled with clay and silt in Dodonis tunnel, Greece (modified after Marinos, 2001).

and limitations in data collection, such as inappropriate models and laboratory testing procedures. Tools to deal with parameter uncertainty in geotechnical engineering include partial factors, quantification using statistical procedures and the observational approach (van der Pouw Kraan, 2014). Reliability analyses, as Langford and Diederichs (2014) performed, can also be carried out to assess the uncertainty in ground response and what support is required using a probabilistic approach. Acceptability criteria must be defined to establish the failure conditions for the project.

A hazard can be defined as a situation that has the potential to cause harm to life, property, the environment or finances. The main potential sources for tunnelling hazards are ground and groundwater interaction with the construction, contaminants, existing structures and human errors in construction technique. These can lead to tunnel collapse or distortion, water inflow and external effects to people, the environment, property and finances. Hazards during tunnelling usually result from insufficient information about the ground or a failure to comprehend the geology in engineering terms (Eskesen et al. 2004).

Risk can be defined as the probability of a hazard occurring in a given timeframe and the magnitude of the resultant consequences. These hazards have the potential to cause harm to life, property, the environment or finances. The main consequences from karst are delays and cost overruns, as any unforeseen conditions can lead to water inrush and potential for collapse, whilst any unanticipated cavities would need preventative measures employed, such as grouting of cavities and the surrounding rock, impervious liners, drainage and dewatering. The cost of such techniques is considerable and typically requires a cost/benefit analysis to be carried out, i.e. balance between accepted risk and reducing costs (Eskesen et al. 2004).

Tunnels strongly interact with the surrounding rock mass, acting as a self-supporting system. The nature of the rock mass sources aleatoric uncertainty and large variation in parameters. Therefore, risk assessment in project management is vital for such projects. Minor delays in construction and maintenance of the works can have major financial implications. The risk in tunnelling is always significant due to high costs, limited design flexibility and ensuring safety during construction.

2.3 *Geotechnical Baseline Reports (GBRs)*

GBRs are a widely used framework for tunnelling projects that contractually define the anticipated ground conditions (Hatem, 1998). They can be used for risk allocation, bid preparation and solving financial and technical disputes during construction (The International Tunnelling Insurance Group (ITIG), 2006). Any anticipated ground conditions stated in the GBR (i.e. in the baseline statements) are the financial responsibility of the contractor, whilst any encountered conditions that exceed the baseline statements shift responsibility onto the owner. This framework can be used during the tendering process to reduce project bid costs, provide a reference baseline for bid evaluation and enable rational financial contingency allocation and during the construction phase as a dispute resolution mechanism. A GBR must be adapted for each project but needs to be objective with no or very little space for interpretation and may typically contain conditions on man-made structures, contamination, geological features, groundwater, etc. Developing accurate or representative numerical baselines can be challenging because of the extensive variability in most geologic formations. This usually requires expertise in interpreting the site investigation, as constraints in project budget and access, limit the extent of data available. Site investigation for karst tunnels will usually include geophysical surveys, groundwater monitoring and borings and the contractors need to be aware of what is factual and what is interpretative (van der Pouw Kraan, 2014).

2.4 *Risk evaluation*

Qualitative and semi-quantitative methods include risk registers, fault tree analysis, event tree analysis and bowtie diagrams. Risk registers are produced for most geotechnical engineering projects, which identify the likelihood and consequences of potential risk events and what controls and mitigation measures are to be put in place. The likelihoods are specified over a given time period and each risk is given a qualitative or semi-quantitative risk rating. Quantitative risk assessment in geotechnical engineering usually involves probabilistic analysis to quantify the uncertainty in input parameters. These methods may include Monte Carlo simulations or discrete sampling in the form of numerical modelling. (Eskesen et al. 2004).

3 METHODOLOGY

The primary aim of the proposed methodology is: a) to create an approach to assess the spatial distribution of karst cavities probabilistically and b) to assess what implications this may have when tunnelling (full-face excavation) through such geological medium based on a typical geotechnical baseline statement that can be set prior to the tendering stage of the tunnelling project, such as "There is a 10% chance of encountering a sediment-filled cavity, with a maximum size of $2m^3$". This baseline statement means that in maximum 10m of a typical 100 m section of the tunnel alignment, karst cavities with a maximum size of $2m^3$ filled with sediments will be encountered. Regardless of the actual geological conditions to be encountered, this baseline statement defines what the contractor bases its bid on and what the owner has to pay.

This baseline statement defines for contractual purposes that the cavity size range has an upper bound of $2m^3$. During construction, cavities of appreciable size within or near tunnel alignments may be dewatered, grouted or filled, so that the rock mass integrity is not compromised.

Figure 2 summarizes the procedure used to gather the results of this project. A 2D model is first created to assess the viability of the modelling procedure and is later adapted to the 3D model, upon which, a true representation of the entire tunnel alignment is assessed.

The model has initially been simplified to 2D, plane strain conditions in RS2 for the tunnel cross section and voids. Karst cavities, modelled as circular in 2D and idealised as spherical in 3D, would assume a cylindrical shape trending into the z-axis in 2D. This means that only the largest cross section of the void would be considered in the 2D plane.

3.1 *Material properties*

The initial model assumed the rock mass was elastic and isotropic. Rock mass parameters are based on Mohr-Coulomb criterion evaluated from the following types of limestone: Slightly

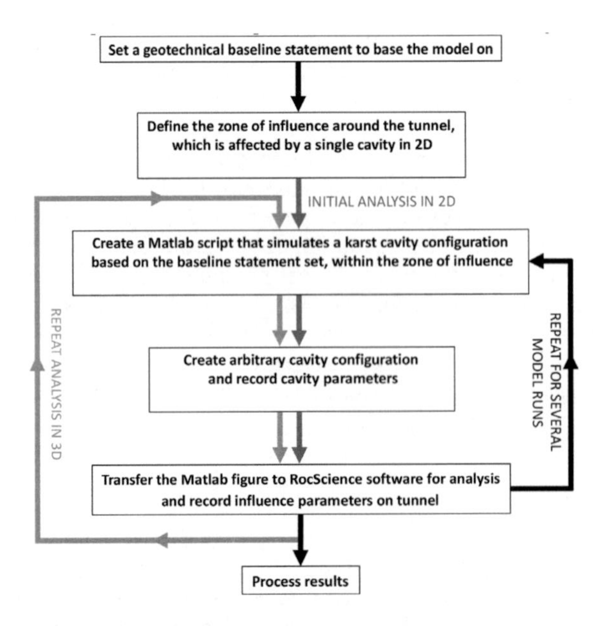

Figure 2. Flowchart of the methodology used in the analysis presented herein.

Table 1. A summary of the geotechnical parameters of HWL and SWL, *the UCS of HWL is undetermined from the samples used as they were too weak, (simplified from Karagkounis et al. 2016)

Rock Unit	Highly Weathered Limestone	Slightly Weathered Limestone
Unit Weight (kN/m^3)	19	22
UCS (MPa)	-	6 - 20
Young's Modulus (GPa)	0.15	1
Poisson's Ratio	0.33	0.33
Permeability (m/s)	$1.6*10^{-6} – 1.7*10^{-4}$	$3.1\&10^{-6} – 3.0*10^{-4}$
Cohesion (kPa)	20	200
Friction Angle (°)	21	26
RMR Range	10 - 30	30 - 50

Table 2. Liner properties used in the 2D base model.

Thickness (m)	UCS (MPa)	Young's Modulus (GPa)	Poisson's Ratio	Tensile Strength (MPa)
0.3	35	30	0.15	3

Weathered Limestone (SWL) is assumed for the rock mass and Highly Weathered Limestone (HWL) parameters are used on the sediment fill of karst voids. HWL represents the common derivation of infill for karst cavities, which is usually a weakened, more permeable version of the source rock or soil above. Table 1 details some of the parameters selected for this analysis.

The liners used are unreinforced concrete and modelled as elastic, with default properties of concrete (Table 2).

3.2 *Boundary conditions and geometry of the model*

The analysis considered the full face excavation of a 6m diameter tunnel, located at 25m depth.

The boundary conditions are set by a 70x50 box, with the tunnel located at the center and all edges restrained, except for the top boundary to represent the surface level (Figure 3). This boundary was determined to be sufficient for analysis, without being computational inefficient. Hydraulic conditions are set using a steady state groundwater analysis. The groundwater level was assumed at ground surface (0m hydraulic head applied to the top/surface boundary) and a zero-pressure boundary around the excavation to simulate free water discharge into the tunnel.

3.3 *Zone of influence*

The base model is adapted to determine the zone of influence for this analysis. This zone can be defined as the distance from the tunnel, where an individual 2m^3 cavity does not affect the

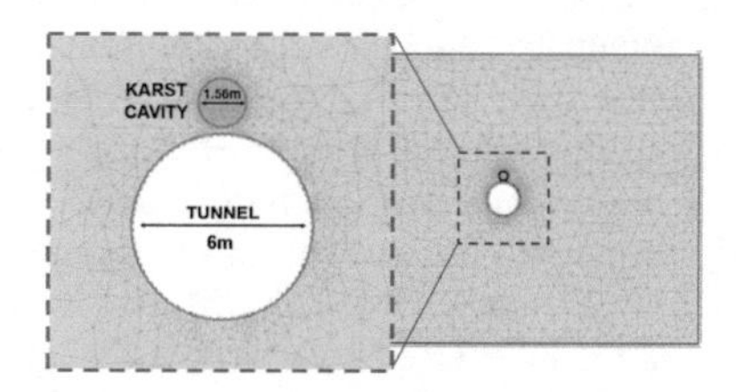

Figure 3. RS2 Model geometry model of the 6m tunnel excavation (white) and 1.56 m karst cavity (the equivalent of 2m^3 in 2D).

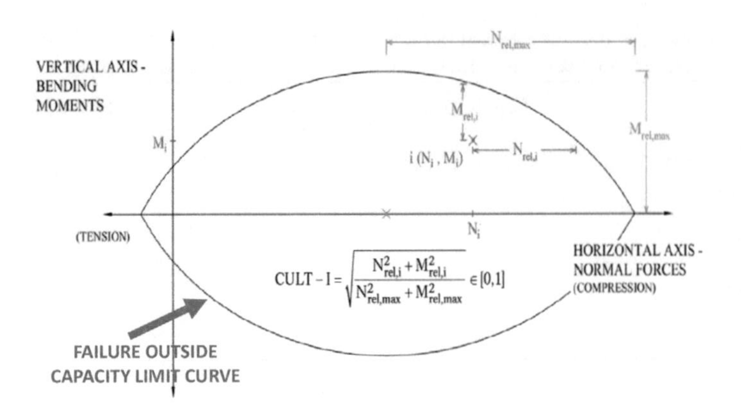

Figure 4. A capacity limit curve, showing the components needed to calculate CULT-I (modified after Spyridis et al. 2016).

tunnel. The variable chosen for analysis to define the effect on the tunnel is the index for capacity utilization of linings in tunnels (CULT-I). CULT-I provides an index for comparison of varying conditions, when considering capacity limit curves of tunnel liners. Figure 4 shows a capacity limit curve, which indicates the values of axial force and bending moment that a liner is 'stable'. Values within the curve indicate a 'stable' lining (i.e. there is reserve in the material utilization), whilst failure occurs outside the envelope (i.e. the material has utilized its full capacity). The limit curve used to find CULT-I in this analysis was specified by the Eurocode 2 EN 1992-1 (EN 1992-1-2, 2004).

The CULT-I ranges from 0 to 1. A zero value indicates that the axial force/bending moment pair of values lies outside the envelope, hence the full material capacity is utilized and exceeded and the liner is considered to have failed and be 'unstable'. Positive values (from zero towards 1) indicate that the axial force/bending moment pair of values lies inside the envelope, there is material capacity to be further utilized and the liner is considered 'stable'. It is important to note that in a framework of optimization and probabilistic approach, the objective is to have a 'stable' liner for all the possible scenarios/combinations, including material and loading variability, with the maximum capacity utilization possible.

3.4 *From 2D to 3D*

Matlab is used to generate the random distribution of karst voids, for subsequent transferal to a suitable format compatible with RocScience software for analysis in 2D analysis. The model aims to quantify the karst cavity distribution properties encountered within the zone of influence of the tunnel. An appropriate boundary is selected so that computation time would not be significant but sufficient for the cavities to be randomly distributed within the zone. The higher bound is the area that is equivalent to the radius of a $2m^3$ cavity (r = 0.78). A random area is first selected using a truncated Gaussian distribution, with -/+3 standard deviation set at 0.1 and 1.92. This is later adapted to a linear distribution, with a linear distribution that generates fewer circles with a higher average area than the Gaussian, with more circles and a higher concentration of average sized circles. For the 2D model, cavities are plotted until 10% of the total area is occupied by the circles (Figure 5.a). Although the set baseline statement (i.e. *there is a 10% chance of encountering a sediment-filled cavity, with a maximum size of* $2m^3$) is to be interpreted in the 3D tunnel alignment, the same assumption was adopted in the 2D for the karst cavities distribution. The plots from the Matlab script are then saved as a scalable vector graphic (SVG), which is converted to a drawing exchange format (DXF) and rescaled in CAD software. This CAD file is then imported to RS2, where analysis of the model is undertaken. The tunnel excavation is then placed on top of the karst distribution, which removes any voids within that area. This is representative of conditions of a tunnel that has been constructed without any ground relaxation (wished-in-place), which would likely be minimal (or its elastic portion would take place) in a competent limestone rock mass and face pressure balance. This procedure is repeated ten times by running the Matlab script again to produce a new cavity configuration, which was transferred to RS2 for analysis (Figure 5.b). The difference in CULT-I values consisting of all the analyses were then plotted as statistical functions for analysis.

The 3D model has the same basic outline as the 2D analysis, except in 3D the entire tunnel alignment can be simulated and the baseline statement can be realistically represented. In 3D analysis the tunnel now becomes a cylinder and karst cavities can be modelled as spheres. In a typical 100m tunnel alignment (Figure 5.c), there may be several combinations of encountering karst cavities that would fulfil the 10% chance of encountering karst. The combinations chosen in this analysis were 1x10m, 2x5m, 4x2.5m and 10x1m sections of the tunnel alignment that encounters karst cavities. Figure 5.c conveys a 10m section occupied with karst, within a 100m tunnel alignment. The different lengths of 10m, 5m and 2.5m that represent a section of the tunnel alignment that encounters cavities are shown in Figure 5.d. In numerical analysis the model is meshed (Figure 5.d) and then is imported into RS3. The model used the same default liner properties as the 2D and applied to the surface of the cylindrical excavation. The steady state groundwater conditions are the same, with zero-pressure applied to the entire tunnel surface. The boundaries are restrained on all faces, edges and vertices, with the exception of the surface face, representing ground conditions. The same gravity loading as the 2D model is applied. The 3D model is then computed and 11 query lines (Figure 5.e) are evenly spaced 1m apart, starting at each end of the 10m section of karst.

This procedure is repeated 10 times using different alignment lengths and with every model having a completely randomized cavity configuration. The results for the differences in CULT-I values of each alignment length are then plotted as statistical functions.

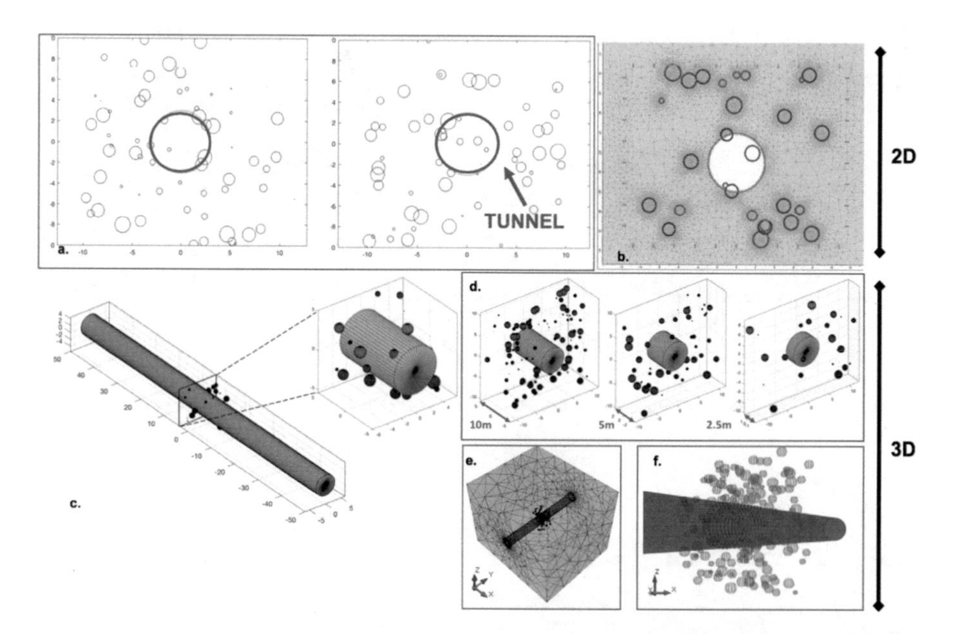

Figure 5. Examples of different karst cavity configurations (black circles) and tunnel (blue circle) in 2D; b. RS2 model showing the implemented cavity configuration (orange circles) from the DXF file; c. a conceptual 100m 3D tunnel alignment with a close-up of a 10m section occupied by karst; d. different lengths of the tunnel alignment (cylinder shape) encountering karst cavities (spheres); e. the mesh of the external boundaries (box) with a 3D tunnel and a cluster of karst cavities in the center; f. an angled view of the tunnel (brown), with 11 query lines around the excavation, surrounded by karst cavities (yellow).

4 NUMERICAL RESULTS

4.1 *2D Analysis*

Figure 6 shows the frequency distribution of difference in CULT-I values. The positive values (blue) indicate an increase in difference in CULT-I and a 'more' stable lining (with reduced lining utilization) whereas the negative values (orange) represent a decrease and a 'less' stable lining (but higher capacity utilization). It is reiterated here that in a framework of optimization and probabilistic approach, the objective is to have a 'stable' liner for all the possible scenarios/combinations, including material and loading variability, with the maximum capacity utilization possible.

From the graph, it can be noted that the maximum positive difference value is less than the negative maximum difference value. The distribution of negative values shows a high frequency of small negative changes, with a decreasing trend towards the larger negative values. The negative values do not reach -100% change (indicating a CULT-I equal to zero), meaning that liner failure does not occur in any of the 2D results. The distribution does not have a discernible trend that could be inferred.

4.2 *3D Analysis*

In general, the contour of axial force on the liner shows low values on the tunnel walls and higher values at the crown and bottom. The cavities in contact with the tunnel then seemed to locally affect the axial force on the liner. High values are seen on the direction parallel to the tunnel alignment on the cavity edges, whilst low values occur perpendicular to that (Figure 7.a).

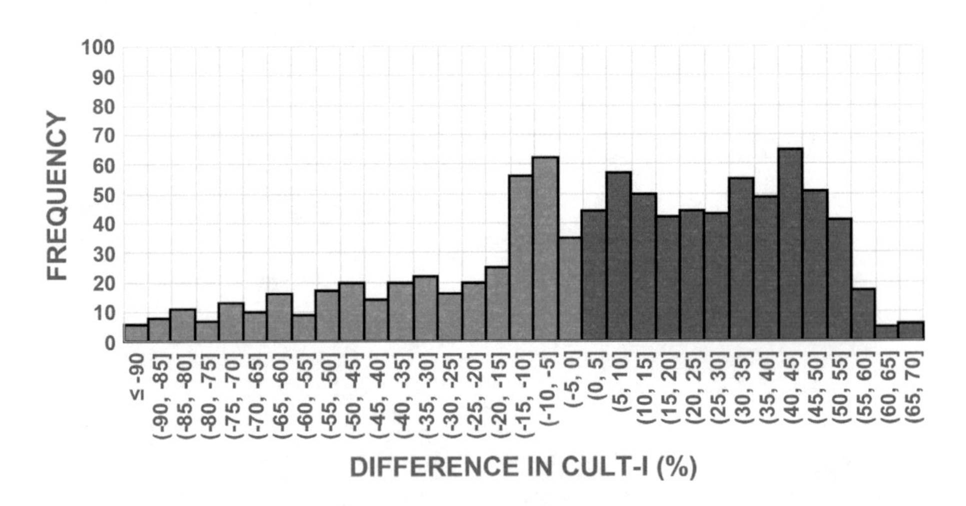

Figure 6. Frequency distribution of the difference in CULT-I values derived from the 2D analysis.

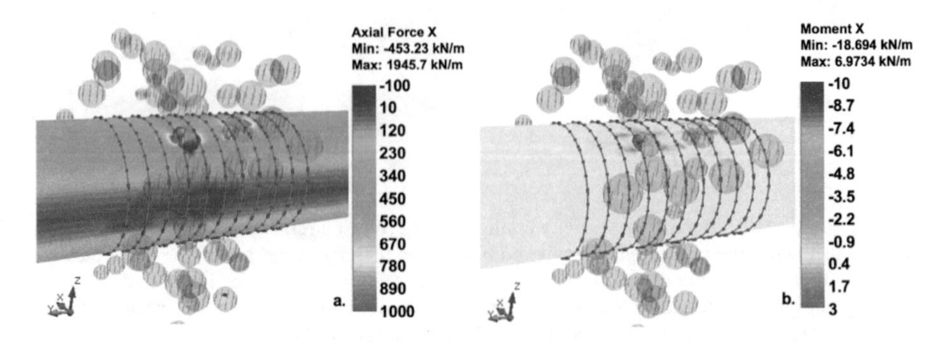

Figure 7. RS3 results: a. Axial force and b. Bending moment values of the tunnel liner.

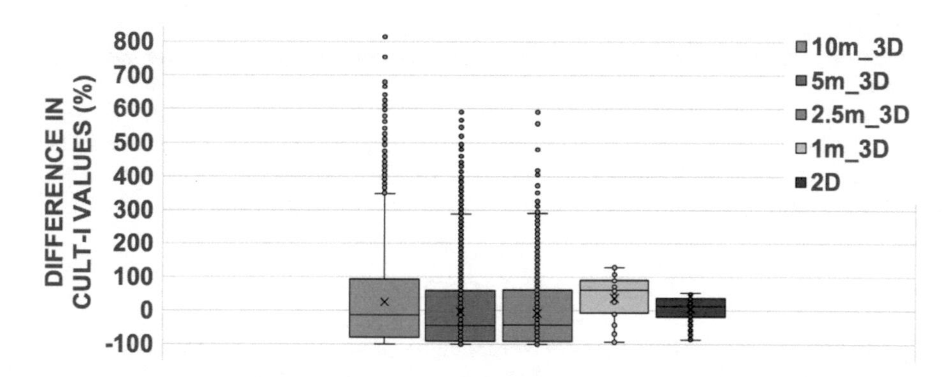

Figure 8. Difference in CULT-I values in the 3D analysis. Negative changes (orange) indicate a decreased CULT-I, positive changes (blue) indicate an increased CULT- I and difference in CULT-I values of 100% (red) indicate lining failure. The frequency refers to the recorded value of equally spaced nodes across the 3D excavation liner surface for all the analyses.

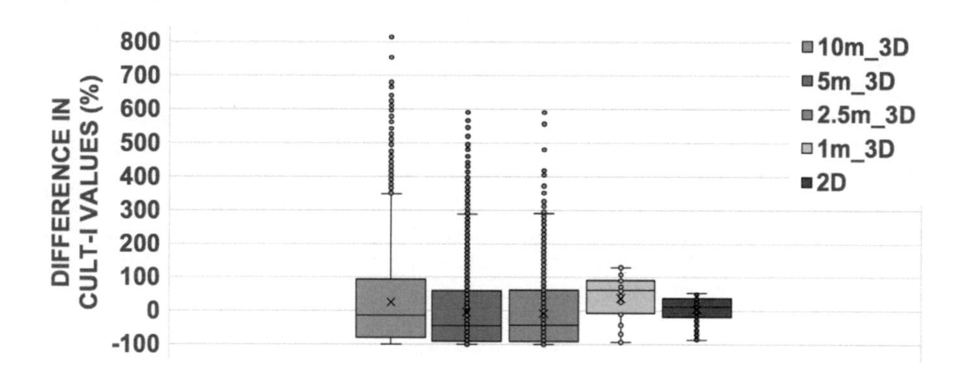

Figure 9. Box plot of the difference in CULT-I values (%) for different alignment lengths.

The bending moment contours show a slightly different pattern that have medium to low values at the sides of the cavity that are perpendicular and parallel to the tunnel alignment. Between the low values on each side, the bending moment is high, which can be seen as the four red areas around the cavity in Figure 7.b.

For a lower CULT-I value to occur, the liner needs to have a higher bending moment and lower axial force, although, there is more influence from a change in axial force. Therefore, the cavity sides that are perpendicular in direction (showing a low axial force) and the cavity corners (red contours on Figure 7.b) produce a lower CULT-I value.

In the 3D analysis (for the 10 m tunnel alignment), there are several CULT-I values that reach a difference equal to 100%, meaning that the liner is outside the capacity curve limit and would fail under the combined effect of the axial force and bending moments (Figure 8).

The majority of data points are negative, with a large peak in frequency in the bin containing -100% to -75% difference in CULT-I values. The values steadily decline from there, with no discernible change between negative and positive values. All values greater than 500% %) were grouped into an upper bin for clarity. The analysis using different alignment lengths of the baseline statement (10m, 5m, 2.5 and 1m) is simplified as box plot, including the 2D model for comparison. The box plot for CULT-I values show a similar distribution for 10m, 5m and 2.5m lengths, although, the 10m values have a larger range (Figure 9).

The 1m length has a smaller overall extent and interquartile range, although, the negative values have a similar range to the other alignments. The median of the 1m values shifts the interquartile range up and indicates that the majority of the data points are positive, meaning that there are more values with an increase in CULT-I values. The 2D values have a smaller range, with more values that are positive, like the 1m alignment length.

The 10m model presents the highest values of CULT-I and so demonstrates the worst expected conditions. This means that this would be the most applicable for projects because the model presents the most conservative estimate for the baseline statement. Any other combination of alignment lengths would produce a lower likelihood of higher values occurring.

5 DISCUSSION AND CONCLUDING REMARKS

The applicability of the proposed model to tunnelling projects is limited to the baseline statement set. The methodology allows the Matlab script to be easily adjusted for varying cavity parameters and probabilities of karst occurring.

The histograms produced using the 3D model for CULT-I values could be applied as probability density functions to assess the risk of certain conditions occurring around the excavation affecting the lining, considering the baseline statement. The greater frequency of negative values in the 3D analysis completely alters the distribution seen in the 2D version, as expected. The 10m, 5m and 2.5m alignment lengths in the 3D analysis yield relatively comparable results, however with significant differences in computational effort.

The reliability of the results is dependent on the accuracy of the model and to the extent that it is representative of the actual conditions, as well as on the number of model runs. The cavities are modelled as spheres, which presents a shape without irregularities that may not accurately portray the effect of irregular void morphologies.

The use of the results as probability density functions is limited by the amount of analyses completed because the model may produce random discrepancies, which become less pronounced when more model runs are carried out. However, the results could still be used for assessment of the expected risks that could occur.

REFERENCES

Eskesen, S.D., P. Tengborg, J. Kampmann, and T.H. Veicherts. 2004. Guidelines for tunnelling risk management: International Tunnelling Association, Working Group No. 2. *Tunnelling and Underground Space Technology* 19: 217–237.

Ford, D. and Williams, P. (2013). *Karst hydrogeology and geomorphology*. Hoboken, N.J.: Wiley, pp.1–208.

Hatem, D.J. 1998. Geotechnical baselines: professional liability implications. *Tunnelling and Underground Space Technology* 13: 143–1150.

Karagkounis, N., Latapie B., Sayers, K., and Mulinti S.D., 2016. Geology and geotechnical evaluation of Doha rock formations, *Geotechnical Research*, 3(3), 119–136. http://dx.doi.org/10.1680/jgere.16.00010

Langford, J.C., and M.S. Diederichs. 2014. Support design for excavations in brittle rock using a Global Response Surface Method. *Rock Mechanics and Rock Engineering*. DOI 10.1007/s00603-014-0567-z.

Marinos, P. (2001). *Tunnelling And Mining In Karstic Terrane; An* Engineering Challenge. Geotechnical & Environmental Applications of Karst Geology & Hydrology, Beck & Herring (eds).

Paraskevopoulou, C., Benardos, A., 2013. Assessing the construction cost of Greek transportation tunnel projects. *Tunnelling and Underground Space Technology* 38, 497–505.

Song, K.I., Cho, GC., Chang, S.B., 2012. Identification, remediation, and analysis of karst sinkholes in the longest railroad tunnel in South Korea, *Engineering Geology Journal*, 135–136, 92–105. https://doi.org/10.1016/j.enggeo.2012.02.018

Spyridis, P., Konstantis, S. and Gakis, A. 2016. Performance indicator of tunnel linings under geotechnical uncertainty. *Geomechanics and Tunnelling*, 9(2), pp.158–164.

International Tunnelling Insurance Group. 2006. A code of practice for risk management of tunnel works.

Van der Pouw Kraan, M. 2014. *Rockmass Behavioural Uncertainty: Implications For Hard Rock Tunnel Geotechnical Baseline Reports*. MScThesis. Queen's University Kingston, Ontario, Canada.

Waltham, A. and Fookes, P. 2003. Engineering classification of karst ground conditions. *Quarterly Journal of Engineering Geology and Hydrogeology*, 36(2), pp.101–118.

Tunnels and Underground Cities: Engineering and Innovation meet Archaeology, Architecture and Art, Volume 3: Geological and geotechnical knowledge and requirements for project implementation – Peila, Viggiani & Celestino (Eds)
 ISBN 978-0-367-46583-4

A new face vulnerability index for mechanized tunneling in subsoil

M. Zare Naghadehi
University of Nevada, Reno, NV, USA

A. Alimardani Lavasan
Ruhr-Universität Bochum, Bochum, Germany

ABSTRACT: In this paper, we address the problem of tunnel face stability in multiple types of ground with a holistic view taking into account a number of the most influencing parameters. We incorporate a recently modified objective systems approach to this topic and a Face Vulnerability Index (FVI) is presented for the first time to assess the stability conditions of tunnels in subsoil. A comprehensive worldwide database of mechanized tunneling case histories is established for this purpose, and the interactions among the parameters are objectively coded by using a soft computing method's capabilities. The results –the FVI predictions- are compared with a number of well-known analytical methods and the actual applied face-support pressures. A good agreement between predictions and measurements has been found that proves the field applicability of the new index to a great extent, which has led to the suggestion of design graphs for future applications.

1 INTRODUCTION

Tunnel boring machines (TBMs) are presently utilized in wide scale in underground construction and tunneling (Jalali and Zare Naghadehi 2013). Slurry shield (fluid support) and earth pressure balance (EPB) (earth support) machines work on the principle that ground deformations can be significantly reduced if the tunnel face is excavated while being supported. Therefore, if the applied pressure can maintain support to the tunnel face, ground displacements can be minimized. Accurate prediction of ground movements associated with face pressures in soft ground is essential to ensure efficient construction and protection of adjacent structures and utilities from damage. The face-support pressure must stabilize the soil at the tunnel front. A number of researchers have studied failure mechanisms of the tunnel face and formulated methods to calculate the minimum required support pressure by analytical or empirical means (see, e.g., Krause 1987, Anagnostou and Kovari 1994, 1996, Jancsecz and Steiner 1994, Kolymbas 2005, Mollon *et al.* 2011, Pan and Dias 2018). Numerical modeling has also extensively been used to investigate tunnel face stability (see, e.g., Ibrahim *et al.* 2015, Zhang *et al.* 2015, Huang *et al.* 2018), in addition to the experimental (laboratory) investigations by physical modeling and centrifuge test models.

In this work, the problem of tunnel face stability in multiple types of the ground is addressed with a holistic view of the problem taking into account a number of influencing parameters. We incorporate a recently modified objective systems approach to this topic for the first time. The Rock Engineering Systems (RES) is a powerful approach to tackle complicated problems, which was first introduced by Hudson (1992). The RES approach has been extensively used in mining and geotechnical engineering analyses (Castaldini *et al.* 1998, Ferentinou and Sakellariou 2007, Budetta *et al.* 2008, KhaloKakaie and Zare Naghadehi 2010, Zare Naghadehi *et al.* 2011, 2013, Ferentinou *et al.* 2012, Khalokakaie and Zare Naghadehi 2012, Ferentinou and Fakir 2018, Khorasani *et al.* 2018). In the RES approach, the interactions between pairs of parameters of the system are presented in an "interaction matrix," whose off-diagonal boxes require to be quantified, in a process called "coding the matrix." Several coding methods have been proposed for this purpose, with the most common being the (deterministic) "expert semi-quantitative" (ESQ) coding method. Despite a recent contribution towards a probabilistic coding of the matrix by Zare Naghadehi *et al.* (2011), the core part of this analysis (i.e. quantification/coding of interactions) is still done "manually" (either deterministically or probabilistically) by "experts" even in the most recent applications of the methodology.

For this reason, the objectivity is in question, and the validity of results based on such coding would be debatable. Motivated by this issue and to minimize the subjectivity due to the human-coding of the interaction matrix in the context of the RES methodology, some efforts have been put on the principle of soft computing capabilities to modify coding by some researchers (Yang and Zhang 1998, Zare Naghadehi and Jimenez 2015, Ferentinou and Fakir 2018). They have utilized Artificial Neural Networks (ANN) to determine the effects of parameters of the system on each other and as a result, to obtain the coding values of the interaction matrix using databases developed for the problems of their studies.

In this research, an extensive worldwide database of mechanized tunneling observations (that we have compiled for the task) is developed to code the interaction matrix in a fully coupled manner on the basis of mentioned objective coding methodology. By coding the interaction matrix of the problem, a new indicative index is proposed to assess the face stability in soft ground mechanized tunneling. Predictions by selected existing analytical methods as well as real observations from the projects are utilized to link the newly developed index to the required face-support pressure of the machine in order to make the index more useful in the future applications.

2 METHODOLOGY

2.1 *Rock Engineering Systems (RES)*

A comprehensive understanding of the entire geo-engineering problem includes all the mechanisms, the parameters and the interactions between them. The rock engineering systems (RES) approach (Hudson 1992, Hudson and Harrison 1992) aims to provide such coherent and general knowledge of complex engineering projects. It also provides a framework from which the complete design procedure can be evaluated, leading to "optimal results" in such engineering projects. The interactions between parameters in the RES approach are represented using an "interaction matrix." The influence of each individual factor on other factors is included at the corresponding off-diagonal box of the matrix, with the intention that the (i,j)-th element represents the influence of parameter i on parameter j as the matrix is generally asymmetric. Technically, no limitation is assumed for the number of factors that may be included in an interaction matrix. A matrix coding value is assigned to each interaction mechanism in the matrix, and by summing the coding values in the row and column through each parameter, "cause" (C) and "effect" (E) coordinates can be computed, allowing each parameter's interaction intensity and dominance to be established (Hudson 1992). The cause-effect plot is also a helpful tool to understand the role of each factor within the project. For instance, levels of interactivity can be used to identify parameters to be kept under control, as their variation is likely to significantly change the system behavior (Mazzoccola and Hudson 1996). As an example of such application, one can differentiate between geotechnical and geometrical properties of a mechanized shield tunneling project if the levels of interactivity are determined and hence the sensitivity of the system to multiple sets of parameters is distinctly highlighted (which should be considered to be kept under control). Such an understanding is essential in conjunction with information about which interactions are beneficial for engineering (and hence should be enhanced) and about which interactions are detrimental for engineering (and accordingly should be inhibited) (Zhao *et al.* 2018).

There are various methods for the selection of the factors and for coding (quantification) their interactions. For instance, in the conventional expert semi-quantitative (ESQ) coding approach, only one unique numerical values (code) is assigned to quantify the influence of a parameter on the others in the matrix, based on the opinion of "expert(s)." Similarly, the "fully coupled model" (FCM) was proposed by Jiao and Hudson (1995) based on the graph theory that assumes all parameters of the system relate to each other in varying degrees of influence. In their work, they considered all the mechanism pathways in the system by the use of partial derivatives (rather than simple derivatives) and finally proposed a global interaction matrix (GIM) rather than the binary one (BIM). However, each component of such GIM is still determined based on the viewpoint of experts, reducing objectivity in solving the problem. To entirely avoid the subjectivity problem, Yang and Zhang (1998) proposed coding the interaction matrix using an artificial neural network (ANN) to model the complex non-linear mapping from system inputs to system outputs.

In this paper, we follow up on the recommendations to use the soft computing capabilities in coding the interaction matrix, and build on the previous works and develop an ANN to code the interaction matrix objectively. The ANN used herein is based on a standard backpropagation (BP) architecture, which, due to its simplicity and applicability, is a robust design process consisting of fully interconnected layers of processing units, incorporated with the well-known sigmoid activation function. For the full details of the calculation process and mathematics, the reader is referred to Zare Naghadehi *et al.* (2013) in which the step by step process of objective interaction matrix coding using the ANN as well as the meaning of them in the framework of the RES methodology have been addressed in detail.

2.2 *The Face Vulnerability Index (FVI)*

The interaction matrix can be coded once the network has been automatically trained using the abovementioned methodology for the available database of mechanized tunneling cases. Then, a vulnerability index for tunnel face can be computed with the use of the values of the parameters considered and their corresponding "weights"; this index will be called the face vulnerability index (FVI). Hudson (1992) has proposed a method for assigning a weight to each parameter which has been applied to many RES applications in the literature. To define such weights, we start by computing the "cause" (C) and "effect" (E) of each parameter in the system. For the $i-th$ factor, we have:

$$C_i = \sum_n I_{mn}(\text{with } m \equiv i);$$

$$E_i = \sum_m I_{mn}(\text{with } n \equiv i). \tag{1}$$

where I_{mn} represents the components of the RES interaction matrix. In other words, C_i and E_i are the sum of the $i-th$ row and column in the interaction matrix, respectively.

Then, the influence weights for each parameter can be computed according to the following equation (Mazzoccola and Hudson 1996):

$$a_i(\%) = \left(\frac{1}{MR_i} \times \frac{(C_i + E_i)}{\sum_j (C_j + E_j)}\right) \times 100 \tag{2}$$

where MR_i is the maximum (input) rating for parameter i that has to be set to one for all parameters considered in this work; i.e., $MR_i \equiv 1.\forall i$ since it is aimed to classify the parameters of the system in a normalized uniform pattern between 0 and 1.

After determination of the parameters weights, a_i, for all considered parameters, the aimed index for a given tunnel (or site) can be computed as:

$$FVI = \sum_i a_i \times R_i \quad (3)$$

where the a_i values are the corresponding weights of the classification system (they are therefore constant for a trained system with the available database) and the R_i values (with) are the ratings assigned to each input parameter considered for our tunnel face stability problem (they are therefore specific for that tunnel or site). In fact, each parameter of the system is assigned different R_i values at different data points (i.e., tunnel sections, here), where the corresponding weights of the parameters, a_i, are kept constant all over the analyses. It is to be noted that R_i is chosen from a classification table of parameters in which the values vary based upon their physical and geometrical conditions at any single data point. The discussion of the parameters as well as the classification table is presented in the next section.

The FVI represents instability potential of a tunnel which ranges from 0 to 100, and (as it will be shown below) higher FVI values indicate more critical face stability conditions (and hence more face-support pressure required).

3 APPLICATION

3.1 *Database of tunneling case histories*

The previously developed methods for tunnel face stability analysis consider different sets of parameters in their calculations dependent upon the amount of simplification taken into account in their methodologies. In the present study, keeping one of the exceptional capabilities of the systems approach in mind, there has been no limitation in the number of parameters that can enter the analyses. However, three criteria were considered in the selection of them including the highest influence, the ease of access (availability), and the least overlap among the parameters. Based on the above and also the parameters of interest of the other researchers, seven influencing parameters were short-listed in the present study as follows:

- Soil Unit Weight (γ)
- Cohesion of Soil (c)
- Friction Angle of Soil (φ)
- Elastic (Young's) Modulus of Soil (E)
- Tunnel Diameter (D)
- Tunnel Overburden (H)
- Water Table (WT)

It should be stated that there is a debate on the influence of the "Elastic modulus of soil" on the face stability of tunnel while it is believed that the elastic modulus mainly governs the geotechnical problems that deal with soil deformability rather than stability. However, the deformability of the soil indirectly alters the stability of a geotechnical system as an excessive plastic strain in the system leads to localization and is usually recognized as instability.

A database of 36 worldwide mechanized tunneling case histories in subsoil consisting of 78 monitoring sections was developed to conduct the analyses in this research. The database has been compiled including general information and geometrical facts related to each project, physical and mechanical data of the soil layers below the surface such as soil unit weight, cohesion, friction angle and elastic modulus for all monitoring sections. It also includes tunnel diameter, overburden, water table level and the machines' data. The actual applied face pressure values are also available for some of the projects within the database.

3.2 *Classification of the system parameters*

In order to start the application process of the systems methodology to the problem, it is required to classify the selected parameters within the system taking their influence on the

Table 1. Parameters classification with intervals equal to 0.1.

Rating value	Unit weight (kN/m³)	Cohesion (kPa)	Friction angle (°)	Young's Modulus (MPa)	Tunnel diameter (m)	Overburden	Water table
0.0	$\gamma<13$	$c\geq 35$	$\varphi\geq 40$	$E\geq 120$	$D<3$	$H\leq 2/3D$	$WT\geq H+D$
0.1	$13\leq\gamma<14$	$30\leq c<35$	$38\leq\varphi<40$	$100\leq E<120$	$3\leq D<4$	$2/3D<H\leq D$	$H+2/3D\leq WT<H+D$
0.2	$14\leq\gamma<15$	$27\leq c<30$	$35\leq\varphi<38$	$80\leq E<100$	$4\leq D<5$	$D<H\leq 11/4D$	$H+1/2D\leq WT<H+2/3D$
0.3	$15\leq\gamma<16$	$25\leq c<27$	$32\leq\varphi<35$	$70\leq E<80$	$5\leq D<6$	$11/4D<H\leq 11/2D$	$H+1/3D\leq WT<H+1/2D$
0.4	$16\leq\gamma<17$	$20\leq c<25$	$30\leq\varphi<32$	$60\leq E<70$	$6\leq D<7$	$11/2D<H\leq 13/4D$	$H\leq WT<H+1/3D$
0.5	$17\leq\gamma<18$	$15\leq c<20$	$28\leq\varphi<30$	$50\leq E<60$	$7\leq D<8$	$13/4D<H\leq 2D$	$3/4H\leq WT<H$
0.6	$18\leq\gamma<19$	$12\leq c<15$	$26\leq\varphi<28$	$40\leq E<50$	$8\leq D<9$	$2D<H\leq 21/4D$	$2/3H\leq WT<3/4H$
0.7	$19\leq\gamma<20$	$10\leq c<12$	$24\leq\varphi<26$	$30\leq E<40$	$9\leq D<10$	$21/4D<H\leq 21/2D$	$1/2H\leq WT<2/3H$
0.8	$20\leq\gamma<21$	$5\leq c<10$	$22\leq\varphi<24$	$20\leq E<30$	$10\leq D<11$	$21/2D<H\leq 23/4D$	$1/3H\leq WT<1/2H$
0.9	$21\leq\gamma<22$	$2\leq c<5$	$20\leq\varphi<22$	$10\leq E<20$	$11\leq D<10$	$23/4D<H\leq 3D$	$1/4H\leq WT<1/3H$
1.0	$\gamma\geq 22$	$c<2$	$\varphi<20$	$E<10$	$D\geq 12$	$H>3D$	$WT<1/4H$

issue into account. Each parameter chosen for our analysis is classified in physical ranges of variation and the corresponding ratings based on the experience of the authors are assigned to the classes (to be also used in the FVI definition; see Eq. (3)). It should be noted that in order to decrease subjectivity in this stage, the ratings were split into very narrow ranges of variation. In this manner, the value of each parameter is subdivided into 11 classes and that each level is rated with values ranging from 0.0 to 1.0 with uniform intervals equal to 0.1 (the higher the rate, the higher the tunnel face instability potential is likely to be). Table 1 depicts this classification. While using the proposed designation, the parameters are independently classified, and any combinations of the parameters in accordance with their specific rating value shown in Table 3 are possible. Apparently, a particular FVI can be identified with different combinations of the parameters.

3.3 *Objective coding of the interactions*

The appropriate neural network with identical ANN structure was set up in the form of a mirror configuration of input and output nodes. The optimal number of neurons in each layer of both networks was found by employing an optimization algorithm. It produced an optimum architecture for each of the training ANN's in the form of 7-14-14-7, consisting of an input layer (7 neurons), an output layer (7 neurons) and two hidden layers (14 neurons each). The network is then entirely trained using the real data within the database, and the partial derivative of an input node can thus be calculated for each output node. Using these derivatives and the weights of all connections, the comprehensive matrix components can be computed.

This process leads to the objective construction of the interaction matrix shown in Table 2 in which the leading diagonal components denote the coupled effect of each parameter on itself. Since a variable does not affect itself, its partial derivative regarding itself will theoretically ever be equal to 1.0. All the leading diagonal components in our objectively coded matrix resulted from the trained neural networks are very close to 1.0 that illustrates the high reliability of coding effort and the calculations. From this matrix (Table 2), the "Cause," C, the influence of each parameter on the system and the "Effect," E, the influence of the system on each parameter can be directly computed.

Accordingly, the cause and effect factors, as well as influence weights, are determined from a holistic point of view (considering all the interactions among the parameters of the system). The a_i values (the interaction intensity of the parameters in the system on tunnel face stability) are computed using Eq. (2) which are shown in Table 3. The weights of parameters within this table indicate that all considered parameters are important to the problem and unidentifiable parameters are not seen in the list.

Table 2. The objectively coded interaction matrix.

1.000[a]	00.41151	0.452	0.5202	-0.0092	0.0799	0.0017	1.47451	Cause (C)
0.0715	1.000[b]	0.0759	0.1348	0.2597	0.2229	-0.0024	0.7672	
-0.0122	0.0912	1.000[c]	0.7578	0.2911	0.1775	-0.0019	1.3317	
0.0332	0.085	0.2609	1.000[d]	0.1777	0.1234	0.0008	0.681	
0.0075	0.0906	0.0017	0.0027	1.000[e]	0.3754	0.0013	0.4792	
0.3276	0.371	0.1718	0.6397	0.0988	1.000[f]	0.0378	1.6467	
-0.681	0.0934	0.6431	0.5737	0.075	0.0055	1.000[g]	2.0717	
1.133	1.14271	1.6054	2.6289	0.9115	0.9846	0.0459		
				Effect (E)				

*The physical meaning of the boxes: [a]Soil unit weight, [b] Cohesion,[c] Friction angle,[d] Young's modulus,[e] Diameter,[f] Overburden,[g] Water table

Table 3. Influence weights of system's parameters

Parameter	Weight (a_i)
P_1: Soil unit weight	0.154
P_2: Cohesion of soil	0.113
P_3: Friction angle of soil	0.174
P_4: Young's modulus of soil	0.196
P_5: Tunnel diameter	0.082
P_6: Tunnel overburden	0.156
P_7: Water table	0.125

3.4 *Computed results, validation, and suggestions*

With the weights of parameters of the system (a_i values, see Table 3) as well as the parameters ratings available from Table 1 for each data point (R_i values, between 0.0 and 1.0), Eq. (3) can be simply used to compute the corresponding FVI values for all 78 cases within the database. On the other hand, seven different analytical methods, the most common methods which are currently widely utilized in the tunneling industry, are considered for computation of the required face pressure values proposed by Atkinson and Potts (1977), Krause (1987), Chambon and Corte (1994), Jancsecz and Steiner (1994), Anagnostou and Kovari (1996), Kolymbas (2005), ZTV-ING (2012). The correlations between each analytical method's estimations and the FVI predictions were found, showing rather good R-Squared values. The results showed better correlations for the methods proposed by Jancsecz and Steiner (1994), Anagnostou and Kovari (1996), and ZTV-ING (2012) which can be categorized as the methods considering the effect of groundwater level while others do not. This is highlighted in higher values of FVI for which the role of groundwater in tunnel face instability is underlined.

In order to validate the newly proposed index, and to evaluate the field applicability, the actual applied face-support pressure which led to the successful execution of the projects depending on the availability for some of the cases within the database (for 43 out of 78 cases), together with the admissible range of support pressure (and its mean value) are compared to the FVI values. Figure 1 shows these comparisons where the permissible ranges of support pressure are specified with the bounding dashed lines, keeping the points related to the analytical method's predictions within the chart.

Apparently, the lower admissible bound represents the minimum required face pressure while the upper permissible bound mainly indicates the upper limit for a failure phase transformation from face collapse to a blow-out. The coincidence of the ZTV-ING (2012) results at the upper limit in Figure 1 specifies this capability of the proposed solution. As seen in this figure, the range of deviation between lower and upper bounds logically increases when

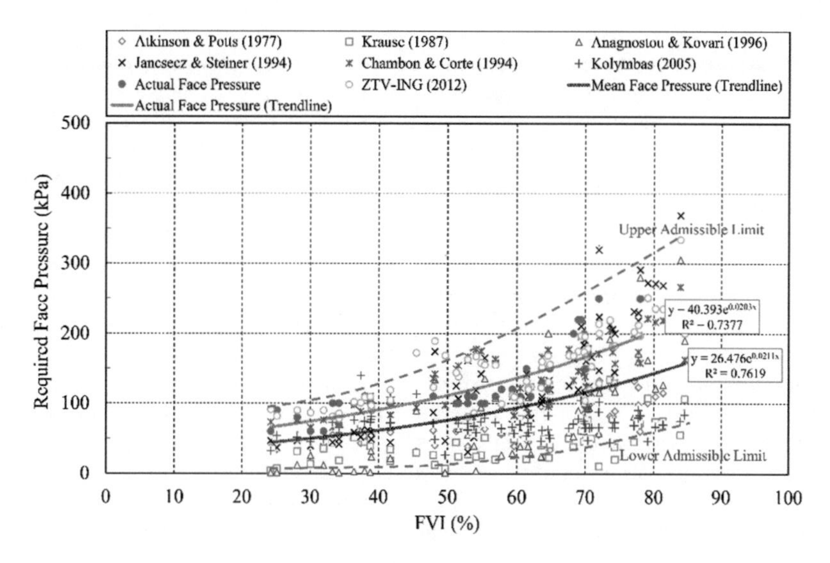

Figure 1. Correlations among analytical methods estimations (average) for required face pressure, actual applied pressure and the FVI predictions.

instability in the face is more likely (higher FVI). As a consequence, it is very reasonable that applying the face-support pressure close to the lower bound induces more ground movements along with the marginal level of safety. Contrariwise, a face pressure in the upper range of appropriate face-support space interval may guarantee fewer deformations while the process is hinged on generating higher slurry or earth pressures which imposes higher costs to the project. However, approaching the upper bound increases the potential of blow-out failure. Furthermore, the allowable range of support-face pressure becomes wider for more critical soil and geometrical conditions (higher FVI) that shows the inability of different analytical methods in offering a unique solution.

The variation of the trend-line that is fitted to the actually applied face pressures is in an excellent match with the mean value of the bound values in the acceptable range. This illustrates the validity of the proposed solution to be employed in order to design the support

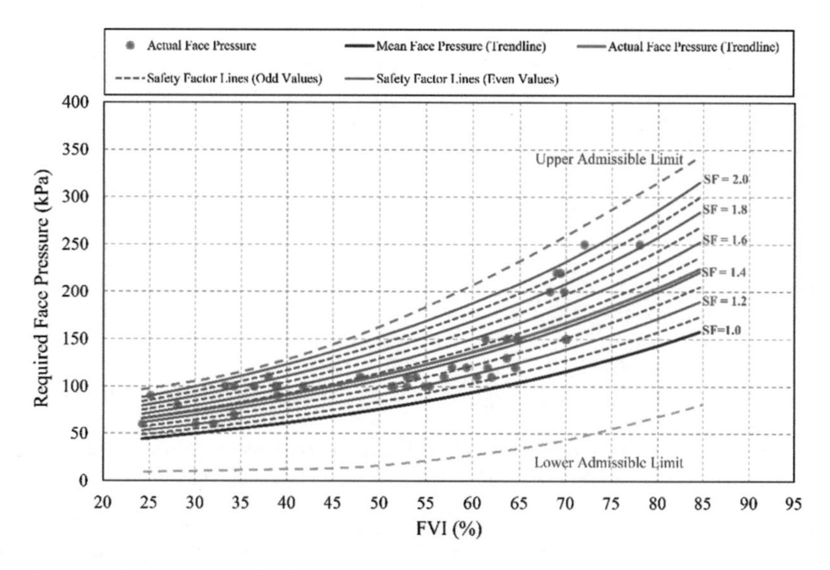

Figure 2. Suggested design graph for face-support pressure.

pressure with different safety factors for real tunneling projects. By assuming the mean value as the most appropriate face-support pressure that is required to stabilize the tunnel face, the difference between the mean value and the actual support pressure can be attributed to the safety factor that is applied to ensure a safe tunnel face stabilizing procedure with minimum ground movements. A comparison between the mean value and the actual values yields to a safety factor between 1.2 and 2.0 in a range of 20<FVI<85.

In order to present the results of current research in the form of guidelines to be utilized in the future, and noting that the field applicability of the suggested index was highly confirmed in the previous sections, a design graph have been prepared in which safety factors are indicated ranging from 1.0 to 2.0. According to this graph (Figure 2), by the calculation of the FVI and assumption of a desired safety factor one can quickly obtain the suggested design face-support pressure value for actual project characteristics. From this graph, it is seen that the growth rate of the fitted exponential curve relating to actual applied pressure is lower compared to all of the safety factor design curves higher than one. Therefore, it is evident that actual applied pressure values conform to different safety factors in various ranges of FVI. Notably, from the design graph, it is seen that the applied values follow an SF almost equal to 1.5 for FVI<40, while it ranges between 1.4 and 1.5 for FVI higher than 40.

4 CONCLUSIONS

The systems approach was adopted to the problem in this paper to present a new face vulnerability index for assessment of the instability conditions of tunnel face in the subsoil and subsequently the required face-support pressure for safe working settings. Seven key parameters affecting tunnel face stability were considered, and soft computing capabilities were utilized along with a worldwide database of mechanized tunneling case histories to determine the complex interactions that exist between parameters of the system. The results showed a strong linkage between FVI values and predictions from the existing analytical methods of face stability analysis. Moreover, field observations were found to be in good agreement with the newly proposed FVI. These altogether proved successful field applicability of the suggested methodology for the future applications. The design graphs were finally presented to be utilized for actual purposes.

The utilization of ANN-based coded interaction matrix for the predictions was highlighted in this research. The objectivity highly increases by the elimination of the manual intervention to determine interactions among the parameters. In fact, we have trusted in real case histories which provide an active link to reality.

In the future, the FVI is to be validated to more extent using a set of different real cases, therefore demonstrating the status of its applicability in other tunnel project sites. Finally, it is evident that the reliability of required face-support pressure predictions based on FVI could be improved as the database is extended.

REFERENCES

Anagnostou, G. and Kovari, K., 1994. The face stability of slurry-shield driven tunnels. *Tunnelling and Underground Space Technology*, 9 (2), 165–174.

Anagnostou, G. and Kovari, K., 1996. Face stability conditions with earth-pressure balanced shields. *Tunnelling and Underground Space Technology*, 11 (2), 165–173.

Atkinson, J.H. and Potts, D.M., 1977. Stability of a shallow circular tunnel in cohesionless soil. *Geotechnique*, 27 (2), 203–215.

Budetta, P., Santo, A., and Vivenzio, F., 2008. Landslide hazard mapping along the coastline of the Cilento region (Italy) by means of a GIS-based parameter rating approach. *Geomorphology*, 94, 340–352.

Castaldini, D., Genevois, R., Panizza, M., Puccinelli, A., Berti, M., and Simoni, A., 1998. An integrated approach for analyzing earthquake-induced surface effects: a case study from the Northern Apennins, Italy. *Journal of Geodynamics*, 26 (2–4), 413–441.

Chambon, P. and Corte, J.F., 1994. Shallow tunnels in cohesionless soil: stability of tunnel face. *Journal of Geotechnical Engineering*, 120 (7), 1148–1165.
Ferentinou, M. and Fakir, M., 2018. Integrating Rock Engineering Systems device and artificial neural networks to predict stability conditions in an open pit. *Engineering Geology*.
Ferentinou, M., Hasiotis, T., and Sakellariou, M., 2012. Application of computational intelligence tools for the analysis of marine geotechnical properties in the head of Zakynthos canyon, Greece. *Computers & Geosciences*, 40, 166–174.
Ferentinou, M. and Sakellariou, M., 2007. Computational intelligence tools for the prediction of slope performance. *Computers and Geotechnics*, 34, 362–384.
Huang, M., Li, S., Yu, J., and Tan, J.Q.W., 2018. Continuous field based upper bound analysis for three-dimensional tunnel face stability in undrained clay. *Computers and Geotechnics*, 94, 207–213.
Hudson, J.A., 1992. *Rock Engineering Systems, Theory and Practice*. Chichester: Ellis Horwood Ltd.
Hudson, J.A. and Harrison, J.P., 1992. A new approach to studying complete rock engineering problems. *Quarterly Journal of Engineering Geology and Hydrogeology*, 25, 93–105.
Ibrahim, E., Soubra, A.H., Mollon, G., Raphael, W., Dias, D., and Reda, A., 2015. Three-dimensional face stability analysis of pressurized tunnels driven in a multilayered purely frictional medium. *Tunnelling and Underground Space Technology*, 49, 18–34.
Jalali, S.M.E. and Zare Naghadehi, M., 2013. Development of a new laboratory apparatus for the examination of the rotary-percussive penetration in tunnel boring machines. *Tunnelling and Underground Space Technology*, 33, 88–97.
Jancsecz, S. and Steiner, W., 1994. Face support for a large mix-shield in heterogeneous ground conditions. *In*: *Tunnelling'94*. 531–550.
Jiao, Y. and Hudson, J.A., 1995. The fully-coupled model for rock engineering systems. *International Journal of Rock Mechanics and Mining Sciences*, 32 (5), 491–512.
Khalokakaie, R. and Zare Naghadehi, M., 2012. Ranking the rock slope instability potential using the Interaction Matrix (IM) technique; a case study in Iran. *Arabian Journal of Geosciences*, 5 (2).
KhaloKakaie, R. and Zare Naghadehi, M., 2010. Ranking the rock slope instability potential using the Interaction Matrix (IM) technique; a case study in Iran. *Arabian Journal of Geosciences*, 5 (2), 263–273.
Khorasani, E., Zare Naghadehi, M., Jimenez, R., Tarigh Azali, S., Jalali, S.-M.E., and Zare, S., 2018. Performance analysis of tunnel boring machine by probabilistic systems approach. *Proceedings of the Institution of Civil Engineers – Geotechnical Engineering*, 1–56.
Kolymbas, D., 2005. *Tunnelling and Tunnel Mechanics: A Rational Approach to Tunnelling*. Berlin Heidelberg: Springer-Verlag.
Krause, T., 1987. Schildvortrieb mit Flussigkeits- und erdgestutzter Ortsbrust. Technischen Universitat Carolo-Wilhelmina.
Mazzoccola, D.F. and Hudson, J.A., 1996. A comprehensive method of rock mass characterization for indicating natural slope instability. *Quarterly Journal of Engineering Geology*, 29, 37–56.
Mollon, G., Dias, D., and Soubra, A.H., 2011. Rotational failure mechanisms for the face stability analysis of tunnels driven by a pressurized shield. *International Journal for Numerical and Analytical Methods in Geomechanics*, 35 (12), 1363–1388.
Pan, Q. and Dias, D., 2018. Three dimensional face stability of a tunnel in weak rock masses subjected to seepage forces. *Tunnelling and Underground Space Technology*, 71, 555–566.
Yang, Y. and Zhang, Q., 1998. The application of neural networks to rock engineering systems (RES). *International Journal of Rock Mechanics and Mining Sciences*, 35 (6), 727–745.
Zare Naghadehi, M. and Jimenez, R., 2015. On the Development of a Slope Instability Index for Open-Pit Mines using an Improved Systems Approach. *ISRM Regional Symposium - EUROCK 2015*.
Zare Naghadehi, M., Jimenez, R., KhaloKakaie, R., and Jalali, S.-M.E., 2013. A new open-pit mine slope instability index defined using the improved rock engineering systems approach. *International Journal of Rock Mechanics and Mining Sciences*, 61, 1–14.
Zare Naghadehi, M., Jimenez, R., KhaloKakaie, R., and Jalali, S., 2011. A probabilistic systems methodology to analyze the importance of factors affecting the stability of rock slopes. *Engineering Geology*, 118, 82–92.
Zhang, C., Han, K., and Zhang, D., 2015. Face stability analysis of shallow circular tunnels in cohesive-frictional soils. *Tunnelling and Underground Space Technology*, 50, 345–357.
Zhao, C., Lavasan, A.A., Holter, R., and Schanz, T., 2018. Mechanized tunneling induced building settlements and design of optimal monitoring strategies based on sensitivity field. *Computers and Geotechnics*, 97, 246–260.
ZTV-ING, 2012. *Zusätzliche Technische Vertragsbedingungen und Richtlinien für Ingenieurbauten. Teil-5: Tunnelbau-Baudurchführung*.

Tunnels and Underground Cities: Engineering and Innovation meet Archaeology, Architecture and Art, Volume 3: Geological and geotechnical knowledge and requirements for project implementation – Peila, Viggiani & Celestino (Eds)
 ISBN 978-0-367-46583-4

Settlements of immersed tunnel on soft ground: A case study

X. Zhang & W. Broere
Geo-Engineering Section, Delft University of Technology, Delft, The Netherlands

ABSTRACT: This paper focuses on the settlement analysis of immersed tunnel on soft ground. Even under careful design, immersed tunnel may have problems of excessive settlement and the resulting concrete cracking or structure leakage during operation period. Yongjiang Tunnel, the first immersed tunnel on soft ground in mainland China, is taken as a case study. The monitored settlement data after a 16-year service is first displayed and analyzed, then a 2D numerical model is built to simulate the ground settlement deformation from tunnel construction to long-term tunnel operation. The effects of back-silting on the stiffness of the foundation layer and further on the settlement is quantitatively analyzed, and the advices on settlement control for immersed tunnel on soft ground are provided.

1 INTRODUCTION

1.1 *Settlement of immersed tunnel*

Immersed tunnels are built under waterways and they are usually better than other crossings like bridges or bored tunnels. Tunnels are superior to bridges mainly in that they do not disturb ship navigation, especially for busy water channels. Bored tunnels usually have a minimum buried depth for safety and thus bored tunnel are usually designed to a certain depth below the river bed, which can increase the buried length of the whole tunnel. Immersed tunnel consists of prefabricated elements (usually around 100 meters long) which are connected underwater with special rubber gaskets. The immersed tunnel structure can be placed directly on the trench excavated on the shallow river bed, and no minimum buried depth is needed, unlike bored tunnel, so when at the same construction site, the total length of an immersed tunnel generally can be shorter than that of a bored tunnel. Also, immersed tunnels are usually factory-prefabricated element by element and then flowed to the immersion site. They have less joints than bored tunnel, which reduce the water leakage risk; what's more, critical sub-projects of immersed tunnel construction, like trench excavation and element fabrication can go simultaneously, and this can save much time and reduce project cost (Lunniss & Baber, 2013).

Immersed tunnels are mostly constructed under canals and waterways, especially where no water navigation interference is allowed. There are more than 200 immersed tunnels in the world, with most of them in the North America, Europe and Asia. For example, in the Netherlands alone, there are more than 30 immersed tunnels. More immersed tunnel projects, including the 6.0km Hongkong-Zhuhai-Macao Bridge Tunnel, the 18km Fehmarnbelt fixed road and rail link, et al are being constructed or designed for fixed links or underpasses beneath waterways (Hu, 2015; Pedersen, 2018)

Generally, immersed tunnel construction will firstly remove part of soil when doing the trench excavation, and the tunnel body is generally lighter than the soil it replaced, so it is easy to infer that pressure on foundation should be within a small level as not to cause large settlement. However, many immersed tunnels have suffered excessive settlement, which is, at least, much larger than the anticipated value in the primary project design. For some projects on soft ground, serious settlement even has caused troubles for normal operation. Some immersed tunnels on soft ground have suffered significant differential settlement and even cause subsequent problems such as leakage, concrete cracking or even damage of joint waterproofing Gasket.

For example, Kiltunnel in Netherlands suffered a serious differential settlement at element joints, and this further caused leakage inside the tunnel and the leaked water freezed to ice in winter, which risk the traffic safety. More information on immersed tunnel settlement are provided by Grantz (Grantz, 2001).

The settlement of immersed tunnel is mainly related to geological conditions, construction quality and design methods. Due to the complexity and uncertainties from project design and construction, immersed tunnel settlement is mostly studied from field measurement and qualitative analysis. For example, Grantz (Grantz, 2001a,b) summarized the immersed tunnel settlement issues, including the potential factors which cause excessive settlement, the effects of settlement on the tunnel structure and provide settlement data of 15 immersed tunnel projects. Settlement problems are more or less can be explained qualitatively by the reasons listed, which include sub-soil conditions, foundation treatment methods, tunnel section, siltation, et al. It should be noted that settlement is inevitable, or inherent, since it is impossible or cost-effective to make soft ground an absolutely rigid body without any deformation, but technically possible to keep settlement within a safety limit. Some researchers have collected settlement data of more than 20 immersed tunnels (Shao, 2003), with most conclusions from (Grantz, 2001b), and concluded that settlement of immersed tunnel are affected by many complex reasons. When founded on soft ground, immersed tunnel is more likely to settle excessively, and special foundation treatment is often needed. For example, the foundation of High-Speed Railway immersed tunnel should be well treated. To mitigate the risk of excess settlement on soft ground in Netherlands, the subsoil was firstly pre-loaded on top with a 6-m thick sand layer, and this accelerated the consolidation rate of soft clay and improve the strength of soil. The final settlement value is within design limit (Mortier, 2013).

In this paper, long-term settlement of Yongjiang immersed tunnel is quantitatively studied. This tunnel is the first tunnel on soft ground in mainland China that constructed with immersion method, and it served as an experimental trial for future immersed tunnel construction. The rest of the paper firstly summarizes the main factors which cause excessive settlement, and then Yongjiang tunnel, in Ningbo, China, is taken as a case study. This tunnel was opened to traffic in 1995 and after a more-then-decade service, very significant settlement occurs, which seriously affects the structure performance. In the case study, settlement of Yongjiang tunnel after a 16-year of service is analyzed, then potential reasons which results in such high settlement are analyzed. Secondly, a numerical model is built to simulate the deformation behavior of immersed tunnel from construction to long-term service period. The key construction steps including trench excavation, gravel pavement, tunnel element placement, backfilling, back-silting of river bed in operation period, are considered.

2 POTENTIAL TRIGGERS FOR IMMERSED TUNNEL SETTLEMENT

2.1 *Uniform settlement or differential settlement*

Uniform settlement, which usually refers to the settlement of the structure as a whole rigid body. Structure loading on elastic foundation or ground will trigger settlement definitely, since the stiffness of the soil medium is limited, though sometimes very large. But uniform settlement generally does not affect the structure safety much if within a certain limit. For immersed tunnel, uniform settlement of tunnel body will not cause rotation of tunnel elements and no concrete cracking, but however, almost all the tunnel experience differential settlement.

Differential settlement means different settlement values of the tunnel body, this usually leads to the element rotation, joint opening, cyclic compression and expansion of rubber gasket, and internal force of the tunnel body. what's more, the subsequent problems like concrete cracking, leakage, et al, will deteriorate the performance of immersed tunnels (Grantz, 2001a).

It should be noted that because foundation stiffness is limited, settlement of immersed tunnel is unavoidable since the pressure on underlying stratum. Anticipated settlement within safety margin is not troublesome to structure safety. But What we are interested is the unexpected excessive settlement which harms the structure significantly.

As pointed out by (Grantz, 2001a), the factors related to immersed tunnel settlement (or, excessive settlement) can be summarized as the follows:

1) Sub-soil conditions. Generally, consolidated sandy layer has a lower settlement value compared with compressible clayey soil, and the latter also needs a longer time to reach a stable final settlement.
2) Siltation. Siltation may accumulate on the trench surface and may cause serious differential settlement, this is especially severe for immersed tunnel founded by sand-jetting or sand-flow method.
3) Method of tunnel foundation construction. Immersed tunnel is generally founded on sand layer (by sand-jetting or sand-flow after the element placement) or screeded gravel bed (before element placement), such as the Øresund tunnel, the Hongkong-Zhuhai-Macao bridge tunnel, et al. both of the methods are widely used and have different settlement control ability.
4) Surcharge. This may include the backfill on top of the tunnel and the increasing heavy traffic loading which exceeds the designed target. Back-silting on the river bed usually causes significantly large settlement, and periodic dredging of the tunnel location has been performed in many tunnel projects.
5) Trench dredging methods. Commonly, hydraulic cutter-head suction dredges and/or clamshell bucket dredges are used to excavate the tunnel trench. The latter may tend to leave a more irregular bottom with larger voids that can take longer for the foundation material to fill and stabilize.
6) Tunnel geometry. Tunnel with octagonal section has a narrower contact width than its projected plan width, while rectangular section has the same contact width as the section width. Large contact area usually means lower foundation pressure and hence a lower settlement.
7) Large tidal variation. Under certain circumstances, large amplitude tidal variation may also cause settlement of sands by their gradual compaction due to daily oscillations in pore pressure. If covered by a clay layer that slows the relief of pore pressures, the upper layers of sand may cause oscillation of the supporting ground.

Generally, final settlement of immersed tunnel is affected by multiple factors and analyzing the effects of each factor are usually too complicated. However, some basic design calculation and modelling, plus a good assurance of construction quality control will help to reduce the structure damage from excessive settlement.

3 SETTLEMENT ANALYSIS OF YONGJIANG IMMERSED TUNNEL

3.1 *Project introduction*

Yongjiang tunnel, which is in Ningbo, China, is located on very soft clayey ground. This tunnel crosses under Yongjiang River, and was the first immersed tunnel on soft ground in mainland China, the construction of Yongjiang tunnel started in June, 1987, and it was opened to traffic in 1995. Longitudinally, Yongjiang tunnel is 1019.97m in total, including the cast-in-place approaches on both sides by cut and cover method, and central immersed section(420m). The immersed tunnel consists of 5 elements (85m×3+80m×2) which are connected by immersion joints. The prefabricated tunnel section has a rectangular cross-section with 11.9m wide and 7.65m high, with dual lanes plus a 1.25m wide side-way for operation inspection (Xie, 2014). It should be noted that Yongjiang Tunnel has a small cross-section, but this project served as a trial for immersed tunnel practices in mainland China. The experience gained in construction of this small tunnel is of great help for the future immersed tunnel engineering practice in mainland China.

Yongjiang tunnel was founded on grouted cementitious material, underlaid by a coarse gravel layer. Firstly, the trench was excavated and then a gravel bed is formed by dumping the coarse gravel on the trench bottom. Then tunnel element is placed on the temporary support and then gap between the element and the gravel layer (40cm) is backfilled with grouting materials (com-posed of cement, fly-ash, bentonite, fine sand and chemical admixture). And the

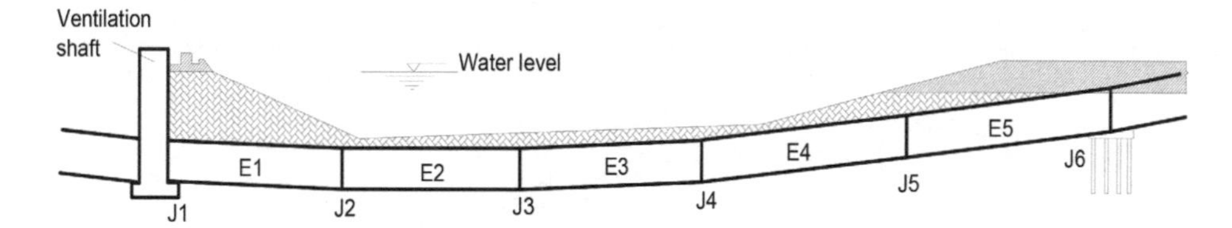

Figure 1. Longitudinal profile of Yongjiang Tunel.

element is fully released to set on the grouted bed. The temporary jacket below water is removed. Then backfill the gap around the tunnel element until to the natural river bed grade. The backfill acts as protective layer as well as ballast against buoyancy.

3.2 *Geological conditions*

The underlaying ground of Yongjiang Tunnel mainly consists of muck soil, mucky clay and medium-sized sand from top down. The muck soil has a thickness ranging from 10–13m, and mucky clay has a thickness of about 5–6m, then underlaid by sand layers interbedding with muck and clay. The muck soil is highly compressible and with low strength, and standard penetration test show a value of 1–2; the tunnel is designed to be placed within saturated muck soil layer, as it was calculated, in the tunnel design, that pressure on foundation bed was very low.

After 11-years of service, Yongjiang Tunnel generally works well, no vital structure damage occurred. But some problems arised, including tunnel structure cracking and water leakage, differential settlement between elements, damage of pavement, reinforcement corrosion on the wall at the approach section, et al. From October 2007 to March, 2008, the tunnel was closed for a major maintenance repair. The repair work mainly includes the grouting at some leaking points to seal the leakage, repairmen of the pavement, strengthen the pillar by expanding the section. As the immersion joint E5 settle significantly, the Gina and omega gasket are inspected, the cover on the wall outside the immersion joint is removed and the joint gap is measured. This can infer the compression status of Gina gasket and asses the waterproofness of immersion joints. And a structure health monitoring network is formed at the finishing of tunnel repair (Li, 2011).

3.3 *Settlement data analysis*

The tunnel structure settlement is monitored since its open to traffic. Vertical settlement is measured by hydraulic static leveler with fixed bolts on the tunnel wall. Until now, the tunnel has undergone a significant settlement and this caused concrete cracking. The Figure 2 shows the tunnel settlement profile at joints locations. After 16-year of operation, the tunnel suffers a serious settlement, with the maximum settlement magnitude about 86mm at immersion joint 5 between Element 4 and 5. While joint 4 at Element 3 and 4 reached a maximum about 56mm (Xie, 2014). This settlement level is well beyond the anticipated value in project design stage.

According to the administration of tunnel operation, excessive settlement of this immersed tunnel may be attributed to the following reasons (Xie, 2014):

1) the underlying soil is extremely soft, highly plastic and compressible, which needs a long time to consolidate under loading.
2) the surcharge from heavy river bed back-silting. Yongjiang Tunnel is located at a delta area and suffered serious siltation, the siltation accumulates at top of the tunnel element and added to the loading, which further causes consolidation of sub-soil and hence an increased settlement.
3) the change of tides causes a varying in the porous pressure in the soil stratum, and lead to the tunnel structure fluctuation.

But the problem is, to what extend did the back-silting surcharge add to the total settlement? Is it the problems of siltation during tunnel immersion, say trench excavation, that

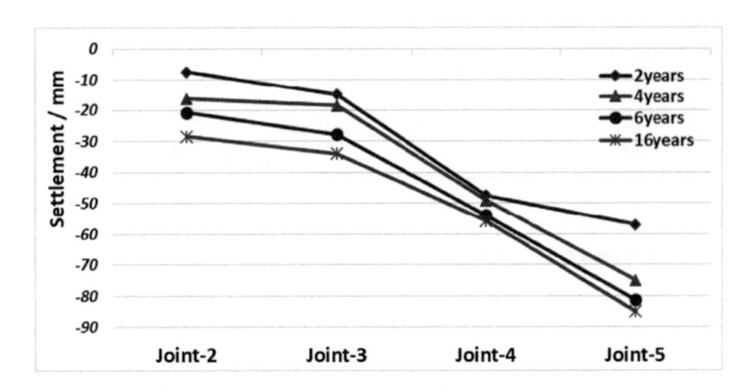

Figure 2. Joint settlement of Yongjiang Tunnel.

cause the most settlement, or rather the surcharge loading from back-silting at operation stage? This calls for a careful study.

3.4 *Numerical simulation*

In this section, a 2D numerical simulation model is built in PlaxisV8.5 to calculate settlement of immersed tunnel through the whole stage, from trench dredging, underwater gravel bed formation, tunnel element placement, backfill to natural river bed grade, back-silting in operation stage. Also, sensitivity analysis of stiffness of the bed on the settlement is analyzed.

During the immersed tunnel construction, underlaid soils are disturbed by a "unloading-reloading" process. This is because trench dredging usually removes the shallow subsoil and rebound occurs at the trench bottom, hence the unloading effects, then coarse gravel is dumped to form a relatively stiff bed, which adds a partial loading on the sub-soil and cause settlement. For tunnel with grouted bed, the element is released on the grouted bed layer and this adds to some loading on the sub-soil; then backfilling is performed, and this further increases loads on the sub-soil as well, as referred to be a reloading process. The back-silting at service stage also adds to the loading which results in long-term settlement, periodic dredging operation will reduce the loading, which may lead to a small rebound of the subsoil. During the unloading of trench excavation, negative porous pressure occurs and the disposition of this negative pressure may last for months, which is accompanied by the trench bottom rebound.

For engineering practices, we hope the subsoil to consolidate fast and most of the total settlement occurs within construction rather than operational period. For sandy soil, usually consolidation occurs constantly and secondary settlement in long-term is expected to be low. But for clayey ground, things are usually quite different, sometimes foundation treatment, such as pre-loading or soil improvement, is performed to accelerate consolidation, this is especially for clayey ground. If we want to consider the final settlement in the operation stage more accurately, long-term soil consolidation is necessary to take into account, rather than only based on elastic soil body.

3.4.1 *Key modelling parameters*

A 2D numerical model is built to simulate the settlement of tunnel structure from trench excavation to backfill, and back-silting during operation stage. The underlying strata analyzed include a muck layer, mucky clay and sandy layer. The trench depth is set as 10m below river bed level, slop of the trench is 1:3 from bottom to 1:4 and then changed to 1:7 till the natural river bed. The tunnel element is designed to lay in the muck layer, since the tunnel bottom pressure was considered to be low in the preliminary design. Here the tunnel structure has an absolutely higher rigidity compared with sub-strata, so tunnel structure deformation is not simulated, only the vertical settlement of tunnel bottom is considered.

The ground strata properties used for numerical modelling is shown in Table 1, since elastic modulus for the Hardening Soil (H-S) model here were not available from site investigation directly, they are set to be 2 times the compression modulus obtained from a confined consolidation test in the investigation document. There is a 1.5m-deep gap between the tunnel top and natural river bed, and backfilling is assumed to reach the natural river bed.

Table 1. Mechanical parameters of sub-strata (based on Xie, 2014).

Stratum	Thickness (m)	density ($g*cm^{-3}$)	Yang's modulus (Mpa)	Cohension (kPa)	Internal friction angle (°)
Muck layer	12	1.81	9.7	19	13
Mucky clay	11	1.72	13.5	23	16.1
Medium-sized sand	15	1.87	50	8	30

According to the foundation design, the 1m-thick bed layer under the tunnel consists of a 0.6m-thick screeded gravel bed and a 0.4-m thick grouted material, due to the lack of the properties of the grouted material, the 1-m thick composite layer is simulated as a single gravel layer, with the property parameters adopted from the Hongkong-Zhuhai-Macau Bridge Tunnel project. The static water pressure is taken as static surface loading on the river bed and on the tunnel trench. The simulation step starts from trench dredging, gravel bed formation, tunnel element placement and backfilling (to natural river bed level). In operation period, different back-silting cases on the river bed is simulated as well, the settlement caused by back-silting is analyzed.

3.5 *Settlement result analysis of tunnel*

Figure 3 shows the 2D simulation model in PlaxisV8.5, the modelled cross-section is a typical geological section under the tunnel element 4, and three ground layers are considered, i.e muck layer, mucky layer and sandy layer.

Simulation shows the trench bottom rebounds to a value of 63.8mm when excavation is finished, and this significant rebound is mainly due to the high compressibility of the first muck layer. Usually on soft foundation, rebound of deep excavation is significant, and this rebound can offset the subsequent settlement in tunnel immersion. Gravel dumping cause a loading on the trench bottom, and this cause a settlement of -6.21mm, which is relatively small. The tunnel element placement triggers a settlement of -33.62mm, and this settlement increases to -57.47mm when the trench is backfilled to original river bed level. Note that the absolute settlement of the tunnel bottom is +6.32mm if taking account of the rebound at trench excavation. The results show the tunnel settlement should not be large, at least not so large as to cause a risk of concrete

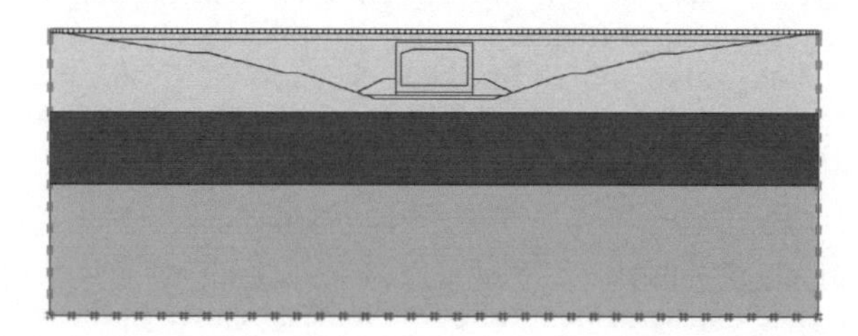

Figure 3. 2D Numerical model of tunnel.

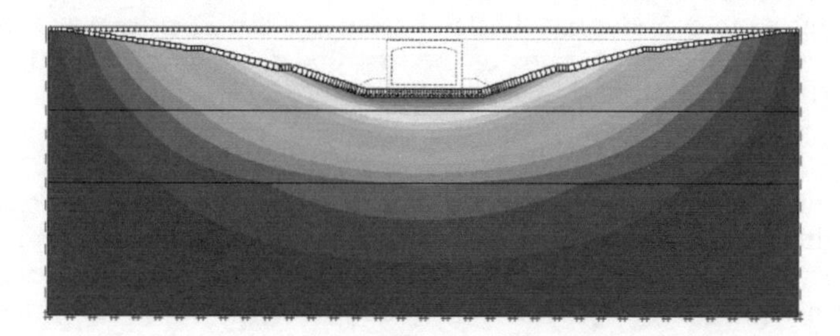

Figure 4. Rebound of immersed tunnel trench bottom due to unloading (maximum as 63.8mm).

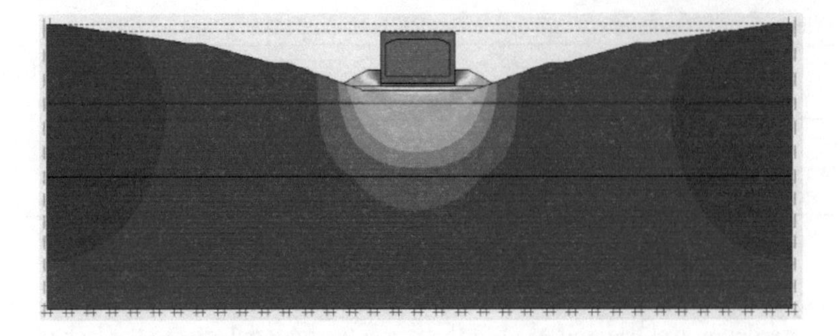

Figure 5. Settlement of tunnel when placed on trench (maximum as 57.6mm).

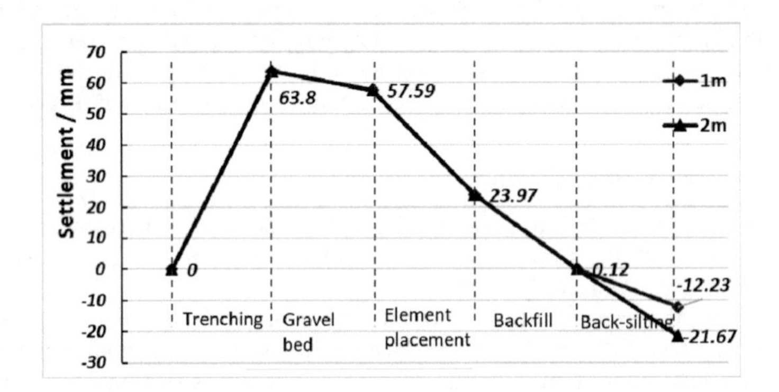

Figure 6. Vertical displacement of tunnel element (positive indicates rebound).

cracking. However, the field monitoring data show there is a surprisingly high settlement value, which is quite beyond expectation. The potential reasons should be quantitatively assessed.

In order to evaluate the potential effects of back-silting on tunnel settlement, different silting loading is considered. Here loading from a 1m-thick and 2m-thick back-silting layer is considered, the newly resulting settlement under 1m-thick back-silting is about -12.23mm, while the corresponding value for a 2m-thick scenario is -21.67mm, that is to say, a unit-meter back-silting generally causes about 10mm settlement of the tunnel. Considering the back-silting, the settlement will increase significantly.

3.5.1 *Effects of bed stiffness*

According to the numerical simulation results, the back-silting alone is not less likely to cause a significant or even shocking settlement (above 80mm) as shown in the field settlement measurement. Since the underwater construction work will bear uncertainties in the foundation quality, and the grouting work control affects the bed stiffness much, as a soft grouted layer will undermine the stiffness of the bed layer right beneath the tunnel bottom. Here a sensitivity analysis is conducted to consider the effects of reduced stiffness on tunnel settlement. In the sensitivity analysis, the elastic modulus of the first underlying layer beneath tunnel bottom is set as 25%, 50% and 75% of the design value in the previous model. And the settlement is recalculated, as shown in Figure 7.

The sensitivity analysis shows that reduction in bed stiffness will cause a much larger settlement compared with the ideally hard bed condition. For example, when the elastic modulus changed to a half, the settlement of the tunnel at backfilling reaches 31.36mm, and this further increases to about 69mm when considered the possible back-silting(2m), this is much closer to the field measured settlement. If considering the reduction of the bed stiffness a step further, say to 25%, the settlement increases much more significantly. Compared with data in Figure 6, it is reasonable to deduce that excessively tunnel settlement is more likely to be, or mainly, caused by the low

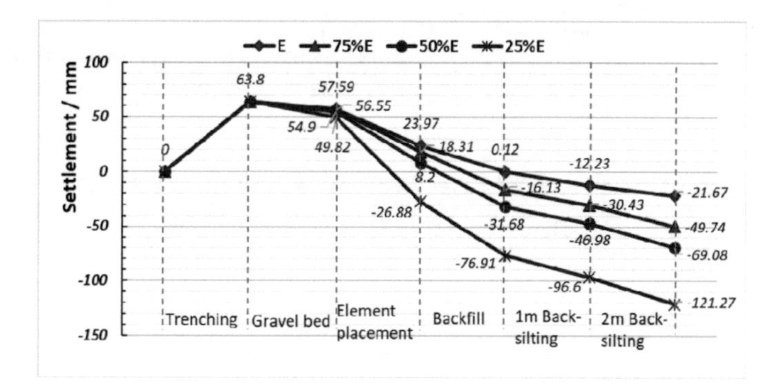

Figure 7. Tunnel settlement under changed bed stiffness.

stiffness of the bed, and this is probably resulting from siltation during trench excavation and element immersion process, hence the importance of trench siltation removal.

4 CONCLUSION

This paper analyzed the settlement of immersed tunnel on soft ground. The settlement of immersed tunnel is mainly related to geological conditions, construction quality and design methods. Long-term settlement of Yongjiang immersed tunnel is quantitatively studied. Settlement of Yongjiang Tunnel after a long-time service is analyzed, then potential reasons which results in such high settlement are analyzed. Secondly, a numerical model is built to simulate the deformation behavior of immersed tunnel from construction to long-term service period. The key construction steps including trench excavation, gravel pavement, tunnel element placement, backfilling, back-silting of river bed in operation period, are considered. Result shows that though back-silting on river bed will significantly increase the settlement due to surcharge effects, the settlement caused by back-silting alone is not likely to reach the measured value. But sensitivity analysis shows bed stiffness reduction causes even larger effects on settlement, which affects the long-term settlement more. The excessive settlement of Yongjiang Tunnel is more likely to be attributed to the bed stiffness reduction, hence siltation control in trench excavation is important for tunnel settlement control.

REFERENCE

Hu Z., Xie Y. & Wang J. 2015. Challenges and strategies involved in designing and constructing a 6 km immersed tunnel: A case study of the Hong Kong–Zhuhai–Macao Bridge. Tunnelling and Underground Space Technology 50: 171–177.

Li W., Wu D., Guo X. & Gao X. 2011. Overhaul Design and Construction of Ningbo Yongjiang Immersed Tube Tunnel. MODERN TUNNELLING TECHNOLOGY 48(1): 82–89.

Lunniss, R. & Baber J. 2013. Immersed Tunnels. New York: CRC Press.

Mortier H & Hakkaart C. J. 2013. Settlements of HSL Immersed Tunnels. Geotechnical Aspects of Underground Construction in Soft Ground: Proceedings of the 5th International Symposium TC28: the Netherlands: 259–264

Grantz, W. C. 2001a. Immersed tunnel settlements. Part 1: nature of settlements. Tunnelling and Un-derground Space Technology 16(3): 195–201.

Grantz, W. C. 2001b. Immersed tunnel settlements: Part 2: case histories. Tunnelling and Underground Space Technology 16(3): 203–210.

Pedersen, S. K. & S. Brøndum. 2018. Fehmarnbelt fixed link: the world's longest road and rail immersed tunnel. Pro. of the ICE - Civil Engineering 171:17–23.

Shao J. 2003. Study on Prediction and Control for Settlements of Immersed Tunnels. Tongji University. Doctor thesis.

Xie X., Wang P., Li Y., Niu J. & Qin H. 2014. Monitoring data and finite element analysis of long term settlement of Yongjiang immersed tunnel. Rock and Soil Mechanics 35(8): 2314–2324.

Tunnels and Underground Cities: Engineering and Innovation meet Archaeology, Architecture and Art, Volume 3: Geological and geotechnical knowledge and requirements for project implementation – Peila, Viggiani & Celestino (Eds)

Design of shaking table tests on atrium-style subway station models under seismic excitations

Z.M. Zhang
Department of Geotechnical Engineering, Tongji University, Shanghai, China
Department of Civil, Architectural & Environmental Engineering, University of Naples Federico II, Naples, Italy

Y. Yuan
State Key Laboratory of Disaster Reduction in Civil Engineering, Tongji University, Shanghai, China

E. Bilotta
Department of Civil, Architectural & Environmental Engineering, University of Naples Federico II, Naples, Italy

H.T. Yu
Key Laboratory of Geotechnical and Underground Engineering of Ministry of Education, Tongji University, Shanghai, China

H.L. Zhao
Department of Civil Engineering, Shanghai University, Shanghai, China

ABSTRACT: In view of the good performance in natural lighting, more and more atrium-style subway stations are used in urban subway systems. Some characteristics of such stations are ex-tremely different from traditional frame subway stations, such as no columns in waiting hall floor and a lot of beams instead of ceiling slab and middle slab. These discontinuous components in the axial direction of the subway lines may lead to special lateral defor-mation while experiencing an earthquake. Shaking table tests are essential to acquire the seismic aspects about atrium-style subway station. This study introduces the design scheme for soil-atrium-style subway station model and verification for free-field model. The in-situ soil is modeled using artificial soil, which is a mixture of sawdust and sand with a mass ratio of sawdust to sand 1:2.5. For the subway station structure, galvanized steel wire and micro concrete were used to model the prototypical reinforcement and con-crete. A flexible-wall container, whose diameter and height were 3 m and 1.6 m, respec-tively, was adopted in the test. The results of free-field test showed that the flexible-wall container presented a shear-type deformation mode under dynamic excitation. The test data proved to be reliable and useful to validate the test scheme of soil-structure model.

1 INTRODUCTION

With the rapid increasing of subways, the safety of underground subway stations has become more important than ever. Especially, after the Hyogoken-Nambu earthquake of January 17, 1995 in Japan, which caused utter devastation of the Daikai subway station and severe damage to some other stations, seismic design of underground structures has drawn a lot of attention. Modern underground subway stations are developing towards long span, large cross section and deeper structural form, which make the seismic performance of many special subway stations unclear (Chen et al. 2014). As an example of these stations, more and more atrium-style subway stations are used in urban subway systems. This kind of station is well-known for its good performance in natural lighting. Singapore and Chinese cities like

Shanghai and Shenzhen have constructed many atrium-style subway stations. Many shaking table tests have been already described in literature on subway station models. However, the previous studies focused on traditional single-story or multiple-story box-frame stations. The scaled model tests of Daikai subway station revealed that the lack of load carrying capacity against shear at the center column caused the collapse (Iwatate, et al. 2000). The effect of change in pore water pressure during and after both main shock and aftershock were also investigated (Chen et al. 2013). Tests on a three-arch type subway station indicated that the columns were more dangerous and the liquefaction difference around the station could lead to remarkable spatial effects of structural strains (Chen et al. 2015). Tests on a three-story box-frame station have shown that interior columns can be the weakest part of the structure and also highlighted the influence of excess pore pressure that should not be ignored for underground structure design (Chen et al. 2015). The tests on a six-story box-frame station indicated that pulse-like ground motion increased dynamic responses of the structure and ground due to its inherent rich low-frequency component and high energy (Chen et al. 2016). A series of shaking table tests on underground subway station in loess ground were also performed to study the dynamic responses of soil-station model (Quan et al. 2016). The shaking table test of a large and shallow metro station with a Y-column revealed that the strains at the branching point of the column and in the middle of lower Y-column were the largest (Wu. et al. 2015). However, until now little is known about the seismic performance of atrium-style subway stations and no shaking table test has been carried out before to study it. The author had conducted some numerical analyses to predict the dynamic responses of atrium-style subway stations ((Zhang et al., 2017; Zhang et al., 2018). Based on these works, the design of a shaking table test campaign on atrium-style subway station models was conducted. The following sections describe such activity.

2 TEST DESIGN

2.1 *Test apparatus and similitude ratio design*

The shaking table tests are conducted using the MTS Company shaking table facility. The table size is 4 m×4 m. Time histories of acceleration can be applied in three directions, having two horizontal and one vertical component. The frequency of input motion can range from 0.1 to 50 Hz. The maximum allowable accelerations are 1.2 g and 0.8 g for horizontal directions and 0.7 g for vertical direction with a mass of 15 t.

The cylindrical flexible-wall container has diameter and height equal to 3 m and 1.8 m, respectively. After Lok (Riemer et al. 1996) compared this kind of flexible container with a rigid-wall box and a box with inclined rigid walls in the numerical analyses, Meymand (1998) had conducted the first shaking table test using this kind of flexible container. Both the numerical analyses and shaking table tests clearly demonstrated the advantage of the flexible-wall container in replicating the prototype response. Chen (2001) and Li (2002) improved the design for the above container and made a new one. But the key materials like rubber and reinforced steel mesh had been renewed for this experimental campaign. In view of this fact, the new container was still validated in the test, which would be shown in the following part. The rubber of flexible container is designed to have the similar shear stiffness to the model soil, in order to minimize any soil-container interactions.

The prototype of the model structure is an atrium-style subway station with dimensions 21.34 m×17.19 m (width×height). The whole subway line has two different cross-sections. The total length of the station is 347 m. Two end parts are rectangular box-type subway station with 177 m and 71 m in length and 21.74 m in width. The middle part is an atrium-style subway station with 99 m in length and 21.34 m in width. The atrium-style subway station has two floors. The first one is a waiting hall floor and the second one is a platform floor. There are arched roof slabs on waiting hall floor. The roof slabs are designed to have seven atrium-style openings with dimensions 9.65 m×11 m and two openings with dimensions 7.45 m×11m (9.65 m and 7.45 m were along the axial direction of the station, while 11 m was perpendicular to the axial direction). There is one transverse beam with dimensions 1.5 m×1 m (width×height) between every two openings in the roof slab plane. This kind of design was intended to meet the requirements of introducing above ground natural light. In view of a lot of openings and less overlaying soil, the waiting hall floor was designed without columns. The middle slab is designed to have four atrium-style openings with dimensions 9.3 m×18.74 m. There is one transverse beam with dimensions 3.5 m×0.8 m (width×height) between every two openings in the middle slab plane. Instead of conventional square column, thin-walled columns, hidden in metro platform screen doors, were adopted. The thin-walled columns have size 3 m×0.4 m (width×height). Cut and cover construction was adopted for the whole subway station. Figure 1 shows the photographs of the atrium-style subway station on-site.

For large-scale shaking table tests, it is not possible or extremely difficult to strictly satisfy the similitude ratio for both the underground structure model and soil model simultaneously. The common practice is to focus on the most important part in the test and to relax the requirements for other part. For example, during the shaking table test of the surface building structures, a common practice is adopting seismically effective mass by a series of masses concentrated at the floor levels (lumped masses). These masses are only seismically effective but not structurally equivalent. In this case the seismically effective mass can be decoupled from the density of the structurally effective material, which relaxes the dimensional requirement (Moncarz and Krawinkler, 1981). In this test, several fundamental physical quantities need to be choosen and other physical quantities can be deduced according to Buckingham π Theorem. For soil model, the length, acceleration, density and dynamic shear modulus were choosen as the fundamental physical quantities. Following the recent studies by Yan et al. (2016), the similitude relation concerning the above quantities could be obtained from the classic dynamic equation. It could be expressed as:

$$S_{G_d}/(S_l S_\rho)=S_a \tag{1}$$

Where S_l, S_ρ, S_a and S_{G_d} are the similitude ratio of length, acceleration, density and dynamic shear modulus, respectively. In this test, S_l and S_a were taken as 1/30 and 1, respectively. The similitude ratios for other physical quantities were listed in Table 1.

2.2 *Model soil and model structure*

The artificial model soil of sawdust and sand is adopted in this test. It has been studied and used by a lot of researchers, e.g. Shang et al. (2006), Yan et al (2016) and Chen et al (2016). Based on a large number of tests on specimens with different proportions of sawdust to sand, the optimal proportion was determined to be 1:2.5. According to the above similitude ratio for the soil, satisfying the similarity of the dynamic behavior like dynamic shear modulus *vs.* dynamic shear strain was most important. Figure 2 showed the G/G_{max} -γ curves and λ-γ curves of both prototype and model soil, where G, G_{max}, λ, γ are dynamic shear modulus,

Figure 1. Photographs of atrium-style subway station on-site.

Table 1. Similitude ratios of model structure and model soil.

Type	Property	Relation	Similitude ratio	
			Model structure	Model soil
Geometry property	Length l	S_l	1/30	1/30
	Linear displacement r	$S_r = S_l$	1/30	1/30
Material property	Elastic modulus E	S_E	0.42	0.033
	Equivalent density ρ	$S_\rho = S_E S_l^{-1} S_a^{-1}$	12.6	0.52
	Shear modulus G	S_G	—	0.033
Dynamic property	Mass m	$S_m = S_\rho S_l^3$	4.4×10^{-4}	1.2×10^{-6}
	Acceleration a	S_a	1	1
	Duration t	$S_t = S_E^{-\frac{1}{2}} S_l S_\rho^{\frac{1}{2}}$	0.1826	0.1826
	Frequency ω	$S_\omega = 1/S_t$	5.48	5.48
	Dynamic stress σ	$S_\sigma = S_l S_a S_\rho$	0.42	0.033
	Dynamic strain ε	S_ε	1	1

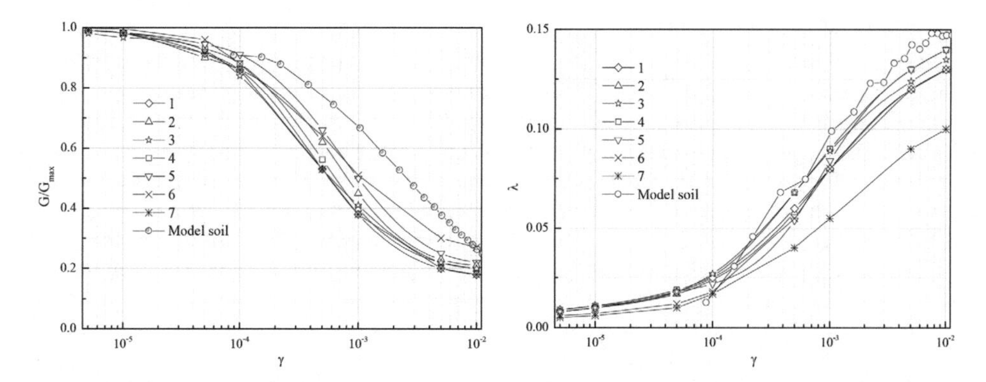

Figure 2. Normalized shear modulus curves and damping ratio curves of model soil and site soil.

maximum dynamic shear modulus, damping ratio and dynamic shear strain, respectively. In Figure 2, curves from 1 to 7 correspond to prototype soil layers of increasing depth.

During preparation, the soil was placed into the container layer by layer. An electronic scale was connected to the crane hook to control the weight of each layer. Four rulers were placed inside the container and around it to control the height of each layer. For each layer, artificial compaction was also conducted.

The prototype station was a cast-in-place reinforced concrete structure. The strength grades of concrete and steel rebar used in the station are C35 and HRB400, respectively (GB50010, National Standard of PR China, 2010). Galvanized steel wire and micro-concrete were used to simulate the prototype steel rebar and concrete, respectively. The designed strength is based on the similitude ratio. Several tests were performed to obtain the optimum ratio for micro-concrete. The mass ratios for micro-concrete were determined as cement : coarse sand : lime : water = 1:5:0.64:1.18. The compressive yield strength and elastic modulus of such a micro-concrete were measured in uniaxial compression test as 10.68 MPa and 1.32×10^4 MPa, respectively. The diameter of the galvanized steel wire has two sizes of 0.7 mm and 1.2 mm. The number and layout of the galvanized steel wires in structural components were based on the similarity principle of bending stiffness of structural components. The detail of the model station dimensions is shown in Figure 3. After the construction, all the model was painted with white coating to be easily observed.

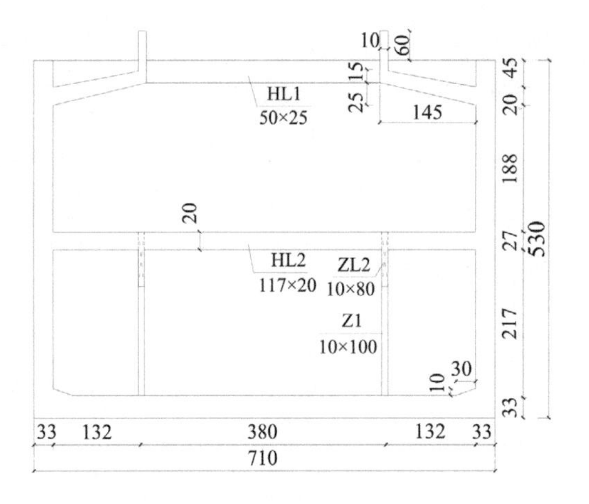

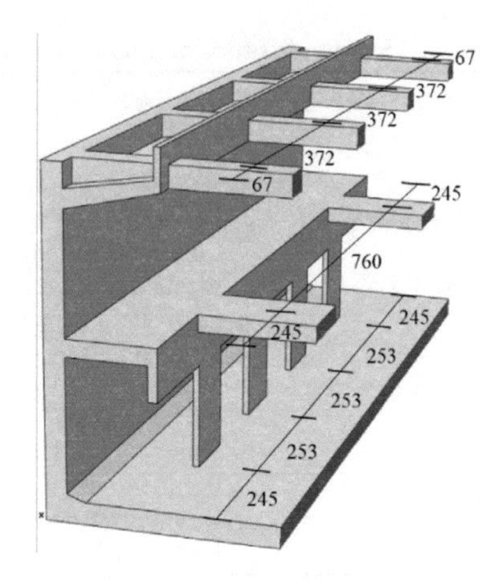

Figure 3. Dimensions of the model station (unit: mm).

2.3 *Loading method and test cases*

In order to study the seismic behavior of the ground and atrium-style subway station, two different types of input motions are adopted. One is sine wavelets with frequency 1–33 Hz and another one is earthquake records including El Centro, Kobe, Loma Preita and Shanghai artificial. The original acceleration histories and Fourier spectra of these input motions are shown in Figure 4. All the peak accelerations of the above input motions were adjusted to around 0.1 g. In addition, the peak acceleration of Loma Preita ground motion was scaled up to 0.3 g, 0.5 g, 0.7 g, 1.0 g and 1.5 g to investigate the seismic response of the model under different shaking intensity. Different input motions were successively imposed at the base of the model. The principle of incremental frequency and intensity was adopted when imposing different input motions. The white noise excitation was also imposed during some cases to check the change in vibration

characteristics of the soil-station system. Table 2 gives the test cases of soil-station model. The test cases of free-field model are shown as no. 1–19 in Table 2, while a white noise case is attached to the end, namely WN3. For space saving the full table for free-field model is not presented.

2.4 *Layouts of sensors*

In total, 36 accelerometers, 6 displacement meters, 32 soil pressure gauges and 27 strain gauges were deployed in the model (Figure 5). Time histories of acceleration, displacement of station model and container, dynamic normal soil stress on the sidewall of the model station and strain of model station were recorded. The symbols used to represent the sensors are as follows: A for accelerometers, D for displacement transducers, P for soil pressure gauges and S for strain gauges. The layouts of sensors shown in Figure 5 were all based on the early finite element analyses (Zhang et al., 2017; Zhang et al., 2018). Accelerometers A3, A5, A7, A10 and A14 were arranged to investigate the propagation of seismic waves in the model ground along the depth. Accelerometers A6, A8, A9, A11 and A12 were arranged to investigate the ground accelerations adjacent to the sidewall of the model station; accelerometers A13-A16 were arranged to check the boundary effects; accelerometers A1s, A4s and A7s were used to record the accelerations of each floor; accelerometers Az7s and Az8s were used to check the rocking mode of vibration. Here the subscript 's' and 'z' denote 'station model' and 'vertical direction'. Soil pressure gauges PL1-PL11 and PR1-PR11 were arranged to investigate the distribution of dynamic earth pressures on left and right sidewalls, respectively. In addition, Figure 6 presents the layouts of accelerometers for free-field model. In total, 16 accelerometers were adopted. The purpose of free field is to verify the boundary effect and compare the acceleration responses with the corresponding soil-station model.

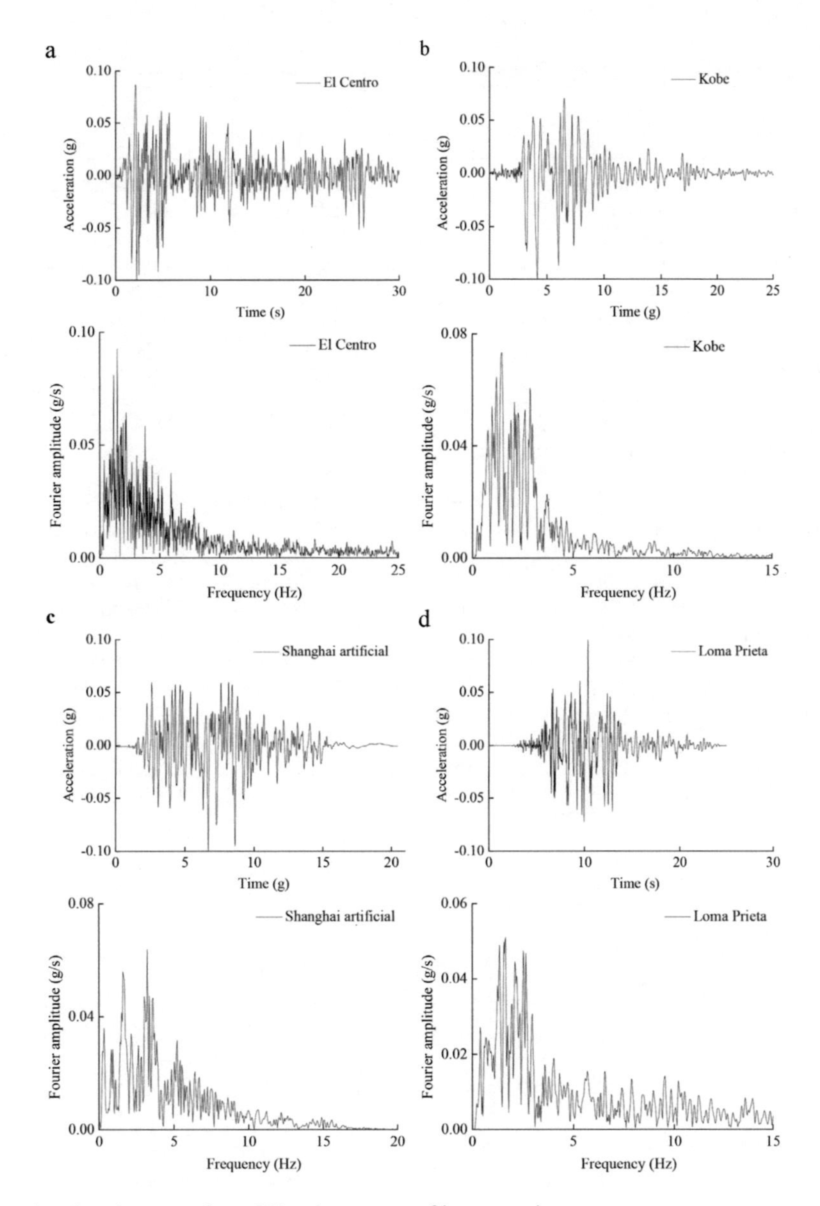

Figure 4. Acceleration records and Fourier spectra of input motions.

Table 2. Test cases for the shaking table tests.

No	Cases	Input waveform	X (g)	Z (g)	No	Cases	Input waveform	X (g)	Z (g)
1	WN1	White noise	0.07	0	24	LP-x0.1-z0.25	Loma Prieta	0.1	0.25
2	SH-x0.1	Shanghai artificial	0.1	0	25	LP -x0.3	Loma Prieta	0.3	0
3	EC-x0.1	El Centro	0.1	0	26	LP -x0.5	Loma Prieta	0.5	0
4	K-x0.1	Kobe	0.1	0	27	5SIN-z0.2	5Hz SINE	0	0.2
5	LP-x0.1	Loma Prieta	0.1	0	28	10SIN-z0.2	10Hz SINE	0	0.2
6	WN2	White noise	0.07	0	29	15SIN-z0.2	15Hz SINE	0	0.2
7	1SIN-x0.1	SINE	0.1	0	30	20SIN-z0.2	20Hz SINE	0	0.2
8	3SIN-x0.1	3Hz SINE	0.1	0	31	5SIN-x0.2	5Hz SINE	0.2	0
9	5SIN-x0.1	5Hz SINE	0.1	0	32	5SIN-x0.3	5Hz SINE	0.3	0
10	9SIN-x0.1	9Hz SINE	0.1	0	33	5SIN-x0.4	5Hz SINE	0.4	0
11	10SIN-x0.1	10Hz SINE	0.1	0	34	WN3	White noise	0.07	0
12	—	—	—	—	35	8SIN-x0.5	8Hz SINE	0.5	0
13	15SIN-x0.1	15Hz SINE	0.1	0	36	WN4	White noise	0.07	0
14	18SIN-x0.1	18Hz SINE	0.1	0	37	N-x0.2-z0.17	Northridge	0.2	0.17
15	21SIN-x0.1	21Hz SINE	0.1	0	38	N-x0.3-z0.28	Northridge	0.3	0.28
16	24SIN-x0.1	24Hz SINE	0.1	0	39	N-x0.4-z0.34	Northridge	0.4	0.34
17	27SIN-x0.1	27Hz SINE	0.1	0	40	WN5	White noise	0.07	0
18	30SIN-x0.1	30Hz SINE	0.1	0	41	LP-x0.7	Loma Prieta	0.7	0
19	33SIN-x0.1	33Hz SINE	0.1	0	42	WN6	White noise	0.07	0
20	LP-x0.1-z0.05	Loma Prieta	0.1	0.05	43	LP-x1.0	Loma Prieta	1.0	0
21	LP-x0.1-z0.1	Loma Prieta	0.1	0.1	44	WN7	White noise	0.07	0
22	LP-x0.1-z0.15	Loma Prieta	0.1	0.15	45	LP-x1.5	Loma Prieta	1.5	0
23	LP-x0.1-z0.2	Loma Prieta	0.1	0.2	46	WN8	White noise	0.07	0

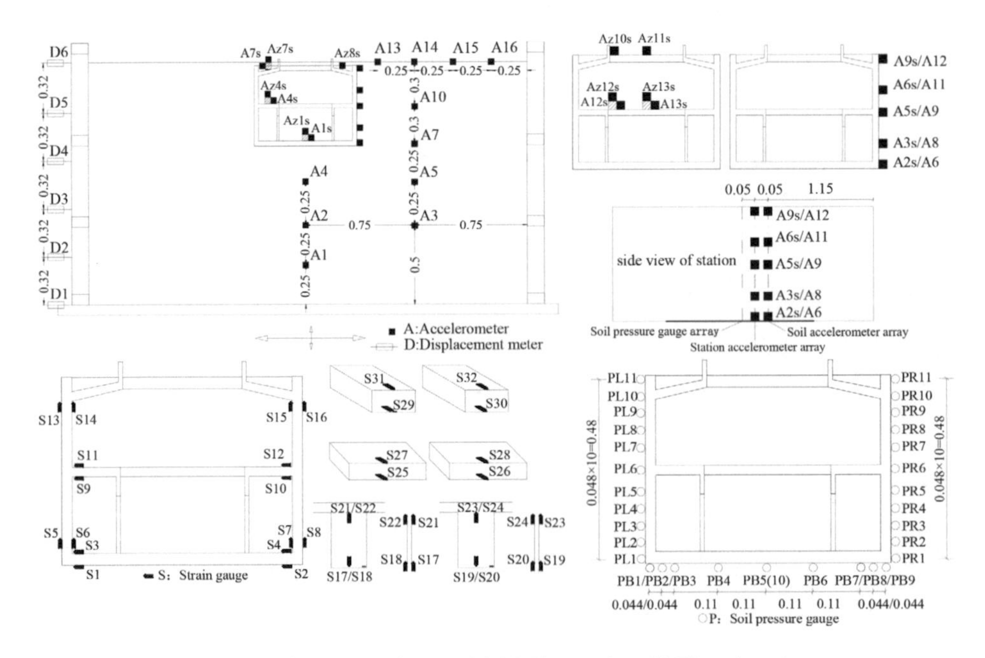

Figure 5. Layouts of sensors for soil-station model (a) Planar view (b) Elevation view.

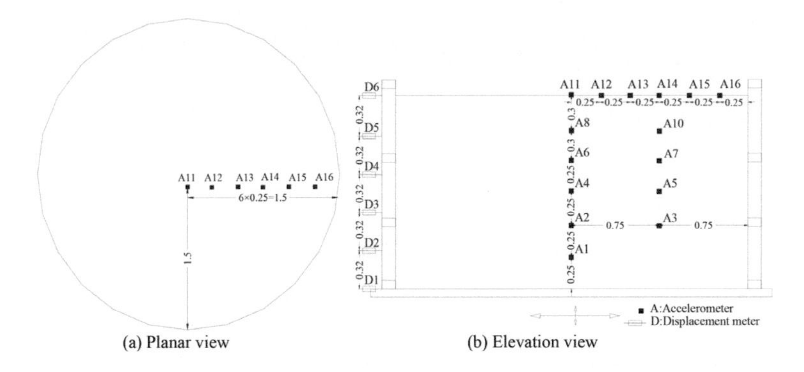

Figure 6. Layouts of sensors for free field model.

3 TEST VERIFICATION

3.1 *Lateral displacement of container*

In the early stage of the experimental campaign, the attention was focused to validate the dynamic deformation mode of the soil-station system in the flexible container. Figure 7 shows the instantaneous lateral displacements of the container (it represents the soil deformation) under the harmonic ground motions of 1 Hz and 3 Hz when the peak acceleration is 0.1 g. In Figure 7, T symbol denoted the period of the corresponding ground motion. At different moments, the distribution of lateral displacement along the height is very similar. Overall, the flexible container that the test has adopted presented a shear-type deformation mode.

3.2 *Boundary effects*

To verify the boundary effects in the flexible container, the accelerometers were arranged on the ground surface at different distances from the container boundary, as shown in Figure 6. Figure 8 shows the peak accelerations measured at these positions in free-field tests. It is easy to find that these peak accelerations were very similar, which indicates that the influence of boundaries is negligible.

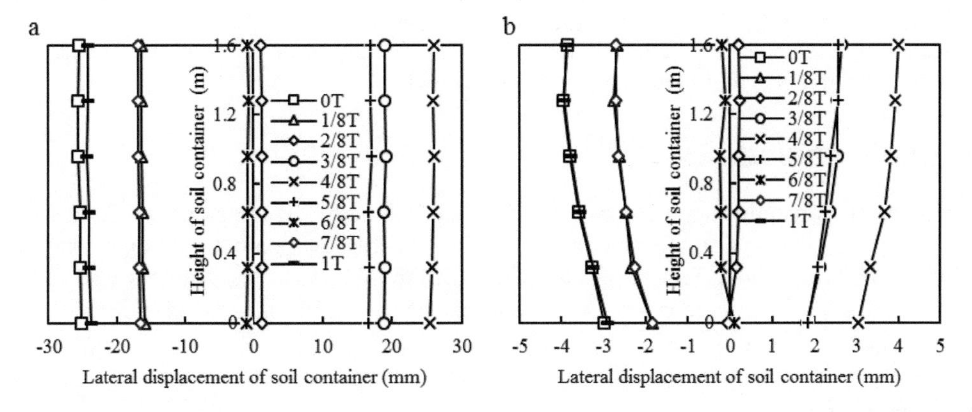

Figure 7. Lateral displacements of container at different steps of one period under sine wavelets with frequency (a) 1 Hz and (b) 3 Hz.

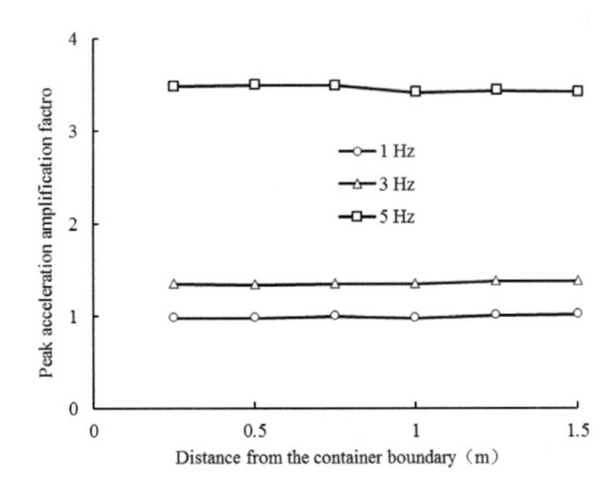

Figure 8. Peak accelerations of ground surface in free field tests.

4 SUMMARY

The paper presented the design and first results of an experimental campaign that is ongoing using the shaking table facility at the State Key Laboratory for Disaster Reduction in Civil Engineering, Tongji University (Shanghai) to estimate the seismic responses of the atrium-style subway station model. Based on current existing laboratory conditions, a shaking table with the size of 4 m×4 m and a cylindrical flexible-wall container with the size of 3 m×1.8 m (diameter×height) was adopted. The model station and model ground were both designed based on similitude ratios according to Buckingham π Theorem. Galvanized steel wire and micro-concrete were used to simulate the prototype steel rebar and concrete. An artificial model soil of sawdust and sand with mass proportions 1:2.5 (sawdust:sand) was adopted in the test. The comparison of G/G_{max} -γ curves and λ-γ curves of both prototype and model soil showed good similarity. Loading method and test programme were also decided according to the test purposes. The layouts of sensors were all based on the early finite element analyses. In order to get required data, enough accelerometers, displacement tranducers, soil pressure gauges and strain gauges were arranged in the model station and model ground.

For the purpose of this paper, the test results of the free-field model, including instantaneous lateral displacements of container and ground surface accelerations, were presented. Results showed that the flexible container presented a shear-type deformation mode and the influences of the verticalboundary on the problem is negligible.

Additional research is ongoing where the dynamic responses of both the atrium-style subway station model and model ground are investigated.

ACKNOWLEDGEMENTS

The research is mainly supported by National Key Research and Development Plan, China (2017YFC1500700) and State Key Laboratory of Disaster Reduction in Civil Engineering, the Ministry of Science and Technology of China (SLDRCE15–02). The Authors also like to acknowledge the anonymous reviewers who played a significant role in improving the manuscript.

REFERENCES

Chen, Z.Y., Chen, W., Bian, G.Q. 2014. Seismic performance upgrading for underground structures by introducing shear panel dampers, *Advances in Structural Engineering* 17(9): 1343–1357.

Chen G.X., Wang, Z.H., Zuo, X., Du X.L., Gao, H.M. 2013. Shaking table test on the seismic failure characteristics of a subway station structure on liquefiable ground, *Earthquake Engineering and Structural Dynamics* 42(10): 1489–1507.

Chen S., Chen G.X., Qi, C.Z., Du X.L., Wang, Z.H. 2015a. A shaking table-based experimental study of seismic response of three-arch type's underground subway station in liquefiable ground, *Rock and Soil Mechanics* 36(7): 1899–1914.

Chen G.X., Chen S., Zuo, X., Du X.L., Qi, C.Z., Wang, Z.H. 2015b. Shaking table tests and numerical simulations on a subway structure in soft soil, *Soil Dynamics and Earthquake Engineering* 76(2015): 13–18.

Chen, Z.Y., Chen, W., Li, Y.Y., Yuan, Y. 2016. Shaking table test of a multi-story subway station under pulse-like ground motions, *Soil Dynamics and Earthquake Engineering* 82(2016): 111–122.

Chen, Y.Q. 2001. Shaking table testing of dynamic soil-structure interaction system, Ph.D Thesis, Tongji University, Shanghai.

GB50010. Code for design of concrete structure. Beijing: China Architecture & Building Press; 2010.

Iwatate, T., Kobayashi, Y., Kusu, H., Rin, K. 2000. Investigation and shaking table tests of subway structures of the Hyogoken-Nanbu earthquake, *Proceeding of the 12 WCEE, Auckland, 30 January – 4 February 2000*. Rotterdam: Balkema.

Li, P.Z. 2002. Shaking table testing and computational simulation analysis of dynamic soil-structure interaction system, Ph.D Thesis, Tongji University, Shanghai.

Meymand, P.J. 1998. Shaking Table Scale Model Tests of Nonlinear Soil-Pile-Superstructure Interaction in Soft Clay, Ph.D Thesis, University of California, Berkeley.

Moncarz P.D. & Krawinkler H. 1981. Theory and application of experimental model analysis in earthquake engineering, Report No. 50, the John A. Blume Earthquake Engineering Center, Stanford University, Stanford CA.

Quan, D.Z., Wang, Y.H., Ye, D., Jing, Y.L., Chen S. 2016. Shaking table test study on subway station built in loess area, *China Civil Engineering Journal* 49(11): 79–90.

Riemer, M. & Meymand, P. 1996. 1-g modeling of seismic soil-pile-superstructure interaction in soft clay, *proc. 4th Caltrans seismic research workshop, Sacramento, July 1996*.

Shang, S.P., Liu, F.C., Lu, H.X., Du, Y.X. 2006. Design and experimental study of a model soil used for shaking table test, *Earthquake Engineering and Engineering Vibration* 26(4): 199–204.

Wu, B.L., Tao, L.J., Li, J.D., Li, S.L. 2015. Shaking table experiment for a large, shallow metro station with a Y-column structure, *Modern Tunnelling Technology* 52(6) : 92–98.

Yan, X., Yuan, J.Y., Yu, H.T., Bobet, A., Yuan, Y. 2016. Multi-point shaking table test design for long tunnels under non-uniform seismic loading, *Tunnelling and Underground Space Technology* 59 (2016): 114–126.

Zhang, Z.M., Yu, H.T., Yuan, Y., Zhao, H.L. 2017. 3D numerical simulation of seismic characteristics of atrium-style metro station, *the 3rd international conference on performance based design in earthquake geotechnical engineering, Vancouver, 16–19 July 2017*.

Zhang, Z.M., Yuan, Y., Bilotta, E., Yu, H.T., Zhao, H.L. 2018. Dynamic soil normal stresses on side wall of a subway station, *the sixth international symposium on life-cycle civil engineering, Ghent, 28 to 31 October 2018*.

Tunnels and Underground Cities: Engineering and Innovation meet Archaeology, Architecture and Art, Volume 2: Environment sustainability in underground construction – Peila, Viggiani & Celestino (Eds)
© 2020 Taylor & Francis Group, London, ISBN 978-0-367-46579-7

Author Index